FUNDAMENTAL ASPECTS OF STRUCTURAL ALLOY DESIGN

BATTELLE INSTITUTE
MATERIALS SCIENCE COLLOQUIA

Published by Plenum Press

1972: Interatomic Potentials and Simulation of Lattice Defects
Edited by Pierre C. Gehlen, Joe R. Beeler, Jr., and Robert I. Jaffee

1973: Deformation and Fracture of High Polymers
Edited by H. Henning Kausch, John A. Hassell, and Robert I. Jaffee

1974: Defects and Transport in Oxides
Edited by Martin S. Seltzer and Robert I. Jaffee

1975: The Physical Basis for Heterogeneous Catalysis
Edited by Edmund Drauglis and Robert I. Jaffee

1977: Fundamental Aspects of Structural Alloy Design
Edited by Robert I. Jaffee and Benjamin A. Wilcox

FUNDAMENTAL ASPECTS OF STRUCTURAL ALLOY DESIGN

Edited by

ROBERT I. JAFFEE

Fossil Fuel and Advanced Systems Division
Electric Power Research Institute
Palo Alto, California

BENJAMIN A. WILCOX

Materials Science Department
Battelle Memorial Institute, Columbus Laboratories
Columbus, Ohio

BATTELLE INSTITUTE
MATERIALS SCIENCE COLLOQUIA

Seattle, Washington—Harrison
Hot Springs, British Columbia
September 15–19, 1975

R.I. Jaffee, Chairman

PLENUM PRESS ● New York and London

Library of Congress Cataloging in Publication Data

Battelle Institute Materials Science Colloquia, 10th, Seattle and Harrison Hot Springs, B.C., 1975.
    Fundamental aspects of structural alloy design.

    Includes indexes.
    1. Alloys—Congresses. I. Jaffee, Robert Isaac, 1917-   II. Wilcox, Benjamin A. III. Battelle Memorial Institute, Columbus, Ohio. IV. Title.
TN689.2.B37 1975   669'.9   77-22519
    ISBN 0-306-31080-5

©1977 Plenum Press, New York
A Division of Plenum Publishing Corporation
227 West 17th Street, New York, N.Y. 10011

Printed in the United States of America

To **PROFESSOR JOHN R. LOW, JR.**

for his contributions to the art and science of alloy behavior, especially that of steels.

PARTICIPANTS

T. H. ALDEN *University of British Columbia, Vancouver, B.C., Canada*

G. S. ANSELL *Rensselaer Polytechnic Institute, Troy, New York, U.S.A.*

M. F. ASHBY *Cambridge University, Cambridge, England*

J. M. BATCH *Battelle, Columbus Laboratories, Columbus, Ohio, U.S.A.*

R. T. BEGLEY *Westinghouse Electric Corporation, Pittsburgh, Pennsylvania, U.S.A.*

A. L. BEMENT *Massachusetts Institute of Technology, Cambridge, Massachusetts, U.S.A.*

S. H. BUSH *Battelle, Pacific Northwest Laboratories, Richland, Washington, U.S.A.*

J. D. CAMPBELL *University of Oxford, Oxford, England*

A. H. CLAUER *Battelle, Columbus Laboratories, Columbus, Ohio, U.S.A.*

L. A. DAVIS *Allied Chemical Corporation, Morristown, New Jersey, U.S.A.*

R. F. DECKER *International Nickel Company, New York, New York, U.S.A.*

J. D. EMBURY *McMaster University, Hamilton, Ontario, Canada*

B.R.T. FROST *Argonne National Laboratory, Argonne, Illinois, U.S.A.*

H. J. FROST *Cambridge University, Cambridge, England*

P. HAASEN *Universität Göttingen, Göttingen, Germany*

G. T. HAHN *Battelle, Columbus Laboratories, Columbus, Ohio, U.S.A.*

J. P. HIRTH *The Ohio State University, Columbus, Ohio, U.S.A.*

E. HORNBOGEN *Ruhr-Universität Bochum, Bochum, Germany*

vii

R. I. JAFFEE *Electric Power Research Institute, Palo Alto, California, U.S.A.*

B. H. KEAR *Pratt & Whitney Aircraft, Middletown, Connecticut, U.S.A.*

I. S. LEVY *Battelle, Pacific Northwest Laboratories, Richland, Washington, U.S.A.*

J. R. LOW, JR. *Carnegie-Mellon University, Pittsburgh, Pennsylvania, U.S.A.*

F. A. McCLINTOCK *Massachusetts Institute of Technology, Cambridge, Massachusetts, U.S.A.*

C. J. McMAHON, JR. *University of Pennsylvania, Philadelphia, Pennsylvania, U.S.A.*

W. S. OWEN *Massachusetts Institute of Technology, Cambridge, Massachusetts, U.S.A.*

H. W. PAXTON *U.S. Steel Corporation, Pittsburgh, Pennsylvania, U.S.A.*

F. S. PETTIT *Pratt & Whitney Aircraft, Middletown, Connecticut, U.S.A.*

T. F. SANDERS *Aluminum Company of America, Alcoa Center, Pennsylvania, U.S.A.*

O. D. SHERBY *Stanford University, Stanford, California, U.S.A.*

P. G. SHEWMON *The Ohio State University, Columbus, Ohio, U.S.A.*

R. F. SIMENZ *Lockheed Aircraft Corporation, Burbank, California, U.S.A.*

C. T. SIMS *General Electric Company, Schenectady, New York, U.S.A.*

J. E. SLATER *Battelle, Columbus Laboratories, Columbus, Ohio, U.S.A.*

J. STRINGER *University of Liverpool, Liverpool, England*

N. P. SUH *Massachusetts Institute of Technology, Cambridge, Massachusetts, U.S.A.*

G. THOMAS *University of California, Berkeley, California, U.S.A.*

J. K. TIEN *Columbia University, New York, New York, U.S.A.*

F. L. VERSNYDER *Pratt & Whitney Aircraft, East Hartford, Connecticut, U.S.A.*

B. A. WILCOX *Battelle, Columbus Laboratories, Columbus, Ohio, U.S.A.*

I. G. WRIGHT *Battelle, Columbus Laboratories, Columbus, Ohio, U.S.A.*

V. F. ZACKAY *University of California, Berkeley, California, U.S.A.*

PREFACE

FUNDAMENTAL ASPECTS OF STRUCTURAL ALLOY DESIGN is the proceedings of the tenth Battelle Colloquium in the Materials Sciences, held in Seattle, Washington, and Harrison Hot Springs, B.C., September 15-19, 1975. The theme of the conference was the emerging science of alloy design. Although the relationships of properties of alloys to their composition and structure have long been a dominant theme in physical metallurgy, it is only recently that metallurgists have turned their attention from the analytical, *post hoc* study of the structure-property relationship to the synthesis approach of alloy design.

As usual in the Battelle colloquia, the first day started with a group of introductory lectures presented by leaders in the field, each emphasizing his personal approach to the problem. This provided a historical perspective for the colloquium. These papers, together with the banquet address of Professor J. R. Low, Jr., who was honored at the colloquium, comprise the introductory section of these proceedings.

Alloy design is generally specific to a given application. Thus, the needs in alloy design in a number of important applications, gas turbines, electrical-power-generation equipment, airframes, pressure vessels, and nuclear applications were presented in a group of papers. An agenda discussion on "Needs in Alloy Design" followed. These papers give the external constraints on alloy design applications, and criteria for mechanical, physical, and chemical properties for which the alloys must be designed.

The individual elements of alloy design then were discussed. These included the properties of solid solutions, transformations, precipitates and dispersions, grain boundaries, interfaces, and textures. Metallurgists generally deal with crystalline materials. A new element available to them is absence of crystallinity in amorphous metal glasses, covered by a separate paper. An agenda discussion covered the roles and limitations of these elements in the processes of alloy design.

The extent to which metallurgists are able to design alloys from scientific principles was addressed through a set of examples, including cases where the limiting design property is mechanical and where the limitation is based on chemical properties. Processing, of course, is an integral part of alloy design. Achievable properties often are limited by the ability of the metallurgist to achieve a desired microstructure. Many of the papers in the colloquium covered advances in processing as a means of achieving sought-for microstructures.

As is the usual custom at Battelle colloquia, one of the distinguished participants was honored at the farewell banquet and by inscribing the proceedings in his name. At Harrison, we honored Professor John R. Low, Jr., of Carnegie-Mellon University for his work on the mechanical behavior of steels. Professor Low's biographical remarks about his career in metallurgy are included in the front matter of these proceedings.

It is our sad duty to report that the Battelle Colloquia in the Materials Sciences concluded with this, the tenth colloquium. We wish to express sincere thanks to the management of Battelle Memorial Institute who supported the colloquia throughout these 10 years—in particular, Dr. B. D. Thomas, retired President of Battelle Memorial Institute, who personally encouraged the initiation of the colloquia series, and Dr. S. L. Fawcett, current President, for his continued support. Dr. F. J. Milford, the original Director of Research in the Physical Sciences of Battelle Institute, and his successor, Dr. C. W. Kern, provided the institutional aegis under which the colloquia were held. We are grateful to Dr. John Batch, Director of the Columbus Laboratories, Dr. E. L. Alpen, then Director of the Pacific Northwest Laboratories, and their staffs for support in the conduct of the colloquium. Julie Swor was in charge of arrangements in Seattle and Harrison Hot Springs; secretarial support was ably and charmingly provided by Darlene Weaver and Jodie Marshall. Forrest Nelson did an excellent job of audio-visual support, and generally was very helpful in the smooth functioning of the colloquium. Edna Jaffee was in charge of the Ladies Program. To all of these, we extend our grateful thanks.

The Organizing Committee is indebted to many people for their advice on the choice of subjects and participants for the colloquium and, particularly, would like to thank G. S. Ansell, A. L. Bement, M. Cohen, P. C. Gehlen, J. J. Gilman, B. H. Kear, F. A. McClintock, E. R. Parker, H. W.

Paxton, J. C. Shyne, M. O. Speidel, R. W. Staehle, J. Stringer, G. Thomas, J. K. Tien, and V. F. Zackay.

Lastly, one of us (R.I.J.) would like to add his personal thanks to all of the colleagues and scientists worldwide who participated in the ten Battelle Institute Colloquia in the Materials Sciences. It has been a great honor and personal pleasure for him to have associated with so many creative and distinguished people in organizing and conducting the series. The proceedings of the colloquia covered all aspects of materials and provide a lasting record of the status of the science of materials over a 10-year period, 1966-1975. In addition, the Battelle colloquia demonstrated a new format for the conduct of an effective scientific conference: the bringing together of a set of leading contributors in a particular aspect of a scientific field, the presentation of the historical background by distinguished contributors, presentation of important new work by current leaders, and the assessment of unresolved issues and routes to future progress through structured agenda discussions. It has been gratifying to note that this format has been adopted in many other scientific meetings, as a sort of legacy of the Battelle colloquia.

The Organizing Committee

R. I. Jaffee, Chairman
S. H. Bush
A. H. Clauer
G. T. Hahn
J. P. Hirth
I. S. Levy
B. A. Wilcox

AUTOBIOGRAPHICAL REMARKS—JOHN R. LOW, JR.

The initial choice which led to my becoming a metallurgist occurred about halfway through my undergraduate program. I had started this program in chemical engineering. However, my home environment was closely connected with the steel industry, and I was persuaded that metallurgy was a better career from both the point of view of economic opportunities and the number of interesting technical problems in the field.

Upon graduation from Purdue University in 1931, I was awarded a graduate fellowship to continue studies in metallurgy at Carnegie Institute of Technology. This fellowship was never taken up because economic conditions in the country were then deteriorating rapidly, and it became apparent that my earnings would be needed in the family. At that time graduate fellows had no net earnings beyond the funds needed to support themselves. Consequently, I began work in various steel mills as a metallurgist and continued in that capacity (with some time lost in the worst of the depression) until about 1935, when I moved into steel mill production as a wire mill foreman. A foreman's pay scale was then about double that of a metallurgist. This work proved very interesting with many problems of technical interest which merely stimulated my curiosity and a desire for time to study, with no opportunity to gratify either. In addition, between 1936 and 1937, steel plant unions were organized and it became apparent that if I stayed in steel mill production supervision, a very large fraction of my time would be occupied in labor relations problems rather than the technical problems which interested me.

I spent some time during the next year investigating research-position possibilities at various laboratories, including Battelle-Columbus, and was convinced by the people to whom I talked that it was not possible to make a career in research without additional training to the level of the doctorate.

Accordingly, I applied for admission to several of the better known universities with graduate programs in metallurgy, including Carnegie and the Massachusetts Institute of Technology. In the meantime, my wife and I attempted to accumulate enough money to support us for a year since no fellowship support was available. Because of our strained financial circumstances, and for no other reason, we finally chose Carnegie Tech since its tuition was slightly lower than that of MIT. Regardless of the reason, the choice was the correct one, since at that time R. F. Mehl and his colleagues were putting together the best program of graduate training in metallurgy available anywhere. In addition, I began work with Maxwell Gensamer, who was doing pioneering work in attempting to understand the mechanical behavior of metals, with emphasis on microstructure-properties relations, an area that has been my principal interest ever since.

After four very stimulating and interesting years at Carnegie Tech, I began teaching and research at the Pennsylvania State College, and remained there until early 1944, by which time all of the full-time metallurgy students had disappeared from the campus. I then became a part of the group which E. C. Bain at the Carnegie-Illinois Steel Company was organizing with a view to integrating all of the research and development activities of the United States Steel Company into a research and development center, which has since been established at Monroeville, near Pittsburgh. Initially, I was responsible for the area of sheet-metal forming and later became involved in fracture research through the problem of catastrophic failure in welded merchant vessels. Both of these problem areas continued to be of major interest when I returned to Penn State late in 1946, by which time I was partially incapacitated by the air pollution from which Pittsburgh was suffering at the time.

I stayed at Penn State until 1948, during which time we carried out investigations on fracture mechanisms in mild steel and the measurement of formability in low-carbon sheet steels for automobile body stampings. This latter investigation, purely by empirical correlations, led to the demonstration of the importance of crystallographic texture in such modes of sheet-metal forming. The empirical correlation initially was between press performance and plastic-strain ratios in the tension test. Toward the end of this period, it became apparent that the sluggish bureaucracy of the state educational system would not be able to keep up with the rate of inflation, and I accepted a position with J. H. Hollomon, who was beginning his synthesis of the General Electric Research Laboratory group in metallurgy and materials science.

The next 18 years, all spent in the climate for research, established and nurtured by Herb Hollomon, were as satisfying and pleasant as one could wish for and must have been unique for an industrially supported laboratory. One had complete freedom in choosing his research area, the only stipulation being that he become a leader in his chosen field of research. In addition, Hollomon had another attribute not apparent to those outside the group: he possessed considerable insight into the needs of various individuals to make them more effective investigators, and much of his time was spent rearranging working conditions and associations to this end.

In the early 1950's, I was employed at the Knolls Atomic Laboratory in charge of a group studying radiation damage to materials. In 1953, I transferred to the General Electric Research Laboratory. I first completed some work on grain-size effects on cleavage fracture in iron begun at KAPL. From this work, it became apparent that the movement of crystal dislocations was intimately involved in cleavage fracture and methods of ex- perimentally observing dislocations were becoming available. With R. W. Guard, I then set about using etch-pit methods to study the slip process in single and polycrystalline iron; and later, with Dale F. Stein, to determine dislocation velocities in 3 Si-Fe as a function of applied stress at several temperatures. By 1958, thin-foil transmission electron microscopy had ad- vanced to the point that Anne M. Turkalo and I were able to investigate the dislocation-multiplication process and slip-band formation in more detail than was possible by etch-pitting techniques.

In the early 1960's, as one might have expected, conditions began to change toward the more normal industrial situation of requiring demonstra- tion of a direct return on research investment, and, in addition, there was a nationwide disillusionment with the profitability of fundamental research. In addition, my own functions were becoming more closely related to teaching than to research and I decided to return to teaching for the balance of my professional career.

The opportunity to do this was made possible by H. W. Paxton, then at Carnegie Institute of Technology, and in 1967, I transferred to that uni- versity. For the past 8 years, I have been teaching undergraduate and graduate courses in mechanical metallurgy and working with graduate students in their doctoral research programs, largely on various aspects of the structural control of fracture in metals.

CONTENTS

xvii

Part One

INTRODUCTORY LECTURES

MECHANICAL PROPERTIES OF SOLID SOLUTIONS

P. Haasen

Institut für Metallphysik
Universitat Göttingen
Göttingen, W. Germany

ABSTRACT

Dislocation theory is used to predict the yield stress of a single-phase alloy from the characteristics of its solute atoms and its microstructure. For fcc dilute solutions, the hardening laws of Fleischer-Friedel and of Labusch are reviewed. Particular attention is paid to the problems of superposition of different hardening mechanisms and of the effects of multiple solute additions. The temperature-independent "plateau" of the yield stress is explained as a dynamic phenomenon. Bcc metals solution harden at intermediate temperatures, most likely by the influence of solute atoms on kink motion. The solution softening observed at low temperatures in this structure is less clearly understood. For polycrystals, a new theory by Suzuki for the Hall-Petch relation is discussed. Work hardening of solid solutions does not differ much from that of the pure metals except for stage III and beyond, where dynamic recovery by cross slip (and by climb) is influenced by stacking-fault energy (SFE) and thus by solute concentration. The SFE is also one of the parameters determining the distribution of slip on various planes and this, in turn, largely influences the fatigue resistance of solid solutions.

1 INTRODUCTION

Metals are strengthened by cold working and by alloying. Here we are concerned with solution and work hardening at not too high temperatures. How far is dislocation theory able to predict the strength of a single-phase alloy from its composition and its microstructure? The model we start with is a single crystal containing a statistical distribution of solute atoms which are unable to move. We want to calculate the minimum shear stress τ_c which drives the dislocation over an arbitrary distance through this alloy. The problem has been solved to various degrees of approximation[1-4] for a dilute alloy, i.e., one whose mean solute spacing is larger than the range of solute-dislocation interaction, and for not too fast dislocations for which inertial effects, as treated by Granato[5], can be neglected. Experimentally, strain rate, and not dislocation velocity, is the parameter which is externally controlled. The proportionality factor $N_M \cdot b$, i.e., mobile dislocation density times Burgers vector, is at present only empirically known and is not at all constant. We return to this point below.

Considering the static balance of forces on a dislocation in a dilute solid solution at T=0, Fleischer[2] assumes a well-defined spacing, L, of obstacles against which a dislocation is pressed under a stress τ. In the critical state the touched obstacles interact, by their maximum (repulsive or attractive) force f_o, with the dislocation and the balance reads

$$\tau_c b \cdot L_c = f_o \quad . \tag{1}$$

The spacing L_c has been calculated by Friedel[6] to depend on the stress as well as on the dislocation line tension E_L as

$$L_c^3 = 6E_L / \tau_c\, b\, C_F \quad , \tag{2}$$

where C_F is the areal concentration of solute atoms. By eliminating L_c from Eqs. (1) and (2), one gets

$$\tau_c b = f_o^{3/2}\, C_F^{1/2} / \sqrt{6\, E_L} \tag{3}$$

for the yield stress at t = 0. f_o is estimated by Fleischer to be caused by two elastic interaction mechanisms between a dislocation and a solute atom: (a) due to its size misfit, measured by the change in lattice parameter with atomic concentration, $\delta = (1/b)(db/dc)$, the solute atom is attracted or repelled (parelastically) by the hydrostatic stress component of the dislocation (in case of anisotropic misfit, other stress components also interact with the strain field of the solute atom[7]); (b) as the solute atom may be a soft or hard spot in the lattice as measured by the modulus change it produces, $\eta = (1/\mu)(d\mu/dc)$, it locally changes the dislocation energy and thus interacts (dielastically) with the dislocation. With $f_{op} \approx \mu b^2 \delta$ and $f_{od} \approx \mu b^2 \eta/\alpha$, where $\alpha \approx 10$, Fleischer's theory yields a total f_o proportional to $(\eta \pm \alpha\delta)$, where the sign depends on whether the solute is at the compression or tension side of the dislocation. This ambiguity of sign is removed in Labusch's theory[4], as is the assumption of point obstacles corresponding to a high degree of dilution such that solute atoms interact with a dislocation at full force or not at all.

2 A THEORY OF SOLUTION HARDENING[4,8]

In a not-so-dilute alloy with obstacles of a finite range of interaction (w), the unit length of dislocation is exposed to a whole spectrum, $\rho(f)df$, of interaction forces (f) of positive and negative signs. Due only to its flexibility, the dislocation feels a nonzero net force, $\int \rho(f)fdf$. Due to the unique force-distance profile f(y) by which a solute atom interacts (par- or dielastically) with the dislocation, the net interaction force can also be written in terms of a spectrum, $\rho(y)dy$, of dislocation-solute spacings (y). For the critical state in which the dislocation is just able to move freely, the balance of forces reads

$$\tau_c b = \int \rho_c(y)f(y)dy \quad . \tag{4}$$

Here ρ_c is the corresponding critical distribution of spacings (ρ later written without the index c). It is calculated straightforwardly by Labusch[4] to be

$$\rho(y) = C_F[1+f'(y) \, G(o)] \text{ for } \rho \geqslant 0, \, \rho = 0 \text{ otherwise.} \tag{5}$$

Here $2G(o) = (E_L \alpha)^{-1/2}$ is the elastic compliance of the dislocation in front of an obstacle characterized by a curvature, $\alpha = \int \rho(y)f''(y)dy$, of the interaction potential. The expression for $\rho(y)$ is interpreted easily considering Fig. 1, a f(y) profile of an obstacle at which the equilibrium positions for a dislocation of mean position 1 are y_1 (1) and y_2 (1). The difference, $1-y_i \equiv f(y) \cdot G(o)$, is shown by the secant construction. Dislocation segments are most frequently found at those positions near the obstacle where the slope f'(y) is largest, but their number decreases as the dislocation line or the obstacle

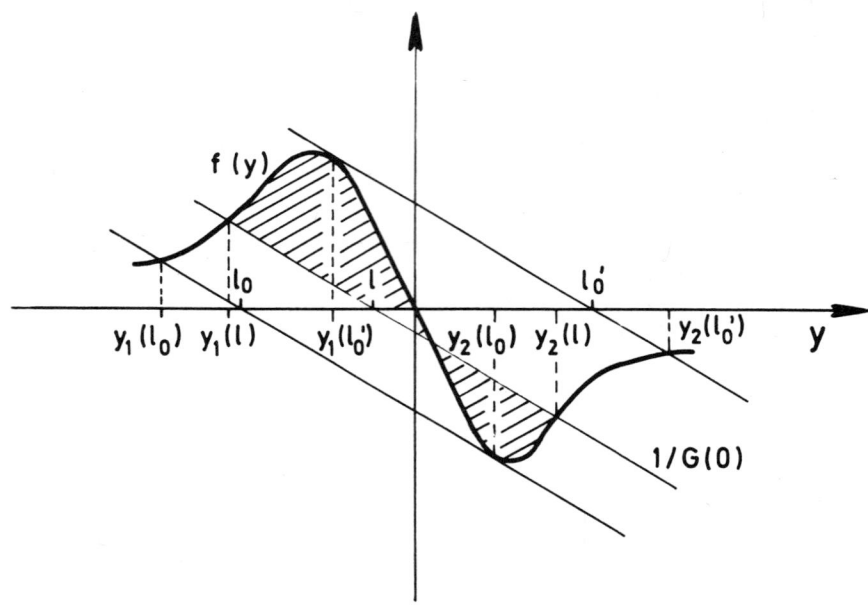

Fig. 1. Force-distance diagram of obstacle in slip plane and graphical determination of equilibrium positions.

becomes "stiffer". At infinite distance from the dislocation, $\rho(\pm\infty) = C_F$, the average solute concentration (cm^{-2}). Note that the distribution function, Fig. 2, has a gap, i.e., there are no equilibrium positions between y_1 (l_0') and $y_2(l_0')$, and that the gap is asymmetric with respect to the origin, so the integral [Eq. (4)] yields a finite value.

If one neglects the "one" in Eq. (5), the integration of Eq. (4) leads to

$$\tau_c b = \frac{C_F}{2\sqrt{E_L\alpha}} \, f_0^2/2 \quad , \tag{6}$$

with

$$\alpha^{3/2} = \frac{C_F}{2\sqrt{E_L}} \int f' \, df = \frac{C_F f_0^2}{2\sqrt{E_L}w} \cdot I$$

and

$$I = \int \frac{d(f/f_0)}{d(y/w)} \, d(f/f_0), \text{ a numerical constant;}$$

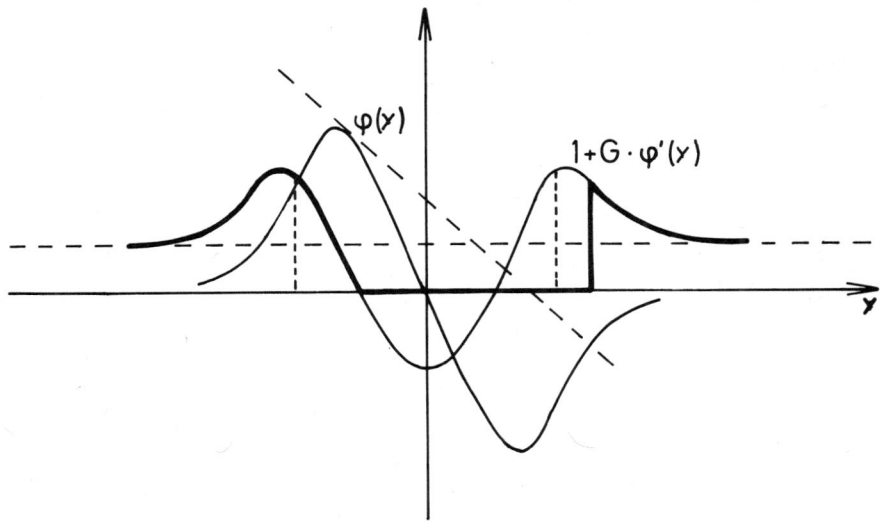

Fig. 2. Distribution function of dislocation-obstacle spacings. (Read f for φ.)

therefore

$$\tau_c b \approx \frac{C_F^{2/3} f_o^{4/3} w^{1/3}}{E_L^{1/3} 2^{5/3}} \qquad (7)$$

except for a numerical factor of order one. It can be shown[4] that the theories of Mott and Nabarro[1] and of Riddhagni and Asimow[3] lead to the same result if some of their assumptions are modified or taken for granted.

Labusch's solution-hardening theory leads to a dependence on C_F, f_o, E_L not too different from Fleischer's, Eq. (3), although it contains the range of interaction (w). On the other hand, it clearly answers the questions with regard to the superposition of various hardening mechanisms and even of different solute atoms: one introduces a distribution function for each species, above and below the slip plane, at various spacings from the slip plane, or for different solutes. Then

$$\tau_c b = \Sigma \int f_i \rho_i dy, \; \rho_i = C_{Fi} \left[1 + f_i'(y) \, G(o) \right]$$

$$\alpha = \Sigma \int f_i' \rho_i dy \qquad (8)$$

with the result

$$\tau_c b = \frac{w^{1/3}}{E_L^{1/3} 2^{5/3}} \left[\sum C_{Fi} f_{oi}^2 \right]^{2/3} = \left(\sum \tau_{ci}^{3/2} \right)^{2/3} \cdot b \ . \tag{9}$$

In particular, for atoms above and below the slip plane acting di- and parelastically on an edge dislocation in proportion to $\eta - \alpha\delta$ and $\eta + \alpha\delta$, respectively, the correct combined interaction parameter is, according to Eq. (9),

$$\epsilon_L = \eta^2 + \alpha^2 \delta^2 \ . \tag{10}$$

This is in fact borne out by experiments, e.g., on copper-alloy single crystals, Fig. 3. Experiments on ternary CuSiGe single crystals confirm Eq. (9); see Fig. 4.

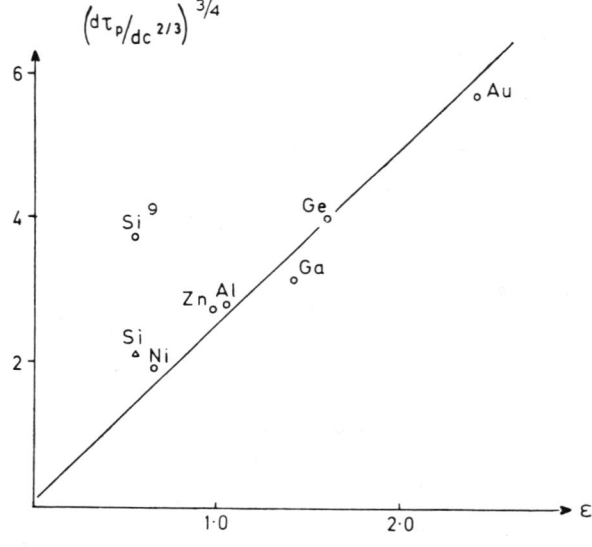

Fig. 3. Specific solution hardening versus Labusch solute parameter for copper-alloy crystals ($Si^\triangle$ Friedrichs, Si° Evans and Flanagan).

3 TEMPERATURE-INDEPENDENT SOLUTION HARDENING

These measurements (Fig. 4) were done in the so-called plateau range of temperatures, for copper alloys between 400 and 550 K, where the CRSS τ_o is rather independent of T and ϵ.

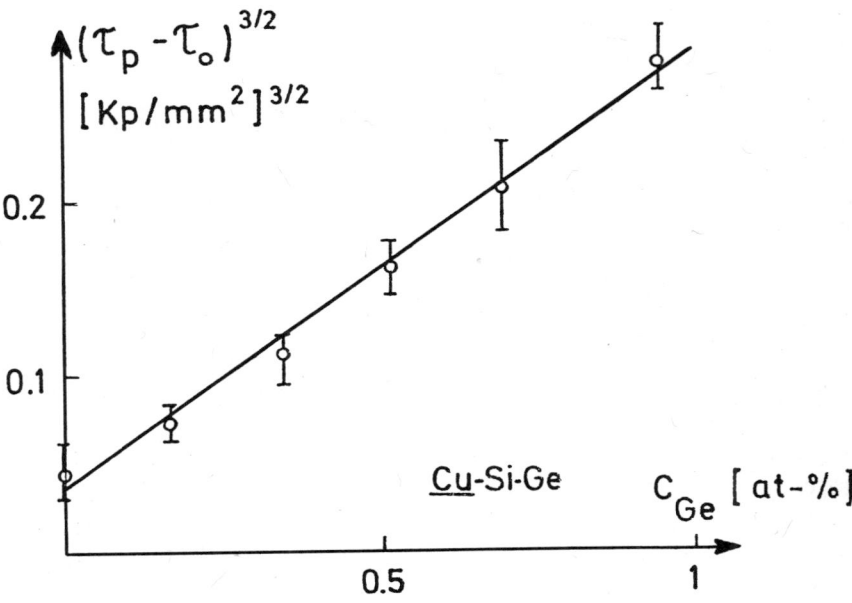

Fig. 4. Plateau stress τ_p of ternary copper alloys with 1 percent silicon versus germanium content (J. Friedrichs, thesis Göttingen); τ_o is the CRSS of pure copper.

This is not easy to understand, as thermally activated dislocation motion over single solute atoms should lead to a continuous and rather rapid decrease of τ_o with increasing T. In a not-too-dilute solid solution there will be statistical fluctuations in the solute arrangement, leading to groups of m obstacles of strength ($\sqrt{m \cdot f_o}$). Labusch[4] has shown that τ_c, Eq. (7), is independent of m, as $C_{Fm} \approx C_F/m$ and w is independent of m. There are statistically too few groups of a large enough m to be thermally insurmountable and to explain the plateau. On the other hand, Labusch et at.[8] discovered a possible dynamic cause of the plateau in the following: thermal activation helps the dislocation to get from y_1 to y_2 in Fig. 1 in proportion to $\nu_o \cdot \exp(-E_+/kT)$, where E_+ is the shaded area for an average position 1. This changes the density of dislocation elements at y_1 by

$$\frac{\partial \rho(y_1)}{\partial t} = v \frac{\partial}{\partial y_1}\left(\rho \cdot \frac{dy}{dl}\bigg/_{y_1}\right) + \nu_o\left[-e^{-E_+/kT}\rho(y_1) + e^{-E_-/kT}\rho(y_2)\frac{dy_2}{dy_1}\right] \quad (11)$$

A similar differential equation holds for y_2. In the stationary state for a dislocation moving with constant velocity v, the left sides of these equations must be zero, which allows $\rho(y_1)$ to be calculated from Eq. (11) and τ_c from Eq. (7). The computation yields

$$\tau_c(T,v) = \tau_{co}\ (T{=}0,v) \cdot \Phi(\Theta,\omega)\ , \tag{12}$$

where

$$\Theta = \frac{kT\beta}{\sqrt{\tau_{co}\cdot E_L b w^3}}\ ,\ \omega = \frac{v}{v_o w}\ ,\ \beta = \text{a number} > 1,\text{ and}$$

Φ is plotted in Fig. 5. For dislocation velocities not too small compared with the limiting velocity $v_o\cdot w \approx 10^{10}\cdot 5\cdot 10^{-8}$ cm/sec $= 5$ m/sec, the flow stress remains high and about constant over a considerable range of temperatures. The physical reason for this plateau is the finite limit that the transition rates between the states y_1 and y_2 assume at high temperatures. [Actually, in this case, thermal activation theory no longer applies for $\exp(-E/kT) \rightarrow 1$]. A finite overall velocity of the dislocation can then be maintained only at a finite stress.

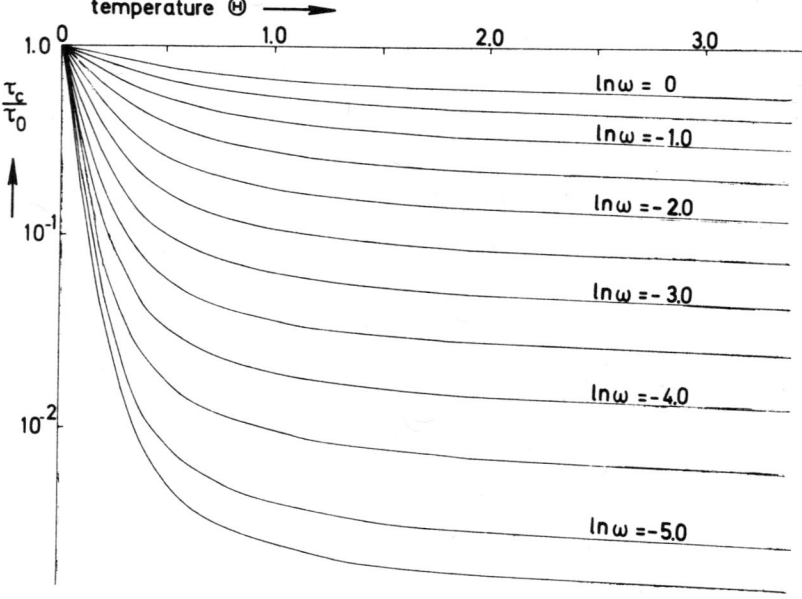

Fig. 5. Calculated stress τ_c relative to that at T=0 versus reduced temperature Θ for various reduced dislocation velocities ω.

This plateau stress should decrease linearly with v for $v \ll v_0 \cdot w$. Direct velocity measurements on Cu-1% Ge single crystals at 300 K by the double-etch technique[9] are in fact compatible with this prediction. The measured velocity near the yield stress is $v \approx 1$ m/sec compared with 10 m/sec for pure copper.[10] According to Fig. 5, this gives a reasonable $\tau_c(\Theta = 1)/\tau_{co} \approx 0.2$. In qualitative agreement with the proposed *dynamic* origin of the plateau stress, Ahearn et al.[11] find the *static* friction stress that stabilizes a dislocation pileup in a Cu-3%Al crystal to be only 16 percent of the flow stress under which the pileup was produced at a strain rate of $4 \cdot 10^{-5}$ sec^{-1}. Also the proposed normalization of the temperature dependence of the CRSS τ_0 of all solutions in terms of the one materials parameter τ_{co} appears to be well obeyed in practice.[12] The measured plateau stress is, however, rather independent of strain rate, whereas the theory expects it to be proportional to dislocation velocity. As indicated in the Introduction, the ratio of the two, $\gamma/v = N_M b$, should be investigated. For the dilute CuGe alloy, one obtains at the yield point in typical tensile test

$$N_M = \frac{\gamma}{bv} = \frac{10^{-4}}{2.5 \cdot 10^{-8} \cdot 40} = 100 \text{ cm}^{-2} \quad . \tag{13}$$

The total dislocation density in the crystal was, however, between 10^4 and 10^5 cm^{-2}, and etching confirmed that less than 1 percent of the dislocations had moved during a stress pulse.[9] Evidently, the crystal reacts to a change in strain rate γ largely by a change in N_M rather than by changing v. This would also make the interpretation of stress-relaxation tests in terms of a slowing down of dislocations rather doubtful.[8] The question remains: What then defines a yield stress if the v-τ relation is linear? One answer, dislocation multiplication, does not show up as a characteristic of τ_0 in the etch-pit observations. The present search is for still larger solute groupings occurring by statistical fluctuations at spacings of several thousand A which must be overcome in order for the large-scale dislocation motion characteristic for yield to occur (R. Labusch, unpublished).

4 SOLUTION HARDENING IN BCC ALLOYS

The deformation of bcc solid solutions differs considerably from that discussed above for fcc alloys: the yield stress of iron alloys for example is less temperature dependent than that of pure iron.[13] At low temperatures there is solution softening, $d\tau_0/dc < 0$, while at room temperature the solute addition hardens the metal. There are strong deviations from Schmid's law, especially in bcc alloys. These effects are attributed to a particular core structure of the bcc screw dislocation which forces it to move by the nucleation and propagation of kinks in a Peierls potential. H. Suzuki[14] discusses the effect of solute atoms on these processes as follows (see also Sato and Meshii[15]). The interaction energy of a kink in a screw dislocation and a substitutional solute atom is rather small because the kink is narrow and the screw does not to first order interact with a center of dilatation but mostly by the modulus defect. [Suzuki uses an effective interaction parameter $2\epsilon_s \approx \eta + 7\delta$ instead of Eq. (10)]. The energy of the dislocation changes proportional to ϵ_s and to the number of solute atoms in its core, which in turn fluctuates at random as the kink moves. For high temperatures where kink motion should be rate controlling, Suzuki calculates (with a numerical factor α)

$$\tau_o b \approx \frac{E_L^2 \epsilon_s^2 c}{\alpha kT} ,\tag{14}$$

in good agreement with experimental data.[13] At low temperatures and small solute concentrations, kink formation should determine the strain rate and this process can be assisted by the interaction with solute atoms. Suzuki calculates for this case a decrease of the Peierls stress

$$\tau_o = \tau_{PN} - \beta E_L^{1/2} \epsilon_s^{1/2} c^{1/4}\tag{15}$$

(with a constant β). The combination of both mechanisms leads to a dependence of yield stress on solute concentration and temperature, as shown in Fig. 6 for the case of FeGe alloys.[13]

A critical test for the hypothesis of solution softening caused by preferred kink nucleation at solute atoms should be its application to alloys with the diamond cubic structure where one is sure that kink nucleation determines plastic deformation.[16] Siethoff[17] has therefore investigated Si-Ge solid solutions with $c_{Ge} = 0.1 \ldots 2.3$ percent in compression at constant strain rate. While the activation energy remained unchanged from that known for double-kink formation in pure silicon, the yield stress *increased* in proportion to $\sqrt{c_{Ge}}$ as predicted by Eq. (3). This finding throws doubt on the model of preferred kink formation at dislocations. Gibala and co-workers[18], on the other hand, have presented strong experimental arguments in favor of a scavenging theory of solution softening. They have shown that pure niobium-oxygen alloys solution harden at all concentrations, while the addition of 50 ppm nitrogen introduces solution softening when the first 200 ppm oxygen is added. Solute interactions evidently reduce the concentration, effectiveness, or distribution of activatable solute obstacles. Sethi and Gibala[19] have recently found an even larger softening effect in niobium whose surface was coated with a 650 A thick Nb_2O_5 film. The authors postulate that this film acts as an effective source of mobile nonscrew dislocations which then carry the deformation.

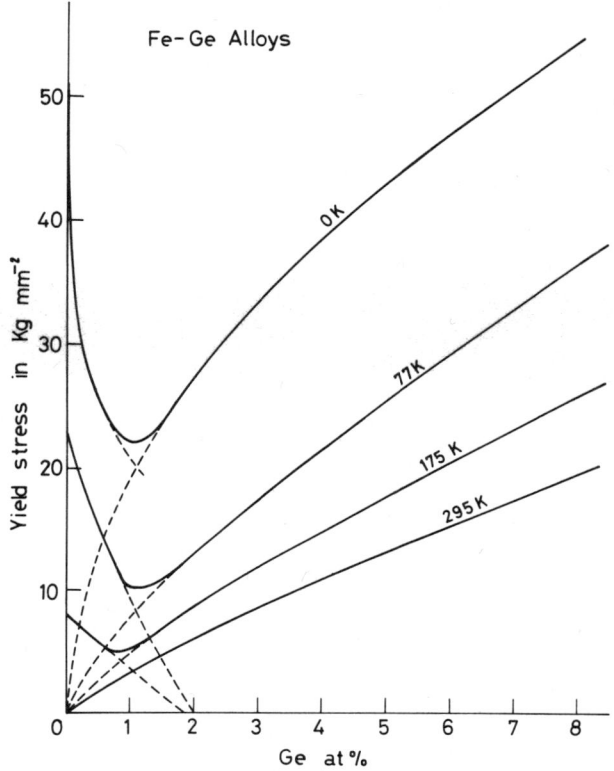

Fig. 6. Calculated yield stress of FeGe alloy single crystals versus germanium concentration.[14]

5 THE HALL-PETCH RELATION IN ALLOYS

The yield stress of a polycrystal depends on its grain size, d, in many metals and alloys as

$$\sigma_o = \sigma_c + k_y d^{-1/2} \quad . \tag{16}$$

The standard theory explains this in terms of the stress concentration of a dislocation pileup which activates a new source in the neighboring grain. $k_y = m_{12} \cdot \tau_s \sqrt{\lambda}$ then is proportional to a stress τ_s necessary to unpin this source which is located at a distance λ from the head of the pileup (m_{12} is an orientation factor). Suzuki and Nakanishi[20] find it difficult to explain their measurements of σ_o ($d^{-1/2}$) on CuAl and CuNi alloys in terms of this k_y theory for the following reasons: (a) the large difference in concentration dependence exhibited for the two kinds of solutes (Fig. 7) does not follow from the theory of "Suzuki locking" by segregation to stacking faults; (b) for a grain size of 10 μm and a grown-in dislocation density of 10^7 cm^{-2} in recrystallized fcc alloys, it seems difficult to find a suitable dislocation to operate as a source in the second grain. The situation is more favorable in these respects in prestrained or quenched mild steel.[21]

As an alternative, Suzuki et al.[20] worked out the theory for the passage of a dislocation through a grain boundary which leaves misfit in the boundary (Fig. 8): when $\mathbf{b}_1$ and $\mathbf{b}_2$ are the Burgers vectors in the two grains, a dislocation $\mathbf{b}_3 = \mathbf{b}_1 - \mathbf{b}_2$ and a grain-boundary ledge are left after the passage. These must be removed before further dislocations can pass through the boundary. Suzuki et al. assume that the component $b_{3\perp}$ perpendicular to the boundary is removed by diffusion along the grain boundary, while the parallel component $b_{3\parallel}$ and the ledge are eliminated by grain-boundary slip or grain-boundary movement over a fraction of an atomic distance; see Fig. 9. The latter processes are hindered by solute atoms segregated to the boundary by either of two effects: solute with a negative modulus defect η

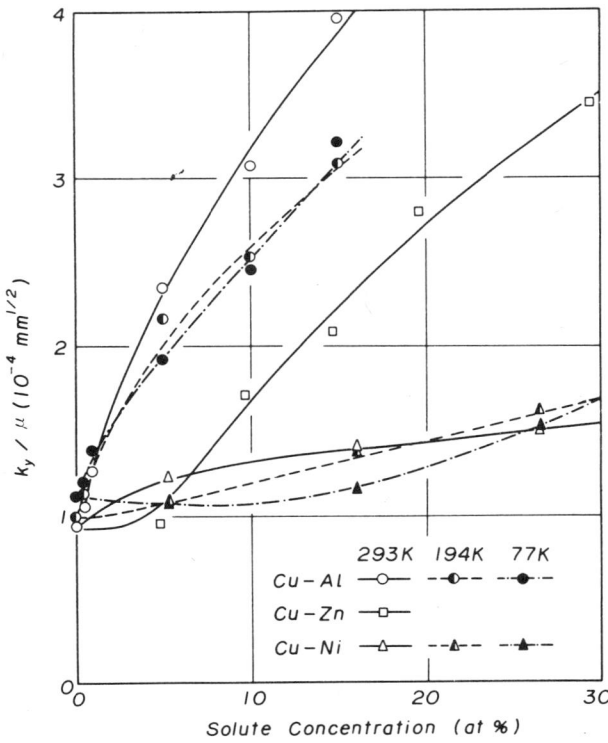

Fig. 7. Concentration dependence of k_y/μ in copper alloys.[20]

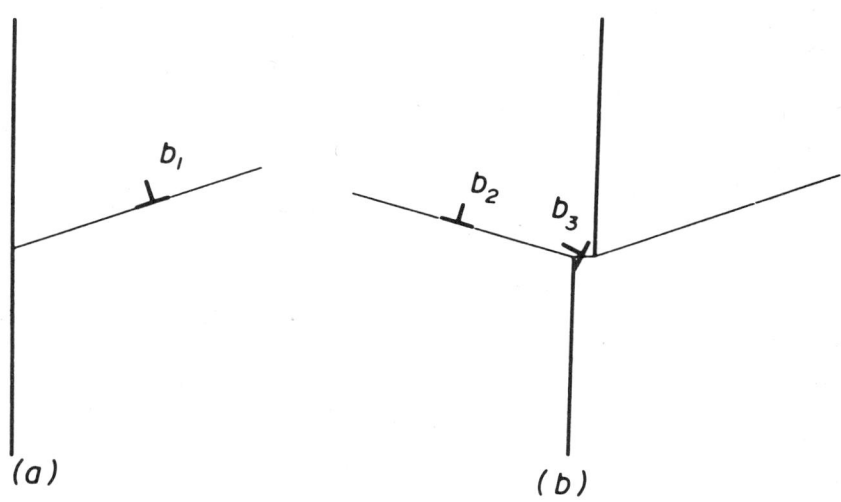

Fig. 8. Penetration of dislocation through grain boundary.

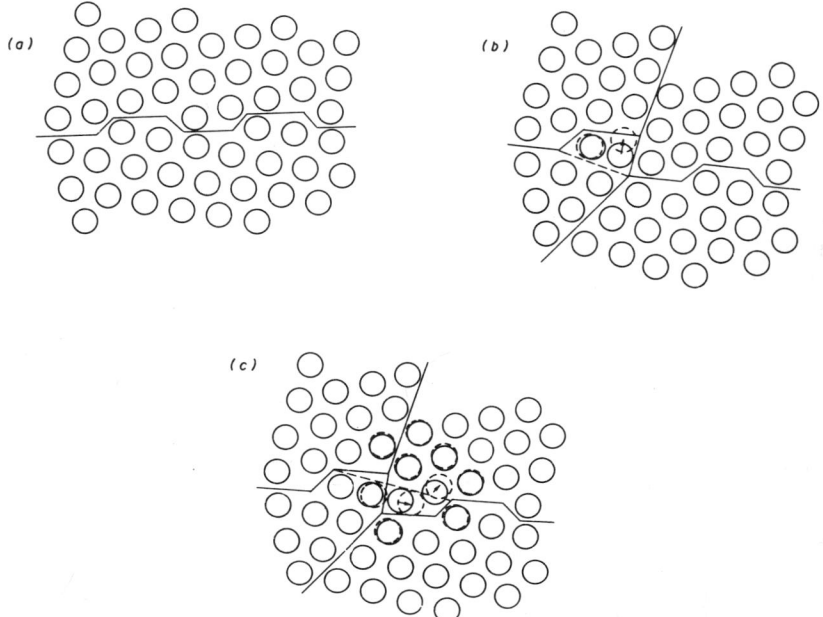

Fig. 9. Rearrangement in grain boundary eliminates ledge.[20]

will segregate to the boundary, while misfitting solute atoms (size misfit δ) will occupy certain favorable sites in the boundary. The latter may convert into unfavorable sites by motion of the grain boundary, thereby hindering this motion and the elimination of $b_3\perp$ and the ledge. The stress necessary to pass a dislocation through a boundary in the alloy is higher than that needed in the pure metal.

The calculation of this stress first involves (1) calculation of the equilibrium solute concentrations c_1 and c_2 at favorable and unfavorable sites of the boundary, respectively, due to δ and η effects; (2) the calculation of the forces $f_\perp$, $f_\parallel$ between the components of b_3 and those of b_2 which have to be overcome to separate these two dislocations; (3) the force f_γ to create the ledge. The forces $f_\parallel$ and f_γ are affected by segregation so that

$$f_{tot} \approx f_\perp + (c_1-c_2)^2 \, (f_\parallel + f_\gamma) \quad . \tag{17}$$

At yield this force must be overcome by that of the applied stress on the leading dislocation of a pileup in order to penetrate the boundary (averaged over all orientations)

$$\overline{n(\tau_o-\tau_c)} \cdot b = \frac{\overline{(\tau_o-\tau_c)}^2}{2\pi A} \, d \cdot b = f_{tot} \tag{18}$$

or

$$k_y = M \cdot \overline{(\tau_o-\tau_c)} \, d^{1/2} \approx M \cdot (f_{tot} \cdot 2\pi A/b)^{1/2} \quad ,$$

with $A = \mu b^2/2\pi(1-\nu)$ and the Taylor factor $M = 3.1$. Suzuki et al. include a temperature dependence of k_y by considering that the leading dislocation does not have to be straight as it penetrates the boundary but can first throw out a loop with the help of thermal activation. The result of the calculation by Eqs. (17) and (18) appears to fit the experimental results rather well (Fig. 10). Evidently, in order to obtain a small k_y, the amount of grain-boundary segregation of the solute must be small, as is the case for nickel in copper.

6 WORK HARDENING IN SOLID SOLUTIONS

The stress-strain behavior of solid solutions is not basically different from that of the pure metals. In fcc single crystals, the flat easy glide stage I is somewhat longer. (In polycrystals, stage I is missing altogether.) The steep stage II has about the same slope, $d\tau/d\gamma \approx \mu/300$ in all fcc single crystals. (In polycrystals, the initial slope $d\sigma/d\epsilon = \Theta_o$ is somewhat larger than

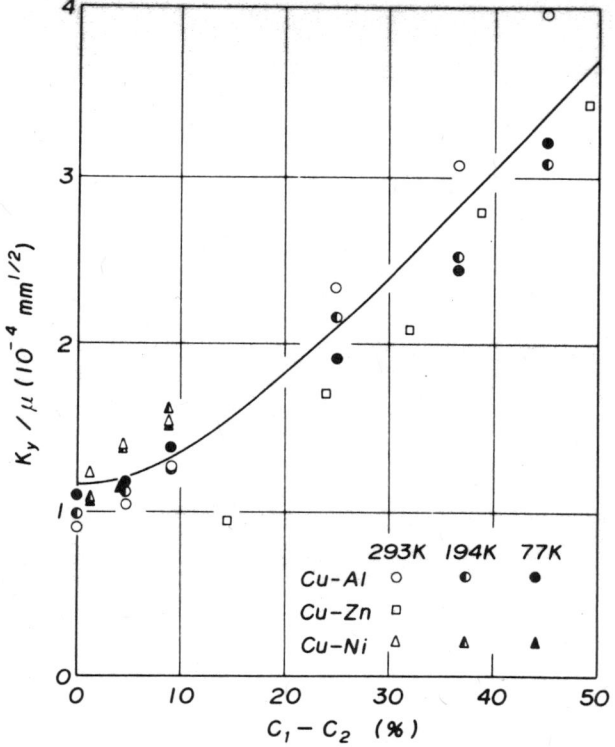

Fig. 10. k_y/μ versus solute segregation (c_1-c_2).[20]

$M^2 d\tau/d\gamma \approx \mu/30$.) A drastic change in work hardening with the addition of solute atoms often occurs in the onset stress τ_{III} of the third stage, in which the work hardening rate again decreases. According to Schoeck and Seeger[22] and Haasen[23], the stress at the beginning of stage III is related to stacking-fault energy γ_s by

$$\ln\!\left(\tau_{III}(T)/\tau_{III}(0)\right) \approx \frac{\beta\gamma_s kT}{\mu^2 b^4}\ln \dot{\epsilon}/\dot{\epsilon}_o \quad . \qquad (19)$$

The γ_s of the Hume-Rothery α solid solutions is a unique function of the electron concentration and approaches zero at the phase boundary with the cph ζ or bcc β phases.[23] This makes $\tau_{III}(T)$ approach its maximum value $\tau_{III}(0) \approx 10^{-3}\ \mu$ and stage II to extend the longest (Fig. 11). Kocks[24] noticed that for fcc polycrystals and at intermediate temperatures the work hardening curves tended (exponentially with strain) towards a saturation stress σ_s

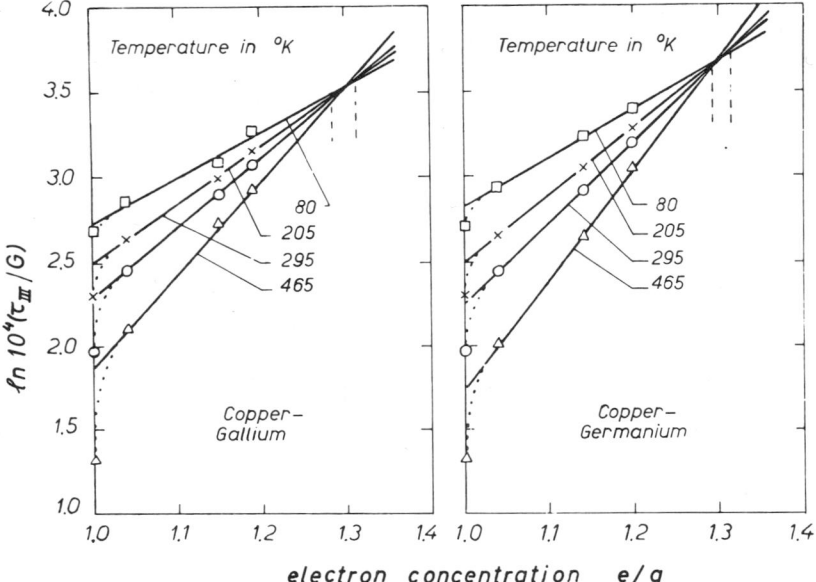

Fig. 11. Reduced stress at beginning of stage III versus electron concentration e/a (G is the shear modulus).

which he finds to be proportional to τ_{III}, Eq. (19). (This saturation has nothing to do with the necking instability in a tensile test.)

The exponential approach towards saturation corresponds to a linear decrease in work hardening with stress

$$\Theta = \Theta_o(1-\sigma/\sigma_s) \quad , \tag{20}$$

and this is interpreted by Kocks as a consequence of dynamic recovery: a decrease in dislocation density by cross slip occurs proportional to the number of times a potential recovery site is contacted by a moving dislocation, i.e., proportional to strain (and not to time as in static recovery). The attainment of a saturation stress in a constant-strain-rate test corresponds to stationary creep under constant stress, and the creep rate deriving from Eq. (19) depends on stress as

$$\dot{\epsilon} = \dot{\epsilon}_o[\tau/\tau_{III}(0)]^n \quad , \quad n = \mu^2 b^4 / \beta \gamma_s kT \quad . \tag{21}$$

Kocks shows that values $n = 4 \ldots 5$ result from this in the range $T/T_m = 0.7 \ldots 0.85$, as are actually observed. Such a power law for $\epsilon(\sigma)$ for high-temperature creep is normally explained[25] by dislocation climb (i.e., static recovery) yielding

$$\epsilon = \epsilon_o(\sigma/\mu)^n(\gamma_s/\mu b)^3 \exp(-Q_D/kT) \quad , \tag{22}$$

where Q_D is the activation energy of diffusion. From this, another saturation stress follows by inversion:

$$\ln \sigma_s/\mu = (Q_D/nkT) + \ln [(\dot\epsilon/\dot\epsilon_o)(\mu b/\gamma_s)^3] \quad . \tag{23}$$

Such a dependence has not been tested so far in comparison with Eq. (19) except for germanium.[26]

In bcc metals and solid solutions, the mobility of screw dislocations determines plastic deformation as we have discussed above. Because of their special core structure, cross slip of screws in bcc crystals should not be much more difficult than their glide. Thus it is no surprise that no stage II is observed in these materials and that stage I turns right into stage III, i.e., that the curvature $d^2\tau/d\gamma^2$ changes from positive to negative values (Šesták and Seeger[27]). In bcc alloys, stage I is more extended, as is also known for fcc solid solutions. A parabolic stage 0 often precedes stage I in which non-screw dislocations are supposed to move until they are exhausted.[28]

7 FATIGUE IN SOLID SOLUTIONS

For the nucleation of a fatigue crack, the distribution of slip throughout the material is very important: coarse slip on relatively few planes leads to a fatigue resistance lower than that induced by fine, more homogeneously distributed slip. This is understood, e.g., by the model of P. Neumann[29], in which a secondary slip plane is set into action at the root of a coarse slip step at the surface during the tensile phase of a push-pull test. If the two slip steps are not completely reversed during the compression phase (and this seems to depend on environmental conditions), they appear again in the next tensile phase and alternate in producing further slip steps (Fig. 12). This mechanism of fatigue-crack formation depends explicitly on the coarseness of slip, in qualitative agreement with many observations on the influence of slip distribution on (high cycle) fatigue. In pure fcc metals, the stress to produce failure after 10^6 cycles, τ_{10^6}, is found to be related to the stress τ_{III}; dynamic recovery by cross slip is in fact strongly coarsening slip.[30] On solution alloying, τ_{10^6} becomes much lower than τ_{III} and approaches the stress τ_{II} at the end of easy glide. The slip lines in the α Hume-Rothery alloys are indeed much coarser than those in pure metals.[31] This is partly an effect of a decreasing SFE which makes for more planar disloca-

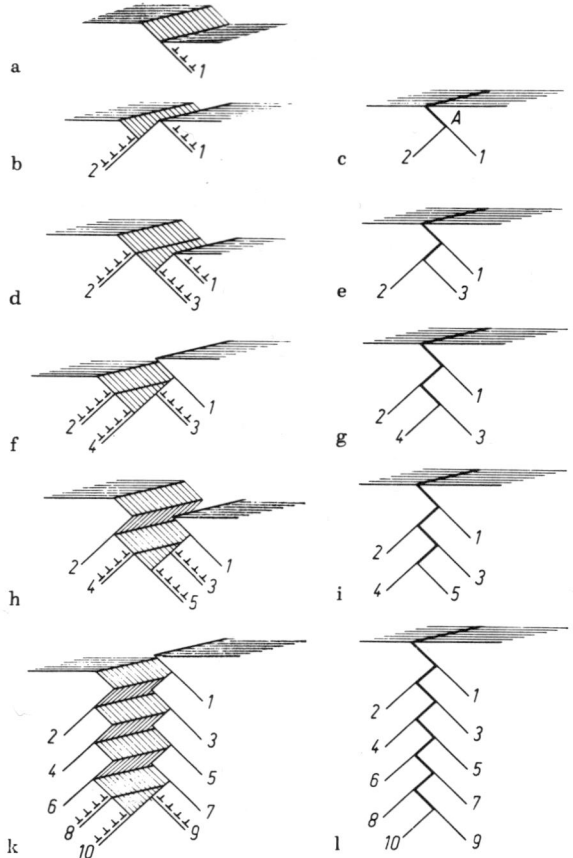

Fig. 12. Model of P. Neumann[29] for crack formation on a coarse surface step in symmetric push-pull; tensile phases left, unloading right.

tion movement. Also, in the more concentrated alloys, short-range order might be destroyed by the first-moving dislocations and this produces an easy glide path for the following ones.

In bcc metals and alloys, the asymmetry of slip resistance on $\{112\}$ planes itself leads to a coarse slip distribution. The crystal uses different planes for tensile and compressive phases of a fatigue cycle, at least in surface grains. This effect has been demonstrated to lead to severe shape changes in symmetric push-pull tests of niobium and α-iron single crystals.[32,33] Cracks then nucleate in this rough surface topography. In polycrystalline α-iron, a surface roughness was observed[33] which had the periodicity of the surface grains and not that of the slip bands as in fcc alloys. The slip asymmetry on $\{112\}$ planes changes with temperature and alloying.[28] It would be most interesting to correlate this with fatigue behavior.

REFERENCES

1. Mott, N. F., and Nabarro, F.R.N., *Report on Strength of Solids,* Phys. Soc., London, England (1950), p. 1.
2. Fleischer, R. L., *The Strengthening of Metals,* D. Peckner (Ed.), Reinhold, London, England (1964), p. 93.
3. Riddhagni, B. R., and Asimow, R. M., *J. Appl. Phys.,* **39**, 4144, 5169 (1968).
4. Labusch, R., *Phys. Stat. Sol.,* **41**, 659 (1970); *Acta Met.,* **20**, 917 (1972).
5. Granato, A. V., *Phys. Rev.,* **B4**, 2196 (1971).
6. Friedel, J., *Dislocations,* Pergamon, Oxford, England (1964), p. 224.
7. Cochardt, A. W., Schock, G., and Wiedersich, H., *Acta Met.,* **3**, 533 (1955).
8. Labusch, R., Grange, G., Ahearn, J., and Haasen, P., *Rate Processes in Plastic Deformation of Materials,* J.C.M. Li and A. K. Mukherjee (Eds.), ASM, Metals Park, Ohio (1975), p. 26.
9. Donch, J., Haasen, P., and Labusch, R., *Nachr. Akad. Wiss. Gott.,* No. 2 (1974).
10. Jassby, K. M., and Vreeland, T., *Phil. Mag.,* **21**, 1147 (1970).
11. Ahearn, J. S., Haasen, P., and Labusch, R., *Scripta Met.,* **8**, 387 (1974).
12. Basinski, Z. S., Foxall, R. A., and Pascual, R., *Scripta Met.,* **6**, 807 (1972).
13. Takeuchi, S., *J. Phys. Soc. Japan,* **27**, 929 (1969).
14. Suzuki, H., *Nachr. Akad. Wiss. Gott.,* No. 6 (1971); *Rate Processes in Plastic Deformation of Materials,* ASM, Metals Park, Ohio (1975), p. 47.
15. Sato, A., and Meshii, M., *Acta Met.,* **21**, 753 (1973).
16. Alexander, H., and Haasen, P., *Solid State Physics,* Vol. 22, F. Seitz (Ed.), Academic Press, New York (1968), p. 27.
17. Siethoff, H., *Mat. Sci. Engr.,* **4**, 155 (1969).
18. Ulitchny, M. G., and Gibala, R., *J. Less-Common Metals,* **33**, 105 (1973).
19. Sethi, V. K., and Gibala, R., *Scripta Met.,* **9**, 527 (1975).
20. Suzuki, H., and Nakanishi, K., *Trans. Jap. Inst. Met.,* **15**, 435 (1974); **16**, 17 (1975).
21. Cottrell, A. H., *The Relation between Structure and Mechanical Properties of Metals,* HMSO, London, England (1963), p. 456.
22. Schock, G., and Seeger, A., *Report of Bristol Conference on Defects in Crystalline Solids,* Phys. Soc., London, England (1955), p. 340.
23. Haasen, P., *Physical Metallurgy,* R. W. Cahn (Ed.), North Holland, Amsterdam, Netherlands (1970), p. 1011.
24. Kocks, U. F., *Micromechanical Modelling of Flow and Fracture,* ASME, Troy, New York (June, 1975).
25. Mohamed, F. A., and Langdon, T. G., *J. Appl. Phys.,* **45**, 1965 (1974).
26. Alexander, H., and Haasen, P., *Acta Met.,* **9**, 1001 (1961).
27. Sestak, B., and Seeger, A., *The Microstructure and Design of Alloys,* Inst. of Met. and Iron and Steel Inst., London, England (1973), p. 563.
28. Taylor, G., Vesely, D., and Christian, J. W., *loc. cit.,* p. 1.
29. Neumann, P., *Acta Met.,* **17**, 1219 (1969).
30. Rudolph, G., Haasen, P., Mordike, B. L., and Neumann, P., *Proceedings of First International Conference on Fracture,* Sendai, Japan (1966), p. 501.
31. Wilsdorf, H.G.F., and Fourie, J. T., *Acta Met.,* **4**, 271 (1956).
32. Neumann, R., *Z. Metallk.,* **66**, 26 (1975).
33. Mughrabi, H., and Wuthrich, Ch., *Phil. Mag.,* **33**, 963 (1976).

DISCUSSION on Paper by P. Haasen

McMAHON: In the copper-base alloys examined by Suzuki, the planarity
of slip varies greatly with the alloy content, being wavy in Cu-Ni and
quite planar in Cu-Al. It appears that Suzuki's analysis of k_y, which considers solute effects at grain boundaries, does not take that feature into
account.

HAASEN: This is correct. Suzuki assumes for the initial stage of deformation that slip is planar in both alloys.

HORNBOGEN: Suzuki's treatment of the effect of solute elements on the
Petch-Hall parameter k_y results in a continuous increase with solute concentration. Bcc interstitial alloys show an upper limit of the k_y value. Can
this be due to a change in mechanism of the dislocation-grain boundary
interaction if the boundary is hardened to a certain extent?

HAASEN: It is quite possible that the mechanism of slip transfer into the
next grain changes as the boundary becomes "hardened". On the other
hand, Suzuki's theory predicts a continuous increase only with the
difference in solute concentrations at favorable and unfavorable boundary
sites.

McCLINTOCK: What progress has been made in, and what are the
prospects for, predicting (1) the stress-strain relations for low-cycle fatigue
(Bauschinger effect) and (2) the slip line length (limiting the applicability
of classical continuum plasticity) in such solution-hardened alloys?

HAASEN: (1) While there has been progress, as indicated in the paper, in
modelling stress-strain relations in unidirectional tests as there has been in
high-cycle fatigue testing, I am afraid that this is not the case for low-
cycle fatigue. Understanding of the Bauschinger effect depends on a better
modelling of the work-hardened state in its stability to stress reversal. (2)
There is a certain understanding of the length of slip lines in terms of the
interaction of dislocations in different slip systems. Slip lines are longer in
the coarse-slipping solute solutions.

SHEWMON: Have you treated at all the cases of *mobile* solutes, e.g., carbon in iron? Can you do a first-order treatment of the effect of solute
clustering or short-range order—phenomena we know are present in most
or all alloys?

HAASEN: H. Gleiter has tried to explain the plateau in terms of dynamic pinning of dislocations by mobile solute atoms. While such an effect might occur at the high-temperature end of the plateau, it does not explain the plateau characteristics in general. (Further work in this direction has been done by Schwink in Braunschweig). As far as other-than-random solute distributions are concerned, one could treat, e.g., solute clusters as a new entity which is distributed statistically itself. Its hardening effect could then be calculated as described in the paper.

CAMPBELL: What is the physical significance of the limiting dislocation velocity of 5 m/sec which is used in the nondimensional parameter ω?

HAASEN: The limiting velocity is determined by the vibration frequency of the dislocation in front of an obstacle times the range of the solute/dislocation interaction. While the latter is of the order of several lattice parameters, the former is about 10^{10} sec^{-1} according to several recent estimates. This frequency corresponds to a cooperative movement of the atoms making up a dislocation and in fact limits its velocity overcoming solute obstacles.

TIEN: It seems to me from your excellent lecture that there exists a wealth of data on the stress-strain behavior of model, chemistry-controlled solid solutions. You emphasized the extraction of hardening and, to a certain extent, work-hardening understanding from this bank of experimental results. Are there now attempts to extract higher strain and especially fracture and ductility understanding from these same systems? In fact, were most of the original stress-strain tests even carried to fracture and the data stored, or do we indeed have to start from scratch?

HAASEN: I am afraid that most of the mentioned stress-strain tests on solid solutions were not carried up to fracture. The cited paper by U. F. Kocks and related work by Mecking and Luecke in Aachen attempt to provide understanding of higher strain behavior of these materials.

SHERBY: I would like to direct my comments to the two plateaus shown in the figure of τ_y versus temperature. It seems to me that the low-temperature plateau should perhaps be viewed, more generally, as a peak rather than a plateau. I have in mind the classical example of the peak flow stress observed in mild steels. In this instance, it is often considered that the peak observed is due to moving interstitial carbon atoms. Thus, I would like to suggest that another model to explain the first plateau region seen in fcc solid solutions could center on the increasing importance of substitutional solute mobility with increasing temperature as the plateau region is approached.

HAASEN: See my comments on a similar question asked by P. Shewmon.

SHERBY: I was quite intrigued to note the presence of the second plateau which looks like it appears up to T_m as you have sketched it on the board. We have not noted such plateaus in the deformation of polycrystalline solid-solution alloys (e.g., Al-Mg solid-solution alloys). In the latter case, the yield strength (σ_y)-temperature curve is quite readily calculated from known creep theories for solid solutions which accurately predict the experimentally observed decrease in σ_y with increase in temperature. Is it possible that the difference in behavior is due to complex slip processes in polycrystals versus single slip in single crystals?

HAASEN: It is likely that in single crystals the CRSS (divided by the appropriate shear modulus) does not drop continuously up to the melting temperature as it does in polycrystals where grain-boundary sliding may be important.

HIRTH: The form of the Labusch equation presentation by Professor Haasen, $\tau_c b = \int f \rho(f) df$, is of interest with respect to the limiting form of a rigid dislocation. Symmetry alone suggests that long-range effects cancel out, so that only local (near-core) interaction effects influence flow; with a random solute concentration, this should give a linear dependence of τ_c or c and this is borne out, for example, by the near-core Suzuki hardening and the core short-range-order models.

It would appear that in many distribution models (i.e., those excluding very short-range interaction effects r^{-n} with n large), the integral on the right side of the equation would be linear in c. This implies that dependences of τ_c on c for flexible versions of the *same* interaction models arises from flexibility via the f and c dependence of L_c. I wonder if this sort of development has been tested and whether a linear dependence of τ_c on c ever emerges from the statistical analysis.

HAASEN: I agree with Professor Hirth that the flexibility of the dislocation is of first importance for the theory of obstacle hardening. Labusch's equation leads to $\tau_c \sim C^{2/3}/E_L^{1/3}$ where E_L is the dislocation line tension. A linear concentration (c) dependence, however, never occurred. Only in the limiting case of point obstacles (Fleischer's equation) does a distance L_c of touched obstacles enter the theory, and this increases as E_L increases. Again the concentration dependence of τ_c is parabolic. The superposition law for the hardening effect of different solutes (i) in this case becomes $\Sigma C_i f_i^2 / (\Sigma C_i f_i)^{1/2}$ (f_i maximum solute interaction force). [See Friedrichs and Haasen, *Phil. Mag.*, **31**, 863 (1975).]. τ_c cannot be expressed here in terms of the binary flow stresses τ_{ci} only.

JAFFEE: The usefulness of a theoretical treatment to an experimentalist working on alloy design is not the a priori calculation of composition and structure, but merely the identification of the important variables and their qualitative effect—e.g., beneficial or detrimental—on some desired property. Soundly based theoretical models are very useful in setting up experimental programs aimed at designing useful alloys, whether or not they quantitatively predict the effects of structural variables on mechanical properties. Also, it is very helpful to know how various hardening mechanisms superimpose theoretically.

HAASEN: I agree.

DEFORMATION-MECHANISM MAPS FOR PURE IRON, TWO AUSTENITIC STAINLESS STEELS, AND A LOW-ALLOY FERRITIC STEEL

H. J. Frost and M. F. Ashby

University Engineering Department
Trumpington Street,
Cambridge, England

ABSTRACT

The construction of deformation-mechanism maps for ferrous alloys is described. Maps for pure iron, 316 and 304 stainless steel, and a ferritic 1 percent Cr-Mo-V steel are presented, and their use is illustrated. The maps are constructed, as far as possible, from model-based constitutive laws which have been fitted to experimental data. They attempt to combine the understanding of the fundamentals of dislocation mechanics and diffusion theory with the observation of the yielding and creep of commercial steels. In this way we retain some of the predictive power that an understanding of fundamentals permits, while giving a good description of the observed behavior of the alloys.

1 INTRODUCTION

If a structure is to be built of a steel, the designer must ensure that the strain appearing during the structure's lifetime is acceptably small. Standard methods permit this at low temperatures, when many materials can be thought of as having a well-defined yield strength below which no flow occurs. But as the temperature is raised, other mechanisms of flow become operative, for some of which only limited data are available. The designer may have to extrapolate from laboratory tests, speeded up to obtain data in a reasonable time span. At this stage, the designer needs some understanding of the fundamental aspects of plasticity in alloys; he can extrapolate properly only if he can identify the dominant mechanism of flow in the structure, and has at his disposal a constitutive equation which properly describes its rate. His problem is made more difficult when the mechanism which dominates in the laboratory test (which may last less than a year) is not dominant in the structure (which may be designed for more than 25 years of use).

A deformation-mechanism map for the steel helps to solve this problem. As shown in Fig. 1, it is a diagram, with axes of stress and temperature, divided into fields within each of which a particular mechanism of flow is dominant—that is, it permits faster flow than does any competing mechanism. Superimposed on the fields are contours of constant strain rate, so that the maps can be used to give a rough idea of the strain rate in the structure. More precisely, the maps relate stress, strain rate, and temperature; if any two are specified, the map can be used to determine the third and to identify the mechanism of flow relating them. There are, of course, other ways of displaying the same information; one other is discussed in the text.

Maps like the one described above have been constructed for a number of pure fcc, bcc, and hcp metals, and for certain nonmetals.[1-4] This paper describes their construction for pure iron and for three commercial steels: AISI 316 and 304 stainless steels (see Table I), and a precipitation-hardened 1 percent Cr-Mo-V steel (see Section V). Such maps are constructed from rate equations, one describing each mechanism of flow, which are adjusted to fit experimental data. An understanding of flow in pure iron is fundamental to the understanding of steels, though the resulting map is complicated by the two allotropic transformations and the magnetic transformation. The two stainless steels are simpler, having a single, non-magnetic, structure. The ferritic steel, like iron, is complicated by a phase transformation, and by a new variable: the state of its heat treatment.

An example of the use of the maps in engineering design is presented.

The maps presented here are the best that we can do with the present understanding of flow and with the data available to us, but they are far

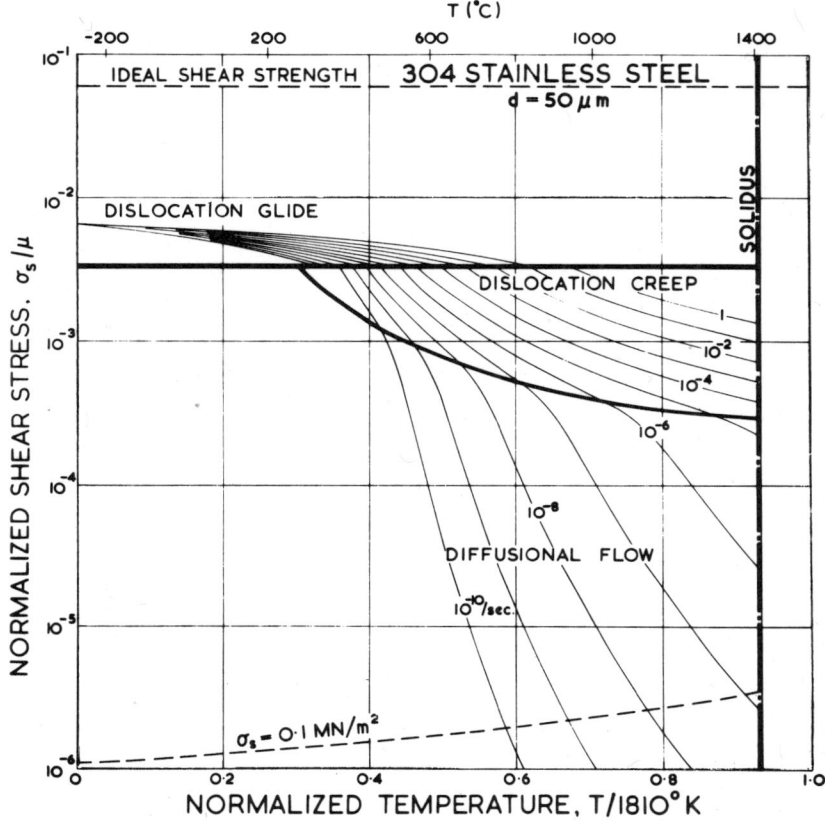

Fig. 1. Map for 304 stainless steel with a grain size of 50 μm.

from perfect or complete. Both the equations in Section 2 and the maps constructed from them must be regarded as first approximations only. They may be used to guide design, but should not be used as the sole basis of it.

2 CONSTRUCTION OF THE MAPS

2.1 The Macroscopic Variables

The maps relate three macroscopic variables: stress, strain rate, and temperature. Commonly, stress and temperature are the independent variables, and it makes sense to use these as the axes of the plot (as in Fig. 1); but any pair can be used, the third then being displayed as a set of contours on the map (examples are shown in Section 4).

Structures are commonly subjected to multiaxial stress states. We have used an equivalent shear stress

$$\sigma_s = \left\{ 1/6[(\sigma_1 - \sigma_2)^2 + (\sigma_2 - \sigma_3)^2 + (\sigma_3 - \sigma_1)^2] \right\}^{1/2} \qquad (1)$$

and an equivalent shear strain rate

$$\dot{\gamma} = \left\{ 6\,[(\dot{\epsilon}_1 - \dot{\epsilon}_2)^2 + (\dot{\epsilon}_2 - \dot{\epsilon}_3)^2 + (\dot{\epsilon}_3 - \dot{\epsilon}_1)^2] \right\}^{1/2}, \qquad (2)$$

where σ_1, σ_2, and σ_3 are the principal strains and $\dot{\epsilon}_1$, $\dot{\epsilon}_2$, and $\dot{\epsilon}_3$ are the principal strain rates. In pure shear, σ_s and $\dot{\gamma}$ become identical with the shear stress and shear strain rate; the maps can be thought of as directly describing the results of such tests. In simple tension (the way in which most creep data is obtained),

$$\sigma_s = \frac{\sigma_1}{\sqrt{3}} \text{ and } \dot{\gamma} = \sqrt{3}\,\dot{\epsilon}_1 \quad . \qquad (3)$$

The maps can be used directly to describe uniaxial tension either by using Eq. 3, or by shifting the stress scale upward by a distance $\sqrt{3}$ and by relabelling all contours of strain rate as $\sqrt{3}$ times the indicated value of strain rate.

Fundamental similarities and differences among materials are brought out by the use of normalized variables: stress divided by shear modulus (σ/μ) and homologous temperature (T/T_M), where T_M is the melting point of a pure element or compound, or some appropriate temperature—say, the solidus for an alloy. The absolute values of stress and temperature are also given on the map: the centigrade scale across the top gives the temperature, and the stress calibration line, labelled $\sigma_s = 0.1$ MN/m^2, near the bottom allows absolute shear stress to be read. This line is curved because the normalizing parameter—the modulus—varies with temperature. The line itself shows where the stress is 0.1 MN/m^2; by displacing it upward by factors of 10, using the scale on the abscissa, lines for $\sigma_s = 1$, 10, and 100 MN/m^2 can be drawn.

2.2 The Rate Equations

The maps are constructed from rate equations: relations between $\dot{\gamma}$, σ_s, T, and internal-structure parameters which depend on grain size, precipitate size and spacing, dislocation density, and arrangement, and so on. They take the form

$$\dot{\gamma} = f(\sigma_s, T, S_i) \quad , \tag{4}$$

where S_i are the i internal parameters.

We now list the rate equations for each mechanism in turn. Some differ in detail from those used in the first paper on this topic[1], and all are discussed from the viewpoint of their application to iron and steels.

(a) **Collapse**. The ideal shear strength defines a stress level above which flow of a defect-free material (or one in which all defects are pinned) becomes catastrophic: its crystal structure becomes mechanically unstable. This is described by

$$\dot{\gamma}_1 = \infty \text{ when } \sigma_s > \tau_{\text{IDEAL}}$$

$$\dot{\gamma}_1 = 0 \text{ when } \sigma_s \leqslant \tau_{\text{IDEAL}} \quad . \tag{5}$$

For ferrite and δ-iron[5], we have used $\tau_{\text{IDEAL}} = 0.11\ \mu$; for austenite[6], $\tau_{\text{IDEAL}} = 0.061\ \mu$. It defines an absolute upper bound for the strength of the steel.

(b) **Dislocation Glide**. Few steels have strengths that approach this upper bound, because an alternative mechanism of flow—that permitted by the conservative motion of dislocations—is available. In austenite, and in disperson-strengthened ferrite, this motion is obstacled-limited: it is the interaction of mobile dislocations with other dislocations, with solute or precipitates, or with grain boundaries, which determines the rate of flow. By contrast, flow in pure ferrite at low temperatures (below about 0 C) is limited not by obstacles of this sort, but by the interaction of the mobile dislocations with the atomic structure itself. This Peierls-resistance- or lattice-resistance-limited glide appears as a characteristic feature of the maps for pure iron (see Section 3) and for other bcc metals[2], but does not appear on maps for austenitic stainless steels and is almost entirely swamped by obstacle strengthening in the low-alloy ferritic steel.

The detailed physical understanding of flow by glide is incomplete, in that we have only a poor picture of how the structure (particularly the dislocation structure) evolves with time or strain. But, if the structure is assumed to be constant, all glide mechanisms can be described by an equation of the form[7]

$$\dot{\gamma} = \dot{\gamma}_o \exp - \frac{\Delta F}{kT} \left\{ 1 - \left(\frac{\sigma_s}{\hat{\tau}} \right)^p \right\}^q \ ,$$

where ΔF is the activation energy for glide, k is Boltzmann's constant. T is the absolute temperature, and $\hat{\tau}$ is, to an adequate approximation, the flow stress at 0 K. The preexponential $\dot{\gamma}_o$ is less important than the exponential term. When ΔF is large (as it is for obstacle-controlled glide in austenite or in dispersion-strengthened ferrite), $\dot{\gamma}_o$ can be assumed to be constant, and equal to 10^6/second. When ΔF is smaller (as it is for lattice-resistance-controlled glide in ferrite), the variation of $\dot{\gamma}_o$ with stress becomes important, and it is a better approximation[7] (which we shall use) to set

$$\dot{\gamma}_o = 10^{11} \left(\frac{\sigma_s}{\mu} \right)^2 / \sec \ .$$

The powers p and q depend on the details of the shape and distribution of the obstacles, or the details of the lattice resistance. They are bounded by the limits

$$0 \leqslant p \leqslant 1$$

$$1 \leqslant q \leqslant 2 \ ;$$

p and q determine the curvature of the strain-rate contours in the glide field. When ΔF is large, their choice makes very little difference: we have set both equal to unity. But when ΔF is smaller, the choice becomes important. We have found[2,7] that p = 3/4, p = 4/3 gives the best fit to data for tungsten, and we shall use it here.

Accordingly, we use the following rate equations for glide, taking the one that leads to the lower strain rate as limiting the actual glide process:

$$\dot{\gamma}_2 = 10^6 \exp - \frac{\Delta F}{kT} \left(1 - \frac{\sigma_s}{\hat{\tau}} \right) \tag{6}$$

(Obstacle-limited flow in ferrite and austenite)

$$\dot{\gamma}_3 = 10^{11} \left(\frac{\sigma_s}{\mu} \right) \exp - \frac{\Delta F_p}{kT} \left(1 - \left(\frac{\sigma_s}{\hat{\tau}_p} \right)^{3/4} \right)^{4/3} \tag{7}$$

(Lattice-resistance-limited flow in ferrite),

where the subscript p has been used in the second equation to show that it refers to the Peierls resistance.

As mentioned above, ΔF depends on the strength of an individual obstacle. We have found that the data (discussed in detail in the Appendix) are best described by a large value of ΔF for the stainless steels of 0.5 μb^3, and for the dispersion-strengthened ferrite of 2 μb^3. This leads to a flow stress in the glide region which is nearly independent of temperature and which is not sensitive to the exact value of ΔF: it depends almost entirely on $\hat{\tau}$ (described below). But for lattice-resistance-limited flow in pure ferrite, the precise magnitude of ΔF is more important; we have determined it by fitting experimental data to Eq. (7), and found a value of 0.1 μb^3.

$\hat{\tau}$ (or $\hat{\tau}_p$) is the flow stress in the absence of thermal activation. For a strong dispersion, it is about $\mu b / l$ (where l is the obstacle spacing); for a work-hardened material, it is about 0.3 $\mu b \sqrt{\rho}$ (where ρ is the dislocation density). We obtained it by extrapolating data for the normalized flow stress to 0 K. To measurements made on single crystals, a Taylor factor must be applied to the extrapolated value of $\hat{\tau}$ or $\hat{\tau}_p$ before it, and any other single-crystal flow stresses, can be used to construct the sort of maps shown here. For ferrite, a factor $\overline{M}$ = 2.9 $\pm$ 5 percent converts the critical resolved shear stress to a polycrystal, tensile flow stress[8]; for austenite, the factor is 3.06. Dividing this by $\sqrt{3}$ to convert to equivalent flow stress of a polycrystal in shear, we obtain the correction factors 1.67 and 1.77, which we have used to convert single-crystal data to the equivalent polycrystal shear strength.

(c) **Power-Law Creep**. Above 0.3 T_m ($\approx$300 C), dislocations in iron and steel acquire a new degree of freedom: they can climb as well as glide. If a gliding dislocation is held up by discrete obstacles, climb may release it, allowing it to glide to the next set of obstacles where the process is repeated. The glide step in its motion is responsible for almost all the strain, although its average velocity is determined by the climb step. We refer to mechanisms which are based on this climb-plus-glide sequence as climb-controlled creep.[9]

The important feature which distinguishes these mechanisms from those discussed in earlier sections is that the rate-controlling process, at an atomic level, is the diffusive motion of single ions or vacancies to or from the climbing dislocations, rather than the activated glide of the dislocation itself. Above about 0.4 T_M ($\approx$450 C), lattice diffusion is the dominant mode of mass transport. Models for power-law creep[9] in this regime lead to the rate equation

$$\dot{\gamma} = A_s \frac{D \mu b}{kT} \left(\frac{\sigma_s}{\mu} \right)^n , \tag{8}$$

where D is the lattice-diffusion coefficient, and A_s and n are dimensionless constants.* Over a limited range of stress, up to about 10^{-3} μ, experiments

---

*Equation (8) is usually written in terms of tensile stress and strain rate. Our constant A_s (which relates shear stress to shear strain rate) is related to the equivalent constant A which appears in tensile forms of this equation by $A_s = (\sqrt{3})^{n+1}$ A.

are well described by this equation. But none of the models convincingly explain the observed values of n and A_s which we have treated as material constants, to be set by fitting Eq. (8) to experimental data for iron and for each steel.

This was the only climb-creep equation incorporated in the first report.[1] We have found it incapable of explaining certain experimental data for creep in pure fcc and bcc metals.[2] To do so it is necessary to assume that transport of matter by dislocation-core diffusion contributes significantly to the overall diffusive transport of matter, and—under certain circumstances—becomes the dominant transport mechanism. Robinson and Sherby[10] suggested that this might explain the lower activation energy for creep at lower temperatures. We have incorporated the contribution of core diffusion by defining an effective coefficient[10,11]:

$$D_{eff} = D_v (1 - f_c) + D_c f_c , \qquad (9)$$

where D_c is the core diffusion coefficient and f_c is the fraction of atom sites associated with dislocation cores. Its value is determined by the dislocation density, ρ, as $f_c = a_c\rho$, where a_c is the cross-sectional area of the dislocation core in which fast diffusion is taking place.

Experimentally, it is possible to measure only the quantity a_cD_c. In general, D_c is about equal to D_b, the grain-boundary diffusion coefficient, if a_c is taken as about $5b^2$. By using the common experimental observation[12] that $\rho \approx 0.1 \ (\sigma_s/\mu b)^2$, the effective diffusion coefficient becomes:

$$D_{eff} = D_v \left[1 + \frac{10a_c}{b^2} (\sigma_s/\mu)^2 \frac{D_c}{D_v} \right] \qquad (10)$$

which, when inserted into Eq. (8), yields a rate equation for dislocation creep which includes the contribution of core diffusion:

$$\dot{\gamma}_4 = \frac{A_s D_{eff}\mu b}{kT} (\sigma_s/\mu)^n . \qquad (11)$$

Equation (11) is really two rate equations. At high temperatures and low stresses, lattice diffusion is dominant; we have called the resulting field high-temperature creep (H.T. creep). At lower temperatures, or higher stresses, core diffusion becomes dominant, and the strain rate varies as σ_s^{n+2} instead of σ_s^n; this field appears on the maps as low-temperature creep (L.T. creep).

Modified in this way, the power-law creep equation describes creep over a wide range of temperature in a number of fcc and bcc metals[2], including pure iron (see Section 3). In alloys, however, the contribution of core diffusion seems to be suppressed. Equation (11) predicts a large low-temperature creep field for stainless steels, yet the data reviewed in Section 4 do not indicate such a field. This seems to be the result of the solid solution; a similar effect is found in nickel-rich Ni-Cr alloys.[3] Two factors may account for the effect. First, the solid solution may lower the dislocation-core diffusion coefficient. We cannot verify this because we have no direct data on core diffusion for isolated dislocations in the alloy. Second, the dislocation density for the alloy may be lower than predicted by Eq. (9); measurements of Challenger and Moteff[13] give some indication of this, and it is consistent with the theory behind Eq. (9). Because of these factors, we have suppressed the contribution of dislocation-core diffusion in the maps for austenitic stainless steels, but have included it for iron and 1 percent Cr-Mo-V steel.

With this additional reservation, Eq. (11) adequately describes steady-state creep at stresses up to about $10^{-3}\ \mu$. Above this stress, the power-law starts to break down; the strain rate exceeds that predicted by the equation. The process is evidently a transition from climb-controlled power-law creep to the glide-controlled flow of Eq. (6). It is scarcely a problem with pure iron, but alloying raises the strength to such a level that for 316 stainless, and the 1 percent Cr-Mo-V ferritic steels, many experimental data lie in the region of power-law breakdown (though most engineering applications do not).

Although there is no firm theoretical model for this regime, there have been various attempts to develop empirical equations (see Jonas, Sellars, and Tegart[14]). Since we have not included these in our treatment of steels, they are not discussed herein.

(d) Diffusional Flow. A stress can induce a diffusional flux of matter through or around the surfaces of grains in iron or steel. This trans- or circum-granular flux leads to strain, provided it is coupled with sliding displacements in the plane of the boundaries themselves. The simple models of the process[15-20] assume that it is diffusion controlled. They are in substantial agreement in predicting a rate equation: if the contributions of both lattice and grain-boundary diffusion are included, the result is:

$$\dot\gamma_s = \frac{42\sigma_s\Omega}{kTd^2}\ D_{\text{eff}}\ ,\tag{12}$$

where

$$D_{\text{eff}} = D_v\left\{1 + \frac{\pi\delta}{d}\frac{D_B}{D_v}\right\}\ .\tag{13}$$

Like the equation for climb-controlled creep, it is really two equations. At high temperatures, when lattice diffusion controls the rate, the flow is known as Nabarro-Herring creep. At lower temperatures, when grain-boundary diffusion takes over, it is often called Coble creep. Equation (12) is based on a simplified model: it neglects the kinetics involved in detaching vacancies from grain-boundary sites and then reattaching them, which may be particularly important for the dispersion-strengthened ferrite.[21,22] Lacking adequate data, we have not included this in the maps shown here.

(e) **Other Mechanisms of Flow**. Mechanical twinning occurs in pure ferrite at low temperatures. Dynamic recrystallization leads to accelerated creep at high temperatures and strain rates. Various transient and anelastic mechanisms of flow can give limited amounts of strain under appropriate conditions. Superplastic flow may be a distinct mechanism, though we believe it to be a type of diffusional flow. None of these is included in the maps presented in this paper, although they could be if sufficiently good models and data were available.

(f) **Construction of the Diagrams**. The problem of the superposition of flow mechanisms is discussed elsewhere.[2,3,7] We have combined the rate equations discussed above to give the net strain rate:

$$\dot{\gamma} = \dot{\gamma}_1 + [\text{greatest of } \dot{\gamma}_4 \text{ (and least of } \dot{\gamma}_2 \text{ and } \dot{\gamma}_3)] + \dot{\gamma}_5 \quad , \qquad (14)$$

where $\dot{\gamma}_4$ and $\dot{\gamma}_5$ each describe the sum of two additive mechanisms. A mechanism is dominant if it contributes more to $\dot{\gamma}$ than does any other mechanism. A field boundary is then defined as the set of values of σ_s and T at which a change of dominant mechanism occurs. The contours of constant strain rate are obtained by (numerically) solving Eq. (14) for σ_s as a function of T at constant strain rate. This and the other type of map shown here are most conveniently plotted directly by a simple computer program. The data required to do this are presented in Table I, and are documented in the Appendix.

3 PURE IRON

The map for pure polycrystalline iron with a grain size of 0.1 mm is presented in Fig. 2. It has three sections, corresponding to the three allotropic forms of iron: α, γ, and δ. The α region shows a dislocation-glide field, in much of which a lattice resistance limits the strain rate; a power-law creep field, with core diffusion important at the lower temperature end; and a field of diffusional flow, with boundary diffusion important at the low-temperature end (to the left of the vertical broken line).

The creep fields are truncated at 910 C by the phase change to austenite, where the field boundaries and the strain-rate contours suffer a

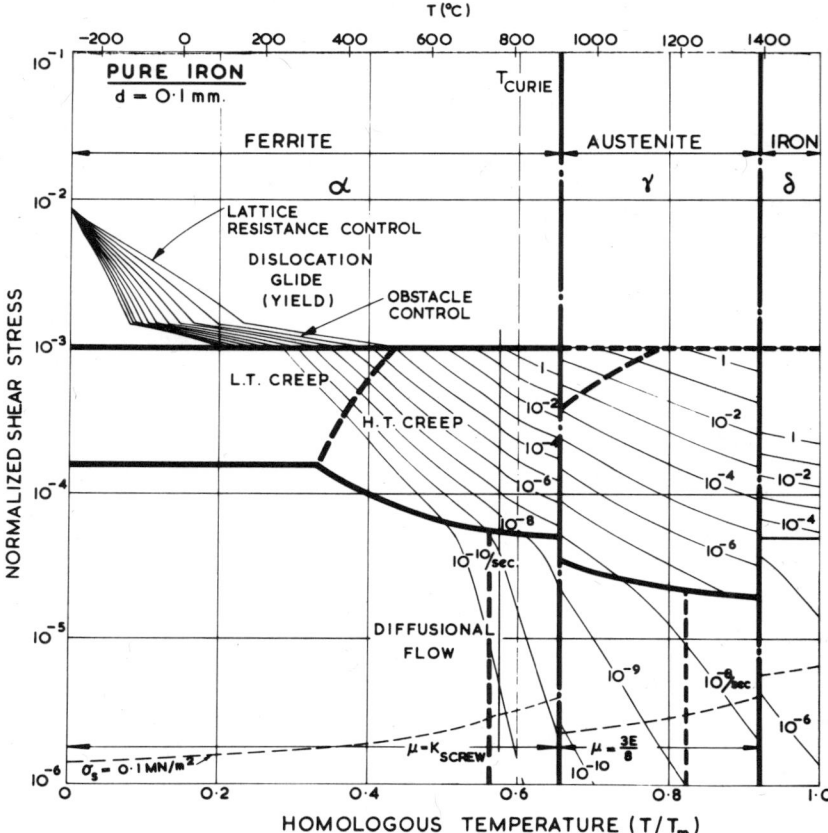

Fig. 2. Map for pure iron with a grain size of 100 μm.

sharp discontinuity. This is because all the physical properties (lattice parameter, Burgers vector, modulus, diffusion coefficients, etc.) change sharply here and again at 1390 C—the most important change being that of a factor of 100 in the diffusion coefficient.

Austenite shows a power-law creep and (for this particular grain size) two diffusional flow fields: boundary-diffusion controlled flow to the left of the broken line and lattice-diffusion controlled flow to the right. Delta-iron shows power-law creep and flow controlled by lattice diffusion only. Flow in iron is further complicated by the loss of ferromagnetism at the Curie temperature, 770 C, which is marked on the map; its effects are discussed below.

The parameters used to construct this map are given in Table I, and are documented in the Appendix. An overview of the considerable body of experimental data from which these parameters were derived is presented in Figs. 3 (which is drawn on the same scale as Fig. 2 and can be superimposed

Table I. Parameters Used in Constructing Creep Maps

	Units	α-Fe	γFe	δFe	1% CrMo Steel	304 Stainless steel	316 Stainless Steel
$\Omega^{(a)}$	10^{-29} m³	1.18	1.21	1.18	1.18	1.21	1.21
$b^{(a)}$	10^{-10} m	2.48	2.58	2.48	2.48	2.58	2.58
T range	°K	0-1184	1184-1665	1665-1810	0-1753	0-1680	0-1680
$\mu_c^{(c)}$	10^4 MN/m²	Ferro 6.4 | Para 6.92	8.1	3.9	8.1	8.1	8.1
$\frac{1}{\mu_0}\frac{\partial\mu}{\partial T}$	10^{-4}/°K	4.5 | 7.23	5.05	4.0	6.0	4.7	4.7
D_{0v}	10^{-4} m²/sec	2.0 | 1.9	0.18	1.9	Ferro 2.0 | Para 1.9	0.37	0.37
Q_v	kJ/mole	251.1 | 239.4	269.9	239.4	251.1 | 239.4	279.6	279.6
δD_{0B}	10^{-13} m³/sec	1.12 × 10³	0.75	11.2	11.2	2	2
Q_B	kJ/mole	173.7	159.0	173.7	173.7	167.4	167.4
$a_c D_{0c}$	10^{-23} m⁴/sec	1	1	1	0.1	—	—
Q_c	kJ/mole	173.7	159.0	173.7	173.7	—	—
n	—	6.9	4.5	6.9	6.0	7.5	7.9
$A^{(d)}$	—	7 × 10¹³	4.3 × 10⁵	7.0 × 10¹³	1.1 × 10⁴	1.5 × 10¹²	1.0 × 10¹⁰
$\dot\gamma_p$	sec⁻¹	10¹¹	—	—	10¹¹	—	—
ΔF_k	10^{-20} joules	10.0	—	—	10.0	—	—
$\hat\tau_p$	10^2 MN/m²	6.4	—	—	6.4	—	—
$\dot\gamma_0$	sec⁻¹	10⁶	—	—	10⁶	10⁶	10⁶
ΔF	—	0.5 μb³	—	—	2.0 μb³	0.5 μb³	0.5 μb³
$\hat\tau/\mu$	—	1.653 × 10⁻³	—	—	6.2 × 10⁻³	6.45 × 10⁻³	6.45 × 10⁻³

(a) Temperature dependence of b and Ω is neglected.

(b) All maps are normalized to 1810K = T_m for pure δ-Fe.

(c) $\mu = \mu_0[1 - (T-300)\frac{1}{\mu_0}\frac{\partial\mu}{\partial T}]$ for all except ferromagnetic α-Fe.

(d) These values are given for the tensile stress—tensile strain-rate equation. For the shear equation we use $A_s = (\sqrt{3})^{n+1} A$.

on it) and 4. The symbols show the range of stress and temperature each of the listed authors studied. Each number against a symbol is the $\log_{10}$ of the strain rate; this allows a direct comparison between the experimental data of Fig. 3 and our description of them in Fig. 2. Broken lines link observations of a single author, and indicate that intermediate data points, not shown on the figure, exist.

This data plot shows that pure ferrite is well investigated, both at low and high temperatures. Data exist in all creep fields, including that of diffusional flow (see below). The sharp discontinuity in strain rate is apparent at the transition to austenite, which, however, has been less intensively studied than ferrite. There are no creep data for δ-iron.

4 TYPE 304 AND 316 STAINLESS STEELS

Figures 1, 5a, and 6 show maps for AISI 304 and 316 stainless steel for two grain sizes (50 μm and 200 μm), which illustrate how grain size changes them. Both steels are fcc solid solutions over their entire solid range; this and the suppression by the alloying of low temperature, core-diffusion-controlled creep make the diagrams particularly simple.

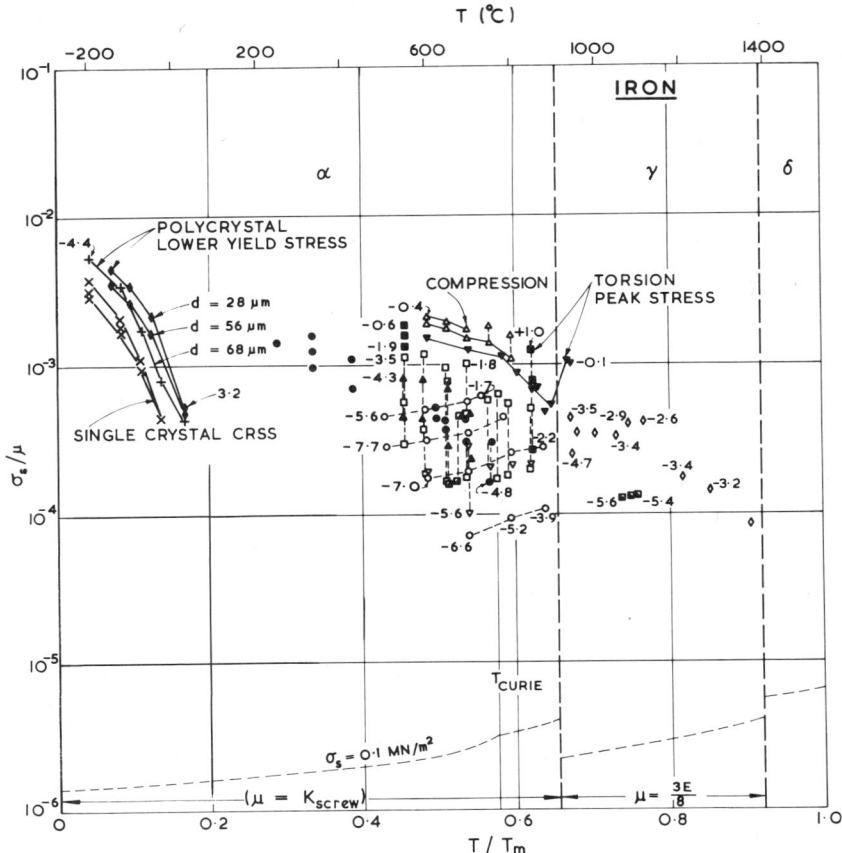

Fig. 3. Experimental data for pure iron.

● Ishida, Y., Cheng, C.-Y., and Dorn, J. E., *Trans. AIME,* **236**, 964 (1966).

○ Karashima, S., Oikawa, H., and Watanabe, T., *Acta Met.,* **14**, 791 (1966).

■ Glover, G., and Sellars, C. M., *Met. Trans.,* **4**, 765 (1973).

□ Čadek, J., Pahutova, M., Ciha, J., and Hostinsky, T., *Acta Met.,* **17**, 804 (1969).

▲ Davies, P. W., and Williams, K. R., *Acta Met.,* **17**, 897 (1969).

△ Uvira, J. L., and Jonas, J. J., *Trans. AIME,* **242**, 1619 (1968).

◪ Murty, K. L., Gold, M., and Ruoff, A. L., *J. Appl. Phys.,* **41**, 4917 (1970).

▽ Watanabe, T., and Karashima, S., *Trans. J.I.M.,* **11**, 159 (1970).

▼ Keane, D. M., Sellars, C. M., and Tegart, W.J.McG., in *Deformation Under Hot Working Conditions,* The Iron and Steel Institute, London, England, Special Report 108 (1968), p. 21.

+ Spitzig, W. A., and Leslie, W. C., *Acta Met.,* **19**, 1143 (1971).

✕ Spitzig, W. A., and Keh, A. S., *Acta Met.,* **18**, 1021 (1970).

◆ Chist, B. W., and Smith, G. V., *Mem. Sci. Rev. Met.,* **65**, 207 (1968).

◇ Feltham, P., *Proc. Phys. Soc. London,* **B66**, 865 (1953).

◣ Hough, R. R., and Rolls, R., *Met. Sci. J.,* **5**, 206 (1971).

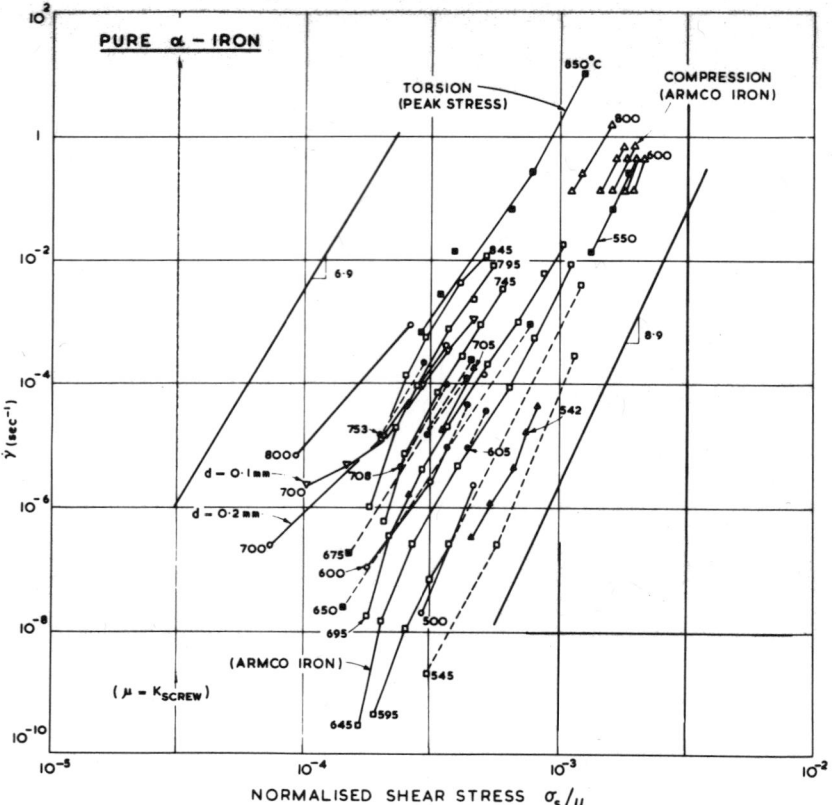

Fig. 4. High-temperature experimental data for pure α-iron.

● Ishida, Y., Cheng, C.-Y., and Dorn, J. E., *Trans. AIME,* **236**, 964 (1966).

○ Karashima, S., Oikawa, H., and Watanabe, T., *Acta Met.,* **14**, 791 (1966).

■ Glover, G., and Sellars, C. M., *Met. Trans.,* **4**, 765 (1973).

□ Čadek, J., Pahutova, M., Ciha, J., and Hostinsky, T., *Acta Met.,* **17**, 804 (1969).

▲ Davies, P. W., and Williams, K. R., *Acta Met.,* **17**, 897 (1969).

△ Uvira, J. L., and Jonas, J. J., *Trans. AIME,* **242**, 1619 (1968).

◪ Murty, K. L., Gold, M., and Ruoff, A. L., *J. Appl. Phys.,* **41**, 4917 (1970).

▽ Watanabe, T., and Karashima, S., *Trans. J.I.M.,* **11**, 159 (1970).

The specifications for these two alloys allow a certain latitude of composition. The maps are based on published creep data, derived from three batches of AISI 316 and two of AISI 304; their compositions are listed in Table II. Both alloys are available in a low-carbon specification designated 316 L and 304 L; these contain less than 0.03 percent carbon.

The parameters used to construct the two sets of maps are given in Table I, and are documented in the Appendix. The data on plasticity and creep in the two steels are more limited than for pure iron, so they have been plotted directly onto Figs. 5 and 6. (Plotting these data onto a map for

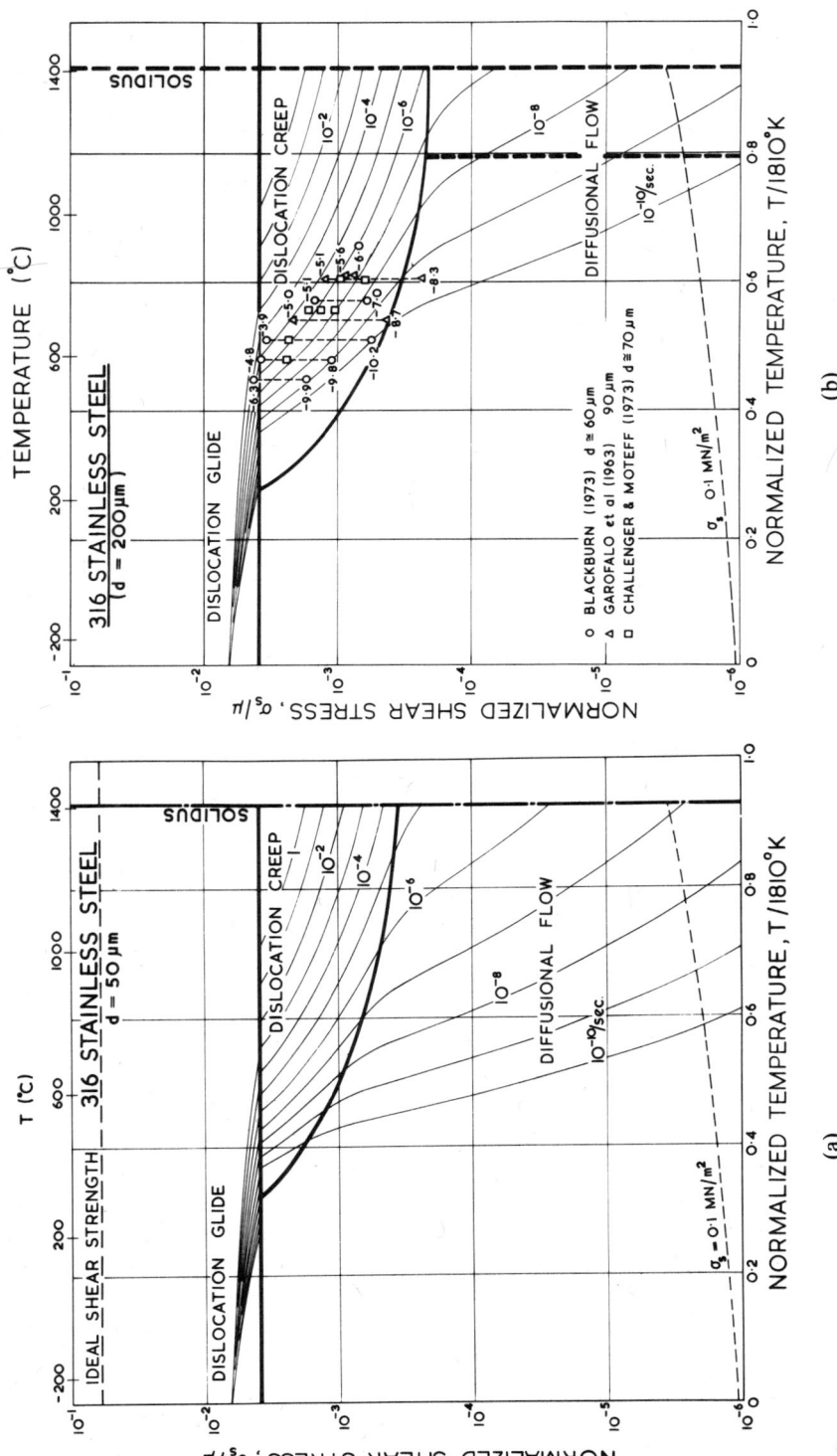

Fig. 5. (a) 316 stainless steel with a grain size of 50 μm; (b) 316 stainless steel with a grain size of 200 μm, showing experimental creep data (for other grain sizes).

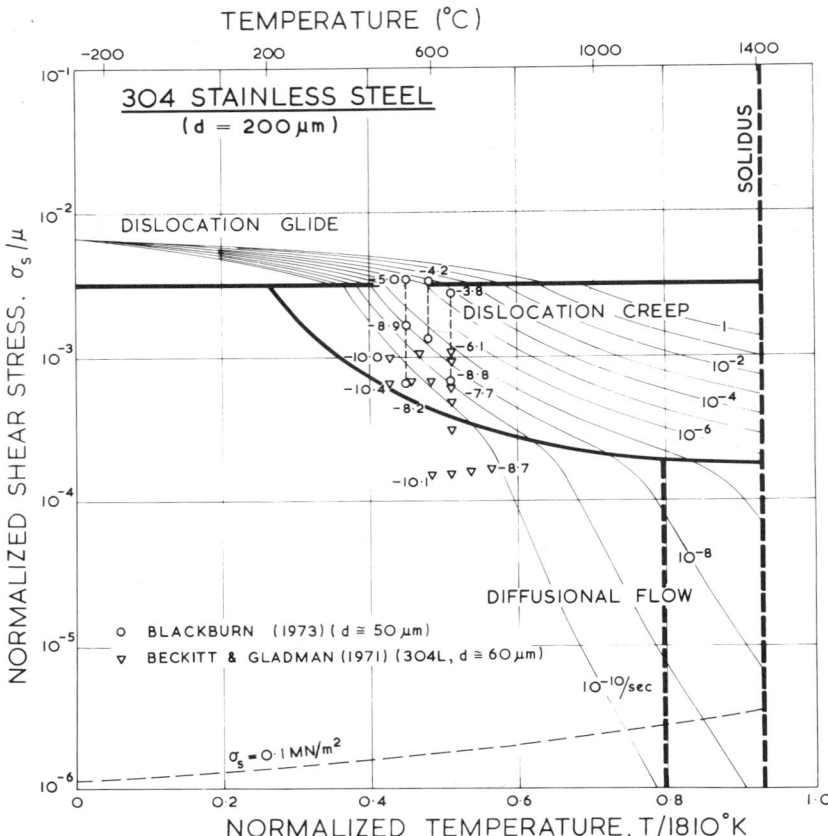

Fig. 6. 304 stainless steel with a grain size of 200 μm, showing experimental creep data (for other grain sizes).

a grain size of 200 μm was an arbitrary choice; the specimens from which the data came had grain sizes of 50 to 100 μm.) As before, the symbols identify the investigator, the numbers are the $\log_{10}$ of the observed shear strain rate, and broken lines link groups of measurements by a single investigator and indicate that more data within the range exist.

Figures 7 and 8 show an alternative form of map: one with axes of strain rate and normalized stress, a more conventional format for the presentation of creep data. The data are shown in these figures also, which have been constructed with a grain size (100 μm) close to that of the specimens from which the data were obtained, and provide a good way of checking the position of field boundaries against experiment.

Though limited in number, the data cover a wide range of stress and (at least for 304) extend well into the diffusional-creep field. The homologous temperature scale has been normalized with respect to the melting point of pure iron; the temperature in centigrade is shown across the top of the maps.

Table II. Stainless Steel Compositions (Weight Percent)

	Cr	Ni	Mo	C	Si	Mn	P	S	N	Ti	Al	B
AISI 316												
Nominal 316, Parr & Hanson (1965)	16-18	10-14	2.0-3.0	0.08 max	1.0 max	2.0 max	0.045 max	0.03 max				
Blackburn (1972)	17.8	13.6	2.4	0.05	0.4	1.7	0.01	0.02	0.04	0.003	0.026	<0.005
Garofalo et al. (1963)	18	11.4	2.15	0.07	0.38	1.94	0.01	0.006	0.05		0.003	
Challenger & Moteff (1973)	18.16	13.60	2.47	0.086	0.52	1.73	0.01	0.006	0.05	<0.005	<0.005	<0.0005
AISI 304												
Nominal 304, Parr & Hanson (1965)	18-20	8.0-12.0		0.08 max	1.0 max	2.0 max	0.045 max	0.03 max				
Blackburn (1972)	18.4	9.7	0.1	0.07	0.5	0.9	0.02	0.01	0.05	0.013		0.0014
Beckitt & Gladman (1971), (304L)	17.7	9.99		0.015	0.31	0.57						

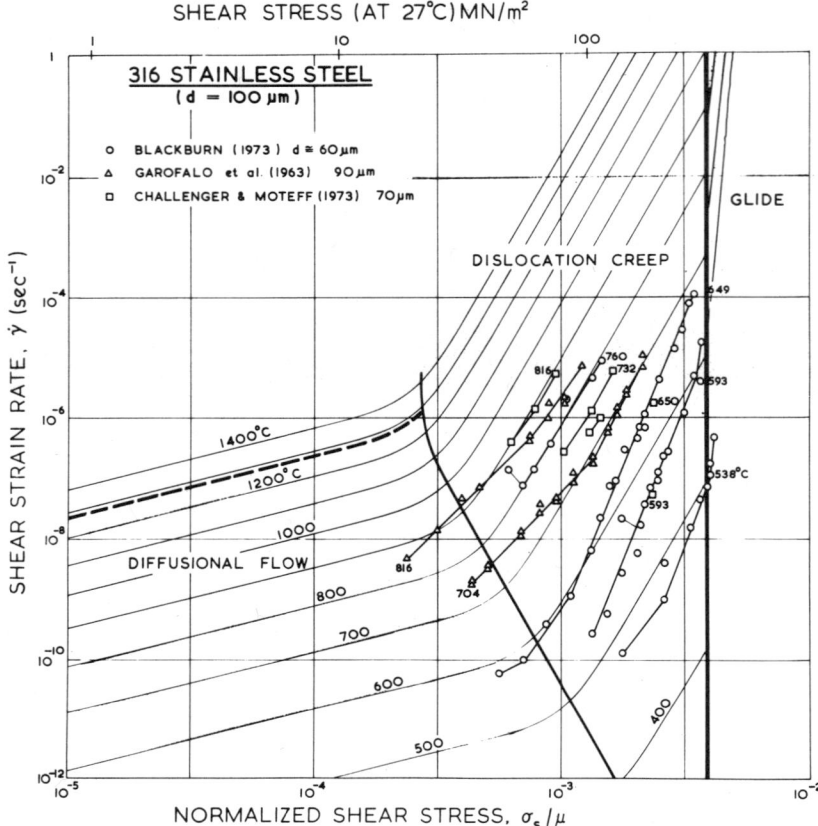

Fig. 7. Map for 316 stainless steel showing contours of constant temperature on axes of strain rate versus normalized stress.

5 A FERRITIC LOW-ALLOY STEEL

A map for a 1 percent Cr-Mo-V steel with a grain size of 50 μm is presented in Fig. 9. This is a heat-treatable steel which derives its strength and creep resistance from a fine dispersion of carbide particles. This introduces a new variable; a map can only describe the steel in one condition of heat treatment and, strictly speaking, only in its virgin state, since aging continues in service. In practice, however, the steel is used in one of a few standard states and is limited to temperatures below about 600 C because of oxidation. Aging in service is of practical importance only above a critical temperature (about 550 C).

The map shown in Fig. 9 describes a particular batch of 1 percent Cr-Mo-V steel, the subject of a joint CEGB–University Collaborative Project.[23] Its composition in weight percent is as follows: C 0.24, Si 0.29, Mn 0.64, Ni 0.21, Cr 1.02, Mo 0.57, V 0.29, S 0.10, and P 0.16. All tests were carried out

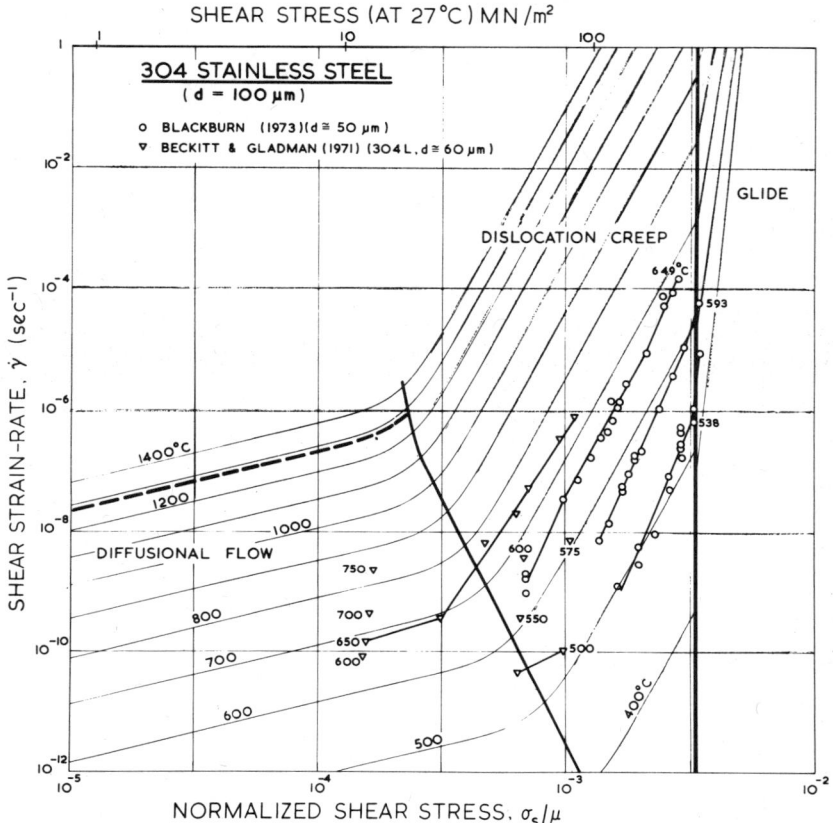

Fig. 8. Map for 304 stainless steel showing contours of constant temperature on axes of strain rate versus normalized stress.

on forged material after a standard heat treatment* designed to give a fine-grain tempered bainite containing a carbide dispersion.

The alloy starts to transform to austenite at about 700 C, and is never used structurally above this temperature, where its strength would be little better than that of pure austenite. For completeness we have shown, in broken lines, the contours and field boundaries for pure austenite (though this ignores the small solution-hardening effect that the alloying elements would exert in the γ phase). The map emphasizes the high flow stress and creep resistance of the dispersion-strengthened ferrite. The alloying suppresses the bcc δ-phase, and therefore it does not appear on the map.

*The heat treatment was as follows: soak at 1000 C; furnace cool to 690 C and hold for 70 hours; air cool. Reheat and soak in a salt bath at 975 C; quench into a second salt bath at 450 C; air cool. Reheat to 700 ± 3 C for 20 hours.

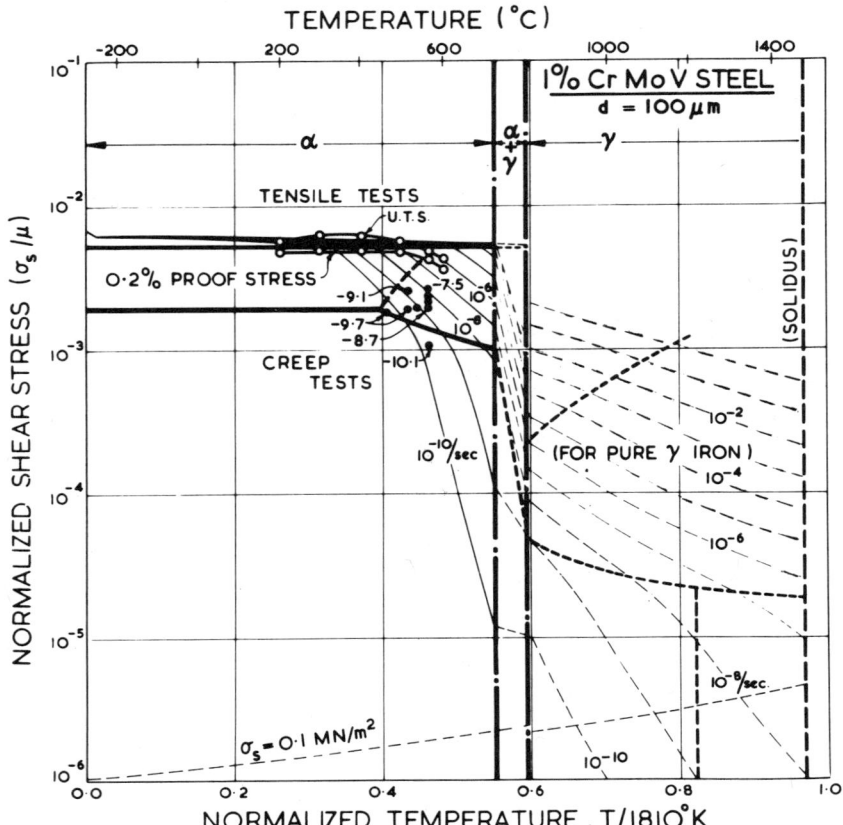

Fig. 9. One percent Cr-Mo-V steel with grain size of 100 μm.

The parameters used to construct the map are given in Table I and are documented in the Appendix. The mechanical data are limited and are plotted in Fig. 9. As before, the numbers are $\log_{10}$ of the observed strain rates. Open circles indicate tensile tests; closed circles indicate creep tests.

The map is a tolerable fit to the data, and is drawn for a grain size close to that of the alloys from which the data came. It shows that the diffusional flow field lies just below the area covered by the data (one point may lie in the field), and suggests that diffusional flow might be the dominant mechanism in any application of the alloy above 450 C. (There is evidence[21,22,24,25] that a dispersion of particles—carbides for example—may suppress diffusional flow; whether they do so in this steel is not known.)

6 APPLICATIONS OF THE MAPS

Design against creep requires, first, that the creep strain appearing during the life of the structure be acceptably small, and, second, that the rupture life exceed (by a suitable safety factor) the design life. To achieve the first, the designer must have at his disposal a constitutive law describing the rate of creep of an element of the structure. We have seen that the constitutive laws describing the various mechanisms of flow differ; if the designer uses the wrong one, the strains he calculates could be in error by several orders of magnitude. The choice is further complicated by the fact, obvious from Figs. 5, 7, and 8, that the laboratory test data may characterize a mechanism which, in long service application, may not be the dominant one.

The following hypothetical case study illustrates these points, and shows how the maps help the designer select the right constitutive law, calculate the approximate strain in elements of the structure, and suggest sensible changes in the metallurgical treatment of the material he uses.

6.1 Case Study: The Creep of Components of a Fast Nuclear Reactor

A much simplified schematic of a section through a hypothetical fast reactor is presented in Fig. 10. Liquid sodium contained in a pressure vessel is pumped through the core and heat exchangers. We shall consider the creep of three components: the pressure vessel, the pipes leading from the pumps to the core, and the reactor skirt. We shall assume that all three are to be made of 316 stainless steel with a grain size of 50 μm.

The pressure vessel operates at the input temperature of the sodium coolant: 390 ± 30 C. Since all the components are suspended from the roof of the reactor, the loading of the vessel is entirely due to the weight of sodium it contains, to buoyancy forces, and to the small pressure of helium gas which covers the sodium. The stresses in the vessel are highest in the sidewall near the bottom, about 8 meters below the sodium surface, where the pressure due to the sodium alone is 0.074 MN/m^2. To this we add the contribution of the overpressure of inert gas (0.007MN/m^2), giving a total of 0.08 MN/m^2. If we take the wall thickness of the pressure vessel* to be 0.0125 meter and its radius to be 6 meters, the axial and circumferential stresses in it are readily calculated. Using Eq. (1) we find the maximum equivalent shear stress to be $\bar{\sigma}_s = 19.3\text{MN}/\text{m}^2$.

---

*The dimensions and pressures given here are hypothetical but are similar to those in current fast-reactor design.

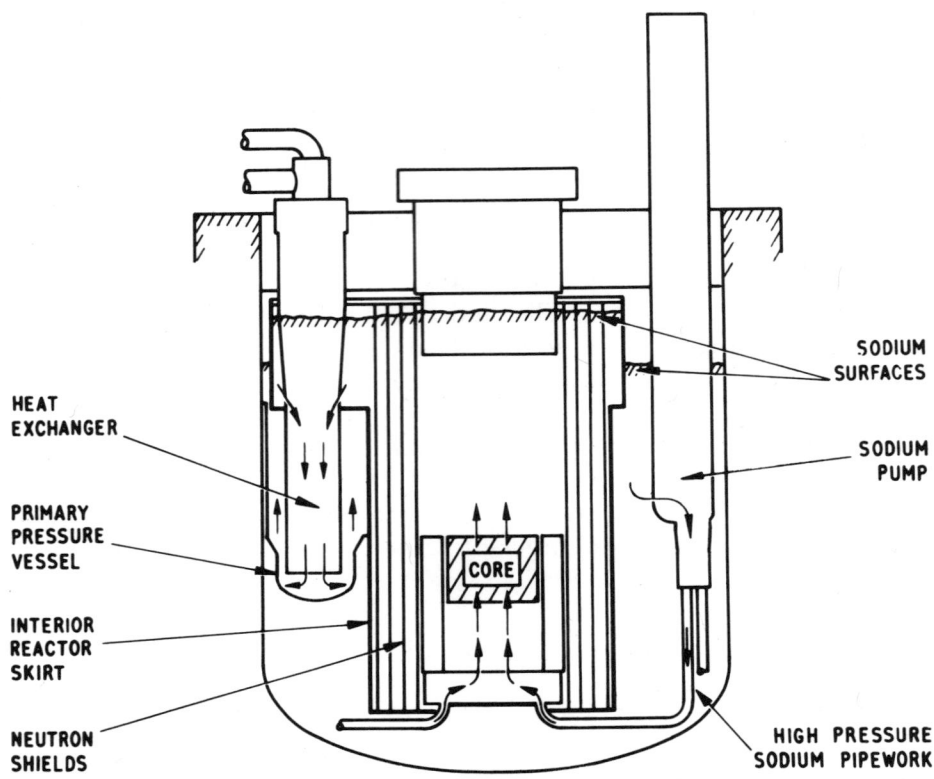

Fig. 10. A hypothetical fast reactor, showing the pressure vessel, the pressure piping, and the reactor skirt.

At the top of the pressure-vessel sidewall, the equivalent shear stress is lower ($\bar{\sigma}$=12.5 MN/m^2). It is also lower in the lower hemispherical part of the pressure vessel because of its shape.

The input sodium pipes, too, operate at 390 ± 30 C. The stresses in them are due mainly to the pressure difference between the sodium inside and outside a pipe, though there is a small additional contribution (which we shall neglect) at a bend in the pipe due to inertial forces set up by the rapidly flowing sodium. If we assume a pipe radius of 0.125 meter, a wall thickness of 0.005 meter, and an internal pressure of 0.83 MN/m^2, then the equivalent shear stress, $\bar{\sigma}_s$, is 10 MN/m^2.

The interior reactor skirt is a cylinder, about 6 meters in diameter, containing the core and associated structure. At least part of it is exposed to the hot sodium leaving the core at a temperature of 580 ± 30 C. It is stressed because of the difference of 2 meters in the sodium level inside and outside the skirt, this difference driving the flow through the heat exchangers; the consequent pressure difference is about 0.02 MN/m^2. If, as before, we take hypothetical values for the skirt radius of 3 meters and a wall thickness of 0.005 meter, we can calculate equivalent shear stress in the skirt as 6.3 MN/m².

This information is plotted on the appropriate map*, Fig. 11. (We have allowed a ±20 percent variation in stress, and a ±30 C variation in temperature, about the values given in the text.) The range in which all laboratory data on AISI 316 were obtained is shown as a shaded box on the map.

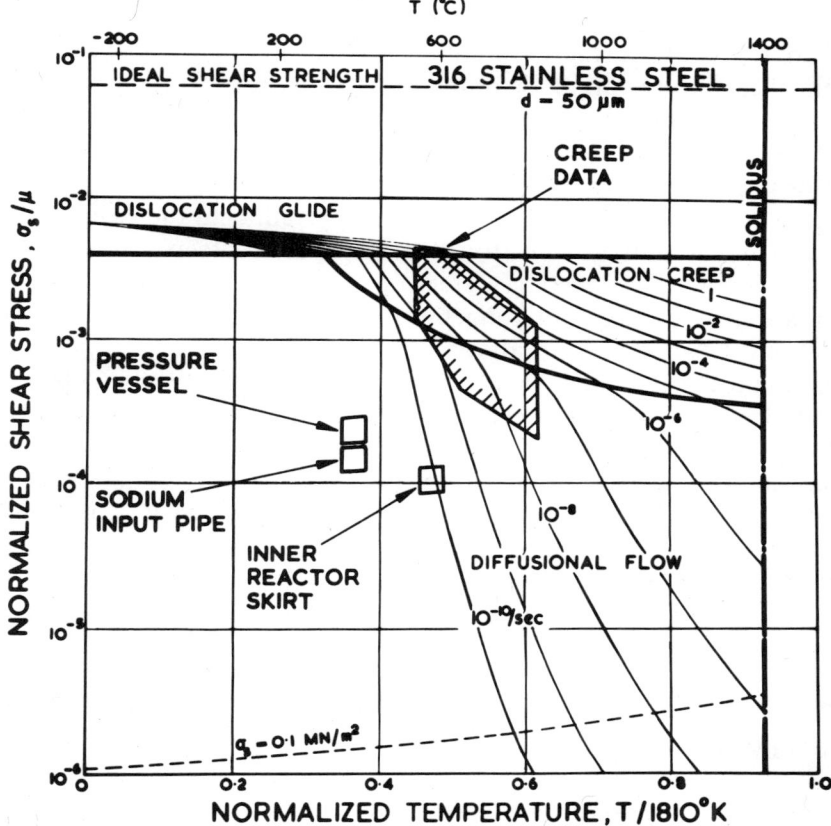

Fig. 11. Map for 315 stainless steel with a grain size of 50 μm, showing the operating conditions of the three reactor components; all three deform by Coble creep.

*These components are not exposed to a heavy flux. If they were, a map incorporating radiation-induced or radiation-enhanced creep would be required.

6.2. Conclusions of the Case Study

All three reactor components discussed above lie well inside the diffusional flow field, in the regime controlled by grain-boundary diffusion. Although our theoretical understanding of this deformation mechanism is fairly sound and we can therefore predict its rate with some confidence, it should be noted that few laboratory data lie in this field and that none lie near the range of service operation of the components. The laboratory data which are available to us are of almost no help in predicting the components steady creep rates.

If the admissible creep strain in any component of the structure were limited to 1 percent during its lifetime (200,000 hours), then the maximum tolerable shear strain rate is of order 2×10^{-11}/second. The pressure vessel and the sodium input pipes are well below this limit. The inner reactor skirt, however, would creep at a rate higher than this if our assumptions about dimensions and pressures were correct.

Increasing the skirt thickness and thereby reducing the stress by a factor of two would not remedy this. An acceptable creep rate in the skirt would be obtained by selecting a steel of the same composition but with a larger grain size. Merely doubling the grain size (to 100 μm) reduces the creep rate by a factor of 8, to about 10^{-11}/second.

The main conclusion is that diffusional flow (particularly Coble creep) is likely to be the dominant mechanism of deformation in the internal parts of fast reactors constructed of 316 stainless steel. Although the theoretical treatment of this mechanism in pure metals is probably adequate, we do not fully understand the effects of alloying on its rate. And it is very poorly studied experimentally; the boundary diffusion coefficients needed to evaluate the theories, for example, are not well known. Appropriate experiments are urgently needed, aimed at the formulation of a reliable constitutive law for this mechanism.

ACKNOWLEDGMENTS

This work was supported by the Science Research Council under Contract Number B1165. We wish to acknowledge helpful discussions with Mr. C.H.A. Townley and Mr. D. A. Wood.

APPENDIX: DOCUMENTATION OF DATA

A-1 Crystallographic and Thermal Data

The crystallographic data (Burgers vector, b; atomic volume, Ω) are from Taylor and Kagle.[26] Their change with temperature, except at a phase transition, is so slight that we have ignored it. Melting points, T_M, and phase-transition temperatures are from Hansen.[27]

A-2 Moduli

The modulus at 300 K (μ_o) and its temperature dependence $1/\mu_o$ ($\partial\mu_o/\partial T$) present special problems. For α-iron we have set μ_o equal to K_{screw}, the modulus which describes the energy of a $1/2\langle 111\rangle$ screw dislocation, which we calculated from the expression given by Hirth and Lothe[28] using the single-crystal constants of Dever.[29] Because of the magnetic transformation at 770 C, the modulus changes with temperature in a nonlinear way. Within the ferromagnetic region (T < 1043 K), we used a modulus given by

$$\mu = \mu_o\,[1 - \frac{1}{\mu_o}\frac{\partial\mu}{\partial T}\,(T-300)] - \underbrace{K_1\,(T-573)^2 - \underbrace{K_o\,(T-923)^2}_{T\,>\,923\ K}}_{T\,>\,573\ K}$$

$$K_1 = 3.2 \times 10^{-2}\ MN/m^2/(K)^2$$

$$K_2 = 2.424 \times 10^{-1}\ MN/m^2/(K)^2\quad.$$

We could find no single-crystal data for γ-iron, and were forced to calculate μ_o and its temperature dependence from $3/7$ E, taking Kosters[29b] measurements of Young's modulus, E. The modulus for α-iron also presents a problem. A linear extrapolation from paramagnetic α-iron goes to zero at a temperature below the melting point. We have therefore extrapolated the polycrystalline modulus of Fe-3.1% Si (as proposed by Lytton[30]) and corrected this to give a screw-dislocation modulus by multiplying it by the ratio of K_{screw} (α-iron, 912 C) to μ(Fe-3.1% Si, 912 C).

The moduli for the stainless steels are based on the data of Blackburn[31]; that for the 1 percent Cr-Mo-V steel is from the CEGB-University Study[23].

A-3 Diffusion Coefficients

Data for the three phases of iron are assembled in Figs. A-1 and A-2. They are dominated by the dramatic discontinuity in D_v and δD_B at the phase transition. A subtler influence is that of the magnetic transformation: diffusion in α-iron does not follow an Arrhenius law but is about three times faster above 770 C than expected from extrapolation from below

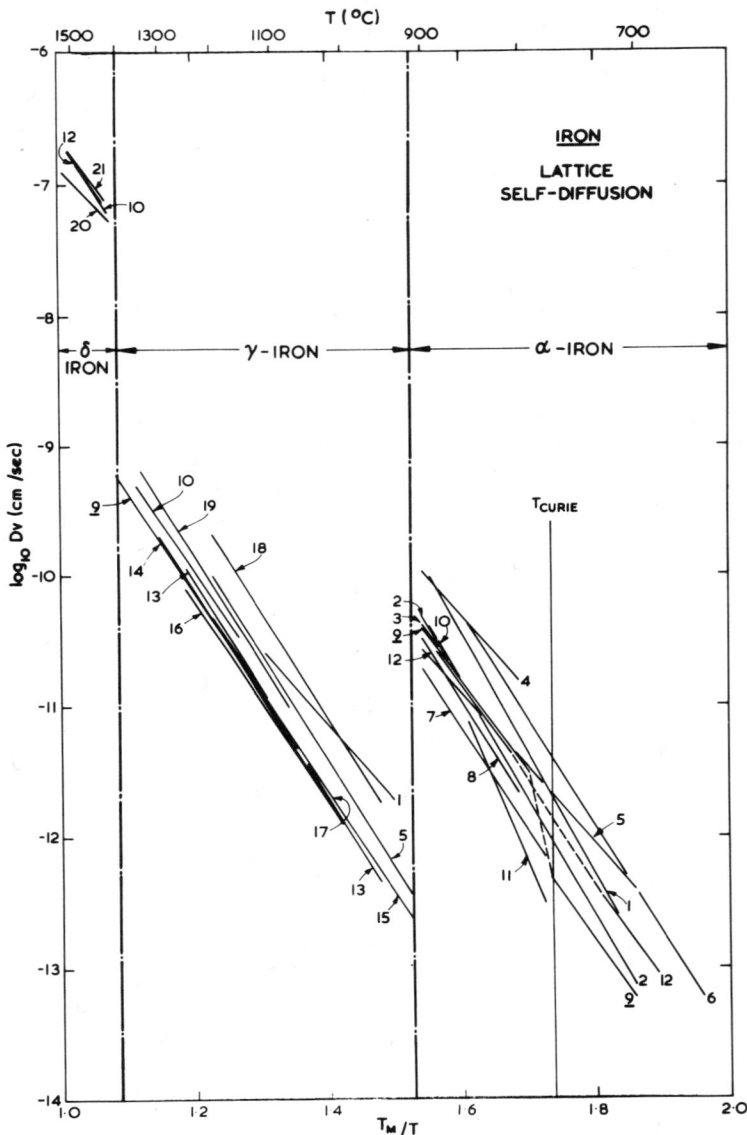

Fig. A-1. Lattice self-diffusion measurements in pure iron.

Legend for Fig. A-1

1 Birchenall, C. E., and Mehl, R. F., *J. Appl. Phys.*, **19**, 217 (1948).
2 Birchenall, C. E., and Mehl, R. F., *Trans. Met. Soc. AIME*, **188**, 144 (1950).
3 Buffington, F. S., Bakalar, I. D., and Cohen, M., *Physics of Powder Metallurgy*, McGraw-Hill, New York, N.Y. (1951), p. 92.
4 Gruzin, P. L., *Probl. Metalloved. i Fiz. Met.*, **3**, 201 (1952).
5 Zhukhovitskii, A. A., and Geodakyan, V. A., *Primenenie Radioaktivn. Izotopov v Metallurg. Sb.*, **34**, 267 (1955); AEC-tr-3100, *Uses of Radioactive Isotopes in Metallurgy Symposium XXXIV*, Pt. 2, p. 52.
6 Golikov, V. M., and Borisov, V. T., *Problems of Metals and Physics of Metals, 4th Symposium* (1955), Consultants Bureau, New York, N.Y. (1957); AEC-tr-2924 (1958).
7 Leymonie, C., and Lacombe, P., *Rev. Met. (Paris)*, **55**, 524 (1958); *Compt. Rend.*, **245**, 1922 (1957); *Metaux (Corrosion-Ind.)*, **34**, 457 (1959); *Metaux (Corrosion-Ind.)*, **35**, 45 (1960).
8 Borg, R. J., and Birchenall, C. E., *Trans. Met. Soc. AIME*, **218**, 980 (1960).
9 Buffington, F. S., Hirano, K., and Cohen, M., *Acta Met.*, **9** (5), 434 (1961).
10 Graham, D., and Tomlin, D. H., *Phil. Mag.*, **8**, 1581 (1963).
11 Amonenko, V. M., Blinkin, A. M., and Ivantsov, I. G., *Fiz. Metal. i Metalloved.*, **17** (1), 56 (1964); *Phys. Metals Metallog. (USSR)*, (English transl.), **17** (1), 54 (1966).
12 James, D. W., and Leak, G. M., *Phil. Mag.*, **14**, 701 (1966).
13 Gruzin, P. L., *Izv. Akad. Nauk SSSR Otd. Tekhn. Nauk*, **3**, 383 (1953).
14 Mead, H. W., and Birchenall, C. E., *Trans. Met. Soc. AIME*, **206**, 1336 (1956).
15 Bokshtein, S. Z., Kishkin, S. T., and Moroz, L. M., *Metalloved. i Term. Obrabotka Metal.*, **2**, 2 (1957).
16 Gertsriken, S. D., and Pryanishnikov, M. P., *Vopr. Fiz. Met. i Metalloved., Sb. Nauchn. Rabot. Inst. Metallofiz. Akad. Nauk Ukr. SSR*, **9**, 147 (1958).
17 Bokshtein, S. Z., Kishkin, S. T., and Moroz, L. M., *Investigation of the Structure of Metals by Radioactive Isotope Methods*, State Publishing House of the Ministry of Defence Industry, Moscow (1959); AEC-tr-4505 (1961).
18 Bogdanov, N. A., *Russ. Met. Fuels* (English transl.), **2**, 61 (1962).
19 Sparke, B., James, D. W., and Leak, G. M., *J. Iron Steel Inst. (London)*, **203** (2), 152 (1965).
20 Staffansson, L. I., and Birchenall, C. E., AFOSR-733 (1961).
21 Borg, R. J., Lai, D.Y.F., and Krikorian, O., *Acta Met.*, **11** (8), 867 (1963).

770 C (Buffington et al.[32], whose data we have used; recently confirmed by Kucera et al.[33]). The activation energy for diffusion well above and below 770 C is about equal, but data obtained around 770 C show an anomalously high activation energy for diffusion, and for any diffusion-controlled process, such as creep.[34,35] For the lattice diffusion coefficient we have used a smoothed transition from the paramagnetic to ferromagnetic state given by

$$D_v = f \ D_{v_{Para}} + (1-f) \ D_{v_{Ferro}}$$

$$f = \frac{1}{2} + \frac{{}^1/{}_2(T-1043°)}{(\ |(T-1043°)|\ + 20.0)}$$

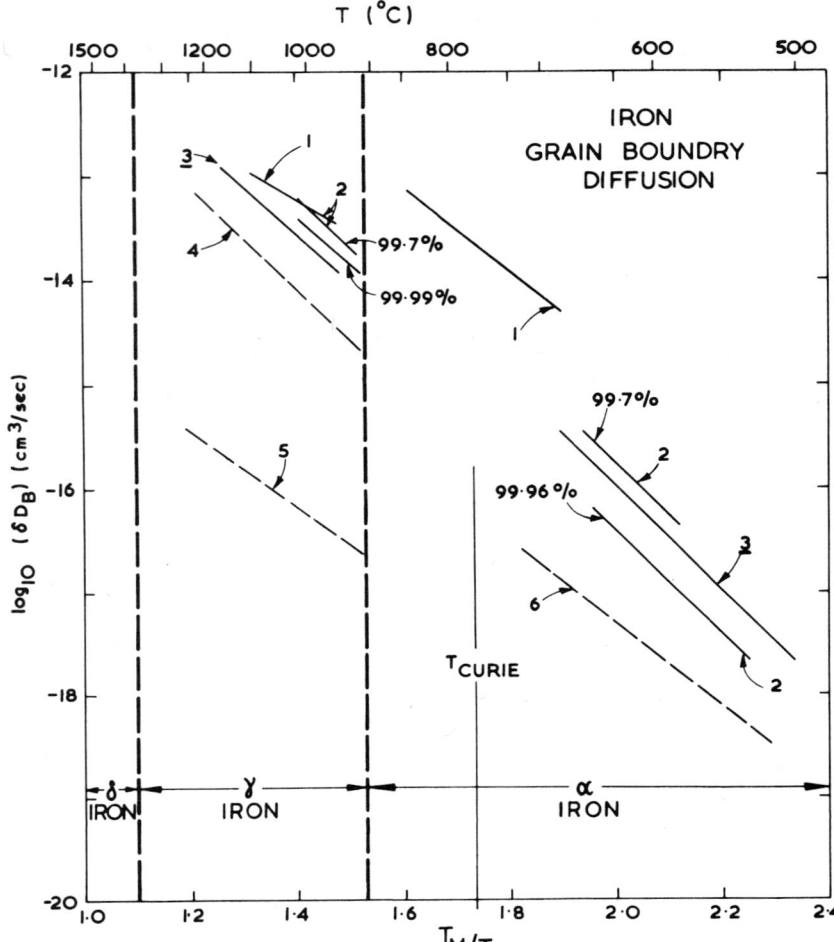

Fig. A-2. Grain-boundary self-diffusion measurements in pure iron.

1 Borisov, V. T., Golikov, V. M., and Scherbedinskiy, G. V., *Fiz. Metal. Metalloved.*, **17** (6), 881 (1964).

2 Guiraldeng, P., and Lacombe, P., *Acta Met.*, **13**, 51 (1965).

3 James, D. W., and Leak, G. M., *Phil. Mag.*, **12**, 491 (1965).

4 Lacombe, P., Guiraldeng, P., and Leymonie, C., *Radioisotopes in the Physical Science Industries,* IAEA, Vienna (1963), p. 179.

5 Bokshteyn, S. Z., Kishkin, S. T., and Moroz, L. M., *Radioactive Studies of Metal Surfaces,* Moscow, U.S.S.R. (1959).

6 Gertsriken, S. D., and Pryanishnikov, M., *Issled. po Zharoprochn. Splavam,* **4**, 123 (1959).

For lattice diffusion in γ-iron, we have used the data of Buffington et al.[32], which typify measurements for this phase (Fig. A-1). For δ-iron and the 1 percent Cr-Mo-V steel, we use the data for α-iron, extrapolated where necessary. Though alloying elements do change the diffusion coefficients for steady-state mass transport (see below), the effect in this steel is slight.

Boundary-diffusion coefficients (Fig. A-2) for all phases of iron and for the 1 percent Cr-Mo-V steel are from James and Leake.[36] Core-diffusion coefficients are not known; these are based on a normalized average of those for bcc and fcc metals.

Diffusion in the stainless steels is well documented. The most relevant study, that of Perkins et al.[37], gives both lattice and boundary diffusion for iron, chromium, and nickel in an alloy with a composition close to that of the two steels. Strictly, the proper diffusion coefficient for steady-state mass transport in a ternary alloy of components A, B, and C is

$$D = \frac{D_A D_B D_C}{D_A D_B \chi_C + D_B D_C \chi_A + D_A D_C \chi_B}$$

where χ_A is the atom fraction, and D_A is the tracer diffusion coefficient of component A, and so on. While Perkins' data are sufficiently complete to allow this D to be calculated, the three major components in the steel diffuse at so nearly equal rates that it is simplest to take the coefficient for the principal component, iron. This choice gives an adequate description of the creep data (see below).

Boundary diffusion, too, should properly be described by a combined diffusion coefficient like that above, the atom fractions being those describing the (possibly segregated) boundary region. Again, we have taken a single coefficient, with an activation energy that is intermediate to those of the three components (167.4 kJ/mole among the quoted values of 150.7 ± 9.6 for chromium, 177 ± 17.2 for iron, and 133.9 ± 8.0 for nickel), and with a preexponential that places the coefficient in the center of the scattered data.

A-4 Power-Law Creep

The power-law creep parameters for α-iron and δ-iron (exponent n, preexponent A, data for core-diffusion $a_c D_c$) were obtained by fitting Eq. (11) to the data shown in Fig. 4.

For high-temperature creep we have used the value n = 6.9 of Ishida et al.[34], which describes all the data reasonably well. Except near the Curie temperature, the activation energy for high-temperature creep is close enough to that for lattice diffusion for the two to be equated; near the Curie temperature, the discrepancy is properly incorporated into our map through our use of a nonlinear temperature dependence of the modulus.

Although there is little evidence for a low-temperature creep field for ferrite, our experience in constructing maps for other metals[2] indicates that this core-diffusion-controlled field is usually present. We took a core-diffusion coefficient based on the grain-boundary[36] for ferrite. Doing this results in the small field shown in Fig. 2, and puts almost all the data of Fig. 3 into the high-temperature creep field—a result consistent with the data.

There are fewer creep data for γ-iron than for α-iron (Fig. 3). Most are due to Feltham[38] whose data we have used. In using the map in Fig. 2, it should be remembered that the stress is normalized with respect to a different modulus above and below the transformation temperature:

$$\mu(\alpha\text{-Fe, 912 C}) = 2.5 \times 10^4 \text{ MN/m}^2$$

$$\mu(\gamma\text{-Fe, 912 C}) = 4.5 \times 10^4 \text{ MN/m}^2$$

This normalization conceals the true magnitude of the discontinuity of strain rate at 912 C: it is larger than it appears from Fig. 2. The difference in normalizations is shown by the dashed line of constant shear stress $\sigma_s = 0.1$ MN/m^2. The strain-rate contours should be displaced accordingly to give the discontinuity in γ at constant σ_s.

The power-law creep parameters for the stainless steels were derived from the sets of data presented in Figs. 7 and 8. These figures show that a simple power law is not a particularly good description, since many of the data are obtained at high stresses, where the power law starts to break down. For 315 stainless steel we used a value n = 7.0, intermediate between the high-temperature, low-stress data of Garofalo et al.[39] and the low-temperature, high-stress data of Blackburn[31]; over the range they investigated, n varies from 4 to 10. Garofalo himself fits his data to a (sinh)^n law which is more flexible than a power law but has little fundamental backing at present. Further, an activation energy for creep, derived by appropriately replotting the data in Fig. 7, is somewhat higher than that for diffusion. We chose the preexponential constant A, so that the maps coincide with the data of Garofalo et al. at 732 C and of Blackburn at 650 C, but because of the discrepancy between the activation energies for creep and for diffusion, and the indifferent fit of the data to a power law, this creep field must be regarded as a first approximation only.

The creep of 304 stainless steel is somewhat better described by a power law (Fig. 8). The observed activation energies for creep and for diffusion are very similar for this alloy, and with n = 7.5, a single power law describes most of Blackburn's[31] data. The alloy studied by Beckitt and Gladman[40] was 304 L—the low-carbon form of 304 stainless steel. It creeps about an order of magnitude faster than 304 stainless steel and has a slightly lower yield strength.

For both 304 and 316 steels, the reproducibility of creep data between different experimenters (using slightly different compositions) is reasonably good—better than for creep of pure metals. This evidently results from the solid-solution alloying which swamps out the effects of trace impurities. The inhibition of recrystallization (at the temperatures used) due to the scattered carbide precipitates also contributes to this.

The data for power-law creep in 1 percent Cr-Mo-V are barely adequate for the construction of a map. We used constant-stress data from "Piece 2" of the CEGB—University Program.[23] "Piece 4" crept more slowly, by a factor of about three, emphasizing that, unlike the stainless steels, heat-treatable alloys such as this one are very sensitive to slight variations in composition of heat treatment. It is clear from the observed activation energy for creep (60 to 90 kcal/mole), which is close to that for lattice diffusion, that the core-diffusion-controlled creep is suppressed in this alloy. We have suppressed it by lowering the effective core-diffusion coefficient by a factor of 10; all the data then lie in the high-temperature creep field.

A-5 Glide Parameters

The glide parameters for α-iron (ΔF_k, $\hat{\tau}_p$, ΔF, and $\hat{\tau}$) were determined by fitting Eqs. (6) and (7) to the data shown in Fig. 3. These best-fit parameters describe flow at constant structure—a particular state of heat treatment and work hardening. Lattice resistance and obstacle resistance are treated as alternatives, giving the sharp kink in the strain-rate contours in the glide region; in reality, the transition between them is a smooth one. There are a great deal of data on the low-temperature flow of pure iron, and we have not attempted to examine all of it.

For both 304 and 316 steels, the dislocation-glide field is based on Eq. (6), with $\Delta F = 0.5 \ \mu b^3$ and $\hat{\tau} = 6.45 \times 10^{-3} \ \mu$. These values give yield stress at room temperature equal to that quoted by Parr and Hanson.[41] A higher value of $\hat{\tau}$ could be used to describe a higher level of work hardening.

The data for dislocation glide obtained by the CEGB—University Program[23] is best fitted by Eq. (6) with a large value of $\Delta F = 2 \ \mu b^3$, which implies strong obstacles (presumably the carbides) and a flow stress that, when normalized, is insensitive to change in temperature or strain rate (it is probably the Orowan stress for carbide dispersion).

REFERENCES

1. Ashby, M. F., *Acta Met.,* **20**, 887 (1972).
2. Frost, H. J., and Ashby, M. F., *Second Report on Deformation Mechanism Maps*, Harvard University Technical Report (October, 1973).
3. Frost, H. J., Ph.D. Thesis, Harvard University, February, 1974.

4. Ashby, M. F., and Frost, H. J., in *Rate Processes in Plastic Deformation of Materials,* American Society for Metals, Metals Park, Ohio (1975).
5. MacKenzie, J. K., Ph.D. Thesis, Bristol University, 1959.
6. Tyson, W. R., *Phil. Mag.,* **14**, 925 (1966).
7. Kocks, U. F., Argon, A. S., and Ashby, M. F., *Prog. in Mater. Sci.,* **19**, 1 (1975).
8. Kocks, U. F., *Met. Trans.,* **1**, 1121 (1970).
9. Weertman, J., *Trans. AIME,* **227**, 1475 (1963).
10. Robinson, S. L., and Sherby, O. D., *Acta Met.,* **17**, 109 (1969).
11. Hart, E. W., *Acta Met.,* **5**, 597 (1957).
12. Vandervoort, R. R., *Met. Trans.,* **1**, 857 (1970).
13. Challenger, K. D., and Moteff, J., *Met. Trans.,* **4**, 749 (1973).
14. Jonas, J. J., Sellars, C. M., and Tegart, W.J.McG., *Met. Rev.,* **14**, 1 (1969).
15. Nabarro, F.R.N., *Report on a Conference on the Strength of Metals,* Physical Society, London, England (1948).
16. Herring, C., *J. Appl. Phys.,* **21**, 437 (1950).
17. Coble, R. L., *J. Appl. Phys.,* **34**, 1679 (1963).
18. Lifshitz, L. M., *Soviet Phys. JETP,* **17**, 909 (1963).
19. Gibbs, G. B., *Mem. Sci. Rev. Met.,* **62**, 781 (1965).
20. Raj, J., and Ashby, M. F., *Met. Trans.,* **2**, 113 (1971).
21. Ashby, M. F., *Scripta Met.,* **3**, 837 (1969).
22. Ashby, M. F., *Surface Sci.,* **31**, 498 (1972b).
23. CEGB—University Collaborative Project on 1 Cr-Mo-V Steel; details available from the CEGB Research Laboratories, Berkeley, Gloucester, England.
24. Burton, B., *Mater. Sci. Eng.,* **10**, 9 (1972).
25. Burton, B., *Mater. Sci. Eng.,* **11**, 337 (1973).
26. Taylor, A., and Kagle, B. J., *Crystallographic Data on Metal and Alloy Structures,* Dover Publications, Inc. (1963).
27. Hansen, M., *Constitution of Binary Alloys,* McGraw-Hill, New York, N.Y. (1958).
28. Hirth, J. P., and Lothe, J., *Theory of Dislocations,* McGraw-Hill, New York, N.Y. (1968).
29. Dever, D. J., *J. Appl. Phys.,* **43**, 3293 (1972); Koster, W. Z., *Metallkde.,* **9**, 1 (1948).
30. Lytton, J. L., *J. Appl. Phys.,* **35**, 2397 (1964).
31. Blackburn, L. D., *The Generation of Isochronous Stress-Strain Curves,* Paper presented at ASME Winter Annual Meeting, New York, November, 1972.
32. Buffington, F. S., Hirano, K., and Cohen, N., *Acta Met.,* **9**, 434 (1961).
33. Kucera, J., Million, B., Ruzickova, J., Foldyna, J., and Jakabova, A., *Acta Met.,* **22**, 135 (1974).
34. Ishida, Y., Cheng, C.-Y., and Dorn, J. E., *Trans. AIME,* **236**, 964 (1966).
35. Čadek, J., Pahutova, M., Cĭha, K., and Hostinsky, T., *Acta Met.,* **17**, 803 (1969).
36. James, D. W., and Leak, G. M., *Phil. Mag.,* **12**, 491 (1965).
37. Perkins, R. A., Padgett, R. A., Jr., and Tunali, N. K., *Met. Trans.,* **4**, 2535 (1974).
38. Feltham, P., *Proc. Phys. Soc. London,* **B66**, 865 (1953).
39. Garofalo, F., Richmond, C., Domis, W. F., and von Germminger, F., in *Joint International Conference on Creep,* Institute of Mechanical Engineers, London, England (1963), pp. 1–31.
40. Beckitt, F. R., and Gladman, T., Special Steels Department Report No. PROD/PM/6041/1/71A (September 16, 1971).
41. Parr, J. G., and Hanson, A., *An Introduction to Stainless Steel,* American Society for Metals, Metals Park, Ohio (1965), p. 54.

DISCUSSION on Paper by M. F. Ashby:

THOMAS: Can you predict from your model and maps what directions the deformation rates will go by manipulating metallurgical parameters? Specifically, for example, in austenitic stainless steels the stacking fault energy can be raised by raising the Ni/Cr ratio (which may be beneficial for reducing transgranular stress corrosion) yet one might then expect cross slip and climb to be easier. In this way can we *design* a composition (SFE) for a stainless steel for your pressure-vessel example? Or are SFE, etc., minor parameters compared to grain size?

Another point I would like to make is that austenitic steels such as 316 are short-range ordered and therefore the dislocation-solute interaction will vary with temperature (with degree of order). Will this ordering effect also be an important parameter for your mapping work?

ASHBY: A single map cannot predict the effects of changing alloy content, or of manipulating metallurgical parameters. These effects can be described if the rate equations used for the maps correctly include these parameters. The only manipulative parameters that are specifically included in our rate equations are the grain size for diffusional flow and the obstacle density for dislocation glide. The effect of stacking-fault energy is implicitly included in the dislocation-creep equation through the constant A_s. It is well known that, in fcc metals, raising the stacking-fault energy raises the creep rate at a given normalized stress and normalized temperature. For example, see C. R. Barrett and O. D. Sherby, *Trans-AIME,* **233**, 1116 (1965) for pure metals, or B. A. Wilcox and A. H. Clauer, *Met. Sci. J.,* **3**, 26 (1969) for Ni-Cr and Ni-Cr-ThO$_2$ alloys. Short-range order should also affect the parameters that describe power-law creep and glide.

McMAHON: I understood you to say that you use the fastest deformation process to construct the deformation maps. Is this correct, and if so, what is the effect of superimposing, say, the fastest two or three processes?

ASHBY: The mechanism fields are determined by which mechanism operates fastest. The strain-rate contours, however, are calculated in a more complicated manner. The diffusional flow and dislocation creep mechanisms are always superimposed. On a logarithmic scale, one or the other dominates, except near the boundary where the contours are smoothed. Dislocation glide is taken as an alternative mechanism to both dislocation creep and diffusional flow. This is done because the glide field is describing yield behavior which is different from steady-state behavior.

CAMPBELL: Were all the data presented in the deformation maps obtained from tests at constant stress and temperature? Can you comment on the problem of determining the deformation under conditions of variable stress or temperature?

ASHBY: All the data presented were quoted as being at constant temperature. Most of the creep tests were carried out at constant load, which is a good approximation to constant stress. In practice, it is more important to determine the deformation under conditions of variable stress and temperature. Although it is possible to describe transient behavior empirically, it is currently very difficult to predict the accumulative effects of many changes in stress and temperature. We believe there remains a lack of predictive models based on physical models.

McCLINTOCK: What are the prospects for reducing the Coble creep by alloy design other than grain size, perhaps by contaminating or plugging the grain boundaries?

ASHBY: There is some evidence that diffusional flow (Coble creep) can be reduced by inhibiting the detachment and attachment of vacancies to the grain boundaries. Particles in the boundaries cause a threshold stress below which no creep occurs, although this has only been measured at high temperatures and very low stresses (of order $\sigma_s \simeq 10^{-5} \mu$). Impurities segregated to the grain boundaries may also have some effect. More work is needed to study these effects more precisely.

BUSH: In your table of significant basic parameters and your maps for such systems as AISI 304 and 316, one set of parameters leading to unanticipated failures are the environmental factors, e.g., environments leading to stress corrosion or corrosion fatigue. Have you considered three-dimensional maps with quantitative parameters related to environmental conditions? The data would be invaluable to the designer in considering stress thresholds for various levels of contamination. To me, such is important in the context of your LMFBR example because one-half to two-thirds of the failures in LMFBR components have been environmentally induced.

ASHBY: No, we have not constructed any maps with environmental conditions varied in the third dimension. It should be possible if the effects of environmental conditions can be quantified over a wide range of stress and temperature. We have not uncovered and analyzed the necessary information. With regard to particular failures, the deformation maps can be used only to predict failures by excessive plastic flow. Creep rupture must be considered separately, although it is generally related to plastic flow. The effect of the environment on creep rupture or crack formation would not necessarily be the same as its effect on the creep rate.

HAASEN: In what way does the mean diffusion distance enter the core diffusion components of your "effective diffusion coefficient" for climb?

ASHBY: The mean diffusion distance for core diffusion-controlled dislocation creep may be thought of as entering the rate equation in the same manner that it enters the lattice diffusion-controlled rate equation, i.e., through the stress to the nth power.

EMBURY: Although for obvious reasons you have commented on the low $\dot{\epsilon}$ range, it would seem that you could also exploit the deformation maps at the high $\dot{\epsilon}$ end. Here the input would be $\dot{\epsilon}$, δy, and temperature, and you could predict deformation load, i.e., mill load or extrusion pressure. This area should be feasible for aluminum and iron; it should also yield the influence of grain size and adiabatic effects in the high $\dot{\epsilon}$ range. Do you have any comments on this possibility?

ASHBY: It would be possible to construct such maps. At high stresses and strain rates, the power-law stress dependence breaks down to an exponential stress dependence. This is commonly described by using $[\sinh(\propto\sigma)]^n$ in place of σ^n. If the total strain in the process were small, then it would also be necessary to take account of transient effects. We do not have sufficient information to describe the effects of grain size and adiabatic heating in this regime, but it should be possible.

MAYER: In analyzing some of the prior data on grain-size effects, is there enough information on differences in texture of polycrystalline materials that might have an influence on the deformation map? Also, it appears that grain-boundary structure (for a given grain size) might be sufficiently alterable, e.g., by different solidification conditions, to make a large change in deformation mode or in the stress required to deform a structure, e.g.,

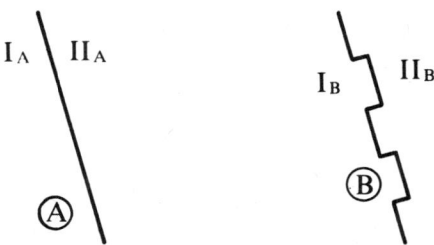

if a heat treatment can develop the structure of B in a grain boundary, such a boundary would clearly be more resistant to G.B. sliding than that of A.

ASHBY: For pure fcc and bcc metals with equiaxed grain structures, texture does not appear to be particularly important to the power-law creep rate. For instance, Barrett el al. [*Trans. A.I.M.E.,* **239**, 170 (19760] found only a factor of two difference in creep rate between textured and randomly-oriented copper.

In HCP metals, in which not all slip planes are equivalent, texturing has a stronger effect. For example, Edwards et al. [*Scripta Met.,* **8**, 475 (1974)] find a strong effect of texture on the creep of zinc. The anisotropy of textured HCP metals at low temperatures is well known.

WILCOX: Do you have a working constitutive equation for grain-boundary sliding? If so, have you applied it to the steels you discussed in this paper?

ASHBY: Yes, there is a constitutive equation which includes the effect of grain-boundary sliding, though we have not included it here. We do not believe that sliding can operate as a deformation mechanism on its own. It is always necessary to make some accommodation of the strain concentration at the edge of the boundary, i.e., the triple point. This can be done in the following ways:

1. Elastically—which can not produce sliding as a steady-state mechanism.

2. By diffusional fluxes, which necessarily leads to a model essentially equivalent to Nabarro-Herring or Coble creep.

3. By concentrated plastic deformation within the grains. This has been modelled by F. L. Crossman and M. F. Ashby, *Acta. Met.,* **23**, 425 (1975).

In their model, the dislocation creep field is divided into two: at high stresses, there is very little boundary sliding; at low stresses, the boundaries slide freely. The sliding results in slightly enhanced strain rates but the same stress dependence. This division is determined by the local resistance to boundary sliding caused by ledges or particles which must be overcome by localized diffusion.

4. By intergranular cracking and void growth. We do not yet have a rate equation to described this process. Hydrostatic pressure should reduce this mechanism of accommodation and thereby reduce the apparent grain-boundary sliding.

CLAUER: You just mentioned that the dislocation-glide-controlled region could actually be split into two areas, one having a small amount of grain-boundary sliding and another showing a large amount of it. This brings up an interesting possibility for mapping creep ductility including the effects of a superposed hydrostatic pressure, as in some components of nuclear reactors. Greenwood and co-workers have suggested that the enhanced creep ductility observed during creep under hydrostatic pressure results from an inhibition of grain-boundary sliding. Thus, in this regime, the stress-rupture behavior might be described in terms of a phenomenological or theoretical description of grain-boundary sliding including the influence of pressures.

ASHBY: This question was answered in the reply to the question by Wilcox.

CLAUER: As you are aware, there are many time-dependent microstructural changes that occur during creep at high temperatures which will influence the creep rate. For example, in your deformation map of the precipitation-hardened Cr-Mo-V steel the time dependence of aging of the precipitate will influence the creep strength and this effect would superpose in your deformation map. Have you given any thought as to how you would include these important effects into your deformation mapping technique?

ASHBY: We have not made any attempt to include the effects of microstructural changes during creep. The effect of grain growth could easily be included in the diffusional flow-rate equation. The effects of aging of the precipitates would be harder to include. It would require determining the dislocation creep parameters at various particle sizes and densities, and then expressing the size and density as functions of time. The map would no longer represent steady-state behavior, and the initial conditions would have to be carefully specified.

HAASEN: Dislocation motion over localized obstacles does not always lead to the exponential velocity-stress and -temperature relation you postulate. Our calculation (Labusch et al., see Eq. (5) in Haasen's paper, this Colloquium Proceedings) of thermally-activated dislocation motion over solute atoms rather predicts a *linear* velocity-stress law in the plateau range. The strain rate—stress relation in this range is very difficult to assess theoretically.

SHERBY: I note, with interest, that the deformation maps you have exhibit large regions where Nabarro-Herring or Coble creep control the deformation process. The example given for the case of low-stress creep of lead was very interesting and nicely illustrates the utility of such deformation maps. It would seem to me that these results suggest the need to understand better all the conditions controlling diffusional creep, and perhaps, more importantly, to learn how to slow down this creep process. I have in mind, for example, that many dispersion-hardened materials containing from grains (e.g., some powder-metallurgy products) do not exhibit diffusional creep at low stresses even though constitutive equations predict that such a creep process is formed over slip (dislocation) creep. Nabarro-Herring or Coble creep has been verified mostly for the case of metallic foils and wires. Much less evidence appears available for such a type of flow in bulk specimens; for example, long-time stress-rupture data for stainless steels have not revealed stress exponents in the order of unity. Your important conclusions would suggest that further experiments are very much in order to determine the possible importance of diffusional creep in commercial ferrite and austenitic steels.

FROST: (1) You stated that diffusional creep had not been observed in Type 316 SS. Under a high neutron flux, creep has been observed at stresses where it is not observed out of the flux. This is probably diffusional creep and must be used in the design of core components.

(2) In designing reactor vessels and pipework one uses creep-fatigue or cumulative damage data rather than simple creep; i.e., one has to go to one more stage of complexity in order to get to the practical case of reactor design.

SIMS: The creep-property mapping tool you have demonstrated has, I believe, the potential of wide application in both alloy development and understanding and correction analysis of alloy failure. Importantly, it provides an understandable tool usable by alloy designers and mechanical-design engineers. While not yet perfected for direct use, perhaps, it seems to me that modification further toward practicality is quite possible.

However, in order to accomplish the latter and also to bring about the important step of moving the tool into the confidence range and thus use by alloy designers will require attention to the process of effective technology transfer. I believe publication alone is insufficient. Specific attention must be paid by those concerned with industrial-equipment needs to bring such a tool to its full fruition.

Incidentally, I do not believe it is necessary to include considerations such as oxidation and corrosion requirements in this particular tool. Such parameters, along with others such as cost, availability, and specific nuclear behavior will be readily fed into applications of such a tool by alloy and engineering designers.

HIRTH: With regard to your information/design flow chart, I have a comment with respect to the glide-plane process to Taylor factor transition. Including averaging of single-glide plane constitutive equations over five (or more) glide deformation systems, one finds that the apparent activation area can differ from the glide-plane activation area by a factor of as much as 20. This implies, of course, also shifts in the apparent stress exponent n. Such an effect could be one of the biggest difficulties in carrying single-dislocation mechanisms to design constitutive equations; however, with computer averaging, such averaging can be done to provide guidelines.

It is interesting in this regard that single to polycrystal shifts for a given mechanism cause shifts in n. Even for single crystals, averaging over secondary systems (which is observed in easy glide) can cause shifts in n. The mechanisms for which no shift in n would be expected are those *linear* in dependence on n. In this connection, it is interesting that polycrystal deformation with n = 1 is observed, in contrast to the other mechanisms. [Reference: J. P. Hirth, in *Rate Processes in the Plastic Deformation of Materials,* American Society for Metals, Metals Park, Ohio, (1975).]

THE MECHANICAL PROPERTIES OF
HIGH-STRENGTH, LOW-ALLOY STEELS

J. D. Embury, J. D. Evensen, and A. Filipovic

Department of Metallurgy and Materials Science
McMaster University, Hamilton,
Ontario, Canada

ABSTRACT

This paper deals with the parameters needed to specify the structural behavior of high-strength, low-alloy (HSLA) steels and the influence of microstructural constituents on these parameters. The development of acicular, nonpolygonal ferrite microstructures through controlled rolling is an important aspect of HSLA design. The microstructural aspects of cleavage failure, ductile rupture, and delamination and of Bauschinger effects on flow stress arising from residual stresses from fabrication are considered.

67

1 INTRODUCTION

Much progress has been made in the past decade in the development of models which describe quantitatively the strengthening contributions from solid solutions, precipitate phases, and dislocation substructure and texture. However, the development of engineering materials demands the use both of combinations of strengthening mechanisms and a more sophisticated understanding of the relationship of microstructure and toughness. Thus, for actual engineering materials, the relationships between structure and properties currently available tend to be somewhat empirical, and furthermore the critical properties may be somewhat more complex than the specification of yield strength or strain to failure. Modern structural steels provide some interesting and challenging examples of the role of physical metallurgy in relation to alloy design. It is the purpose of this paper to consider both the definition of the parameters that are needed to specify the behavior of structural steels and the influence of microstructural constituents, and thus alloy design, on these parameters.

The development of improved structural steels with yield strengths in the range 60 to 80 ksi for line-pipe and other applications has been one of the most important metallurgical endeavors of the past decade. The results of these developments have brought into focus the need to understand the complex phase transformations occurring during continuous cooling of thermomechanically processed low-alloy steels and the need to correlate the critical parameters governing the fracture behavior with the existence and distribution of various microstructural constituents. There have been a number of definitive reviews[1-4] concerning the development of high-strength, low-alloy (HSLA) steels, and the present work does not consider the detailed chemistry or processing of the various HSLA steels. However, it is important to outline the major metallurgical factors which have been utilized in the development of the current categories of HSLA steels and to consider the difficulties of producing a quantitative description of the mechanical properties of these materials. The properties of HSLA steels are determined by the following factors.

(a) *The use of alloying elements such as columbium, vanadium, and titanium which promote both grain refinement of the austenite and precipitation hardening in the ferrite.* It is important at this point to realize that one major barrier to the quantitative understanding of the property-strength relationships is the ability to predict the detailed chemistry of the alloy additions, e.g., the volume fraction of phases precipitated in the austenite and ferrite. In part, this arises from the complex chemistry of these elements in the presence of both carbon and nitrogen[5] and from the lack of suitable activity data in complex solutions.

(b) *The ability to process the austenite phase by hot deformation in a controlled manner involving both the degree and temperature of deformation and the subsequent rate of cooling.* Much effort[67] has been expended on the optimization of rolling and cooling schedules and, in some cases, the range of deformation temperatures has been extended into the $\gamma + \alpha$ region.

(c) *The trend toward improvement of weldability by reducing the carbon content, in addition to control of the alloy content.* Inherent in this practice is the need to understand and control deoxidation and reoxidation reactions in the steelmaking operation to ensure an acceptable oxide-inclusion level, particularly with the use of rare-earth alloy additions.[8]

(d) *The demand for structural materials with isotropic response which has led to the practice of inclusion control using both desulfurization practices and sulfide-shape modifications.* The objective of these practices has been to improve a variety of factors such as through-thickness ductility, cold formability, and resistance to in-plane delamination during welding.

The basic metallurgical variables outlined above can be used to provide HSLA steels by a variety of processing routes. Baird and Preston[4] have recently very clearly outlined the basic structure and property relationships found in ferrite-pearlite steels, controlled rolled steels, and low-carbon steels containing acicular ferrite.

Despite much effort, both at the fundamental research level and in the optimization of both chemistry and processing routes, there remain a number of basic structure-property relationships which need to be clearly defined. In attempting to outline an approach to these problems, let us first consider the engineering properties demanded of the HSLA steels.

For line-pipe applications, it is self-evident that both toughness and high yield strength are required. However, as shown in Fig. 1, the demand is not simply for simultaneously improved toughness and yield properties, but the attainment of these properties in larger section sizes. This latter aspect presents difficulties with steels rolled at low finishing temperature, since this may produce marked variations in structure throughout the section as shown in Fig. 2. The gradients in structure are complex and lead to variation in grain size, in the amount of nonferritic constituents, and in texture throughout the section.[9,10] Thus, local parameters such as the critical cleavage stress are difficult to relate to a unique structural feature such as grain size. Further, the variations in structure arising from differences in transformation history may produce considerable residual stresses in the as-rolled structure which complicate the resultant mechanical behavior. A further demand on structural materials is the provision of isotropic properties which arise from the inclusion shape and distribution. In an effort to control the anisotropy of properties, rare-earth additions are usually made and some property such as the short-transverse ductility may be specified in addition to the usual fracture-resistance parameters. It will be shown later

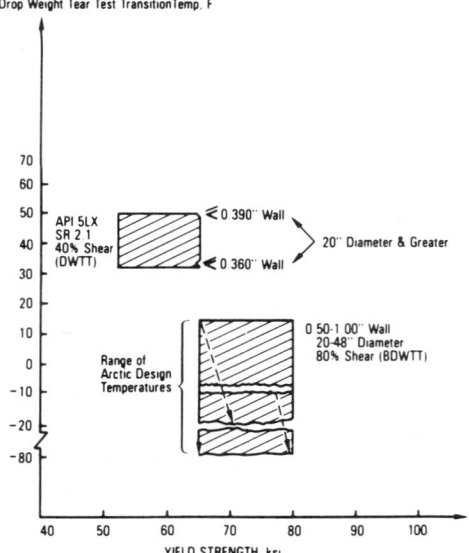

Fig. 1. Combinations of strength and fracture resistance (drop weight tear test, DWTT) needed for structural steels for use in Arctic pipelines, together with the changes in dimensions of the structures.

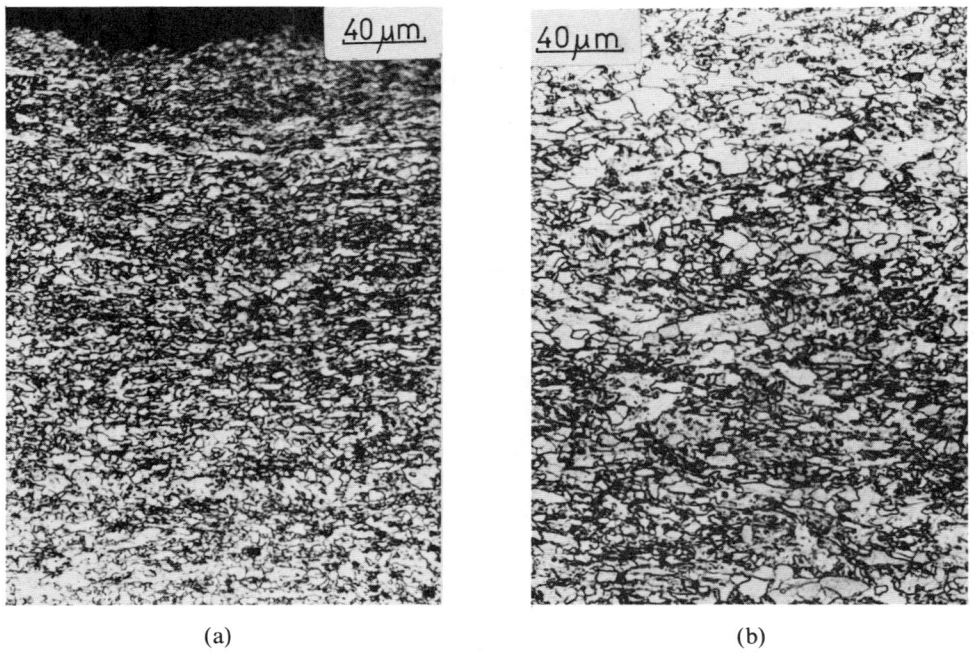

(a)   (b)

Fig. 2. Variation in microstructure from plate surface (a) to center of plate (b) (taken from steel B).

that for high-strength hot-rolled structural steels, one basic engineering difficulty is not simply the attainment of adequate toughness but the attainment of acceptable isotropy of fracture resistance.

A subtle aspect of the mechanical response which is of increasing importance is the form of the initial yield curve in the steel.[11,12] Many fabrication processes such as the U-O-E pipemaking process subject material to strain reversals (Fig. 3) which may lead to strength reductions via the Bauschinger effect.[13,14] Thus, it is important for the physical metallurgist to be able to predict the influence of microstructural constituents not only on the level of the yield stress but also on features such as the magnitude of the Bauschinger effect and the initial shape of the stress-strain curve.

From the considerations above it is clear that currently available descriptions which relate the yield strength or ductile-to-brittle transition temperature to features such as the grain size, composition, and degree of precipitation hardening will have limited value in providing a complete description of the range of properties required of the HSLA steels. Thus, it is of value to consider origin and description of the microstructural features of these materials as well as the influence of the various microstructural constituents on properties such as the resistance to cleavage failure, the short-transverse ductility, the form of yielding, and the behavior in reverse flow.

It is clear at the outset that the present paper will not provide a definitive description of the structure-property relations which are required. Hopefully, it will, however, provide some guidelines for the future consideration of microstructural control and alloy design needed in these most important structural materials.

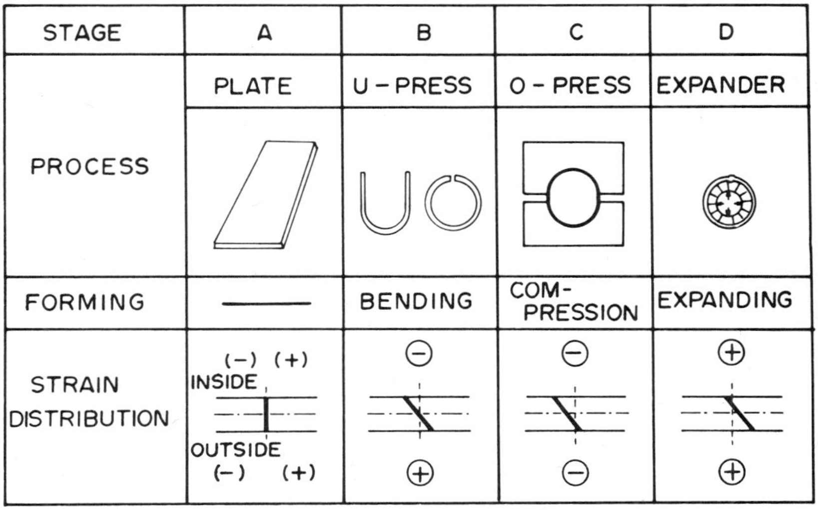

STAGE	A	B	C	D
	PLATE	U – PRESS	O – PRESS	EXPANDER
PROCESS				
FORMING	—	BENDING	COM-PRESSION	EXPANDING
STRAIN DISTRIBUTION	(−) (+) INSIDE, OUTSIDE (−) (+)	⊖ ⊕	⊖ ⊖	⊕ ⊕

Fig. 3. Processing steps and associated strain distributions developed in U-O-E pipemaking processes.

2 MICROSTRUCTURAL ASPECTS OF HSLA STEELS

To provide an adequate description of the microstructural features of structural steels it is necessary to consider in detail the effect of both the thermal-mechanical history and the composition on the transformation of austenite under continuous cooling conditions. Irvine[15] has given a very lucid comparison of the role of alloying elements such as columbium, vanadium, and titanium in modifying the properties of structural steels. The complexity of the microstructures arises from the fact that in order to improve conventional ferrite-pearlite steels for use either in the as-rolled or normalized condition, the grain size must be refined. A material with a grain size of 5 μm (ASTM 12) and carbon content in the range 0.12 to 0.16 weight percent possesses a yield strength of the order of 50 ksi and an impact transition temperature of -50 C.[16] Although the yield strength can be improved by precipitation hardening or dislocation substructure, the most satisfactory method is to suppress the transformation to polygonal ferrite and facilitate the transformation to a nonpolygonal or acicular structure. Examples of polygonal and nonpolygonal structures are presented in Fig. 4 for several steels discussed in this paper, the compositions and processing of which are given in Table I. The mechanical properties of these steels are given in Table II.

In principle, the continuous cooling transformations can be predicted with the aid of a diagram such as that in Fig. 5 which indicates the regimes in which polygonal and nonpolygonal ferrite are produced. However, the processing of alloy austenites in which either solutes or small precipitates retard recrystallization is complex. The grain size, shape, texture, and stored energy of the austenite all influence the kinetics of the transformation to ferrite, as does the amount of alloy element remaining in solid solution. Thus, in many Mn-Mo-Cb steels, mixed microstructures are produced because at low finishing temperatures the transformation to polygonal ferrite is accelerated by strain.[17] In this type of material, it is essential that the scale of the polygonal ferrite be refined so as to avoid poor impact properties resulting from the influence of a small areal fraction of large ferrite grains. This is usually achieved by finish rolling in the range 700 to 770 C.

The objective of controlling rolling practice and the addition of alloy elements is to produce a fine-grained ferritic transformation product. The scale of the ferrite can be considered in terms of the nucleation and growth models outlined by Shewmon.[18] As the ferrite start temperature is decreased, the rate of nucleation rises and the grain size is refined. Also, at greater degrees of undercooling more free energy is available to create plate and needle morphologies. Hence, the faster growing acicular products dominate the structure at lower transformation temperatures. Although there has been much debate concerning the nomenclature of the acicular ferrite, a number

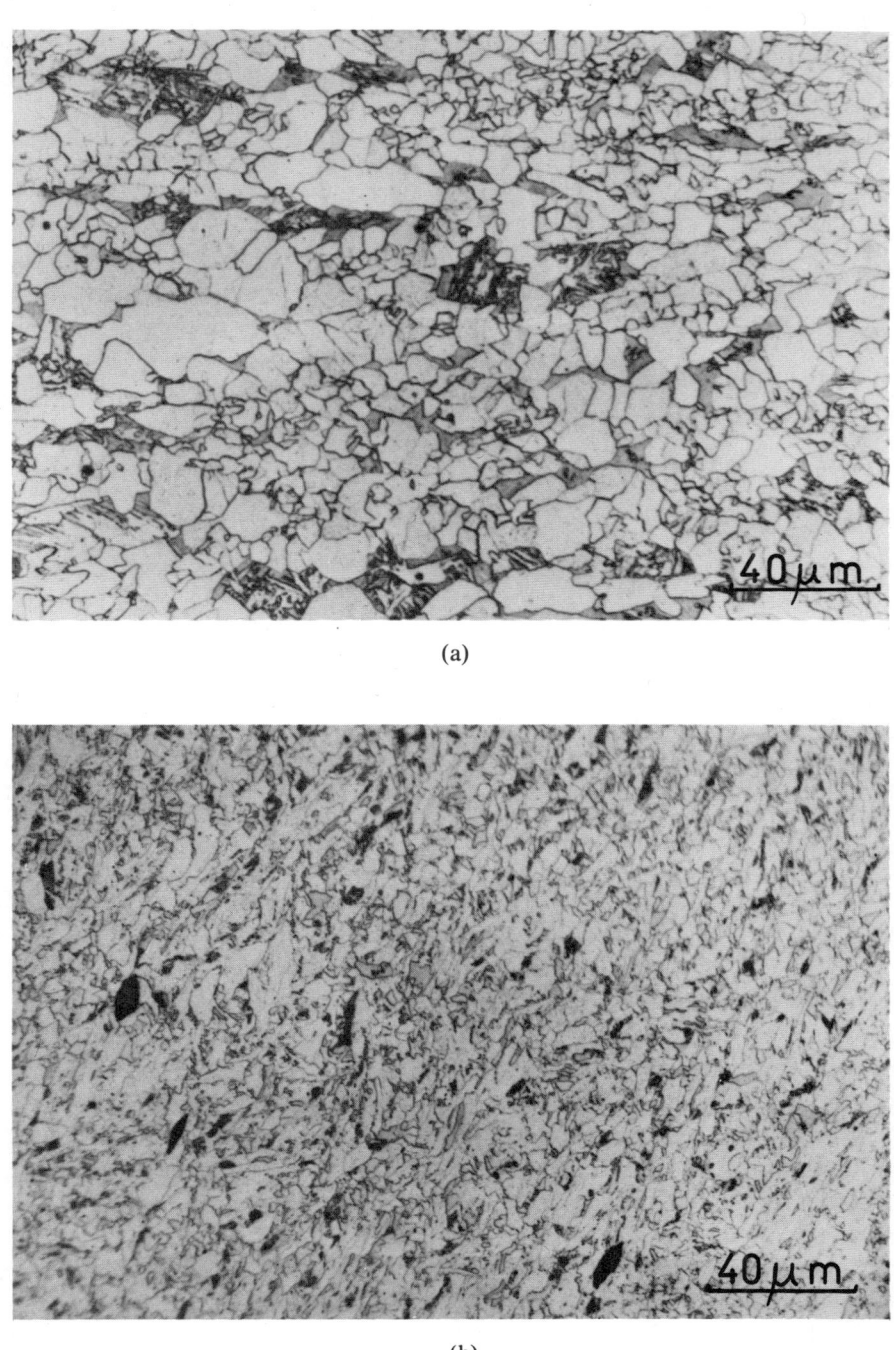

(a)

(b)

Fig. 4. Typical microstructures for (a) polygonal ferrite from steel C and (b) acicular ferrite from steel D.

Table I. Composition and Processing of Steels

Alloy	C	Mn	Mo	Si	Ni	Sb	S	Ce	Al	Final Plate Thickness, inch	Processing
A	0.04	1.80	0.33	0.08	—	0.06	0.012	—	0.109	3/4	Deformed 80% below 900 C; finished at 780 C
B	0.05	2.07	0.56	0.31	0.22	0.067	0.007	0.004	0.039	3/4	Deformed 80% below 900 C; finished at 780 C
C	0.063	1.71	0.14	—	—	0.10	0.001	NA	0.06	5/8	Soaked at 1300 C; hot rolled and finished at 780 C
D	0.063	1.6	0.25	0.06	—	0.05	0.023	—	0.02	3/4	Hot rolled and finished at 870 C

NA — not available.

Table II. Mechanical Properties of Steels

Alloy	Yield Stress, ksi	Tensile Stress, ksi	True Fracture Stress, ksi	Fracture Strain Measured for Longitudinal Samples	Work Hardening Exponent n
A	72.3	90	199	1.15	NA
B	79.0	109	217	0.98	NA
C	71.0	103.3	180	0.85	0.14
D	67.1	95.4	187	1.08	0.13

NA — not available.

of salient features remain undefined. The nature of the interphase interfaces is obscure and few data are available concerning the kinetics of interface motion. These parameters are of relevance in attempting to explain both the high dislocation density of the acicular product (Fig. 6) and the very marked variations in dislocation density which are observed in a given sample.[16] The dislocation density appears to make a substantial contribution to the overall strength and thus studies of its origin and stability are germane to the optimization of the technology of HSLA steels.

The acicular ferrite is composed of groups of parallel ferrite laths arranged in colonies (Fig. 6) and most of the evidence gathered so far indicates that the properties are determined by the scale of the colonies rather than the lath size.[19,20] This may reflect either the small lattice misorientations which exist between individual laths, or some form of segregation or precipitation on the scale of the colony size. The microstructure of the

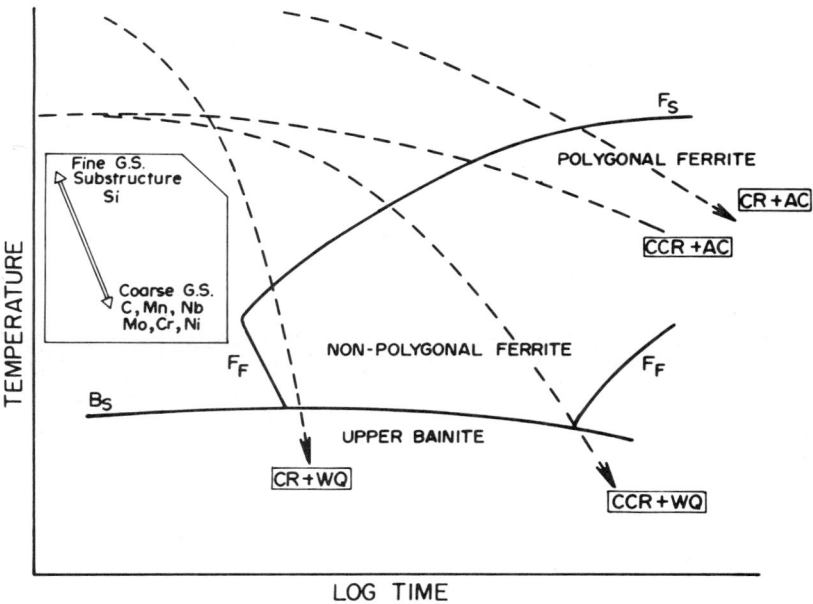

Fig. 5. CCT diagram for low-carbon steel[66]; broken lines indicate transformation paths for the different rolling schedules: CCR, conventional controlled rolling; CR, continuous rolling; AC, air cool; WQ, water quench.

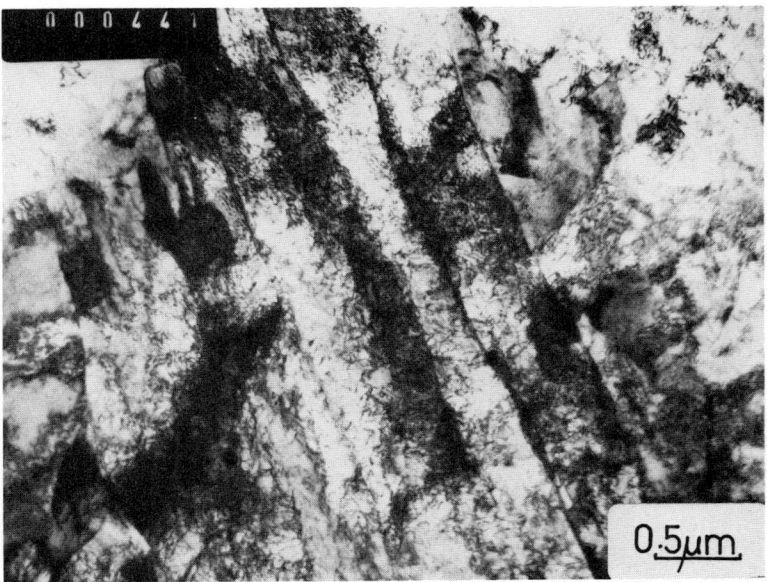

Fig. 6. Region of acicular ferrite consisting of narrow laths containing a high dislocation density.

acicular ferrite is too complex to permit a quantitative estimate of either the dislocation density or the extent of precipitation by transmission electron microscopy.

In addition to the morphology and scale of the ferrite, attention must be given to the redistribution of carbon during transformation and the formation of nonferritic products. These may arise in the form of carbides or complex transformation products such as retained austenite or martensite formed by carbon rejection in the presence of alloying elements such as manganese. The carbides may be in the form of small alloy carbides (Fig. 7) formed either by precipitation at the transforming interface[21] or in the ferrite. These carbides make a contribution to the strengthening which may be of the order of 10 to 15 ksi. In addition to the discrete alloy carbides, cementite may be formed at the ferrite boundaries (Fig. 8). The scale of the carbide phase varies with the temperature of transformation, but in some cases the carbide may form either continuous films or colinear arrays of particles which serve to act as preferred fracture paths in a manner analogous to the carbides formed in upper bainitic structures.[22]

In addition to the formation of carbide, the rejection of carbon from the growing ferrite may result in the formation of regions of retained

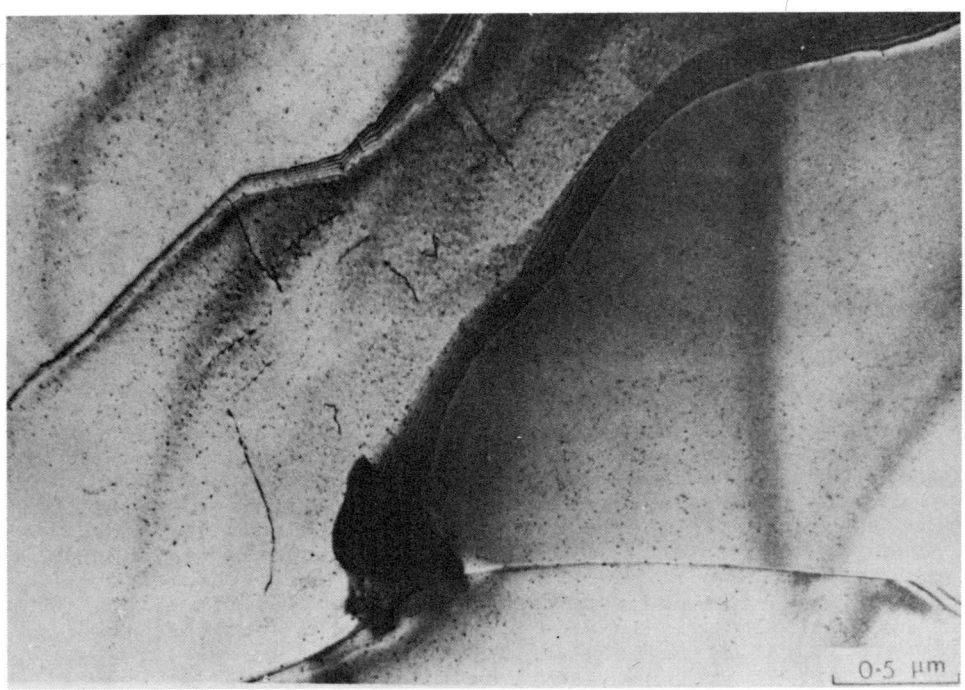

0.5 μm

Fig. 7. Small columbium carbides precipitated in a ferrite matrix.[66]

Fig. 8. Cementite particles (indicated by arrow) precipitated on ferrite grain boundaries (steel C).

austenite or martensite, as outlined by Biss et al.[23] The martensite constituent is often twinned (Fig. 9), indicating that the local carbon level probably exceeds 0.5 weight percent. However, the constituent may also be autotempered in some cases[16] and thus it is difficult to generalize on its detailed structure. The amount and distribution of the austenite/martensite constituent depend both on the thermal history and kinetics of transformation and on factors such as the homogeneity of the distribution of the substitutional alloying elements. To date, no detailed and definitive study of the occurrence of this constituent and its relationship to the ferritic structure has been produced. This is a serious omission because, as shown later, the austenite/martensite constituent plays an important role in determining the yielding characteristics and energy absorption of HSLA steels.

Recently, much effort has been devoted to studies of the role of texture in structural steels and the definitive work of Bramfit and Marder[7], and Webster et al.[24] have clearly indicated the importance of this aspect of the microstructure. The texture of the resultant ferrite influences both the cleavage fracture behavior and the plastic anisotropy and formability of the material.[25] The texture may arise both from transforming a pretextured austenite and from hot rolling the transformed ferrite.[26,27] The influence of finishing temperature on the texture is illustrated in Fig. 10. In addition to

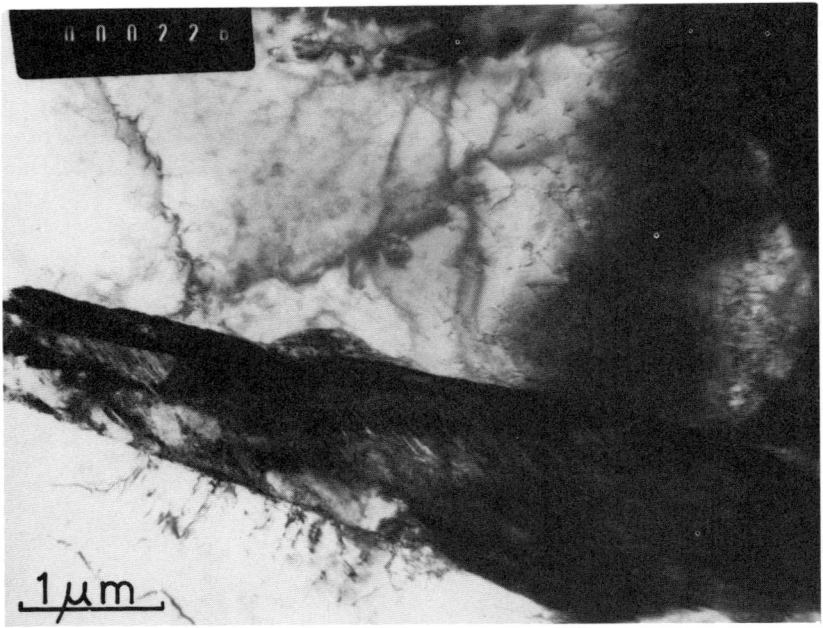

Fig. 9. Twinning in martensite needle (steel C).

the crystallographic aspects of the ferrite, the morphology of the prior austenite may determine the shape and distribution of the transformation products. This in turn may exert a marked influence on the mechanical anisotropy of the final product.

It can be seen from the brief survey of microstructural features presented above that a number of microstructural variables contribute to the strength and fracture resistance of structural steels. It is difficult to present a quantitative model of the strengthening without recourse to considering how the various factors such as grain size, dislocation density, and precipitation hardening can be combined. To date, the majority of attempts to analyze the strengthening mechanisms have assumed linearly additive constituents. However, this linear additivity assumes that a dislocation encounters obstacles which are in effect a few strong obstacles interspersed among many weak ones[28], which is not really justified for the mechanisms under consideration here. However, from a pragmatic viewpoint it is important to consider how the influence of the basic production variables can be rationalized. A number of empirical studies have indicated that factors such as transformation temperature, rolling temperature (Fig. 11), and composition (Fig. 12) can be related to the strength of the finished steel by qualitatively considering that reduction in transformation temperature occurs with increased hardenability. This is reflected in three important

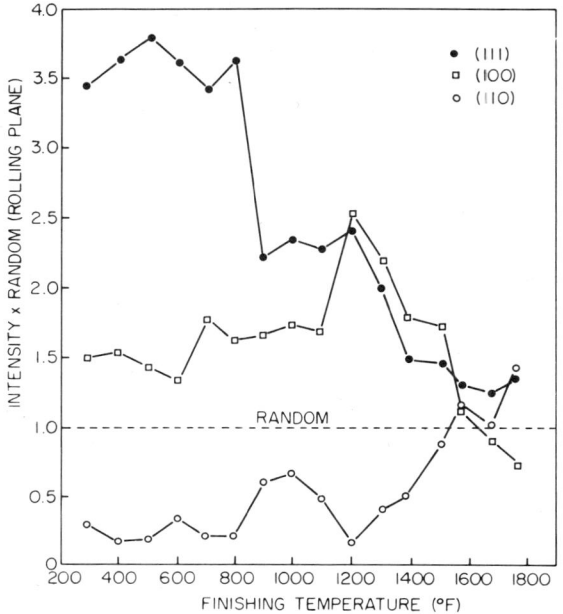

Fig. 10. Influence of finishing temperature on the relative intensities of three crystallographic planes measured parallel to the rolling plane.[7]

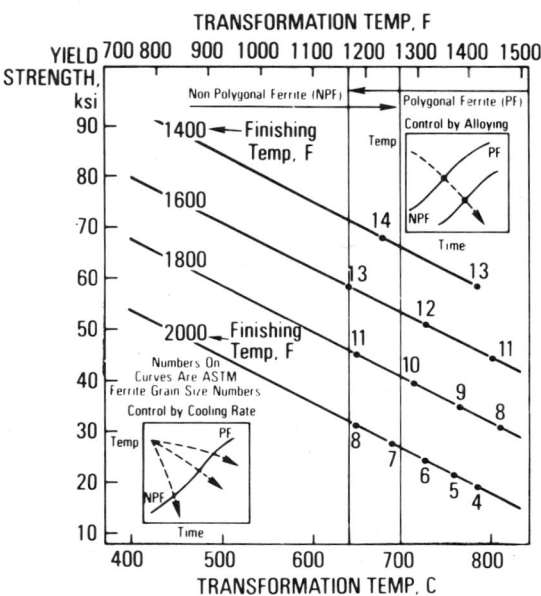

Fig. 11. Relationship between yield strength and transformation temperature for different finishing temperatures.[1]

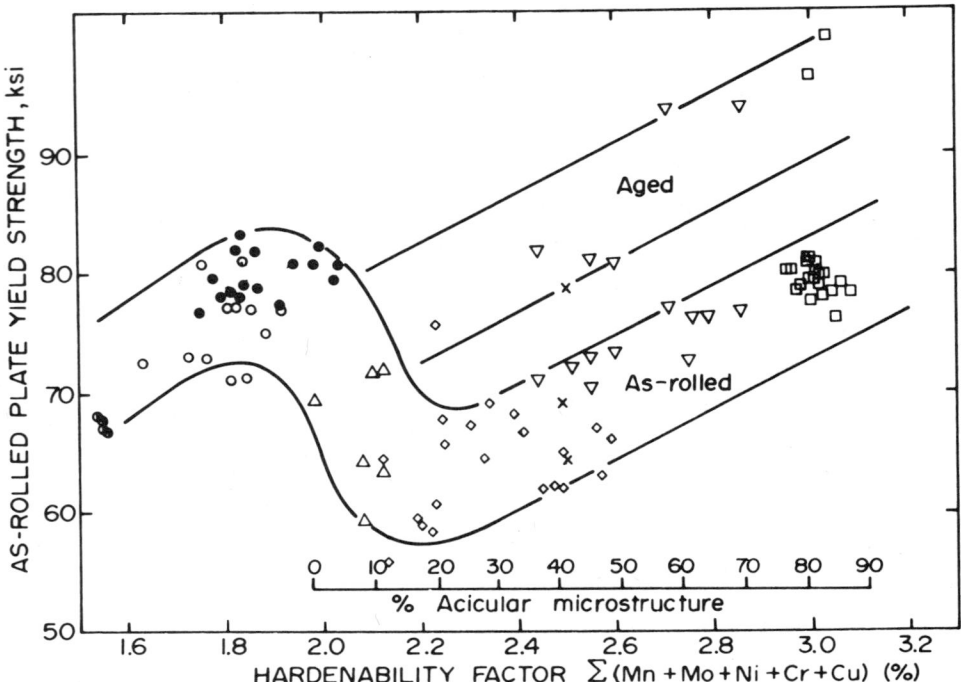

Fig. 12. Effect of hardenability, expressed as the sum of alloying elements, on the yield strength of HSLA steels in both the hot-rolled condition[10] and after aging to promote precipitation hardening.

microstructural changes: (a) refinement of the ferrite grain size, (b) increase in dislocation density, and (c) suppression of alloy carbide precipitation, all of which influence the overall strength level.

This gives at least a qualitative method of rationalizing the relationship of strength and microstructure. The fracture properties are more difficult to rationalize because of the scale of events on which fracture occurs and the important influence that heterogeneous features of the microstructure exert on the nucleation of fracture. Thus it is pertinent to consider the various fracture modes and the influence of various microstructural constituents on nucleation and propagation of fracture.

3 FRACTURE PROCESSES IN HSLA STEELS

For structural steels in which extensive plastic yielding occurs during failure, the concepts of fracture mechanics are difficult to apply and the fracture resistance must be specified by reference to a number of fracture modes. Of concern is the occurrence of cleavage failure and thus the specification of the ductile-to-brittle transition temperature, the total energy-absorption capacity, and the maximum Charpy value of shelf energy. Also, as discussed earlier, the anisotropy of fracture resistance is important, particularly in the occurrence of delamination failures in the plane of rolling.[29] Thus, for clarity, cleavage failure, ductile rupture, and delamination are considered separately below, along with their relationship to the detailed microstructure.

In attempting to produce an overview of the various processes of fracture, it is of value to draw information from a range of steels containing a variety of microstructural constituents. Thus the steels used in the current work are listed in Tables I and II, together with a brief description of their processing history and properties. Some typical microstructural features are exemplified by the composite micrographs in Fig. 13.

3.1 Cleavage Failure

At cryogenic temperatures, the condition for cleavage is usually described in terms of a critical cleavage stress σ_f which is considered independent of temperature and strain rate if failure is induced by slip.[30] For a blunt notch the stress state of the adjacent region can be described by a system of exponential spiral slip lines[31] such that the tensile stress σ_{yy} at X below a notch of radius ρ is

$$\sigma_{yy} = 2\tau_y[1 + \ln(1 + X/\rho)] \quad , \tag{1}$$

or for a notch of included angle θ,

$$\sigma_{yy}^{max} = 2\tau_y\left(1 + \frac{\pi}{2} - \frac{\theta}{2}\right) \quad , \tag{2}$$

where τ_y is the yield stress in shear.

If we use the experimental criterion that at a given temperature, T_{GY}, the fracture event is coincident with general yielding, we can calculate the value of σ_f. By varying the notch angle θ, the value of T_{GY} can be varied and the temperature independence of the critical cleavage stress σ_f established.[32] It remains to define what microstructural features determine the value of σ_f and also the scale or range over which σ_f must be attained. The latter aspect has been considered in detail by Rice and Johnson[33] and by Ritchie, Knott, and Rice[34], who concluded that the maximum tensile stress must exceed σ_f over distances on the order of two grain diameters.

In regard to the microstructural dependence of σ_f, a variety of models (Smith[35], Almond et al.[36], and Lindley et al.[37]) have been developed which

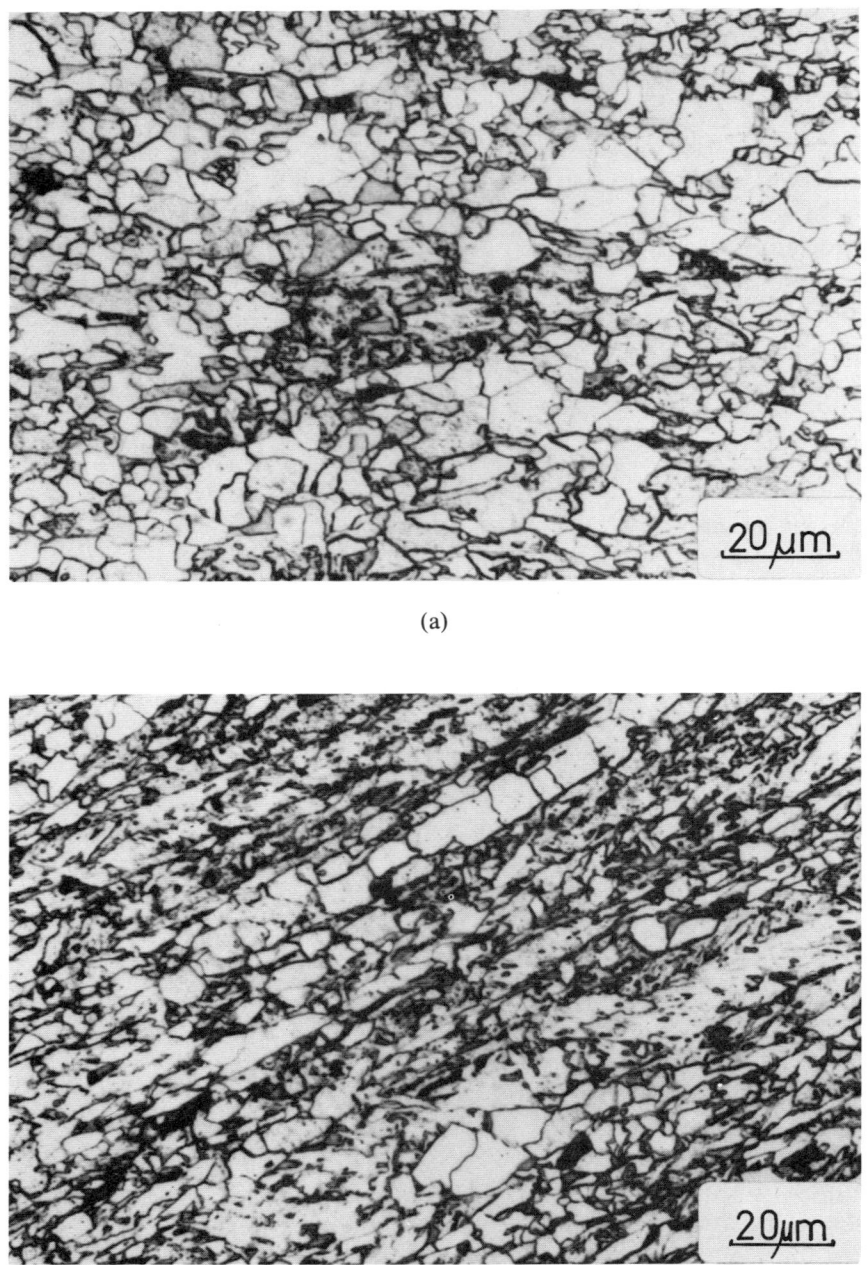

(a)

(b)

Fig. 13. Typical microstructural features of steels A, B, C and D; steels A and C consist of polygonal ferrite plus M/A constituent; steel B consists of both polygonal and acicular ferrite in bands together with M/A constituent; steel D is an acicular steel containing M/A constituent.

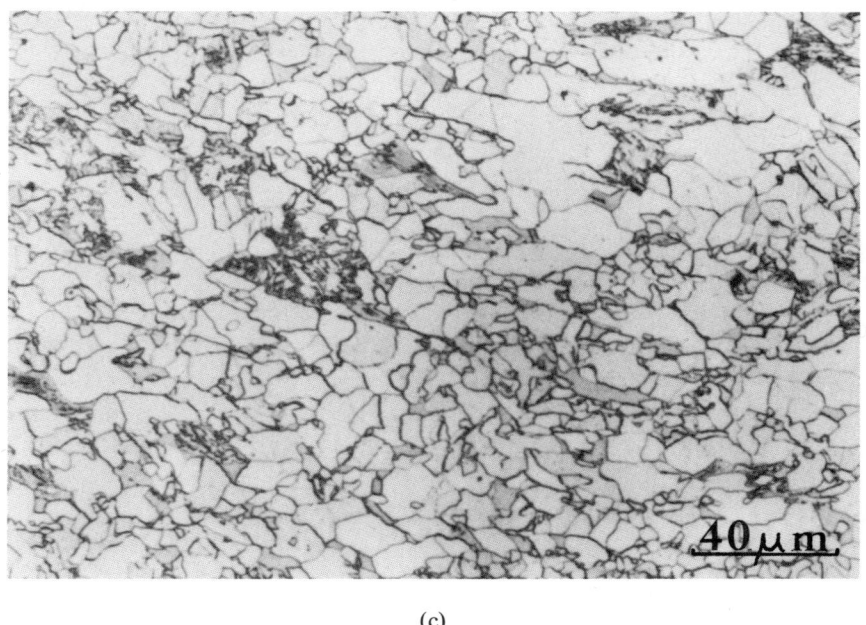

(c)

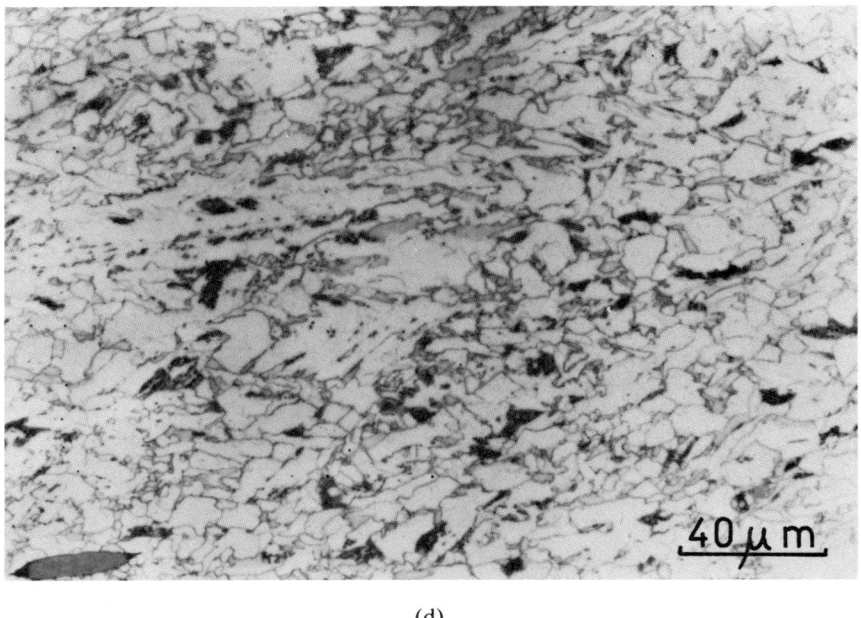

(d)

Fig. 13 (continued)

consider the influence of both the grain size d and the carbide thickness c on the magnitude of σ_f. A typical formulation leads to a fracture stress σ_f of the form

$$\sigma_f = \left(\frac{k^2 d}{4c^2} + \frac{8\mu\gamma}{\pi(1-\nu)c}\right)^{1/2} - \frac{kd^{1/2}}{2c} \quad, \tag{3}$$

where k is the slope of the Hall-Petch equation for uniaxial tension, μ is the shear modulus, γ is the surface energy, and ν is Poisson's ratio.

This type of equation has been used to explain the effect of carbide refinement in quench aging[38] and the influence of carbide thickness on the toughness of alloy steels[39]. In the context of the HSLA steels possessing complex microstructures, the existing models for predicting σ_f assume uniform grain size and no textural effects and take no account of the influence of the transformation stress adjacent to the martensitic or retained-austenite regions of the structure.

To determine the applicability of the critical-cleavage-stress concepts to HSLA steels, notch bend tests were performed on steels A and B (Table I). The results are shown in Figs. 14a and 14b, which indicate that the values of T_{GY} are very low, of the order of -170 C, and the values of σ_f are on the order of 350 ksi for both materials. To determine the critical events occurring during cleavage, the form of the nonpropagating microcracks below the notch and the detailed form of the fracture path (Figs. 15 and 16) were studied. It is apparent that in the regions of polygonal ferrite the non-propagating cracks correspond to the ferrite grain size and the magnitude of the critical cleavage stress is of the order of that expected for a material with grains 5 microns in diameter containing carbides 1-micron thick. However, in the acicular regions, the fracture path is dictated by the scale of the bundles of ferrite laths rather than by the individual lath size, and in many instances failure is initiated by cleavage of the complex nonferritic products.

These observations suggest that some care should be exercised in regard to the definition and control of the cryogenic properties of acicular steels. The scale of the bundles of ferrite laths and the size and distribution of the nonferritic transformation products are very dependent on deformation history, finishing temperature, cooling rate, and section size. Thus, the values of σ_f produced in such structures may exhibit considerable variation through the section, particularly for thick sections. To illustrate this, let us consider the predictions of fracture toughness K_{Ic} based on the values of σ_f and the yield stress σ_y. Following Malkin and Tetelman[40], we can write

$$K_{Ic} = 2.89\, \sigma_y \left[\exp\left(\frac{\sigma_f}{\sigma_y} - 1\right) - 1\right]^{1/2} \rho_0^{1/2} \tag{4}$$

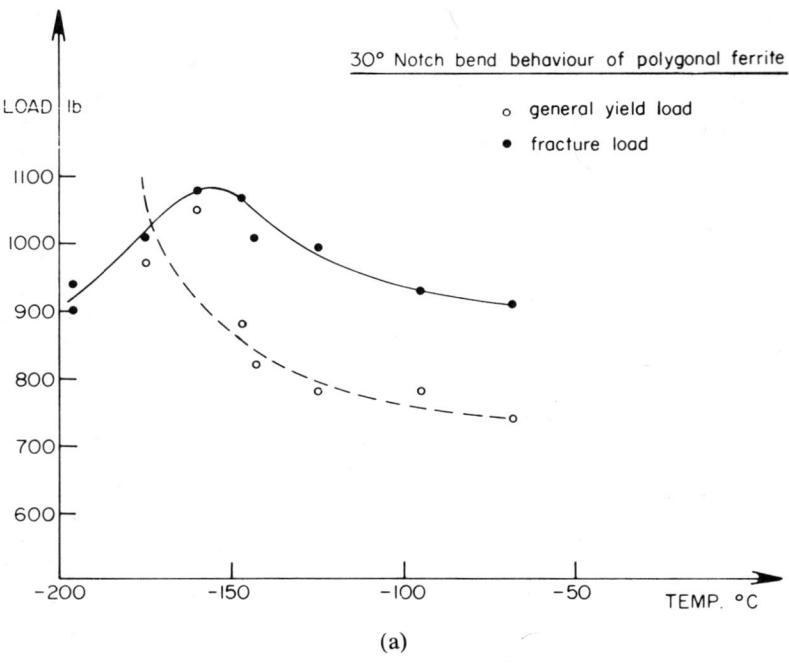

(a)

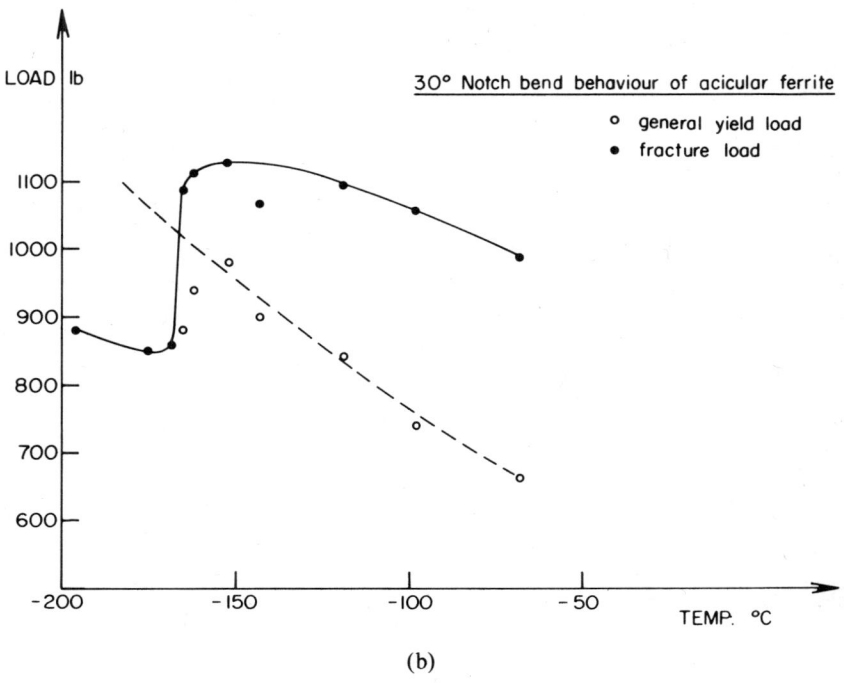

(b)

Fig. 14. Yield and fracture loads in slow bend tests for (a) steel A and (b) steel B.

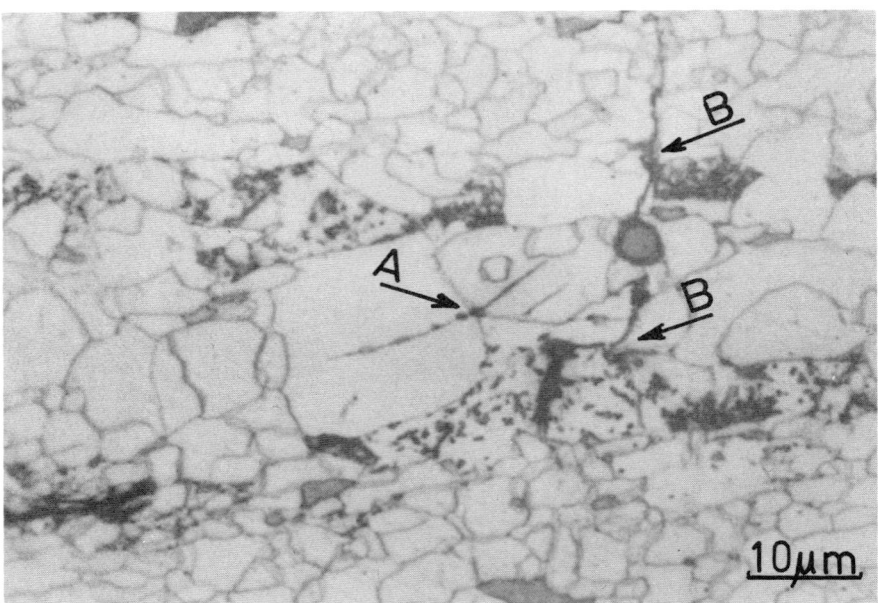

Fig. 15. Cracks initiated below root of notch in steel A; cracks originate from grain-boundary carbides (A) and nonferritic phases (B).

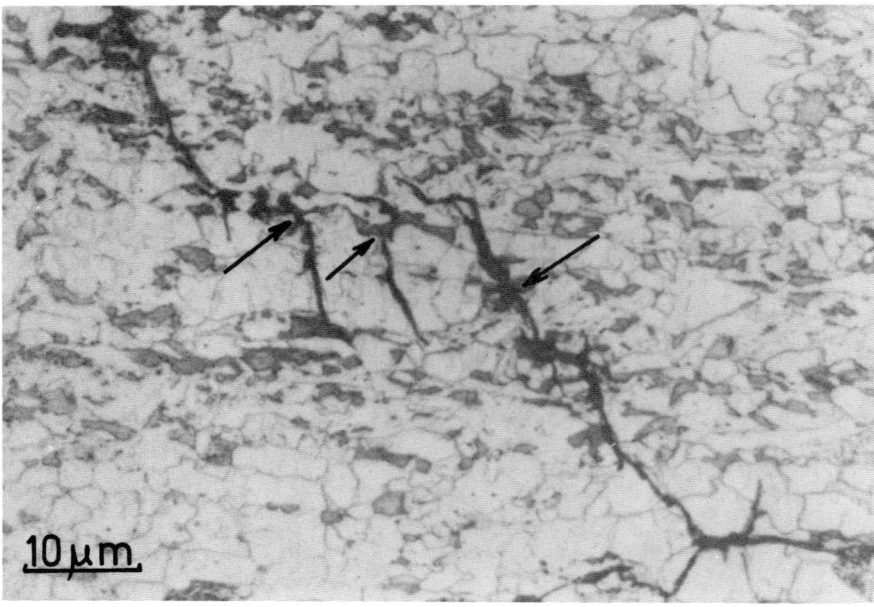

Fig. 16. Cleavage fracture in steel B; arrows indicate cracks in M/A constituent responsible for crack nucleation in the ferrite matrix.

or

$$K_{Ic} = 0.13 \ \sigma_y \left[\exp \left(\frac{\sigma_f}{\sigma_y} - 1 \right) - 1 \right]^{1/2} , \qquad (5)$$

where ρ_o is the limiting notch radius of the order of $-.002$ inch. The predicted toughness of the acicular steel at -170 C with σ_f equal to 350 ksi and σ_y equal to 130 ksi is 36 ksi$\sqrt{\text{in}}$., which is of the same order as that calculated for current reactor-grade steels. However, if a region in which coarse particles of nonferritic transformation product or coarse polygonal grains are considered to be on the order of 10 microns, the value of σ_f could fall to 250 ksi and the resultant K_{Ic} would be reduced to 20 ksi$\sqrt{\text{in}}$.

Thus the nature and scale of the transformation products and the scale of the bundles of ferrite laths in the acicular regions appear to exert a major influence on the criterion for cleavage failure and thus the expected low-temperature toughness. The major problem appears to be not the attainment of fine-scale structures but the ability to reproduce these structures in large section sizes for a variety of processing and cooling conditions.

3.2 Ductile Failure

The process of ductile failure by void initiation and growth at inclusions has been well documented.[41-43] The process of void initiation is difficult to quantify both in terms of the experimental problem of detecting the onset of voiding and the theoretical problem concerning the local competition between continued plasticity and voiding at the particle-matrix interface. Recent work by Argon and co-workers[44] indicates that the process of void initiation can be quantitatively described in terms of the attainment of critical stress at the matrix-particle interface. The strains to cause voiding, ϵ_n, are often in excess of one-half the true strain to failure. However, for closely spaced inclusions, the local plastic fields surrounding the inclusions interact and reduce the strain needed for void nucleation. This has important consequences when the cumulative strains for nucleation and growth are considered, because void initiation occurs most readily at the closely spaced inclusions and thus the total strain depends not only on the volume fraction of inclusions but on the distribution of particle spacings.

The problem of void growth has two important aspects. It can be assumed in a simple geometric model[43] (Fig. 17) that void growth depends on the volume fraction of inclusions V_f, and an expression for the growth strain ϵ_g can be developed in terms of V_f. However, the distribution of strain envisaged in the void-linking process has an important influence. Hahn and others[45] have described processes of strain localization between particles

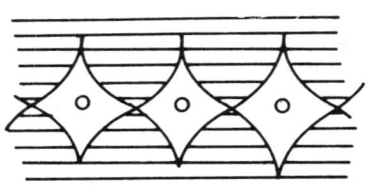

$$2r_0(1 + \varepsilon_g) = r_0 \left(\sqrt{\frac{2\pi}{3f}} - \sqrt{\frac{8}{3}} \right)$$

$$\varepsilon_r = \ln \frac{A_o}{A_f}$$

$$= \ln (1 + \varepsilon_g + \varepsilon_n)$$

$$= \ln \left(\sqrt{\frac{\pi}{6f}} - \sqrt{\frac{2}{3}} + \varepsilon_o \right)$$

Fig. 17. Onset of local necking and void linkup.

which they term "superband" formation. This process is intimately concerned with the extent of ductility in plane strain. In plane strain, in addition to the process of void nucleation and growth, we must consider the cumulative events of strain localization, shear crack initiation, and shear crack growth to a critical size. These events are shown schematically in Fig. 18.

The onset of localized shear is sensitive to both the local work hardening rate and the strain rate sensitivity.[46] The occurrence of negative strain-rate sensitivity leads to both localization of shear and marked reductions in the expected strain for void growth.[47] In structural steels, the reduction of local work hardening capacity either by imposed prestrains or by virtue of the conditioning due to low finishing temperature results in void linkage by shear with low crack-opening-displacement values. In HSLA steel, two features are important which reflect the processing and the nature of the transformation products. The regions of hard constituent consisting of mixed martensite and austenite act essentially as inclusions, i.e., they remain rigid and cause the development of local-plastic-strain gradients. This is important because, despite the low nonmetallic inclusion level, i.e., the amount of sulfides, oxides, etc., a large areal fraction of hard transformation product can severely limit the ductility. Hence, the shelf energy should decrease with increasing amounts of nonferritic transformation product (Fig. 19). In addition to the influence of the hard constituents, the ferrite matrix often possesses a very high dislocation density owing to the low transformation temperature and possibly the deformation of the ferrite during the rolling process. Thus the local work hardening capacity of the matrix is reduced and void coalescence can occur by shear linkage with low

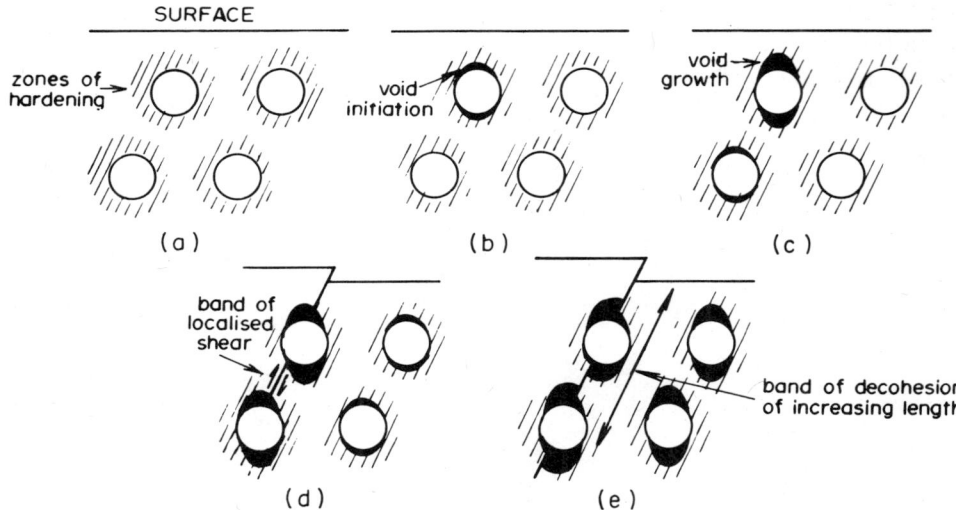

Fig. 18. Cumulative events of strain localization, formation of shear bands, and nucleation of shear crack in relation to the processes of void nucleation and growth.

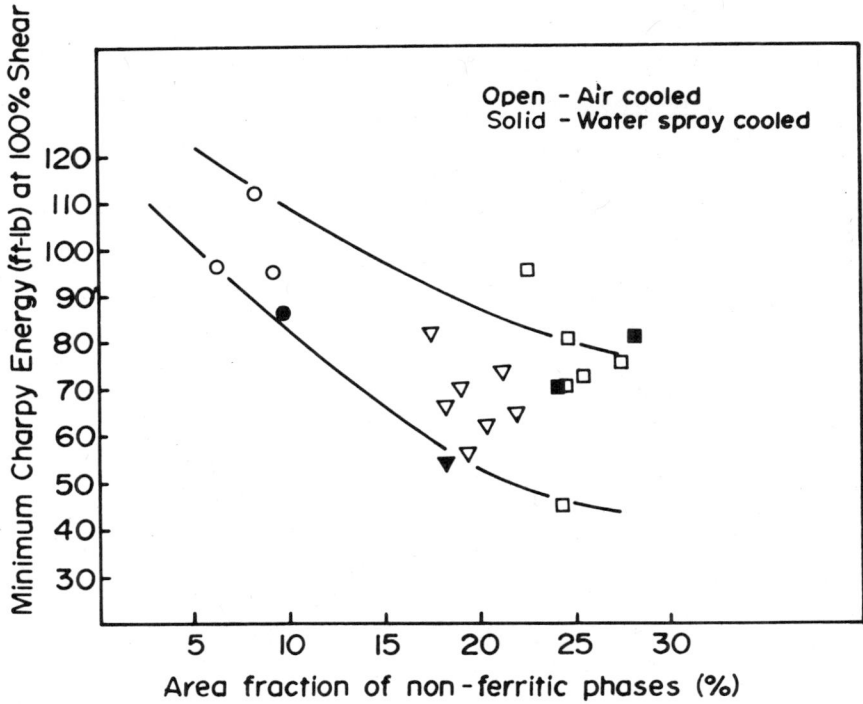

Fig. 19. Relationship between Charpy V-notch shelf energy and volume fraction of non-ferritic phases.[10]

ductility and small crack-opening displacement. This process can be seen in the failure of both tensile samples and samples subject to stretch bend testing, as shown in Fig. 20.

Fig. 20. Voids nucleated at nonferritic constituents and the development of shear bands between the voids (arrowed region); the local shear strains are of the order of unity for crack growth.

In a complex test, such as the stretch bend test, there exists a strain gradient through the thickness. The sequence of events appears to be that void initiation occurs at the interface between the hard transformation products and the ferrite at the region adjacent to the outer surface. Voids grow and link by localized shear along the loci of planes of velocity discontinuity. Similarly, in tensile failures soon after necking, the sample fails along shear zones within the neck at overall strains which result in low values of the true strain to failure, e.g., a comparison of the strains to failure between steels C and D indicates that steel C fails at a lower strain despite the low sulfur content and inclusion modification used in steel C.

This observation has important implications in regard to the toughness of HSLA steels and their resistance to rapid tear failures. Much of the data on fracture resistance measured by both the Charpy shelf energy and by fracture-toughness parameters such as K_{Ic} suggested that for many structural materials there is a correlation between fracture resistance and strain to failure[48,45], as shown in Fig. 21. Thus the presence of the microstructural factors discussed above and the tendency toward localized shear may reduce the fracture resistance. The extent of this deterioration depends on the distribution and volume fraction of the transformation product and on the state of stress which determines the strain to failure. Clausing[49], Ostermann[50], and others have shown the importance of measuring the strain to

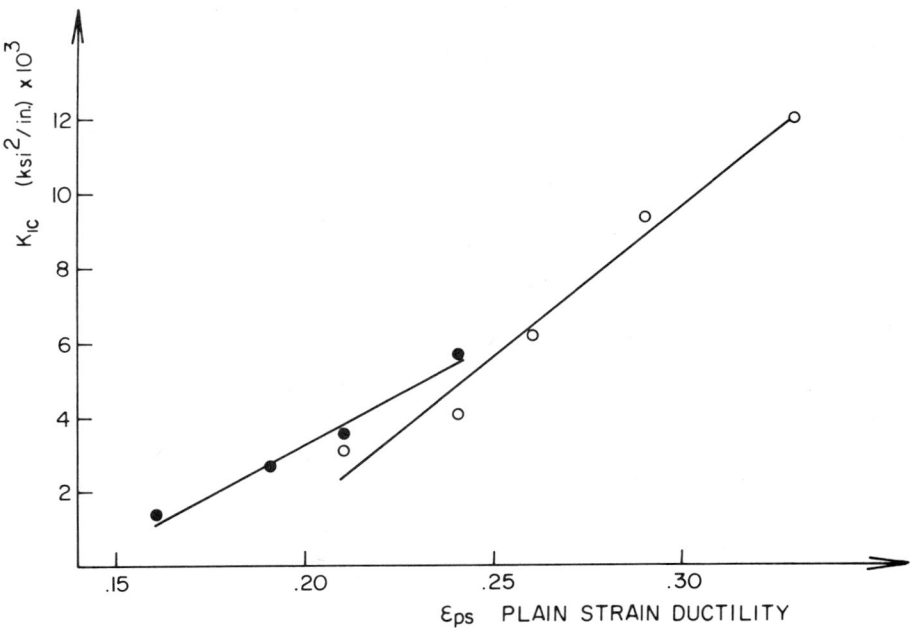

Fig. 21. Correlation between fracture toughness and plane-strain ductility.[45,49]

failure under nominal plane-strain conditions and the correlation between toughness and plane-strain ductility. This concept can be extended to map the strain to failure as a function of the state of stress. If we draw an analogue with the yield surface and consider the net strain history in any stress state, we can proceed as follows. The work of deformation $d\omega$ can be expressed for any stress state as

$$d\omega = \sigma_{ij} \cdot d\epsilon_{ij} \quad .$$
(6)

If we consider stress and strain as vectors (rather than tensor quantities), then the work done is the product $\sigma \cdot d\epsilon$

$$d\omega = \sigma \cdot d\epsilon = \sigma_1 d\epsilon_1 + \sigma_2 d\epsilon_2 + \sigma_3 d\epsilon_3 \quad .$$
(7)

The general concept of the maximization of external work leads to the condition that the incremental strain vector is normal to the yield surface.[51] Thus, if we consider the measured strain to failure in a given stress state ϵ_f*, we can determine the extent to which the yield surface can be extended in stress space to become a failure surface, assuming that the normality rule applies throughout the strain path. This surface can in effect be viewed as the surface of maximum plastic potential defined in such a way that by measuring the strain to failure and using an associated flow rate, we can follow the effective stress—effective strain relation to define the extent to which the yield surface, i.e., initial surface of plastic potential, can be expanded. This is shown schematically in Fig. 22 and an example of this construction for steel D is shown in Fig. 23. Although this type of construction has not been used in a quantitative sense, it does allow a comparison to be made between materials such that one can consider the influence of microstructural features on the fracture strain for a variety of stress states. This may be of value in considering the forming potential of HSLA steels, where components are subject to complex strain histories involving a variety and sequence of stress states.

3.3 Short-Transverse Properties

As stated earlier, the successful utilization of high-strength structural steels in many formed and welded structures depends on the attainment of satisfactory through-thickness properties. Recently, many workers[29,52,53] have reported the occurrence of delamination failure parallel to the rolling plane. This type of failure can occur as splitting in Charpy samples, as delamination parallel to the tensile axis in uniaxial tensile tests, and in full-scale tests of sections of pipeline. At present, it is difficult to discuss the influence of this type of delamination failure on factors such as the overall

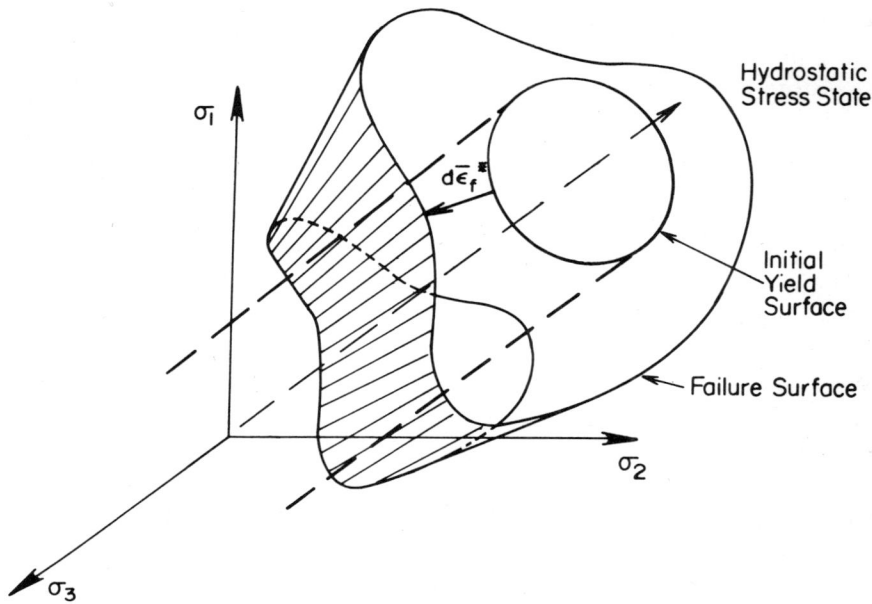

Fig. 22. Schematic showing how the initial yield surface can be expanded to a failure surface; the displacement in stress space is indicated by the effective strain vector $\bar{\epsilon}_i^*$.

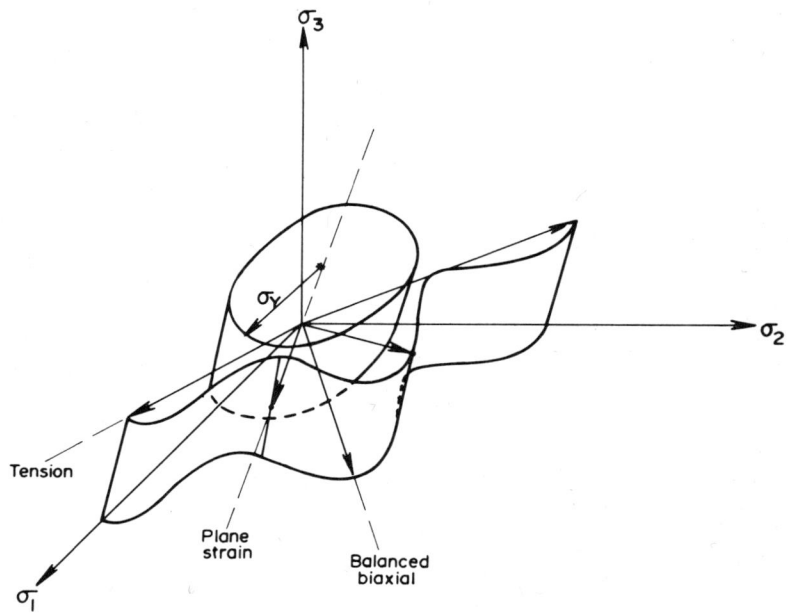

Fig. 23. Failure surface developed for steel D tested in uniaxial tension, plain strain, and
balanced biaxial-strain conditions.

fracture resistance in large-scale tests because of the scatter in the available data. One important feature may be that delamination causes changes in the stress state ahead of the advancing crack. As the energy absorbed per unit area depends on the state of stress and thus the local ligament thickness, the rate of energy consumption can vary, depending on the scale on which delamination occurs. This is also reflected in tests where the energy absorption is measured. For example, the transition from ductile to brittle fracture in a Charpy test will take place over a broad temperature interval because of the change in effective specimen geometry when delamination occurs.

Several models have been proposed in order to explain the mechanism of delamination fracture, assuming that planes of weakness exist parallel to the rolling plane. The origin of these planes of weakness can be either crystallographic anisotropy or mechanical fibering. Miyoshi et al.[52] have suggested that delamination fracture is in fact a cleavage failure in material with a strong (100)-texture component. This texture component has been observed in materials subjected to rolling in the ferrite range (Coleman et al.[54]). Mechanical fibering is noncrystallographic in origin and can be described as alignment of matrix discontinuities from mechanical processing. The most commonly observed discontinuities are inclusions, segregation, and nonferritic transformation products. In mild steel, delamination fracture along MnS stringers is often observed. Recently, delamination along sheets of cerium oxides resulting from unsatisfactory steel-melting practice has been found, Fig. 24. Segregation of impurities to austenite grain boundaries might provide a path for easy fracture propagation, but experimental evidence for this mode is lacking.

Hero et al.[29] and others[53] have shown how the delamination failure propagates along ferrite grain boundaries in experimental steels containing few inclusions (Fig. 25). This mechanism can arise in banded microstructures or due to the presence of carbides on ferrite grain boundaries. Decohesion of the ferrite-carbide interfaces and the ferrite boundaries gives rise to linear arrays of voids when subjected to large strains, as in the neck of a tensile specimen (Fig. 26). In establishing criteria for delamination, it must be recognized that not only can a variety of microstructural features provide planes of weakness, but the criteria for delamination may involve the attainment of either a critical stress or critical strain condition.

A quantitative comparison of these conditions can be established by comparing the delamination process using short-transverse tensile specimens and a Bridgeman analysis on longitudinal tensile specimens. From the Bridgeman calculation[55], the transverse stress in the neck can be expressed as

$$\sigma_{rr} = F \ln \left(1 + \frac{a}{2R}\right) , \qquad (8)$$

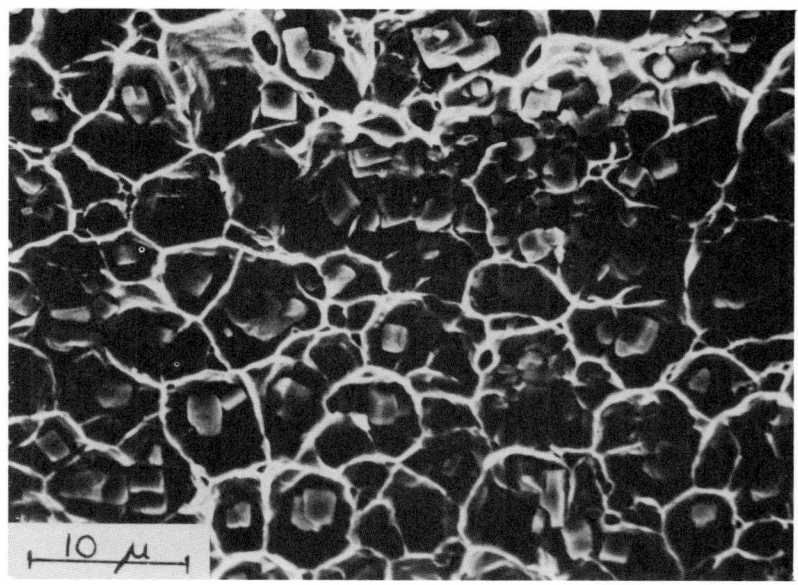

Fig. 24. Oxides of aluminum and cerium on fracture surface of through-thickness tensile specimen from steel A.

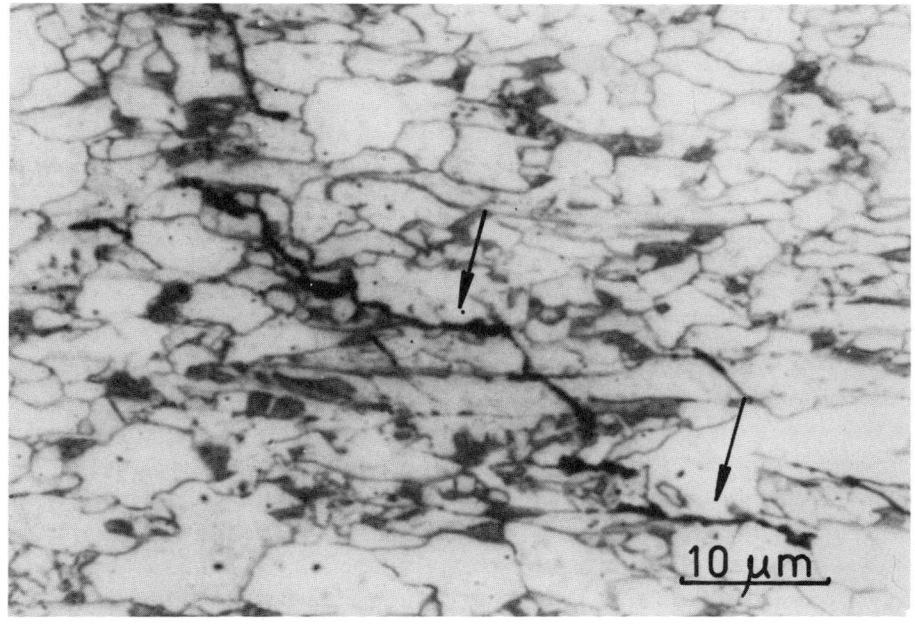

Fig. 25. Delamination along ferrite grain boundaries in steel B (grain-boundary cracks are indicated by arrows).

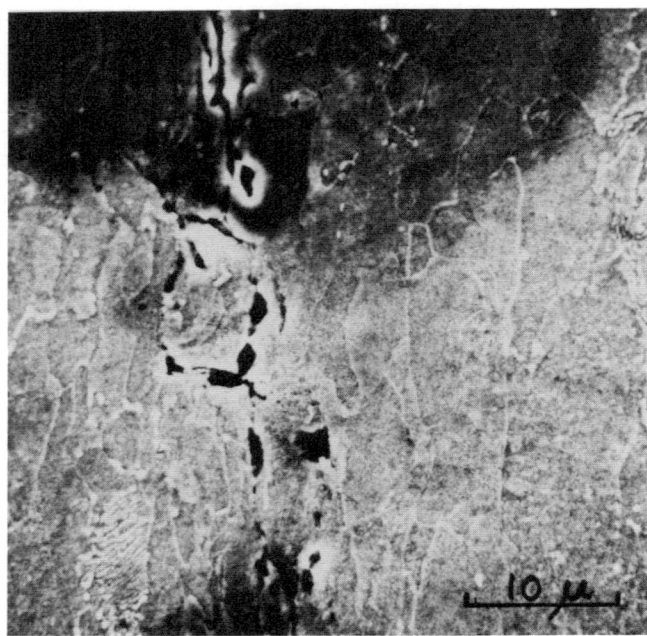

Fig. 26. Linear array of voids in the neck of a longitudinal tensile specimen (from steel A).

where F is the flow stress of the material, a is the radius of the load-bearing area, and R the radius of curvature of the neck. Delamination stresses (in the transverse direction) of the order of 50 to 65 ksi have been found in materials with 75-ksi yield strength at strains of the order of 1.0. Subjecting the same materials to short-transverse tensile tests yielded failure stresses of the order of the yield stress over a range of temperatures, the failures occurring at very small strains. Investigation of fracture surfaces shows that failure in the short-transverse tensile direction is associated with large colonies of inclusions (cerium oxides), whereas delamination in longitudinal specimens is connected to tearing of ferrite boundaries. Thus, the delamination due to grain-boundary failure appears to involve the decohesion of boundaries after they are subject to large strains, and occurs at stress levels lower than those needed for failure at inclusions. This is in agreement with data from Groom and Knott[53] who found that the tendency to delamination increases with increasing amount of prestrain in the material before testing.

   It is clear from both the data available in the literature and the results reported above that there is no unique microstructural cause for delamination failure. However, the sensitivity of the onset of delamination to microstructural features such as texture and the distribution of carbides at grain boundaries, as well as the degree of local plastic strain, poses serious problems in regard to quality control. If the occurrence of delamination severely reduces the fracture resistance of structures, such as pressure vessels or pipelines, and increases the tendency to in-plane delamination at welded joints, then as-rolled materials may be difficult to optimize and the tendency to delamination may have to be reduced by a separate heat treatment such as normalizing.[29]

4 STRAIN REVERSALS IN HSLA STEELS

In fabricating operations such as pipemaking and pipe expansion, the material is subjected to a strain reversal. This induces a Bauschinger loss, i.e., the flow stress on reversal is lower than the flow stress in the original direction of straining. A number of authors have attempted to quantify the magnitude of the Bauschinger effect and to relate this both to the microstructure and to the detailed strain cycle[14,56-59], e.g., a comparison of the yield strengths measured on the original plate, ring expansion tests on the finished pipe, and cold-flattened samples obtained from the pipe. However, before attempting to model the Bauschinger losses sustained in complex fabrication processes, some attention must be given to the origin of the flow-stress differences in relation to the microstructure of HSLA steels. Also, it is important in complex deformation processes to distinguish changes in flow stress which arise from the storage of elastic back stresses due to the presence of rigid nonyielding phases and those which arise from a change in the loading path of the structure.

The structural portion of the Bauschinger effect can be considered to arise from the presence of an elastic back stress which aids reverse flow. The back stress may be generated by the original prestrain or by a phase transformation.[60] For simple, well-characterized two-phase materials, Brown and Stobbs[61] have produced a model to relate the magnitude of the Bauschinger effect to the scale of the microstructural features, e.g., the size and volume fraction of second-phase particles and the extent of the prestrain. However, the material may contain patterns of residual stress due to fabrication, e.g., nonuniform plasticity, or due to local phase transformations or texture gradients which also affect the flow process. It is important at small plastic strains to recognize that a distribution of residual elastic stress affects not only the magnitude of the flow stress but also the form of the flow curve and the detailed transition from elastic to fully plastic behavior.

Following the model of Friedel[62], we can consider a material containing a distribution of back stresses $-\sigma_B$ to $+\sigma_B$. For a material which initially has a sharp yield point at σ_C, the inclusion of a distribution of back stresses modifies the yield behavior as shown in Fig. 27, so that the initial flow curve is essentially parabolic in form. A similar model can be developed by considering a dislocation model in which a distribution of source lengths is included. Thus, the removal of the initial yield point and the production of a

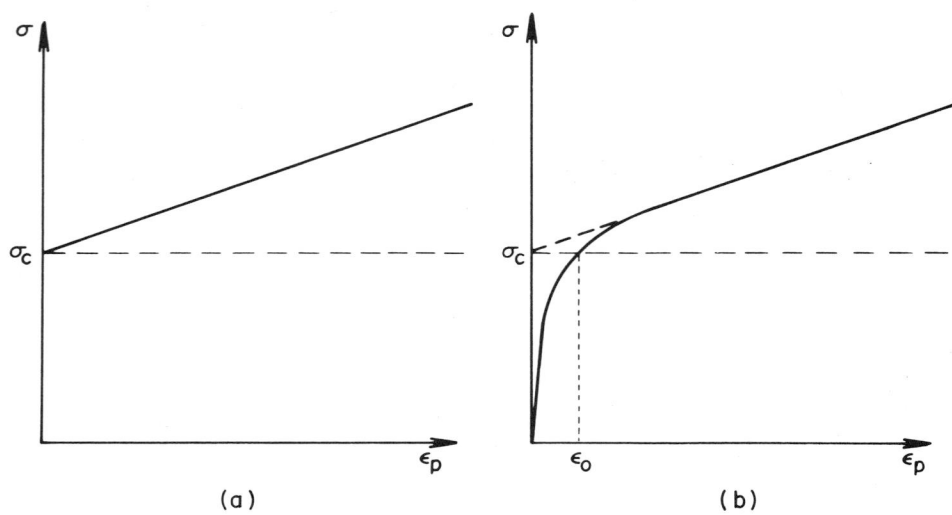

Fig. 27. Effect of internal stress on the slope of the yield curve: (a) stress free material, (b) material containing internal stresses.

rounded stress-strain curve with a high initial hardening rate which is observed[11] in many acicular ferrites may reflect either the influence of a pattern of back stresses due to the presence of the austenite/martensite transformation products or the high dislocation density of the nonpolygonal ferrite.

Currently, there is no theoretical treatment which treats fully the magnitude of the forward and reverse flow stresses and the associated forms of the stress-strain curves in a quantitative manner. However, we can develop an approximate model following the model of Brown and Stobbs[61] and others[63]. If the material has an initial flow stress σ_0, let the forward flow stress be σ_F at a given strain and the reverse flow stress be σ_R at the same strain. Let the work hardening be composed of a forest hardening term σ_{for} which is assumed symmetrical in its action with respect to forward and reverse strain and an elastic back stress σ_B which aids reverse flow.

$$\sigma_F = \sigma_0 + \sigma_{for} + \sigma_B \tag{9}$$

$$\sigma_R = \sigma_0 + \sigma_{for} - \sigma_B \quad . \tag{10}$$

Thus we can define a parameter

$$\frac{\sigma_F - \sigma_R}{\sigma_F - \sigma_0} = \frac{2\sigma_B}{\sigma_F - \sigma_0} \tag{11}$$

which we will term the Bauschinger effect parameter (BEP) and which is a *qualitative* measure of the fraction of total work hardening due to the elastic back stress. It should be noted that the factor BEP is not an indication of the local value of σ_B in the vicinity of some plastic inhomogeneity such as a second-phase particle. The magnitude and extent of the local back stresses are difficult to estimate without use of either detailed fitting of forward and reverse stress-strain curves, which is difficult in ferrous systems, or by reference to some supplementary observation such as X-ray determinations of lattice parameters.

Given the limitations of the approach, we can consider a two-phase material containing a volume fraction V_f of rigid inclusions of radius r subject to a strain ϵ. Following Brown and Stobbs[61] we can express σ_{for} and σ_B as

$$\sigma_{for} = \sqrt{3\alpha\mu} \; V_f^{1/2} \left(\frac{8b'\epsilon}{\pi r}\right)^{1/2} \tag{12}$$

$$\sigma_B = \left(\frac{8\pi b}{\epsilon r \alpha^2}\right)^{1/8} \alpha\mu V_f \left(\frac{8b\epsilon}{\pi r}\right)^{1/2} \tag{13}$$

Thus $(BEP)^{-1} = C_2 + C_1 f^{-1/2}$

and $\dfrac{\sigma_B}{\sigma_{for}} \simeq C_3 f^{1/2}$

Hence, the term BEP should be essentially independent of forward strain and the ratio of the elastic back stress to the forest hardening terms is proportional to the square root of the volume fraction of hard particles. These relationships have been tested by Ibrahim[64] for a range of carbon steels, and the results are shown in Figs. 28 and 29. In addition, in Fig. 29, it can be seen that for some HSLA steels, the hard nonferritic constituents appear to act in a manner qualitatively similar to that of rigid precipitates, as exemplified by the carbides (circles in Fig. 29). Thus, despite the differences in the scale of the flow processes, it appears that the approximate magnitude of the effects in reverse flow can be predicted in terms of the volume fraction of nonferritic components. It is of value to note that the reverse flow parameter BEP defined in Eq. (11) is of value in regard to commercial steels. The magnitude of the flow-stress losses due to strain reversals decreases with increasing initial hardening rate, i.e., a steep initial rate of hardening compensates the buildup of elastic back stresses which is the essential feature emphasized by the manufacturers of acicular steels.[12]

Thus far, it has been tacitly assumed that the back stresses considered above arise from prestrain. However, similar arguments could be advanced in regard to back stresses generated by transformation strains.

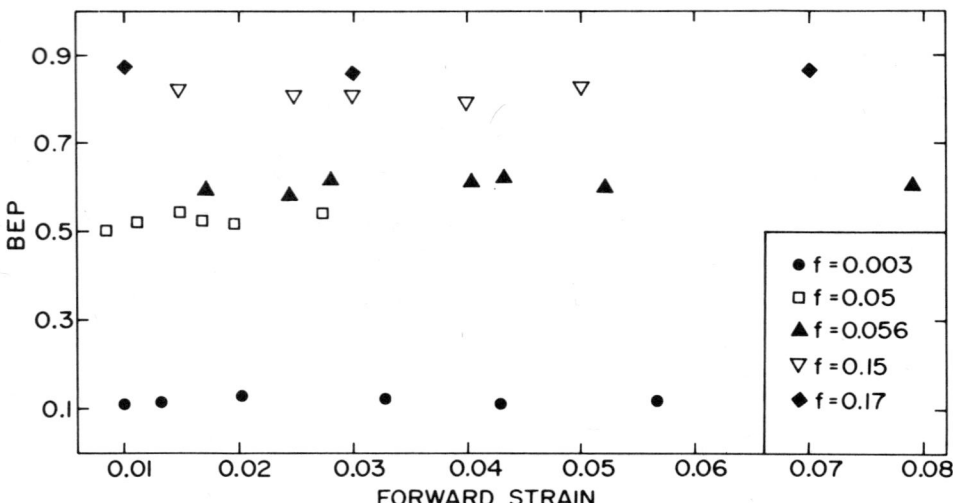

Fig. 28. Correlation between the magnitude of the Bauschinger effect (BEP) and forward strain for different volume fraction of second-phase particles.[65]

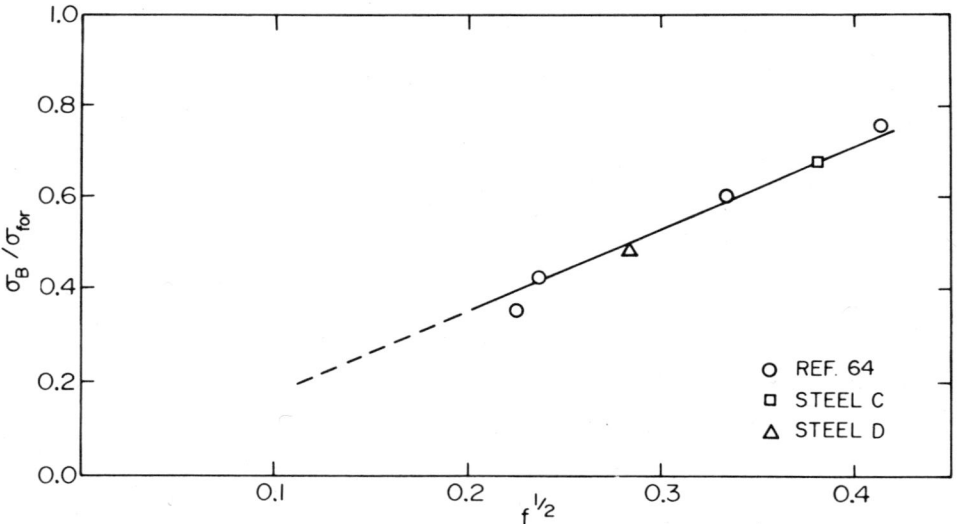

Fig. 29. Ratio between back stress and forest hardening as a function of volume fraction of hard phases.

If we consider a volume fraction of material V_f in which a transformation strain ϵ_T occurs, then, in the absence of plastic relaxation around the regions, the change in elastic energy ΔF_{el} can be expressed as

$$\Delta F_{ei} \simeq \mu V_f\, \epsilon_T^2 \quad,$$

where μ is the shear modulus and the upper limit of the back stress σ_B will be of the order of

$$\sigma_B = 2\mu\, V_f\, \epsilon_T \quad.$$

Thus we would expect, qualitatively, that as the volume fraction of the transforming phase, e.g., the austenite martensite constituent, increases, the yield curve becomes more rounded and the magnitude of the Bauschinger effect should increase, which is in accord with the evidence presented above. However, more work is needed to make a more quantitative approach to the problem. The important point raised here is that at small reverse strains in HSLA steels, one must consider the interaction between the stress system generated by inhomogeneous yielding and that due to localized transformation. This is emphasized by the experimental observations that the magnitude of the Bauschinger effect recorded in HSLA steels containing quantities of the austenite/martensite constituent is dependent on *the initial sense* of prestrain and on heat treatment.[57,65]

5 CONCLUSIONS

From the evidence presented in the preceding sections, it can be seen that modern structural steels exhibit a number of interesting features in regard to alloy design. Many of the important features of their behavior, such as the nucleation of cleavage cracks, their ductility and behavior under reverse strain condition, are closely connected to heterogeneities such as the presence of nonferritic phases. These constituents differ from the conventional definition of inclusions in that their volume fraction and distribution is extremely sensitive not only to alloy chemistry, but also to the thermal-mechanical history, cooling rate, and section size. Further, the detailed characterization of these important constituents is currently obscure and they represent a potentially major problem in regard to quality control and uniformity of behavior in the HSLA steels.

In considering fracture processes, it is clear that in descriptions of ductile failure, the strain to failure is sensitive not only to the amount of microstructural inhomogeneities and the operative stress state, but also to the distribution of strain. The amelioration of strain localization may prove as important as improved alloy cleanliness in providing adequate ductility for high-strength structural materials when considered in terms of their response under a variety or sequence of stress states.

The preliminary descriptions of failure surfaces advanced in this work are not rigorous in dealing with the strain history of a given element in terms of sequential loading paths. They do, however, provide a mechanism for comparing material in regard to identifying both stress states and microstructural conditions in which competitive modes of failure must be considered.

The behavior in reverse-strain operations suggests that detailed consideration be given to the interaction of transformation stresses and those arising from plastic inhomogeneity. Further, the crude description advanced in Section 4 may provide a basis for linking the microstructural features to a more general description of mechanical anisotropy.

ACKNOWLEDGMENTS

The authors wish to acknowledge the research support received from the National Research Council of Canada and Noranda Research Centre. In addition, the provision of research fellowships by the Steel Company of Canada and N.T.N.F. (Norway) is gratefully acknowledged.

REFERENCES

1. Gray, J. M., "Metallurgy of HSLA Pipeline Steels", Molybdenum Corp. of America, Report No. 7201 (1972).
2. Fukuda, M., Hashimoto, T., and Kunishige, K., *The Sumitomo Search*, p. 8 (May, 1973).
3. Irvine, K. J., Gladman, T., Orr, J., and Pickering, F. B., *J.I.S.I.*, **208**, 717 (1970).
4. Baird, J. D., and Preston, R. R., *Processing and Properties of Low Carbon Steel*, J. M. Gray (Ed.), AIME Metals Society Conference Proceedings, New York (1973), p. 1.
5. Johansen, T. H., Christensen, N., and Augland, B., *Trans. Met. Soc. AIME*, **239**, 1651 (1967).
6. Dewsnap, R. F., and Frost, M. G., *The Controlled Processing of a Low Carbon Niobium Steel Suggested for Low Temperature Linepipe Applications*, British Steel Corporation, Report No. MG/36/72 (1972).
7. Bramfitt, B. L., and Marder, A. R., *Processing and Properties of Low Carbon Steel*, J. M. Gray (Ed.), AIME Metals Society Conference Proceedings, New York (1973), p. 191.
8. Lu, W.-K., and McLean, A., *Metals and Materials*, **8**, 452 (1974).
9. Thompson, T., and Baker, J. M., *Aust. J. of Metallurgy*, **14**, 84 (1969).
10. McCutcheon, D. B., Trumper, T. W., and Embury, J. D., *Proc. J. Internat. de Siderurgie*, Paris, France, October 4, 1974.
11. Mihelich, J. L., Paper presented at CIMM Conference, Edmonton, Canada, 1975.
12. Civallero, M., Parrini, C., and Pizzimenti, N., *Proc. Microalloying 75*, Washington (1975).

13. Nakajima, K., Mizutani, W., Kikuma, T., and Matumoto, H., *Trans. I.S.I.J.*, **15**, 1 (1975).
14. Jamieson, R. M., and Hood, J. E., *J.I.S.I.*, **209**, 46 (1971).
15. Irvine, K. J., *Symposium on Low Alloy High Strength Steels*, published by The Metallurgy Companies, Dusseldorf (1970).
16. Woodhead, J. H., and Whiteman, J. A., *Processing and Properties of Low Carbon Steel*, J. M. Gray (Ed.), AIME Metals Society Conference Proceedings, New York (1973), p. 145.
17. Coldren, A. P., Cryderman, R. L., and Smith, Y. E., loc. cit., p. 163.
18. Shewmon, P. G., *Transformations in Metals*, McGraw-Hill, New York (1969), p. 210.
19. Cooper, K., and Embury, J. D., *Can. Met. Quart.*, **14**, 69 (1975).
20. Matsuda, S., Inoue, T., Mimura, H., and Okumura, Y., *Toward Improved Ductility and Toughness*, Climax Molybdenum Co. (1970), p. 45.
21. Davenport, A. T., Berry, F. G., and Honeycombe, R.W.K., *Met. Sci. J.*, **2**, 104 (1968).
22. Pickering, F. B., *Symposium: Transformation and Hardenability in Steels*, Climax Molybdenum Co. (1967), p. 109.
23. Biss, V., and Cryderman, R. L., *Met. Trans.*, **2**, 2267 (1971).
24. Webster, T. H., Dillamore, I. L., and Smallman, R. E., *Met. Sci. J.*, **5**, 74 (1971).
25. Davidson, R. M., and Dendel, D. R., Paper presented at CIMM Conference, Quebec City, Canada (1973).
26. Whiteley, R., and Wise, J., *Flat Rolled Products*, Vol. III, AIME (1962).
27. Richards, P. N., *J.I.S.I.*, **207**, 1333 (1969).
28. Forman, A.J.E., and Makin, M. J., *Can. J. Phys.*, **45**, 511 (1967).
29. Hero, H., Evensen, J. D., and Embury, J. D., *Can. Met. Quart.*, **14**, 117 (1975).
30. Knott, J. F., *Fundamentals of Fracture Mechanics*, John Wiley & Sons, New York (1973).
31. Hill, R., *Mathematical Theory of Plasticity*, Oxford, London (1950).
32. Knott, J. F., *J.I.E.I.*, **204**, 104 (1966).
33. Rice, J. R., and Johnson, M. A., *Inelastic Behaviour of Solids*, M. F. Kanninen, W. Adler, A. Rosenfield, and R. Jaffee (Eds.), McGraw-Hill, New York (1970), p. 641.
34. Ritchie, R. O., Knott, J. F., and Rice, J. R., *J. Mech. Phys. Solids*, **21**, 395 (1973).
35. Smith, E., *Proceedings of Conference on Physical Basis of Yield and Fracture*, p. 36, Institute of Physics, Physical Society, Oxford, England (1966).
36. Almond, E. A., Timbres, D. H., and Embury, J. D., *Fracture 1969*, Proceedings of Second International Conference on Fracture, Chapman and Hall, London, England (1969).
37. Lindley, T. C., Oates, G., and Richards, C. E., *Acta Met.*, **18**, 1127 (1970).
38. Almond, E. A., Timbres, D. H., and Embury, J. D., *Can. Met. Quart.*, **8**, 51 (1969).
39. Gerberich, W. W., and Guest, P. J., *Int. J. Fract. Mech.*, **6**, 98 (1970).
40. Malkin, J., and Tetelman, A. S., *Eng. Fract. Mech.*, **3**, 151 (1971).
41. Gurland, J., and Plateau, J., *Trans. ASM*, **56**, 442 (1963).
42. McClintock, F. A., *Ductility*, American Society for Metals, Metals Park, Ohio (1968), p. 255.
43. Brown, L. M., and Embury, J. D., *Proceedings of Third International Conference on Strength of Metals and Alloys*, Cambridge (1973), p. 164.
44. Argon, A. S., Im, J., and Safoglu, R., Paper presented at Third International Conference on Fracture, Munich, W. Germany (1973).
45. Hahn, G. T., Barnes, C. R., and Rosenfield, A. R., Report to Aerospace Research Labs, Air Force Systems Command; Contract AF 33615-71-C-1915, Battelle Memorial Institute, Columbus, Ohio (1975).
46. Chung, N., Embury, J. D., Evensen, J. D., Hoagland, R. G., and Sargent, C. M., Submitted for publication in *Acta Met.*

47. Thomason, P. F., *Met. Sci. J.*, **3**, 139 (1969).
48. Kula, E. B., and Anctil, A. A., *J. Materials*, **4**, 817 (1969).
49. Clausing, D. P., *Int. J. Fract. Mech.*, **6**, 71 (1970).
50. Osterman, F., *Abhangigkeit der Bruchzahigkeit von plastischen Werkstoffkenngrossen bei hochfesten Aluminium-Legierungen*, Internal Report, Vereinigte Aluminium Werke AG, Bonn, W. Germany (1974).
51. Backhofen, W. A., *Deformation Processing*, Addison-Wesley (1972).
52. Miyoshi, E., Fukuda, M., Iwanaga, H., and Okazawa, T., *Proc. Inst. Gas. Eng. Conf., Crack Propagation in Pipelines*, Newcastle (1974).
53. Groom, J.D.G., and Knott, J. F., *Met. Sci. J.*, **9**, 390 (1975).
54. Coleman, T., Dulieu, D., and Gouch, A., *Proceedings of Third International Conference on Strength of Metals and Alloys*, Cambridge (1973), p. 70.
55. Bridgeman, P. W., *Studies in Large Plastic Flow and Fracture*, McGraw-Hill, New York (1952).
56. Kishi, T., and Tanabe, T., *J. Mech. Phys. Solids*, **21**, 303 (1973).
57. Sejnoha, R., and Palacios, F., *J.I.S.I.*, **211**, 778 (1973).
58. Wilson, D. V., and Konnan, Y. A., *Acta Met.*, **12**, 617 (1964).
59. Kishi, T., and Gokyu, I., *Met. Trans.*, **4**, 390 (1973).
60. Eshelby, J. D., *Internal Stresses and Fatigue in Metals*, G. M. Rassweiler and W. L. Grube (Eds.), Elsevier, Amsterdam, Netherlands (1959), p. 41.
61. Brown, L. M., and Stobbs, W. M., *Phil. Mag.*, **23**, 1201 (1971).
62. Friedel, J., *Internal Stresses and Fatigue in Metals*, G. M. Rassweiler and W. L. Grube (Eds.), Elsevier, Amsterdam, Netherlands (1959), p. 220.
63. Ibrahim, N., and Embury, J. D., *Mat. Sci. Eng.*, **19**, 147 (1975).
64. Ibrahim, N., Ph.D. Thesis, McMaster University, Hamilton, Ontario, Canada (1974).
65. Filipovic, A., unpublished results.
66. Boyd, J. D., Canada Centre for Mineral and Energy Technology, Report No. ERP/P-MRL-75-13(R) (June, 1975).

DISCUSSION on Paper by J. D. Embury

LOW: What leads to strain localization? Is it a change in work-hardening rate or is it induced by the mechanical situation in the material with growing voids approaching each other?

When strain localization does occur what is the failure mechanism— void sheet formation or mode III crack propagation? If void sheet, what void nucleating constituent is responsible?

EMBURY: Let me try and deal with the two points separately because I consider that they are separate issues. First, the strain localization does not require voids to be present in the material. It can occur when the *local values* of parameters such as strain hardening or strain-rate sensitivity reach critical values. We have recently examined this in detail for aluminum alloys (*Acta Met.*, 1975, in press) and it is clear that by taking cognizance of the local values of these parameters, two situations may be defined: (a) strain localization due to geometrical instability, e.g., tensile necking; and (b) localization due to material instability, e.g., catastrophic shear prior to necking.

In regard to the mechanism of growth, both the possibilities you suggest may occur. However, the strains required for linkage of the void sheets may be very small under conditions such as discontinuous yielding. Also, the local strains are so large that void nucleation can occur at very small precipitates or at grain boundaries. In general, I think it is better to consider a mode II extension and try and model the stress equilibria between the growing crack and the matrix in various idealized plastic gradients.

BEMENT: To what extent is banding a problem in low-alloy, Mn-Mo steels? Do you encounter a problem with temper brittleness if periodic heterogeneities occur in manganese and molybdenum composition due to poor ingot forging control?

EMBURY: The problem is not one of temper embrittlement due to periodic heterogeneities, but one of periodic variations in hardenability. The segregation of manganese and molybdenum will certainly influence the extent and distribution of the nonferritic constituents and thus influence the mechanical properties of the material. However, in general, these factors reflect the rolling practice and cooling rate rather than shortcomings in ingot quality.

BUSH: Work of Stan Rolfe and others on the "effective utilization of yield strength" may lead to code (ASME) changes where permissible design stresses will be much closer to yield than now permitted. This approach should reduce tonnages of steel for your pipeline example. More specifically, the increase in design stress may deemphasize alloy design aimed at higher yield strength and shift emphasis to other properties such as upper shelf energy or nil-ductility temperature. Would you see such a code change as a significant factor influencing the direction of future alloy R&D?

EMBURY: It will indeed influence future R&D and lead, I think, to a more rational use of properties such as shelf energy. It is clear that the control of desulfurization reaction, inclusion control, and deoxidation can be improved to raise the shelf energy. Also, there is an urgent need to provide an economic optimization of balancing alloy development versus inclusion control in terms of net increases in shelf energy. The data presented clearly indicate that transformation products may be equally as deleterious as inclusions in terms of ductility.

HORNBOGEN: The question is, what are the conditions under which delamination occurs along grain boundaries of ferrite? There may be the possibility that dynamically recrystallizing grains form filmlike carbide precipitates during continuous cooling because large areas of noncoincidence structure exist.

EMBURY: That is possible; however, the recent data by Groom and Knott (*Met. Sci. J.*, July, 1975) indicate that the delamination can be induced by prestrain. This means that boundaries are not inherently brittle but that their ability to transmit plasticity ceases after local imposed strains which are quite large, e.g., 0.4 to 0.7. Also, in general, the austenite will not be recrystallized in HSLA steels. Their solute redistribution by dynamic recrystallization is unlikely in either the austenite or ferrite.

HIRTH: With respect to the generation of a failure surface from a yield surface, it is of interest to connect to Ashby's discussion of mechanistic constitutive equations. For quasi-reversible processes, it is reasonable to expect the usual normality rule to apply: i.e., that the *net* strain is normal to the yield surface. However, with stress-activated or thermally activated processes, one would expect the activation strain to follow the normality, but, to the extent that it differed from the activation strain, the net strain not to follow it. This would give a further complication in generating the failure surface.

EMBURY: It is true that the assumption of normality of the net strain rather than incremental strains is a simplification. In principle, if the strain path on loading were known, one could still transform into stress space; however, for the present we have assumed that the normality rule exhibits the data in the form of a failure surface. Comparisons of these surfaces yield data which permit *both* the microstructural features and stress which limit ductility to be delineated.

LEVY: Relative to your yield surface to "failure surface" plastic potential, have you examined the use of uniform strain instead of failure strain? The latter includes the inconsistencies associated with propagation of cracks after attaining maximum (ultimate) stress, while the former involves strain only through the formation of the cracks.

EMBURY: One can define the failure surfaces on the basis of both uniform and total strains. In either case, there are shortcomings concerning the use of an effective stress–effective strain relationship to map the surface in stress space and the assumption that the limit strain is path-independent. A simple two-dimensional mapping can be accomplished by mapping the forming-limit curves in stress space.

KEAR: Certain nickel-base superalloy components are now routinely manufactured by powder-metallurgy techniques. This is to overcome problems associated with the presence of gross heterogeneities, both chemical and microstructural, in hot-deformation-processed ingot material. Although such an approach would not be suitable for making steel pipe, it seems to me that the new techniques of continuous casting, e.g., spray rolling, would be satisfactory, and probably can be made cost effective. Have you considered possible practical methods of high rate/gradient solidification to minimize chemical segregation and microstructural variations in the processing of HSLA steels?

EMBURY: While it is true that the scale of inclusion can be controlled in rapid solidification processes, one should exercise some caution in extrapolation to structural steels. The need to process the material by hot rolling in order to simultaneously shape the product, process the austenite, and control the form of the transformation product means that the detailed physical metallurgy is circumscribed by this process route in current steels.

PROGRESS IN FERROUS-ALLOY DESIGN

V. F. Zackay and E. R. Parker

*Department of Materials Science and
Engineering, College of Engineering
University of California, Berkeley, California*

ABSTRACT

Attempts to develop alloys with thermally stable intermetallic compounds, rather than alloy carbides, as the strengthening dispersoid in ferritic steels have failed, largely because of the embrittlement resulting from precipitation of the intermetallic compound in grain boundaries during heat treatment. Recent work has shown that this problem can be overcome by the appropriate use of the α-to-γ phase transformation in iron-base alloys strengthened by the Laves phase, $TaFe_2$. The new alloys have creep strengths at 650 C that are superior to most of the commercially available ferritic steels and are comparable in creep strength to the 300-grade austenitic steels.

Promising combinations of strength, ductility, toughness, ductile-to-brittle transition temperatures, and work-hardening rates can be obtained in

109

nickel-free alloys of iron and manganese, when the manganese content is in the range 16 to 20 percent. Several complex ternary and quaternary alloys were also investigated and some of the room- and cryogenic-temperature properties of these steels were found to be comparable and, in some instances, superior to those of existing commercial steels.

The deleterious or beneficial effects of chemical and structural features on the strength and toughness of ultrahigh-strength steels was demonstrated by appropriate control of composition and microstructure. In early work, the microstructural control was obtained by varying the austenitizing temperature. Extremely high (1100 C to 1200 C) austenitizing temperatures were shown to confer combinations of strength and fracture toughness (K_{IC}) that were far superior to those obtained by conventional heat treatment. In later work, similar properties were obtained by changes in composition and the use of more practical heat treatments.

1 INTRODUCTION

Associated with a rapidly expanding sophisticated technology is the ever-present need for new materials. This need imposes an extraordinary demand on the materials scientists. Almost all of the structural steels used during the past several decades were "designed" by a combination of empirical testing, collective experience, and accidental discovery. Currently used structural steels are, for the most part, either the same ones used in the post-World War II period or they are incrementally improved versions of the older types.

Compensating somewhat for the lack of innovation in alloy design is the substantial progress that has been made in the accumulation of basic knowledge. Great progress has also been made in the process- and quality-control aspects of large-scale metal production. In a similar vein, availability of the new high-resolution instruments, useful for identifying and characterizing defect structure and microstructure, have taken much of the "guess work" out of materials-science research. Such developments have set the stage for what should be a golden age of opportunity for materials science and materials engineering.

For the past 15 years, the authors, their students, and professional colleagues have largely concentrated their research efforts on steels. Their objective has been to synthesize alloys, some of which might serve to illustrate certain solid-state phenomena or principles of potential value in the alloy design of the future, as was done with the TRIP steels[1] or, in other cases, to generate model alloys which, with appropriate modifications by steel metallurgists, would (hopefully) take their place in the technology.

In the latter category, there are three steels whose composition, processing, structure, and properties are briefly described below. These are a ferritic alloy which is strengthened by an intermetallic compound rather than by alloy carbides; a metastable austenitic steel exhibiting the TRIP mechanism with a composition based on the binary Fe-Mn system; and an isothermally heat treated ultrahigh-strength steel with properties that are dependent on the volume fraction, morphology, and stability of retained austenite. In each instance, an attempt has been made to synthesize steels with some properties that are superior to those of existing commercial steels of the same general type.

2 DISCUSSION

2.1 A New Type of Ferritic Alloy for Moderately Elevated Temperature Service

Both ferritic and austenitic iron-base steels are commonly used for moderately elevated temperature service, i.e., up to 800 C. At service temperatures of about 600 C and above, the higher strength austenitic grades are used in preference to the ferritic ones. For structural applications involving service temperatures above 800 C, the nickel- or cobalt-base superalloys are preferred. The temperature limitation of the lower alloyed ferritic steels is a serious one and it severely limits their service in many applications in the petrochemical, nuclear, and aircraft industries.

Most commercially available creep-resistant materials derive their strength from a uniform dispersion of a second phase which hinders metal flow. In creep-resistant ferritic steels, the dispersion almost always consists of one or more types of alloy carbides; whereas, in the nickel-base superalloys, the dispersion consists mainly of a large volume fraction of the more thermally stable intermetallic compound, $Ni_3(Al,Ti)$. The composition, morphology, and crystal structure of carbides dispersed in the ferritic-steel matrix tend to change slowly at elevated temperatures with a concomitant change in creep rate.[2] An example of this phenomenon occurs in the well-known ferritic steel having the composition 0.1% C-1% Cr-0.5% Mo. After heat treatment, the carbide is of the M_3C-type. With long-time exposure to elevated temperatures and/or stress, the carbides M_7C_3, M_2C, and M_6C sequentially form. These structural and compositional variations and the concomitant changes in the matrix are accompanied by a decrease in creep resistance.[3] It is desirable, therefore, to eliminate carbides as the strengthening dispersoid in creep-resistant steels and to devise alternative means of enhancing elevated-temperature creep strength. The use of stable intermetallic compounds is an obvious possibility; several are known to provide significant strengthening when dispersed in iron.[4]

The concept of using intermetallic compounds, rather than carbides, for enhancing the elevated-temperature strength of ferritic alloys is decades old, but progress towards a realization of this concept has been slow. A major deterrent has been the tendency of such compounds to precipitate at grain boundaries and thereby cause severe intergranular embrittlement. Clearly, means have to be devised to eliminate or nullify the effects of the brittle grain-boundary network if the potential of these systems is to be realized.

Early studies on dislocation-precipitate interactions at room temperature suggested that alloys based on the Fe-Ta system would, if properly heat treated, have a high volume fraction of a stable intermetallic compound of the Laves type, viz., $TaFe_2$.[5,6] In the same investigation, a technique to alter the morphology of the grain-boundary brittle film by the use of certain phase transformations was devised. This subject is discussed in the following section.

It was found by Jones et al.[6] that the grain-boundary precipitate of the Laves phase could be spheroidized by heat treating the alloys in the $(\gamma + TaFe_2)$ phase field (this heat treatment is hereafter referred to as the spheroidizing treatment). Scanning electron micrographs which illustrate the process of spheroidization in an Fe-2 at. % Ta alloy are presented in Fig. 1. It is seen that the optimum time was 10 minutes.[6] The morphologies of the precipitates within the grains before and after spheroidization in an Fe-Ta-Cr alloy are compared in Fig. 2, (a) and (b).[2]

On cooling to room temperature, the gamma matrix transformed to alpha by a mechanism believed to be similar to the massive transformation observed in several iron alloys.[7] The resulting grain boundaries were very ·irregular, as shown in Fig. 3. The microstructure thus consisted of a uniform dispersion of about 3 vol % of almost spherical particles of $TaFe_2$ in a matrix of the α solid solution. The spheroidizing treatment resulted in a considerable enhancement of the room-temperature elongation. The fracture surfaces of the ductile and brittle alloys are shown in the scanning electron micrographs of Fig. 4, (a) and (b). A room-temperature yield strength of 40,800 psi and an elongation of 30 percent were observed for the Fe-1 at. % Ta alloys spheroidized for 10 minutes at 1100 C, followed by aging for 1 hour at 700 C. The high thermal stability of this system suggested that it might provide a good basis for a moderately elevated temperature ferritic alloy. Subsequent elevated-temperature tensile tests indicated that a large fraction of the room-temperature yield strength was retained up to a test temperature of about 600 C.[6] Above this temperature, the yield strength dropped rapidly.

Two major limitations of the binary Fe-Ta alloys were (1) the high-temperature heat treatments that were required and (2) the inadequate oxidation resistance at the expected service temperature. At 700 C in air, the weight gain of the Fe-1Ta alloy was about 30 percent of that of pure iron.[8] It is well known that chromium additions to iron (below 7 percent) lower both the $\delta \rightarrow \gamma$ and $\gamma \rightarrow \alpha$ phase-transformation temperatures[2] and that the oxidation of Fe-Cr alloys decreases significantly with chromium additions

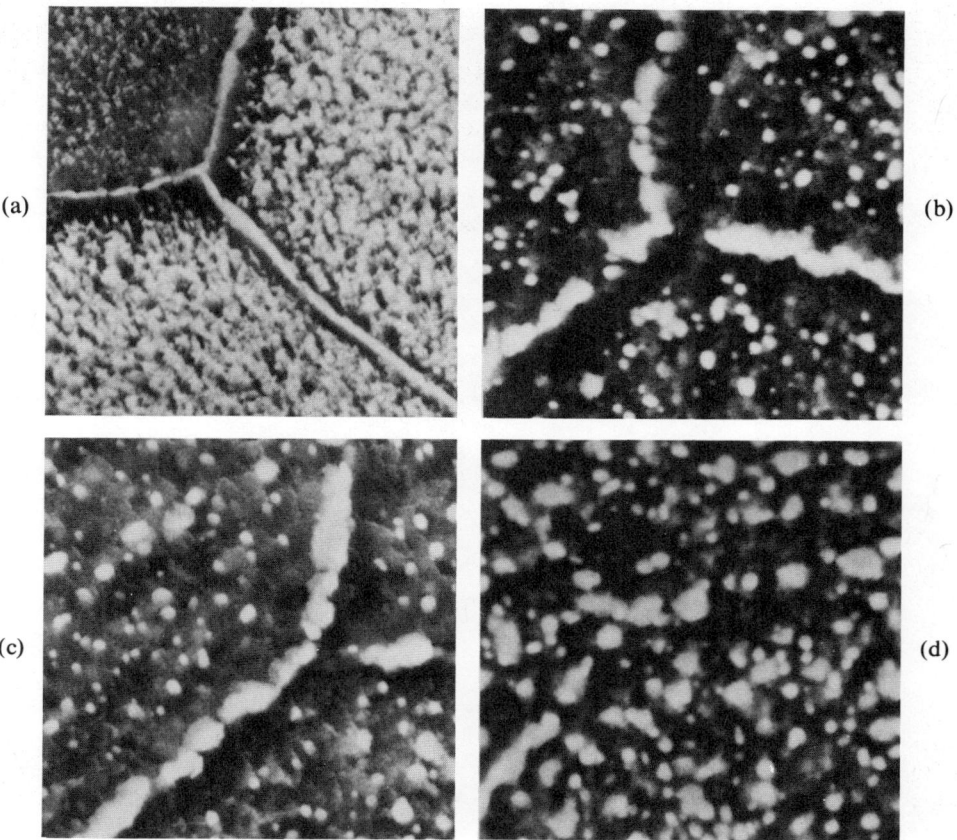

Fig. 1. Spheroidization of grain-boundary Laves phase in an Fe-2 at. % Ta alloy: (a) aged condition, (b) 4 min at 1100 C, (c) 12 min at 1100 C, and (d) 52 min at 1100 C.

of about 7 percent. The factors discussed above led to the development of a ternary Fe-Ta-Cr alloy, Ta7Cr (iron, 1 at. % tantalum, 7 at. % chromium). A more detailed study of the effect of chromium additions on the phase transformation temperatures of Fe-Ta alloys has been reported elsewhere.[9] The creep and stress-rupture properties of the Ta7Cr alloy were investigated extensively.[2] It was shown that the rupture strength of alloy Ta7Cr was higher than those of the ferritic steels containing 5 to 12 wt % chromium that were strengthened by dispersions of chromium and molybdenum carbides. However, the rupture strength was lower than those of the ferritic steels containing significant amounts of vanadium and tungsten as solid-solution strengtheners and carbide formers.

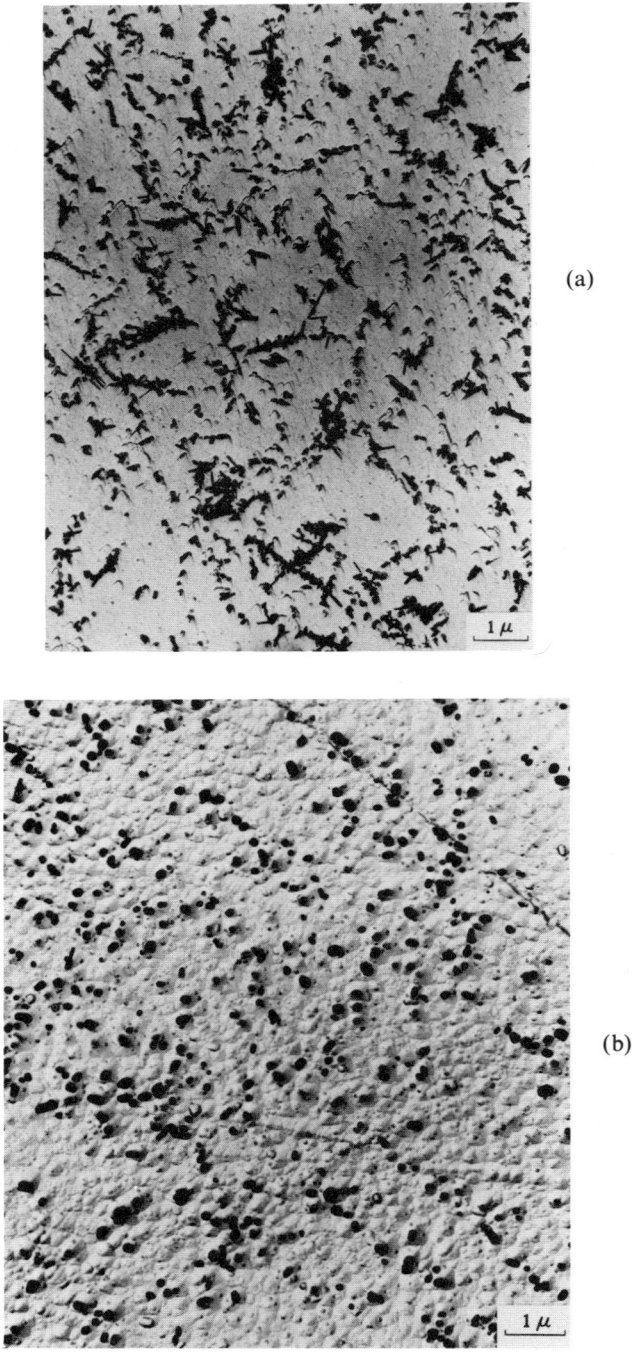

Fig. 2. Carbon replicas showing precipitate morphology (a) before spheroidization and (b) after spheroidization in an Fe-Ta-Cr alloy.

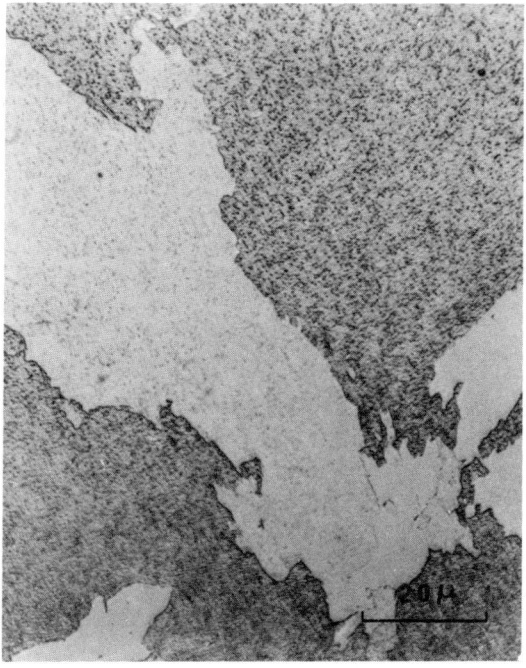

Fig. 3. Microstructure of an Fe-Ta-Cr alloy spheroidized at 1100 C and cooled to room temperature.

The study of tensile and creep properties and the examination of structural features of specimens of alloy Ta7Cr before and after creep testing indicated that partial recovery of dislocation substructure occurred in the alloy. This was particularly evident in specimens exposed for a long time (1809 hours) at 1100 F (593 C) and for a relatively short time (119 hours) at 1200 F (649 C), as shown in Fig. 5, (a), (b), and (c). Molybdenum is a well-known effective solid solution strengthener at high temperatures for both ferritic and austenitic steels.[10-12] The addition of 0.5 at. % molybdenum to alloy Ta7Cr led to a substantial improvement in elevated-temperature properties. Table I gives the nominal composition of the alloys used in these investigations.

An effective heat treatment of the Fe-Ta-Cr-Mo alloy was established in studies of the relationships between the creep properties (rupture time and creep rates), microstructures, and the associated heat treatments.[13] From these studies, the following heat treatment evolved: solution treat at 1350 C for 1 hour, quench in hot water, age at 700 C for 1 hour, air cool, spheroidize at 1100 C for 2 hours, and air cool.

Table II gives the 1000-hour rupture stress and the stress for a creep rate of 10^{-4} %/hr at 1100 F (593 C) for the Fe-Ta-Cr-Mo alloy heat treated in this manner, compared with the best of the commercial ferritic steels

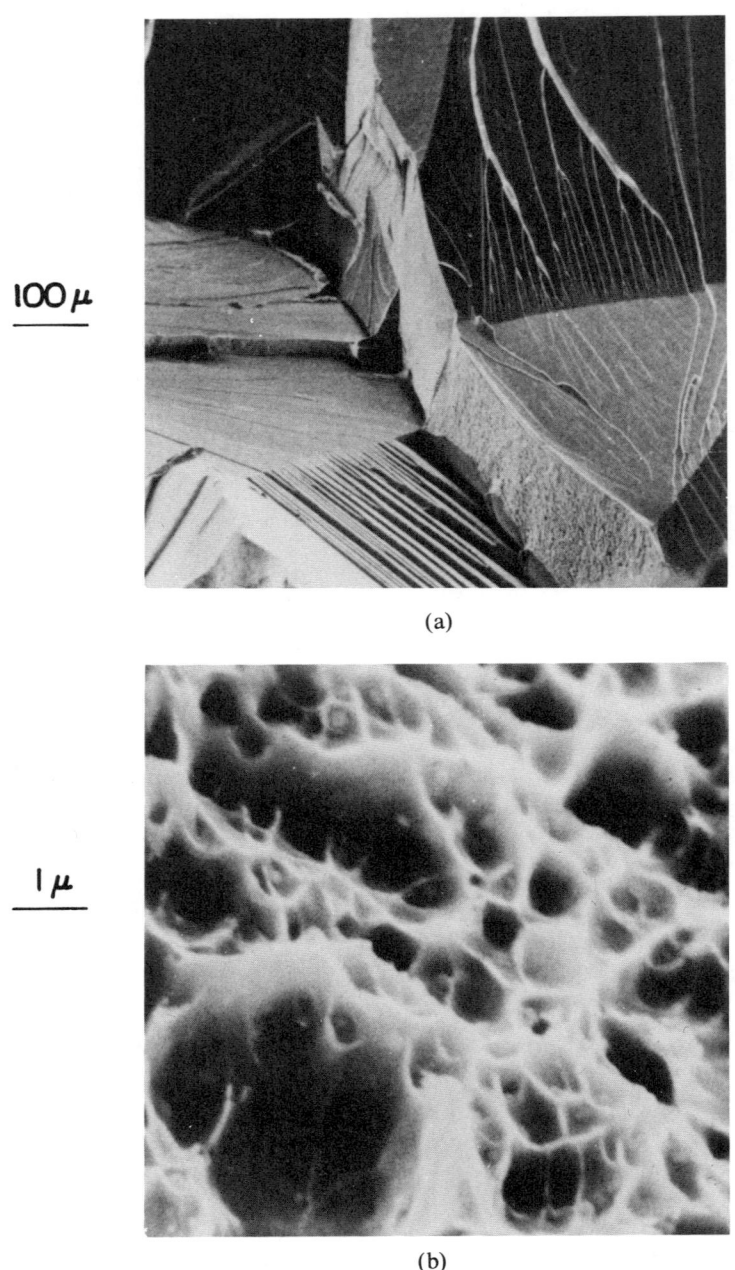

100 μ

(a)

1 μ

(b)

Fig. 4. Typical fracture surfaces of Fe-Ta alloys: (a) aged condition, (b)spheroidized condition.

(a)

(b)

(c)

Fig. 5. Thin-foil specimens of an Fe-Ta-Cr alloy: (a) before creep testing, (b) after testing to rupture at 1100 F (593 C) and 15,000 psi (1809 hr), (c) after testing to rupture at 1200 F (649 C) and 11,000 psi (119 hr).

Table I. Compositions of Fe-Ta-Cr and Fe-Ta-Cr-Mo Alloys

| | Compositions | | | | | | | |
| | Atomic Percent | | | | Weight Percent | | | |
Alloy	Ta	Cr	Mo	Fe	Ta	Cr	Mo	Fe
Fe-Ta-Cr	1	7	—	Bal	3.18	6.4	—	Bal
Fe-Ta-Cr-Mo	1	7	0.5	Bal	3.17	6.38	0.84	Bal

available as well as some austenitic steels. The creep strength (in terms of stress for a creep rate of 10^{-4} %/hr) of the Fe-Ta-Cr-Mo alloy was superior to that of all the other steels shown in the table. However, the 1000-hour rupture strengths of the 0.3C-1Cr-1Mo-0.25V steel and the 422, 316, 321, and 347 stainless steels were higher than that of the Fe-Ta-Cr-Mo alloy. The same properties at 1200 F (649 C) are also given in Table III. The 1000-hour rupture stresses for the 422 ferritic stainless steel and all of the austenitic steels were either comparable to or higher than that of the Fe-Ta-Cr-Mo alloy. However, the creep strength of the Fe-Ta-Cr-Mo alloy was higher than that of all the steels except the 316 and 347 stainless steels.

The microstructure of the Fe-Ta-Cr-Mo alloy consisted of spherical particles of a Laves phase in both the grain-boundary areas and within the grains. In an ancillary study of this alloy, it was learned that the decomposition of the delta (δ) phase followed classical "C"-curve kinetics.[13] The δ phase can be decomposed into the two phases, γ and $TaFe_2$, in two ways. One is by quenching directly from the single-phase δ region into the two-phase γ+$TaFe_2$ and transforming isothermally (termed "reaction on cooling"). The second is by quenching fast enough to retain δ at room temperature, followed by reheating the retained δ phase to temperatures in the γ+$TaFe_2$ region. This latter transformation is termed "reaction on heating". Although the exact kinetics for the cooling and heating reactions were slightly different, the "C" curves obtained were similar in nature. The kinetics of the reaction on cooling were shown to be similar to those of eutectoid transformations occurring in steel and some nonferrous alloys.[14,15] The shape of the "C" curve is shown in Fig. 6. The reaction on cooling gives rise to a lamellar type of structure and that on heating, a spheroidized one, as shown in Fig. 7, (a) and (b).

Preliminary creep tests[16] on the specimens with the lamellar-type structure at 1200 F (649 C) have indicated that the specimens with the spheroidized structure have better creep properties, as shown in Table II. However, specimens with the lamellar structure exhibited slightly better properties for shorter rupture times. This behavior is shown in Fig. 8. Long-time (greater than 400 hours) exposure at 650 C could result in a breakdown of the lamellar structure leading to the observed results. Preliminary tests

Table II. 1000-Hour Rupture Stress and the Stress for Creep Rate of 10^{-4} %/Hr for Alloy Fe-Ta-Cr-Mo and Some Commercial Ferritic and Austenitic Steels

Alloy	1000-Hour Rupture Stress, psi	Stress for Creep Rate of 10^{-4} %/Hr, psi
Test Temperature 1100 F (593 C)		
Fe-Ta-Cr-Mo (spheroidized)	29,000	25,000
0.3C-1Cr-1Mo-0.25V steel	36,000	(a)
0.15C-9Cr-1Mo steel	14,100	6,300
403,410 stainless steel	10,000	4,200
Greek Ascaloy	22,000	(a)
422 stainless steel	33,800	19,000
304 stainless steel	22,600	11,500
304L stainless steel	23,000	9,700
309 stainless steel	28,000	11,600
310 stainless steel	25,000	13,000
316 stainless steel	33,000	18,200
316L stainless steel	26,900	14,100
321 stainless steel	29,500	17,100
347 stainless steel	34,000	23,000
Test Temperature 1200 F (649 C)		
Fe-Ta-Cr-Mo (spheroidized)	14,000	10,000
Fe-Ta-Cr-Mo (lamellar)	13,000	9,000
0.3C-1Cr-1Mo-0.25V steel	(b)	(b)
403,410 stainless steel	4,900	2,000
Greek Ascaloy	11,000	(b)
422 stainless steel	15,000	(b)
304 stainless steel	15,100	6,900
304L stainless steel	15,800	6,800
309 stainless steel	20,000	8,000
310 stainless steel	17,000	8,600
316 stainless steel	24,700	12,700
316L stainless steel	17,300	8,200
321 stainless steel	18,000	9,700
347 stainless steel	22,000	16,500

(a) No values quoted for 1100 F.
(b) No values quoted for 1200 F.
Source: Data for the ferritic steels were obtained mostly from Ref. 18, and those for the other steels from *Engineer's Guide to High Temperature Materials*, F. J. Clauss (Ed.), Addison-Wesley Publishing Co., Inc., Reading, Massachusetts (1971), Chapter 4, pp. 86-128.

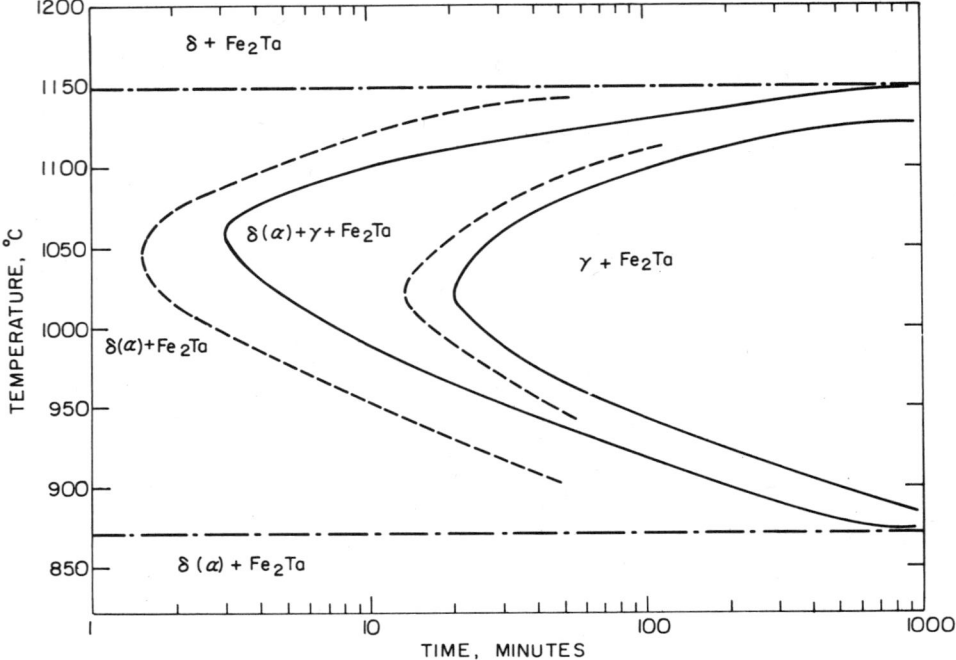

Fig. 6. Time-temperature-transformation curves for the heating (solid curves) and cooling (broken curves) reactions in an Fe-Ta-Cr-Mo alloy.

have also indicated that the higher the spheroidization temperature, the better were the elevated-temperature properties, as shown in Table III.[17] This behavior can be rationalized on the premise that the solubility of tantalum and molybdenum in γ increases with increasing temperature, thereby causing greater solid-solution strengthening.

These preliminary structure-property results suggest that the original objective of this program may be met in the near future, viz., that "model" ferritic alloys can be designed with a useful temperature range exceeding the present limit of ferritic steels (about 600 C) and with properties that are comparable to those of the well-known iron-base 18-8 type of stainless steels.

2.2 Austenitic Alloys for Room and Cryogenic Service Based on the Fe-Mn System

The Fe-Cr-Ni austenitic stainless steels have occupied a unique position in metals technology for more than half a century. Their popularity among designers, engineers, and manufacturers is obviously well deserved. They are

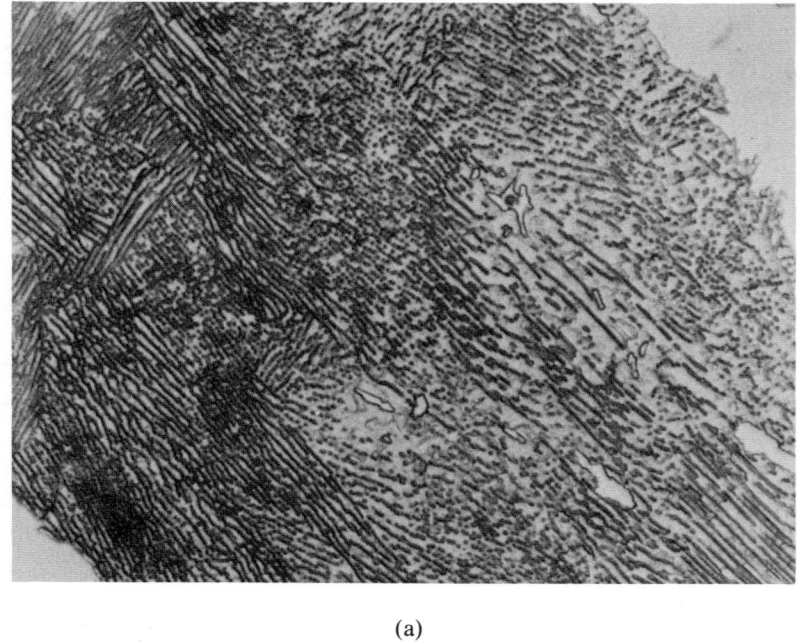

(a)

(b)

Fig. 7. Alloy Fe-Ta-Cr-Mo transformed by holding at 1020 C for (a) 10 min in the cooling reaction and (b) 15 min in the heating reaction.

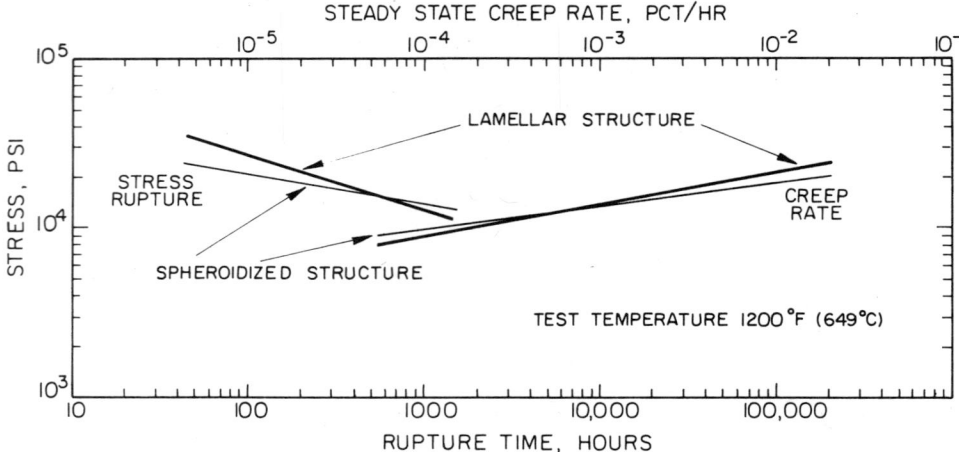

Fig. 8. Stress versus rupture time and versus steady-state creep rate for alloy Fe-Ta-Cr-Mo at a test temperature of 1200 F (649 C) for the lamellar and spheroidized microstructures.

Table III. **Effect of Spheroidization Temperature on the Rupture Time of Fe-Ta-Cr-Mo Alloy, Tested at 1200 F (649 C) and 17,000 psi[a]**

Spheroidization Temp, C	Rupture Time, hr
960	30
1020	68
1100	326

(a) The heat treatment preceding the spheroidization treatment consisted of a solution treat at 1350 C for 1 hour followed by a hot-water quench. The specimens were then aged at 700 C for 1 hour and air cooled to room temperature. Time at the spheroidization temperature was 2 hours.

fabricable, weldable, corrosion resistant, and pleasing in appearance and, as a consequence, are perhaps the most versatile of all commercially available steels. There would be technological advantages, however, in having austenitic steels of different chemical properties but similar mechanical properties. For example, in certain severe chemical, radiation, or stress environments, complex solid-state reactions will occur at grain boundaries and may lead eventually to structural failure.[18] Also, it is well known that many nickel-containing alloys are especially susceptible to sulfur attack at elevated temperatures. This limitation is particularly severe in coal-gasification plants. Lastly, the element nickel has been, on occasion, in short supply and it would seem desirable to have available substitute nickel-free alloys whose properties are comparable to those of the Fe-Cr-Ni system. The search for alternatives to the austenitic Fe-Cr-Ni alloys is an old one and the literature is replete with references to promising candidates. In the past, interest has usually centered on the partial or complete substitution of manganese for nickel and other stronger-austenitizing elements, such as carbon and nitrogen.

The authors' interest in this subject stems from their studies of the TRIP-steels.[19] The presence of diffusionless transformations can lead to high (and predictable) levels of uniform elongation, increased rates of work hardening (ultimate-to-yield ratios of two or more), and fracture toughness (at the 200,000-psi yield-strength level) far in excess of that of any other known family of alloys. It is well known that binary Fe-Mn alloys exhibit stress- and strain-induced transformations.[20,21] For this reason, it seemed logical to use them as a base upon which a family of alloys could be designed with desirable combinations of chemical and mechanical properties.

The desirable mechanical properties of the austenitic Fe-Cr-Ni alloys are a reflection of their favorable crystal structure. The FCC crystal structure precludes, under normal circumstances, failure by cleavage, and it virtually ensures high ductility. The number of elements which stabilize the FCC structure of iron is small. For example, the element manganese, although an austenite former like nickel, is a relatively weak one. Consequently, it has often been used only as a partial substitute for nickel. Alternatively, nickel-free alloys have been designed and used which, in addition to the manganese, depend on relatively high levels of the strong-austenitizing elements, carbon and nitrogen. Both approaches have been exploited in the technology and have proven to be practical, especially in times of material emergencies.

In a recent paper, Holden et al.[22] have suggested a somewhat different attack on the problem. They investigated the influence of manganese on the crystal structures and associated mechanical properties of high-manganese Fe-Mn alloys with manganese contents as high as 40 percent. Holden et al. demonstrated clearly that unusual combinations of strength, ductility, and toughness could be obtained by controlling the level of manganese and without the aid of other austenite formers or stabilizers. Our early studies, described in the following section, corroborate Holden's results.

A part of the iron-rich end of the phase diagram of the Fe-Mn system is shown in Fig. 9. Several features of this diagram are of direct relevance to the ensuing discussion. First, crystal structures that are sequentially encountered with increasing manganese content are the body-centered cubic, the hexagonal (or epsilon phase), and the face-centered cubic. Second, the transformations are characteristically sluggish and they occur at relatively low temperatures, especially at the higher manganese contents. The epsilon phase is a crystallographic intermediate between the body-centered and the face-centered cubic phases. The epsilon phase can transform to the body-centered cubic phase by a stress- or a strain-induced transformation. The face-centered cubic phase is believed to transform in a similar manner by the sequence $\gamma \rightarrow \epsilon \rightarrow \alpha$.[23] The extremely low stacking-fault energy of the higher manganese Fe-Mn alloys is undoubtedly a major factor in the formation of the epsilon phase.

The tensile properties of Fe-Mn alloys can be readily correlated with their crystal structures and the transformations not thermally activated. For example, the elongation at room temperature, as might be expected, remains more or less constant with increasing manganese content until about 12 percent manganese, beyond which it begins to rise rapidly, as shown in Fig. 10. The change in elongation is associated with the presence of the epsilon phase and its strain-induced transformation to the α phase.[23] Incidentally, these changes in crystal structure can be followed during the tensile test by means of simultaneous magnetic saturation measurements.[24] The elongation continues to rise with increasing manganese content up to about 16 percent, the composition marking the beginning of the austenite phase. In mechanical-property tests conducted at liquid-nitrogen temperature, the influence of the manganese content on the diffusionless phase transformations, and on the elongation, is even more pronounced.

The yield and ultimate strengths exhibit a similar dependence on manganese content (and the associated crystal structure), as shown in Fig. 11. In this case, a stress-induced rather than a strain-induced transformation is operative, as indicated by the abrupt decrease in yield and ultimate strengths between 12 and 16 percent manganese. Similarly, the ductile-to-brittle transition temperature (DBTT) decreases with increasing manganese (above about 8 percent) as shown in Figs. 12 and 13. Rather surprisingly, manganese appears to be more effective than nickel in decreasing the DBTT of iron binary alloys, as shown in Fig. 13. These early corroborative results strongly suggested that further studies were warranted on more complex versions of these alloys.

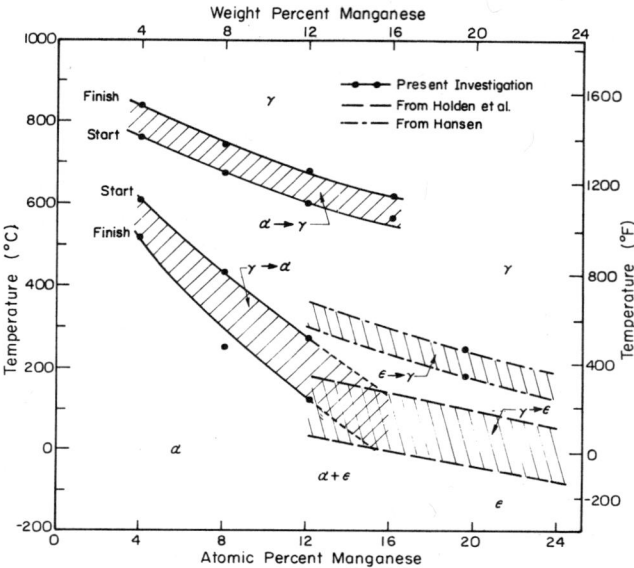

Fig. 9. Phase-transformation temperatures for Fe-Mn alloys.

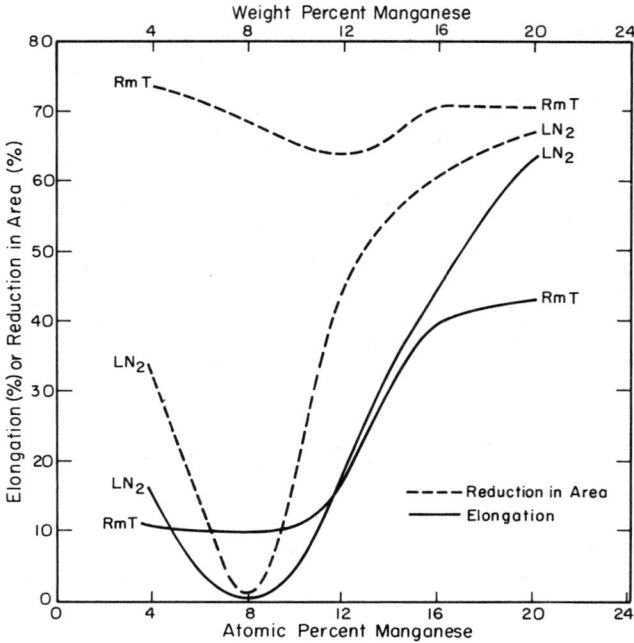

Fig. 10. Elongation and reduction in area at 25 C and −196 C versus manganese content for Fe-Mn alloys.

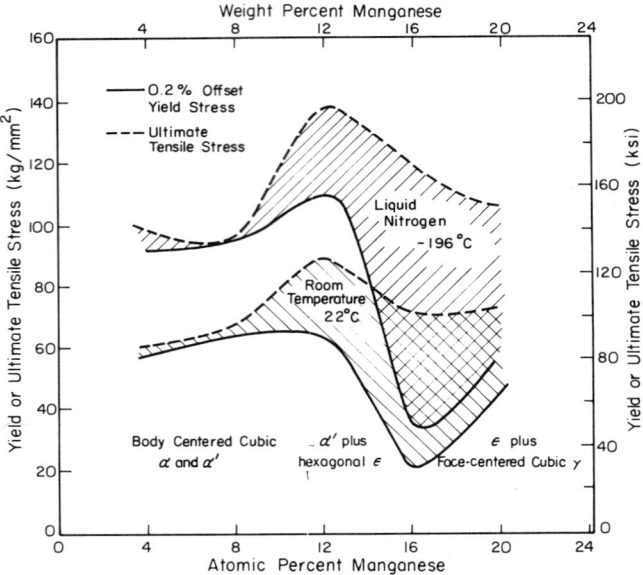

Fig. 11. Yield and ultimate tensile strengths at 25 C and −196 C versus manganese content for Fe-Mn alloys.

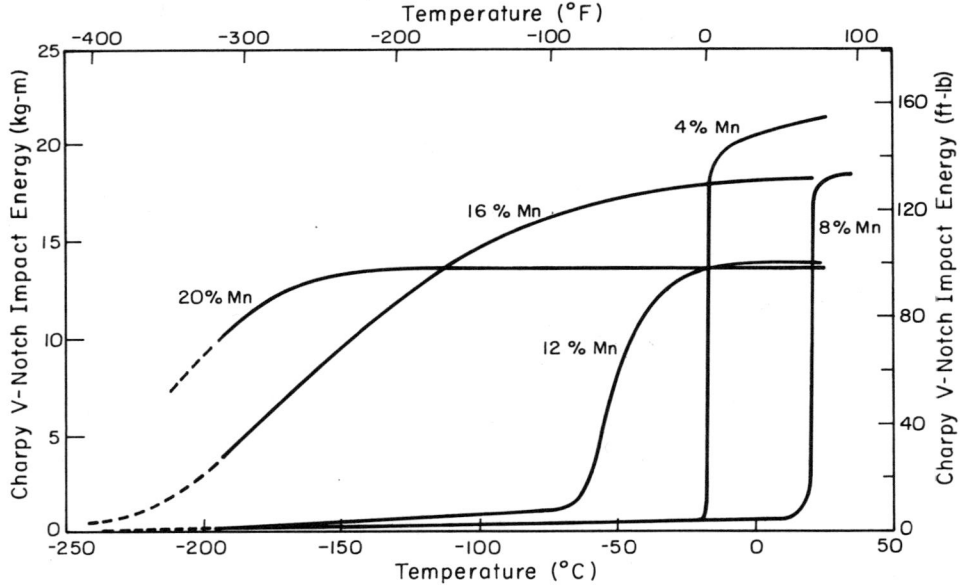

Fig. 12. Charpy V-notch impact toughness versus testing temperature for Fe-Mn alloys without chromium.

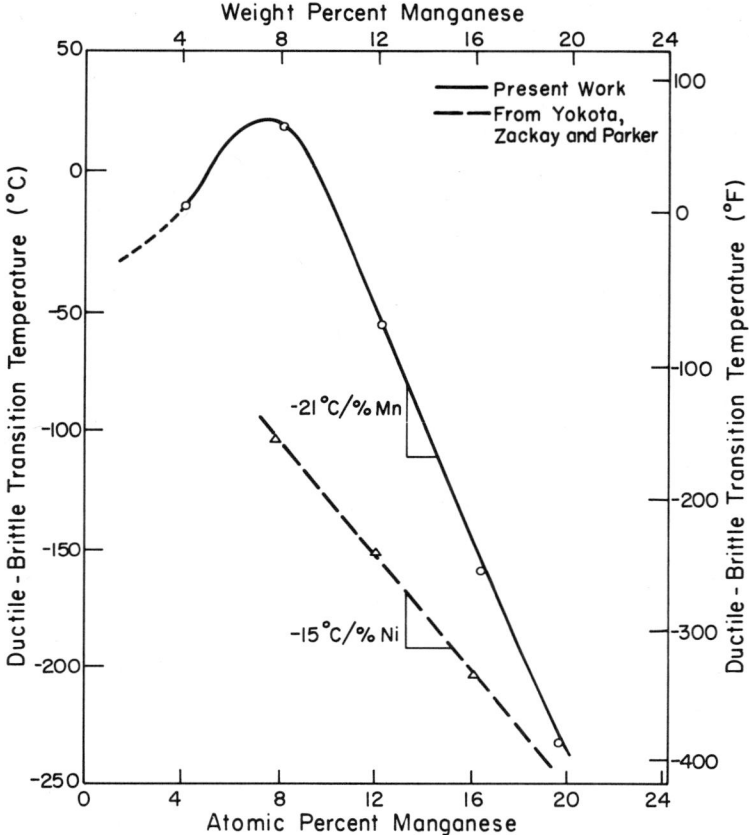

Fig. 13. Ductile-brittle transition temperature (DBTT) versus manganese content for Fe-Mn alloys; also plotted are DBTT versus nickel content for noninterstitial Fe-Ni alloys.

Recent studies have focused on the design of alloys with mechanical properties that are more or less equivalent to those of the familiar Fe-Cr-Ni alloys.[25] In the practical range of manganese content, i.e., 16 to 20 percent, the yield strengths of the binary alloys are unacceptably low because of the stress-induced phase transformation. Furthermore, the oxidation and the corrosion resistance of the binary alloys are lower than those of the Fe-Cr-Ni alloys. For these reasons, ternary and more complex alloys have been made with the objective of overcoming or minimizing these limitations. The compositions of some of the alloys studied are shown in Table IV and their associated properties are shown in Table V. In all the alloys, chromium was added to enhance the corrosion and oxidation resistance.

Table IV. Compositions of Ternary and More Complex Fe-Mn Alloys

Alloy Designation	Composition, wt %[a]							
	Mn	Cr	Si	Mo	N	C	Al	Ti
2	16	18	—	—	0.130	—	—	—
6	16	18	—	1.99	0.180	—	—	—
16	16	13	1.44	—	—	0.20	0.10	—
18	20	13	1.46	—	—	0.22	0.10	—
23	20	13	1.42	—	0.130	—	—	—
24	20	18	1.42	—	0.130	—	—	—
32	18	13	—	—	—	—	0.10	0.15

(a) Balance Fe in all cases.

Table V. Tensile and Charpy V-Notch Properties of Complex Fe-Mn Alloys

Alloy Designation[a]	Tensile Properties				Charpy V-Notch Properties	
	Yield Strength, ksi	Ultimate Strength, ksi	Elong., %	Reduction in Area, %	Room Temp, ft-lb	DBTT, C[b]
1	47	122	67	70	145	174
2	54	105	69	76	122	−151
6	55	106	70	74	177	−169
16	48	162	65	55	128	−82
18	48	125	64	63	128	−116
23	45	106	77	79	206	−189
24	62	97	60	78	180	−95
32	45	92	51	82	168	−196

(a) Alloys 1, 2, 16, 18, 23, and 24 were austenitized at 1000 C, alloy 6 was austenitized at 1200 C, and alloy 32 was austenitized at 900 C. The austenitizing time was 2 hours and all the alloys were ice-brine quenched.
(b) The DBTT is defined as the temperature of one-half the upper shelf energy.

Carbon and nitrogen additions were made to the Fe-Mn-Cr steels to stabilize the austenite and thereby lower the M_s and M_d temperature. In so doing, the stress-induced transformation was lowered to below room temperature.[19] Molybdenum and silicon were added to improve the corrosion resistance; molybdenum improves the general corrosion and pitting resistance and silicon improves the oxidation resistance when added to austenitic stainless steels.[26] Aluminum and titanium were added as solid-state gettering agents.

In an earlier investigation[23] it was shown that 8 percent chromium additions to an Fe-16% Mn alloy nearly doubled the yield strength at approximately the same DBTT. However, chromium additions to a 20 percent manganese alloy did not significantly change the yield strength or the DBTT. Since about 12 percent of chromium is needed for good corrosion resistance and 18 percent is used in the 18-8 stainless steels, these additions were made to the Fe-Mn alloys which contained silicon and nitrogen (alloys 23 and 24). The tensile and Charpy V-notch properties at room temperature of these two alloys are compared in Table V. Increasing chromium from 13 to 18 percent increased the yield strength but decreased the ultimate strength, thereby reducing the extent of strain hardening. The increase in chromium also resulted in a decrease in the Charpy V-notch impact energy and an increase in the DBTT.

The effect of manganese in lowering the DBTT is clear even in these complex alloys by a comparison of alloys 16 and 18. However, the higher manganese alloy appears to have a reduced ultimate strength and has the same Charpy V-notch energy at room temperature. Alloys 16 and 32 exhibited excellent tensile and impact properties. The former had molybdenum and nitrogen additions.

A comparison of the mechanical properties of the 300 series of the austenitic Fe-Cr-Ni alloys with those described in the preceding section is informative. In general, the yield and ultimate strengths of the ternary (and more complex) Fe-Mn alloys are higher. In some cases, this difference in strength is almost 50 percent. The ductility of the complex Fe-Mn alloys is equal to or greater than that of the standard austenitic Fe-Cr-Ni grades—the values of elongation and reduction in area often exceeding 70 percent (see data for alloys 2, 6, 23, and 24). Lastly, while complex Fe-Mn alloys do have DBTT's and the Fe-Cr-Ni alloys do not, the DBTT's are quite low and the upper shelf energies are high in the case of the former. The existence of a DBTT in these steels reflects, of course, the stress- or strain-induced transformation of metastable austenite to martensite. Steels based on the Fe-Mn system will probably suffer somewhat in comparison with those of the Fe-Cr-Ni system with respect to overall corrosion and oxidation resistance. In addition, there are many features that have yet to be determined, such as producibility (steel-making aspects), castability, weldability, final cost, and market need. All of these factors must be determined by industrial metallurgists before the potential, if any, of these steels can be established. However, encouraged by these early promising results, our group is continuing studies of the corrosion behavior of these steels in various liquid and gaseous environments, the structural changes occurring during elastic and plastic flow at room and cryogenic temperatures, and of design studies directed toward the development of more versatile alloys.

2.3 New Concepts in the Design of High-Strength Steels

A widely held opinion among metallurgists and mechanical engineers is that toughness necessarily decreases with an increase in strength or with an increase in section thickness. This view appears to be substantiated by theory, reinforced by physical intuition, and supported by countless experiments in the laboratory. The weight of this combined evidence has strongly influenced the course of both the design of steels and the design of large steel structures.

Much of the effort in the design of ultrahigh-strength steels has consisted of the identification and characterization of the chemical and structural factors which degrade the toughness of existing (or modified) commercially available steels. Another approach, less often taken, consists of attempts to create steels whose composition and practical heat treatments would be entirely different from those presently used. Both these approaches have been followed in the authors' laboratory and a description of this work is summarized in the following section.

The prevalence of opinion regarding the previously mentioned correlation between toughness and section size has profoundly influenced the design of structures with heavy sections; viz., designers have tended to choose low-strength steels of high toughness and thick section sizes rather than higher strength steels of lower toughness and thinner section size. Generally, the low-strength steels have a low-alloy content and are characterized by relatively low hardenability. The unfortunate overall result of this combined design and metallurgical-engineering approach has often been the fabrication of large engineering structures made of steels having microstructures that varied with section thickness in an unpredictable manner. Attempts to obviate this problem by the use of more highly alloyed steels of greater hardenability were often negated by cost considerations. Clearly, a need exists for high-strength steels of low-to-medium alloy content and of considerable toughness which can be used safely in engineering structures utilizing large section sizes. This problem is perhaps one of the most important presently confronting the materials scientist and engineer. Some preliminary attempts being made in the authors' laboratory to solve this problem are summarized below.

1. The Role of Undissolved Brittle Microconstituents

Elucidation of the role played by microstructure in controlling the mechanical properties of ultrahigh-strength steels has been, and continues to be, a major part of our research effort. Significant progress has been made in a number of laboratories in identifying and characterizing several factors (chemical and microstructural) which influence mechanical behavior. Some of these are: (a) soluble and insoluble impurities; (b) the size, morphology, and volume fraction of weak (ferrite) and brittle (carbide) microconstituents; (c) twinned versus dislocated martensite substructures; and (d) the morphology, stability, and volume fraction of retained austenite. The deleterious or beneficial effects of these chemical and microstructural features on the strength and toughness of several laboratory and commercial heats of ultrahigh-strength steels was demonstrated by appropriate control of composition and microstructure.[27-29] In early work, the microstructural control was obtained by varying the austenitizing temperature.[30]

A critical part in the heat treatment of any steel is the austenitizing step. During austenitization, many compositional and structural changes occur and these profoundly influence the final microstructure and its associated mechanical properties. For example, low austenitizing temperatures favor a small austenite grain size but may leave a large fraction of brittle, undissolved carbides. Conversely, high austenitizing temperatures lead to dissolution of a great proportion of such carbides but cause a concomitant increase in austenite grain size. In an effort to evaluate relationships among austenitizing temperature, volume fraction of undissolved carbides, fracture toughness, and strength, several laboratory-type secondary hardening steels were studied.[31] Their compositions are given in Table VI.

The effect of varying the austenitizing temperature on the room-temperature yield strength and the fracture toughness of as-quenched 0.30C-5Mo and 0.41C-5Mo steels is shown in Fig. 14. The increase in fracture toughness with austenitizing temperature for the lower carbon steel was particularly striking. Metallographic examination of these high-molybdenum steels indicated that extensive solution of alloy carbides occurred above a critical austenitizing temperature. The fracture toughness was observed to

Table VI. Chemical Compositions of C-Mo Alloys

Alloy	Composition, wt %[a]					
Designation	C	Mo	Mn	S	P	Ni
0.32C-2Mo	0.32	1.96	0.65	0.005	0.007	—
0.30C-5Mo	0.30	5.03	0.60	0.005	0.008	—
0.41C-5Mo	0.41	4.93	0.51	0.005	0.007	—
0.35C-1Mo-3Ni	0.35	0.95	0.61	0.005	0.007	3.1

(a) Sn,Sb < 0.002 percent, As < 0.005 percent, and Si < 0.02 percent in all steels.

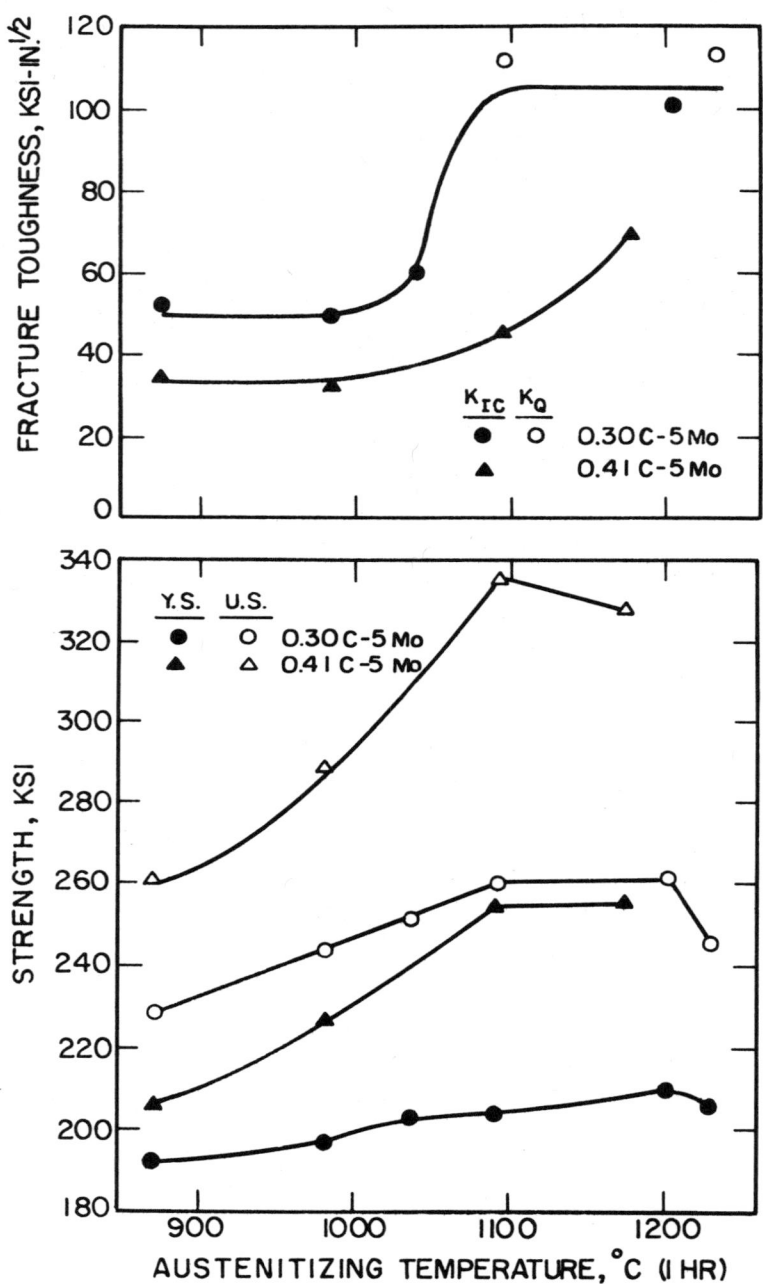

Fig. 14. Influence of austenitizing temperature on room-temperature fracture toughness, yield strength, and ultimate strength of 0.30 C-5 Mo and 0.41C-5 Mo steels.

increase appreciably in the same temperature range. Associated with the increased solution of carbides and the improved fracture toughness was, as expected, a pronounced increase in grain size. Similar experiments were performed with another set of lower alloy steels in which complete dissolution of carbides could be expected at austenitizing temperatures as low as 870 C. In this instance, the fracture toughness was relatively independent of the austenitizing temperature (or grain size) as shown in Fig. 15.

Transmission electron microscope studies, using carbon replicas, indicated that an austenitizing temperature of 870 C left undissolved carbides approximately 0.05 micron in diameter for a 0.32C-2Mo steel and 1 to 3 microns in diameter for the 0.30C-5Mo steel, as shown in Figs. 16 and 17, respectively. The deleterious effects of the larger carbides was reflected by the values of the ratio of plane-strain fracture toughness to yield strength for the two steels in the as-quenched conditions. These were 0.42 for the 0.32C-2Mo steel and 0.27 for the 0.30C-5Mo steel. Furthermore, the ratios were equal for the 0.32C-2Mo steel austenitized at 870 C and the 0.30C-5Mo steel austenitized at 1200 C. These results strongly indicated that the presence of hard, brittle, undissolved particles above a certain critical size could lead to a significant degradation of the fracture toughness of alloy steels. The results also suggest that the fracture toughness, as conventionally measured, appeared to be insensitive to large variations in prior austenite grain size for those alloy steels in which the matrix was free of undissolved carbides of sizes above a critical value.

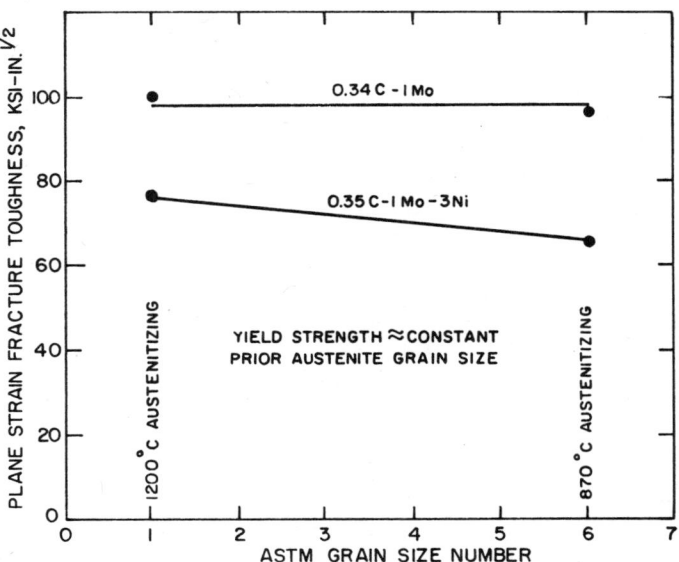

Fig. 15. Room temperature plane-strain fracture toughness versus prior austenite grain size for as-quenched 0.34 C-1 Mo and 0.35 C-1 Mo-3 Ni steels.

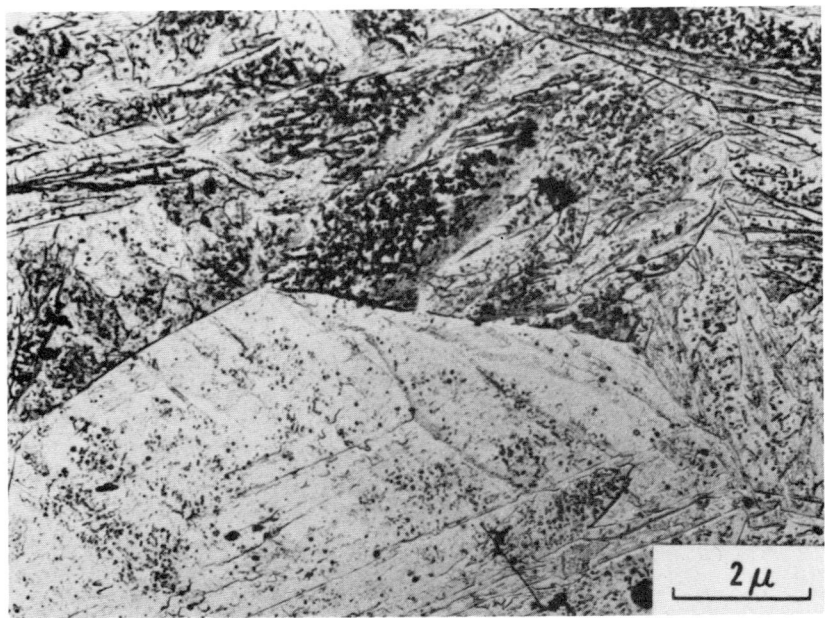

Fig. 16. Carbon replica showing fine undissolved carbides in as-quenched 0.32 C-2 Mo steel, austenitized at 870 C.

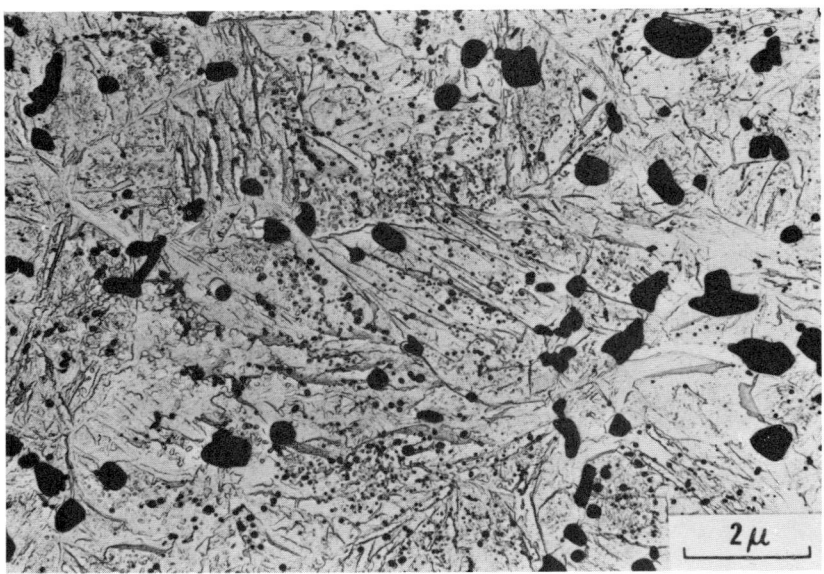

Fig. 17. Carbon replica showing 1 to 3-micron size undissolved carbides in as-quenched 0.30 C-5 Mo steel, austenitized at 870 C.

2. *Alloy Composition and the Kinetics of
Decomposition of Austenite*

The alloying elements in steel and the solid-state reactions which steels undergo determine the structure and, therefore, the properties of the alloys. The complex nature of these reactions is not yet fully understood, but these reactions are important in the economic design of superior steels. For example, even if the individual influences of all the common alloying elements on the thermodynamics and kinetics of the principal isothermal reactions of steels were known, the information would be insufficient for the effective design of complex steels. Commercial steels contain many interacting elements whose total effect is often different from that predicted on the basis of the effects of individual elements alone. A real need exists, therefore, for a technique that will permit the rapid determination of the single *and* the combined effects of all the common alloying elements on the solid state metallurgical reactions of importance in the heat treatment of steel. A knowledge of the initiation (incubation) and finish times of these reactions, as well as the associated kinetics, is particularly important. Progress made in the authors' laboratory in developing such a technique is described below.

The rapid method of studying isothermal reactions in steels consists of quenching the steel sample from the austenitizing temperature to a subcritical temperature in an isothermal bath and holding it within the magnetic field of an inductor coil. The increase in permeability accompanying austenite decomposition increases the inductance of the coil, and this changes the resonant frequency of the circuit. An automatic continuous recording of the corresponding period provides a convenient and accurate method for following the austenite decomposition. Quantitative information on austenite decomposition kinetics can be obtained within 2 seconds after the start of quenching.[32,33] An example illustrating some of the preliminary results obtained by this method is presented below.

In the recently-reported T-T-T diagrams for AISI 4340 steel, the bainite range is shown as a smooth C-shaped curve, Fig. 18(a). Investigations by the new method showed significant differences in the shape and character of the lower bainite region. As shown in Fig. 18(b), the incubation period for the formation of lower bainite decreased at temperatures just above the M_s and gave an S-shaped curve for the bainite reaction above the M_s. Acceleration of austenite decomposition at temperatures just above M_s has been observed previously.[34-36]

That part of the lower bainite curve which extended below the M_s had a C-shape, as shown in Fig. 18(b). Immediately below the M_s temperature, the bainite reaction begins almost immediately after the end of the martensite reaction. The rapid onset of this reaction at temperatures just below the M_s is well established in the literature and is associated with the increased

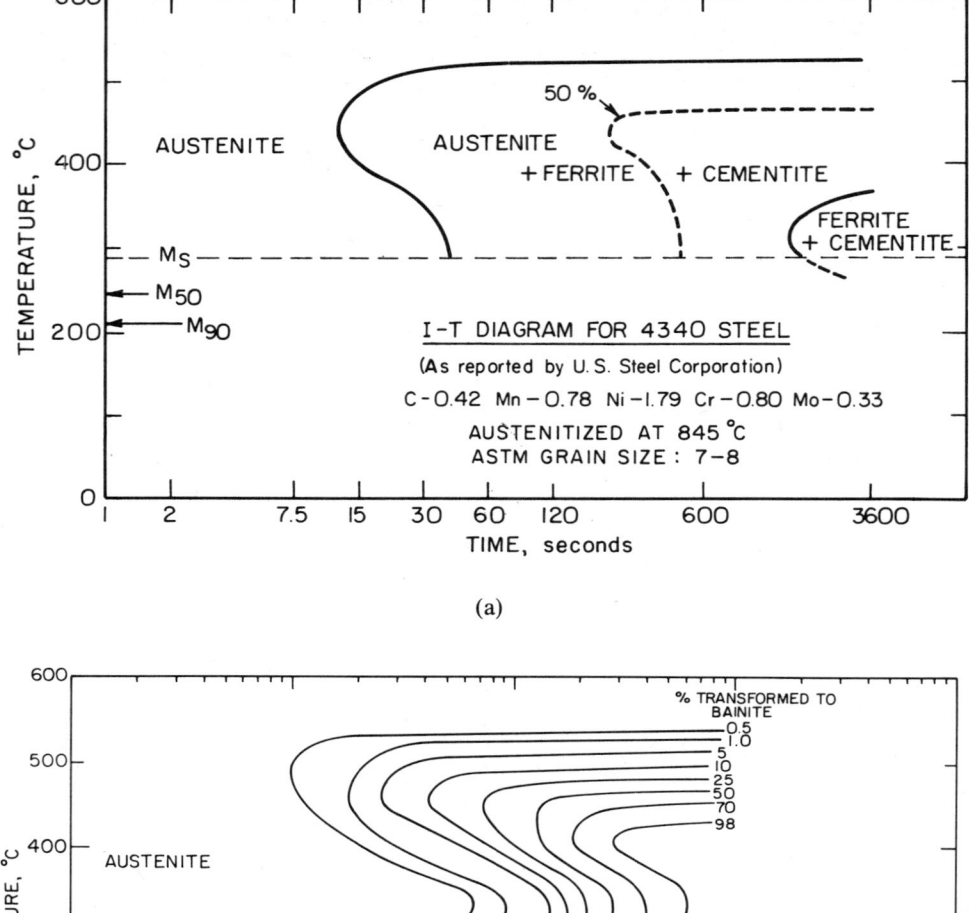

(a)

(b)

Fig. 18. T-T-T diagrams of AISI 4340 steel: (a) as reported in the literature and (b) as determined by the new magnetic-permeability technique. (Only the bainite and martensite ranges are shown.)

nucleation of the bainite by the strain effects associated with the austenite-to-martensite transformation. The decreased rate of bainite formation at temperatures well below the M_s is apparently due to the lower diffusion rates of carbon.

Silicon is known to affect the kinetics of tempering[37-40], and the mechanical properties of tempered martensite steels[41,42]. The reasons for this behavior are not known, but speculation has centered both on the retardation of the nucleation of iron-carbide by silicon[38,42], and its well-known influence on the thermodynamic activity of carbon[43]. These observations suggested that silicon might have a significant influence on the isothermal formation of lower bainite from metastable austenite.

The low-temperature portion of a T-T-T diagram of AISI 4340 steel with the 3.0 percent silicon addition is shown in Fig. 19. (The standard alloy contains about 0.25 percent silicon.) The effect of the silicon addition on the kinetics of transformation is striking, as shown in the T-T-T diagrams in Figs. 18(b) and 19. The time for isothermal transformation of 50 percent of the austenite is about 1200 seconds for the silicon-modified 4340 steel, as compared with a few hundred seconds for the standard 4340 steel. The morphology of the retained austenite in the silicon-modified steel is clearly seen in the bright field-dark field pair of electron micrographs of Fig. 20. The dark field picture was taken using an austenite reflection. The retained austenite was present in the form of films decorating the lath boundaries of ferrite, as shown in A, with some austenite regions with a thickness equal to the lath size of bainitic ferrite, as seen at B. A pool of austenite can also be observed at C. It was possible to vary the amount of retained austenite over a wide range by changing the temperatures and times of isothermal holding.

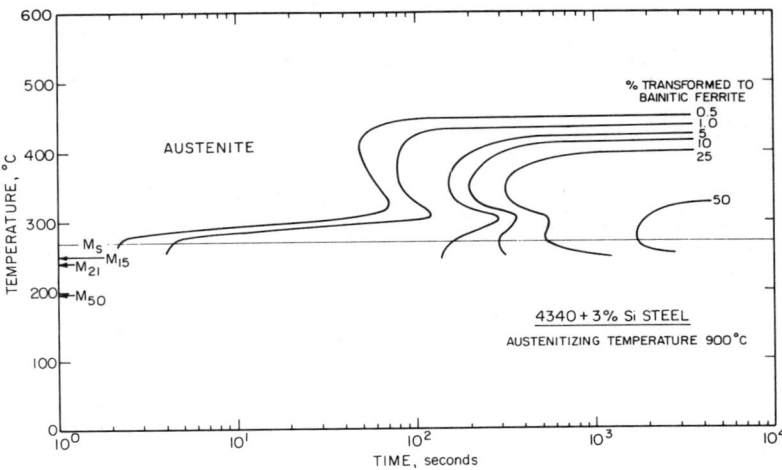

Fig. 19. T-T-T diagram of AISI 4340 steel modified with a 3.0 percent silicon addition. (Only the bainite and martensite ranges are shown.)

(a)

(b)

Fig. 20. Thin foils obtained from 3 percent silicon-modified 4340 steel: (a) bright field and (b) dark field of an austenite spot; steel austenitized at 900 C and isothermally transformed at 307 C. Films of austenite decorate the lath boundaries of ferrite at A, with austenite regions of thickness equal to the lath size of bainitic ferrite at B, and a pool of austenite at C.

3. Some New Directions

In the introductory remarks to this section, reference was made to the two parts of the alloy design efforts: (1) the identification and subsequent minimization of the deleterious microconstituents (relative to fracture toughness) in commercial steels and, (2) the synthesis of entirely new steels. Much progress has been made by the metallurgical fraternity in achieving the objectives of the first part, viz., the understanding and improvement of current materials. These objectives have been met by the collective efforts of the process metallurgist, the physical metallurgist, and the materials scientist. Progress toward achieving the objectives of the more advanced aspects of alloy design, viz., the creation of entirely new alloys, is slow but there are many encouraging signs.

In early work[30], the microstructural control was obtained by varying the austenitizing temperature of AISI 4130, 4330, and 4340 steels. Extremely high (1100 C to 1200 C) austenitizing temperatures were shown to confer combinations of strength and fracture toughness (K_{Ic}) that were far superior to those obtained by conventional heat treatments. Transmission electron microscopy of thin foils of specimens austenitized at 1200 C revealed extensive networks of retained austenite, as shown in Fig. 21. In some areas of the specimen, almost every martensite plate or lath was practically surrounded by austenite in the form of films 100 to 200 A thick. Attempts to establish either the metastable-austenite stability with respect to stress or strain or its influence on the mechanical properties were not successful. These studies clearly demonstrated the need for methods by which the volume fraction and stability could be varied over a wide range.

In our recent work, we were encouraged by the possibilities of incorporating, in new steels, retained austenite of the appropriate volume fraction, morphology, and stability to match the requirements of the desired mechanical properties. To date, virtually all of the research has been on modified commercial steels. The reasons are twofold: (1) a large background of information is available on these steels and they are available in relatively high purity form and (2) the accumulation of the required kinetic data on the decomposition of metastable austenite of simple and complex steels has been more difficult to obtain than first anticipated.

In the previous section, the unusual microstructures of the silicon-modified and isothermally treated AISI 4340 steel were described. These steels have a combination of fracture toughness and strength approximating that of the highly alloyed maraging steels.[44] To date, it has not been possible to establish quantitatively the influence of the retained austenite present in these steels on the fracture toughness. The fracture toughness appears to be directly related to the strength and indirectly to the volume fraction of the stable retained austenite. In an effort to resolve this problem, heat

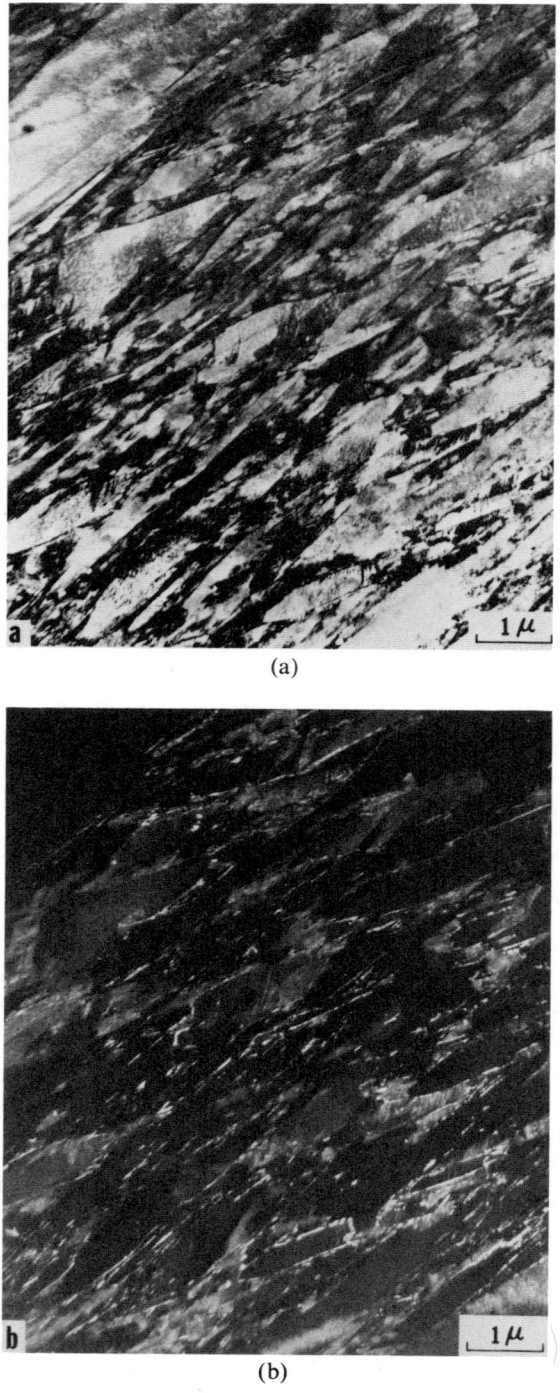

(a)

(b)

Fig. 21. As-quenched AISI 4340 steel: (a) bright field (b) dark field of an austenite spot for the 1200 C→870 C austenitized specimen.

treatments have been devised, with the aid of dilatometric measurements, to "condition" the retained austenite such that it decomposes in the tensile and fracture tests by a strain-induced transformation. The effect of the strain-induced decomposition on the tensile properties is unequivocal, as shown in Table VII for a series of silicon-modified AISI 4340 specimens. The elongation increases with increasing amounts of prior-conditioned retained austenite. The elongations shown are all relatively lower (by about a factor of two) than those obtained with ASTM standard specimens because of the special type of tensile specimen that was required for the magnetic measurement of retained austenite. These preliminary data indicate that practical modifications of composition and heat treatment of medium alloy steels can be made which will result in steels with yield strengths of about 200,000 psi and elongations of over 25 percent, as measured with standard specimens.

Some preliminary fracture-toughness data are available on high-silicon, low-alloy ultrahigh-strength steel (modified AISI 4330) given an isothermal heat treatment similar to that described in the above paragraph. This steel has a total alloy content of less than 5 percent. Two percent silicon was added to the base 4330 composition, making a total of 2.28 percent. A heat treatment was devised to condition the retained austenite (by tempering at 350 C) after air cooling from an isothermal hold temperature of 300 C. The tensile and fracture-toughness values were as follows: yield and ultimate tensile strengths, 199,000 and 244,000 psi, respectively; elongation, 11 percent; and fracture toughness (K_{Ic}), 96 ksi-in.$^{1/2}$. Similar properties can be obtained in this alloy by the use of a high austenitizing temperature (1200 C), followed by a drastic quench in iced brine and finally a low tempering treatment at 200 C. However, it is more desirable to use the isothermal holding treatment than the high austenitizing treatment. The combination of fracture toughness and strength for both treatments is about equal to that of the highly alloyed (over 30 percent) maraging steel at the same yield-strength level.

Table VII. Mechanical Properties of AISI 4340 Steels[a]
Modified With Silicon Additions

Silicon Addition to AISI 4340 Steel, %	Yield Strength (0.2% Offset), ksi	Ultimate Strength, ksi	Elongation, percent	Retained Austenite[b], percent
1.0	199	228	6.5	7
2.0	204	237	8.7	13
3.0	200	237	13.0	16

(a) The specimens were austenitized at 900 C (1 hour), isothermally transformed at 300 C for 1 hour and air cooled. They were tempered at 350 C for 1 hour and air cooled to room temperature.

(b) Determined by a magnetic saturation method.

The practical disadvantages of the high austenitizing treatment are obvious. Perhaps less obvious is the fact that such heat treatments can lead to complicated fracture behavior. For example, in 4340 and similar low-alloy steels[30,45-47] high-temperature austenitizing results in increased fracture toughness (K_{Ic}) values but decreased Charpy impact energies, when compared with conventional austenitizing at 870 C. A complete analysis explaining this behavior has recently been developed in this laboratory on the basis of the influence of notch root radius on toughness.[47] A consequence of this behavior is that, although the toughness of high-temperature austenitized steels is superior when rated by fracture mechanics (K_{Ic}), their performance in service in the presence of blunt notches is likely to be inferior.

The promising results obtained in the silicon-modified low alloy steels given isothermal heat treatments (rather than quench and tempering) has stimulated research on the problem of obtaining high toughness in thick sections. It has been ascertained that the natural air-cooling rates (from the austenitizing temperature) of thick sections, comparable to the wall thickness of large pressure vessels, will result in the production of microstructures in high-silicon ultrahigh-strength steels similar to those produced by isothermal holding (for example, large amounts of retained austenite can be obtained in 300 M which contains 1.65 percent silicon).[48] The correlationships among composition, microstructure, cooling rates, section size, and properties are currently being studied. The object of this study is to develop low-alloy high-strength steels (100,000 to 150,000-psi yield strength) with high fracture toughness in thick sections.

In the foregoing discussion, it has been shown that the tensile and fracture properties of low- and medium-alloy steels can be made equivalent to those of the best of the ultrahigh-strength steels, viz., the maraging steels, by several variations in composition and heat treatment. These compositional variations and modified heat treatments appear to be practical ones and are within the capabilities of current steel technology.

ACKNOWLEDGMENT

The authors thank the following individuals for kindly granting permission to use unpublished data: L. Thompson, G. Kohn, and R. Horn. They also thank Dr. Ritchie for his contribution on the problem of the observed discrepancy between Charpy and fracture-toughness data. The authors are especially grateful to Mr. M. Bhat for his critical review of the manuscript and his many reliable suggestions. This work was supported by the U.S. Energy Research and Development Administration.

REFERENCES

1. Bhandarkar, D., Zackay, V. V., and Parker, E.R., *Met. Trans.*,3, 2619 (1972).
2. Bhandarkar, M. D., D. Eng. Thesis, LBL-1858, Lawrence Berkeley Laboratory, Berkeley, California.
3. Toft, L. H., and Marsden, R. A., *Iron Steel Inst. Special Report*, No. 70, 276 (1961).
4. Decker, R. F., and Floreen, S., *Precipitation from Iron-Base Alloys*, G. R. Speich and J. B. Clarke (Eds.), AIME Metals Society Conference Proceedings, Gordon and Breach Science Publishers, New York (1965), Vol. 28, p. 69.
5. Jones, R. H., Zackay, V. F., and Parker, E. R., *Met. Trans.*, 3, 2835 (1972).
6. Jones, R. H., Parker, E. R., and Zackay, V. F., *Proceedings of Fifth International Materials Symposium*, Berkeley, California, September, 1971, p. 829.
7. Massalski, T. B., *Phase Transformations*, American Society for Metals, Metals Park, Ohio (1970), p. 433.
8. Baghdasarian, A., unpublished work.
9. Jin, S., M.S. Thesis, LBL-443, Lawrence Berkeley Laboratory, Berkeley, California.
10. Austin, C. R., St. John, C. R., and Lindsay, R. W., *Trans. Met. Soc. AIME*, 162, 84 (1945).
11. Parker, E. R., *Proc. ASTM*, 60, 1 (1960).
12. Archer, R. S., Briggs, J. Z., and Loeb, C. M., Jr. (Eds.), *Molybdenum*, Climax Molybdenum Company, New York (1970), p. 163.
13. Bhat, M. S., M.S. Thesis, LBL-2277, Lawrence Berkeley Laboratory, Berkeley, California.
14. Christian, J. W., *The Theory of Transformations in Metals and Alloys*, Pergamon Press, Oxford, England (1965), p. 670.
15. Spencer, C. W., and Mack, D. J., *Decomposition of Austenite by Diffusional Processes*, V. F. Zackay and H. I. Aaronson (Eds.), Interscience Publishers, New York (1962), p. 549.
16. Bhat, M. S., unpublished work.
17. Bhat, M. S., Singh, S., and Bhandarkar, D., *Inorganic Materials Research Division Annual Report, 1974*, LBL-3530, Lawrence Berkeley Laboratory, Berkeley, California.
18. Lyman, T. (Ed.), *Metals Handbook*, Vol. 1, American Society for Metals, Metals Park, Ohio, p. 409.
19. Zackay, V. F., Bhandarkar, M. D., and Parker, E. R., paper presented at the 21st Sagamore Army Materials Research Conference on Advances in Deformation Processing, Raquette, N.Y., August 13-16, 1974; also, LBL-2775, Lawrence Berkeley Laboratory, Berkeley, California.
20. White, C. H., and Honeycombe, R.W.R., *J.I.S.I.*, 200, 457 (1962).
21. Reed, R. P., and Guntner, C. J., *Trans. Met. Soc. AIME*, 230, 1713 (1964).
22. Holden, A., Bolten, J. D., and Petty, E. R., *J.I.S.I.*, 209, 721 (1971).
23. Schanfein, M. J., Yokota, M. J., Zackay, V. F., Parker, E. R., and Morris, J. W., Jr., paper presented at ASTM Symposium on Properties of Materials for Liquid Natural Gas Tankage, Boston, May 21-22, 1974; also LBL-2764, Lawrence Berkeley Laboratory, Berkeley, California.
24. de Miramon, B., M.S. Thesis, UCRL-17849, Lawrence Berkeley Laboratory, Berkeley, California.
25. Thompson, L., M.S. Thesis, research in progress, Lawrence Berkeley Laboratory, Berkeley, California.
26. Archer, R. S., Briggs, J. Z., and Loeb, C. M., Jr. (Eds.), *Molybdenum*, Climax Molybdenum Company, New York (1970), p. 162.
27. Zackay, V. F., Parker, E. R., Morris, J. W., Jr., and Thomas, G., *Mat. Sci. and Eng.*, 16, 201 (1974).

28. Zackay, V. F., Parker, E. R., and Wood, W. E., *Proceedings of Third International Conference on the Strength of Metals and Alloys*, Cambridge, England, August 20–25, 1973, Vol. 1, p. 175.
29. Zackay, V. F., and Parker, E. R., *Alloy Design*, J. K. Tien and G. S. Ansell (Eds.), Academic Press, New York (1975); also LBL-2782, Lawrence Berkeley Laboratory, Berkeley, California.
30. Wood, W. E., Parker, E. R., and Zackay, V. F., LBL-1474, Lawrence Berkeley Laboratory, Berkeley, California (1973).
31. Tom, T., Ph.D. Thesis, LBL-1856, Lawrence Berkeley Laboratory, Berkeley, California.
32. Naga Prakash Babu, B., D. Eng. Thesis, LBL-2772, Lawrence Berkeley Laboratory, Berkeley, California.
33. Ericsson, C. E., M.S. Thesis, LBL-2279, Lawrence Berkeley Laboratory, Berkeley, California.
34. Howard, R. T., Jr., and Cohen, M., *Trans. AIME*, **176**, 384 (1948).
35. Schaaber, O., *Trans. AIME*, **203**, 559 (1955).
36. Radcliffe, S. W., and Rollason, E. C., *J.I.S.I.*, **191**, 56 (1959).
37. Allten, A. G., and Payson, P., *Trans. ASM*, **45**, 498 (1953).
38. Owen, W. S., *Trans. ASM*, **46**, 812 (1954).
39. Keh, A. S., and Leslie, W. C., *Materials Science Research*, Vol. 1, Plenum Press, New York (1963), p. 208.
40. Speich, G. R., and Leslie, W. C., *Met. Trans.*, **3**, 1043 (1972).
41. Shih, C. H., Averbach, B. L., and Cohen, M., *Trans. ASM*, **48**, 86 (1958).
42. Alstetter, C. J., Cohen, M., and Averbach, B. L., *Trans. ASM*, **55**, 287 (1962).
43. Wada, T., Wada, N., Elliot, J. F., and Chipman, J., *Met. Trans.*, **3**, 1657 (1972).
44. Kohn, G., Ph.D. Thesis, research in progress, Lawrence Berkeley Laboratory, Berkeley, California.
45. Lai, G. Y., Wood, W. E., Clark, R. A., Zackay, V. F., and Parker, E. R., *Met. Trans.*, **5**, 1663 (1974).
46. Lai, G. Y., Wood, W. E., Parker, E. R., and Zackay, V. F., LBL-2236, Lawrence Berkeley Laboratory, Berkeley, California (1975).
47. Ritchie, R. O., Francis, B., and Server, W. L., *Met. Trans.*, **74**, 831 (1976), Lawrence Berkeley Laboratory, Berkeley, California (1974).
48. Horn, R. M., Ph.D. Thesis, research in progress, Lawrence Berkeley Laboratory, Berkeley, California.

DISCUSSION on Paper by V. F. Zackay

McMAHON: With regard to the use of high austenitizing temperatures to achieve high toughness in 4340-type steels, this appears to be an artifact of the sharp crack toughness test. When a blunt notch (e.g., Charpy V) is used the toughness is found to be degraded by high-temperature austenitizing. This degradation is due to the increased tendency for fracture along prior austenitic grain boundaries as grain size increases. If the grain size is much larger than the plastic zone size of the notch, the effect is masked. This has been observed by a number of groups, but is best documented and analyzed in recent work by R. O. Ritchie (U. Cal., Berkeley, to be published).

ZACKAY: The use of high austenitizing temperatures in the heat treatment of ultrahigh-strength steels causes many microstructural and microchemical changes and the associated changes in mechanical properties. If, *and only if,* the deleterious microstructural and microchemical features of an ultrahigh-strength steel are either eliminated or significantly minimized in this type of heat treatment, then, and only then, will there be an influence of the acuity of the notch in the toughness test. Further, and, far more importantly, if these deleterious features are eliminated by means of other more simple (practical) heat treatments and/or composition modifications which do not involve high austenitizing temperatures (and changes in grain size), then there is *no* effect of notch acuity even though the combinations of toughness and strength are equal to those of the best and most highly alloyed commercial steels currently available, i.e., the maraging steels. The synthesis of new steels with simple heat treatment, and significantly improved properties has been our goal in these studies, an effort now more than eight years old. The statements made above and the physical evidence supporting them are given in our paper and in the references listed therein.

TIEN: Your Fe-Ta-Cr-Mo alloy, which can be strengthened by the non-tantalum dispersoids, appears to be an ideal alloy system for the study of the effect of well controlled strengthening particles on alpha-iron properties. You have touched on such properties as yield and ultimate strength. Have you or are you planning to determine the effects of the particles or particle interface structure on fatigue or sustained load crack growth rates? May I suggest that this may be worth doing, since the results can shed some light, without bringing in other microstructure complications, on the effects of strengthening particles on crack growth behavior. As you know, there is a great deal of confusion on whether particles are beneficial or not.

ZACKAY: We are presently finishing the construction of apparatus which will allow us to answer the critical questions you have raised. Your encouragement will hasten our efforts to complete these studies.

SIMS: What is the long-time stability of your Laves phase, Fe_2Ta? At least in austenitic systems, such as the cobalt superalloy-type matrices, such phases have not shown high promise since they overage at too low a temperature.

ZACKAY: The Laves phase, Fe_2Ta, is, relative to the expected service temperature (maximum 700 C), extremely stable. Long time (several thousand hours) at temperatures as high as 1200 C does not appreciably alter the microscopic appearance of the Laves phase. This is true, especially, of those particles with a spherical shape. Lamellar-type microstructures do, however, tend to break up into discontinuous particles at temperatures above about 1100 C.

SHERBY: There is much need to improve the creep resistance of ferritic steels in the range 600 C, and your new results on an Fe-Ta alloy are most encouraging. The creep behavior of dispersion-hardened materials is not well understood, and we don't fully know what key variables may be manipulated for optimizing their high-temperature properties. I was curious as to the creep characteristics of your Fe-Ta alloy since such knowledge can assist in determining the rate-controlling process in creep. I had in mind, for example, the nature of the creep curve (e.g., amount of primary creep), the stress exponent for creep and the activation energy. In addition, I was curious about the creep ductility at 600-650 C. I ask this because a low ductility might be expected if no strain hardening occurs and if the stress exponent is high (that is, strain-rate sensitivity is low). On a related subject, I was wondering if the authors have considered thermal mechanical processing with the thought of attaining a fine subgrain structure which might then remain stable at warm temperatures from the presence of spherical precipitates of $FeTa_2$.

ZACKAY: With respect to the last question of your comments, i.e., the possibility of attaining a favorable substructure, we found that variations in heat treating procedure involving cold working before aging, and repeated cycling between room temperature and the spheroidizing temperature, did not cause changes in short-time tensile properties of a ternary Fe-Ta-Cr alloy, particularly at test temperatures near 700 C. A substructure developed by cold working the spheroidized Fe-Ta-Cr alloy was not effective in enhancing short-time yield strength at temperatures higher than about 700 C. In spite of these negative results we agree that the possibility remains for a substantial improvement in creep properties of these alloys. We intend to renew our efforts in this direction on the newer and more versatile Fe-Ta-Cr-Mo alloys discussed in our paper.

A stress sensitivity exponent of n = 6.5 and an apparent activation energy of 94,650 cal/mole were determined for the ternary Fe-Ta-Cr alloy. The precise values of creep ductility of these alloys were not available to me at the time of writing these comments, however, the tensile fracture elongation of these alloys in the 600-700 C range was usually between 20 and 30 percent.

MECHANICS IN ALLOY DESIGN

F. A. McClintock

Department of Mechanical Engineering
Massachusetts Institute of Technology
Cambridge, Massachusetts

ABSTRACT

The mechanical designer must effect a compromise between the costs of a large factor of safety, the cost of high hardness or high crack-initiation resistance, and the cost of high resistance to crack propagation, along with any necessary inspection during service. The economics of recycling will ultimately become a factor. Available equations are given relating these mechanical attributes to the microstructure. The primary possibilities for improvement seem to be (a) preferred orientation for increased stiffness, (b) increasing the adherence between the matrix and second-phase particles to delay hole nucleation, and (c) increasing crack meandering and branching to impede crack growth.

The difficulties of calculation suggest that more realistic tests are needed, e.g., of plane-strain ductility and of cracked plates suggestive of various weld defects.

1 INTRODUCTION

The optimum metallurgical design of an alloy will depend on the particular compromise between objectives that the structural or machine designer makes for the job at hand. After reviewing these objectives we turn to estimates of the mechanical behavior from the microstructure to see how the alloy designer can choose the microstructure that will best give the desired compromise.

2 OBJECTIVES IN THE DESIGN OF STRUCTURES, PRESSURE VESSELS, AND MACHINE PARTS

Some of the conflicting objectives that must be compromised in reaching an economical design are:

1. In structures and machines whose failure would involve low costs (relative to the annual income of the producer or owner), it may be most economical (for rare failures) to simply pay off on a guarantee. Here the emphasis is on quality control to reduce initial defects and on high strength for resistance to crack initiation in normally unnotched parts, or on low material cost for a large factor of safety.

2. If the cost of failure is high compared to the annual business income, and if the costs of overdesign are low, the designer can simply use a large factor of safety. If the costs of overdesign are high, it may be more economical to design on the basis of allowable crack growth between inspections. This involves the concepts of maximum undetected crack size, acoustic emission, crack growth rate, and critical crack size for fracture. Alternatively or simultaneously, the designer can use fail-safe design, where the fracture of any one element will not lead to loss of the entire structure or machine.

3. Parts that are subject to uncertain overloads for just a few cycles, especially those containing stress or strain concentrations, require ductility. Sometimes the product must remain functional after overloads. Under other circumstances, as in automobile collisions, the structure may be regarded as sacrificial. In intermediate cases, repairability may be a dominant factor.

4. Ductility may also be required because of stresses arising from heat treating, welding, forming, or assembling.

5. The forming, heat treating, and assembly processes themselves must be economical relative to the overall cost.

6. Consumer products should be made of easily recyclable materials, whereas long-life parts, such as for central-station and air-transport turbines, can be recycled as alloys.

These various objectives are all reflected in the desire for minimum annual cost to the customer, to society, and to future generations for the service provided. Specific examples of the properties needed to meet these different objectives will now be given.

3 BRITTLE GRAIN-BOUNDARY FRACTURE

Although the idealized situation of purely brittle grain-boundary fracture is not attained in metallic alloys, it provides a reasonably well worked out example of some of the compromises that must be reached.[1,2] To study the strength of parts under the interactions of internal flaws, a numerical model was first developed. The grain structure was idealized as a regular hexagonal plane array of grains whose strengths were randomly assigned from an extreme value distribution of the third kind, where they vary at some power $m-1$ of the amount with which the stress S exceeds some lower limit S_L. The model assumes an elastically homogeneous and isotropic material. The program computes the externally applied stress S necessary for fracture of the weakest grain boundary relative to its locally applied stress, for each cracked segment in turn. If contiguous crack propagation is more critical than renucleation, the lowest grain-boundary toughness relative to the local stress-intensity factor is found. The critical segment is assumed cracked, a new elastic solution is found, and so on. The ratios f_{K0} and f_{KL} deal with the relative toughness and strength of grain boundaries.

As expected, microcrack initiation, at S_{1st}, occurs sooner with increasing coefficient of variation and with increasing size of part. On the other hand, microcracking occurs at a lower fraction of the maximum strength S_{max} with increasing coefficient of variation and with increasing size of part. Hence, the design of parts with highest strength must be compromised in order to obtain a greater possibility of crack detection by acoustic emission, for example.

An approximate analytical theory was developed to provide more insight into the numerical results. Space does not allow statement of the equations here, but a comparison between the numerical and analytic calculations is shown in Fig. 1, in which the number of crack segments is plotted versus the applied stress. The maximum stress S_{max} is estimated from the probability of getting a number of contiguous crack segments N_c which would allow the average critical-stress-intensity factor to be reached, after which further crack extension requires progressively lower applied stresses. Once the maximum stress is reached, the increase in total number of crack segments is made up almost entirely of the added number of segments along the main crack.

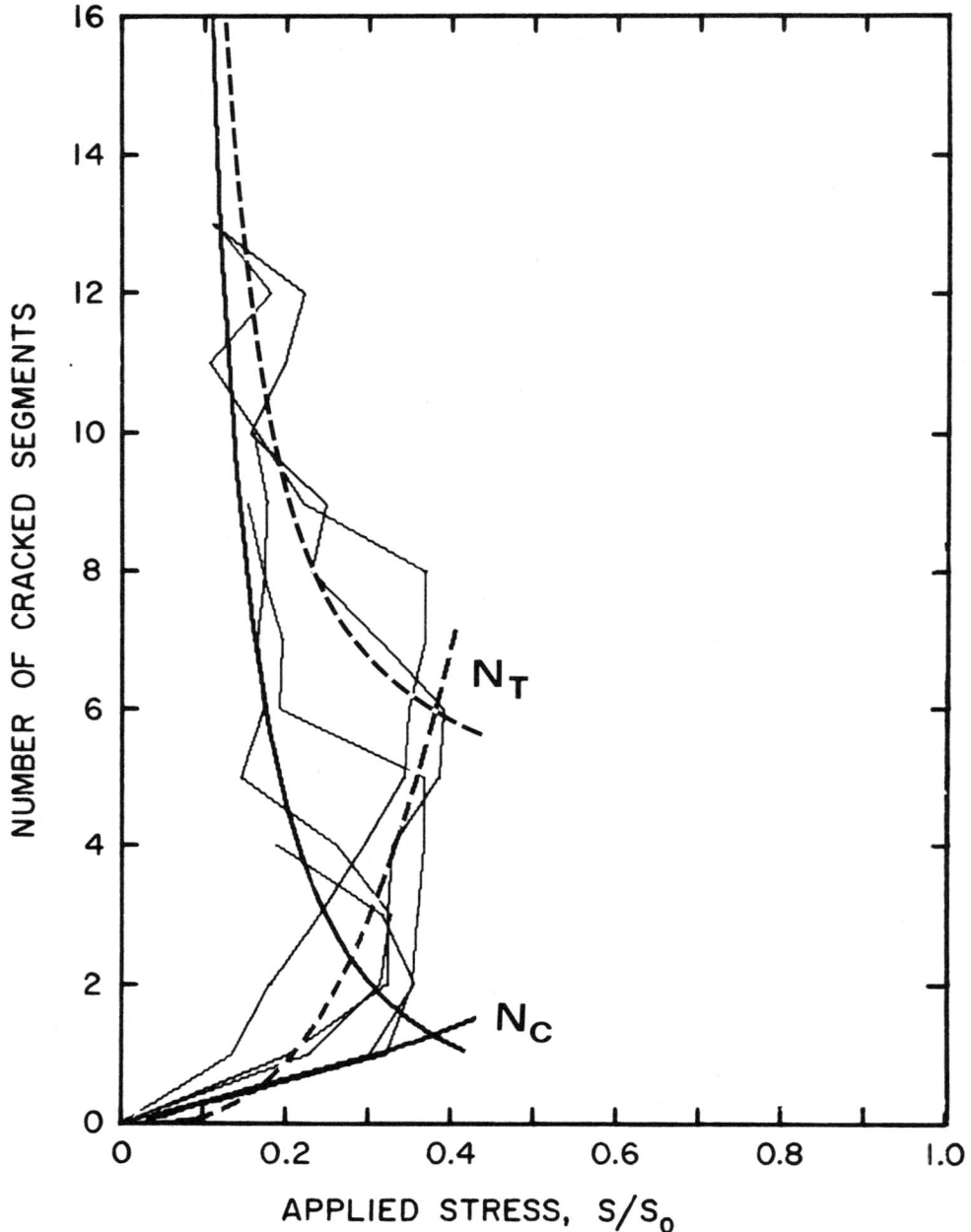

Fig. 1. Total cracked segments N_t and maximum number of contiguous cracked segments N_c versus applied stress from the analytic and numerical models of brittle grain-boundary cracking, for $m = 3$, $S_L/S_0 = 0$, $f_K = 0.77$, and $N = 111$.

The results for variety of part sizes are shown in Fig. 2, indicating the progressively larger spread between the stress for first fracture S_{1st} and the stress at maximum load S_{max} as the part size is increased, for two probabilities Φ of fracture of the part.

For crack growth, the numerical model shows that a coefficient of variation in grain-boundary strengths of 0.15 can increase the macroscopic fracture toughness by a factor of approximately 2 over that for a homogeneous material. The increase in crack toughness is attained at the expense of a small loss in strength.

The introduction of anisotropy in grain-boundary strengths tends to halt propagation by causing delamination perpendicular to the crack. To increase the crack toughness by such anisotropy appears to require that grain boundaries parallel to the crack must be at least three or four times stronger than those of other orientations. Crack toughness in other directions would be correspondingly reduced.

4 ELASTIC STIFFNESS

The elastic stiffness is essentially related to the atomic binding forces. In fact, for many materials the ratio of bulk modulus B to binding energy per unit volume, ρu, falls in the range of 2 to 4, while those properties individually vary by as much as a factor of 10.[3] The only real possibilities for improving stiffness seem to be in improving other properties of an inherently stiff element, such as beryllium, or using metal-matrix graphite-fiber-reinforced composites at the expense of serious anisotropy in other properties, or on a more modest scale, economically producing steel with a <111> orientation in the direction of the working stress.

Conversely, for using elasticity to understand microstresses in materials, some 40 references involving inclusions, corners, average elastic constants, and cracks in three dimensions, in dissimilar materials or in arrays, have been assembled.[4]

5 RESISTANCE TO PLASTIC FLOW

The derivation of flow strength from the underlying metallurgical structure is better treated by many other papers in this book and elsewhere. There are only two points to mention here.

First, in low-cycle fatigue under an arbitrary history of strain, temperature, and time, the ordinary concept of a stress-strain relation will probably be insufficient. Rather, it appears desirable to use first a constitutive equation, giving the (tensor) dependence of the stress on the strain

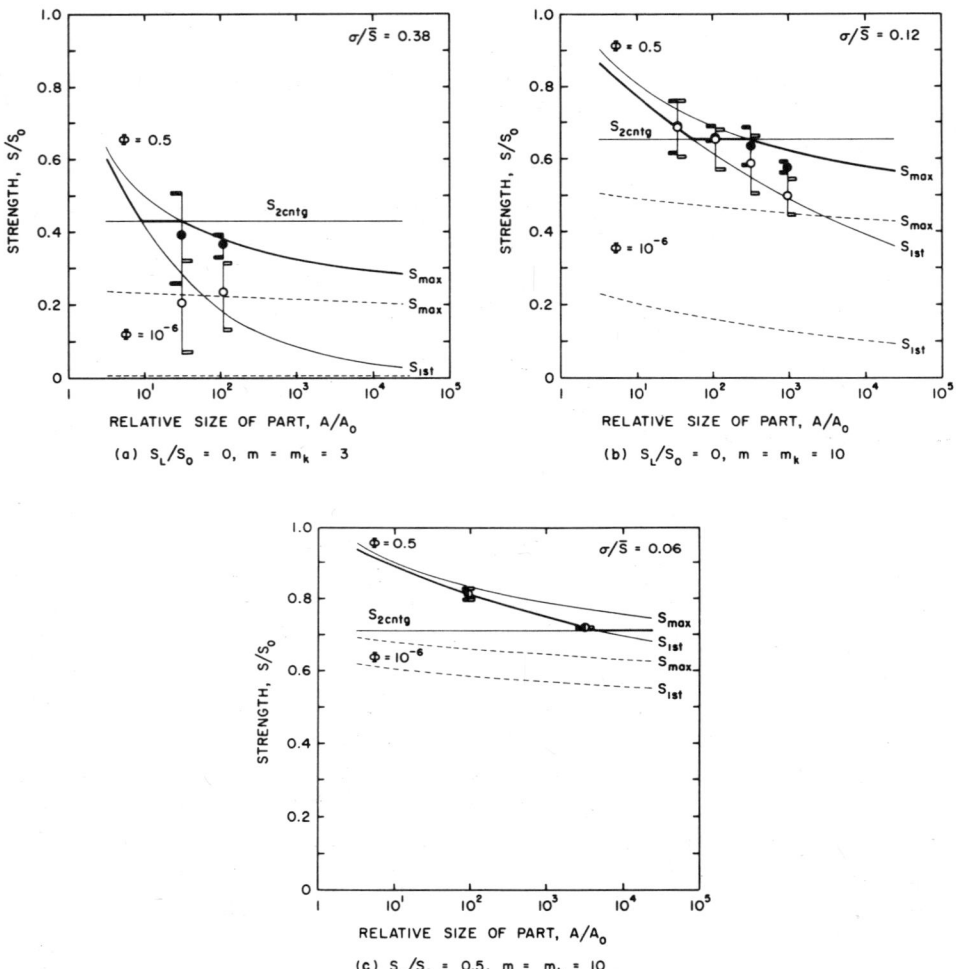

Fig. 2. Statistical strength as a function of desired fracture probability Φ and relative part size A/A_o for various strength distributions with exponent m, and lower limit S_L/S_0, $f_{K0} = f_{KL} = 0.77$. (o = first cracking, S_{1st}, ● = instability, and S_{max}. $S_{2\ cntg}$ is stress for second microcrack if contiguous.) (a) $S_L/S_0 = 0$, $m = m_k = 3$, $\sigma/\overline{S} = 0.38$; (b) $S_L/S_0 = 0$, $m = m_k = 10$, $\sigma/\overline{S} = 0.12$; and (c) $S_L/S_0 = 0.5$, $m = m_k = 10$, $\sigma/S = 0.06$.

rate, on the temperature, and on internal parameters describing the essence of the structure:

$$\sigma = \sigma\,(\dot{\epsilon},\ T,\ P_1,\ P_2,\ \ldots)$$ (1)

and then introduce process equations giving the rate of change of internal parameters with their present values, the temperature, and the strain rate:

$$\dot{P}_i = \dot{P}_i\,(P_i,\ T,\ \dot{\epsilon})\quad .$$ (2)

It is almost certain that no one characterization will be simple and accurate enough to describe a very broad range of alloys and conditions. In principle, the computer could use any such relations; in practice, with a finite element containing as many as 30 points and with (tensorial) structure parameters, stress components, strains, and (vector) displacements at each point, 500 data can easily be required for each element. When one considers that structural analyses often involve thousands of elements, it becomes clear that economy and accuracy in describing the constitutive and process equations will be very important. For a long time to come we shall not be able to simply turn a problem over to the computer without considerable thought and preprocessing, especially since the rate of increase of computer power does not seem to be changing much from a doubling every other year, and costs do not seem to be decreasing much.

Second, in connection with flow strength, one should bear in mind that the very concepts of stress and strain are no longer useful where there are extremely high stress or strain changes over the mean free path of a dislocation. Such situations occur near surfaces or hard grain boundaries. Inclusion of these effects would require introducing strain gradients and property gradients into Eqs. (1) and (2). Some speculation along these lines has been included in the couple stress, micropolar, or nonlocal theories of elasticity. Very little has yet been done at the plastic scale, where these effects extend over distances of as much as microns, in contrast to the elastic case where the corresponding distances are of the order of atomic dimensions. Fourie[5] has reviewed the evidence for the existence of a soft layer near a surface. One might compare the depth of this layer with the interaction or pinning distance of the dislocation. In either case, a typical distance is given in terms of the shear modulus of elasticity G, the Burgers vector b, and the yield strength in shear k, by

$$\ell = G\,b/k\quad .$$ (3)

Applying Eq. (3) to Fourie's data using the bulk strength indicates that the softened layer was larger than ℓ by a factor of 100 to 1000. Presumably, if one used the strength of the softened layer, a better agreement would be found. Applying the same factor to a high-strength steel with $G/k = 100$ gives a characteristic distance of the order of 2 to 20 microns, comparable to dimensions and spacings of the holes which give rise to ductile fracture. Such softening effects are therefore real possibilities. Suh[6] has found it possible to intentionally introduce such a soft layer at the scale of microns to reduce wear coefficients to the order of 10^{-8} or below.

6 DUCTILE FRACTURE

The absence of solutions for the mechanics of ductile fracture is a major impediment to making quantitative predictions about the effects of changes in the design of alloys, but some results can be given.

6.1 Hole Nucleation

Weak boundaries between inclusions and matrix may cause holes to nucleate at nearly negligible stress. In aluminum alloys, holes are often present from earlier processing. On the other hand, holes may also nucleate later as the matrix strain hardens.

Argon and co-workers[7] have shown that for equiaxed inclusions and a variety of solutions ranging from linearly hardening to nonhardening, including elastic-plastic numerical results, the peak interface stress occurs in line with the direction of maximum applied stress (point A of Fig. 3). In terms of the mean normal applied stress σ and the shear flow strength of the matrix k, the interface stress σ_{rr} falls in the range

$$\sigma_{rr} = \sigma + (1 \text{ to } 2) \, k \quad , \tag{4}$$

The range of coefficients covers all degrees of strain hardening and might also vary with relative elastic moduli. Taking a factor of 1.73, Argon and Im[8] found that for copper chromium in copper, titanium carbide in VM-300, and Fe_3C in 1045 steel, the interface stresses for nucleation of holes from inclusions were of the order of 0.008 E to 0.009 E. One would very much like to have a more accurate idea of the ideal interfacial strengths to know how much room for improvement there is. For a perfect close-packed lattice the strength is likely to be no less than 0.05 E, according to Kelly.[9] McClintock and O'Day[10] have shown that for soap-bubble rafts simulating the interatomic forces between metal atoms, high-angle grain boundaries have a cohesive strength of about half that value, leaving only about another factor of 3 to 5, to the observed interfacial strengths [depending on the factor of Eq. (4)].

If the separation occurs 30 degrees or so from the pole of the inclusion,

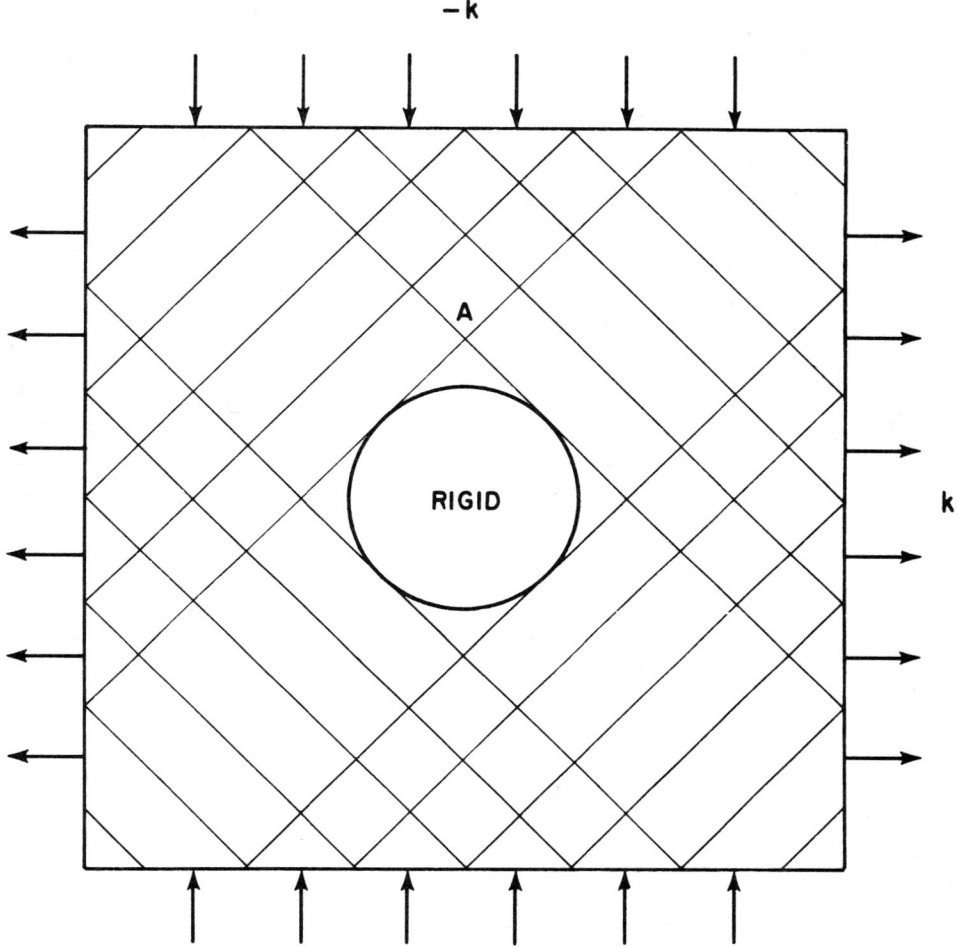

Fig. 3. Slip lines around a rigid inclusion.

the interface stress would be less, but the shear stress would be greater, and the shear strain might contribute microstresses due to inhomogeneous dislocation structure.

The regions of highest strain are away from the inclusion (where the slip lines of Fig. 3 first cross), and fracture has been observed there by Gurland[11] and others, as well as in fatigue by Lankford and Kusenberger[12]. More work remains to be done, but in any event, it is clear that in front of a crack, where the maximum normal stress reaches $(2 + \pi)$ times the shear flow strength, $Y/\sqrt{3}$, the stresses in alloys with $Y/E = .01$ would reach $E/30$ and any interface would break. Notch toughness could not be attained by a strong interface. The question is how else?

6.2 Growth of Holes, Including Internal Pressure

Because of the interest in radiation damage, it may be worth including the effects of internal pressure. First consider pure equiaxial tension, σ, along with an internal pressure p. For spherical holes in an infinite medium with constant yield strength Y, Poisson's ratio ν, and modulus of elasticity E, containing holes under internal pressure p and subject to an applied equiaxial tension σ at infinity, the condition for continued growth of a hole is (e.g., Hill[13])

$$(p + \sigma)/Y = (2/3) (1 + \ln[E/(3(1 - \nu)Y)]) \quad . \tag{5}$$

Similarly, for cylindrical holes under plane strain,

$$(p + \sigma)/Y = (1/\sqrt{3}) (1 + \ln[E\sqrt{3}/(5 - 4\nu)Y]) \quad . \tag{6}$$

For Y/E from .00125 to .01, the values of $(p + \sigma)/Y$ range from 4.6 to 3.2 for spherical holes and from 4 to 2.8 for cylindrical ones. (This is some justification for the common assumption of plane strain.) Note that the mean normal stress already present in front of a crack provides most of the action needed, even without internal pressure or the softening effect of a surrounding plastic strain field:

$$\sigma = (1 + \pi)Y/\sqrt{3} = 2.4 \ Y \quad . \tag{7}$$

With strain hardening at a constant rate according to

$$Y(\bar{\epsilon}^p) = Y + Y'\bar{\epsilon}^p \quad , \tag{8}$$

Hill[13] has also given the following equations for the growth of spherical and cylindrical holes in an incompressible material:

$$(p + \sigma)/Y = (2/3) (1 + \ln(2E/3Y)) + 2\pi^2 Y'/27Y \quad , \tag{9}$$

$$(p + \sigma)/Y = (1/\sqrt{3}) (1 + \ln(E/\sqrt{3}Y)) + \pi^2 Y'/18Y \quad . \tag{10}$$

In steels, the rate of strain hardening is less with increasing strain and for steels of higher strength. Taking Y'/Y of 0.6 to 0.2 as Y/E goes from .0025 to .01 gives $(p + \sigma)/Y = 4.8$ to 3.6 and 4.0 to 3.0 for spherical and cylindrical holes, respectively, not a very large effect resulting from strain hardening.

If the holes are spaced closely enough together, full plasticity intervenes before the conditions for expanding them in an infinite medium are attained. For holes currently of diameter 2a spaced $2\ell_0$ on centers, the conditions for growth of spherical and cylindrical holes in nonhardening material are, respectively,

$$(p + \sigma)/Y = 2 \ln (\ell_0/a) \quad , \tag{11}$$

$$(p + \sigma)/Y = \ln (\ell_0/a) \quad . \tag{12}$$

For spherical holes in steel with $E/Y = 100$, full plasticity is attained with a spacing ratio $\ell_0/a = 5$; for the softer material with $E/Y = 800$, the spacing ratio is 10, corresponding to a volume fraction of 0.001. With cylindrical holes, as might be encountered around strings of inclusions, the fully plastic limit intervenes much sooner. For example, with a spacing ratio $\ell_0/a = 10$, Eq. (12) indicates the stress ratio of $(\sigma + p)/Y = 2.66$, which is lower than that for elastic-plastic growth in steels of any Y/E. Once the fully plastic limit is reached, the hole can grow under decreasing pressure.

The above analyses of critical conditions for hole growth have been made on the assumption that the stress at infinity (that is between the holes) is pure equiaxial (hydrostatic) tension. Hole growth can occur far more readily when the material is already undergoing macroscopic plastic deformation, and the internal pressure or applied triaxial stress need only bias the strain. An approximate equation, exact for zero or linear hardening and transverse strain and axial shortening of cylindrical holes, has been obtained by McClintock.[14] The rate of change of semiminor axis a of holes spaced $2\ell_a$ apart, subject to applied stress of σ_a^∞ and σ_b^∞ and internal pressure p, in a material with an equivalent flow stress Y undergoing an equivalent strain increment of $\bar{\epsilon}^\infty$ is

$$d \ln(a/\ell_a)/d\bar{\epsilon}^\infty = \sinh[(1-n) (\sigma_a^\infty + \sigma_b^\infty + 2p)/(2Y/\sqrt{3})]/(1-n) \quad . \tag{13}$$

Note that in the limit as the strain-hardening exponent n approaches 0 the hole growth rate becomes exponential with the applied stress, whereas for linear hardening, n = 1, the hole growth rate per unit applied-strain increment varies linearly with the applied stress.

To visualize the importance of variables in plastic hole growth, consider the applied strain necessary to attain necessary hole growth ratios, as shown in Table I. The growth ratios to fracture are, if anything, less than the initial relative hole spacing ℓ_a/a because of the localization of plastic flow to be discussed below. The volume fraction of inclusions would be of the order of the initial value of $(a/\ell_a)^3$. Typical high-strength alloys have strain-hardening exponents of 0.1 or less. Several conclusions can be drawn from Table I:

Table I. Strains for Hole Growth Ratios

[As functions of strain-hardening exponent n and ratio
of applied transverse stress to plane-strain
yield strength, $(\sigma_a^\infty + \sigma_b^\infty)/(2Y/\sqrt{3})$]

Transverse Stress, $(\sigma_a^\infty + \sigma_b^\infty)/(2Y/\sqrt{3})$	1.5			$1.5 + \pi/2$			$1.5 + \pi$		
Flank-to-Flank Angle 2ϕ	180 deg			90 deg			0 deg		
Exponent n	0	0.1	0.2	0	0.1	0.2	0	0.1	0.2
Growth Ratio $(a/\ell_a)/(a/\ell_b)_0$									
2.5	0.430	0.458	0.486	0.085	0.104	0.127	0.018	0.025	0.036
5	0.756	0.805	0.853	0.150	0.183	0.222	0.031	0.044	0.063
10	1.081	1.152	1.220	0.214	0.262	0.318	0.044	0.064	0.090
20	1.407	1.499	1.588	0.279	0.341	0.414	0.058	0.083	0.117

1. The effect of triaxiality is very large, corresponding to a factor of 5 reduction in ductility going from unnotched plane strain to opposing 90-degree grooves, and another factor of 5, again to a sharp crack, regardless of inclusion spacing or strain hardening.

2. In the usual range for structural alloys, the effect of strain hardening is small at low triaxiality, but a factor of 2 at high triaxiality. Actually, the benefits of high strain hardening are even greater, for more hole growth would occur before localization.

3. While the benefits of increased hole spacing are pronounced for small ratios, the benefit is less with increasing purity, because the effect shows up as a logarithm in Eq. (13), and reducing the volume fraction of holes from .001 to .000001, an impractical feat metallurgically, would only increase the allowable strain by a factor of 2.

Note that the effect of internal pressure p is to accelerate the hole growth rate, just as would externally applied triaxial tension. External pressure does not suppress crack growth entirely, but rather leads to elongated holes squeezed out laterally from the inclusions as shown, for example, by French and Weinrich[15] and by Tipnis and Cook[16]. Such processing would weaken the metal under subsequent transverse tension.

6.3 Localization of Hole Growth

A metal containing widely separated holes or cracks at first deforms rather homogeneously. The effect of the porosity on the macroscopic properties remains small. At some stage, however, there arises the possibility of localized flow in which the applied deformation is concentrated in a thin region whose thickness is of the order of the spacing between the holes. Practically, this marks the end of the useful ductility of the part, because the subsequent extension to fracture of the entire part is only of the order of the hole spacing. Nagpal et al.[17] have given necessary and sufficient conditions for the localization of plastic flow of such dilatational material:

Plastic flow localizes at a point when there exists (a) a surface through that point at some orientation that gives no distortion in the plane of the surface, and (b) some field of relative motion across a thin layer parallel to the surface that does not require further deformation in the material on either side of the layer.

The first condition requires plane strain, with the plane normal to the deforming layer. The second condition requires demonstrating a flow field in which localized plastic flow can occur between the holes in the layer more easily than general plastic flow of the bulk material, and the growth of holes decreases the load more rapidly than strain hardening increases it.

These conditions for localization of dilatational plastic flow are analogous to the formation of a neck in a round tensile bar, or in a sheet where the necking region is a strip whose width is of the order of the sheet thickness, at an oblique angle to the tensile axis that does not have any normal strain along it, so the sheet can remain rigid, away from the neck. In plane strain of nonhardening material, the slip lines provide surfaces of incipient localization. Any holes along one of these lines will lead to localization. A more confined mode of localization is shown in Fig. 4. For this mode to occur, the limit load exerted by the decohering region must be less than that for the distributed flow field. If we assume that the latter limit load is reduced in proportion to the linear relative hole spacing, then from the traction exerted by the logarithmic spirals of the localized mode, this further localization begins when

$$2k \ln (\ell_a/a) = [2k(1 + \pi/2 - \theta) + p] (1 - a/\ell_a) \tag{14}$$

Note how the required hole size, on a linear basis, turns out to be only 0.0985 for $\theta = 0$, 0.1908 for $\theta = 30°$, both with no internal pressure. (These results correct an earlier error by the author.[17])

With strain hardening, the conditions will depend essentially on the inhomogeneous strain distribution in the part. Numerical analysis is needed, but no results are known at this time.

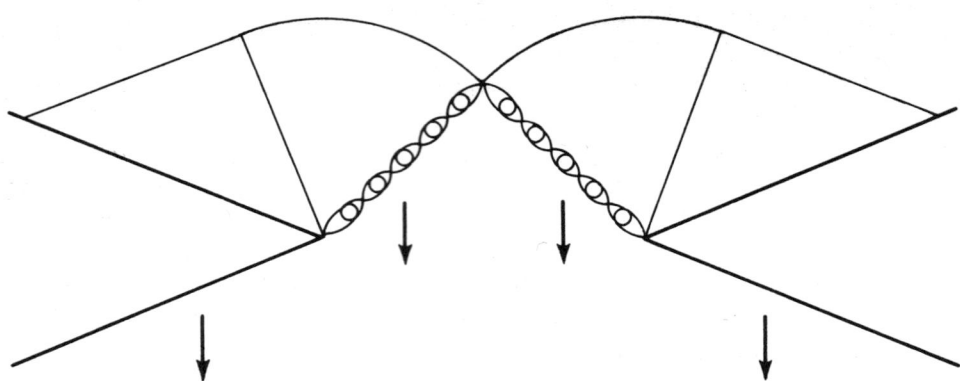

Fig. 4. Secondary localization in plane strain.

In the neck of a tensile specimen there is initially no plane strain, so the localization criterion cannot be met. Rather, plane strain must gradually develop as the material becomes progressively more porous. Again, no results are known.

6.4 Final Separation

Whatever the porosity at which localization of flow has occurred, the traction falls off with displacement at an initial rate of nearly 100 percent reduction in traction for a displacement of only 0.2 times the hole spacing.[17]

The work of final separation, integrated from the traction-displacement result, is approximately

$$W = (Y/\sqrt{3})\, \ell\, (1 - a_0/0.6\,\ell_a) \quad \text{for} \quad a_0/\ell_a < 0.5 \ . \tag{15}$$

Equation (15) also applies to flow localized to depleted grain boundaries.

6.5 Initial Crack Growth

In typical high-strength alloys, crack propagation occurs primarily by the interaction of the crack with one hole at a time.[18] For the initial growth, assume that the first hole will join with the crack when the opening of the crack flanks one tip radius away from the crack tip is equal to the spacing between holes:

$$\text{CTOD} = 2\ell \ . \tag{16}$$

(For a nonhardening plastic crack running in from one side of a plate, the actual displacement can be worked out from the deformation field of Garr, Lee, and Wang.[19] The assumption used here is likely to be a mild overestimate of the required displacement.)

Conditions at the tip of a crack are related to the macroscopic geometry and loads by fracture mechanics. If the body were linearly elastic, the stress field in the neighborhood of the crack tip (and hence the conditions for fracture) could be characterized by a single scalar parameter, the stress-intensity factor K. For a crack of length 2c in an infinite solid subject to a stress σ^{∞} across the crack,

$$K_I = \sigma^{\infty} \sqrt{\pi c} \quad . \tag{17}$$

Values for other loads and geometries are tabulated, for example, by Tada et al.[20] The resulting stress and displacement fields are given there, or, for example, by Rice.[21] But structural alloys are usually chosen to have plastic flow. For nonhardening plastic deformation, Levy et al.[22] have found numerically that the plastic zone extends mostly above and below the crack tip, to a radius from the tip of nearly

$$r_Y = \frac{1}{2\pi}\left(\frac{K_I}{Y}\right)^2 \quad . \tag{18}$$

It is the critical value of this plastic zone size, r_{Yc}, not K_{1c} itself, that should be tabulated as a material property characterizing crack toughness because its dimensions are those of length and are much more easy to visualize than $FL^{-3/2}$. In particular:

1. The designer can directly compare r_{Yc} with the dimensions of his part. If the calculated r_{Yc} is greater than the ligament dimension, the part will carry its fully plastic limit load before fracture (although fracture may still reduce the extension of the part).

2. r_{Yc} gives a rough measure of the maximum allowable half-length of a crack that can withstand working stresses up to a large fraction of the yield strength. Specifically, up to about half the yield strength, Eqs. (17) and (18) combine to give

$$c_c = \frac{1}{\pi}\left(\frac{K_{Ic}}{\sigma_c}\right)^2 = \frac{r_{Yc}}{2}\left(\frac{Y}{\sigma_c}\right)^2 \quad . \tag{19}$$

For instance, for a maraging steel with a reasonably tough-sounding value of $K_{Ic} = 90$ MN/m$^{3/2}$ (83,000 lb/in.$^{3/2}$) and a yield strength of 1806 MN/m^2 (262,000 psi), the critical plastic-zone size is only 0.39 mm, and cracks down to about this size would have to be found and eliminated, unless one adopts fail-safe design or is willing to pay off on failures.

3. The plane strain r_{Yc} can be compared directly to the specimen thickness. The plane-strain assumption is valid if the part if thicker than about 6 times the plane-strain critical plastic-zone radius.

4. If r_{Yc} approaches the ligament dimension, nonlinear elastic fracture mechanics (the J-integral) must be used.

5. For nonhardening materials, the flank-to-flank crack-tip opening displacement can be readily pictured in terms of the plastic-zone radius by[22]

$$CTOD = 2.7 \; r_Y \; Y/E \quad . \tag{20}$$

Now return to estimating the fracture toughness at initial crack growth, r_{Yi}, from the inclusion spacing ℓ assuming $CTOD = 2\ell$. For a nonhardening material, Eq. (20) gives

$$r_{Yi} = 0.37 \; (2\ell) \; (Y/E) \quad . \tag{21}$$

For a hardening material, the near tip stress, strain, and displacement fields can be given in terms of the J-integral, which can be thought of as a scalar, like K_1. These fields hold for a nonlinear, elastic, incompressible stress-strain relation of the form

$$\sigma = \bar{\sigma}_1 \; \epsilon^n \quad . \tag{22}$$

If Eq. (22) represents the total strain, the nonlinear-elastic or deformation-theory plasticity gives the same result as the more physical incremental plasticity. It is not valid for crack growth, wherein the ratios of strain increments change as the crack passes by. For most alloys, the exponent n is of the order of 0.05 to 0.25. The constant $\bar{\sigma}_1$ can be thought of as the equivalent stress at unit strain. The nonlinear stress and strain fields also depend on a numerical coefficient I_n that varies slowly with n[23] within 2 percent of

$$I_n = 10.3 \; \sqrt{0.13 + n} - 4.8n \quad . \tag{23}$$

In terms of these quantities and normalized stress, strain, and displacement functions which depend only on angle, the Hutchinson[24], Rice and Rosengren[25] (HRR) fields are

$$\frac{\sigma_{ij}(r_1\theta)}{\bar{\sigma}} = \left(\frac{J}{\bar{\sigma}_1 I_n r} \right)^{n/(n+1)} \bar{\sigma}_{ij}(\theta) \quad ,$$

$$\epsilon_{ij}(r,\theta) = \left(\frac{J}{\bar{\sigma}_1 I_n r} \right)^{1/(n+1)} \bar{\epsilon}_{ij}(\theta) \quad , \tag{24}$$

$$\frac{u_i(r,\theta)}{r} = \left(\frac{J}{\bar{\sigma}_i I_n r}\right)^{1/(n+1)} \bar{u}_i(\theta) \quad .$$

At the flank of a crack, the normalized displacement function is unity. The desired flank-to-flank opening where $r = u$ is then

$$\text{CTOD} = \frac{2 J}{I_n \bar{\sigma}_1} \quad . \tag{25}$$

Equation (24) gives perhaps the most direct physical insight into the meaning of J. Since the stress-strain Eq. (22) is not at all valid for crack growth, the interpretation of J as a strain-energy release rate is not physically very useful, even though it provided the insight for the original derivation.

When the plastic zone is extending nearly all across the specimen, J must be evaluated numerically for the particular structure and loading. When there is a surrounding linear elastic stress singularity, J can be found from K:

$$J = K_I^2 (1 - \nu^2)/E \quad . \tag{26}$$

In terms of the nonhardening plastic-zone radius,

$$J = 2\pi r_Y Y^2 (1 - \nu^2)/E \quad . \tag{27}$$

Finally, we can relate the critical plastic-zone size for crack initiation to the spacing between hole nuclei by combining Eqs. (16), (25), and (27):

$$r_{Yi} = \ell \frac{I_n}{2\pi} \frac{E\bar{\sigma}_1}{Y^2 (1 - \nu^2)}, \text{ with } \nu = 0.5 \quad . \tag{28}$$

For a nonhardening material this reduces to $r_{Yi} = 0.39(2\ell)E/Y$, compared to $0.37(2\ell)E/Y$ obtained from Eq. (21). For a hardening material with $n = 0.1$, $Y/E = 0.004$, $Y/\sigma_1 = 0.174$, $r_{Yi} = 0.822(2\ell)E/Y$, indicating a rather strong effect of strain hardening.

6.6 Crack Instability

It is still not clear whether or not a plane-strain Mode I crack is stable after the first hole has joined the crack. There are no numerical solutions, and not much prospect of any reasonably exact ones being obtained soon, taking into account geometry changes, the incremental flow rule, leaving strained material behind, strain hardening, and residual stresses. Perhaps some insight can be obtained from the Mode III solutions (shear parallel to the leading edge of the crack). The most unrealistic feature is that in Mode

III the plastic zone extends mostly in front of the crack, whereas for initial Mode I loading it runs off to the side. This discrepancy is likely to be reduced by strain hardening and residual stress, but experiments still frequently show zigzagging of a Mode I crack, and a fully plastic solution seems to confirm it for growth, although there is not general agreement.

With these caveats, the Mode I analogue of Mode III instability is given in terms of the fracture strain (perhaps including any needed to initiate holes and only the further strain to the point of localization)[26]:

$$r_{Yc} = 2\ell \exp\left[\sqrt{2\,\epsilon_f^p/(Y/E) + 1} - 1\right] . \tag{29}$$

The analogue for initiation is:

$$r_{Yi} = 2\ell\ (\epsilon_f^p/(Y/E) + 1) . \tag{30}$$

The estimates from Eqs. (28) and (30) for initiation and Eq. (29) for instability are given in Table II for the three aluminum alloys and the steel alloy studied by Joyce[18], where the hole growth ratio was taken roughly from his scanning electron micrographs. The primary difference between the alloys is in the yield strain Y/E. The displacement initiation criterion fails to predict the large observed differences between instability in alloys. The nonhardening equation for instability from strain is low, perhaps because crack blunting and hole growth reduce the triaxiality, and also because it does not account for strain hardening.

Table II. Initial and Critical Plastic Zone Radii

Alloy	Y, MN/m^2	E, MN/m^2	n	2ℓ, m	$\dfrac{(\ell_a/a)_s}{(\ell_a/a)_f}$	Initiation, r_{Yi} Strain, Eq. (28) mm	Displac., Eq. (30) mm	Instability, r_{Yc} Strain, Eq. (29) mm	Observed mm
7075-T6	500	72,000	0.10	12 E-6	3	1.42	0.064	0.1	1.34
				16 E-6	4	1.89	0.104	0.189	
2024-T4	390	72,000	0.12	12 E-6	3	1.82	0.084	0.163	3.39
				40 E-6	3	6.07	0.281	0.544	
6061-T6	281	72,000	0.09	12 E-6	6	2.53	0.159	0.689	11.
1117 steel	510	200,000	0.07–0.11	4 E-6 20 E-6	5	1.89–6.44	0.332–0.377	2.14–3.13	>25.

How can changing the alloy change the crack toughness? Both Eqs. (28) and (29) indicate that increasing the hole spacing has a direct effect. Reducing the yield strength is effective in either case. Only if instability governs is the fracture strain directly effective, and this is due primarily to increasing strain hardening. With a constant yield strength, increasing the strain hardening will decrease the spacing between holes as more inclusions break, or break away from the matrix. The limit to improvement is probably no more than a factor of two.

What else can be done? Crack branching seems to be the best prospect. Uniaxial reinforcement leads to transverse weakness. Can short, multidirectional, multisized, interwoven reinforcements be introduced by powder metallurgy so that different parts of the crack will run in different directions, bringing the higher toughnesses of Modes II and III into play? Analysis is not likely to provide an answer; experiment seems called for.

For the more ductile materials, Joyce[18] found that crack blunting and sliding off became more important, and differed between singly and doubly grooved specimens. The strain hardening was apparently insufficient for the HRR fields scaled by the J-integral to dominate, and the differing stress and strain fields of Fig. 5 were felt clear to the tip. A second example of the fully plastic states of stress is the transition temperature for cracked steel plates, where a specimen with two opposing cracks exhibited a transition temperature at least 30 C above that for a similar plate with a crack in just one side, or for a Charpy specimen.[27]

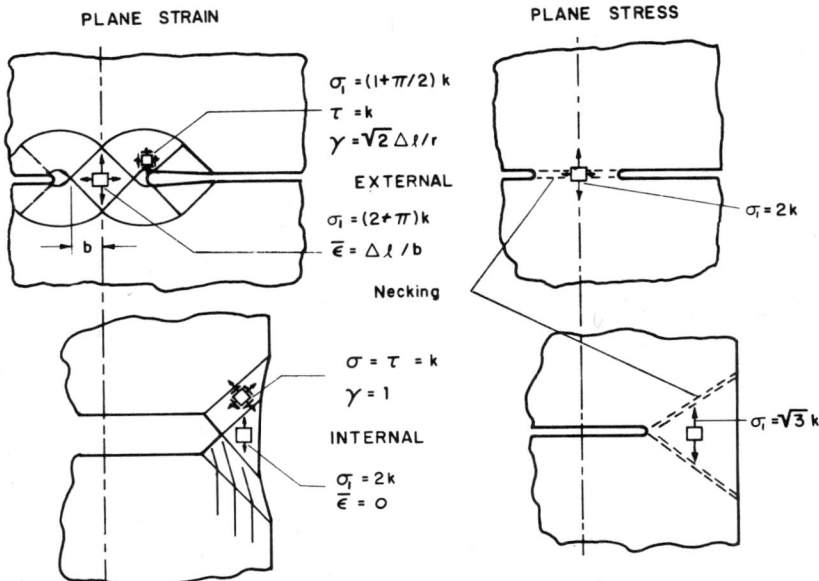

PLANE STRAIN

PLANE STRESS

$\sigma_i = (1 + \pi/2)\,k$
$\tau = k$
$\gamma = \sqrt{2}\,\Delta l/r$

EXTERNAL

$\sigma_i = (2 + \pi)k$
$\bar{\epsilon} = \Delta l / b$

Necking

$\sigma = \tau = k$
$\gamma = 1$

INTERNAL

$\sigma_i = 2k$
$\bar{\epsilon} = 0$

$\sigma_i = 2k$

$\sigma_i = \sqrt{3}\,k$

Fig. 5. Fully plastic flow fields in tension.

7 FATIGUE

While mechanisms for the formation of fatigue cracks in unnotched specimens have been thoroughly studied in relatively pure materials as well as in certain of the high-strength alloys, more information will hopefully be available at this meeting. It still does not appear possible to predict the unnotched fatigue curve quantitatively from a knowledge of the microstructure of the material. Furthermore, under loading conditions where the principal components of stress rotate, as in rolling contact, wear, or erosion, there is no satisfactory empirical correlation to predict crack initiation from tests under uniaxial stress at constant load amplitude. We are working on one that is based on normal stress and shear strain on each orientation in the material. The criterion for initiation is then based on the plane with the most sensitive orientation. This criterion is in contrast to describing the state of stress in terms of principal components, which in effect averages the orientation first and then applies a fracture criterion.

Once the crack has initiated, its growth should be predictable from mechanics if the striation markings are relatively uniform. Since the yield strength doubles on load reversal (aside from the Bauschinger effect), equating crack growth rate to crack-tip-opening displacement gives, from Eq. (20),

$$\frac{d\ell}{dN} = \frac{2.7}{2\pi} \left(\frac{\Delta K}{2Y}\right)^2 \frac{2Y}{E} \cdot \tag{31}$$

For the glassy metal of Davis[28] with $Y = 245$ kg/mm^2, $E = 13.5 \times 10^3$ kg/mm^2, Eq. (31) gives

$$dc/dN \text{ (mm/cycle)} = 6.5 \times 10^{-8} \Delta K^2 \quad, \tag{32}$$

compared with his value of $2 \times 10^{-8} \Delta K^{2.2}$. It would be interesting to see how much using a version of the HRR displacements [Eq. (25)], with a strain-hardening coefficient to match any Bauschinger effect, would improve the estimate.

The tentative nature of the comparison of CTOD and dc/dn is indicated by studies of fatigue-crack propagation in steels with tensile strengths of from 450 to 1400 MN/m^2 by Barsom.[29] To give the benefit of the doubt, base the correlation on tensile strength rather than yield strength to allow for cyclic hardening. Then the crack growth rates should have varied by a factor of 3. In fact, Barsom found a variation of only 25 percent. Expressed another way, the observed crack growth rates were smaller than those predicted from Eq. (30) by factors of from 13 for the lowest strength steel to 2 for the highest strength. Fractographic studies should be made of these specimens to see what contributes to this unexpectedly large resistance to fatigue crack growth. There is other evidence[30-32] that areas of increased impurity content, or crack roughening by stress-corrosion cracking, also sometimes exhibit lower crack growth rates. Here, as in monotonic loading, the introduction of irregular structure might further increase the crack resistance, while probably decreasing the resistance to crack initiation.

8 CREEP

The difficulties of giving mechanics solutions for ductile fracture increase correspondingly when time effects must be taken into account. There is a possibility that numerical solutions will become easier, while at the same time requiring more computer time and being more expensive. Here again, the compromises between monotonic ductility, yield strength, and fracture resistance must be made. For instance, the keying of grain boundaries which increases creep strength may markedly reduce the fracture toughness, not only at low temperatures but also at service temperatures.

9 WEAR AND EROSION

Correlating equations for various wear mechanisms have been available for some time, but it now appears that subsurface fatigue and progressive deformation play an important role even in ordinary wear.[6] The requirements of high hardness and minimum sites for crack nucleation are similar to those for other modes of failure, but the desire for easy formation of a soft, micron thick film may be counter to the fatigue-strength requirement except where the film is repeatedly smoothed over.

10 RECOMMENDATIONS FOR FRACTURE RESISTANCE

The effects of changes in several characteristics of alloys on their fracture behavior are qualitatively summarized in Table III. The effects are given for small changes relative to typical high-strength alloys. For instance,

1. Increasing the volume fraction of inclusions with the same inclusion spacing has a size effect on the strain for hole nucleation. For small inclusions with no interaction, the strain for a given hole growth is unaffected. It would have a fairly deleterious effect on the strain to localization, ϵ_ℓ, and there would be only a slight effect on crack growth because the strain appears under the logarithm.

2. Increasing the spacing while holding the same volume fraction is essentially a size effect, and holes would be expected to nucleate sooner with larger, geometrically similar inclusion arrays. Equations (29) and (30) indicate linear increases in critical plastic zone size for crack initiation and in-

Table III. Effects of Variables on Fracture Behavior Relative to Typical High-Strength Alloys

Fracture Behavior	Alloy Variable				
	Higher Volume Fraction f, Same Spacing ℓ	Larger Spacing ℓ, Same Volume Fraction f	Higher σ_{rr}	Larger n	Higher Y
Strain for stages in unnotched crack formation					
Hole nucleation, ϵ_i	0−	0−	+	−	−
Hole growth strain, $\dfrac{d\epsilon}{d(a/\ell)}$	0	0	0	−	0
Localization, ϵ_ℓ	−	0	+	+	0
Plastic zone for crack growth					
Initiation, r_{yi}	0−	++	0+	+	—
Instability, r_{yc}	0−	++	0+	+	—

stability, with increasing geometrically similar inclusion size and spacing. It should be emphasized that this beneficial effect will continue only until cracking nucleates at intermediate structural features such as grain boundaries or precipitate particles.

3. A higher interface bond strength, σ_{rr}, should be beneficial through increasing the strain to hole nucleation in unnotched specimens and through decreasing the hole density, which will increase the strain to localization. At high triaxiality, however, so many holes will be nucleated that the effect on crack growth is expected to be small.

4. A higher strain-hardening exponent n will raise the flow stress more rapidly and nucleate holes sooner, but they will grow more slowly, which will improve the unnotched strain to crack formation and the plastic-zone sizes at initial growth and instability.

5. Increasing the yield strength Y will decrease the strain for hole nucleation, not further affect the strains for hole growth and localization to form cracks in unnotched parts. It will have a very deleterious effect on the plastic-zone sizes for crack initiation and instability.

The above discussion indicates that the primary candidate for improvements in unnotched ductility is an increase in the bond strength of inclusions with the matrix, provided the inclusions themselves do not fracture. For high-strength alloys there does not seem to be any great prospect for increasing the toughness against crack initiation for instability except through a change in the structure of the inclusions to promote crack branching and meandering as illustrated in Fig. 6. Whether such a structure is practically obtainable is an open question.

The small critical plastic-zone sizes or crack lengths indicate that further increases in yield strength without marked improvements in toughness are not likely to be of practical value.

The only feasible mechanism for increasing the modulus of elasticity, where stiffness is of importance, lies in preferred orientation for steels, but will be of little help in aluminum, which is nearly isotropic.

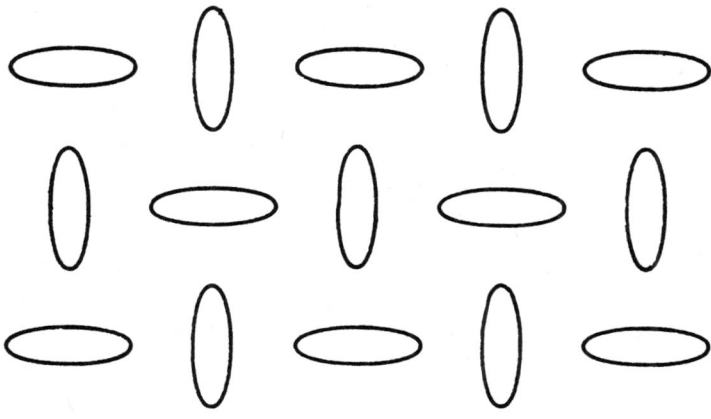

Fig. 6. Inclusion configuration to promote toughness by crack branching and meandering.

ACKNOWLEDGMENT

The support of the Army Research Office through Contract DAHG04-72-C-0043 in the preparation of this paper is gratefully acknowledged.

REFERENCES

1. McClintock, F. A., and Zaverl, F., in *Proceedings of 11th Annual Meeting of the Society of Engineering Science*, G. J. Dvorak (Ed.), Duke University, Durham, North Carolina, 1974, pp. 6–7.
2. McClintock, F. A., and Zaverl, F., submitted to *Int. J. Rock Mech. and Mining Sciences* (1976).
3. McClintock, F. A., and Argon, A. S. (Eds.), *Mechanical Behavior of Materials*, Addison-Wesley, Reading, Massachusetts (1966), p. 83.
4. McClintock, F. A., Paper presented at ARPA Materials Research Conference, University of Michigan, Ann Arbor, Michigan, 1973.
5. Fourie, J. T., in *Corrosion Fatigue*, O. F. Devereaux, A. J. McEvily, and R. W. Staehle (Eds.), National Association Corrosion Engineers (1972), pp. 164–175.
6. Suh, N. P., Paper presented at Battelle Colloquium on Fundamental Aspects of Structural Alloy Design, Seattle, Washington, September, 1975.

7. Argon, A. S., Im, J., and Safoglu, R., *Met. Trans.,* **6A**, 825 (1975).
8. Argon, A. S., and Im, J., *Met. Trans.,* **6A**, 839 (1975).
9. Kelly, A., *Strong Solids,* Oxford University Press (1966), p. 5.
10. McClintock, F. A., and O'Day, W. R., Jr., *Proceedings of First International Conference on Fracture,* T. Yokobori, et al. (Eds.), Japanese Society for Strength and Fracture, Sendai (1966), Vol. 1, p. 75.
11. Gurland, J., *Acta Met.,* **20**, 735 (1972).
12. Lankford, J., and Kusenberger, F. N., *Met. Trans.,* **4**, 553 (1973).
13. Hill, R., *The Mathematical Theory of Plasticity,* Clarendon Press, Oxford (1950).
14. McClintock, F. A., International Symposium on Fracture Mechanics, Noordhoff Publishing, Groningen (1968), pp. 101–130.
15. French, I. E., and Weinrich, P. F., *Met. Trans.,* **6**, 1165 (1975).
16. Tipnis, V. A., and Cook, N. H., *J. Basic Eng., Trans. A.S.M.E.,* **89**, 533 (1967).
17. Nagpal, V., McClintock, F. A., Berg, C. A., and Subudhi, M., *Foundations of Plasticity,* A. Sawczuk (Ed.), Noordhoff Publishing, Leyden (1973), pp. 365–385.
18. Joyce, J. A., Sc.D. Thesis, Department of Mechanical Engineering, M.I.T., Cambridge, Massachusetts (1974).
19. Garr, L., Lee, E. H., and Wang, A. J., *J. Appl. Mech.,* **23**, 56 (1956).
20. Tada, H., Paris, P. C., and Irwin, G. R., *The Stress Analysis of Cracks Handbook,* Del Research Corporation, Hellertown, Pennsylvania (1973).
21. Rice, J. R., *Fracture,* H. Liebowitz (Ed.), Academic Press, New York, N.Y., (1968), Vol. II, pp. 191–311.
22. Levy, N., Marcal, P. V., Ostergren, W. J., and Rice, J. R., *Int. J. Fracture Mechanics,* **7**, 143 (1971).
23. Shih, C. F., *Fracture Analysis, ASTM STP,* No. 560, 187–210 (1974).
24. Hutchinson, J. W., *J. Mech. Phys. Solids,* **16** 13 (1968).
25. Rice, J. R., and Rosengren, G. F., *J. Mech. Phys. Solids,* **16**, 1 (1968).
26. McClintock, F. A., *Fracture,* H. Liebowitz (Ed.), Academic Press, New York, N.Y. (1971), Vol. III, p. 43.
27. Marcolini, R., Master's Thesis, Departments of Ocean Engineering and Mechanical Engineering, M.I.T., Cambridge, Massachusetts (1975).
28. Davis, L. A., Paper presented at Battelle Colloquium on Fundamental Aspects of Structural Alloy Design, Seattle, Washington, September, 1975.
29. Barsom, J. M., *J. Eng. Indust. Trans. ASME,* **93B**, 1190 (1971).
30. McClintock, F. A., *ASTM STP, No. 415,* 170–171 (1967).
31. El Soudani, S. M., and Pelloux, R. M., *Met. Trans.,* **4**, 519 (1973).
32. Chu, H. P., and Wacker, G. A., *J. Basic Eng.,* **91D**, 565 (1969).

DISCUSSION on Paper by F. A. McClintock

SANDERS: In the data presented there was a reference to the deleterious effect of the base-aluminum-metal purity on the fracture behavior of 7xxx alloys. It is difficult to speculate on the effect of purity when there are a number of interactions occurring. First, we must decide on what is meant by purity. If by purity we mean removal of all elements except zinc, magnesium, and copper, we may very well be addressing another issue. Chromium and manganese are intentionally added in controlled amounts to 7xxx alloys to suppress recrystallization and grain growth through the dispersed phase of $Al_{12}Mg_2Cr$ or $Al_{20}Mn_3Cu_2$. By taking out the chromium or the manganese 7xxx alloy we may be just describing the effect of degree of recrystallization and grain size on the fracture behavior.

On the other hand, if purity is associated with the level of iron and silicon, that is another issue. These elements are in liquid solution during the melting of the charge; they combine with other elements and separate during ingot solidification as a coarse constituent particle ranging in size up to approximately 30 μm in the longest direction. These particles may be partially broken up during fabrication, but cannot be taken into solid solution. Investigations at Alcoa [J. A. Nock and H. Y. Hunsicker, *J. Metals,* **15**, 216–224 (1963)] have shown that by decreasing the iron and silicon content, properties such as fracture toughness are increased.

LOW: Please clarify why one might, by eliminating the larger void nucleating particles, encounter a condition of low ductility.

JAFFEE: The detrimental effects of small void nucleation spacing and purity of single-phase alloys on fracture toughness suggest the design of high-purity, high-strength composites with wide spacing of the elements. This is contrary to the usual scheme of promoting small spacing of elements. It has been found that toughness of composites is promoted by reducing the interelement adhesion, in line with your concepts.

LEVY: That large inclusion spacings improves crack toughness is not surprising to superalloy developers for they have learned that, for example, in Nimonic-type alloys, discrete carbide particles at grain boundaries provide improved rupture strength relative to either a thin, continuous film or no carbide precipitation at all.

McCLINTOCK: I appreciate the detailed comments of Dr. Sanders about the 7000 series aluminum alloys, and agree that in many cases the higher purity has led to greater crack toughness. I am only warning that such will not always be the case. An explanation may answer the questions raised by Drs. Low and Jaffee, and at the same time be consistent with the further examples in high-temperature alloys and composites pointed out by Drs. Levy and Jaffee.

My remarks on the surprising deleterious effect of purity come from experimental evidence in the case of fatigue-crack growth and speculation about ultimate limits (not yet attained) for fracture toughness. My recollection of ref. 30 was wrong; it does indicate a higher fatigue-crack growth rate for the 7178 alloy with the lower iron content, apparently because of the accelerating effect of inclusions. Reference 31 indicates a similar effect at higher growth rates, but a lower growth rate for the commercial material, in contrast to material with half the volume fraction of inclusions. I have seen similar results in unpublished work by others on the effect of inclusions on fatigue-crack growth in T1 steel. The reason for the effect appears to be illustrated by ref. 32, in which a rougher crack gives a higher critical stress-intensity factor on subsequent loading. Here, the roughness is caused by corrosion cracking, but one would expect similar effects when the roughness is caused by inclusions with spacings of the order of the plastic zone size.

In regard to the ultimate limits attainable for critical stress-intensity factor under monotonic loading, I agree that current experiments indicate that the greater the freedom from inclusions, the greater is the toughness. As pointed out in the written paper, however, I fear that this process will reach a limit. With alloys whose yield strain is the order of 1 percent, and with triaxiality in front of a crack of the order of 3 times uniaxial yield, the state of stress is so close to the ideal cohesive strength that even clean grain boundaries or dislocations would nucleate fracture. Assume that there are no other sources of fracture. Then, by dimensional analysis alone, as well as by Eq. (29) or Fig. 4, the finer the structure is, the smaller the critical plastic-zone size or required crack opening and the smaller the critical stress-intensity factor. So far, we have been saved by the fact that the strains to nucleate fracture kept rising; with increasing yield strengths, the strain to fracture will sooner or later reach a limit and then decrease. To be sure, we should keep on cleaning up the alloys until we reach this limit.

On the other hand, the increased toughness due to intentional inclusions for crack branching will be at the expense of lower resistance to crack initiation and to transverse ductility; a compromise will be involved. The inclusion pattern of Fig. 6 might help, especially with a multimodal distribution of sizes. Even here, the fatigue-initiation strength would be reduced by the inclusions.

RADIATION ENVIRONMENTS AND ALLOY BEHAVIOR

P. G. Shewmon

National Science Foundation
Washington, D.C.

ABSTRACT

Water reactors, breeder reactors, and fusion reactors will play an expanding role in our society for at least the next few centuries. Each requires that materials perform reliably under intense irradiation for years or decades. Several new phenomena occurring when alloys are exposed to such an environment have been identified, e.g., radiation creep, swelling, hardening, helium effects, and embrittlement. All but the helium effects result from the precipitation and annealing of the point defects continuously produced by fast neutrons. The current theories concerning these phenomena are described, and the requirements for future alloy design are outlined.

1 INTRODUCTION

To obtain useful heat from nuclear fission or fusion, the high-velocity atomic particles produced must plow through the fuel and structural members of the reactor, converting kinetic energy into sensible heat. A good candidate material for such a reactor must maintain its useful mechanical properties and dimensional stability in such an environment for one or several years. This means that each atom will have been knocked from its initial lattice position on the order of ten times during the life of the structural piece (in the fuel, 1000 times). In view of this internal churning, it is not surprising that the metal behaves differently than it does outside the reactor—although, in reality, the behavior is not drastically different.

This paper catalogues the main property differences that develop in the intense radiation of a nuclear reactor and summarizes the theoretical ideas that have been put forward to explain them. Corrosion effects, the other main criterion for selecting reactor core alloys, are not discussed. Since the radiation effects are caused by the formation and/or annealing out of point defects, the temperature of irradiation plays a central role. All defect effects wash out above $1/2\ T_m$ because of thermally activated annealing. All reactor materials operate above $1/4\ T_m$ where interstitials and vacancies are mobile, but how mobile is not clear. Much experimental work has been done at very low temperatures in an effort to establish the kinetics of interstitial motion. The question is still unresolved. However, for our purposes, questions of crowdion versus dumbbell motion of interstitials are largely irrelevant. An uncertainty in their relative mobilities is an irritant, but it is far from the only qualitative element in the theoretical discussion.

The only damage that does not anneal out above $1/2\ T_m$ is the transmutation products. Helium from (n,α) reactions with transition elements is the most damaging on a per-atom basis because it precipitates as bubbles. However, zirconium development in niobium or silicon in aluminum can also produce significant effects. The effect of these substitutional elements is not novel and will not be discussed herein.

2 RADIATION DAMAGE[1]

A fission neutron is produced with a most probable energy of 1 Mev. At this energy it undergoes a direct collision with a lattice atom about every centimeter along its path. Such a collision between a moving neutron and a lattice atom can transfer up to 1 percent of the neutron energy to the atom. The primary knockon atom undergoes collisions much more frequently, having a mean-free path, for iron, which decreases from about 10^3 A at 10 kev to the interatomic distance at 0.1 kev. The resulting distribution of damage consists of Frenkel pairs whose density increases along the path of the primary knockon and ends in a so-called displacement spike or high concentration of pairs. The initial separation of an atom and the vacancy it came from are also variable. Many are quite close, but a significant minority, say 10 percent, are separated by many interatomic distances. The generally accepted picture of a displacement spike is shown schematically in Fig. 1; this consists of a center with a relatively high concentration of vacancies surrounded by a halo of interstitials.[2] This diffuse cluster of vacancies may condense into a dislocation loop or void. Either way it is too small to study easily under the microscope, but it is large enough to restrain the motion of passing dislocations and thus cause hardening.

The annealing of these point defects begins immediately, with over 99 percent of the kinetic energy of the primary knockon appearing in the lattice as heat in under 1 nanosecond. The close pairs are attracted to each other and some will recombine before this heat is conducted away. If one or both defects are mobile, near pairs combine after a few jumps. The exact number of vacancies which escape the cascade and contribute to the average free vacancy concentration is not known but seems to be only a few—for example 6 out of the 100 to 200 formed in the cascade.[3]

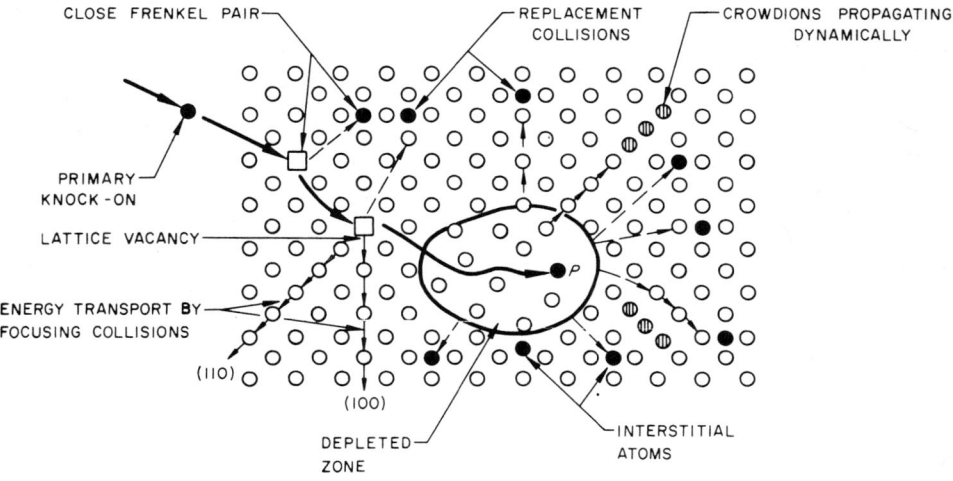

Fig. 1. Schematic representation of radiation damage.[2]

3 POINT-DEFECT ANNEALING

In a reactor environment, vacancies and interstitials are produced in equal numbers and at a roughly constant rate. Over a period of time, a steady state will develop in which the rate of removal equals the rate of production. Any treatments of phenomena which occur in a volume much larger than the dimensions of one cascade resort to a discussion of the average concentration; examples are radiation-enhanced diffusion[4-6], annealing stages[7], radiation-induced voids[8-10], and interstitial clustering[11,12]. Such a treatment is useful and often powerful, but the processes are so many that only a few can be treated at one time. Thus the device is of more pedagogical value than a rigorous tool. The main themes to be developed here are the time and changes that occur in the transient leading up to steady state, and the role impurities can play in trapping the relatively mobile interstitials.

The change in concentration of vacancies (v) or interstitials (i) in a unit volume is the sum of the production rate (K), rate of recombination (αvi), annihilation at sinks ($D_v \rho v$), and the divergence of the flux ($D \Delta^2 v$). Thus

$$\frac{dv}{dt} = K - \alpha vi - D_v \rho v + D_v \Delta^2 v \tag{1}$$

$$\frac{di}{dt} = K - \alpha vi - D_i Z_i \rho i + D_i \Delta^2 i \quad . \tag{2}$$

Here α is a kinetic coefficient determined primarily by the highly mobile interstitial and a recombination volume (about 100 atomic volumes, Ω) within which the two defects spontaneously recombine; ρ is a dislocation density, this being the primary internal sink. $Z_i > 1$ allows for a slightly larger dislocation capture cross section for interstitials than for vacancies.

The alloy is usually assumed to be free from concentration gradients, thus eliminating the last term in Eqs. (1) and (2). This assumption of homogeneity makes the problem tractable and is adequate unless one wishes to treat problems involving a free surface or processes around a grain boundary, as in ref. 6.

Consider first the low-temperature case. Here the vacancies are only barely mobile, while, as always, the interstitials have a much higher mobility.

Qualitatively the behavior of a sample can be described as follows, assuming

$$i = v = 0 \quad \text{at } t = 0.$$

Initially both i and v will rise. If the sink density is low, i and v rise until, after a relaxation time of $(K\alpha)^{-1/2}$, the recombination rate approaches the production rate K, then

$$i = v = (K/\alpha)^{1/2} \tag{3}$$

After a time $t_i = 1/D_i Z_i \rho$, interstitials begin to diffuse to fixed sinks in significant numbers. This plus the continuing rise in v makes i drop, though with the constant product $vi = K/\alpha$. After a time $t_v = 1/D_v \rho$, the diffusion of vacancies to sinks equals that of interstitials. Then

$$D_v \rho v = D_i Z_i \rho i \quad . \tag{4}$$

Thus, $i/v \simeq D_v/D_i \ll 1$. Though the product iv still equals K/α at this new steady state, i and v have shifted to

$$i = \left(\frac{KD_v}{\alpha D_i}\right)^{1/2} \quad .$$

$$v = \left(\frac{KD_i}{\alpha D_v}\right)^{1/2} \quad . \tag{5}$$

Given the rapid diffusion of interstitials, t_i is less than 1 second at $1/4\ T_m$. However, D_v is so low at this temperature that t_v may be days or weeks. At $1/2\ T_m$, t_v has risen to seconds, provided ρ is relatively high, for example $\geqslant 10^9/\mathrm{cm}^2$.

The formation energy of interstitials is several electron volts for copper. Thus the thermal equilibrium concentration, i_o, is very low. It follows that as v and i rise, initially the supersaturation for interstitials i/i_o rises much more rapidly than that for vacancies, v/v_o. Thus homogeneous nucleation of interstitial loops occurs quite early. Brown et al. have postulated that a di-interstitial is a stable nucleus for such a loop and they were thus able to describe the observed high density of interstitial loops found in electron-irradiated graphite.[11] This is a general theory that would apply equally well to interstitials in metals, e.g., Makin.[12] In this model, an interstitial may be (a) annihilated at a vacancy; (b) trapped by another interstitial, thus nucleating a dislocation loop; or (c) annihilated at the edge of an interstitial dislocation loop.

If the vacancies are assumed to be immobile, the time variation of v and i, concentrations of loops N_1, di-interstitials N_o, and the predicted average loop radius $\bar{R}$ are given in Fig. 2. Note that the number of loops quickly saturates though $\bar{R}$ increases as $(\mathrm{time})^{1/3}$. Also $v \sim \theta^{2/3}$, while $i \sim \theta^{-2/3}$, the inverse relationship required. If the mobility of the vacancies is finite, then v and i will ultimately reach the steady-state value given by Eq. (4).

Substitutional impurities whose size differs from the solvent will trap interstitials with a binding energy sufficient to significantly increase the concentration of interstitials available to form di-interstitials and thus nucleate

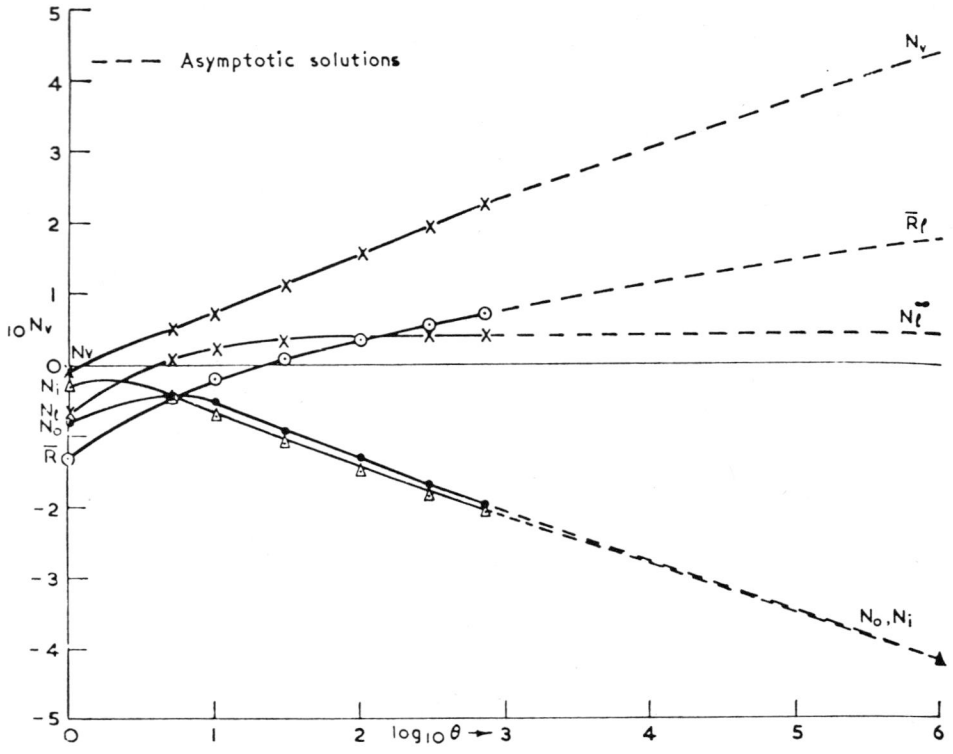

Fig. 2. Development of defect concentration with time under irradiation[11]; N_v, N_i, N_1, N_o are vacancy, interstitial, loop, and di-interstitial concentrations, respectively.

more loops. If the vacancies are mobile, such trapping would also increase the rate of recombination.[10,11]

At higher temperatures, but temperatures such that $v < v_o$, diffusion to fixed sinks is more important than recombination. At steady-state then,

$$v = \frac{K}{D_v \rho} \quad , \qquad i = \frac{K}{D_i Z_i \rho} \quad . \tag{6}$$

The variation of the self-diffusion coefficient D is easily seen from Eq. (6):

$$D = D_v v + D_i i \simeq \frac{2K}{\rho} \quad . \tag{7}$$

In the temperature range where the defects diffuse to fixed sinks, D is proportional to the flux and inversely proportional to the sink density.

When recombination is dominant, noting that α equals a constant times D_i, one obtains

$$D \sim (KD_v)^{1/2} \quad . \tag{8}$$

Thus, at the lowest temperatures, D varies less with flux and decreases with temperature. At still higher temperatures the thermally produced vacancy concentration becomes much larger and may annihilate most of the interstitials.

Figure 3 shows the concentration of radiation-produced defects at steady state. There are three distinct temperature ranges: a low-temperature region in which i and v are large, recombination dominates, and dislocations play no role; a middle range where diffusion to fixed sinks (dislocations) dominates and i or v are inversely proportional to the dislocation or sink density; and a high-temperature range where the defects anneal out promptly, no supersaturation is produced, and $v_o \gg v$ is so great that recombination again dominates. Radiation effects are observed only in the range where defect supersaturation develops, so this high-temperature range is of no further concern here.

Some of the radiation effects we are concerned with, e.g., void growth and creep, depend on the flux of defects to fixed *sinks,* and are unaffected by defects which mutually recombine. Thus, an alloying element that traps interstitials can significantly raise the concentration of interstitials in the alloy, increase the recombination probability, and decrease the flux to sinks.[10] Alloying that slows dislocation recovery and thus keeps the fixed-sink content higher will reduce the supersaturation but will increase the flux to sinks. Here, the net effect on the radiation-produced effect can go either way, depending perhaps on whether nucleation or growth is more important.

4 DUCTILE-BRITTLE TRANSITION IN STEEL

In terms of tonnage, or dollar value, of product, the most important radiation effect on structural materials is probably the effect of fast-neutron irradiation on the ductile-brittle transition temperature (DBTT) of pressure-vessel steel. This is a more complex phenomena than is usually addressed in basic studies of radiation damage. The complexity stems from the rich variety of things that can occur in a heat-treated alloy steel, the complexity of the ductile-brittle transition (DBT), plus the effect of irradiation. As a result of several studies, the problem is now understood in the sense that one can specify the factors which cause the sensitivity to neutron radiation.

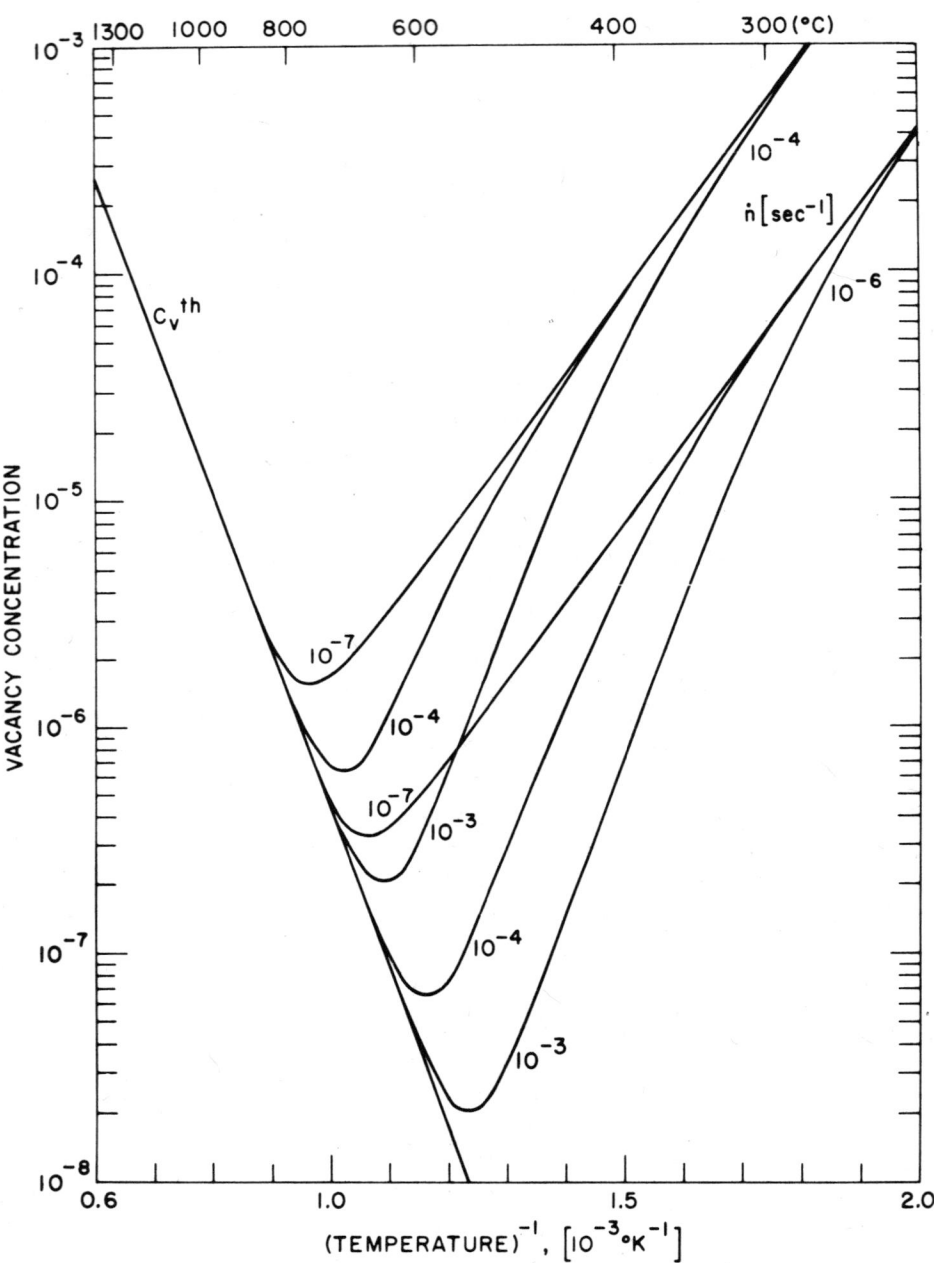

Fig. 3. Steady-state vacancy concentration versus temperature for nickel data; $\dot{n} = 10^{-6}$ sec^{-1} is typical of a fast-reactor core; P is the probability that a site is a sink and is related to the dislocation density P $\simeq 10^{15}\rho$; the values of P shown (10^{-3} to 10^{-7}) thus range from those for annealed to those for severely cold-worked material.

Empirically, the following characteristics have been established[13]:
1. Energetic, not thermal, neutrons are responsible for the effect.
2. The temperature of irradiation has a pronounced influence on radiation embrittlement, embrittlement decreasing with higher temperatures (Fig. 4).
3. For irradiation temperatures and fluence in the range of interest for power reactors, the substitutional impurities copper and phosphorus play a primary role in determining ΔDBTT, e.g., see Fig. 5 for the effect of copper content. Empirically,

$$\Delta DBTT = (F) = -118 + 14{,}800 \ (\%P) + 990 \ (\%Cu) \quad . \qquad (9)$$

4. The transition temperature increase rises as $\log (\phi t)$ in the range of interest, i.e., $10^{18} < \phi t < 10^{20}$ n/cm^2 (Fig. 5).
5. The radiation-induced increase in yield stress parallels that of ΔDBTT, i.e., it rises with increasing copper content and decreases with higher T_{irr} (Fig. 6).
6. Intersitital elements increase ΔDBTT at low temperatures but are unimportant at reactor temperatures, that is, at about 400 to 650 F (204 to 343 C).
7. Fine-grained, quenched-and-tempered steels are less affected by irradiation.

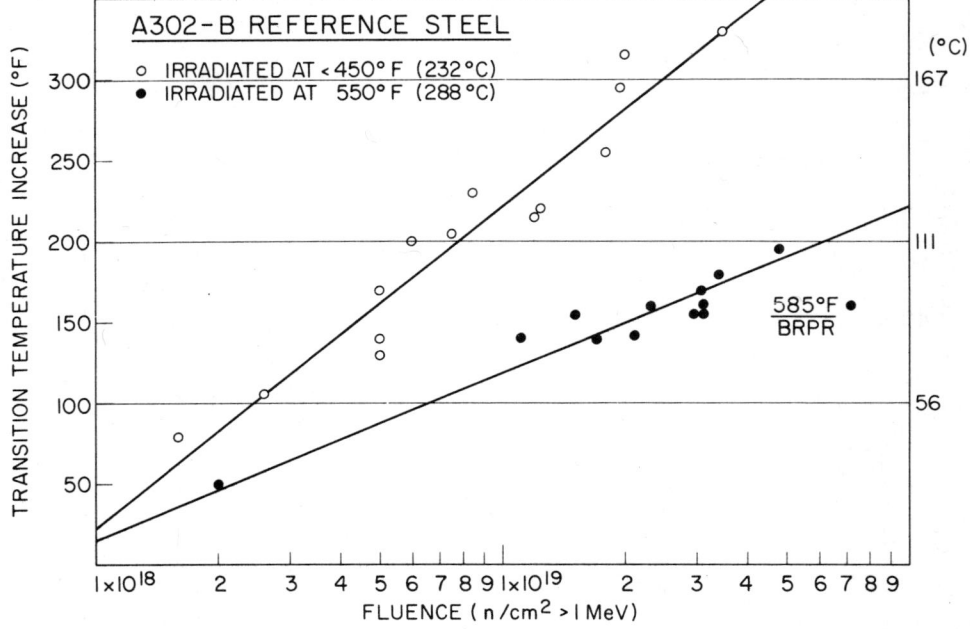

Fig. 4. Effect of temperature of irradiation on transition temperature of A302-B steel.[14]

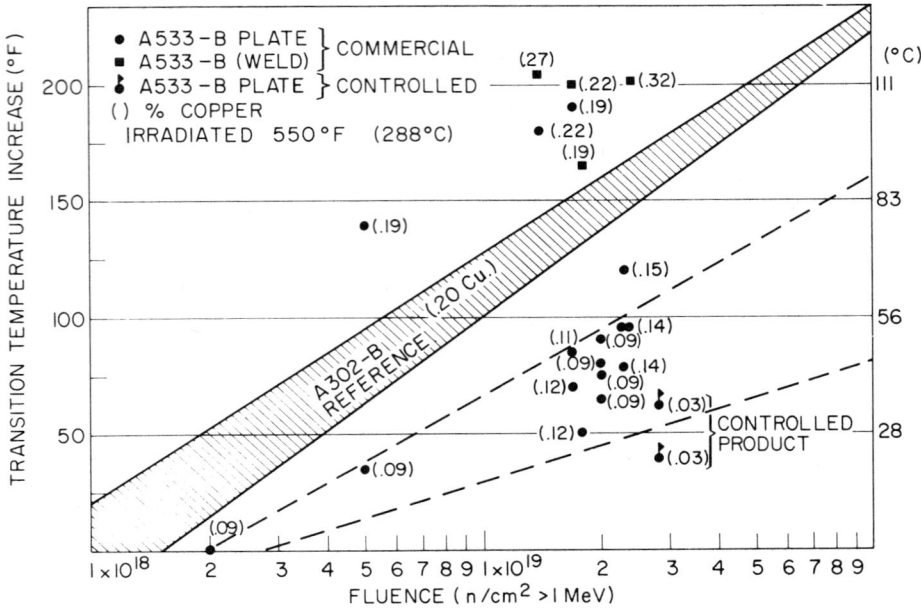

Fig. 5. Radiation-induced increases in the DBTT of commercial steel plates and weldments of Type A533-B steel. The copper content is presented beside each point.[13]

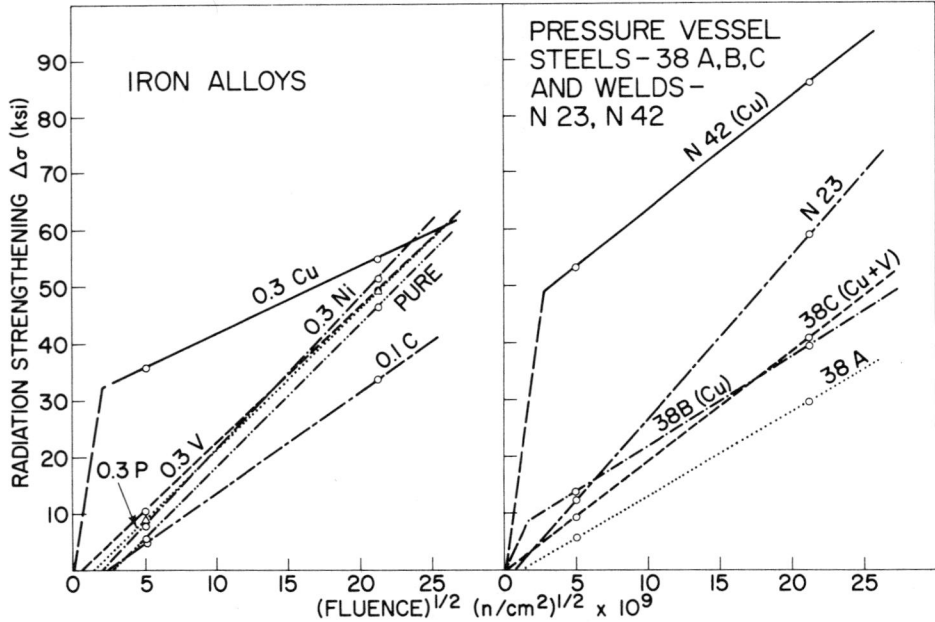

Fig. 6. Radiation strengthening as measured by difference in yield of irradiated and un-irradiated binary alloys of iron, pressure-vessel steels, and welds.[16]

Considerable experimental effort has gone into selecting and developing the best composition and steel-making practice to obtain the lowest DBTT after irradiation. Considerably less has gone into explaining why. Those interested in a review of the extensive test programs, their intercomparison and the rationale for current practice are referred to reviews by Bush[15] and Steele[13].

The radiation effect on the fracture energy of pressure-vessel steel shows up in two, not necessarily related, ways—the shifts in DBTT and the fracture energy above the DBTT, or shelf energy. Radiation lowers the shelf energy and the degradation due to phosphorus or sulfur shows up primarily here. The mechanism is not clear. However, a reasonably clear story has evolved on the effect of copper and its role in raising the DBTT.

The relation between irradiation and rise in transition temperature involves two steps. The rise in DBTT is related to the radiation hardening. This in turn is related to such factors as fluence, impurities, and irradiation temperatures. The relation between hardening and transition temperature is shown schematically in Fig. 7. The cleavage stress, σ_c, varies little with temperature, while for bcc metals, like carbon steel, the yield stress, σ_y, rises rapidly with falling temperature. If the alloy is hardened by irradiation, the increase in flow stress is found to be independent of test temperature. This causes a shift in the ductile-brittle transition. In a study of a low-alloy carbon steel (ASTM A212B), Wechsler et al.[17] measured both the change in flow strength of smooth tensile specimens and the DBTT as measured with Charpy V-notch specimens. Making corrections for strain rate and triaxiality, their principal conclusion was: ". . . that the radiation hardening is in itself adequate to explain the observed increase in the DBTT upon irradiation, based on the Ludwig-Davidenkov concept illustrated in Fig. 1" (our Fig. 7). "This means that it is not necessary to postulate a change in the nature of the fracture process as such or, in particular, a decrease in the cleavage fracture stress." They also note that at the transition temperatures of embrittled steel, the variation in flow stress with temperature is relatively small so that small changes in flow stress can lead to significant changes in the DBTT. This would explain the differences in DBTT from one heat to another of the same alloy or even in different parts of the same heat.

There has been much less study of the atomic processes that give rise to residual-element effects in the irradiation embrittlement of steel. Most of the work done has been at the Naval Research Laboratory. They believed two mechanisms were consistent with the data[18]: (1) a modification of the defect microstructure from impurity-defect interactions, which would increase the yield strength; and (2) embrittlement of an interface due to impurity segregation accelerated by radiation-induced vacancy supersaturation.

The two elements which most influence the DBTT are copper and phosphorus. A study of the radiation-induced hardening of pressure-vessel

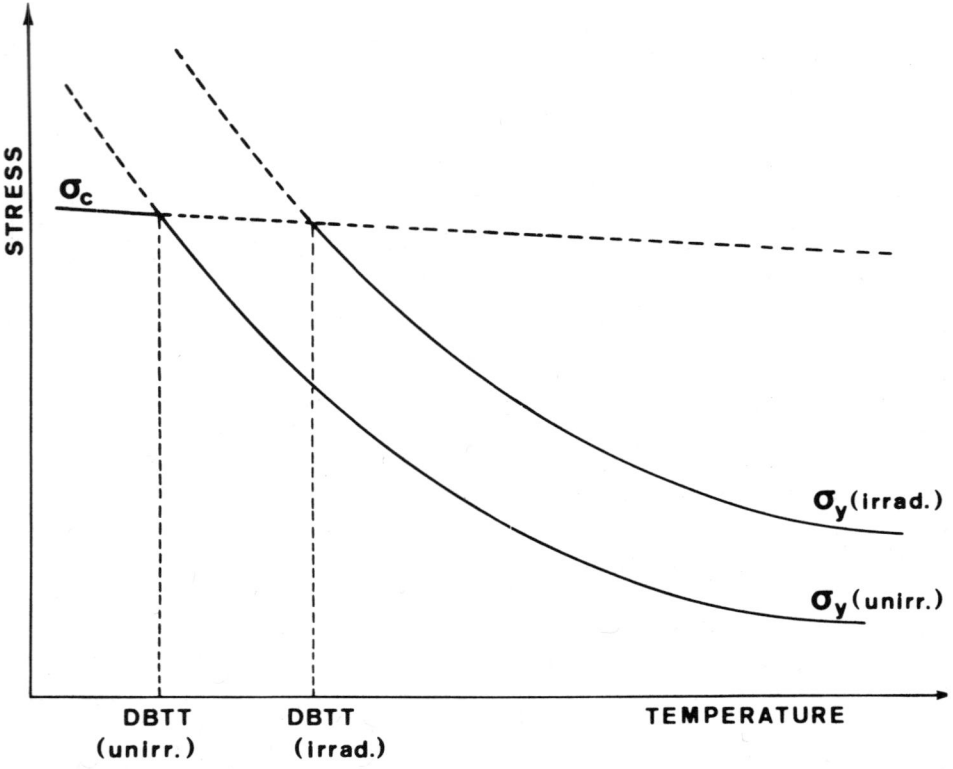

Fig. 7. Schematic illustration of the Ludwig-Davidenkov approach to the relation between strength and ductility; σ_y is yield stress and σ_c is the cleavage fracture stress.[17]

steels, welds, and binary iron alloys at 290 C shows that copper causes a much more rapid radiation strengthening than phosphorus, vanadium, nickel, or carbon (Fig. 6). They conclude that vacancy trapping by copper atoms significantly increases the rate of nucleation of defect aggregation. While copper is not the only factor influencing radiation strengthening in these alloys, it clearly is a major one. The increased flow stress fits in with the higher defect aggregate density and a satisfactory mechanistic argument can be made.[19]

The explanation of the effect of phosphorus is not nearly as satisfactory. Phosphorus can produce temper embrittlement, and interfacial segregation seems to be a plausible mechanism. Unfortunately, efforts to check this hypothesis by Auger spectroscopy on the fracture surfaces of irradiated steel do not show segregation.[20]

5 HELIUM EFFECTS

Helium injected into structural materials is an unavoidable by-product of residence in a nuclear reactor. In a fast reactor, fission neutrons, from (n,α) reactions, produce on the order of 10 ppm of helium in 316 stainless steel over the lifetime of a fuel element. The 14-Mev neutrons of a deuterium-tritium fusion reactor produce about ten times as much helium, per neutron, since the cross section for (n,α) reactions in transition elements rises rapidly with energy. In addition, in a fusion reactor, some of the helium produced by fusion will imbed itself in a wall and come to rest near the surface, i.e., at depths < 1 μm.

Helium, which is essentially insoluble, is important in three different phenomena:

1. At high temperature (>0.5 T_m), the helium collects as bubbles on grain boundaries and significantly reduces creep ductility.
2. It probably plays a role in the nucleation of voids.
3. At the high concentration and shallow depths of helium deposition in the first wall of a fusion reactor, surface blistering and peeling may occur.

At low temperatures, helium in metals can occupy a rich variety of metastable states, so we begin by discussing high temperatures and low concentrations—10 to 100 ppm. Here the helium forms a high density of bubbles whose prssure (P) is balanced by surface tension (γ) according to the equation $P = 2\gamma/r$. For the bubbles to grow to 200 A in diameter requires the removal of roughly 30 host atoms per helium atom by self-diffusion to sinks—usually dislocations. Thus the bubbles form only after anneals at or above 0.5 T_m. The bubbles execute Brownian motion resulting from matrix atoms diffusing over their surface from one region to another. Bubble collision occurs and from the increase in mean bubble radius with time, one can calculate a surface diffusion coefficient (D_s) using an analysis due to Gruber.[20a]

Figure 8 shows values of D_s obtained in this way by Smidt for vanadium, aluminum, and 316 stainless steel.[21] Also plotted are lines due to Gjostein showing D_s which fit most data for clean fcc and bcc metals.[22] The reduction in D_s by 4 to 5 orders of magnitude, determined from bubble motion, is thought to be due to the barrier to surface fluctuations posed by nucleating steps on the faceted bubbles found in these materials and was found earlier in gold.[23]

Smidt finds that atomic helium diffuses only far enough to form small bubbles, there being little or no release of helium from a target injected with helium to a depth of 1 μ, until the annealing temperature approaches 0.9 T_m. If the sample is cold worked, moving dislocations or grain boundaries sweep bubbles and play an important role in clearing the grains of bubbles.

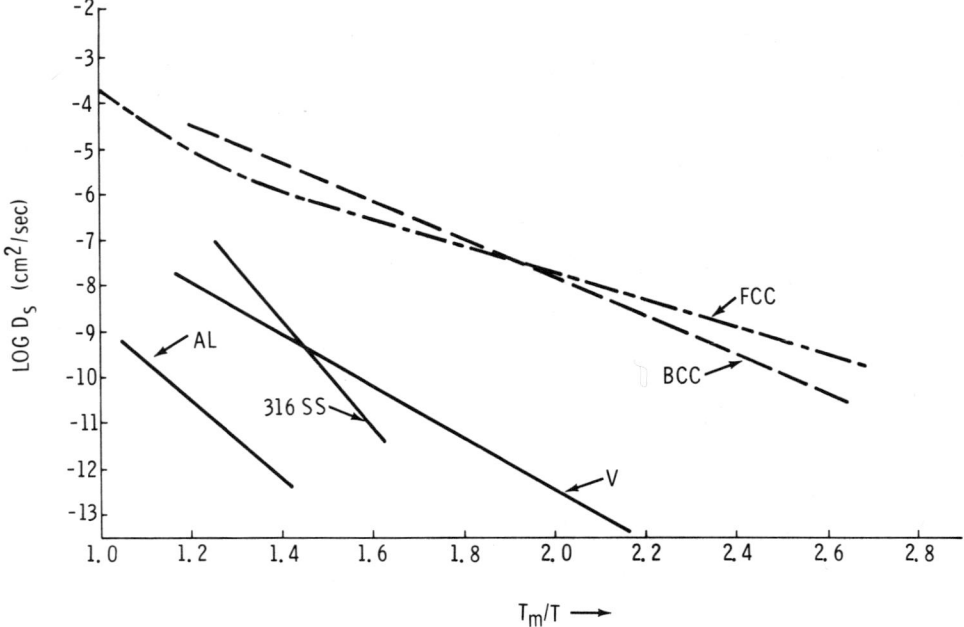

Fig. 8. Surface diffusivities as measured for bcc and fcc metals[22] and values inferred from bubble-coarsening kinetics[21]; the low values from coarsening seems to be the rule.

As little as parts per billion of helium can have an effect on the tensile-test ductility of stainless steel, and the effect increases steadily with helium content (Fig. 9). Here the grain-boundary bubbles serve as nuclei for cracks and markedly decrease the strain before the grain-boundary fracture develops. If the specimen is undergoing creep, the grain boundaries migrate and sweep up bubbles. Also, the bubbles will have a more marked effect on strain to failure than in a tensile test. This is felt to be the main effect of irradiation at higher temperature where all the displacement damage has annealed out.[24]

At low temperatures, the injection of the insoluble helium leads to various metastable states depending on the concentration of traps. In a pure metal, the injected helium atom may come to rest in an interstitial position. If a vacancy is present, there is a significant reduction in energy, i.e., a trap-ping or binding energy, if the interstitial moves into the vacancy. If there are many more helium atoms than vacancies, several helium atoms will bind to one vacancy. If there are more vacancies than helium, a helium atom tends to stabilize divacancy clusters. Also, a host-atom interstitial will replace a helium atom on a substitutional site, and move the helium into an in-terstitial position.[26,27]

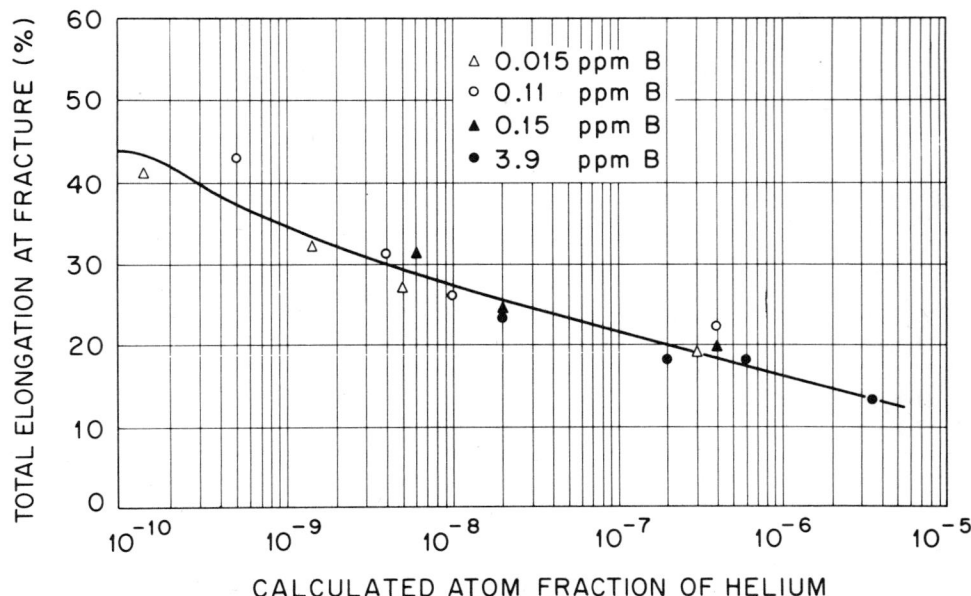

Fig. 9. Elongation in 700 C tensile test for 304 stainless steel containing various helium con-centrations.[24]

Table I lists the energies of motion or binding for various con-figurations. All of the energies are calculated, but for tungsten they agree well with values obtained by Kornelsen in a careful set of experiments with helium injected at low energies and reemitted on continuous heating.[25] The conclusion one draws from this is that the state of helium injected at low temperatures depends sensitively on the relative concentration of interstitials and vacancies, and thus on the relative flow of each of these to sinks and traps.

Impurities would certainly provide additional traps for point defects and helium atoms. For example, helium release from 316 stainless steel is significantly slower on heating than that from pure nickel. There seem to be traps which keep the helium from diffusing out or moving around to form bubbles in the stainless steel until the temperature is significantly higher.[28] Whitmel and Nelson suggest that the more rapid development of bubbles in nickel provides nuclei which give rise to the more rapid swelling of nickel compared with that of 316 SS in a fast-neutron flux.

Detailed kinetic analyses of the role of helium on nucleation of voids[29] indicate that helium probably has a significant effect in-reactor, though probably not with the much higher displacement rates typical of accelerator experiments. The complexities introduced by uncertain properties of the

Table I. Summary of Calculations of Helium in Copper and Tungsten

(All energies in ev.)

	Cu^{26}	W^{27}
Activation Energy for Motion		
Interstitial	0.45	0.24
Mutual Diffusion (Exchange)	2.15	4.69
Binding Energy		
Helium Interstitial to Vacancy, $E^B_{IV.1\ He}$	1.84	4.40
Helium Interstitial to Vacancy Containing 1 He, $E^B_{IV.2\ He}$	0.79[a]	2.90
Helium Interstitial to Vacancy Containing 2 He, $E^B_{IV.3\ He}$	0.57[a]	2.52
Helium Interstitial to Vacancy Containing 3 He, $E^B_{IV.4\ He}$	0.66[a]	2.50
Helium Interstitial to 2 Vacancies Containing 1 He, $E^B_{2V.2\ He}$	1.84[a]	4.42

(a) M. Baskes and W. D. Wilson, personal communication.

traps in stainless steel make it difficult to draw firm conclusions about the behavior of stainless steel from such an analysis.

A different type of problem arises in experiments done to simulate container walls ("first-wall"), beam limiters, diverters, etc., for a fusion reactor which are bombarded with beams of energetic helium ions. Here the penetration range into the metal is on the order of 1 μm and the dose is such that the concentration builds up to 10 atom percent. Obviously, a plane of bubbles forms at a depth roughly equalling the range. These combine to form blisters, but at low and high temperatures, discrete, bulbous blisters form and rupture with their skin remaining intact to attentuate the incoming beam and protect the metal below. At the intermediate temperatures (≈ 0.4 T_m), the entire surface forms one large blister which peels off. Another blister forms beneath it and this in turn peels off (Fig. 10). This repeated peeling gives a pronounced maximum in the "erosion" or metal-loss rate (Fig. 11). Such a loss is undesirable on at least two counts: (1) the flakes would come off into the plasma and be damaging and (2) the reduction in section is intolerable, sputtering rates of 10^{-2} or 10^{-3} atoms per incident helium being an upper limit for reasonable wall life.[30]

Figure 12 shows the rate of helium reemission from a similar study of 316 SS as a function of dose and specimen temperature.[31] Note that the dose to give the first burst of reemission falls as temperature rises. The emission before this first burst may be due to helium diffusion by an interstitial mechanism. The remainder of the helium probably combines with vacancies in clusters, though the vacancy/helium ratio in a "bubble" is unclear. It would seem likely that the ratio is near unity at -170 C. As the concentration of helium rises, some of these overlap. As a developing bubble gets large enough for the helium to act like a gas, the pressure is very high. This

Fig. 10. Surfaces of annealed 304 SS after irradiation at 450 C with 0.5 MeV He$^+$ ions for doses of (a) 0.1 C/cm^2, (b) 0.5 C/cm^2, (c) 1.0 C/cm^2 (C = 6.3 x 10^{18} ions).[30]

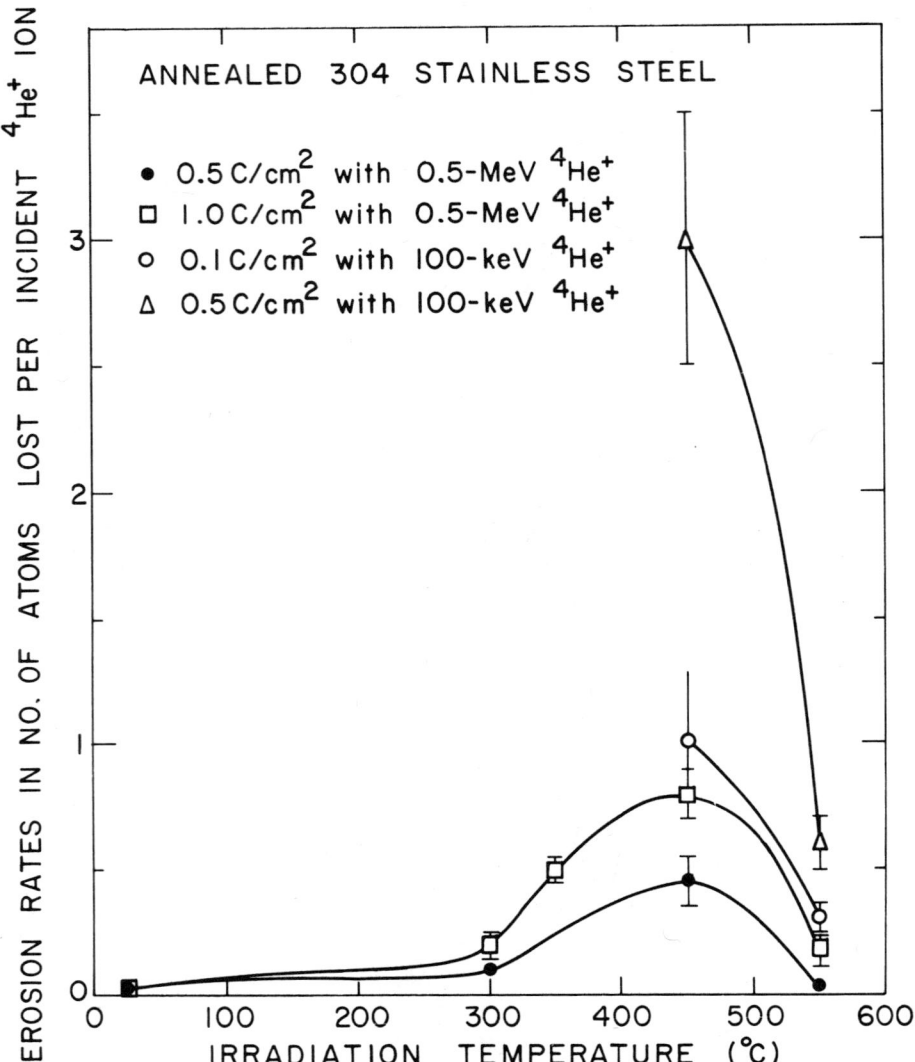

Fig. 11. Erosion rates for annealed 304 SS versus irradiation temperature for varying helium-ion energies and doses.[30]

would then exceed the theoretical strength of the strips of metal between adjacent bubbles (it must, since the strips are much too narrow to contain dislocation sources and display the macroscopic yield). Once this tearing begins, it will propagate until the tear attains a diameter such that the surface pops out, relieving the pressure and tearing at the edge to release the pressure.

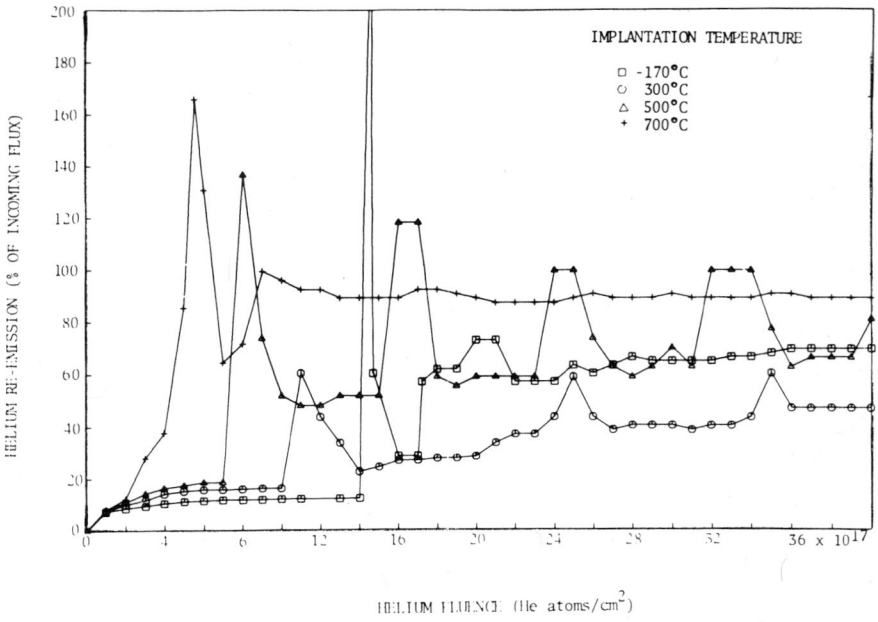

Fig. 12. Helium reemission during implantation versus helium fluence in annealed 316 SS; implantation at the four temperatures indicated.[31]

At higher temperatures, the bubbles will be more distinct and the vacancy/helium ratio somewhat higher. The interesting point here is that now the tear propagates at a pressure which does not bulge up the surface greatly but moves the tear all the way across the beam. At this point, the surface may tear or may get so hot in the center that it ruptures. The unpeeling is probably due to compressive stresses set up in the interior side of the film where more helium comes to rest. The ease of this "tearing" or film "unzipping" increases with temperature, partly because the bubble diameter in the layer beneath the film increases—perhaps also because the thinning of the strips between bubbles may involve stress-assisted surface diffusion.

At higher temperatures ($\geq 0.6\ T_m$), the reemission rate approximates the influx. This is a temperature range where bubble coarsening beneath the surface will proceed readily and self-diffusion is rapid enough so that the bubbles expand to give the gas pressure $2\gamma/r$. A network of interconnecting bubbles probably develops beneath the surface and some break through, connecting this network to the surface. Injected helium atoms will come to rest near an internal surface, and diffusion of substitutional helium proceeds rapidly enough to bring the helium to the gas phase, or to a new small bubble. Either way, no peeling occurs and no new layer of bubbles develops deeper beneath the surface.

All metals studied seem to show this sequence of release mechanisms, with increasing temperature.[30,32] It is not yet clear how or whether alloying will help. Also, the worst peeling comes in around 0.3 to 0.5 T_m, as with swelling. However, studies are still too preliminary to say whether certain base metals or alloying may be used without peeling in the temperature range of the designers' ultimate choice.

Another unknown is just how well these injections run at one energy simulate the first wall. Three questions arise which have yet to be studied:

1. What effect will a spectrum of deposition depths (energies) have?
2. How will the fast neutron damage modify the helium blistering?
3. How well do tests run in hours simulate wall behavior over decades?

The first of these is under study now. The second must probably await a test fusion reactor. Answers to the third must certainly await such a test reactor.

6 VOID FORMATION

The formation of voids in a fast-reactor environment is the motivation for what is currently the most active alloy-development program in the field of nuclear energy. There is a conference on the subject roughly each year, and the effort is going on in several countries.[33-36] Thus, the subject is too important to ignore and too large to treat adequately in a few pages. My accommodation to this conundrum is to list the references most helpful to me, briefly discuss some of the ideas I find most interesting, and leave the body of the discussion to later authors who are closer to current work. (See B.R.T. Frost, Paper IV-4.)

Early studies of void formation showed that swelling occurred in essentially all metals between 1/3 T_m and 1/2 T_m and that the volume increased as (fluence)^n where $n > 1$ but < 2. Also, volume increases of over 10 to 15 percent appeared likely in certain parts of a LMFBR core if the fuel burnup goals were met. Such volume changes and the accompanying core distortions would be unacceptable, so active programs of study and alloy development began.[37]

The most important experimental development has been the simulation of fast neutron damage with heavy ions or electrons in a HVEM. In this way one can get in a day the average displacement per atom (dpa) that takes years in a fast test reactor.[36,38]

Once the temperature is high enough for vacancies to migrate, the basic requirement for void formation is that interstitials are preferentially adsorbed at dislocations, thus building up an excess of vacancies to form voids. As discussed earlier, the only clear way to eliminate swelling is to enhance recombination, though reducing nucleation by lowering the amount of helium production or raising the dislocation density may be helpful.

Clearly, any complete theoretical treatment must deal with nucleation rates and other subtle factors like the relative flow of interstitials and vacancies to dislocations, the relative trapping ability of various solutes or particles for interstitials and vacancies, etc.

It is now clearly established that alloying plays a central role in controlling swelling. Though cold work, impurities, and precipitates all have an effect, solid-solution effects are the most striking and the most difficult to rationalize. Figure 13 shows the swelling of several commercial alloys, including 304 and 316 SS, as induced by 5 Mev nickel ions. It also shows swelling in four Fe-15Cr-XNi alloys. The variation in swelling is two orders of magnitude on increasing nickel content by 20 to 30 percent! Some of these alloys are precipitation hardened (A-286), while some are not (IN 800).[40] This figure would indicate that matrix composition plays a central role and variations in structure or impurity level should start with an inherently low swelling composition. It would seem clear that the probability

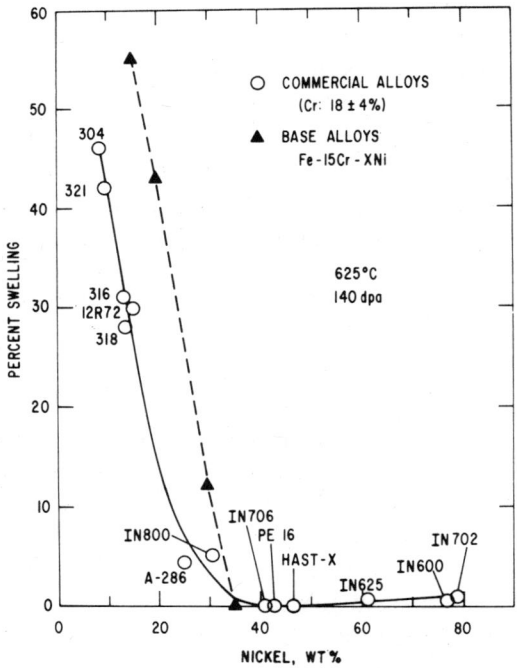

Fig. 13. Swelling versus nickel content of pure Fe-15 Cr-Ni alloys and commercial alloys containing 18 ± 4 wt % chromium; the damage level was 140 dpa by 5 MeV Ni$^+$ ions; the target temperature was 625 C but similar results were observed over the range 575 to 725 C. See ref. 40 for a discussion of alloy compositions and heat treatment.

of vacancy-interstitial mutual annihilation is significantly increased in the higher nickel alloys, but no satisfactory explanation of this composition dependence has been given.

7 RADIATION-ENHANCED CREEP

Interest in radiation-enhanced creep (REC) stems equally from the question of when and how radiation affects creep, and the need to predict the behavior of structural materials in the core of a reactor.

The most striking early evidence of REC was given by Hesketh's work with low stresses at 316 to 77 K.[41] He found that strength before irradiation bore no relation to the in-pile creep rate, and that the creep rate varied *inversely* with temperature. Such creep is certainly not due to thermally activated dislocation motion but to athermal rearrangements in displacement cascades. Those working with engineering materials at reactor temperatures had a harder experimental problem since the stresses are relatively high; and reactor power cycles and control of temperature were essential to clearly establish the differences in creep rate between in-pile and out-of-pile control specimens.

Mechanisms of REC can be divided into those which develop strain by depositing point defects on dislocation loops whose orientation is dictated by the applied stress and those that strain by dislocation glide. Dislocation glide is favored at higher temperatures and stresses.

A tensile stress in a specimen will influence the orientation of planes on which edge-dislocation loops nucleate. In fact, loop formation may not involve significant thermally activated motion of defects but may result from the stress-induced collapse of the depleted (vacancy rich) zones into loops on planes oriented so the stress can do work.[41,42] This leads to a creep rate

$$\epsilon = B\epsilon_e\phi \quad , \tag{10}$$

where ϵ_e is the elastic strain, ϕ the fast-neutron flux, and B a constant. This model stems from the fact that creep occurs at low stresses and very low fractions of T_m. For example, tungsten creeps at 77 K under only a 1000-psi pressure and the REC rate *decreases* with increasing temperature up to 25 C. The absence of dislocation glide is inferred from the fact that all metals and alloys studied seem to creep at the same rate, within an order of magnitude, and "strengthening" the alloy by cold work or precipitation hardening increases the creep rate.

At higher stresses and temperature, dislocation glide is involved. At the low end of this intermediate stress range (near the yield in annealed, unirradiated material), dislocation motion between barriers will be limiting, but at higher stresses, climb over barriers is limiting. At still higher stresses, at or above the postirradiation yield stress, barriers are cut by dislocations and there is no radiation enhancement of the rate.

The nature of the barriers in the intermediate range is a variable element and not always clearly established. At $T < 0.3\ T_m$, the vacancy-rich depleted zones from displacement cascades are formed as fast as they anneal out. The majority of the in-pile creep studies in this intermediate stress range have been performed on cold-worked Zircaloy or similar alloys which are the main core structural material for water reactors and operate there below $0.3\ T_m$. Nichols and Dollins have extensively developed a model based on climb over such a depleted zone barrier.[42,44] Since vacancies as well as interstitials are thought to be mobile in this temperature range, if the flux of both to dislocations is the same, the net climb rate is zero. The flux imbalance needed for dislocation climb might come from the interstitials corresponding to the vacancies left in depleted zones. Or, adequate local climb to overcome barriers might come from fluctuations in the flux causing a net excess of vacancies in some regions and excess interstitials in others—an idea put forward by Gittus[45] and more fully developed by Nichols and Dollins[46].

The situation in zirconium is made more difficult by the anisotropic growth that occurs under fast-neutron irradiation (probably because of the growth of interstitial loops on prism planes). In a polycrystal this anisotropic growth could cause the grains to stress one another, thus reducing the flow stress of the aggregate. Piercy has carefully examined various in-pile experiments on cold-worked Zircaloy and concludes that under the relatively high stress creep conditions of interest in reactor design, radiation-induced growth does not play an important role.[43] (At low stresses, it certainly does.)

The various mechanisms put forward for in-reactor creep of zirconium-base alloys can be summarized with the aid of Figure 14. Here, n is defined by the equation

$$\epsilon \propto \sigma^n \ . \tag{11}$$

At zero stress, radiation-induced growth of the grains in the sample gives rise to a yielding strain of the specimen. As stress rises, these growing dislocation loops tend to align relative to the applied stress and ϵ increases. Dislocation motion inhibited by barriers is dominant in causing strain in the "intermediate stress region", while at high stresses the dislocations cut through essentially all barriers.

Much less work, experimental or analytical, has been done on REC in the stainless steel, which is the core structural material of fast-breeder reactors.[47,48] Walters et al. found that the steady state in-core creep rate of 304 L

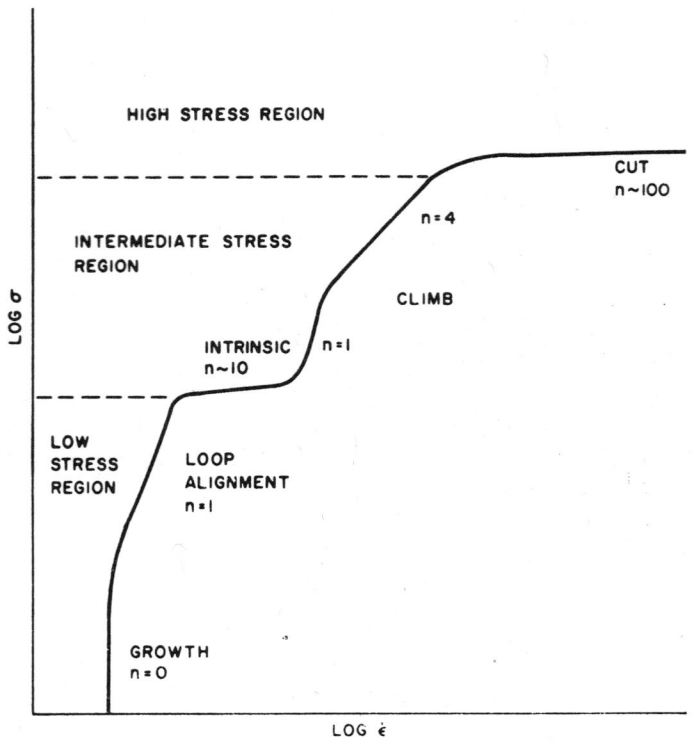

Fig. 14. Schematic representation of irradiation creep behavior of zirconium-base alloys below 0.3 T_m.[44]

was linear in stress from 10 to 27.3 ksi at 380 C and linear in flux.[47] These stresses are below the unirradiated yield stress of the tubes being tested, so are in the low stress region of Fig. 14, and creep is due to the nucleation and growth of oriented dislocation loops.

The swelling of structural members that occurs in a fast reactor causes dimensional changes which interact with, and may be confused with, the shear strains of radiation-enhanced creep. An important engineering reason for interest in REC is the determination of the degree to which it can relax the stresses built up by local variations in the swelling of core components due to variation in flux and temperature across the core. If these stresses are relaxed by REC, the swelling-induced distortion of the core will be greatly reduced.

One of the more interesting ideas in this area is a possible direct proportionality between REC strain and swelling, apparently first put forth by Martin and Poirer.[49] The argument can be developed with detailed assumptions on models[50] or by using mass balance arguments.[9]

To date there are not enough careful data to check out the linearity of this prediction. However, there seems to be a clear correlation between REC rate and swelling. For example, the rank order

$$PE16 < ST316 < ST304L$$

applies to in-pile creep rate and swelling rate (the letters ST designate solution treated).[47,48] Walters finds that in 304 L the swelling is directly proportional to REC strain.[51] This is of interest to those concerned with core warping, but does not help identify the mechanism of creep.

Void growth (swelling) is an indicator of the excess flow of interstitials over that of vacancies to sinks (dislocations). This excess can give rise to creep through loop growth or dislocation climb. Either way, creep rate would be proportional to swelling, though the constant of proportionality would differ. In either case, alloying with an aim to reducing swelling would probably help reduce creep. However, void growth is not essential for creep so the importance of swelling in creep should not be emphasized too much.

8 PRECIPITATE STABILITY

In view of the central role of precipitates in alloy strength, some comment should be made about their role in reactor-core materials. However, corrosion resistance has been of more concern than strength in selecting core materials. Also, radiation-induced hardening and loss of ductility reduce interest in high-strength alloys with their customary lower ductility.

In a reactor core, the stability and size of a second phase are modified by several factors.
1. The diffusion coefficient is increased.
2. Displacement cascades tend to dissolve the precipitate.
3. Nucleation rates are increased.

The increase in diffusivity has been covered above in Eqs. (7) and (8). These equations give an adequate approximation for the solvent. However, they ignore any effect of a vacancy breeze. This would sweep solute either with the vacancies, if there was vacancy-solute binding, or against the vacancy flux, if the solute-vacancy exchange frequency exceeded that of the solvent atoms, and at the low temperatures involved, the diffusivity of certain solutes could be much higher.[52]

Precipitate dissolution by displacement cascades has been treated by Nelson et al.[53] and has been most recently reviewed by Hudson[54]. Several authors have reported that cascades put precipitates into solution, especially at lower temperatures where diffusion and reprecipitation are slow.[54] What Nelson et al. clearly showed with the electron microscope is that coherent, ordered precipitates are dissolved faster than incoherent, disordered precipitates, and that the solute put in solution by irradiation precipitates again as a fine precipitate a few hundred angstroms in diameter. That is, instead of causing coarsening, irradiation causes the precipitate to become finer under conditions that suggest the same would happen in a fast-reactor core.

They suggest two mechanisms of resolution. *Recoil dissolution* is analogous to surface sputtering and involves the mechanical ejection of atoms from the surface into the lattice. The rate of volume change for a spherical particle of radius r is then

$$\frac{dV}{dt}(\text{recoil}) = (4\pi r^2 \alpha K)\Omega \quad, \tag{12}$$

where α is the surface density of atoms, Ω is the atomic volume, and K is the atom fraction displaced per second.

In an ordered precipitate, the cascades create disorder and give rise to *disorder dissolution*. Inside the particle, the lattice will order again. Disorder near the surface region leads to a high-solute surface layer which they say "will lose solute to the surrounding matrix by diffusion". However, they ignore diffusion in their rate equation and represent it by:

$$\frac{dV}{dt}(\text{disorder}) = 4\pi r^2 \Psi K \quad, \tag{13}$$

where Ψ is an average thickness lost per surface atom displaced. Since 10 to 100 more atoms will be disordered by a cascade than permanently displaced, the authors argue that this dissolution mechanism will dominate for any ordered particles.

At steady state, the solute diffusion flux toward the precipitate is given by

$$\frac{dV}{dt} = \frac{3D'cr}{p} \quad, \tag{14}$$

where c and p are the average solute concentration in the lattice and precipitate, respectively, D' is the solute diffusion coefficient in the lattice (Eq. 11 or Eq. 12), and the solute concentration in equilibrium with the particle is zero, or at least much less than c. Finally, for dilute alloys, the average concentration C is

$$C = \frac{4}{3}\pi r^3 pn + c \quad, \tag{15}$$

where n is the number of particles per unit volume. Combining these equations for the disorder dissolution case gives

$$\frac{dr}{dt} = \frac{D'}{r}\left(\frac{3C}{4\pi p} - nr^3\right) - \Psi K \quad .$$ (16)

The equation has the virtue of explaining the observed radiation-induced refinement of coherent precipitates because it says big particles will shrink and little particles will grow. Also, it fits the observation that below some critical temperature, radiation puts particles in solution but diffusion is too slow to give reprecipitation. Conceptually it has the problem of particles simultaneously having a high solubility due to disorder and a small solubility so solute will diffuse to the particle and precipitate. It ignores the problem of nucleation of new particles. A satisfactory treatment would be complicated by the need to treat nucleation as well as particles of varying degrees of disorder (and thus varying tendencies to grow or dissolve). Also, one would expect substantial dragging of solute, at least in the intermediate temperature range where sinks play a dominant role in defect annihilation.

The work on nucleation rates consists of many experimental observations that radiation raises the rate.[54] This may be due to heterogeneities provided by displacement cascades, or to enhanced diffusion with existing heterogeneities. In general, if there is supersaturation with respect to a given phase, radiation may aid its formation. Thus, in stainless steels, carbides[55] and/or ferrite[56] may develop in reactor at temperatures and times where they would not out of pile.

One of the problems inherent in the stainless steels now used in the LMFBR program, e.g., 316, 316L, 304, is that the long-time precipitation sequences out of pile have not been well studied and may differ for various heats.[57] The sequences certainly depend on the degree of initial work, or stress cycles at temperature.[58] Given this, singling out the effect of irradiation is not simple. More importantly, prediction of the behavior of complex commercial alloys in the reactor is uncertain.

REFERENCES

1. Thompson, M. W., *Defects and Radiation Damage in Metals*, Cambridge University Press (1969). [A good text to flesh out the superficial treatment given here.]
2. Nichols, F. A., *Annual Reviews of Materials Science*, R. A. Huggins (Ed.) (1972), Vol. 2, p. 463.
3. Blewitt, T., AEC Sym. Series CONF-710601, J. W. Corbett and L. C. Ianniello (Eds.) (1972), pp. 798–824.
4. Sharp, J. V., AERE Report R-6267 (1969).
5. Dienes, G. J., and Damash, A. C., *J. Appl. Phys.*, **29**, 1713 (1958).

6. Lam, N. Q., Rothman, S. J., and Sizmann, R., *Rad. Effects,* **23**, 53 (1974).
7. Frank, W., and Seeger, A., *Rad. Effects*, **25**, 17–26 (1975).
8. Wiedersich, H., *Rad. Effects*, **12**, 111 (1972).
9. Wiedersich, H., *Proceedings of International Conference on Physical Metallurgy of Reactor Fuel Elements*, Berkley Laboratory of CEGB, U.K. (September, 1973). To be published.
10. Koehler, J. S., *J. Appl. Phys.*, **46**, 2423 (1975).
11. Brown, L. M., Kelly, A., and Mayer, R. M., *Phil. Mag.,* **19**, 721 (1969).
12. Makin, M. J., *Phil. Mag.,* **20**, 1133 (1969).
13. Steele, L. E., IAEA Technical Report Series No. 163, Chapter 6, Vienna (1975).
14. Steele, L. E., *Nuclear Metallurgy,* **16**, 270 (1970).
15. Bush, S. H., *ASTM J. Testing & Eval.,* **435** (1974).
16. Smidt, F. A., Jr., and Watson, H. E., *Met. Trans.,* **3**, 2065 (1972).
17. Wechsler, M. S., Berggren, R. G., Hinkle, N. E., and Stelzman, W. J., *ASTM-STP No. 457*, 242–260 (1969).
18. Smidt, F. A., Jr., and Steele, L. E., NRL Report 7310 (September 1, 1971).
19. Smidt, F. A., Jr., and Sprague, J. A., *ASTM-STP No. 529*, 78 (1973).
20. Smidt, F. A., Jr., Stein, D. F., and Joshi, A., NRL Report 7660 (December 21, 1973).
20a. Gruber, E. E., *J. Appl. Phys.,* **38**, 243 (1967).
21. Smidt, F. A., NRL Memorandum Report 2866, pp. 78–90 (July, 1974).
22. Gjostein, N. A., *Diffusion*, American Society for Metals, Metals Park, Ohio (1973), p. 241.
23. Willertz, L. E., and Shewmon, P. G., *Met. Trans.,* **1**, 2217 (1970).
24. Wier, J. R., *Science,* **156**, 1689 (1967).
25. Kornelsen, E. V., *Rad. Effects,* **13**, 227 (1972).
26. Wilson, W. D., and Bisson, C. L., *Rad. Effects,* **9**, 53 (1973).
27. Wilson, W. D., and Bisson, C. L., *Rad. Effects,* **22**, 63 (1974).
28. Whitmel, D. S., and Nelson, R. S., *Rad. Effects,* **14**, 249 (1972).
29. Wiedersich, H., and Burton, J. J., *J. Nucl. Matl.,* **51**, 287 (1974).
30. Das, S. K., and Kaminsky, M., *J. Nucl. Matl.,* **53**, 115–122 (1974).
31. Bauer, W., and Thomas, G. J., *J. Nucl. Matl.,* **47**, 241–245 (1973).
32. Bauer, W., and Thomas, G. J., *J. Nucl. Matl.,* **53**, 127 (1974).
33. Corbett, J. W., and Ianniello, L. C. (Eds.), AEC Symposium Series CONF-710601 (1971).
34. Pugh, S. F., Loretto, M. H., and Norris, D.I.R. (Eds.), *Voids Formed by Irradiation of Reactor Material*, British Nuclear Energy Society (1971).
35. Nelson, R. S. (Ed.), AERE-R7934 (1975).
36. Norris, D.I.R., *Rad. Effects,* **14**, 1–15 (1972); **15**, 1–22 (1972).
37. Shewmon, P. G., *Science,* **173**, 987 (1971).
38. Johnston, W. G., Rosolowski, J. H., Turkalo, A. M., and Lauritzen, T., *J. Nucl. Matl.,* **47**, 155 (1973); **48**, 330 (1973).
39. Brailsford, A. D., and Bullough, R., *J. Nucl. Matl.,* **44,** 121 (1972). [Probably the best recent effort at a complete theoretical treatment.]
40. Johnston, W. G., Rosolowski, J. H., Turkalo, A. M., and Lauritzen, T., *J. Nucl. Matl.,* **54**, 24–40 (1974).
41. Hesketh, R. V., Report BNL 500.83 (C-52), p. 389 (1968).
42. Nichols, F. A., *J. Nucl. Matl.,* **37**, 59–70 (1970).
43. Piercy, G. R., *J. Nucl. Matl.,* **26**, 18–50 (1968).
44. Dollins, C. C., and Nichols, F. A., *ASTM-STP, No. 551*, p. 229 (1974).
45. Gittus, J. H., *Phil. Mag.,* **25**, 345 (1972).
46. Nichols, F. A., and Dollins, C. C., to be published in *Rad. Effects* (1975).

47. Walters, L. C., Walter, C. M., and Pugacz, M. A., *J. Nucl. Matl.,* **43**, 133 (1972).
48. Mosdale, D., Lewthwaite, G. W., Leet, G. O., and Sloss, W., *Nature,* **224**, 1301 (1969).
49. Martin, G., and Poirer, J. P., *J. Nucl. Matl.,* **39**, 93 (1971).
50. Wolfer, W. G., Foster, J. P., and Garner, F. A., *Nuclear Technology,* **16**, 55–63 (1972).
51. Walters, L. C., and Walter, C. M., Argonne National Laboratory, private communication.
52. Anthony, T., *Diffusion in Solids,* A. S. Nowick and J. J. Burton (Eds.), Chapter 7, Academic Press, New York (1975).
53. Nelson, R. S., Hudson, J. A., and Mazey, D. J., *J. Nucl. Matl.,* **44**, 318–330 (1972).
54. Hudson, J. A., AERE Report R-7943 (1974).
55. Brager, H. R., and Straalsund, J. L., *J. Nucl. Matl.,* **46**, 134 (1973).
56. Keefer, D. W., Pard, A. G., Rhodes, G. G., and Kramer, D., *J. Nucl. Matl.,* **39**, 229 (1971).
57. Stickler, R., and Weiss, B., *Met. Trans.,* **3**, 851 (1972).
58. Diercks, D., Argonne National Laboratory, private communication.

DISCUSSION on Paper by P. G. Shewmon

HIRTH: A comparison of the decrease in ductility with bubble formation in nickel-base alloys under irradiation with those formed in HIP nickel-base powder alloys supports the view on effects of voids suggested by McClintock. The voids mentioned by Shewmon had spacings on the order of $\sim10^3$ A and caused decreases in ductility at volume fractions of 10^{-7}. In HIP powder alloys made by helium atomization, the bubbles with spacings of $\sim10^{-3}$ can form after anneals at ~2200 F, but these cause decreases in ductility only for volume fractions greater than 10^{-2}. This difference is consistent with McClintock's suggestion that decreased void spacing increases the likelihood of ductile fracture propagation.

SIMS: Would you expect any structural effects from helium in construction metals operating in the gas streams of HTGR's—where the helium atmosphere is at about 1000 psi and 1000 C for gas/metal exposures up to design times of 300,000 hours?

SHEWMON: I presume you are thinking of something like helium going into solution at high temperature and precipitating on cooling. I would not expect this because I know of no one who has been able to detect the equilibrium solubility of helium in any metal. It seems to be really insoluble in metals.

FROST: Some advance in the theory of solute drag has been made recently by Bob Johnson of the University of Virginia while at Argonne National Laboratory this past summer. His work applies particularly to migration towards voids.

On the question of blister formation, Manfred Kaminsky at Argonne has injected helium into S.A.P. (sintered aluminum powder) which had interconnected porosity. The helium was mostly released and blistering was minimal. This indicates an avenue for development of blister-resistant alloys for fusion-reactor use.

REMARKS ON ALLOY DESIGN

J. R. Low, Jr.

Carnegie-Mellon University
Pittsburgh, Pennsylvania

Some years ago, it was the custom of large industrial organizations with large research laboratories to have published what were, I believe, called institutional advertisements in various technical and popular journals.

I would like to read to you a portion of the text of one such advertisement which appeared about 20 years ago and is pertinent to the subject matter of this conference.

The headline was *New Alloys for Special Uses* and it read as follows:

"Recently the Research Laboratory was asked to design an alloy to be used in a new type of heating element. In addition to good formability, the alloy had to have a special temperature-resistivity curve not available in any commercial material. Dr. _____ , after less than an hour with pencil and paper, came up with the answer—a new composition and detailed processing instructions. Dr. _____'s success was dramatic evidence of how metallurgy has progressed from an industrial art to a science. Until the last few years, new alloys with prescribed physical, electrical or mechanical properties had to be developed primarily by trial-and-error "cookbook" methods. Dr. _____ and his associates, through their basic studies of atomic arrangement in metals, are shedding new light on the relationship between the structure of alloys and their properties. This research will play an important role in the many areas of our technology where future progress is dependent on improved material."

This achievement must be the all-time record for alloy design and one gets the feeling that we have gone downhill ever since.

As all of you know, alloy design is not quite this easy and, as you will note, the advertising agency does not discuss whether or not the initial concept actually worked when it was tested nor all of the difficulties encountered in making the alloy and its subsequent processing into a useful form, which are only a few of the problems which the alloy designer and producer face. It may be just as well if the initial alloy design is done without too much concern for the production problems, since these problems may well be better solved by others.

There are several aspects to alloy design and reduction to practice, and I would now like to discuss these briefly.

It is important to be able to specify all of the service conditions under which a new material is to perform its function. The design engineer who is to use the material in his design may not be aware of all of the conditions that may affect the performance and may not specify all of the properties necessary for satisfactory performance. A recent example is the attempt to use reinforced polymers for aircraft gas-turbine-compressor blading. The material satisfied the primary requirements of good unidirectional strength in the temperature range of the contemplated application with a considerable reduction in weight and power consumption. It was not until the application had been carried to an advanced stage of development that the material was found not to be able to withstand various forms of impact loading and environmental attack. The consequences were that the design and materials had to be completely revised with unfortunate and serious financial losses. It is small wonder then that the most convincing recommendation for a material that one can make to a design engineer is that the material has a long history of satisfactory performance in a similar application, preferably an identical application.

If we assume that the service conditions can be specified to everyone's satisfaction, it then becomes the task of the alloy designer to invent the optimum material to perform the service function. To do this he uses what information is available to him concerning the effects of composition, microstructure, and environment on behavior. He may also rely on prior records of other materials in general use at the time. In some cases, it may be necessary only to design an alloy that will provide a modest improvement in properties but which may nevertheless result in a large improvement in performance or reliability, e.g., a 50 F increase in the temperature of operation of gas-turbine blading alloys. In cases such as this, it may even be possible to predict that an improvement of this kind will be accomplished at a certain time (say, 3 years hence) and to begin the mechanical design and gamble that the improved material will be available when it is necessary to begin production.

The above approach is possible when there is a long history of materials and performance to work from. There are other cases when so little is known about the effects of new conditions of use that we have to rely much more heavily on fundamental knowledge and theoretical analysis. This was the case, for example, in attempting to select or design materials for nuclear-power generation some 25 years ago. At that time, it was necessary, to take just one aspect of the problem, to predict the interaction of fast neutrons with crystalline solids, the probable kinds of crystalline defects that might be produced and the effects of such defects on chemical, mechanical, or electrical behavior of materials. Initially, there was no alternative but to simply irradiate all candidate materials and examine them after irradiation. The early literature on radiation damage describes irradiation experiments on a broad range of materials and devices including, for example, simple alloys, vacuum tubes, and other devices. Gradually, as we acquired more fundamental understanding, general principles evolved, in part from experiment and in part from advances in theory; but as is to be expected, the conditions of service continue to become more demanding and we are frequently confronted with startling surprises. Nevertheless, some progress has been achieved.

If we consider the complex array of stipulations which may be written down for the design of a single alloy for a single application, it becomes apparent that a single individual is unlikely to be knowledgeable enough in all areas to make the required decisions or to foresee all of the possible complexities that may be encountered. Further, if we consider the explosion in the technical and scientific literature which has been and is occurring, it is highly unlikely that an individual will even be sufficiently aware of the state of current understanding in more than one or two areas of alloy behavior.

This suggests that initially, at least, the planning should be done by a group of specialists working together and contributing their opinions and ideas to a solution. At first blush this sounds like a satisfactory approach, but I am reminded of Bruce Old's article on the statistics of committees. One of his principal conclusions was that if he plotted the effectiveness of a committee versus the number of committee members, his curve showed a maximum at three-quarters of a person. I suggest that it is far better to have one individual responsible for the alloy design, say someone knowledgeable in the structure-properties relations of principal concern. However, his proposed solution should be periodically reviewed by a group of his colleagues specializing in other areas to guard against possible catastrophic oversights.

In the above brief discussion, it has been implicit that we understand quantitatively what a change in composition or microstructure will do to change some aspects of mechanical or chemical behavior. As all of you are aware, much of our knowledge is qualitative and we rely largely on empirical correlations in much of our work. For example, Ashby a few years ago pointed out, in discussing strengthening mechanisms, that the only quantitative microscopic theory having some reliability was Orowan's dispersoid-strengthening model, provided the dispersoid met all the qualifications initially assumed. In the application of dislocation theory to the calculation of behavior, we have evolved fairly sophisticated models, but we are still searching for a method of determining the number of mobile dislocations in the total dislocation population, and it is the mobile dislocation fraction which is important in such calculations.

We, therefore, need a continuing strong research effort aimed at improving our quantitative understanding and description of atomic and microscale phenomena which we now know qualitatively to be involved in alloy behavior. Further, there must be constant interaction between research investigators working to understand the phenomena that govern behavior and alloy designers attempting to optimize materials to meet present and future service requirements. I would urge research investigators to attempt to select their research problems so that new understanding will have the broadest possible influence on the task of the alloy designer. To this end, the research investigators should be aware of the critical areas where a lack of understanding is blocking an improvement in materials behavior. The research investigator seldom suffers from a shortage of phenomena to investigate, his problem is one of selection and separation of the interesting, but trivial, from the just-as-interesting but more important subjects for investigation.

This interaction between research investigators and alloy designers is the principal motivation for the present conference in which we are now participating. I will conclude these brief remarks with the hope that our activities during the next few days will be both enjoyable and fruitful, as I am sure they will be!

_____ *Part Two*

NEEDS IN ALLOY DESIGN

NEW DIRECTIONS IN ALLOY DESIGN FOR GAS TURBINES

F. L. VerSnyder and M. Gell

Pratt & Whitney Aircraft Division
United Technologies Corporation
East Hartford, Connecticut

ABSTRACT

A description is given of alloy trends and requirements for aircraft gas-turbine components for which there exist major alloy-design opportunities. In the near term, emphasis is on reduced cost and improved durability. In the longer term, improved alloy performance becomes more dominant. There is ample opportunity for alloy design, research and development in new materials for gas-turbine engines despite the marked change in the research and development environment.

1 INTRODUCTION

Materials and their processing have historically played a key role in the development of gas-turbine engines. Higher performance alloys and unique processing methods have permitted the designer to achieve significant increases in engine performance, particularly thrust-to-weight ratio. The development and application of directionally solidified (D.S.) superalloy turbine blading and high-strength, lightweight superalloy powder metallurgy disks formed by Gatorizing™ are two examples of this kind of materials innovation. Although the importance of materials and processing is a continuing feature of power-plant development, the state of the art and the economic climate have significantly altered the development approach. Through the 1960's, the great potential of various classes of materials, particularly titanium and cast superalloys, was rapidly exploited to achieve significant improvements in strength and temperature capability. Both the military and commercial engine fields were expanding vigorously and development funding was quite readily available. Cost was not a critical factor. The current development atmosphere differs in several ways. Materials and processing development are now pursued more selectively with considerable attention to requirements, benefits, and cost. This is the case because (1) the traditional metallurgical approaches of alloy development have exhausted much of the apparent potential, (2) increasing design sophistication has made materials development more complex, and (3) new engine programs and development funding are limited.

Our current materials-development philosophy divides programs according to near-term and long-term objectives. The near-term programs are aimed at reduced cost and increased durability. The longer term programs are aimed at improved performance and insuring cost-effective materials utilization. The major alloy-design opportunities exist in materials for the high-pressure compressor and for turbine disks, blades, and vanes. The alloy systems involved are titanium alloys and nickel-base alloys. Not only are they employed in the critical components of advanced engines but in total they represent about 60 percent of the weight of the engines, with nickel-base alloys accounting for approximately 40 percent of the weight.

2 ALLOY RESEARCH/DEVELOPMENT

2.1 Titanium-Base Alloys for Blades and Disks

The introduction of titanium alloys to gas-turbine engines about 20 years ago required considerable invention and development to insure reliability. Since that time, a major effort in titanium metallurgy for aircraft gas turbines has been to advance the temperature capability of titanium alloys. These improvements in titanium-alloy properties have been accomplished through alloying, processing techniques, and heat treatments.

Several years ago, the development of the Ti-6Al-2Sn-4Zr-2Mo alloy (6-2-4-2) processed to an $\alpha+\beta$ microstructure marked a significant advance in titanium-alloy design for elevated-temperature service (Fig. 1). More recently, processing advances provided a transformed β form of the alloy with increased temperature capability. Refinement of the β microstructure has resulted in further improvements in durability (Fig. 2). This alloy (Ti-6-2-4-2) is currently used in some of the more advanced jet engines and in sheet form in developmental aircraft. It is expected that further improvements in temperature capability will be achieved through β processing in the near term and that for the long-term potential, major gains can be achieved using titanium aluminides (Fig. 1).

The short-term, high-temperature, titanium-alloy program has as its goals reduced cost and the replacement of nickel alloys in the high-pressure compressor and the low-pressure turbine. The development program follows essentially the traditional ways of titanium-alloy development, that is, modest changes in alloy processing techniques and heat treatments, especially for disks, with special chemistry modifications for low-pressure-turbine airfoil castings (Fig. 3). An example of processing development is the isothermal beta-forging part of the program. The advantages of beta forging were demonstrated several years ago, including improvements in fracture toughness. The ideal material would simultaneously exhibit high tensile properties, high toughness, resistance to fatigue-crack propagation, and a high level of creep resistance. In practice, it is difficult to gain improvements in all these properties areas, so the alloy-development program must be closely coordinated to the requirements of a particular application.

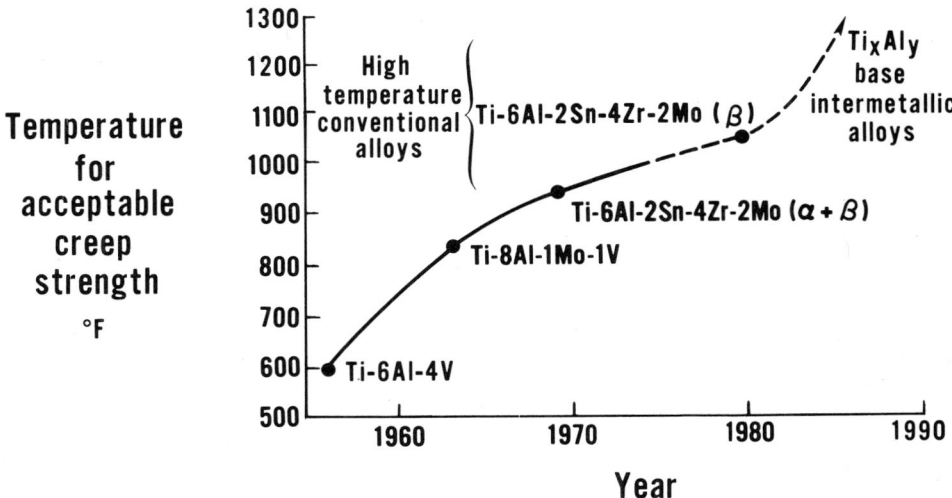

Fig. 1. Titanium-alloy trends.

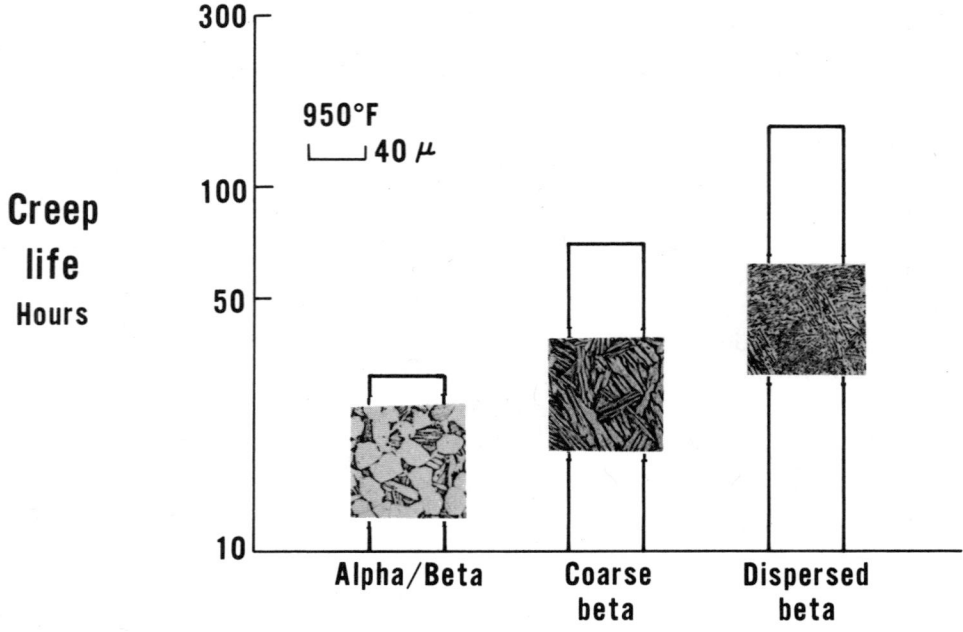

Fig. 2. Creep life related to titanium-alloy microstructure (stress is 35 ksi).

- **Goals**
  - **Reduce cost**
  - **Replace Ni in high compressor and low turbine**

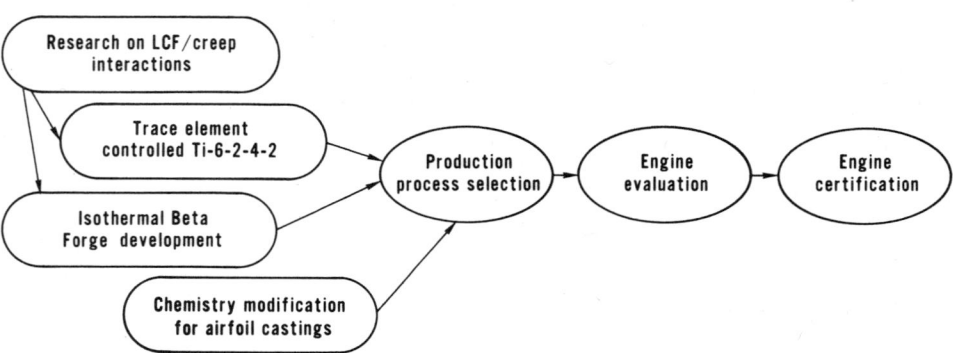

Fig. 3. Short-term high-temperature titanium-alloy program.

The areas for high-temperature titanium-alloy research in the short term are more straightforward and could likely pace advances in development. Critical areas of research include:

- Mechanical Properties
  — Creep Mechanisms
  — Microstructural Effects on Creep/Fatigue Behavior

- Corrosion
  — Transient Thermal-Fluctuation Effects on Stress-Corrosion Cracking
  — Atmospheric Environmental Effects on Low Stress Cracking.

Moving to longer term programs aimed at the 1980's, one finds that the Ti-Al intermetallics offer the most promise for high specific strength (Fig. 4). Exploratory investigations of the Ti-Al intermetallics, Ti-Al (α) and Ti$_3$Al (γ), have indicated outstanding high-temperature strength compared with the conventional titanium alloys, as well as attractive oxidation resistance and hot workability, the latter particularly for Ti$_3$Al. On the basis of these features, these systems offer the only foreseeable avenue to significantly reduce the weight and perhaps the cost of the higher temperature compressor airfoils as well as turbine section components now produced from the intermediate-strength superalloys such as INCO 901 and INCO 713C. Preliminary studies have suggested that the total engine weight for an advanced fighter engine could be reduced by as much as 20 percent through extensive utilization of such materials, particularly in large static structures operating in the 1200 to 1800 F range. Of course, the current

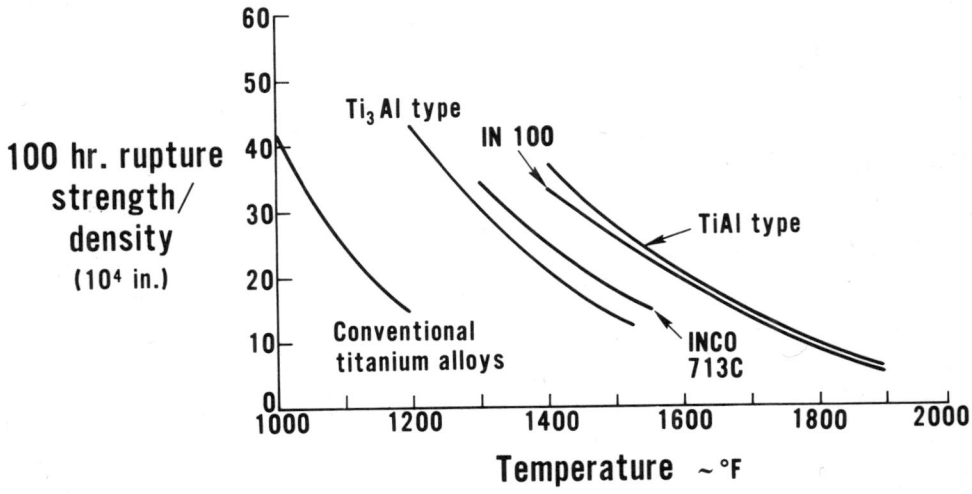

Fig. 4. Ti-Al exhibits high specific strength.

problem with these materials is that they exhibit ductile/brittle transition temperatures well above room temperature. Therefore, in view of the likely applications and the presently observed limitations, areas for research/development in titanium intermetallic alloys are:

- Mechanical Properties
  - Basic Deformation Modes
  - Alloy Influence on Ductility

- Corrosion
  - Oxidation Mechanisms
  - Stress Corrosion
  - Basic Diffusion Studies of Ti-Al-X Ternary Systems.

2.2 Nickel-Base Superalloys for Disks

The trend in turbine disk materials is toward greater strength (Fig. 5). However, at the same time, the final ready-for-assembly disks must be lower in cost. Net or near-net shape processing by direct hot-isostatic compaction of powder is therefore receiving considerable attention. Hot isostatic pressing (HIP) has grown, over the past 15 years, from a laboratory curiosity to a viable process capable of producing large-scale, high-quality, powder-metallurgy products on a production basis. High-pressure equipment has progressed from small, laboratory-size pressure vessels with long turn-around cycles to large-diameter equipment capable of handling large charges in relatively short times. Unique methods have been devised for HIP

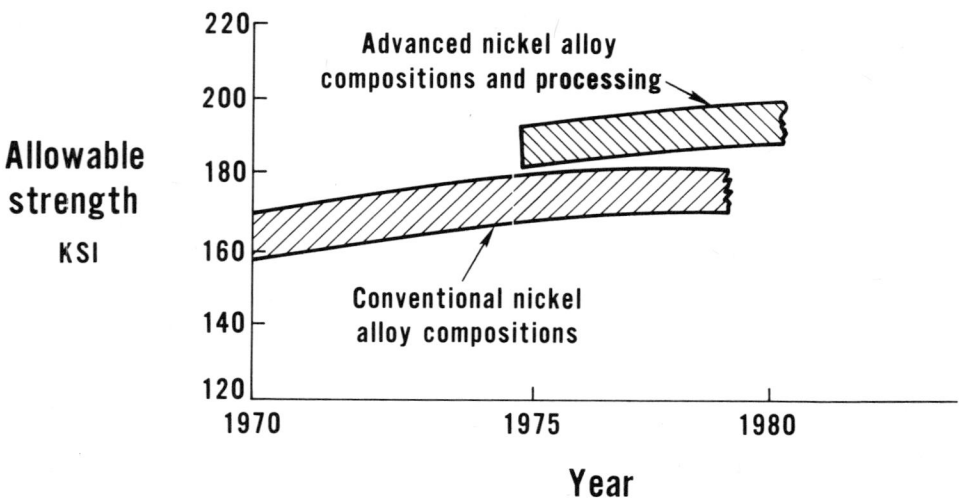

Fig. 5. Nickel-disk-alloy trends.

container fabrication which permit pressing of components to nearly net configuration in relatively low cost, reliable containers. Utilizing these methods, components have been fabricated from several materials with properties equivalent to and sometimes superior to those produced by other methods.

Although cost is a prime consideration for developing this processing approach, high-temperature-alloy, powder-metallurgy development may yield high-strength alloys not attainable by the conventional melt, cast, forge methods of processing. As a result of a high alloying content and the combination of high- and low-melting-point constituents in most nickel-base superalloys, segregation and banding of the cast ingots is often a severe problem. Owing to the presence of complex carbides and a γ' phase that has a solutioning temperature very close to the incipient melting temperature of the alloy, hot working is often impossible. Therefore, alloys which cannot be cast to shape because of inhomogeneities or cannot be wrought subsequent to casting because of brittleness can be utilized in powder form. The powder-metallurgy or HIP product made from atomized powder is inherently superior in this sense to a large casting, since each powder particle is in essence an individual ingot.

In the net-shape processing of superalloy disks, cost savings can arise in several ways (Fig. 6), i.e., from less input material, lower machining cost, and fewer chips, from the same finished disk weight. All this is not to imply that there are no problems to be solved in the development of HIP processing for high-temperature alloys. Two problems illustrative of the difficulties to be overcome are internal defects which range from voids to ceramics introduced from the atomization nozzles. There is also the phenomenon of prior particle boundaries which are associated with the carbon content of the alloy and the morphology of the carbides in the individual powder particles. Consequently, the short-term net-shape-processing program for nickel-base alloys includes as its goals reduced cost and control of the defect characteristics (Fig. 7). The research and development areas for improved nickel disks from powder include:

- Powder Surface Chemistry
- Gas/Metal Desorption Kinetics
- Nonequilibrium Chemistry/Structure Control in Small Particles
- High-Temperature Plasticity/Compaction Prediction Models
- Unique or Improved Methods for Reduced Flaw-Detection Limits.

2.3 Superalloys for Turbine Blades and Vanes

Currently, superalloys make up approximately 40 percent of the weight of advanced gas-turbine engines. Most of these alloys are modifications of a

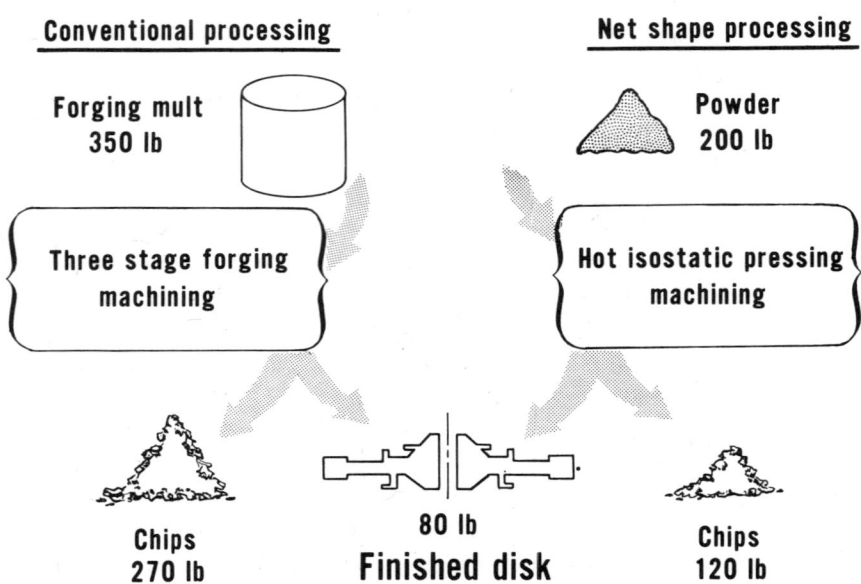

Fig. 6. Net shape processing of superalloy disks.

- **Goals**
  - **Reduce cost**
  - **Control defect characteristics**

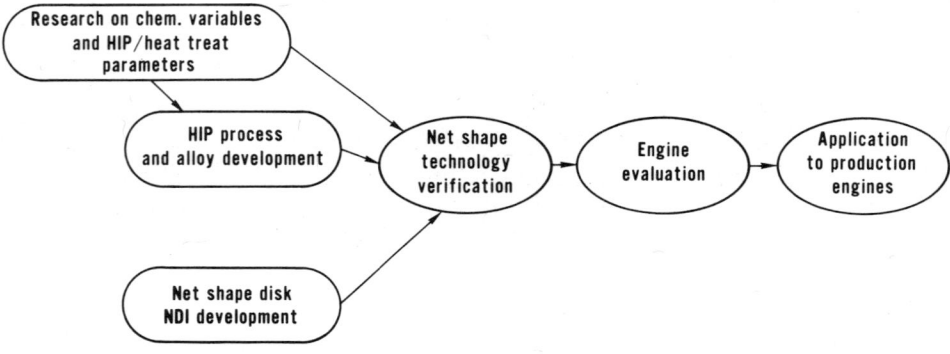

Fig. 7. Short-term nickel-disk-alloy processing program.

Nichrome matrix hardened by γ' but with further additions of cobalt, iron, tungsten, molybdenum, vanadium, titanium, niobium, tantalum, boron, zirconium, carbon, and manganese. Thus, today's superalloys are known for their complexity, with their compositions having been evolved empirically over the past 30 years.

The plan for further development of high-temperature turbine blades and vanes also involves a near-term program and a long-range program. In the near-term program, the emphasis is on low-cost, increased-life, directionally solidified (D.S.) alloys. The longer term strategy involves a single-crystal superalloy program and a D.S.-eutectic program, both of which are justified on the basis of cost/benefit studies. The trends in turbine-airfoil temperature and strength capability required in the years to come are compatible with this strategy (Fig. 8). The advantages of D.S. alloys over conventionally cast alloys have been well publicized in recent years. Basically, one is able to achieve greater thermal-fatigue life, greater rupture life, and greater rupture ductility, all simultaneously. The directionally solidified MAR-M 200 + Hf alloy is now used in a number of advanced production engines, powering both commercial and military aircraft (Fig. 9).

Although the benefits of this D.S. alloy have been amply realized, additional improvements would be desirable. Hafnium was added to MAR-M 200 to prevent grain-boundary cracking of hollow-blade castings during solidification and to provide good postcasting transverse ductility. However, the addition of hafnium had also led to the formation of hafnium containing inclusions which affected casting yields. To improve casting yields, hafnium should be minimized or eliminated and, perhaps, replaced with other grain-boundary-strengthening elements. To identify a substitute for hafnium, an understanding of the local chemistry and microstructure that control grain-

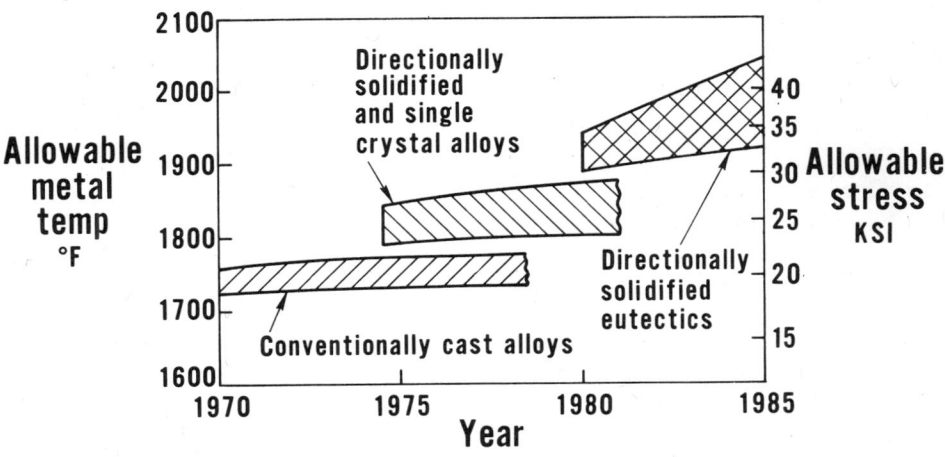

Fig. 8. Turbine-airfoil-alloy trends.

Advantages over conventional castings
- **10X greater thermal fatigue life**
- **4X greater rupture ductility**
- **2X greater rupture life**

Used in advanced production engines

Engine	Aircraft
• JT9D-7F	Boeing 747
• JT9D-59/70	Boeing 747, DC-10
• F100-PW100	F-15, F-16
• TF30-P-100	F111F

Fig. 9. Directionally solidified turbine airfoils.

boundary strength would be useful in conjunction with alloy-modification and heat-treatment studies. Figure 10 summarizes the microstructures and grain-boundary-strength characteristics of two current superalloys, MAR-M 200 and IN 1851, and emphasizes the need for additional understanding of grain-boundary-strengthening behavior.

So, in the short-term superalloy-airfoil program, the goals are reduced cost and improved durability. To achieve these goals, it is necessary to develop alloys with improved castability and hot-corrosion resistance. These facets are included in a current development program (Fig. 11). Consequently, some areas for research for superalloy airfoil materials are:

- Mechanical Properties
  - Alloys/Structures for Maximum Grain-Boundary Strength and Ductility
  - Alloys/Processes for Maximum Strength (1400 to 1700 F)

- Corrosion/Sulfidation of High-Strength Alloys
  - Mechanisms of Sulfidation Attack
  - Role of Alloying Elements
  - Improved Coatings

- Improved Lifetime-Prediction Methods.

The first step in the longer term program involves a single-crystal-superalloy program. Single crystals offer a means of achieving the highest possible creep strength in nickel-base superalloys because of the elimination of grain boundaries, which are primary high-temperature-failure sites.

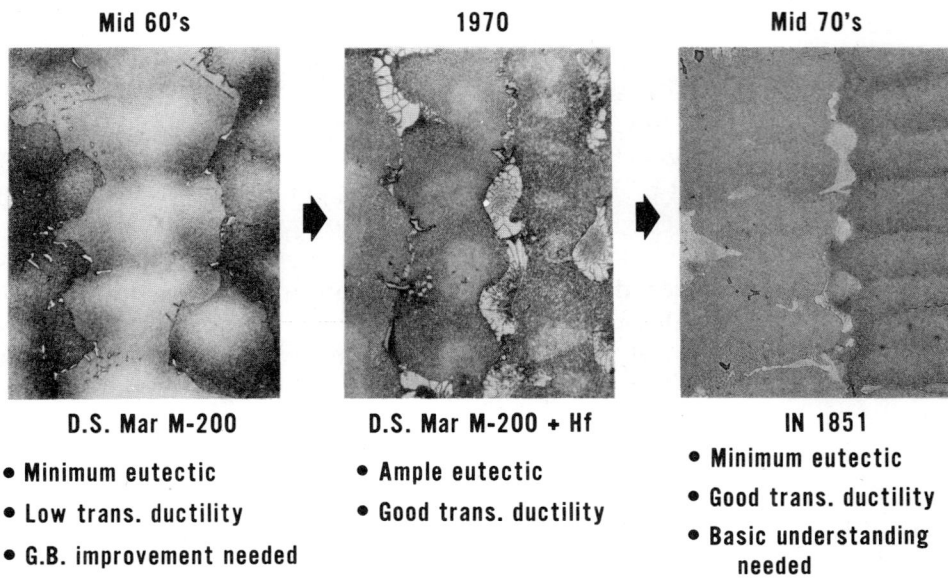

Fig. 10. Grain-boundary characteristics of several directionally solidified alloys.

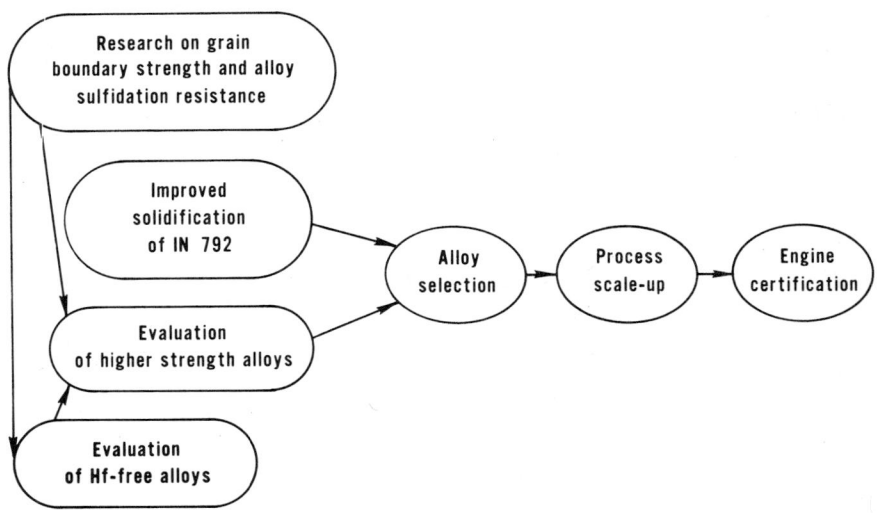

Fig. 11. Short-term superalloy airfoil program.

Single crystals also offer a means of maximizing superalloy melting temperature which is desirable for turbine vanes, by eliminating those alloying elements which are added to improve grain-boundary properties, but which also tend to lower the incipient melting temperature. This compositional freedom available with single crystals provides the best possibility for achieving the optimum combination of superalloy properties, namely, strength, thermal-fatigue resistance, hot-corrosion and oxidation resistance, and castability. Evolving from the initial work on columnar-grain superalloys, a modified directional solidification process was developed to grow superalloy single crystals. This process involves a liquid-tin cooling technique. Single crystal growth is not rate sensitive as there are no grain boundaries to consider. Consequently, with the liquid-tin cooling techniques, single crystal superalloy bars have been grown at rates up to 200 in./hr. Columnar-grain superalloys are normally grown at rates of 8 to 25 in./hr, although a rate of 80 in./hr could be achieved using liquid-tin cooling. Monaloy 444 is a single-crystal superalloy having no carbon, boron, zirconium, or hafnium and it exhibits a useful advance in strength over columnar-grain MAR-M 200 + Hf (Fig. 12). A thin-section-data comparison would show an even greater improvement in creep properties because of the absence of all grain boundaries.

Although single-crystal alloys have been identified for both turbine blades and vanes, considerable additional information is required prior to

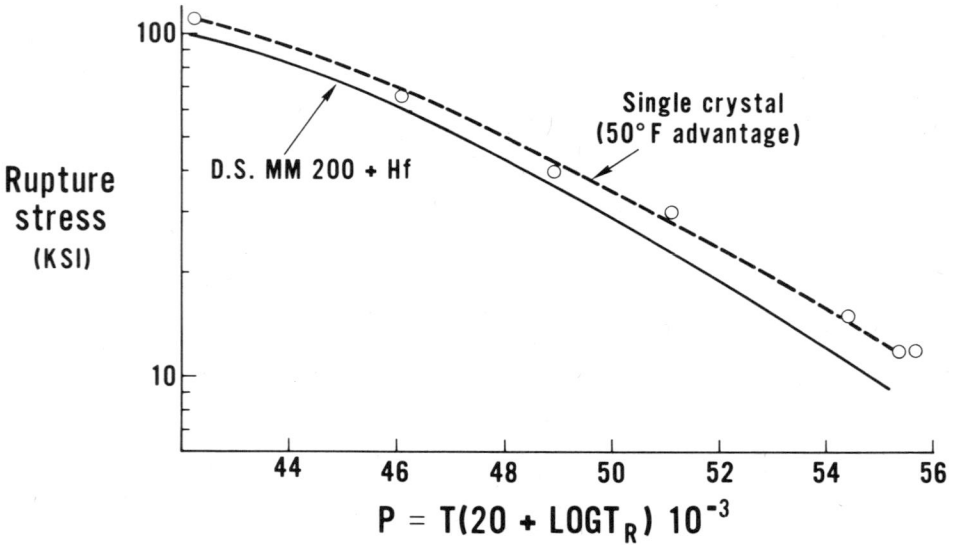

Fig. 12. Comparison of single crystal Monaloy 444 and columnar-grained MAR-M 200 + hafnium.

their introduction into sophisticated, advanced gas-turbine engines. The usual characteristics will have to be evaluated, namely:

- Creep/Thermal-Fatigue Properties
- Oxidation/Sulfidation Resistance
- Microstructural Stability.

A concurrent program for the long term is the development of directionally solidified eutectics. Eutectic alloys reinforced with the intermetallic compound γNi_3Cb, such as $\gamma/\gamma'+\delta$ (composition, weight percent: Ni-20Cb-6Cr-2.5Al) and $\gamma'+\delta$ (composition, weight percent: Ni-23.1Cb-4.4Al), have been identified as possessing many of the desired mechanical-property characteristics for gas-turbine applications. For example, in terms of creep behavior, the $\gamma/\gamma'+\delta$ D.S. eutectic offers a temperature advantage of up to 100 to 150 F (56 to 83 C) over existing advanced superalloys such as D.S. MAR-M 200 + Hf, as indicated in Fig. 13. This higher temperature capability can be translated into significant improvements in thrust, engine efficiency, and component durability, as shown in a recent cost/benefit study. Although the longitudinal mechanical properties of a $\gamma/\gamma'+\delta$ system are outstanding, improvements in off-axis strength and ductility are being sought through minor alloy modifications. Further improvements in longitudinal creep strength may be possible with major alloy additions and processing improvements which permit solidification at higher rates, and therefore produce a finer lamellar spacing.

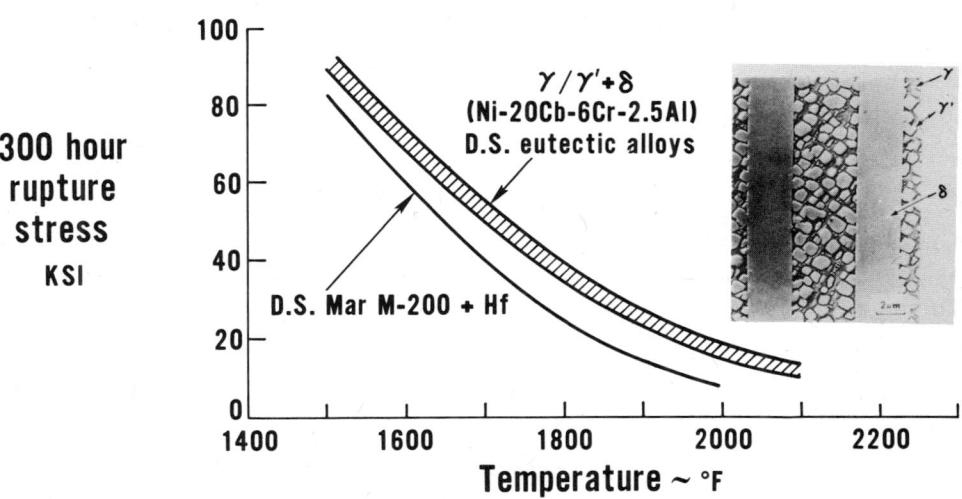

Fig. 13. Directionally solidified eutectic alloy stress-rupture comparison.

In addition to the modification of existing D.S. eutectics, the exploration of new eutectic-alloy systems uniquely conceived to achieve superior high-temperature strength, together with a favorable balance of properties to meet other criteria, is required. The alloy-development process for D.S. eutectics is in its infancy and, because of the potential payoffs, could profitably continue for many years.

Solidification processing is also a key requirement for the successful development of D.S. eutectics. Eutectic alloys must be solidified with a planar solid/liquid interface in order to achieve the desired microstructure and elevated-temperature mechanical properties. The necessary conditions for plane-front solidification of multicomponent alloys have been considered for the case of both dilute single-phase alloys and for eutectic-alloy compositions. These studies have shown that for a given alloy composition and solidification rate, it is necessary to exceed some particular value of temperature gradient (G_L) in the liquid at the solid/liquid interface in order to obtain a plane solidification front. Casting processes with improved temperature gradients are required for reduced-cost processing as well as to achieve a higher level of creep strength through microstructural refinement.

Consequently, the most promising areas for research/development for directionally solidified eutectic airfoil materials are:

- Strength/Ductility
  — Improved Shear Strength
  — Minor Element Effects

- New Alloy Systems

- Processing
  — Higher Temperature Mold and Core Materials
  — Higher Gradient Casting Processes.

3 SUMMARY

In summary then, though the research and development environment has undergone marked changes in the past 5 years, there is ample opportunity for alloy design, research and development in new materials for gas turbine engines. Some of the most significant opportunities are considered to be:

- Mechanical properties and corrosion resistance of high-temperature titanium alloys

- Characterization and ductility improvement of Ti-Al intermetallics

- Surface chemistry, HIP/heat-treat interaction and NDI studies for advanced disks from nickel-base superalloy powders

● Turbine airfoil materials studies including:
  — Corrosion/sulfidation behavior of nickel-base superalloys
  — Grain boundary strength/ductility interactions in D.S. superalloys
  — Alloy design for increased strength of columnar-grain and single-crystal airfoils
  — Optimized eutectic-alloy processing
  — Exploration for new D.S. eutectic-alloy systems.

ACKNOWLEDGMENT

The authors appreciate the assistance of M. J. Donachie in the preparation and review of this paper.

BIBLIOGRAPHY

Bisset, J. W., NASA Rept. No. CR-134701 (October, 1974).

Coyle, J. E., et al., *Metals Engr. Quart.*, **8** (3), 10–15 (1968).

Jaffee, R. I., and Promisel, N. E. (Eds.), *The Science, Technology and Application of Titanium*, Pergamon Press, New York, N.Y. (1970).

Jaffee, R. I., and Burte, H. N. (Eds.), *Titanium—Science and Technology*, Plenum Press, New York, N.Y. (1973).

Breme, J., *Metall.*, **27**, 127–134 (1973).

Bowen, A. W., and Stubbington, C. A., Royal Aircraft Establishment (England), Tech. Rept. 73019 (May, 1973).

Seagle, S. R., et al., *Metals Engr. Quart.*, **15** (1), 48–54 (1975).

Gray, H. R., *Metals Engr. Quart.*, **12** (4), 10–17 (1972).

Gray, H. R., and Johnston, J. R., NASA Rept. No. TMX-3145 (December, 1974).

Blackburn, M. J., Feeney, J. A., and Beck, T. R., in *Advances in Corrosion Science and Technology*, Vol. 3, Plenum Press, New York, N.Y. (1974), pp. 67–292.

Williams, J. C., and Blackburn, M. J., *Ordered Alloys*, Claitor's Press, Baton Rouge, La. (1970), pp. 425–445.

Hayden, S. Z., and Blackburn, M. J., Second Semi-Annual Report for Aerospace Research Laboratories, Wright-Patterson AFB, Contract No. AF-33615-74-C-1140 (15 June 1975).

Burke, J. J., Weiss, V., (Eds.), *Powder Metallurgy for High Performance Applications*, Syracuse University Press, Syracuse, N.Y. (1972).

Hanes, H. D., *op. cit.*, pp. 211–230.

Gessinger, G. H., and Bomford, M. J., *Intl. Met. Reviews*, **19**, 51–76 (1974).

Production of Powder Metallurgy Nickel-Base Superalloy Turbine Disks, Contract in progress at Pratt & Whitney Aircraft, Contract No. AF-33615-74-C-5108.

VerSnyder, F. L., and Shank, M. E., *Mater. Sci. Engr.,* **6**, 213–247 (1970).

Sullivan, C. P., et al., *Cobalt-Base Superalloys—1970,* Cobalt Information Center, Brussels, Belgium (1970).

Sims, C. T., and Hagel, W. C., *The Superalloys,* J. Wiley & Sons, Inc., New York, N.Y. (1972).

Superalloys—Processing, MCIC-72-10, Metals and Ceramics Information Center, Battelle Memorial Institute, Columbus, Ohio (1972).

Sahm, P. R., and Speidel, M. O. (Eds.), *High Temperature Materials in Gas Turbines,* Elsevier, Amsterdam, The Netherlands (1974).

Betteridge, W., and Heslop, J. (Eds.), *The Nimonic Alloys,* Crane, Russak and Co., New York, N.Y. (1974).

Erickson, J. S., et al., in *High Temperature Materials in Gas Turbines,* Elsevier, Amsterdam, The Netherlands (1974), pp. 315–343.

Proceedings of the Conference on In-Situ Composites, Rept. No. NMAB-308, NTIS, Springfield, Va. (January, 1973).

Directionally Solidified Composites, Rept. No. NMAB-301, NTIS, Springfield, Va. (April, 1973).

Thompson, E. R., and Sahm, P. R. (Eds.), *Specialists Meeting on Directionally-Solidified In-Situ Composites,* Rept. No. AGARD-CP-156, NTIS, Springfield, Va. (August, 1974).

Gell, M., and Donachie, M. J., in *Materials on the Move, Proceedings 6th SAMPE Conference,* SAMPE, Asuza, Calif. (1974), pp. 188–195.

DISCUSSION on Paper by F. L. VerSnyder

SHERBY: It is well documented that atom mobility in titanium is very high in contrast to that of other metals of comparable melting temperature and crystal structure. The modulus of titanium is very low (equal to about that for polycrystalline copper). Both these factors are known to influence the creep resistance of metals and undoubtedly contribute to the low strength of titanium alloys at warm temperatures. The results you showed on the high creep strength of Ti-Al alloys are indeed impressive. I was curious as to whether aluminum additions to titanium decreased the overall atom mobility in the alloy (i.e., make atom mobility in titanium behave more normally). Similarly, I wondered whether the modulus of titanium increases with aluminum additions, especially in the range of temperature of interest for improved creep resistance (1000 to 1500 F). It would seem to me that new inroads into improving titanium alloys for high-temperature applications may arise from basic studies on the stress dependence and activation energy for creep of titanium-aluminum alloys. Such studies may help determine the mechanism of creep in TiAl alloys, and this knowledge, in turn, may help to optimize the properties of such alloys.

You made mention that titanium can be made quite creep resistant by beta forging and you showed us that a fine structure was developed as a result of such processing. I was curious as to why such a structure is desirable for high creep resistance. Is it possible that working in the beta region created a well-defined, fine subgrain structure and that this structure, in turn, transforms to a fine alpha structure also consisting of fine subgrains?

JAFFEE: Titanium alloys of the $\alpha+\beta$ type have markedly better creep resistance in the acicular condition as produced by cooling from the beta field into the alpha-beta field. The acicular structure consists of large colonies of alpha platelets of essentially the same orientation. This approximates a large-grained alpha structure with a stable irregular grain boundary, a structure which would be expected to have and does have better creep strength than the fine-grained equiaxed alpha-beta structure produced by working and annealing in the $\alpha\beta$ field.

VERSNYDER: As Sherby has indicated, atomic mobility in titanium is higher than expected based on its melting temperature and crystal structure. Atom mobility seems to relate more to the α to β phase transformation temperature than to the melting point. The Ti_3Al and $TiAl$ alloys do have a higher elastic modulus resulting from the ordered structure, with the elastic modulus of Ti_3Al about equal to 20×10^6 psi at room temperature. However, it is recognized that creep strength does not depend simply on elastic modulus and atomic mobility, but depends also on the nature of the operative slip processes.

The improved creep properties produced by β forging of Ti-6-2-4-2 arise from a number of factors, including dynamic strain aging, microstructural size, and phase-orientation relationships.

TIEN: My understanding of the powder nickel-base alloys is that they are by nature fine grained. If that is correct, by what means are you hoping to enhance their creep resistance?

VERSNYDER: Powdered nickel-base disk alloys are fine grained. However, with maximum disk metal temperatures of less than 1400 F, disks are not creep limited. For commercial-engine applications, improvements in disk low-cycle fatigue life are being sought, whereas for military engines, improved disk materials will require greater yield and ultimate tensile strengths.

SHEWMON: Could you briefly describe the HIP approach to disk manufacture? What pressures and temperatures are involved? What powder sizes are commonly used and how are they formed?

VERSNYDER: The powder used in disk manufacturing can be produced by inert gas atomization and rotating electrode methods. In addition, soluble gas atomization is used for some nickel-base superalloys and the hydride/dehydride process for titanium. The powder, which is about −80 mesh in size, is then placed in a container, outgassed, and consolidated by hot isostatic pressing at temperatures between 2000 and 2250 F and pressures of 5 to 30 ksi. Present development efforts are concentrated on hot isostatically pressing the disk to near final shape so that raw material and machining costs can be minimized.

MAYER: There have been recent examples of failures due to fretting corrosion in nickel-base and titanium-base alloys employed in jet-engine components. Do you have ongoing research on this subject? How great is your concern about this sort of failure mechanism with respect to future alloy development and engine design?

VERSNYDER: While fretting failures have occasionally been observed, it is not considered to be a major source of component failure. As a result, only limited research is being conducted in this area.

BUSH: What are the concerns with regard to nondestructive examination? Is it surface or volumetric since the problem appears to be one of critical flaw size (N_{IC} or J_{IC}), surface or subsurface? Can significant or critical flaw sizes be determined so that the probable reliability or specific NDE technique can be set?

VERSNYDER: The chief concern with respect to nondestructive inspection is related to fatigue crack growth. From a design viewpoint, the lifetime of a part is established on the basis that an initial flaw, whose size is determined by the resolution of the NDI technique employed, not grow to a critical size in the disk lifetime. An NDI technique with improved small-flaw resolution provides for a longer design lifetime, and for the potential use of materials with higher crack growth rates or lower critical crack lengths.

McCLINTOCK: What are the fractography and microstructure of the transverse fractures in the directionally solidified alloys?

VERSNYDER: The transverse fractures of low-hafnium specimens are intergranular. The fracture surface, itself, shows a dendritic appearance corresponding to those dendrites that terminate at a grain boundary.

TIEN: Can you comment on the status of your competition's NiTaC eutectic? You mentioned that the "weak link" in your $\gamma/\gamma'+\delta$ alloy is transverse ductility. What is NiTaC's weak link or links?

VERSNYDER: NiTaC-13 and derivative compositions have undergone intensive development and evaluation in the past few years. Much of the activity, to date, has been on specimen evaluation, but increased effort on turbine-blade evaluation has been initiated. Some of the difficulties with NiTaC-13 lie in the casting area. The alloy has a melting point and a high critical G/R (G is temperature gradient in the liquid and R is solidification rate) that requires a superheat temperature in excess of what can be withstood by current core and mold materials. This high superheat temperature and the high carbon content of the alloy also produce reactivity with mold and core materials that makes cast-to-size turbine airfoils difficult. The volume fraction of the reinforcing phase, and therefore the creep strength, has been found to be a function of the fraction solidified and the strength of a NiTaC-13 blade will, therefore, vary along its length.

SIMS: The question of the state of development of NiTaC—a eutectic alloy of interest to General Electric—was raised. The answer given by Frank VerSnyder was appropriate, and I can add only that its state of developmental progress appears to be similar to that of $\gamma/\gamma'+\delta$—the Pratt & Whitney eutectic of interest.

Very importantly, a major type of equipment, in addition to steam turbines and aircraft gas turbines has failed to be mentioned. These are industrial "heavy-duty" gas turbines of the type manufactured by G.E., Westinghouse, Brown-Boveri, Sulzer Bros., and others. These gas turbines, used for electric-power generation and a variety of other industrial prime-mover power applications are intermediate in size between that of aircraft engines and that of steam turbines. Their latter stage bucketing, for instance, can be 25 to 30 inches long—yet they operate under very approximately the same general conditions as aircraft engines.

From the standpoint of this equipment, I will try to summarize briefly alloy-development needs.

As a general guideline, alloy design and modification will be directed by the need to improve efficiency and to ensure product reliability. For instance, in recent years, contaminants in the fuels have led to an industry-wide problem called "hot corrosion"—characterized by chemical surface degradation of hot-stage superalloy components. This problem is being solved by development of improved alloys and coating and cladding of the hot parts. Other specific needs are for hot-stage alloys with improved thermal-fatigue characteristics and greater strength.

Alloy development or modification is also needed because of the large size of the investment-cast hot parts. This leads to casting problems that can be assisted materially by improved alloys.

ALLOY NEEDS AND DESIGN: THE AIRFRAME

R. F. Simenz

Manager, Materials Engineering
Science and Engineering Division
Lockheed California Company
Burbank, California

M. A. Steinberg

Director of Technology Applications
Lockheed Aircraft Corporation
Burbank, California

ABSTRACT

Materials applications in aircraft are discussed, and design requirements for strength and fatigue resistance are examined to afford an insight as to the problems of selecting materials for structural applications. Examples of materials selection and application are discussed for a new generation/wide-body aircraft. The concepts of fail-safe and damage-tolerant structures are reviewed.

Alloy needs for future aircraft are expected to continue to depend on aluminum, steel, and titanium for the principal components. However, in aluminum, new developments in alloy design are needed that achieve superior fatigue and strength properties combined with immunity to stress-corrosion cracking. Aluminum alloys with enhanced high-temperature strength and stability are also needed. In titanium, future alloys should have improved fabrication characteristics to achieve lower cost structure and better metallurgical controls to provide more reliable fatigue and toughness properties. Future steel-alloy developments must provide more damage tolerant and more stress-corrosion resistant grades without compromising strength properties or fabrication characteristics.

1 INTRODUCTION

In any discussion of transportation systems, and in particular, aircraft systems, one must refer to concepts and systems prior to 1973, and the consequences of the pressure and constraints of economics, technology, and utility brought on by the energy crisis and by changing social values. To these constraints must be added those imposed by EPA, OSHA, the FAA, and other political/regulatory agencies that will be a major consideration in the future design and production of new transportation systems.

The pressures and constraints detailed originally by Stampfer of the Boeing Company are shown in Fig. 1. Our discussion is limited to the technology aspects and, more importantly, to materials and processes and the design criteria imposed on their application. However, with the great emphasis now on improvement in fuel utilization we want to examine what advanced structural materials and design can accomplish to help improve fuel utilization. A projection of this potential is shown in Fig. 2, where all technological improvements possible are projected as a function of time with their corresponding effect on efficiency.

Our discussion attempts to judge the present state of the art in design and materials application, and to project requirements based on these constraints for the future, which may give some insight as to material needs and future alloy requirements.

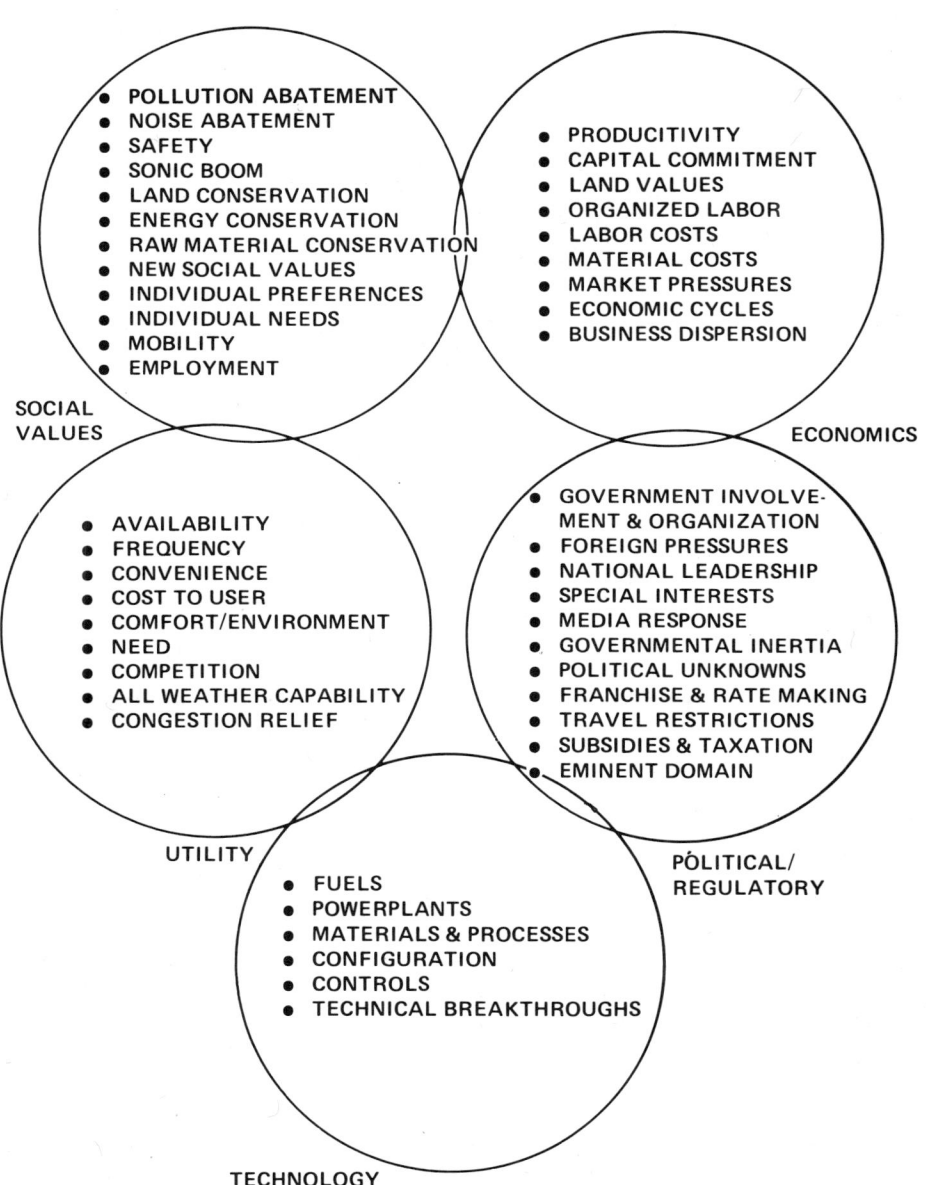

Fig. 1. Pressures and constraints of transportation.

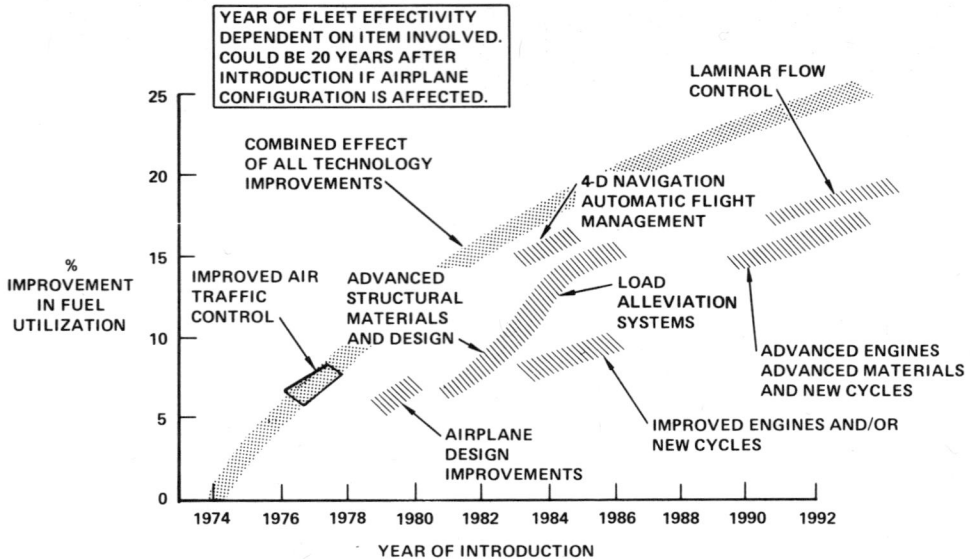

Fig. 2. Effect of technology improvements.

2 AIRCRAFT DESIGN REQUIREMENTS

In discussing materials applications in modern aircraft, an examination of general aircraft design requirements can give us an insight regarding the problems of selecting materials for structural applications. These requirements (Table I) cover the total aerodynamic spectrum through the environment, the safety of flight requirements and, finally, cost. They are described in terms of material behavior, properties, and fabrication needs familiar to all.

3 DESIGN CONSIDERATIONS

Modern aircraft structural design now accounts for five basic mechanisms of structural analysis (Table II). These are as follows:

1.Static strength and stiffness of the undamaged structure provide strength integrity for maximum loads (i.e., material properties which determine minimum weight).

2.Creep deformation and stress-rupture limits are established for parts subject to long-time elevated-temperature exposure.

Table I. General Aircraft Design Requirements

Load	Strength
Vibration	Fatigue, galling, fretting
Temperature	Creep, rupture, oxidation, corrosion, expansion, conductivity
Environment	Corrosion, stress corrosion, vapor pressure
Weight	Density
Rigidity	Modulus of elasticity, ductility
Stability	All
Safety	Elongation, toughness, reliability, uniformity
Economy	Cost-material, manufacturing, maintenance, useful life

Table II. Five Basic Mechanisms of Structural Analysis for Modern Aircraft Structural Design

1. Static strength and stiffness of undamaged structure

2. Creep deformation and stress rupture

3. Residual static strength and stiffness of damaged structure

4. Fatigue of undamaged structure

5. Life of damaged structure

3. Static strength and stiffness of damaged structure are the primary means of providing safety from unexpected and unknown damage from any source (i.e., flutter, loss of control, and excessive or uncontrollable vibration within normal operating limits).

4. Fatigue of undamaged structures is defined as the time to initiate and develop cracks to unacceptable size under cyclic loading and environments encountered in service.

5. Life of damaged structure is an essential requirement of the fail-safe damage-tolerant criteria used in design and can allow for sufficient time for inspection and repair.

Early designs were based, in the main, on ultimate- and yield-strength criteria. Design methodology evolved to include fatigue design (stress-cycling S-N concepts), strain-cycling concepts, and, finally, linear elastic fracture-mechanics procedures and fatigue-crack-growth concepts. The development of each of these design methodologies has provided another tool to aid the designer in:

- Improving structural reliability
- Establishing the design life
- Setting operating stress or strain allowables
- Component sizing and configuring
- Developing viable material, process, raw material, and inspection specifications and controls.

Each of the methodologies is somewhat different but they are all directed at the five stated objectives.

4 FAIL-SAFE OR DAMAGE-TOLERANT STRUCTURE

One of the most far-reaching structural developments of the past 15 years is the philosophy of intentional design for structural integrity under any nominal damage condition. Necessary elements in the design process include establishment of design criteria for:

- Design load levels in the damaged state
- Type and size of assumed damage for design
- Periodic inspections to find and fix any possible damage.

Static limit, ultimate, and fail-safe load levels are indicated along the vertical axis in Fig. 3. Design criteria are that zero or positive margins exist any time in the life of the aircraft. The curve concave upward indicates a typical crack growth in size (vertical axis) versus time (on the horizontal axis). As the crack grows with time, the ultimate strength may decrease according to the upper curve-concave downward. Time to initiate and grow a fatigue crack to minimum inspectable size is indicated in Fig. 3 as (1), and growth to most probable size at discovery may be at (2), allowing enough time for repair before (3) is reached.

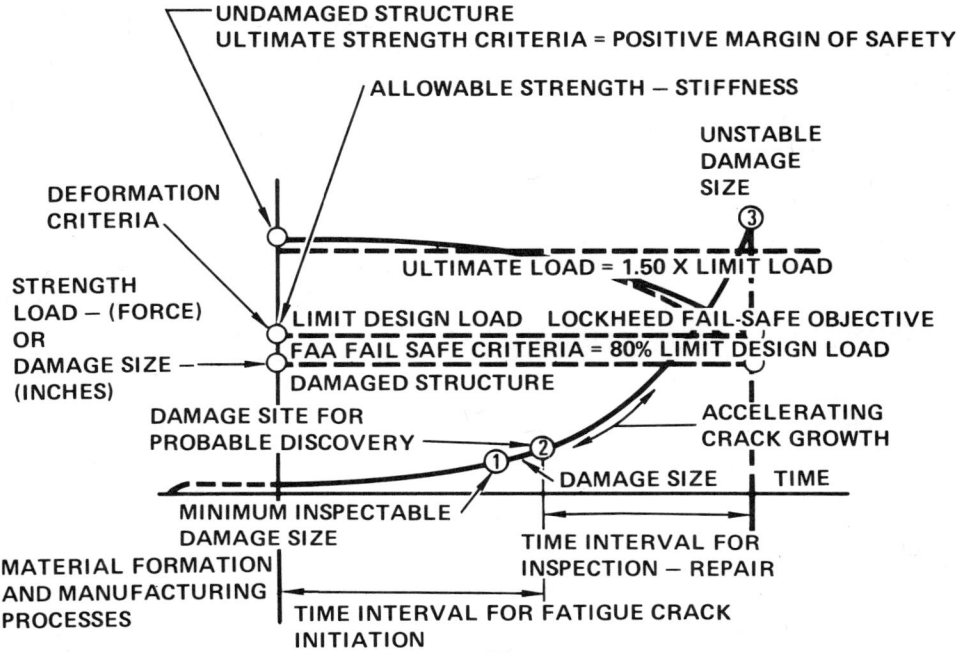

Fig. 3. Relationships of applied load history, strength, and damage size with time.

5 FATIGUE DESIGN CONCEPTS

The approach to fatigue is to design for unlimited life. To achieve this objective, criteria are established to enable the design to meet a predetermined crackfree base lifetime under operational conditions. To meet the fatigue-life requirements, high-quality detail design is essential. Fatigue tests of speciments which have failed in service on past models, along with tests of successful structural elements, confirm the selection of a minimum fatigue quality level and, by this link, tie the projected new design to actual service results on prior models. Development tests are required to demonstrate the achievement of this acceptance standard quality level in addition to demonstrating the long-life requirements. Assuming that all vital areas of the structure are going to meet the established quality standards, fatigue analyses of the structure may proceed and material comparisons, controlling design stress levels, and detail design decisions may be made with confidence.

6 RELATIVE FATIGUE DAMAGE

Joints and other discontinuities are tested under flight-by-flight spectra conditions to ensure meeting the fatigue quality acceptance standard; thus there is confidence that all the requirements for fatigue-resistant design are met. In performing fatigue analyses or tests, all of the significant flight and ground loadings are included, such as taxi, gust, landing impact, and the ground-air-ground cycle. As indicated in the graph (Fig. 4), the ground-air-ground cycle which occurs once per flight produces by far the most significant effect on fatigue life. The number of flights flown is, therefore, much more significant in the evaluation of fatigue damage than the number of hours flown. Many ground-air-ground cycles will be experienced in the short-to-medium range of aircraft operation.

Cyclic loading of cracked material, either constant-amplitude or flight-by-flight spectrum loading, provides crack-growth-rate data up to a desired crack size. Increasing tension load then defines the stability boundary [points (1) and (2), Fig. 3] between slow, stable crack propagation and rapid, unstable failure. Comparison of the stability boundaries of several materials provides a rank-ordering selection criterion—the highest stress for the largest stable crack being the better material.

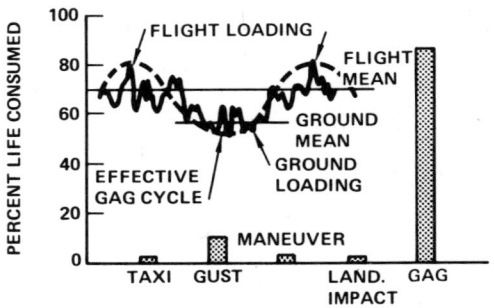

Fig. 4. Relationship between fatigue damage and the flight spectrum showing the major importance of the ground-to-air cycle as it affects fatigue damage.

7 RECENT APPLICATION OF NEW CONCEPTS TO MODERN WIDE-BODIED JET TRANSPORT AIRCRAFT

The introduction of wide-bodied jet transport aircraft was largely prompted by requirements for greatly increased load-carrying capability over previous aircraft with marked improvement in revenues. A third, no-less vital goal was to produce these new advanced technology aircraft with a significant increase in reliability over the previous generation of jet transports.

8 THE L-1011 STRUCTURAL DESIGN REQUIREMENTS

In the design of the L-1011, Lockheed established its own design requirements, augmented by airline requirements. At the onset of the design of the L-1011, the airlines were surveyed to determine the nature of structural deficiencies that they had encountered in their past operation of transport aircraft. The unanimous opinion was that there were two main structural problems—fatigue and corrosion. In general, the first-generation jet transports, while fully qualifying to all structural requirements for strength and safety, have had corrosion and fatigue problems which, throughout the economic lives of the aircraft, have caused flight delays, flight cancellations, long aircraft-out-of-service times, and high cost of repairs. Therefore structural-design objectives were established for the L-1011 which, in addition to providing additional strength, stiffness, and fail-safe characteristics for flight and safety, would also provide *unlimited life* for the primary airframe components in operational service. By this is meant that with normal operation, inspection, maintenance, and repair, it is intended that the ultimate retirement of the airplane, when it occurs, will be for reasons other than structural fatigue, such as wear and tear, economic obsolescence, accidental damage, or other unpredictable factors. An unlimited-life structure can be achieved by the proper choice of materials and processes, design stress levels, detailed design quality, and adequate corrosion protection. This allows more freedom in structural design, even though a rigid weight empty program was established. Of the various actions taken to satisfy the requirements, the commitment to employ extensive metal-to-metal bonding was probably the most important, providing significant advantages for long fatigue life, improved fail-safe capability, and corrosion resistance.

Fatigue-resistant structure is provided by design considerations that include:

- Careful detail design, particularly in such fatigue-prone areas as splices, joints, and major structural intersections and discontinuities
- Use of fatigue-resistant materials
- Extensive use of interference-fit fasteners in primary structures
- Use of adhesive bonding in conjunction with fasteners for fuselage skin splices
- Use of conservative fatigue quality index factors
- Use of low ultimate gross-area tension stresses in all primary structures
- Use of low hoop-tension allowable stresses in fuselage skins
- Extensive fatigue testing of components during design development.

9 MATERIAL SELECTION

Lockheed worked closely with material-manufacturing companies to determine the most suitable materials for the structure. An early decision was to make extensive use of clad aluminum surfaces for improved corrosion resistance and new exfoliation-resistant alloys. Details of materials application are depicted in Fig. 5.

Titanium alloys and steel forgings are used where strength and size make it necessary. Incorporating the latest technology of the aluminum industry, the T73- and T76-temper aluminum alloys were developed to significantly improve the corrosion characteristics of the 7000 aluminum series. Thus, while 7075-T6 is used for many applications of highly loaded structures, 7075-T73 or 7075-T76 is used where resistance to stress or exfoliation corrosion becomes an important consideration. These potential problems are particularly associated with wing skins machined from heavy plate stock, so that 7075-T76 aluminum alloy is used for both the upper and lower wing surfaces of the L-1011.

As proven by tests, 7075 in the new T76 temper has fatigue resistance, crack-propagation resistance, and toughness properties comparable to those of 2024-T3—an aluminum alloy noted for these characteristics—while also possessing strength properties much superior to those of 2024-T3 and only slightly less than those of 7075 in the T6 temper. Additionally, age-stabilized 7075-T76 showed stress-corrosion thresholds for each grain direction higher than those of other aluminum alloys tested. The L-1011 7075-T76 wing skins have an aluminum-zinc-magnesium (7008 alloy) layer on the external surfaces. This 7008 cladding not only provides a more corrosion-resistant surface but, unlike typical soft aluminum cladding, also responds to heat treatment so that it approaches the strength level of the base metal.

The fuselage skins are of sheet-metal gages rather than plate stock as used for the wings. Exfoliation and stress corrosion in any grain direction are not so severe a problem for the fuselage, because of the nature of the structure and its load paths. Therefore, the use of 7075-T76 for fuselage skins does not offer the advantages that it does when used for the wings. Selection of 2024-T3, with pure aluminum cladding for corrosion protection, for the fuselage skins is consistent with its long successful history in this application.

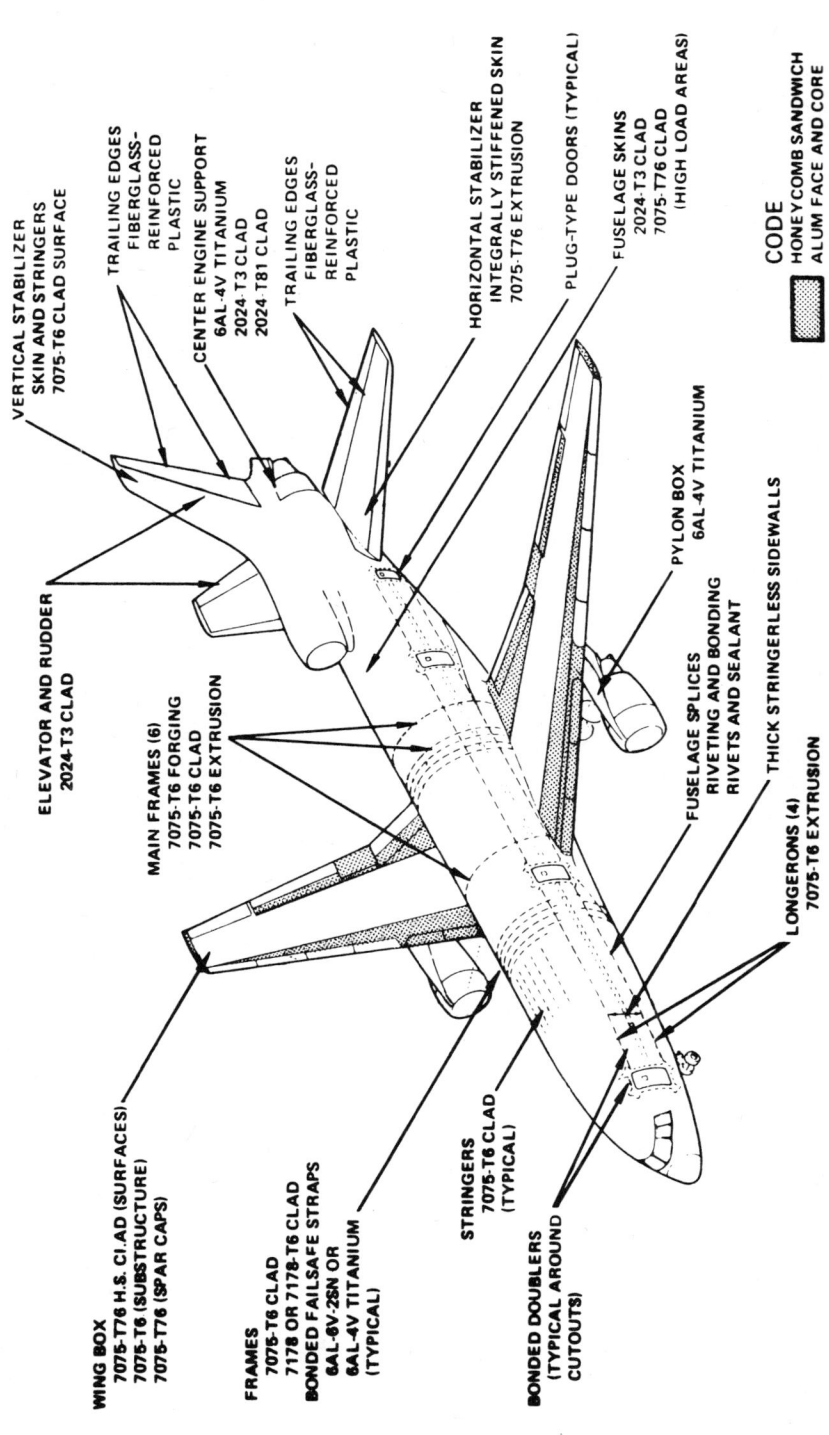

VERTICAL STABILIZER
SKIN AND STRINGERS
7075-T6 CLAD SURFACE

TRAILING EDGES
FIBERGLASS-
REINFORCED
PLASTIC

CENTER ENGINE SUPPORT
6AL-4V TITANIUM
2024-T3 CLAD
2024-T81 CLAD

TRAILING EDGES
FIBERGLASS-
REINFORCED
PLASTIC

HORIZONTAL STABILIZER
INTEGRALLY STIFFENED SKIN
7075-T76 EXTRUSION

PLUG-TYPE DOORS (TYPICAL)

FUSELAGE SKINS
2024-T3 CLAD
7075-T76 CLAD
(HIGH LOAD AREAS)

CODE

HONEYCOMB SANDWICH
ALUM FACE AND CORE

ELEVATOR AND RUDDER
2024-T3 CLAD

MAIN FRAMES (6)
7075-T6 FORGING
7075-T6 CLAD
7075-T6 EXTRUSION

PYLON BOX
6AL-4V TITANIUM

FUSELAGE SPLICES
RIVETING AND BONDING
RIVETS AND SEALANT

THICK STRINGERLESS SIDEWALLS

LONGERONS (4)
7075-T6 EXTRUSION

WING BOX
7075-T76 H.S. CLAD (SURFACES)
7075-T6 (SUBSTRUCTURE)
7075-T76 (SPAR CAPS)

FRAMES
7075-T6 CLAD
7178 OR 7178-T6 CLAD
BONDED FAILSAFE STRAPS
6AL-8V-2SN OR
6AL-4V TITANIUM
(TYPICAL)

STRINGERS
7075-T6 CLAD
(TYPICAL)

BONDED DOUBLERS
(TYPICAL AROUND
CUTOUTS)

Fig. 5. L-1011 basic materials and design features.

10 ALLOY NEEDS FOR FUTURE AIRFRAMES

On the basis of the previous discussion and summary on current airframe design and construction, let us consider alloy needs for future aircraft. Generally, aluminum alloys are expected to remain the principal materials of construction. There will be increased use of composites in competition with aluminum and continued use of steels and titanium for special design applications.

10.1 Aluminum Alloys

The principal trend in aluminum-alloy development for airframes has aimed at improved corrosion and stress-corrosion resistance, and increased fracture toughness. Improvements in these characteristics have generally been accompanied by a reduction in strength properties. This trend is clearly illustrated in Fig. 6 which indicates that the 7178-T6 composition remains the highest strength aluminum alloy available today. First used extensively 25 years ago, unfavorable stress-corrosion and exfoliation experiences have limited its application during the past 10 years. Therefore, a high-priority need exists for a replacement material which provides strength properties equal to or greater than those of 7178, and greatly improved toughness and corrosion-resistance characteristics. Such a product could be used to provide the benefits tabulated below in typical applications on a transport such as the L-1011.

Product and Application	Presently Used Material	Potential Weight Reduction with an Advanced Aluminum Alloy	Approx. Reduction in Annual Fuel Consumption[a], lb
Misc. Forgings (2000 lb)	7075-T73	300	42,000
Extrusions (5000 lb), Horizontal Stabilizer and Beam Caps	7075-T76	400	56,000
Fuselage Floor Supports (3000 lb)	7075-T76	150	21,000
Fuselage Skin (900 lb)	7075-T76	80	11,000
Upper-Wing Skin (6000 lb)	7075-T76	540	70,000

(a) Per aircraft, 3000 hours utilization per year.

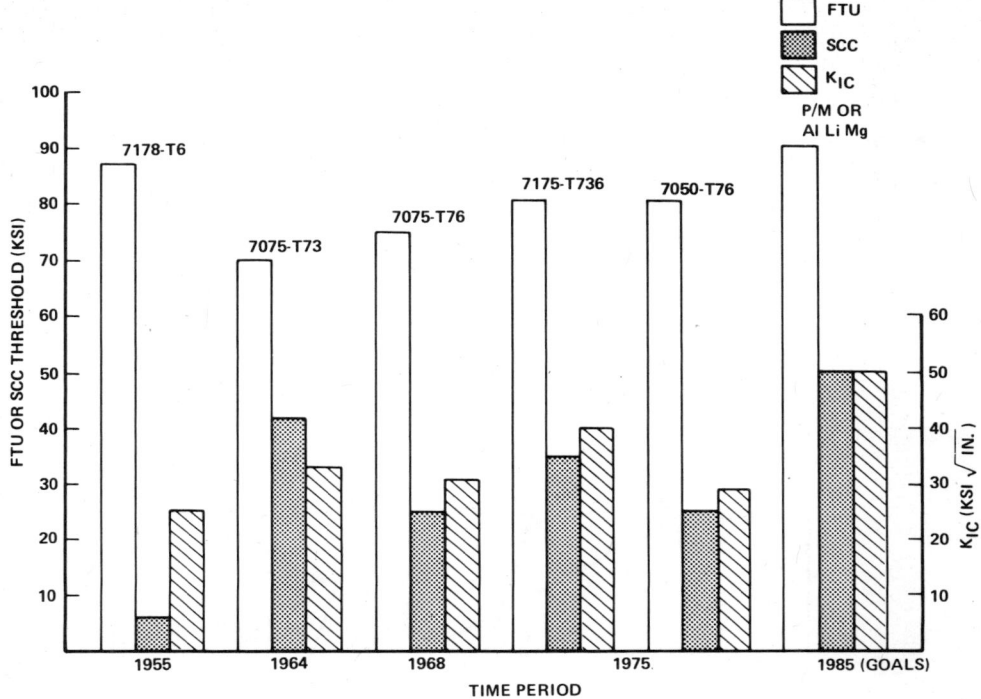

Fig. 6. Aluminum-alloy trends.

Aluminum alloys with improved fatigue and stiffness are also of great interest and would obviously translate into similar weight reductions when used in airframe applications designed to fatigue and stiffness criteria.

Ongoing research and developments in aluminum alloys that hold great interest for potential application in airframe design include the following:

- *Al-Mg-Li.* Alloys in this system have potential to provide high mechanical properties and good resistance to corrosion plus a unique combination of high modulus and low density. Work is under way[1] to obtain an optimized composition of Al-Mg-Li with controlled amounts of chromium, manganese, and zirconium. Minor additions of cadmium and beryllium are being investigated as a means of increasing aging kinetics.

- *Retrogressive Aging.* This process has been shown to substantially improve resistance to stress corrosion in 7000 series alloys while retaining original strength properties. It is believed that the retrogression treatment disperses networks of dislocations responsible for susceptibility to stress corrosion, and also partially redissolves precipitates responsible for the previous hardening. Subsequent reaging at the age-hardening temperature recovers original maximum strength.[2]

- *Powder-Metallurgy Processing.* Results to date (Fig. 7) indicate potential for forgings and extrusions with maximum fracture toughness, stress-corrosion resistance, and exfoliation resistance for a range of specific yield strengths.[3]

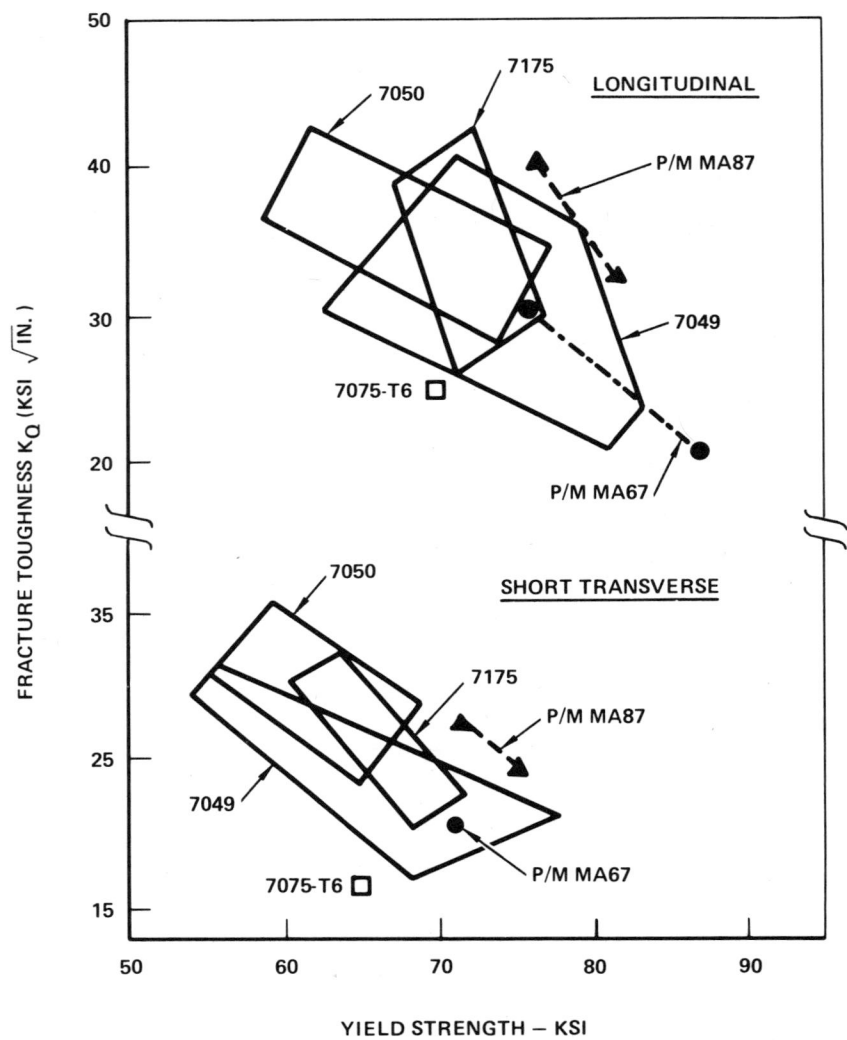

Fig. 7. Comparison of fracture toughness of P/M MA67 and MA87 die forgings to commercial I/M alloy forgings.

• *Controlled Solidification Processes.* Initial results indicate that specialized high-cooling-rate processes such as splat cooling are very effective in controlling solidification structure. High-cooling-rate processes can overcome conventional solidification-process difficulties of segregation of alloying elements, impurities, and excess phases.[4] Very coarse dendrite and grain structure can also be avoided. Splat cooling of 2024 has been shown to completely surpress formation of the brittle complex CuMgFeAl phase. Absence of this phase undoubtedly was a factor in the sevenfold increase found in fatigue life of splat-cooled 2024 when compared with ingot-processed 2024 (Table III).

Table III. Room Temperature Properties of Aluminum Alloy 2024-T4

Product	Yield Strength (0.2% Offset)		Ultimate Tensile Strength		Elongation, %	Reduction in Area, %	Fatigue Life, number of cycles at 30,000 psi (21.2 kg/mm^2)
	Ksi	Kg/mm^2	Ksi	Kg/mm^2			
Ingot bar	40.2	28.5	67.2	48.7	23	37	100,000
Air atomized	42	29.7	70	49.5	25	33	300,000
Splat atomized	47	33.3	76	54	24	34	700,000

10.2 Need for Improvements in Fatigue of Joints

The inherent fatigue strengths of conventional production alloys are all quite similar. While attempts to improve fatigue properties using a metallurgical approach have been successful to some extent, it has yet to be shown that the notched fatigue properties can be improved. Thus, until current R&D to improve fatigue produces a breakthrough which is commercially viable, the airframe designer is very limited in his capability to gain fatigue improvements through alloy selection since notched properties govern design of most components.

However, the aircraft industry has achieved considerable fatigue improvement in aluminum joints through sophisticated approaches to fastener systems and hole conditioning. Adhesive bonding can provide the optimum joint for a long fatigue life but its use is limited by certain production constraints. For applications where bonding is not practical, Fig. 8 illustrates the vast improvement attained in riveted joints by the use of a special rivet configuration. Similarly, Fig. 9 shows the benefits of interference-fit fasteners and hole conditioning.

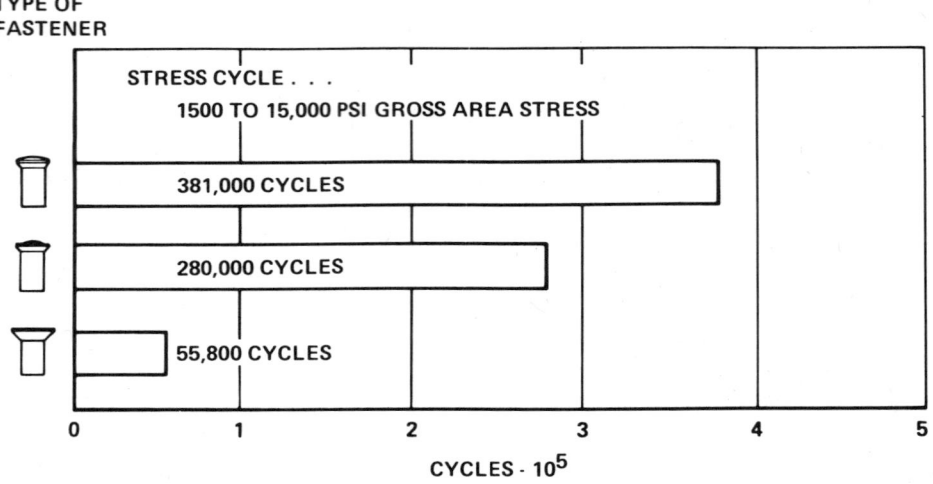

Fig. 8. Constant-amplitude fatigue tests of lap joints.

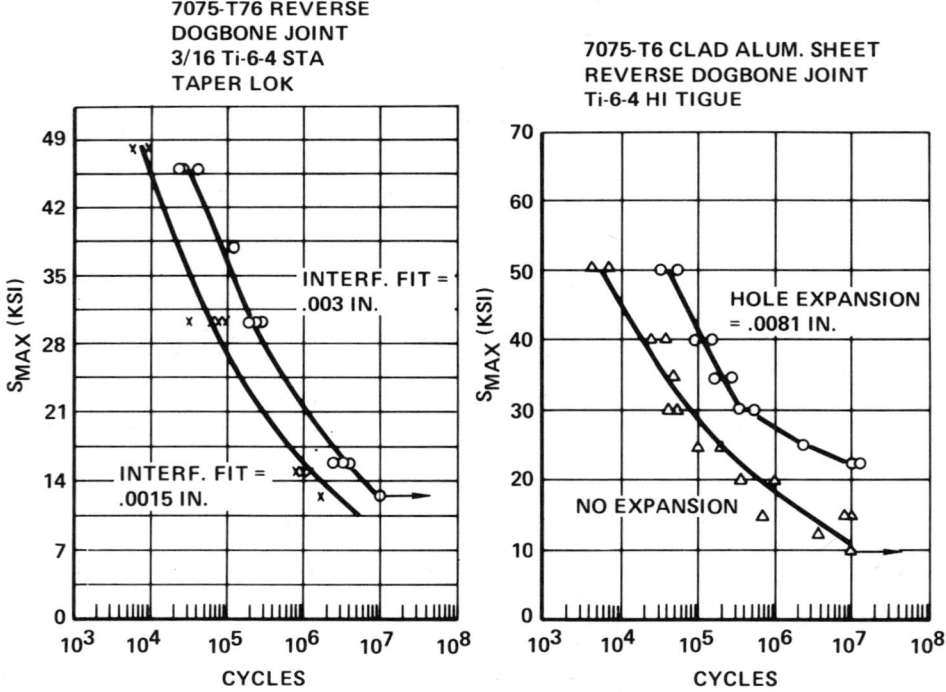

Fig. 9. Effects of amount of interference fit or hole expansion on fatigue life of joints.

All of these techniques for improving the fatigue life of fastened joints largely involve creation of favorable residual stresses in the vicinity of the hole (Fig. 10). It has been found[6] that these procedures result in residual stresses that may be considerably higher than the stress-corrosion threshold stresses for many currently used airframe aluminum alloys (Fig. 11). Thus, further progress in fatigue improvement in fastened joints is impeded by inadequate stress-corrosion resistance of many current alloys in current tempers.

Our technology for improved fatigue is not nearly so developed or successful for complex mechanically fastened joints. For example, predictable fatigue improvement is very difficult in multiple stackups involving differing principal grain orientations, different mill products, and dissimilar metals, including combinations of aluminum alloys, steel, titanium, and composites. The aircraft industry would welcome some attention by the metallurgical community to help gain further fatigue improvement through R&D of fastened joints to complement efforts in basic alloy design aimed at improved fatigue properties. The work by Nam Suh presented at this conference on microstructural and chemical effects in wear and fretting is an excellent example of fundamental research that should prove very helpful in gaining an understanding of mechanisms of failure initiation in complex joints. The potential is obvious for greater progress and more rapid implementation of results through bridging the scientist and engineer approaches in this area of wear and joints.

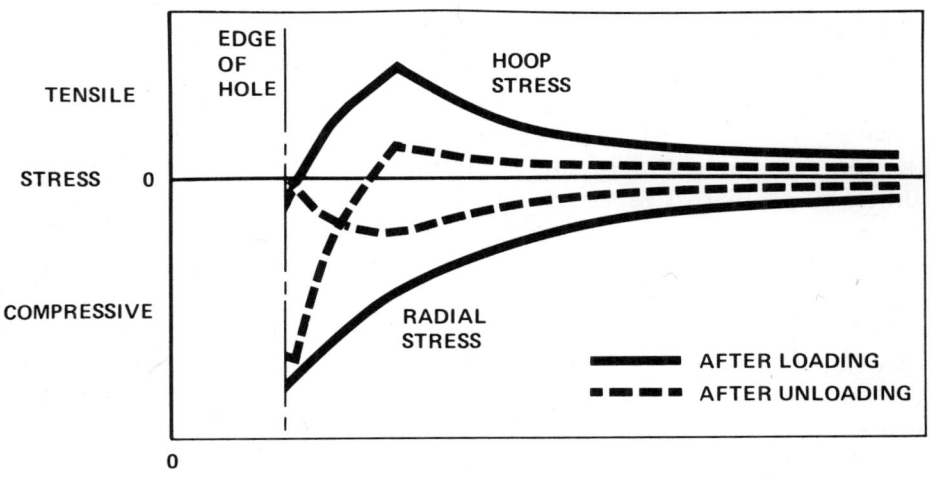

Fig. 10. Stress fields in an aluminum alloy due to hole expansion.

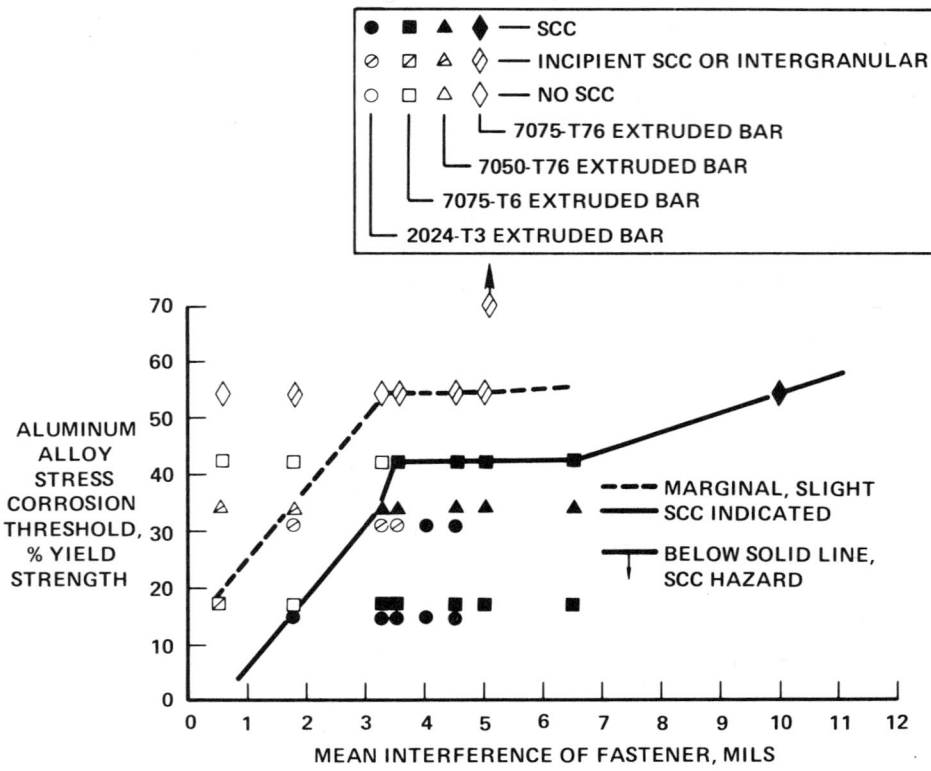

Fig. 11. Estimated stress-corrosion potential for interference fit fasteners in aluminum alloys.

10.3 Airframe Needs for Steel Alloys

Another perspective of alloy needs can be gained by a review of service problems with materials used in current aircraft construction. Failures experienced with current aluminum, titanium, and steel materials most often result from residual stress, stress corrosion, cracking in mechanical fastened joints, misprocessing, poor surface condition, corrosion, or poor detail design which introduces stress concentrations such as sharp corners and abrupt stiffness changes.

Obviously most of these problem sources cannot be directly remedied by providing the aircraft designer with an improved alloy which has been optimized only for metallurgical features concerning microconstituents, morphology, atomic bonds, aging kinetics, and defect densities.

Thus, when speaking of alloy design to meet airframe needs, we have the situation first described by Paul Kuhn, and amplified by Grover[7], which compares scales of observation of structural design and classification of failure, as shown in Table IV.

Table IV. Comparison of Scales of Interest in Structural Design and Alloy Design

| Area of Interest | Scale of Consideration | | Causes of Failure | |
	Unit	Approx. Size, inches	Mechanical	Metallurgical
Preliminary Design	Foot	1 x 10	Overload/improper stress	Material choice
Detailed Design & Development	Inch	1	High stress concentration	Metallurgical choice
Engineering Research	Millimeter	4×10^2	Details of stress concentration, inclusions, internal stress	Inclusions, grain size, internal stress
Metallurgical Research	Micron	4×10^5	Dislocation stress	Dislocation array
Physics Research	Angstrom	4×10^6	Atomic forces	Atomic forces

If we are to make significant progress, greater effort is needed to bridge the spectrum of causes of failure which has the engineer concerned with one half of this scale and the metallurgical scientist concentrating on the other half of the classification scale.

The state of the art in high-strength steels is a case in point. Virtually all large aircraft for the past 20 years have used, and will continue to use, high-heat-treat (260 ksi min) low alloy steels, primarily of the 4340 or 300M grades, because these steels offer the best combination of structural-strength and fatigue-strength efficiency at moderate costs. Successful use of these steels is achieved through precise design practices and controls, and very careful attention to all stages of processing and fabrication. Experience with thousands of HHT steel parts has evolved empirically derived limits on sustained stress levels to avoid stress-corrosion cracking. As shown in Fig. 12, sustained stresses in short transverse grains may typically be limited to only 25 percent of the yield strength to avoid SCC. Clearly, steels capable of much higher thresholds would be very welcome to the aircraft designer.

Similarly, fracture-toughness-related properties, K_{Ic} and K_{Iscc}, of commonly used low-alloy steels show considerable room for improvement (Fig. 13). Figure 13 also indicates the trends in alloy development, which certainly are in the proper direction.

The past 20 years have seen numerous unsuccessful attempts at alloy design to obtain new, improved, high-strength steels to replace the currently used low-alloy steels. One reason for this lack of success has been the failure to adequately consider the importance of the "engineering end of the classification scale" wherein 300M- and 4340-type materials provide capability for readily attaining consistent, high integrity in large parts through highly developed melting, forging, heat-treating, and other practices. Too often laboratory alloy developments have been prematurely

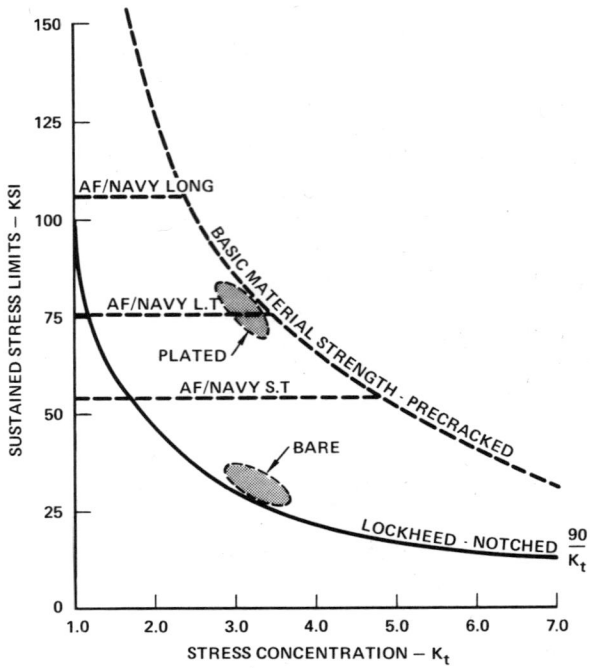

Fig. 12. Corrosive environment sustained stress allowables—high-strength steel.

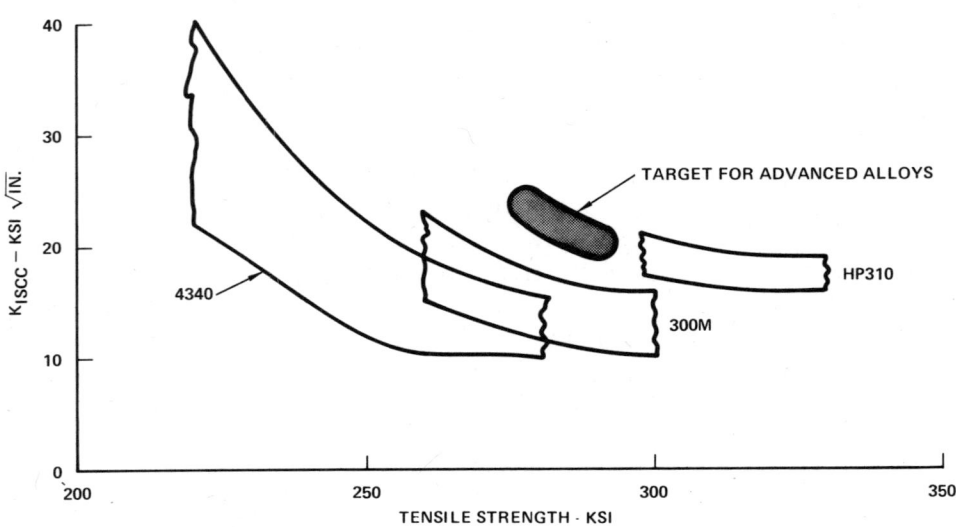

Fig. 13. Threshold stress intensity K_{Iscc} versus strength for high-strength landing-gear alloys.

touted for their "significant breakthroughs" in stress-corrosion-cracking fracture-toughness properties, only to find this improvement has been attained at the expense of such a drastic sacrifice in processing and producibility that it precludes the alloy from ever reaching production status. The aircraft industry is extremely interested in alloy development of improved damage-tolerant and more stress-corrosion-resistant high-strength steels; therefore, we urge those engaged in such alloy development to include producibility criteria in their development parameters such that processibility of new materials at least approaches that of current alloys (see Fig. 14).

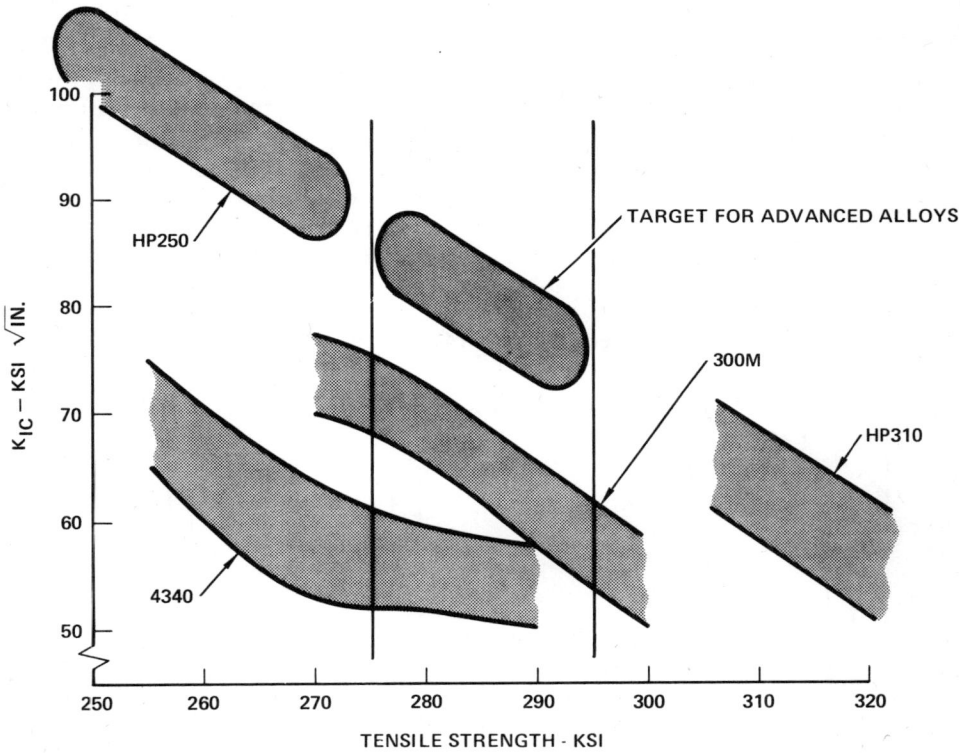

Fig. 14. Fracture toughness versus strength for high-strength landing-gear alloys.

11 CURRENT R&D TRENDS AND
AIRFRAME TITANIUM-ALLOY NEEDS

Improvement in present alloys is being sought through the use of cleaner master alloys and improved melting procedures. Higher strength alloys are being investigated through interstitial hardening of beta and alpha-beta alloys and by developing alloys with modulated microstructures. The raw-material cost of an aluminum fighter is 1 percent of the flyaway cost; for 100 percent titanium, the raw-material cost is 3.5 percent of the flyaway cost. If titanium-raw-material costs were halved, the flyaway cost of an all-titanium fighter aircraft would change by only 1 percent. It is apparent that raw-material cost of titanium is not of major economic importance, and that cost of fabrication is the significant factor. Improved cold-formable and age-hardenable beta alloys have made their appearance. Further improved performance is expected from alloys now in development. These new developments are expected to expand the use of titanium through lowering fabrication costs and increasing the utilization ratio, as seen in Fig. 15.

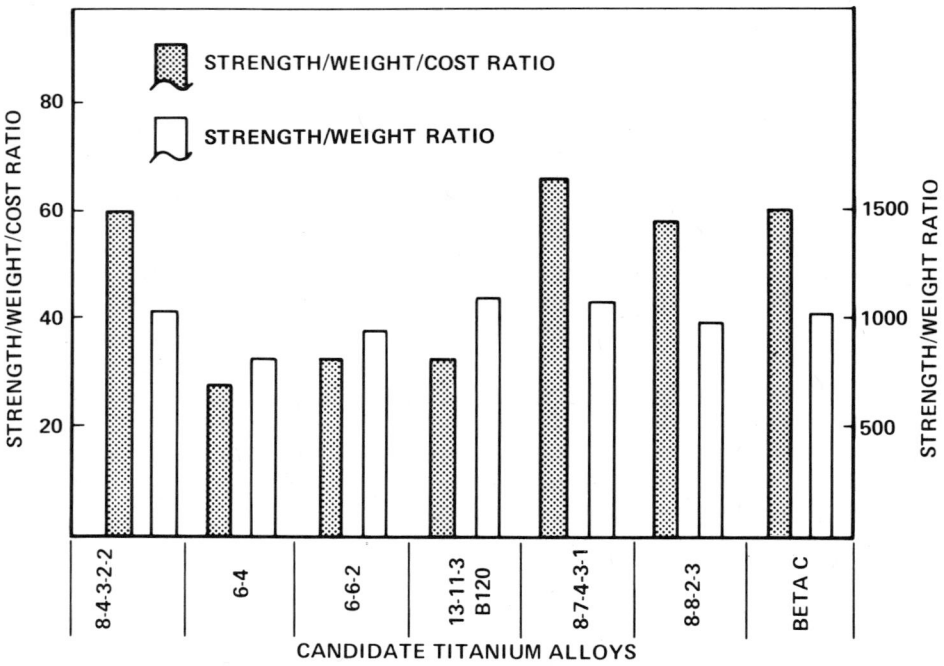

Fig. 15. Strength/weight and strength/weight/cost ratios for candidate titanium-alloy strip products.

It is not only in supersonic aircraft that titanium can be used to advantage in airframes. The use of titanium will increase with the trend to larger commercial aircraft. The longer sections and spars of the larger aircraft have rigidity requirements beyond the capacity of aluminum alloys. Titanium alloys with elastic-modulus values from 50 to 80 percent higher than those of aluminum alloys and possessing increased strength and corrosion resistance represent optimum materials for airframe construction.

The strength of conventionally heat-treated alpha-beta alloys and beta alloys is being substantially increased to the 250-ksi level by texture hardening and thermomechanical treatments, and by producing modulated microstructures in current commercial alloys. Finally, deep-hardenable alpha-beta alloys are being explored as replacements for high-strength steels in landing gear.

12 THE FUTURE OF ALUMINUM ALLOYS

The major structural airframe materials will still be the aluminum alloys in subsonic or slightly supersonic aircraft, because they will be improved until they are the most cost-effective material.[8] It is expected that at least 50 percent of the airframe weight will come from the improved aluminum-alloy systems. The major developments will be in modifications of existing alloys and the improvement in processing and purity of the high-strength aluminum alloys, as well as in thermal-mechanical treatments. These developments may produce an increase in strength to 100 ksi and an increase in stress-corrosion-cracking resistance to about 60 ksi, with a tenfold improvement in fatigue life.

To achieve these improvements, the answers to the following questions must be obtained:

1. What microstructural features control corrosion fatigue and fretting fatigue in Al-Cu and Al-Zn-Mg alloys? How much of the "fatigue problem" in aluminum is actually corrosion fatigue? What are the basic mechanisms of fatigue in precipitation-hardened aluminum alloys?

2. The optimum heat treatments for stress-corrosion-cracking resistance and for achieving fatigue and tensile strength are not compatible in Al-Zn-Mg alloys (such as 7075). How can these be brought closer together through alloying, heat treatment, or thermomechanical treatments?

3. How can the high-temperature strength (up to 300 F) and stability of aluminum alloys be raised?

4. What additional metallurgical and stress-state factors can be explored to achieve increased fatigue performance in complex mechanically fastened joints?

5. What are the relationships of the insoluble phases found in commercial aluminum alloys to fatigue-endurance limits of these alloys? What effect do they have in crack initiation and crack propagation? How are these particles formed in the melt? What are the partition coefficients of the various elements that make up these particles? And what effect do they have on the remainder of those elements that participate in age hardening? Finally, what is the relation of these particles, their size and size distribution, to melting and casting variables such as ingot size, mass of material cast, and rate of cooling? And what happens to them in ingot breakdown and subsequent mill processing?

6. Do we know anything about inclusions and their effect on properties, including fatigue? We know that reducing iron, silicon, and chromium can improve fracture toughness but does this necessarily improve endurance limit? Or is it the size and size distribution that have the major effect?

To answer these questions, we must determine the basic mechanisms of fatigue and corrosion fatigue of precipitation-hardening aluminum alloys. Concurrently, thermomechanical treatments, the use of clean alloys, and surface protection techniques must be evaluated as means for improving the fatigue resistance under various stress and environmental conditions.

Besides improvement in alloy chemistry and microstructure, we are in dire need of better test methods to develop an economical, reliable test method for evaluating crack-growth resistance behavior of thin and transitional-thickness materials. Also, optical and photographic instrumentation principles are needed to enhance data acquisition in the field of fracture-fatigue and corrosion testing.

13 THE FUTURE OF HIGH-STRENGTH STEELS

Since current airframes do employ a substantial weight fraction of steel, any aircraft of the near future will most probably continue to specify steel in the landing gear as well as in a variety of other applications.

The versatility of steels provides a most attractive basis for development that many competitive materials do not possess. Many high-strength and ultrahigh-strength steels are readily formed and joined. A major concern in employing high-strength steels is that nondestructive flaw-detection capability is not compatible with the critical crack sizes associated with the high-strength levels.

If steel alloys are to maintain their current competitive position, alloy development efforts must provide answers to the following concerns:

1. How can useful strength be increased? Are sufficient metallurgical principles available to provide alloys with increased fracture toughness and stress-corrosion resistance without degrading tensile strength? Will such alloys maintain adequate producibility and processability?

2. Have the critical combinations of hydrogen and stress level been determined? Are these conditions compatible with the ability to control hydrogen content?

3. What are the quantitative relationships among machining conditions, amount of untempered martensite, fracture toughness, and fatigue-crack-initiation behavior?

4. Can improved inspection techniques be developed to more readily detect critical flaw sizes in steels at strengths of 260 ksi and above?

14 THE FUTURE OF TITANIUM ALLOYS

To ensure the growth of titanium applications in the decade ahead, a number of problems must still be solved. A few which may indicate the future trends of research and development in the field of titanium metallurgy include the following:

1. What quantitative features of the microstructure and the environment control the fracture and fatigue processes in alpha-beta and beta titanium alloys? What features will improve the uniformity of fatigue and toughness properties?

2. How can we increase the modulus of titanium alloys, particularly in beta alloys? Texture strengthening and alloying with boron are indicated solutions.

3. What are the controlling hot-deformation mechanisms at temperatures above and below the beta transus? How can formability be improved by influencing these mechanisms as through alloying or strain-rate control?

4. What methods are commercially practical for forming net shapes of titanium components to avoid expensive machining and material scrap costs? Conversely, what can be done with titanium scrap to recoup part of this expense?

5. What alloy constitutes the ideal for low mill production costs due to optimum strip-rolling characteristics combined with room-temperature formability and competitive strength/weight ratios with current alpha-beta alloys?

REFERENCES

1. Evancho, J. W., NADC Contract No. N62269-74-C-0438, Alcoa Technical Center (1975).
2. Cina, B., U.S. Patent 3,856,584 (December, 1974).
3. Otto, W. L., AFML Contract F-33615-74-C-5077, Alcoa Laboratories (April, 1975).
4. Grant, N. J., and R. M. Pelloux, *Liquid-Metal Atomization for Hot Working Preforms,* Metals and Ceramics Information Center (October, 1972).
5. Grant, N. J., paper presented at Third NORDIC High Temperature Symposium, M.I.T. Center for Materials Science and Engineering (1972).
6. Kaneko, R. S., and Simenz, R. F., paper presented at ASTM 78th Annual Meeting, Symposium on New Approaches to Stress Corrosion (June, 1975).
7. Grover, H., *Mechanical and Metallurgical Causes of Failure,* Source Book in Failure Analysis, ASTM (October, 1974).
8. Steinberg, M. A., *Materials of the Future-Metals,* Centennial Volume, AIME, New York (1971).

DISCUSSION on Paper by R. F. Simenz

TIEN: Do you see mineral availability (i.e., cost) as a consideration in the forced selection of aluminum alloys over titanium alloys or vice versa within, say, the next decade or so?

SIMENZ: I don't believe cost will force selection of aluminum over titanium to any greater extent than it already does. However, I do see availability of critical alloying elements, such as vanadium, as a constraint in the selection of alloying systems to be used in alloy design of new commercial titanium alloys with improved formability, etc.

Related to your question is the situation which has occurred in aluminum-alloy development. Recent work has shown conclusively that high purity can be very beneficial to fracture toughness, crack-propagation rate, and even stress-corrosion resistance. However, the viability of commercial production is very questionable for high-purity (low iron and silicon) aluminum alloys because of present requirements for large scrap recycling.

THOMAS: I'd like to make three comments:
(1) The relatively low modulus of aluminum limits the potential for obtaining very high strengths (theoretical $\sim G/20$).
(2) The utilization of current high-strength aluminum alloys by precipitation strengthening is limited by the fundamentals of nucleation and growth of the phase transformations which tend to give heterogeneous microstructures (e.g., Al-Mg-Zn type alloys, grain-boundary heterogeneities, and GP zones), especially at maximum age hardening, which can cause high notch sensitivities, so K_{Ic} or plastic-zone size becomes more important than strength.

(3) New, economical, high-strength steels with 300,000-psi tensile strength and 80 ksi $\sqrt{\text{in}}$. K_{Ic} can be made simply by quenching to martensite, even without tempering, e.g., in a simple Fe/4Cr/0.35C alloy. (The data are given in my paper in this Colloquium proceedings.)

SANDERS: I would first like to comment on the recent trends associated with aluminum-alloy development. Initially, aluminum alloys were produced by manipulating microstructures to optimize static strengths. However, other design criteria such as fracture toughness, fatigue-crack growth, and resistance to stress-corrosion cracking have necessitated modifications to the maximum strength microstructures. Generally, to optimize for these important criteria, slight reductions in static strengths were required. It has been the goal of the aluminum-alloy designer to alter the microstructures in such a way that there will be little sacrifice in strength while producing alloys which have high fracture toughness. Also, I would like to comment on the state of the Al-Li-Mg alloy system. As pointed out in the discussion, a combination of high modulus, low density, and medium strength makes this alloy system attractive in providing a potential weight saving in certain aircraft systems. The strengthening mechanisms in a commercially significant temper are associated with the decomposition and precipitation of coherent σ', a Al_3Li precipitate having a structure of the ordered Cu_3Au type, and solid-solution strengthening of the matrix by magnesium. Much work has been done on this sytem in the Soviet Union and has led to the development of the alloy 01420. Work in the United States is being supported by the U.S. Naval Air Systems Command and is being conducted at Alcoa Laboratories. Modulus, density, and static strength have met expectations; however, toughness has been below that of existing commercial aluminum alloys. The work that is anticipated during the next contract period will be directed toward isolating the microstructural feature(s) that control fracture toughness, and using this we hope to produce an alloy that will have acceptable fracture-toughness properties.

MAYER: In terms of microstructural effects, defect structure, etc., will you describe the effects of your improved heat treatments in aluminum alloys that enhance SCC resistance and fatigue resistance (at the same time maintaining strength level), and what your future plans for improvement of such properties along these lines are?

I take it that you do not have much confidence in the reproducibility and/or durability of your adhesive bonds (since you build in much structural redundancy with fasteners). Have you done much work on production aircraft with diffusion bonding (especially with reference to titanium alloys)?

SIMENZ: There are still several points of view regarding the chemical, electrochemical, metallurgical, and microstructural features associated with SCC in Al-Zr-Mg-Cu-Cr. In most instances, however, grain constituent of 7075-T6 is not attacked but the margin along the grain is. This margin is at least partially depleted of copper as compared with the grain. Overaging causes reduction of copper in the grain and grain margins to the same low value, thereby minimizing anodic electrochemical action and contributing to excellent resistance to stress-corrosion cracking.

It has also been observed that the larger grain-boundary precipitate particle size and greater interparticle spacing of overaged material correlates with reduced stress-corrosion susceptibility. Overaging also disperses dislocation networks believed associated with stress-corrosion susceptibility.

I believe the following background on the selection of adhesive bonding for the L-1011 will help answer your question regarding what you termed redundancy in our approach to adhesive bonding. Early in the L-1011 program, we determined that the two most serious problems that have hampered in-service operation by the airlines are fatigue and corrosion. Thus, the goal was set to design the L-1011 Tristar with an airframe durable enough to promise unlimited structural life in normal operational service, and relative freedom from corrosion.

The commitment to employ extensive metal-to-metal bonding was one of the most important actions providing significant advantages for long fatigue life and improved failsafe capability and corrosion resistance. We have complete confidence in our adhesive-bonding system. It has been thoroughly proven in a wide range of accelerated and real-time environmental tests and, in addition, extra protective measures of primer, topcoat, and sealant are used. Service performance has been excellent.

It was a basic design consideration to provide sufficient mechanical fasteners in all riveted and bonded joints to carry 100 percent of the ultimate static load, ignoring any strength contribution from bonding. The presence of full static strength capability is thus assured in one mechanical joining system. The adhesive-strength properties are such that adhesive strength alone also exceeds ultimate joint-strength requirements. The bonding cannot be overloaded and so it is always available to improve fatigue life and to provide an environmental barrier protection for mechanical fasteners and the interior of the joint.

Our work on diffusion bonding has been confined to research and development projects.

NEEDS IN ALLOY DESIGN
FOR NUCLEAR APPLICATIONS

A. L. Bement

Massachusetts Institute of Technology

ABSTRACT

In developing reactor-core alloys for current light-water-reactor systems, which operate at relatively low temperatures, primary consideration has been given to compatibility with the coolant environment, nuclear economy, reasonable strength at the operating temperature, and fabricability to both exacting product specifications and quality control standards. These criteria alone, however, have narrowly limited the choice of candidate alloy systems. Furthermore, thermal stability of the alloy system has not been a serious problem for these reactors, and it has been generally possible to design around radiation-induced changes in properties.

In advanced reactor systems, the alloy design requirements are more stringent because of higher operating temperatures; higher thermal cycles during off-normal reactor operation and shutdown; reduced compatibilities

257

with coolants, fuel, and fission products; higher surface-to-volume ratios; and thinner sections for heat-transfer surfaces. Radiation-induced creep, growth, volume swelling, embrittlement, and atomic interchanges require special attention in the design analysis for these reactor systems.

For most alloys in metastable equilibrium, the thermodynamic relationships are often complex. Neutron radiation can affect alloy reactions through (1) destruction of order, (2) breakup of precipitates, (3) enhanced irradiation, (4) enhanced diffusion, and (5) retarded diffusion. Most efforts to date in alloy development to counter these effects have involved relatively modest changes in the composition, melt practice (residual element control), or thermal-mechanical treatments of commercially available alloys developed for other applications. However, new understandings of alloy behavior in neutron environments show promise of guiding the development of improved alloys which will have reduced susceptibilities to undesirable radiation effects.

1 INTRODUCTION

The development of advanced nuclear power-generating systems is responsive to the general missions of achieving both higher fuel conservation, through an increased fissile-isotope conversion ratio, and higher thermal efficiencies as a result of increased primary-coolant-outlet temperatures. Hence, there is an evolution from light-water "burners" to gas-cooled "advanced converters" and to either liquid-sodium-cooled or gas-cooled "fast breeders". The accelerating development of controlled thermonuclear reactors provides a growing expectation that a replenishable form of energy and/or fissile isotopes will be available shortly after the turn of the century.

This evolutionary path not only places increasingly difficult performance and reliability requirements on structural materials exposed to the radiation environment, but also places increasing economic benefits on the design of improved alloys other than those in widespread commercial use. This is necessary to stress since, with only a few rare exceptions (such as the development of the Zircaloys), the nuclear market has not warranted the expense of an alloy-development program on the part of suppliers to meet specific requirements. Therefore, most product specifications represent relatively minor deviations from established commercial practice. However, the rigors placed on alloy performance by higher service temperatures, particle fluxes, and neutron-energy spectra suggest that this trend may not continue into the future. Therefore, the challenge to the materials scientist and engineer is to carefully define the relationships among steady-state alloy phase constituents, defect structures, changing impurity compositions, mechanical and physical properties, and compatibilities with surrounding environments. But more than that, they must not only translate these interrelationships into requirements for improved product and process specifications but must also develop the data necessary to substantiate design principles for the use of these new materials. The costs of developing these data will be considerably greater than those for nonnuclear systems because of the enormous expense of neutron irradiation and remote testing and the added costs for special test reactors, radiation sources, and handling facilities needed to conduct such large-scale development programs.

2 SYSTEMS AND GENERAL CHARACTERISTICS

Before proceeding with a discussion of materials requirements and problems, it is worthwhile to distinguish among various fission- and fusion-reactor options. Fission reactors can be divided into two general categories on the basis of the conversion ratio: that is, the ratio in the number of fissile atoms produced by transmutation from fertile isotopes to the number of fissile atoms "burned" in fission. Referring to Table I, converters or "burners" have a conversion ratio less than one; therefore, they consume more fuel than they produce. Light-water reactors (boiling- and pressurized-water reactors), which fall under this category, have already achieved a high degree of commercial acceptance and application in both the United States and abroad for electrical power generation. However, the thermal efficiency of these reactors is only 30 to 33 percent as compared with 38 to 41 percent for modern fossil-fueled steam plants. High-temperature, gas-cooled reactors (HTGR's), for which technology is currently being demonstrated with

Table I. Fission Reactors—Technology Options

Reactor Type	Conversion Ratio	Primary/Secondary Heat Removal Media	Nominal Outlet Coolant Temp, C	Thermal Efficiency, %
Fission Converters (Burners)				
Light Water (BWR,PWR)	0.55	H_2O/H_2O	285 (BWR), 320(PWR)	30-33
Gas Cooled (HTGR)	0.6-0.9	He/H_2O	730	40+
Fission Breeders				
Fast Breeders (U^{238}-Pu^{239})				
Sodium Cooled (LMFBR)	1.2-1.4	Na/H_2O	650	38-41
Gas Cooled (GCFR)	1.55	He/H_2O	550	40+
Thermal Breeders (Th^{232}-U^{233})				
Light Water (LWBR)	1.0+	H_2O/H_2O	320(PWR)	30-33
Molten Salt (MSBR)	1.0+	BeF_2-LiF-ThF_4-UF_4/$NaBF_4$-NaF	500	45

the Peachbottom and Fort St. Vrain reactors, have the advantages over light-water reactors of a higher conversion ratio (0.6 to 0.9) and a higher thermal efficiency (greater than 40 percent, depending upon the thermodynamic cycle).

Thermal breeders which operate on a Th^{232}-U^{233} cycle and fast breeders which operate on a U^{238}-Pu^{239} cycle are the two general categories of breeder reactors (having a conversion ratio greater than 1.0). Light-water (LWBR) and molten-salt (MSBR) thermal breeder reactors have a conversion ratio only slightly greater than 1.0. Therefore, while they can sustain their own fuel resupply indefinitely, they cannot provide fuel in any significant quantities for new reactors.

Fast-breeder reactors, by virtue of their higher conversion ratios, can supply sufficient excess fissile material to fuel a new reactor about every 7 years. (Recent experience and analyses for current demonstration fast-breeder reactors reveal that this time period will probably be significantly longer, however.) Therefore, these reactors have the major advantages of not only conserving natural U^{235} reserves but also greatly reducing fuel-enrichment requirements. Sodium-cooled fast-breeder reactors (LMFBR's) are currently in an advanced stage of development and have the earliest date of potential commercial availability (~1986). However, even when fully developed, LMFBR's will have a thermal efficiency not much different from present fossil-fueled plants. Therefore, thermal release to the environment will also be comparable. Gas-cooled fast reactors (GCFR), which are currently under study and early development as an alternative breeder concept, have the potential advantage, when operated under a Brayton cycle, of a substantially improved thermal efficiency over steam supply systems.

The various fusion-reactor schemes under both study and active development are outlined in Table II. These can be categorized as toroidal

**Table II. Fusion Reactors—
Technology Options**

Toroidal Devices

- Diffuse pinch
- Stellarator
- High-β
- Tokamak

Magnetic Mirror Machines

- Minimum B trap
- Direct conversion

High-Density Schemes

- Laser ignition
- Electron-beam ignition

devices, magnetic-mirror machines, and high-density ignition schemes. Among the various options listed, the Tokamak has come nearest (~10 percent) to satisfying the Lawson criterion* and is generally looked upon as the configuration most likely to first achieve a sustained thermonuclear reaction. Advanced Tokamak experiments are currently being constructed with the high hope of demonstrating scientific feasibility of sustained fusion within the next 2 to 5 years.

Three of the various fission- and fusion-reactor options outlined above, namely, the light-water reactors (pressurized-water reactor, PWR, or boiling-water reactor, BWR), the liquid-metal fast-breeder reactor (LMF-BR), and the Tokamak fusion reactor, will suffice to focus attention on the increasingly difficult challenges to design and materials engineers in providing reliable components and systems for the operating environments corresponding to each concept. Therefore, only these three concepts are discussed further.

The critical components and materials used in the fueled core region of light-water reactors are listed in Table III. The Zircaloy fuel cladding has the vital functions of providing a fixed neutronically reactive geometry and spacing for the nuclear fuel (UO_2), defining the heat-transfer surface for thermal-energy removal from the core, and acting as the initial containment of fission products generated during the burnup of the fuel. Because of their low neutron-absorption cross sections, the Zircaloys [Zr-Sn alloys with additions of iron, chromium, and nickel (Zircaloy-2 only)] are the principal

Table III. Critical Components and Materials Used in Core Region of LWR's

Component	Material
Fuel Assembly (Except Fuel)	
Cladding	Zircaloy-2 (Zr-Sn-Fe-Cr-Ni) BWR Zircaloy-4 (Zr-Sn-Fe-Cr) PWR
Fuel-rod spacers, springs	304SS, Zircaloy, Inconel-600 Inconel X-750
End fittings	304SS (cast)
Flow Guides (BWR)	Zircaloy-2
Primary Pressure Vessel	SA533-1-B (C-Mn-Mo-Ni)
Support Rings	SA508-2 (C-Mn-Mo-Ni)
Control Rods	B_4C (BWR), Ag-In-Cd or Hf (PWR)

*A figure of merit which is the product $n\tau$, where n is the plasma density and τ is the confinement time. For a fusion reactor using deuterium and tritium, $n\tau$ must be greater than $\sim 10^{14}$.

structural materials used in the core region of LWR's. Stainless steels and nickel-base alloys which have much higher neutron-absorption cross sections in a thermal-neutron spectrum, are used sparingly for springs and supports, where the stiffness and strength of the Zircaloys are inadequate. All commercial PWR and BWR pressure vessels installed to date have been fabricated from low-alloy ferritic steels (predominantly A533, Class 1, Grade B). The vessels are usually shop fabricated of hot-formed, quenched-and-tempered, heavy-thickness plates, having wall thicknesses ranging from 6 to 14 inches for PWR's and from 5 to 12 inches for BWR's. The vessels for both BWR's and PWR's are designed to withstand working stresses of 1000 lbf/in.^{2} (70.3 kgf/cm^2) and test stresses of 1875 lbf/in.^{2} (131.4 kgf/cm^2) under the respective size and system pressure.[1] The engineering problems inherent in the control and surveillance of neutron-irradiation embrittlement of reactor pressure vessel steels have been recently and comprehensively reviewed by Steele.[1]

The critical core components and principal materials (other than fuel) of an LMFBR are outlined in Table IV. Austenitic stainless steels are used extensively for fuel cladding, fuel-assembly hardware, coolant flow ducts, core-support hardware, pressure vessels, piping, and numerous out-of-core components. This extensive use of stainless steel is necessary because of its excellent corrosion resistance to sodium at the operating temperature and the importance of minimizing the transport of component elements among dissimilar alloys in the primary sodium circuit. Nickel-base alloys are used much less extensively for neutron reflectors, springs, and control-rod devices. Finally, stainless-clad boron carbide pellets (or possibly tantalum) will be used as the neutron absorbers in nuclear control components.

Table IV. Critical Components and Materials Used in Core Region of LMFBR's

Component	Material
Fuel Assembly (Except Fuel)	
Cladding	316SS (reference)
End fittings, spacers	304SS, 316SS
Flow Guides	316SS
Primary Pressure Vessel, Core Support Structure	304SS
Springs	Inconel 600, 625, 718
Control Rods	B$_4$C, Tantalum

Tokamak fusion reactors are in a stage of active invention, and a variety of design concepts involving various blanket configurations, magnet designs, and materials applications are being subjected to detailed technical assessment. The general characteristics of such a reactor have been summarized by Steiner.[2]

"A Tokamak reactor, as currently envisioned, would operate as a pulsed device. The strength of the pulsed field might be as much as 20 percent of the steady state field at the center of the plasma.[3] The pulse duration might be 10 sec to 1000 sec.[4,5] Thus fuel recycling during the pulse may be required."

Among the critical components for such a device, as listed in Table V, are the first vacuum wall, diverters to remove fusion products (helium) and impurities from the plasma, and fuel injectors. The first vacuum wall is perhaps the most critical component, since it must (1) withstand wall erosion caused by bombardment with both highly energetic particles and various forms of electromagnetic radiation emitted from the plasma, (2) be compatible with the coolant, and (3) retain mechanical integrity from the standpoints of radiation-induced creep, volume swelling, and embrittlement (both from radiation hardening and the neutronically generated gases hydrogen and helium).

The principal materials required for the construction and operation of a fusion reactor are as yet tentative because of the early stage of technical development. However, 316 stainlesss steel is a leading contender, with alloys of niobium, vanadium, and molybdenum also being given serious attention. Significant amounts of niobium [as TiNb, Nb_3Sn, or $Nb_3(Al,Ge)$] superconductors and copper stabilizer will also be required for the toroidal field, ohmic heating, and diverter coils. Stainless steel and nickel-base alloys will be required for segments of the reactor toroid, the primary heat-

Table V. Critical Components and Materials Proposed for Early Tokamak Reactors

Components	Material
Vacuum Wall	316SS, Al-Al$_2$O$_3$, Alloys of Nb, V, Mo, Ni
Breeding Zone	Lithium, Li$_2$BeF$_4$, LiAlO$_2$, LiAl, beryllium
Reflector and Spectral Shaper	Graphite
Shield	304SS, B$_4$C, lead
S/C Magnet	304SS, copper, TiNb, Nb$_3$Sn (other)
Other Components	
Limiters	Tungsten alloys
Diverters	Copper, TiNb, Aluminum
Fuel Injectors	Various

removal system, and other power-plant components. Helium, liquid lithium, or fused Li_2BeF_4 salt are alternative candidates for the coolant. Beryllium may be required as both a neutron multiplier and moderator, tungsten (or W-Re alloys) may be required for limiters or other temperature-resistant components, and ceramic electrical and thermal insulators will be required at the fuel injectors and at other locations.

3 SELECTION CRITERIA AND NEUTRONIC CONSIDERATIONS

The general selection criteria for nuclear-reactor structural materials include: (1) cost, (2) availability, (3) fabricability, (4) neutronic properties, (5) adequate strength and ductility, (6) compatibility, (7) alloy stability, (8) wastage due to erosion and corrosion, and (9) physical properties. Of these, items (4) through (9) can vary substantially with the neutron fluence; therefore, the designer and materials engineer must understand in detail the microstructural state and mechanical behavior of structural materials at the various stages of service lifetime. Of the above criteria, neutronic properties and the effects of neutron bombardment set nuclear reactors apart from other energy conversion systems and require special design measures and analyses. In comparing light-water, liquid-metal fast-breeder, and Tokamak fusion reactors one must consider the reaction with matter of neutrons with energies ranging from approximately 0.05 ev to 14.1 Mev.

Calculated spectra for two fast reactors (FFTF and EBR-II) and a conceptual Tokamak reactor (UWMAK-I) are illustrated in Fig. 1.[6] The peak

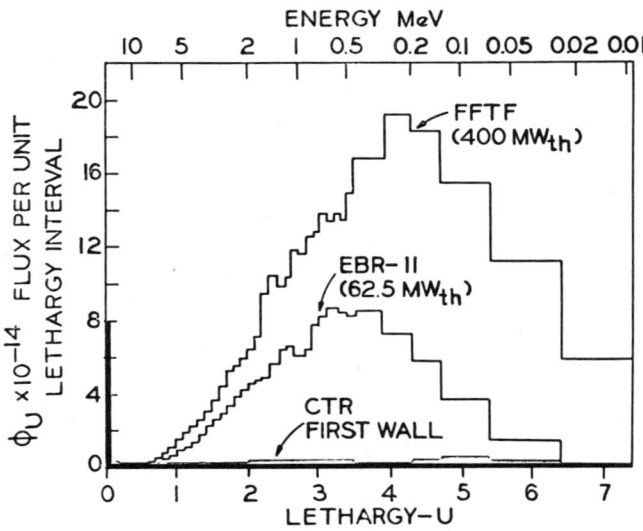

Fig. 1. Calculated neutron spectra for EBR-II, FFTF, and UWMAK-I (5000 MW$_t$).[6]

of the neutron spectra for fast-breeder reactors would occur between 0.1 and 1 Mev, while for a fusion reactor, the first wall would be subjected to a strong 14.1 Mev peak. In comparison, the light-water reactor would exhibit a highly degraded fission spectrum with a high energy peak at about 2 Mev and a dominant component below about 0.7 ev. Except for a few elements of low mass numbers (such as B, Be, etc.) which undergo n,p and n,α reactions with slow neutrons, most elements undergo radiative n,γ capture. However, with fast neutrons having energies of 1 Mev or more, n,p and n,α capture reactions frequently occur more readily than the n,γ reaction. And, with incident neutrons having energies of about 10 Mev and higher, reactions such as n,2n; n,np; n,3n; n,2np; etc., are possible.

With the exception of a resonance region, generally in the range 0.1 to 1 kev, the neutron-absorption cross section of structural materials is inversely proportional to neutron velocity, and at high energies (in the Mev range), it becomes reduced to about the same order as the scattering cross section (of the order of the geometrical nuclear cross section). Consequently, the ranking of candidate materials according to absorption cross section can undergo dramatic changes with increasing neutron energy, as indicated in Table VI. Hence, niobium or molybdenum would exhibit relatively low cross sections in a thermal-neutron spectrum but relatively high cross sections in a fast-neutron spectrum. Conversely, vanadium and cobalt which have undesirably high absorption cross sections in a thermal-neutron spectrum would exhibit relatively low cross sections in a hardened-neutron spectrum.

Table VI. Nuclear Absorption Cross Sections for Candidate Materials

Material	Macroscopic Absorption Cross Section (Σ_a), 10^3 cm^{-1}		
	Thermal (2200 m/sec)	Soft (120 kev)	Hard (305 kev)
Zircaloy-2	10.7	1.5	0.97
Aluminum	13.8	—	—
Niobium	60	13.6[a]	9.5[a]
Molybdenum	160	16.6[a]	8.7[a]
Iron	221	6.0	3.6
304 stainless steel	258	6.4	3.0
Vanadium	360	1.2[a]	0.8[a]
Nickel	422	8.6	5.6
Cobalt	3380	10.9	1.7[a]

(a) Major shift in relative position.

The increasing incidence of neutron-capture reactions with increasing neutron energy creates special problems that must be accounted for by the materials engineer: (1) the increasing generation rate of helium and hydrogen, (2) the transmutation of alloy components to new radioactive isotopes, and (3) the afterheat given off after shutdown by radioactive decay of the isotopes produced. The first two of these problems are of significant concern for fast-breeder and fusion reactors and the third problem is of major consequence for fusion reactors.

Table VII, based upon calculations by Kulcinski et al.[6], gives gas production rates in candidate fusion reactor materials for typical nuclear facilities in the United States. The table reveals that the annual production

Table VII. Calculated Gas Production Rates for Typical Nuclear Facilities

Facility[a]	Atomic Parts per Million per Sec $\times 10^7$					
	316SS[b]	316SS[c]	Nb	Mo	V	SAP[d]
Helium						
Fusion						
CTR (Tokamak)	64	N	7.6	15	18	130
LMFBR						
FFTF	2.5	N	0.53	0.95	0.15	5.5
EBR-II	1.5	N	0.31	0.57	0.097	2.5
Thermal						
ETR	0.16	48	0.15	0.27	0.030	1.3
Hydrogen Isotopes						
Fusion						
CTR (Tokamak)	170		25	30	33	250
LMFBR						
FFTF	133		3.6	1.8	7.7	33
EBR-II	85		2.1	1.1	4.6	16
Thermal						
ETR	43		1.1	0.53	2.5	11

(a) Legend:
    CTR — Controlled Thermonuclear Reactor
    FFTF — Fast Flux Test Facility (Richland, Washington)
    EBR-II — Experimental Breeder Reactor-II (Idaho Falls, Idaho)
    ETR — Experimental Test Reactor (Idaho Falls, Idaho).
(b) Primary reactions only.
(c) Helium (in atomic parts per million per sec $\times 10^7$).
    Contribution from $^{58}Ni(n,\gamma)^{59}Ni(n,\alpha)$ after only 1 year of irradiation.
    N = Negligible, $\ll 0.1 \times 10^{-7}$ appm·sec^{-1}.
(d) Sintered aluminum product.

of helium in 316 stainless steel irradiated in a light-water (ETR), liquid-metal fast-breeder (FFTF), and Tokamak (CTR) reactor would be, respectively, 1.9, 7.8, and 191 ppm, with a significant fraction (~50 percent) of the helium in the light-water reactor after 1 year of irradiation being generated by the B^{10} (n,α) Li^7 reaction with thermal neutrons. The corresponding hydrogen production would be 134, 416, and 529 ppm, respectively. Lesser but significant amounts of these gases are produced in the refractory metals niobium, molybdenum, and vanadium. The presence of these soluble and insoluble gases not only adds to the stored internal energy of materials but also creates major uncertainties in the amount trapped or otherwise occluded at internal defects; the effect of these gases on interfacial, stacking-fault, and free-surface energies; and the size distribution of voids or gas bubbles resulting from these gases.

The fraction of alloy component transmuted to new isotopes as a function of service lifetime adds additional complexity in attempting to understand phase stability and mechanical behavior of engineering alloys in a fusion reactor. For example 13.5 percent of a niobium vacuum wall in a Tokamak subjected to 14.1 Mev neutrons will transmute to zirconium after 20 years lifetime at a wall loading of 10 MW/m^2.[7] Less benign reactions are possible in most candidate alloys. In the case of liquid-metal fast-breeder reactors, the transport and deposition of radioactive isotopes in the primary circuit can result in special maintenance problems.

The afterheat and decay radiation given off by radioactive species generated in a Tokamak first wall can prevent accessibility for significant times after reactor shutdown, as illustrated in Fig. 2.[7] Here it is seen that the afterheat generated in niobium and tantalum can exceed that generated in

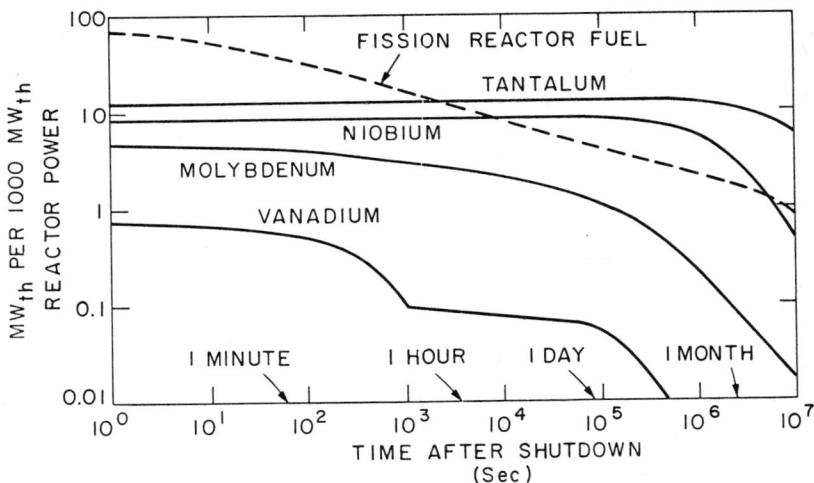

Fig. 2. Afterheat in fusion reactors.[7]

fission-reactor fuel after only a few hours, and the rate of heat production in these metals remains nearly constant out to a month of shutdown time. This undesirable characteristic alone has greatly reduced the interest in niobium or tantalum as candidate first-wall materials relative to that in molybdenum or vanadium.

Elastic fast-neutron/atom interactions will generate displacement spikes at the termini of collision cascades involving primary and higher progeny knock-on atoms. For metals with relatively high atomic masses (greater than ~40) there is a dynamic separation of point defects in the displacement spike due to correlated interstitial collisions, resulting in a vacancy-rich core surrounded by a mantle of dispersed interstitial atoms. Beeler[8] has shown by computer simulation that a primary knock-on atom (PKA) with energy greater than 2 kev will generate a finite average number of large vacancy embryo clusters ($\geqslant$10 vacancies) per displacement spike in α-iron. Beeler[9] has also shown that successive overlaps of many collision cascades in α-iron and copper can result in widely divergent damage states. At one extreme, vacancy clusters formed by previous cascades are dispersed into a nearly uniform distribution of mobile vacancies with low survival probability. At the other extreme, the vacancy clusters not only remain stable but also become enlarged.

In analyzing the kinetics and distribution of defect agglomerates produced by neutron irradiation, it is important to account for the detailed spatial structure of the collision cascade and the thermal stability of vacancy clusters generated during the time period of such cascades. To the extent that stable nuclei or near-critical embryo are formed directly by the displacement-spike process, significant dislocation barriers (and point-defect sinks) in sizable concentrations will exist far in advance of incubation periods predicted from homogeneous nucleation theory. The significance of this consideration can be examined from the data of Table VIII which reveal the variation of PKA damage potential with neutron energy in iron.[10] As the neutron energy increases from 0.5 to 14.1 Mev, 50 percent of the PKA's will have initial energies greater than 10 and 50 to 100 kev, respectively. Furthermore, the number of large vacancy clusters (n $\geqslant$10 vacancies) generated per PKA will increase from 0.35 to 0.8 as the PKA energy increases from 10 to 20 kev. Clusters of 20 to 25 vacancies are predicted with PKA energy of 20 kev, and small microvoids (10^1 to 10^2 vacancies) would be expected for PKA energies greater than 50 kev. It is evident, therefore, that damage correlations based solely on the total number of equivalent displacements per lattice atom will become less valid as the PKA energy spectrum resulting from the incident radiation shifts to higher energy. Thus, even though the calculated maximum displacement rates for a fusion-reactor wall loaded at 1 MW/m^2 are significantly lower than those for a fast-breeder reactor (FFTF) as indicated in Table IX[6], the damage potential of the 14.1 Mev neutron radiation may be considerably higher.

Table VIII. Variation in PKA Damage Potential With Neutron Energy in Iron[10]

Neutron Energy (E_n), Mev	Fraction of PKA's Produced With $E > E_0$, kev			
	F=1%	F=10%	F=20%	F=50%
0.5	31	29	25	10
1.0	67	60	52	26
2.0	135	110	95	45
9.5	600	175	100	50
14.1	870	200-400	100-150	50-100

PKA Energy (E), kev	Large Vacancy Clusters ($n \geqslant 10$) per PKA
5	0.22
10	0.35
20	0.8
	(n_{max} = 20 to 25)
>50	Microvoids

Note: Values within box are average E_n for LMFBR.

Table IX. Calculated Maximum Displacement Rates for Typical Nuclear Reactors[6]

Facility[a]	DPA/Sec × 10^7				
	316SS	Nb	Mo	V	Al
Fusion					
Tokamak (1 MW/m^2)	3.1	2.3	2.6	3.7	5.4
LMFBR					
FFTF (400 MW)	26	16	19	34	49
EBR-II (62.5 MW)	14	9.0	9.6	17	24
Thermal					
ETR (175 MW)	5.1	3.2	3.5	6.6	8.4

(a) Legend:
   CTR — Controlled Thermonuclear Reactor
   FFTF — Fast Flux Test Facility (Richland, Washington)
   EBR-II — Experimental Breeder Reactor-II (Idaho Falls, Idaho)
   ETR — Experimental Test Reactor (Idaho Falls, Idaho).

Another selection criterion which substantially limits the choices of candidate materials for reactor structural applications is compatibility with the coolant environment and, in the case of fuel cladding, compatibility with both the fuel and fission products generated in the fuel. In the cores of light-water reactors, most problems of compatibility center on the Zircaloy fuel cladding and the stainless-steel overlay cladding for the low-alloy steel pressure vessel.

In BWR coolants with significant dissolved oxygen concentrations (two-phase H_2O), the neutron flux can enhance the oxidation of Zircaloy-2 both by radiolytic decomposition of the water and enhanced film transport kinetics.[11] Hydrogen injected into PWR coolant (single-phase H_2O) suppresses radiation-enhanced radiolytic decomposition of the coolant but increases the percentage of corrosion hydrogen picked up by the cladding. Therefore, internal hydride formation and attendant embrittlement must be taken into account in determining cladding-performance limits for PWR fuel. Other sources of cladding attack in light-water reactors include internal hydriding due to residual moisture and hydrocarbon contaminants and stress cracking from either residual fluorine contamination (carried over from UF_6 to UO_2 fuel-reconversion processing or from cladding pickling operations) or from fission-product iodine. Most of these problems are well understood and generally are adequately controlled in practice.

The compatibility problems in liquid-metal fast-breeder reactors are generally of five types: (1) solution corrosion of selective alloy components of stainless steel in the hot leg and deposition of these components in the cold leg, (2) activity gradient migration of alloy components from regions of higher concentration to regions of lower concentration, (3) mechanical erosion and galling at contact points, (4) selective intergranular attack of the stainless-steel fuel cladding by fission-product cesium, and (5) uniform cladding wall wastage by mechanical fuel-cladding interaction and iodine transport. The uniform, steady-state corrosion of stainless steel by sodium can be minimized both by strict control of oxygen content in the sodium and by limiting the nickel content of the stainless steel. However, the selective leaching of nickel and carbon (especially from unstabilized grades of stainless steel) can result in the formation of surface ferrite layers. The rate of cladding attack by fission-product cesium is generally less in hypostoichiometric, high-density fuels.

The compatibility and wastage problems in the vacuum wall of a Tokamak reactor consist primarily of surface vaporization, sputtering, blistering, exfoliation, and other surface phenomena from the following forms of irradiation: 14.1 Mev neutrons, 3.5 Mev α-particles, deuterium and tritium ions, bremsstrahlung radiation, secondary gamma radiation, X-rays, synchrotron radiation, and high-Z ions originating from the wall and energized by the plasma.

The nature and recession rates due to various surface phenomena on the first wall have been recently reviewed[12]; however, one set of calculations will illustrate the relative effects of the various blistering and sputtering phenomena. Figure 3 is a plot of erosion of a 316 stainless-steel wall of UWMAK-I as a function of irradiation time for a neutron loading to the first wall of 1.25 MW/m^2 and operating temperatures from 300 to 500 C.[12] A 90 percent efficient diverter is assumed in the calculations. The data reveal that two-thirds of the wall erosion comes from the high-energy neutrons and the remainder of the wall thinning is due to sputtering and blistering equally. The effect of no diverter would be to increase the thinning rate by a factor of 3.5. It should be noted, however, that the neutron-sputtering yield used in these calculations (S = 0.2) is that derived by Kaminsky and Das[13] for cold-worked niobium. However, it is possible that these high yields are due to unconsolidated fragments remaining on the surface after polishing such that the emission of these fragments will quickly saturate.[14]

One must remember that sputtering from lower energy neutrons (0.1 to 10 Mev) occurs on both sides of the wall and that such sputtering ($\sim 1 \times 10^{-2}$ atoms/neutron) on the back side may significantly affect corrosion mechanisms with the coolant, especially for alloys dependent upon protective films. The mechanisms could be further complicated by hydrogen and tritium transport through the wall and helium generation at the near surface. The combined effects of these contributing factors on coolant compatibility are relatively unexplored and speculative.

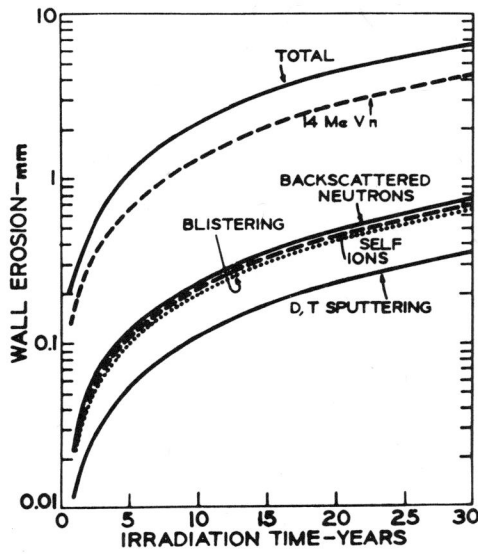

Fig. 3. Wall erosion of UWMAK-I 316SS first wall.[12]

4 EFFECTS OF NEUTRON IRRADIATION

The selection and successful employment of structural materials in nuclear-reactor systems requires accurate models and data for a variety of neutron-irradiation effects on bulk and surface properties. Among the problem areas of interest are (1) radiation hardening, (2) dimensional instability, (3) radiation-enhanced creep, (4) internal gas production and transport, (5) surface phenomena, and (6) alloy stability. As indicated in Table X, the number of these problem areas increases as one progresses from light-water reactors to fast-breeder reactors and in turn to fusion reactors. Since items (1) through (5) are being discussed by others at this conference and the conference theme is alloy design, the comments below are restricted to this problem area.

Of importance to the stability of complex engineering alloys under neutron bombardment are the changes brought about by long-range atomic movement; the disruptive nature of displacement spikes; and the thermodynamic relationships among the constituents under highly non-equilibrium, irreversible conditions. As Damask[15] has summarized, neutron radiation can affect the rates and mechanism of atomic interchanges in alloys through: (1) destruction of order, (2) breakup of precipitates, (3) enhanced nucleation, (4) enhanced diffusion, and (5) retarded diffusion.

Table X. Radiation Problem Areas

Form of Damage	Reactor Type[a]		
	LWR	LMFBR	Tokamak
Radiation Hardening			
Brittle fracture	X		X
Plastic instability	X	X	X
Loss of toughness	X	X	X
Dimensional Instability			
Volume swelling		X	X
Growth	X	X	X
Radiation-Enhanced Creep	X	X	X
Internal Gas Production and Transport			
Helium embrittlement		X	X
Hydrogen and tritium effects		m	X
Alloy Instability	m	X	X
Surface Phenomena			
Blistering			X
Sputtering			X

(a) X — significant; m — minor.

These processes are generally caused by kinetic exchanges between energetic neutrons and atoms or between "knocked-on" atoms and other atoms in the lattice. Atoms can also be transported large distances by cooperative focusing, replacement collisions, and channeling along close-packed lattice directions.

If an alloy is ordered, both the replacement collision and the displacement spike can break up the order. If the alloy contains precipitates, the displacement spikes can break up the precipitates and place the constituents into solution. In a precipitating alloy, the spike-damage regions can serve as nucleation sites for precipitate particles. If the alloy diffuses by the vacancy mechanism, the excess vacancies created by irradiation can enhance the diffusion rate. Finally, if an alloy diffuses by the interstitial mechanism and a binding energy exists between the diffusing species and a lattice vacancy, a retardation of diffusion can occur. Damask[15] has given several examples for each of these irradiation-induced reactions in his review.

The probability and extent to which some of the above reactions will occur depends upon the temporal thermodynamic stability of the alloy and the irradiation temperature. However, in such irreversible systems subjected to huge energy and mass fluxes (represented by the bombarding particles) and to highly supersaturated concentrations of point defects (both vacancies and interstitials), the phase diagram can at best be considered a first-order approximation in predicting which phases will actually be present. If an alloy is unstable, the rate of conversion to a lower free-energy state can be accelerated by either irradiation-enhanced diffusion or nucleation. Atomic mixing by the breakup of either long-range-order, localized segregations, or precipitates can occur at very low temperatures, where thermal diffusion is negligible. However, the reestablishment of equilibrium would require heating the alloy to temperatures where vacancy diffusion is appreciable.

The effectiveness of displacement spikes in nucleating new phases or in dissolving precipitates is subject to restrictions which are not completely understood. For example, Piercy[16] found in the Cu-Co (2 percent) system that only cobalt precipitates with a radius less than 12.4 A and which are completely inside displacement spikes could be dissolved by fission neutron irradiation at 50 C. As the size of the displacement spike increases and spike overlap becomes more frequent with increasing PKA energy, however, both the maximum and average precipitate disrupted by spike resolution would likewise increase.

The kinetics of internal conversion to lower free-energy states in a neutron environment are dependent upon the steady-state concentrations and mobilities of point defects produced by the irradiation. Vacancy and interstitial concentrations in alloys are dependent not only on linear annealing at sinks, but also on trapping at solute and impurity atoms and at defects and interfaces. Potential point-defect sinks in alloys include (1) voids,

(2) unpinned and pinned edge dislocations, (3) partial dislocations, (4) incoherent twin boundaries, (5) incoherent precipitate boundaries, and (6) grain boundaries. Potential trapping and recombination sites include (1) substitutional and interstitial impurities having large misfit strains, (2) coherent precipitate boundaries, (3) coherent twin boundaries, and (4) screw dislocations.

Coupled steady-state equations for point-defect production according to Wiedersich[17] are as follows:

$$\frac{dC_v}{dt} = \dot{n} - (\nu_i a_i + \nu_v a_v)(C_v C_i - C_v^e C_i^e) - \nu_v P_v(C_v - C_v^e) \tag{1}$$

and

$$\frac{dC_i}{dt} = \dot{n} - (\nu_i a_i + \nu_v a_v)(C_v C_i - C_v^e C_i^e) - \nu_i P_i(C_i - C_i^e) \ , \tag{2}$$

where

$\dot{n}$ = radiation-induced defect production rate

ν_i, ν_v = jump frequencies for interstitials and vacancies, respectively

C_v, C_i = atomic fractions of vacancies and interstitials, respectively

C_v^e, C_i^e = atomic fractions of vacancies and interstitials, respectively, at thermal equilibrium

P_v, P_i = the probabilities, respectively, that a vacancy or an interstitial jump will lead to annihilation at a sink

a_v, a_i = the number of nearest-neighbor atomic sites from which spontaneous recombination occurs.

Examination of Eqs. (1) and (2) reveals that the first term on the right side of both equations (generation terms) are identical, as are the second terms (recombination terms), due to the fact that vacancies and interstitials form and recombine pairwise. The last term in both equations (sink annihilation terms) varies among theoretical treatments, primarily because a complete analysis requires detailed expressions not only for vacancy and interstitial interactions with various sinks and traps but also for the time, temperature, and fluence relationships for the sink densities. Sink-sink interactions and sink stability as a function of size and configuration are additional complications.

The vacancy concentration calculated by Wiedersich[17] for nickel as a function of inverse temperature is plotted in Fig. 4 for defect production rates typical of those for a fast-breeder reactor ($\dot{n} \simeq 10^{-6}$ sec^{-1}) and for heavy ion bombardment ($\dot{n} \simeq 10^{-4}$ sec^{-1}).

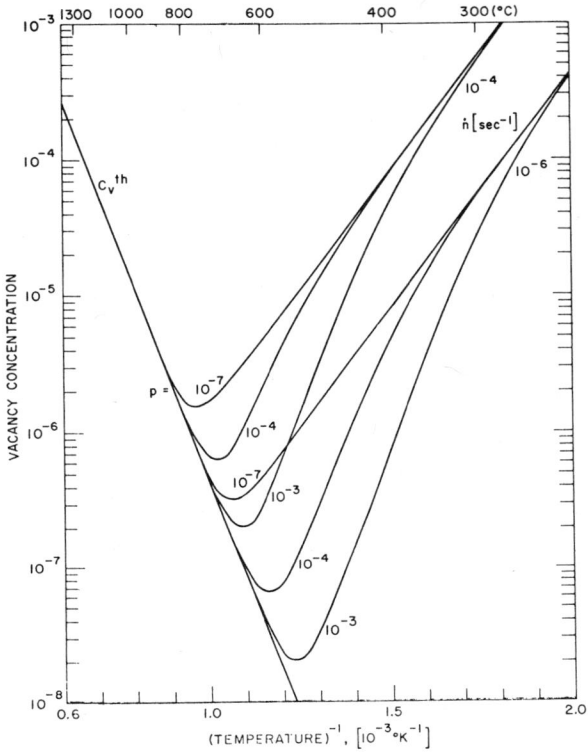

Fig. 4. Steady-state vacancy concentration in nickel.[17]

Three annihilation probabilities are plotted: $p \simeq 10^{-7}$ characteristic of an annealed material; $p \simeq 10^{-4}$ for a cold-worked (10^{11} cm^{-2} dislocations) lightly damaged material; and $p \simeq 10^{-3}$ for a heavily cold-worked or damaged material. The curves in Fig. 4 show that the vacancy concentration decreases markedly with increasing temperature (mainly because of enhanced recombination with increased vacancy mobility), goes through a minimum, and converges to the thermal-equilibrium concentration at high temperatures.

Recent investigations into the control of void swelling in 316 stainless steel[18-20] have demonstrated that alloy additions having significant lattice-misfit strains can substantially reduce the magnitude of swelling through vacancy trapping. Smidt et al.[21] have treated vacancy trapping by modifying the recombination term of Eqs. (1) and (2) to the form $Z_R \, \nu_i C_i (C_v^F + C_v^T)$, where the superscripts refer to free and trapped vacancies and Z_R is the recombination volume. The formation rate of trapped vacancies is given as follows:

$$\frac{dC_v^T}{dt} = (ZC_S - C_v^T)\, \nu_v^F C_v^F - \nu_v^T C_v^T - Z_R \nu_i C_i C_v^T \, , \tag{3}$$

where Z is the trapping volume. The first term in Eq. (3) considers formation of trapped defects at empty trap sites, the second term accounts for thermal detrapping $[\nu_v^T = \nu_v \exp(-E_{diss}/kT)]$, and the third term accounts for the loss of trapped vacancies by recombination. Figure 5 by Smidt et al.[21] shows the free-vacancy concentrations in iron as a function of temperature with a defect production rate of 10^{-6} sec^{-1}, a dislocation density of 10^7 cm^{-2}, a solute concentration of 0.3 at.%, and binding energies (E_B) of 0.2, 0.4, and 0.5 ev. The vacancy formation and migration energies were taken as 1.5 and 1.0 ev, respectively. The rather modest changes in the vacancy supersaturation due to trapping, as shown in Fig. 5, have been found[21] to decrease the void-nucleation rate in iron by a factor of at least 10^{14} and more likely $>10^{21}$ under fast reactor conditions, which is indicative of the strong role alloy additions can play in void-swelling control.

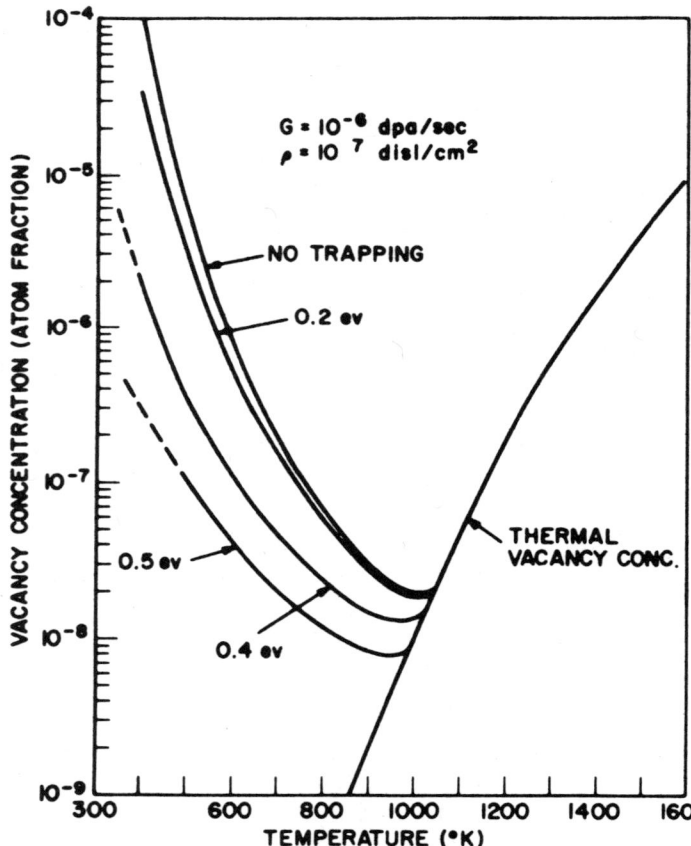

Fig. 5. Steady-state vacancy concentration in iron with and without trapping.[21]

In addition to enhancing vacancy-interstitial recombination, impurity trapping can significantly influence the mobilities of vacancies and interstitials. Likewise, the presence of a point-defect binding energy can both hinder the diffusion of interstitial impurities (such as carbon and nitrogen, in iron) and influence impurity segregation at traps.

According to the treatment of Schilling and Schroeder[20], the flux, the total current (I_S) of vacancies arriving at a void in the absence of traps, can be written as

$$I_S = 4\pi R_S \cdot D_v \cdot C_v \quad , \tag{4}$$

where D_v is the vacancy diffusion constant, C_v is the average vacancy concentration in the crystal, and R_S is the void radius. In the presence of trapping, Eq. (4) would be given by

$$I_S = 4\pi R_S \cdot D_v^+ \cdot C_v^+ \quad , \tag{5}$$

where now D_v^+ is an effective diffusion constant and C_v^+ denotes the total of the free plus temporarily trapped vacancies. The effective diffusion constant D_v^+ can be related to the unhindered diffusion constant D_v by the expression

$$D_v^+ = \frac{D_v}{1 + 4 \, C_t R_t \cdot D_v \cdot \tau} \quad , \tag{6}$$

where C_t is the trap concentration, R_t is the trap radius, and τ is the trapping period, which varies with temperature according to $\tau = \tau_0 \cdot \exp(E_{diss}/kT)$, where E_{diss} is the dissociation energy.

Schilling and Schroeder[20] find that the void-swelling threshold temperature in austenitic stainless steel can be increased by about 100 C by trapping interstitials with binding energies of 1.7 ev or by trapping vacancies with binding energies of 0.4 ev. (The difference in required binding energy is related to the difference in mobilities between the two species). In reviewing experimental studies of interstitial and vacancy trapping in nearly pure metals and binary alloys, Schilling and Schroeder[20] conclude that (1) the binding energies of vacancies to substitutional impurities are of the order of 0.1 to 0.3 ev so that strong trapping effects are not to be expected, (2) vacancies are efficiently trapped by interstitial impurities with binding energies of the order of 0.5 ev, and (3) self-interstitials can be trapped by substitutional impurities up to and above the temperatures where vacancies become mobile with binding energies of the order of 1 ev. So that while trapping effects may not be sufficiently effective to completely suppress swelling, they would nevertheless substantially alter swelling behavior with respect to incubation periods, threshold temperatures, and nucleation rates.

Even at small binding energies, coupled interactions between point defects and solute atoms can have a substantial effect on irradiation-damage processes such as void swelling. Vacancy currents in multicomponent systems have been investigated by Anthony[22] and were found to cause significant differential currents of solute atoms relative to solvent atoms, resulting in a significant solute enrichment or depletion at such vacancy sinks as voids and grain boundaries. Therefore, the use of chemical potential or surface-tension values for a homogeneous solid solution in analyzing void nucleation and growth could lead to significant errors. The relative effects of misfit strains and valency effects on the binding energy between point defects and solutes have not been clearly resolved by theory. However, Anthony[22] finds that even small valency contributions to the binding energy (as low as 0.05 ev) can substantially reduce the vacancy jump frequency away from a substitutional solute atom.

Okamoto and Wiedersich[23] have proposed a "drag" mechanism involving interstitial solute transport. They find that if undersized solutes are more readily accommodated in interstitial sites, the interstitial flux will be composed disproportionately of undersized solutes, resulting in a buildup of these atoms around voids. The segregation that can occur at point-defect sinks can be far in excess of that permitted by the Gibbsian absorption mechanism and can create anomalously high coherency strains of either sign which can be large in comparison with the intrinsic surface energy. In their work on 18-8 austenitic stainless steel containing silicon, they observed an enrichment of subsized nickel and silicon atoms at void surfaces and a depletion of oversized chromium atoms. These segregation effects led to high coherency strains at void surfaces after 0.3 displacements per atom and precipitation on void surfaces at 4.3 displacements per atom.

These results and suggestions by Beeler[10] indicate the following possible consequences of point-defect-coupled segregation effects:

1. The segregation of solute concentration at internal sinks will enhance second-phase nucleation.

2. The accumulation of a nearly impervious second phase layer at voids could severely inhibit assimilation of both gas atoms and vacancies at voids, thereby constraining void growth.

3. The coherency strains developed at sinks could greatly affect interaction forces with migratory point defects and bias absorption and emission rates.

4. The presence of brittle coatings on microvoids could modify blistering behavior either by enhancing fragmentation and sloughing or by permitting early venting of the underlying gas pressure short of sloughing.

A number of studies have shown that the kinetics for precipitate formation in irradiated alloys are much more rapid than in thermally aged alloys. This has been well demonstrated in austenitic stainless steels irradiated in fast-breeder reactors where the temperature threshold for precipitation reactions is reduced about 100 to 150 C by the neutron flux. In early investigations of M316 stainless steel fuel cladding irradiated in the Dounreay Fast Reactor above 500 C, copious $M_{23}C_6$ precipitates were observed at stacking faults. A subsequent investigation by Brager and Kissinger[25] of 304 stainless steel irradiated in the 370 to 425 C temperature range and at fluences above 1×10^{22} n/cm^2 revealed the presence of small (50 to 300 A), coherent, cuboid precipitates distributed uniformly throughout the matrix. These precipitates were identified by both electron and X-ray diffraction to be $M_{23}C_6$ carbides. At higher irradiation temperatures (450 to 600 C), Brager and Kissinger[25] observed acicular precipitates parallel to <111> in addition to the larger cuboid precipitates. These acicular precipitates were substantially found to be a new nonequilibrium phase related to but different from FeCr sigma phase. At a still higher irradiation temperature of 700 C, only large $M_{23}C_6$ carbide precipitates were observed. Bisson[26] also observed acicular precipitates in addition to $M_{23}C_6$ carbides in 316 stainless steel irradiated in the RAPSODIE fast reactor at 550 C at a neutron fluence of from 2 to 3×10^{22} n/cm^2. He observed that both the $M_{23}C_6$ carbides and the acicular precipitates could be made to disappear by a postirradiation anneal at 600 C for 1 hour and were absent in 316 stainless steel irradiated above 550 C. Clearly, one must account for the combined effects of enhanced nucleation, enhanced diffusion, and displacement spike resolution in either predicting or explaining precipitation phenomena in a neutron flux.

The effects of irradiation on precipitate stability have been considered recently by Russell.[27] He finds that the free-energy change, $\Delta G^\circ(n, \chi)$, on forming a precipitate of x solute atoms and n vacancies from a system at solute and vacancy supersaturations S_x and S_v is as follows:

$$\Delta G^0(n, x) = -nkT \ln S_v - xkT \ln S_x$$
$$+ (4\pi)^{1/3} (3\Omega)^{2/3} x^{2/3}$$
$$+ X\Omega e(\delta - n/\chi)^2 / 9(1-\nu) \quad , \tag{7}$$

where Ω is the atomic volume of the precipitate, E and ν are Young's modulus and Poisson's ratio, respectively (assumed the same in both phases), and δ is the fractional difference between the precipitate and matrix atomic volumes. Figure 6 is a schematic representation of the trajectory of a precipitate particle in n,x space with the atomic volume of the particle less than and greater than the matrix under irradiation conditions. If the arrival rate of vacancies (β_v) is greater than that for interstitials (β_i) at the interface, the growth of particle A will be encouraged by the presence of excess defects. Since particle B consumes vacancies to decay, dissolution will be favored by the excess vacancies. It is expected that the excess defects will

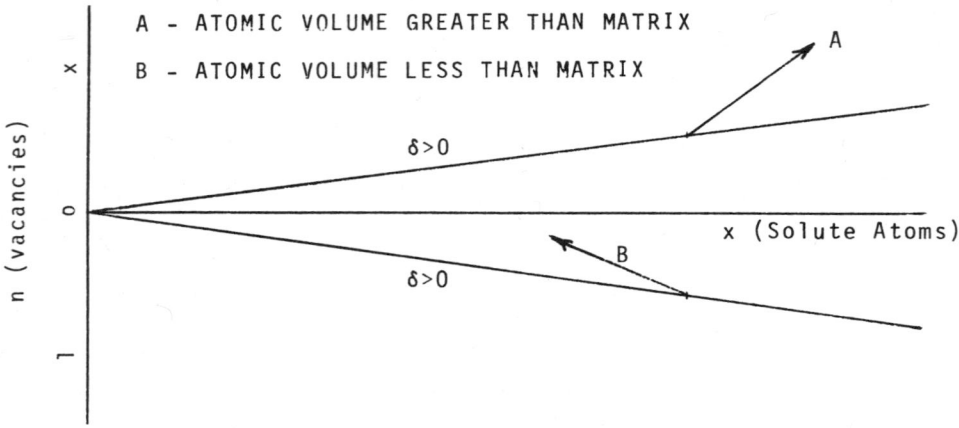

Fig. 6. Schematic representations of precipitate particle trajectories in (n, χ) space.[27]

have a strong stabilizing or destabilizing effect on precipitates as long as $\beta_v \neq \beta_i$ (the usual case). Therefore, excess defects may very well reverse the order of phase stability and lead to nucleation of unusual phases which appear nowhere in an equilibrium phase diagram.

The implications of these and other effects have been treated by Nelson et al.[28] in their study of γ' stability in Nimonic alloy PE16. They considered the following effects: (1) recoil dissolution in which atoms within the precipitate which achieve sufficient energy from a collision cascade can cross the interface and become dissolved in the matrix, (2) disordering dissolution in which the disordering of ordered precipitates by displacement spikes creates localized regions of high solute concentration in the absence of diffusion, and (3) radiation-enhanced vacancy-controlled diffusion in the presence of excess vacancies. Theoretical and experimental investigations with Ni-Al alloys and Nimonic PE16 bombarded with nickel ions revealed the following: (1) precipitate stability under irradiation is a balance between the radiation dissolution of precipitates and their growth by irradiation-enhanced and thermal diffusion; the precipitate radius and concentration changes to maintain the correct solute concentration in dynamic solution; (2) there is a critical temperature below which the dissolution rate exceeds the growth rate and the value of the critical temperature is dose-rate dependent; and (3) during irradiation coarse particles will shrink and small particles will grow until they reach an equilibrium size defined by the secular equilibrium conditions under irradiation.

In the case of Tokamak reactors which may employ a variety of superconducting coils for plasma confinement, ohmic heating, and impurity diversion, it is necessary to know the basic mechanisms by which the critical current density, J_c, and other technically important superconducting properties are changed by fast neutrons. It is known[29] for Nb-Ti alloys that J_c may be increased by at least four mechanisms: (1) plastic deformation, (2) omega-phase precipitation, (3) alpha-phase precipitation, and (4) oxide precipitates. The reaction of the superconductor to fast neutrons will not only depend upon which of these four will dominate J_c, but also the size scale of these structural features: omega phase particles are of the order of 100 A, dislocation subcells are somewhat larger (depending upon the degree of cold work), alpha-phase particles somewhat larger, and oxides possibly largest of all. Irradiation studies[30-32] of Nb-50%Ti wires have shown both increases and decreases in J_c with irradiation. Unfortunately it is not possible from these investigations to determine the relative effects of alloy structural changes and irradiation defect production on the measured changes.

In the case of A15 compounds (Nb_3Sn, V_3Ga, V_3Si, Nb_3Al, etc.), a high degree of atomic ordering is necessary to achieve a high T_c. Neutron irradiation would be expected to induce disorder in these compounds, and, indeed, fluences of 2×10^{19} n/cm^2 or better at 60 C have been found[33] to remove all superconductivity above 4.2 K in Nb_3Sn, Nb_3Al, Nb_3Ga, and Nb_3 (Al,Ge), and probably would do so for the other A15's as well.

The overall implication of the various effects of irradiation on alloy stability described above is that intuitive predictions based upon current alloy theories are likely to be in error and that the error will be greater for conditions of increasing neutron dose and harder neutron spectra. Consequently, commercial alloys optimized for nonnuclear applications and the thermal-mechanical treatments of these alloys will most likely be inappropriate for advanced nuclear-reactor systems. The rapidly expanding program in advanced cladding-alloy development for LMFBR's supports this contention.

5 MECHANICAL DESIGN CONSIDERATIONS

Areas where additional attention is required for the mechanical design of components for LMFBR and Tokamak components subjected to the nuclear environment are radiation creep, creep-rupture, and the effects of thermal and stress cycles on creep-fatigue interaction and void swelling. As yet there are no mechanical equations of state to adequately treat hardening laws under reversed cyclic stresses (isotropic versus kinematic) or for analyzing creep behavior under changing stress (strain hardening versus time hardening).

Although considerable theoretical and experimental attention has been given to understanding irradiation creep behavior, there is much less guidance in alloy design, not to speak of mechanical design, for the accommodation or control of radiation creep than is available for void-swelling behavior; and yet irradiation creep must be accurately accounted for in the following technical areas:

1. *Fuel-Cladding Design (LWR, LMFBR):* (a) inward creep down and collapse under external coolant pressure (PWR); (b) outward creep under internal pressure (PWR, LMFBR); and (c) fuel-clad mechanical interaction and fuel swelling constraint.

2. *Spacer Supports (LWR, LMFBR):* stress relaxation of spacer supports.

3. Core Support Systems (LWR, LMFBR): (a) stress relaxation in bolts and supports (LWR) and (b) relaxation of stresses generated by differential thermal expansion and void swelling (LMFBR).

4. *Flow Guides (BWR, LMFBR):* nonuniform outward creep causing coolant flow redistribution and duct contact.

5. *Outward Plenum Components (LMFBR) and First Wall and Blanket Components (Tokamak):* plastic strain ratchetting, thermal fatigue, and combined creep-fatigue.

While it has generally been observed that irradiation-induced creep exhibits a reduced stress dependence (of order 1 to 2) and a nearly linear flux dependence[34], neither the temperature dependence nor the effects of changing barrier distributions with neutron fluence are well established. Furthermore, the relative contributions of growth (due to dislocation climb without glide, defect loop alignment, and volume change) to dimensional changes are not completely known over the temperature and neutron-flux ranges of interest. The distribution between creep and growth can be important when one recognizes that dimensional changes due to growth can generally be reversed by thermal annealing, can be independent of stress, and may not accrue against plastic strain limits, in contrast with expected creep behavior.

In recent measurements of radiation-induced creep (by proton bombardment) in high-purity nickel by Hendrick[35], as shown in Fig. 7, the temperature dependence exhibited a minimum such that the creep rate was directly dependent upon temperature above 215 C and inversely dependent upon temperature below 215 C ($0.28T_m$) which is near the temperature (277 C) for Stage IV recovery in nickel. In the model for radiation-controlled creep by Harkness et al.[36], the model for climb-controlled creep over dispersed obstacles developed by Ansell and Weertman[37] has been adopted in the form

$$\dot{\epsilon} = \frac{A\sigma^2 L}{\mu^2 bd} (\nu_{climb}) \quad , \tag{8}$$

where

$$\nu_{climb} = 2D_s \left[\frac{\sigma^2 bL}{\mu kT} \right] \text{ for } T > 0.5T_m \tag{9}$$

$$\nu_{climb} = \frac{(Q_i, \text{dis} - Q_v, \text{dis})b^2}{L} \text{ for } T < 0.5T_m \quad , \tag{10}$$

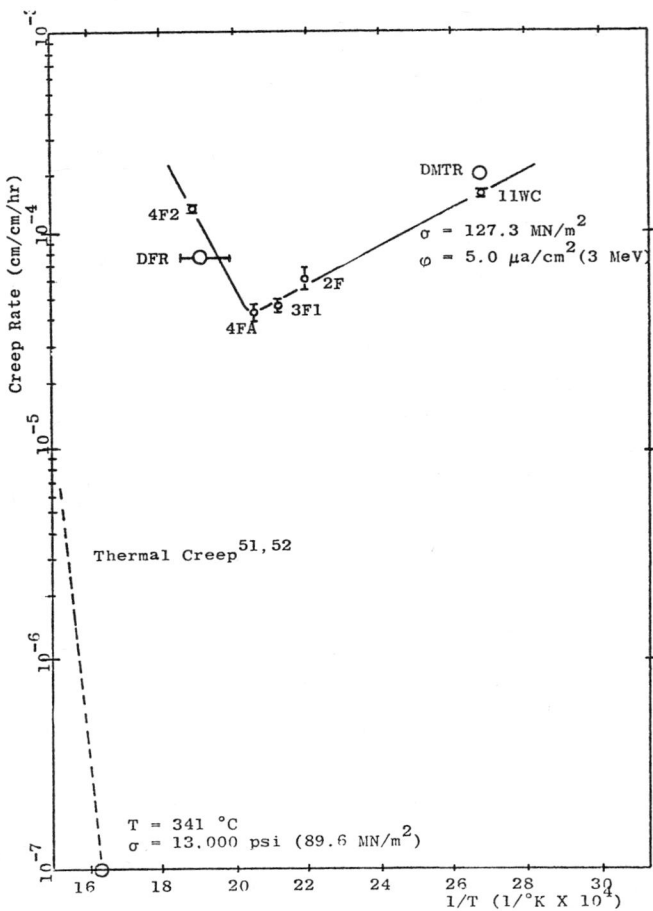

Fig. 7. Temperature dependence of creep in high-purity nickel induced by proton bombardment.[35]

and where D_s is the self-diffusion coefficient, A is a constant, σ is the applied stress, L is the average spacing of the obstacles, μ is the shear modulus, and d is the obstacle height. It is generally expected that the arrival rate of intersititals at edge dislocations will be greater than that for vacancies because of stress biasing effects. Whereas Eq. (10) would predict an inverse temperature dependence due to both lower vacancy mobility and reduced point-defect recombination (higher interstitial concentrations) with lower temperatures, one must also account for volume changes and growth mechanisms which can also vary inversely with temperature.

The implication of radiation creep behavior shown in Fig. 7 is that cryogenically cooled coils and magnet components in Tokamak reactors may be subjected to dimensional changes far in excess of expectations. In the higher temperature regions of LMFBR's and Tokamaks where structural materials are employed between 0.35 and 0.5T$_m$, radiation-enhanced creep can be expected but at a much-reduced activation energy as compared with thermal creep.

Subsequent creep measurements by Hendrick[38] on nickel irradiated with α-particles at 224 C (near the minimum of Fig. 7) reveal that radiation-induced creep under changing stress may exhibit marked departures from either strain-hardening or time-hardening laws. Figure 8 reveals that a stress increase from 287 to 345 MPa following nearly steady-state creep at 287 MPa resulted in a transient creep response far in excess of that predicted by a strain-hardening construction. If the barriers controlling the creep rate under irradiation are defect clusters generated by the irradiation, then unpinning and channeling effects resulting from dislocation-defect interaction at the higher stress level would result in reestablishment of most of the transient response at that stress level.

In applying alloys such as austenitic stainless steel in both elevated-temperature and fast-neutron service, careful attention must be given to the superimposed effects of radiation defect structure, helium generation by n,α capture reactions, and changes in alloy stability on stress-rupture properties. Available information[39] on the effects of fast-reactor irradiation on the creep and stress-rupture behavior of austenitic stainless steel reveal that (1) irradiation defect structures (voids and Frank partial loops) have a marked effect in reducing creep rate, ductility, and rupture life at

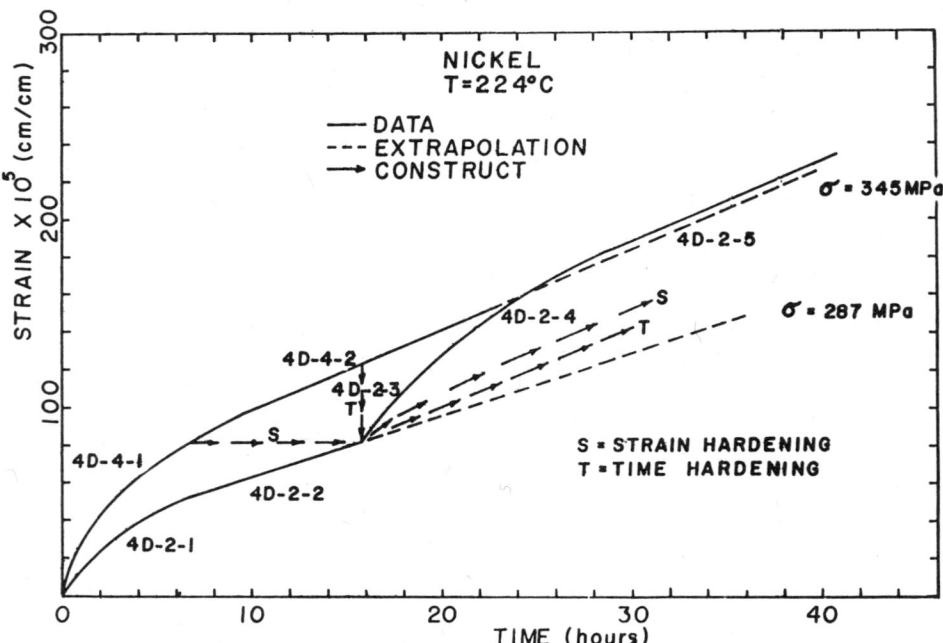

Fig. 8. Irradiation-induced creep behavior following a change in stress level at 224 C.[38]

temperatures below about 650 C; (2) rupture life below about 600 C is reduced markedly by the effects of matrix hardening on ductility; (3) rupture life above 650 C is greatly reduced by the combined effects of irradiation on both increasing creep rates due to second-phase precipitation (reduction in solid-solution strengthening) and reduced ductility due to helium embrittlement; and (4) rupture life in the temperature range 550 to 650 C can be longer than at either lower or higher temperatures owing to compensating effects of reduced creep rates due to matrix hardening, sufficient damage recovery to provide a degree of ductility, and rapid relaxation of stress concentrations at grain-boundary junctions and at second phases within the boundaries. This last point is illustrated in Fig. 9 from the work of Lovell and Barker[40] which shows that the reduction in rupture life for 316 stainless steel irradiated to 1.2×10^{22} n/cm^2 at 480 C is less at 649 C than at either 538 or 760 C.

Perhaps the most challenging design problem for advanced reactor systems is analyzing hold-time effects under conditions where low-cycle fatigue and either stress relaxation or creep are superimposed. In LMFBR's, cyclic plastic strains are generated in structures exposed to the outlet plenum thermal transients. These cyclic plastic strains not only produce fatigue damage in components but also generate residual stresses which contribute to the stress-rupture contribution to the creep-fatigue cumulative

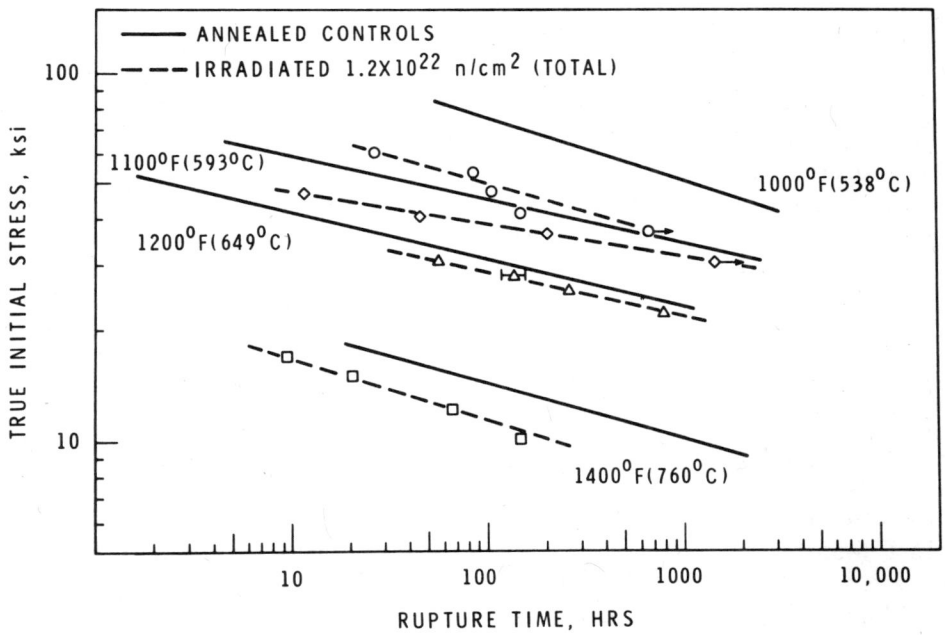

Fig. 9. Effect of EBR-II irradiation on the rupture life of AISI type 316 stainless steel determined in uniaxial tests.[40]

damage generated within the structure.[41] Considerable effort has gone into incorporating design rules and constitutive equations in Section III (Nuclear Service) of the ASME pressure vessel code in recent years to account for relaxation-fatigue and creep-fatigue behavior in austenitic stainless steel under elevated-temperature service. However, there is still considerable uncertainty in setting life fraction coefficients for the various stress regimes and cycling periods. The added complications of both the coolant and neutron environments add substantially to the uncertainty.

Until recently, the combined effects of creep and low-cycle fatigue have been treated according to the fractional-damage concept proposed by Wood.[42] According to this concept, the fatigue-damage term N_c/N_f is assumed to be inversely related to the creep-rupture term t_c/t_f, where N_f is the room-temperature fatigue life at a given strain range, N_c is the elevated-temperature life at the same strain range, t_c is the total time under creep at a given stress level, and t_f is the stress-rupture life at a stress component comparable to that for the hold time applied. Although Wood originally proposed an inverse linear relationship, both positive and negative departures from a linear law have been observed. However, the reduction in fatigue life with an increase in hold time per cycle has generally been attributed to creep at the maximum stress.

Recently, Coffin[43] has proposed the concept of a frequency-modified fatigue life which is based on the equation

$$\nu^k t = C_1 \quad , \tag{11}$$

where ν is the frequency, t is the failure time, and k is a material constant. Equation (11) applies to one temperature and a single strain range; however, the constant k is independent of the plastic strain range. The term C_1 in Eq. (11) can also be expressed as

$$C_1 = N_f \nu^{k-1} \quad , \tag{12}$$

where N_f is called the frequency-modified fatigue life. When $N_f \nu^{k-1}$ is plotted against $\Delta\epsilon_p$ (the plastic-strain range) on logarithmic coordinates for a variety of iron-base and nickel-base alloys, a straight line generally results.

Coffin[44] has demonstrated that a time dependency in fatigue life can be attributed perhaps more to environmental effects than to creep. In a dramatic series of experiments on 304 stainless steel at 430 F (221 C) to 816 F (435 C), as illustrated in Fig. 10, he found that the fatigue behavior in air was vastly different from that in high vacuum (10^{-8} torr).

Metallographic investigations confirmed a marked difference in fracture behavior in the two environments. Samples tested in air fractured intergranularly because of internal oxidation. Localized oxidation at the base

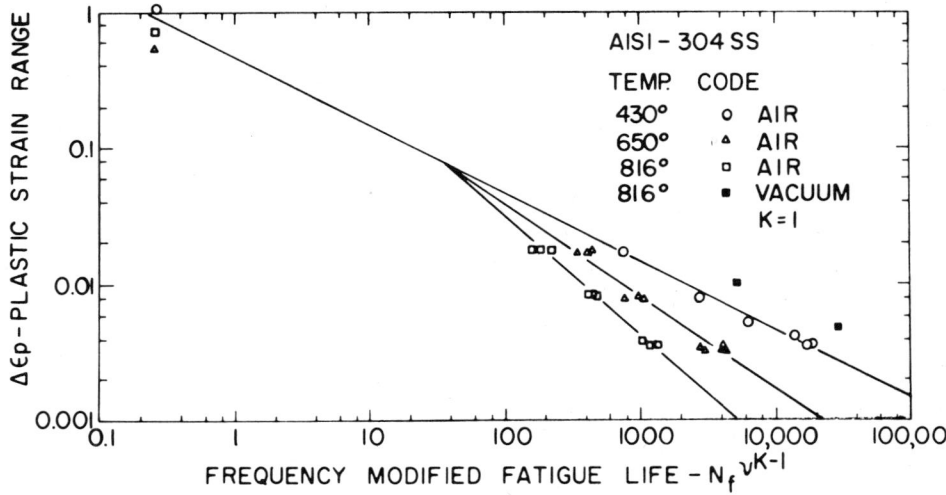

Fig. 10. Frequency-modified fatigue life in AISI 304 stainless steel.[44]

of a fatigue crack was more extensive at lower frequencies, permitting greater crack growth per cycle. In contrast to this behavior, samples tested in vacuum fractured by a transgranular mode, and the frequency effect shown by the air tests disappeared.

The significance of these findings to LMFBR and Tokamak applications arises not only from considerations of the thermal and strain cycling that metallic components will be subjected to in service but also the potential effects of the coolant environment, internally generated gases (helium and hydrogen), and, in the case of Tokamak first walls, injected ions. Design relationships derived from test data in one environment in the laboratory may be substantially in error when applied to the specific service environment.

The effects of damage accumulation and recovery kinetics become added complications when designing a first wall for a Tokamak reactor. Typical incubation and relaxation periods are compared with proposed fusion burn and neutron-off periods in Table XI. As noted from the table, the fusion burn period can be of the order of the steady-state, point-defect-production period but short relative to the damage-saturation period. In actuality, steady-state point-defect production is a temporal concept since the sink density is changing over a time period comparable to that for damage saturation. The neutron-off period is relatively long compared with the average vacancy lifetime but may overlap to some extent with normal damage-recovery periods. The damage-spike relaxation period is that time required for a vacancy-concentration peak generated by a displacement

Table XI. Incubation and Relaxation Periods[a]

Period for:	$\tau = 63\%$ of t_{max}, sec
4. Steady-state point-defect production	10^0-10^3
2. Average vacancy lifetime	10^{-3}-10^0
1. Displacement-spike relaxation period	10^{-4}-10^{-3}
3. Spike injection period within average vacancy migration volume	10^{-3}-10^0
7. Damage saturation period	10^4-10^6
5. Fusion burn period	10^2-10^4
6. Neutron-off period	10^0-10^2
7. Confinement time	10^0-10

(a) Stainless steel, $\tau = 500$ C, $\phi = 2 \times 10^{15}$ n/cm^2 sec.

spike to relax and the displacement-spike injection period is the period between spike overlap in a volume defined by the average migration distance of the vacancy.[45]

The implications of Table XI are that a cyclic neutron flux during a Tokamak duty cycle can substantially alter the kinetics of irradiation-damage phenomena measured during steady operation. Furthermore, when these flux cycles are superimposed upon stress and temperature cycles, unusual ratchetting effects may result.

Recent theoretical calculations by Choi et al.[46] reveal that steady state may never be achieved under the time constants and operating conditions of proposed fusion devices. Preliminary results of transient-void-nucleation calculations for nickel at 500 C are shown in Fig. 11. The procedure was to establish an initial damage rate and watch the void concentrations increase with time. The initial vacancy supersaturation ($S=10^4$) gave copious nucleation, which in 34.5 seconds produced the void densities [$\rho(n)$ versus n, where n is the number of vacancies in the void] labelled t = 0 in Fig. 11. The term $F_i=\beta_i/\beta_v$ is the ratio of interstitial-to-vacancy arrival rates at a neutral sink. At t = 0 (following the initial 34.5-second burn period), the vacancy supersaturation was reduced to 10^3 and the readjustment in void concentrations was followed. The adjustment at the small n is more rapid at short times (t ~ 10 seconds), but even large voids are seen to decompose after 7 × 10^3 seconds. The results of these preliminary calculations indicate that even though the time constant for void decay is much longer than that for void formation (under the assumption of homogeneous nucleation), voids above the critical nucleus size may decay after a sharp reduction in damage rate. Therefore, a transient nucleation formulation must be used to describe void nucleation in CTR materials subjected to a pulsed duty cycle.

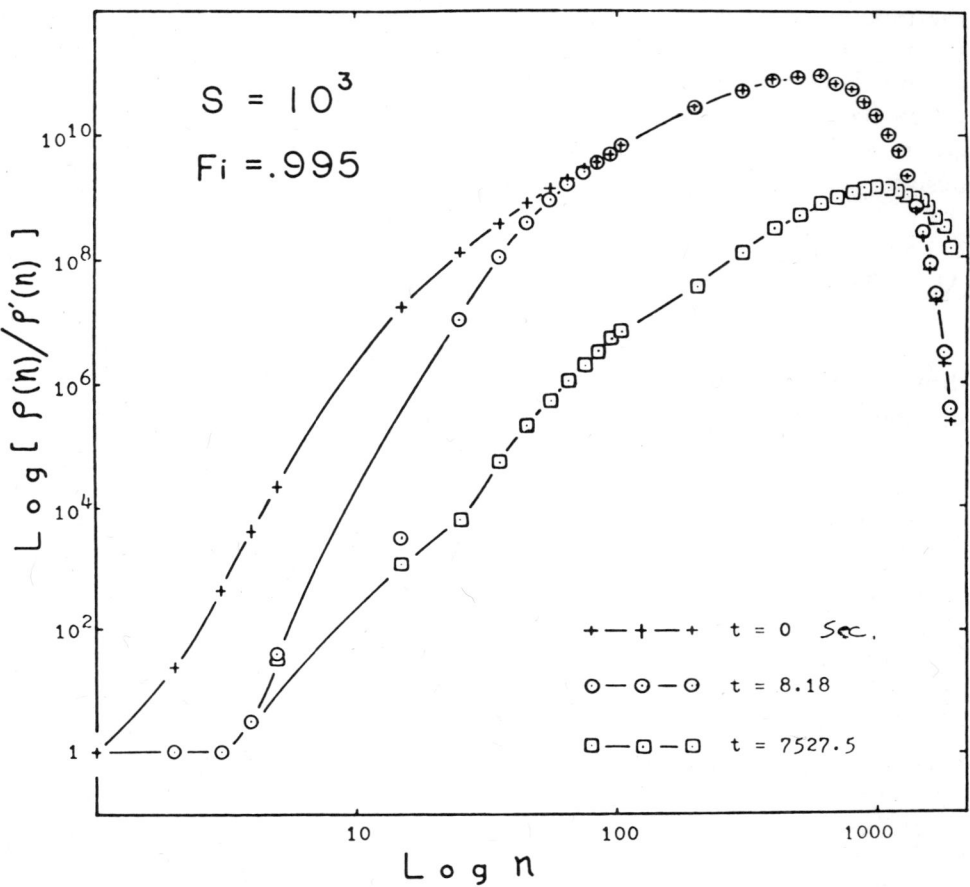

Fig. 11. Void density versus size after an instantaneous change in vacancy supersaturation (S is vacancy supersaturation, F_i is the ratio of interstitial-to-vacancy arrival rates at the voids).[45]

6 CLOSING STATEMENT

Structural alloys subjected to a neutron environment will store significant amounts of internal energy due to atomic displacements, clustered
defect formation, and the accumulation of neutronically generated gases, all
under highly nonequilibrium and irreversible conditions. The manner in
which the alloy system undergoes internal conversion to lower energy states
or arrives at a steady-state condition is highly complex and is subject to considerable uncertainties because of the presence of high supersaturations of
two point-defect species (vacancies and interstitials) and displacement
cascades. These considerations become increasingly important at high
neutron fluxes and neutron energies. Alloy theories based upon classical

thermodynamic relationships and equilibrium considerations are inadequate to treat these problems and have little predictive capability in such environments. Furthermore, design practices and empirical relationships which have stood in good stead for the design of light-water-reactor core components have only limited applicability for LMFBR and Tokamak design.

ACKNOWLEDGMENT

Appreciation is expressed to K. C. Russell and R. M. Rose for helpful discussions concerning nucleation effects and irradiation effects on superconductors, respectively. The sponsorship of ERDA (DCTR) under contract C00-2431 is also gratefully acknowledged.

REFERENCES

1. Steel, L. E., *Neutron Irradiation Embrittlement of Reactor Pressure Vessel Steels*, Technical Report Series No. 163, International Atomic Energy Agency, Vienna, Austria (1975).
2. Steiner, S., Wash-1206, U.S. Atomic Energy Commission (April, 1972).
3. Lubell, M. S., et al., in *Fourth IARA Conference on Plasma Physics and Controlled Nuclear Fusion Research*, Madison, Wisconsin (June, 1971).
4. Golovin, I. N., et al., in *Nuclear Fusion Reactor Conference Proceedings*, Culham (1970).
5. Hubert, P., *loc. cit.*
6. Kulcinski, G. L., Doran, D. G., and Abdou, M. A., in *Proceedings of the 7th ASTM International Symposium on Radiation Effects on Structural Materials*, Gatlinberg, Virginia (June, 1974).
7. Kulcinski, G. L., Personal Communications.
8. Beeler, J. R., Jr., in *Radiation Damage in Reactor Materials*, Vol. II, IAEA, Vienna, Austria (1969), p. 12.
9. Beeler, J. R., Jr., *Lattice Defects and Their Interactions*, Gordon and Breach, New York, N.Y. (1967), p. 621.
10. Beeler, J. R., Jr., *J. Nucl. Mater.*, **53**, 207–212 (1974).
11. Johnson, A. B., Jr., and Irvin, J. E., BNWL-463, Battelle's Northwest Laboratories, Richland, Washington (1967).
12. Kulcinski, G. L., and Emmert, G. A., *J. Nucl. Mater.*, **53**, 31 (1974).
13. Kaminsky, M., and Das, S. K., *J. Nucl. Mater.*, **53**, 162 (1974).
14. Harling, O. K., Private Communication.
15. Damask, A. C., paper presented at *AIME Symposium on Radiation Effects*, Asheville, North Carolina, September 8-10, 1965.
16. Piercy, G. R., *J. Phys. Chem. Solids*, **23**, 463 (1963).
17. Wiedersich, H., in *Second International Conference on Strength of Metals and Alloys*, Vol. II, American Society for Metals, Cleveland, Ohio (1970), p. 784.
18. Leitnacker, J. M., Bloom, E. E., and Stiegler, J. O., *J. Nucl. Mater.*, **49**, 57 (1973).
19. Johnston, W. G., Rosolowski, J. H., Turkalo, A. M., and Lauritzen, T., *J. Nucl. Mater.*, **54**, 24 (1974).

20. Schilling, W., and Schroeder, K., AERE Report R-7934, AERE, Harwell, England (1974).
21. Smidt, F. A., Jr., et al., in *Defects and Defect Clusters in B.C.C. Metals and Their Alloys,* National Bureau of Standards, Gaithersberg, Maryland (1973), p. 341.
22. Anthony, T. R., *Phys. Rev. B,* **2**, 264 (1970).
23. Okamoto, P. R., and Wiedersich, H., AERE Report R-7934, AERE, Harwell, England (1974).
24. Cawthorne, C., Fulton, E. J., Bramman, J. I., Linekar, G.A.B., and Sharpe, R. M., in *Proceedings BMES Void Conference,* British Nuclear Energy Society, Harwell, England (1971), p. 35.
25. Brager, H. R., and Kissinger, H. E., *Trans. Amer. Nucl. Soc.,* **12**, 118 (1969).
26. Bisson, A., in *Proceedings of Karlsruhe Conference on Fast Reactor Fuel and Fuel Elements* (1970), 720.
27. Russell, K. C., in *KTG/BNES Conference on Irradiation Behavior of Fuel Cladding and Core Component Materials,* Karlsruhe, West Germany, December 3–5, 1974.
28. Nelson, R. S., Hudson, J. A., and Mazey, D. J., *J. Nucl. Mater.,* **44**, 318 (1972).
29. Rose, R. M., Personal Communications.
30. Sugisakik, T., Okada, T., and Suita, T., *Rept. Osaka Univ.,* **21**, 385 (1971).
31. Soell, M., Wipf, S. L., and Vogl, G., in *Proceedings of 1972 Applied Superconductivity Conference* (1972), p. 434.
32. Coffey, H. T., Keller, E. L., Patterson, A., and Autler, S. H., *Phys. Rev.,* **158**, 335 (1967).
33. Sweedler, A. R., Schweitzer, D. G., and Webb, G. W., to be published in *Appl. Phys. Lett.*
34. Gilbert, E. R., *Reactor Technology,* **14**, 258 (1971).
35. Hendrick, P. L., *Proton Simulated Irradiation-Induced Creep,* SM Thesis, Massachusetts Institute of Technology (1974).
36. Harkness, S. D., Tesk, J. A., and Li, Che-Yu, *Nucl. Appl. and Tech.,* **9**, 24 (1970).
37. Ansell, G. S., and Weertman, J., *Trans. AIME,* **215**, 838 (1959).
38. Hendrick, P. L., Ph.D. Thesis, Massachusetts Institute of Technology (1975).
39. Bement, A. L., *Adv. in Nucl. Sci.,* Vol. 7, Academic Press, New York, N.Y. (1973).
40. Lovell, A. J., and Barker, R. W., *ASTM STP,* No. 484, 468 (1970).
41. Pennell, W. E., Personal Communications.
42. Wood, D. S., *Welding J. Res. Suppl.,* **45**, 925 (1966).
43. Coffin, L. F., *Proceedings of 2nd International Conference on Fracture,* Chapman and Hall, London, England (1969).
44. Coffin, L. F., GERL Report No. 71-C-108, General Electric Research and Development Center, Schenectady, N.Y. (April, 1971).
45. Li, Che-Yu, Franklin, D. G., and Harkness, S. D., *ASTM STP,* No. 484, 347 (1970).
46. Choi, Y. A., Russell, K. C., and Bement, A. L., in *Technical Progress Report for Period July 1, 1974, to April 30, 1975,* M.I.T. Fusion Technology Program, COO-2431-2 (1975).

DISCUSSION on Paper by A. L. Bement

HAASEN: The critical current of niobium solid solutions first increases but then decreases with increasing neutron dose (at an intermediate magnetic field). The A15 compounds seem to be less irradiation sensitive in their critical currents if I remember correctly the work by Saur and Giessen. Can you confirm this?

BEMENT: For the case of niobium solid solutions, the point-defect clusters formed during the irradiation contribute to the fluxoid pinning, giving an initial increase in the critical current. However, at higher neutron dose, fine omega-phase precipitates can be redissolved due to the dynamic atomic mixing effects of displacement spikes. The concern in this case is replacement of a pinning microstructure that is relatively thermally stable with one which may undergo rapid recovery during warming cycles, either accidentally or intentionally imposed. The A15 compounds, which depend upon a high degree of order for a high T_c, are particularly susceptible to irradiation degradation due to disordering effects by γ recoil, replacement collisions, focusing, and channeling. The effect of irradiation on the critical temperature of A15 compounds may be of greater importance than that on critical current.

THOMAS: In view of the homogeneity and low mismatch in spinodal systems and in view of the fact that even in overaging when the wavelength exceeds ~ 1000 A, stable interface dislocation networks are formed, don't you think some investigations of the role of spinodal alloys under irradiating conditions would be worthwhile? The structures may change or coarsen but maybe the properties would not be too badly lowered. Considering all the alloy systems that have been studied, it would seem worthwhile to start a program on spinodal alloys (including searching for suitable spinodal systems).

BEMENT: Recently, there has evolved a much better understanding of the effectiveness of rather small trapping binding energies in enhancing the point defect recombination and retarding void nucleation and growth kinetics. Your suggestion viewed from these findings has considerable merit and should be pursued. One can expect, however, that the periodicity in spinoidal systems will be quickly destroyed by random atomic displacements and recoils. Nevertheless, the conventional thermodynamic approaches for predicting probable alloy structural states are largely inapplicable for a neutron environment, and direct experimental exploration of spinodal alloys systems is warranted.

McCLINTOCK: What are the best sources of data and combined stress/strain/time/temperature/structure/radiation relations for mechanics analysis of reactor components? Are data, relations, or calculation techniques (or all three) the primary obstacles?

BEMENT: One of the greater limitations in obtaining useful design data for the mechanical analysis of reactor components is the extreme expense and difficulty of conducting even simple tests on radioactive specimens under remote handling conditions. Some notable work on austenitic stainless steels in this area includes the subcritical-crack-growth measurements and analyses by James at Westinghouse-Hanford and Shahinian at Naval Research Laboratories, the analysis of constitutive relationships and failure limits by Blackburn of Westinghouse-Hanford, and combined creep-fatigue measurements by Brinkman of Aerojet Nuclear Co. However, the overall understanding of creep under variable stress, combined creep-fatigue effects, and rupture limits under history-following conditions for irradiated structural materials operating in the elastic-plastic regime is very inadequate for confident mechanical design.

PROBLEMS OF ALLOY DESIGN IN PRESSURE VESSEL STEELS

C. J. McMahon, Jr.

Department of Metallurgy and Materials Science
University of Pennsylvania
Philadelphia, Pa.

ABSTRACT

The fracture resistance of alloy steels at ordinary and elevated temperatures and in aggressive environments is at present limited by tendencies toward intergranular fracture at low stress levels. The theses developed here are that these tendencies result from segregated metalloid impurities and that this problem can be alleviated by microstructural control and by adjustments of the alloy content of the steel.

1 INTRODUCTION

As with alloy design in general, the optimum design of a pressure-vessel steel requires a set of compromises among conflicting requirements. Some of these constraints are illustrated by Table I, in which arrows indicate whether particular compositional or microstructural variables should be increased or decreased to achieve the various requirements. The first four requirements involve compromises to which metallurgists have become accustomed, e.g., strength versus toughness. Thus, it is well known that to achieve maximum toughness, one must use the minimum allowable carbon content, in addition to a fine grain size, a fine carbide distribution, and a low sulfur content. The fifth requirement refers particularly to usage under conditions of high-pressure H_2 at high temperatures, as in petroleum processing systems. Here, stable carbides are required to inhibit the reaction:

$$\underline{H} + M_x C_y \rightarrow \underline{M} + CH_4$$

which produces methane-filled voids in the steel.

Table I. Conflicting Relationships Between Variables and Requirements for Pressure Vessel Steels

Requirements	C	Mn	Si	Cr Mo V	Ni	S	Grain Size	Hardness	P Sb Sn As
Hardenability	↑	↑		↑	↑				
Weldability	↓	↑	↑					↓	
Creep Strength	↑			↑			↑	↑	
Fracture Toughness (TG fracture)	↓	↑			↑	↑	↓	↓	
Resistance to CH₄ Cavitation	↓			↑		↓		↑	
Fracture Toughness (IG fracture)	↓	↓	↓		↓		↓	↓	↓
Stress-Rupture Ductility (IG fracture)	↓	↓	↓				↓	↓	↓
Environment Resistance (IG fracture)	↓	↓	↓			↓	↓	↓	↓

The last three requirements have not been well understood, and this appears to be the area which offers the most return for expended effort. Here, one must deal with the effects of metalloid impurities (mainly, group IV and V elements) in promoting intergranular fracture by:

1. Lowering intergranular cohesion
2. Enhancing intergranular diffusive crack growth at elevated temperatures
3. Promoting intergranular H_2-assisted cracking
4. Promoting intergranular stress-corrosion cracking in aqueous media.

These effects place constraints on the choices of variables to achieve the requirements listed in the upper part of Table I; note that about half of the arrows indicating "increase" in the upper part are reversed in the lower part. Thus, if we could be released from these constraints (e.g., on the use of nickel and manganese) we could improve alloy performance in several critical areas. Hence, we concentrate here on the impurity effects.

2 INTERGRANULAR EMBRITTLEMENT

2.1 Methods of Measurement

The tendency for impurity-induced intergranular embrittlement is generally measured by a blunt-notch test, such as the Charpy V-notch impact test, which yields curves such as in Fig. 1(a). For a given steel, with a fixed carbon content, microstructure, and strength level, the baseline is the left-most curve showing the reduction in fracture energy accompanying the shift in fracture mode from microvoid coalescence to cleavage. As the tendency toward intergranular fracture increases, the curves shift to the right and generally rotate clockwise. Essentially, the transition is from microvoid coalescence after large plastic strains to intergranular cracking with only small plastic strains. The embrittlement can be denoted by: (i) the reduction in fracture energy at a convenient arbitrary temperature, (ii) the shift in "transition temperature" at which some arbitrary fracture energy is reached, or (iii) the shift in "transition temperature" at which some arbitrary fracture appearance is reached. An example of the latter is the 50 percent FATT shown in Fig. 1(b). Such tests are widely used for comparison of materials and as criteria of acceptability, but of course they provide no information about critical flaw sizes in components such as pressure vessels or turbine rotors. For this, a sharp crack test would be necessary, and, in principle, one would measure a shift in G_c (or K_{Ic}) at an arbitrary temperature [as in (i), above] or a shift in "transition temperature" at an arbitrary G_c (or K_{Ic}) level [as in (ii), above]. This is generally not done in embrittlement

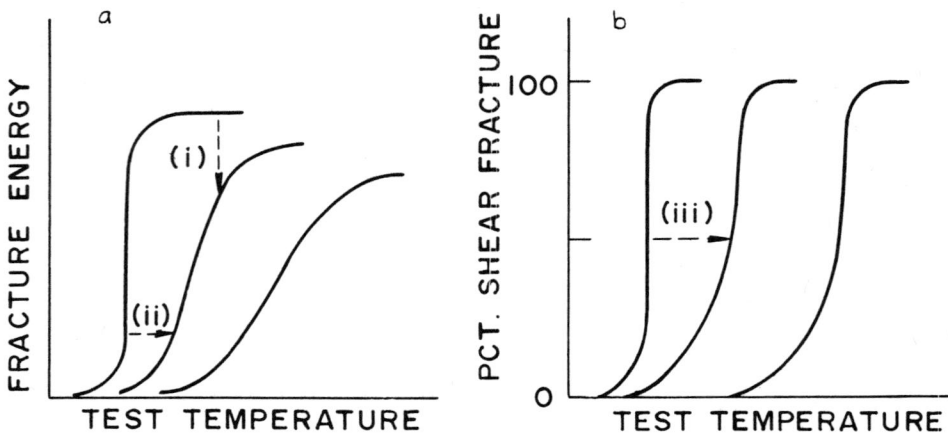

Fig. 1. Various methods for characterizing embrittlement with Charpy V-notch specimens (see text).

studies with medium-strength pressure-vessel and rotor steels because of the specimen size and expense involved.*

Since impurity-induced intergranular embrittlement is normally observed in microstructures of tempered martensite or bainite, it is called temper embrittlement. In an attempt to clarify the nomenclature, it has been suggested[1] that we differentiate between one-step temper embrittlement, which is found in high-strength steels and is measured by method (i), above, and two-step temper embrittlement of medium-strength steels, which is normally measured by method (ii) or (iii), as indicated by Fig. 2. Here, we deal mainly with the latter.

2.2 Nature of the Problem

Optimization of an alloy steel for resistance to temper embrittlement is best carried out in two stages, which can be stated simply as: (1) getting the best possible behavior out of the present steel and (2) improving the type of steel. To accomplish the former, one must recognize that the controlling variables are two kinds:

Chemical: 1. Concentration of a particular impurity in the grain boundaries
                 2. Potency of that impurity in lowering cohesion

---

*Also, sharp-crack-toughness tests can be dangerously misleading in steels which tend toward intergranular fracture, particularly in coarse-grained microstructures.[28,29]

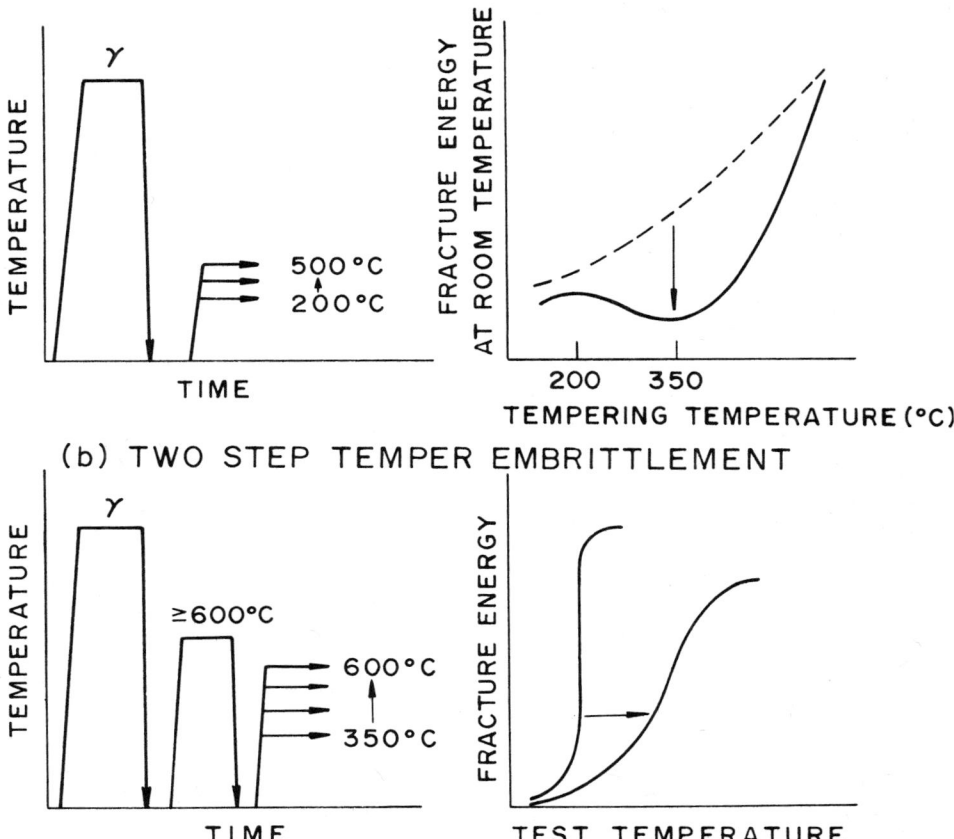

(a) ONE STEP TEMPER EMBRITTLEMENT

(b) TWO STEP TEMPER EMBRITTLEMENT

Fig. 2. Thermal treatments and usual characterization tests for one-step and two-step temper embrittlement.

Microstructural: 3. Hardness
                      4. Grain size

Their effects can be illustrated in the following way: Figure 3 shows how the 2 ft-lb transition temperature [method (ii)] of 1/4-inch circumferentially notched bars of 3.5% Ni-1.7% Cr steel varies with grain-boundary concentration of phosphorus, tin, and antimony in specimens with fixed hardness and grain size.[2,3] Here we can see that the embrittlement is approximately proportional to impurity concentration, as measured by Auger electron spectroscopy (AES), and that antimony and tin are much more potent than phosphorus. The effect of varying only hardness in the same steel is shown in Fig. 4. The grain-size effect is illustrated by Fig. 5, taken from the work

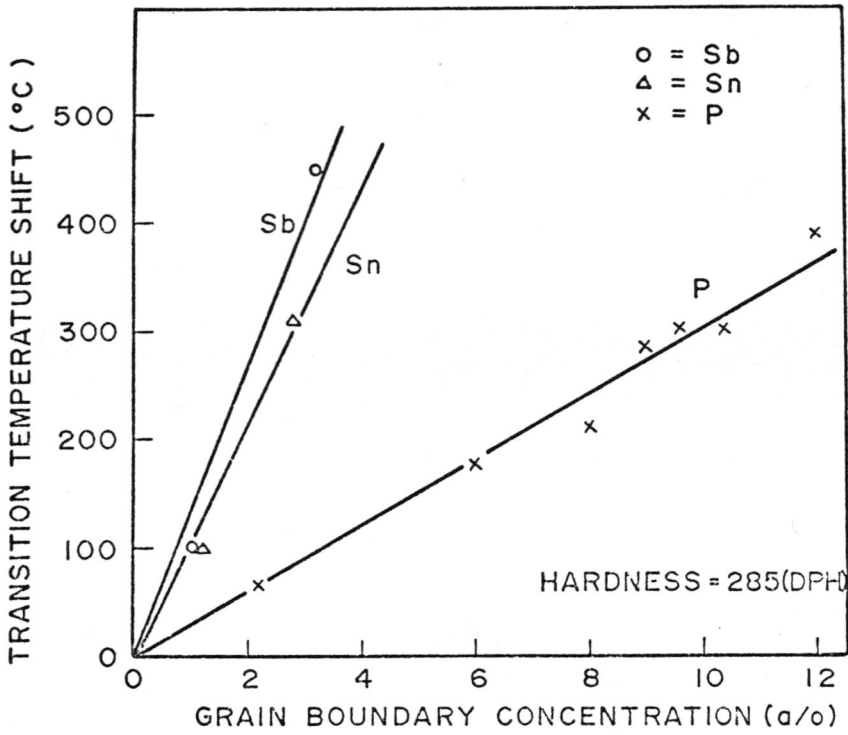

Fig. 3. Embrittlement diagram showing variation of transition temperature with grain-boundary concentration of phosphorus, tin, or antimony at a fixed hardness and grain size.[2,3,29]

of Capus[4] on a 3% Ni-0.75% Cr steel using Charpy specimens. Such plots can, in principle, be used to ensure that a component made from a given heat of steel never attains a transition temperature above a certain value (or a toughness lower than a certain value at some selected temperature). For example, suppose that the steel must be exposed to a particular thermal history. The grain-boundary composition that will result from this can be determined by use of AES on samples given either the same treatment, or treatments for short times at high temperatures to establish the segregation kinetics and the equilibrium segregation level as a function of temperature. Examples of such data for phosphorus in a NiCr steel are shown in Fig. 6. Knowing the maximum intergranular impurity concentration which can be achieved, the hardness and grain-size effects can be measured in samples having that concentration. In the three-dimensional diagram shown in Fig. 7, in order to remain below the maximum allowable transition temperature, shown by the shaded plane, the hardness and grain-size values must fall within the cross-hatched region. Thus, if the grain size cannot be refined below point A, then the hardness must be kept below point B.

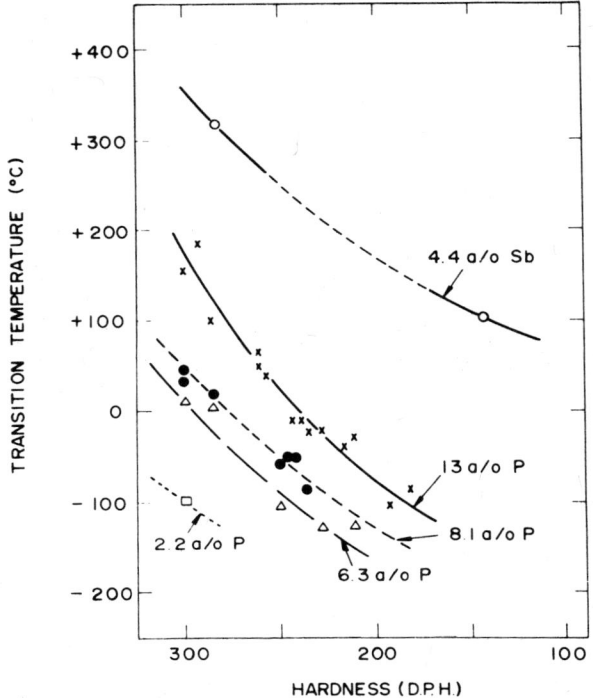

Fig. 4. Embrittlement diagram showing variation of transition temperature with hardness for various grain-boundary concentrations of phosphorus or antimony at a fixed grain size.[2]

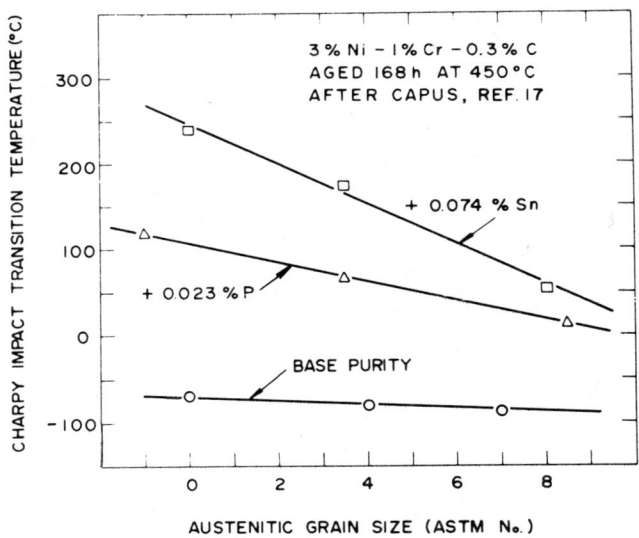

Fig. 5. Embrittlement diagram showing variation of transition temperature with grain size in a Ni-Cr steel with fixed hardness and degree of segregation of phosphorus or tin.[4]

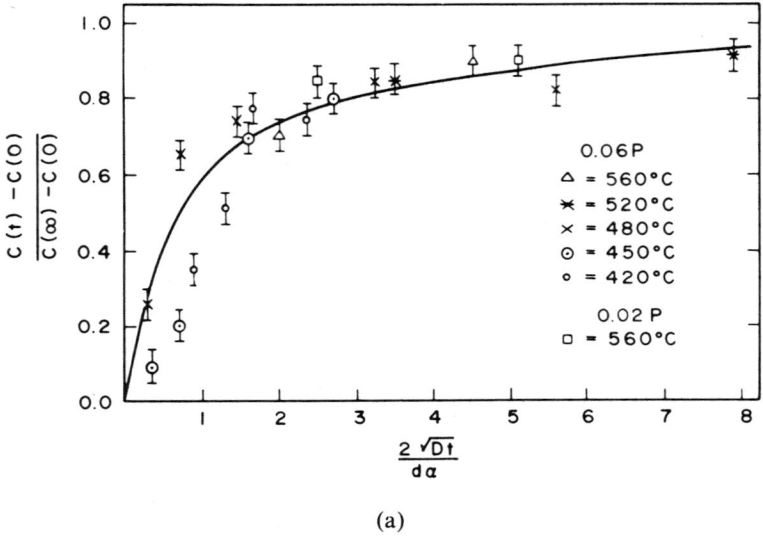

(a)

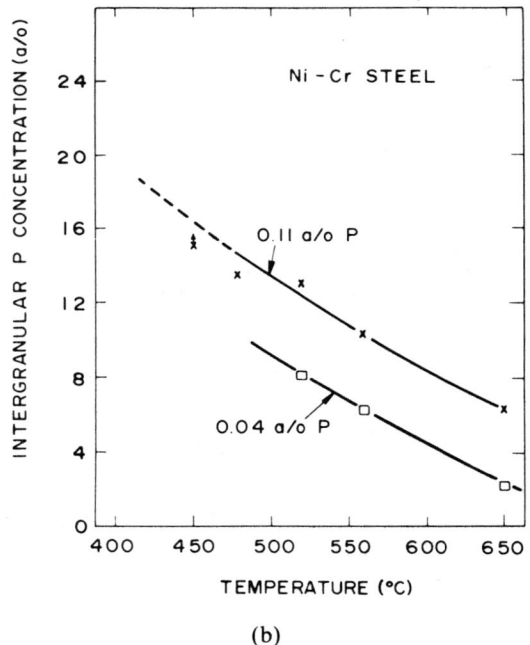

(b)

Fig. 6. (a) Rate of approach to saturation of phosphorus segregation in a Ni-Cr steel, show-ing experimental points and calculated curve. (b) Variation of equilibrium grain-boundary concentration of phosphorus with temperatures for two bulk concentrations 0.06 wt % (0.11 at. %) and 0.02 wt % (0.04 at. %).[2]

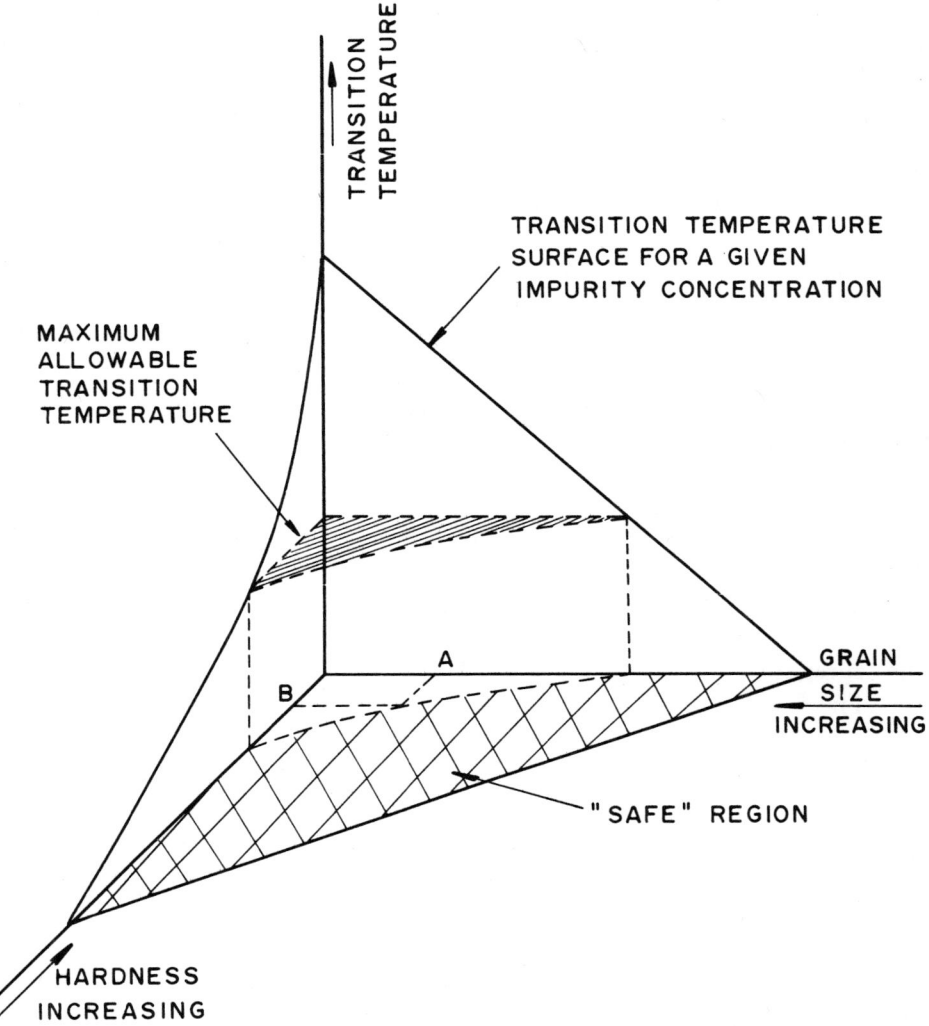

Fig. 7. Three-dimensional embrittlement diagram of the type needed for optimization of resistance to embrittlement (see text).

2.3 Factors Which Control Segregation

The procedure just described would be relatively straightforward in a simple steel, such as the laboratory heats of Ni-Cr steel which have been extensively studied. However, in an actual commercial steel there can be delayed reactions which would complicate the embrittlement kinetics. An example is the case of the Mo-P-C reaction, which is discussed below.

The above approach is essentially a stopgap, and it would be imprac-ticable in many cases, although it would always be useful for rough ap-proximations and for conceptual purposes. The second stage of optimiza-tion will involve fundamental compositional improvements in the various types of alloy steel. Of course, reductions in the residual impurity levels would always be beneficial, but the major improvements will arise from an understanding of the alloy-impurity interactions which control impurity segregation. These are now beginning to be understood, and the current status is as follows.

It is now clear from recent studies of a Ni-Cr steel doped with an-timony[5], phosphorus[2], and tin[3] and of a manganese steel doped with an-timony[6] that the impurity segregation to grain boundaries is an equilibrium phenomenon. The attractive interaction between specific alloy and impurity elements gives rise to cosegregation of these elements. This has been rationalized by the thermodynamic theory of Guttmann[6], who used a regular solution model of a ternary alloy to take account of solute interac-tions. He derived equations which contain an interaction coefficient α, the magnitude of which can be used to assess the tendency for segregation, and for his FeMnSb steel he estimated α from the heat of formation of MnSb to be ~ 20 kcal/g-atom. In this alloy, the embrittlement was accompanied by the segregation of manganese and antimony.

The Guttmann analysis is consistent with our own findings on the em-brittlement of Ni-Cr steel by antimony, tin, and phosphorus. These studies involved the correlation of variations in transition temperature and grain-boundary composition, as measured by AES, which resulted from isother-mal aging. It was found that in the antimony-doped steel there existed a relationship between the concentrations of nickel and antimony in the grain boundaries, as shown in Fig. 8, and that a similar relationship occurred in the tin-doped steels (Fig. 9). This was rationalized in terms of the results of Nageswararao et al.[7] who found that nickel lowers the solubilities of an-timony and tin in α-Fe, as shown in Fig. 10. This occurs because nickel par-titions to the FeSb and Fe_5Sn_3 phases and lowers their free energy. It is in-teresting to note that Guttmann's estimated α value for NiSb is 27 kcal/g-atom; this indicates that the Ni-Sb interaction should produce somewhat more embrittlement than that of Mn-Sb. Figure 10 also indicates that chromium lowers the solubility, and thus the chemical potential[1], of tin and antimony in iron; hence, chromium should also promote segregation of these elements. The AES results for chromium concentration on embrittled grain boundaries are not clearly interpretable because of the presence of chromium-rich carbides. However, we have indirect evidence that chromium is important. This is shown in Fig. 11; the high- and low-carbon NiCr steels exhibit essentially equivalent amounts of antimony segregation when aged at 480 C, where formation of chromium-rich carbides is slow. However, ag-ing at 520 C, where these carbides form much more rapidly, produces much

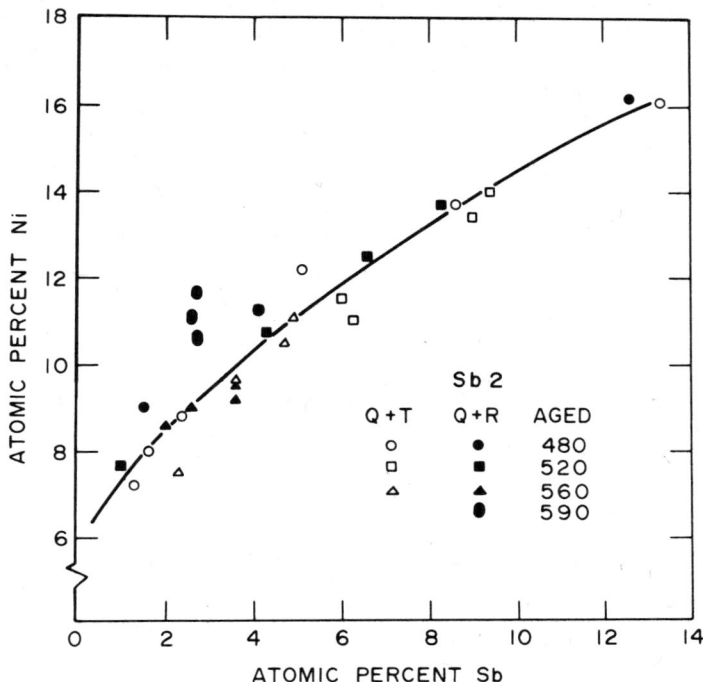

Fig. 8. Relationship between nickel and antimony concentrations in grain boundaries of embrittled Ni-Cr steel.[5]

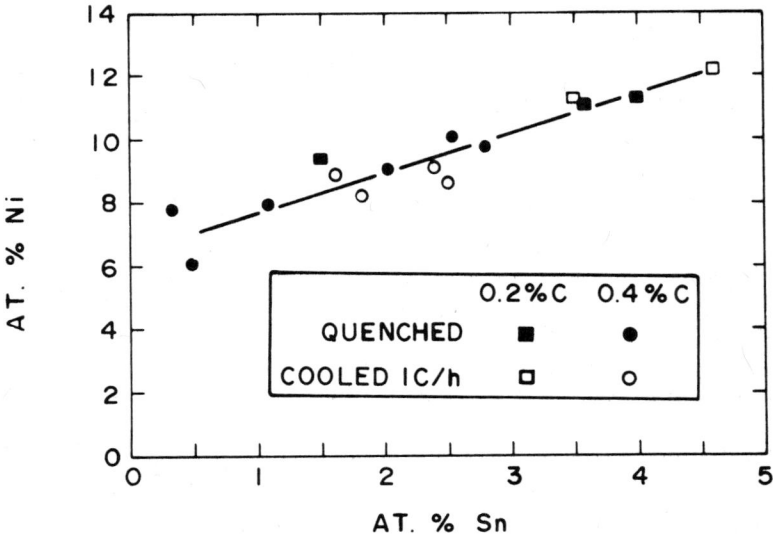

Fig. 9. Relationship between nickel and tin concentrations in grain boundaries of embrittled Ni-Cr steel.[3]

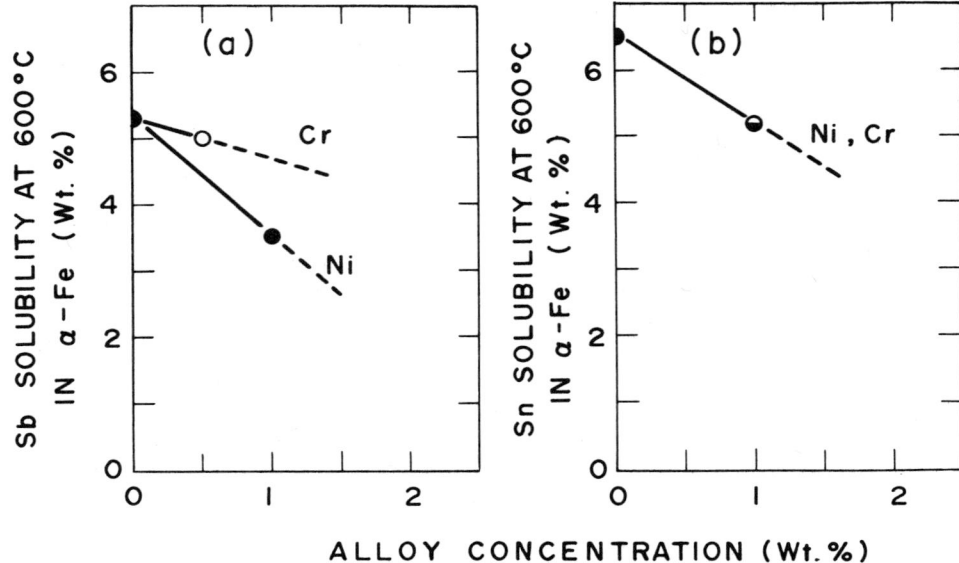

Fig. 10. Effects of nickel and chromium solubilities of antimony and tin in α-Fe.[7]

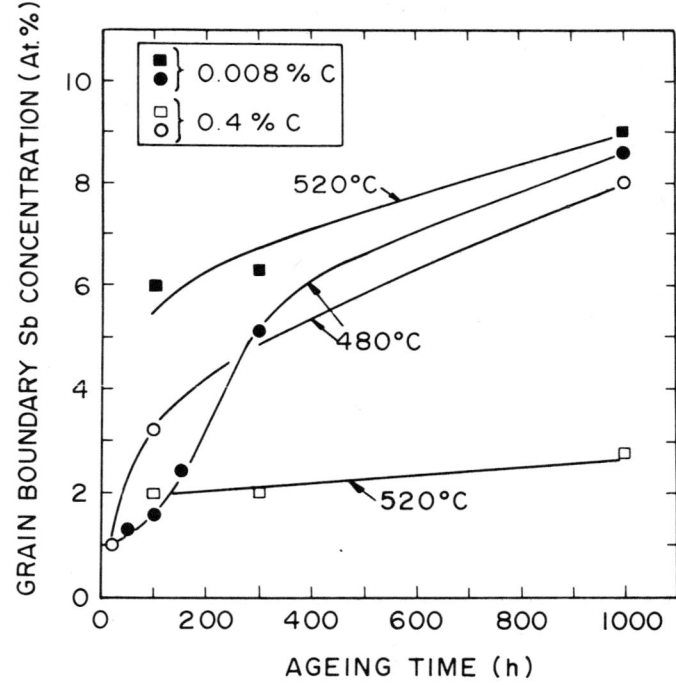

Fig. 11. Influence of aging temperature on the segregation of antimony in high- and low-carbon Ni-Cr steel, showing effect of chromium-depletion in the matrix.[29]

less antimony segregation in the high-carbon steel, in which the matrix is presumed to have been depleted of chromium by the carbide formation.

Carbide formation has also been implicated in the embrittlement reaction in a different way.[8-10] Since nickel is essentially insoluble in carbides, it will be expelled when carbides form in grain boundaries, and the grain-boundary concentration of nickel will thereby be enriched. This will in turn create a gradient in the chemical potential of, say, antimony, which will then diffuse in from the matrix. Hence, carbide formation can hasten the approach to equilibrium, since the segregation process involves only the diffusion of antimony into the free-energy "well" provided by the nickel excess, rather than the cosegregation of both nickel and antimony from the matrix.

A different kind of alloy-impurity interaction occurs when 0.1 percent titanium is added to a low-carbon Ni-Cr steel doped with antimony.[9,11] This is illustrated by Fig. 12 which has several important features:

1. The titanium addition virtually eliminates the embrittlement by antimony, except for a small transient effect.

2. The transient transition-temperature rise in the titanium-treated steel is matched by a similar trend in the antimony concentration at grain boundaries.

3. For a given amount of antimony at the grain boundaries, the titanium-treated steel has a substantially lower transition temperature. Since titanium was also found to be segregated, it appears that the substitution of titanium for iron at the boundaries mitigates the embrittling effect of antimony. This raises the question of the physical basis for embrittlement by groups IV to VI "metalloid" elements. A brief, tentative discussion of this point has been given[11] in terms of the increased electron density contributed by such elements, the possibility that the extra electrons enter the d-band of iron and decrease the d-electron contribution to bonding, and the increase in the density of states at the Fermi level in the d-band when titanium is added to iron. Obviously, this issue needs much more attention.

4. The titanium-treated steel exhibited a large concentration of nickel in the grain boundaries after tempering (1 hour at 625 C) which diminished during aging at 480 C. This suggests that a transient Ni-Ti interaction may occur, but the details are not all clear.

One effect of titanium is to eliminate carbide formation during the aging treatment at 480 C; this presumably would at least alter the kinetics of embrittlement. We have no knowledge yet of a possible strong interaction between titanium and antimony which might cause precipitation in the matrix.

The behavior of phosphorus in the Ni-Cr steel is analogous to that of antimony and tin. The main differences are that phosphorus is a less potent embrittling element, it diffuses much faster, and its interaction with nickel is

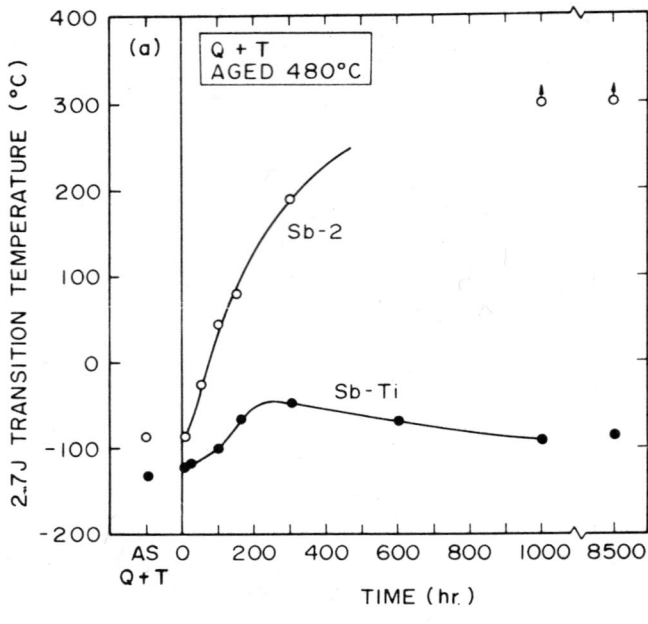

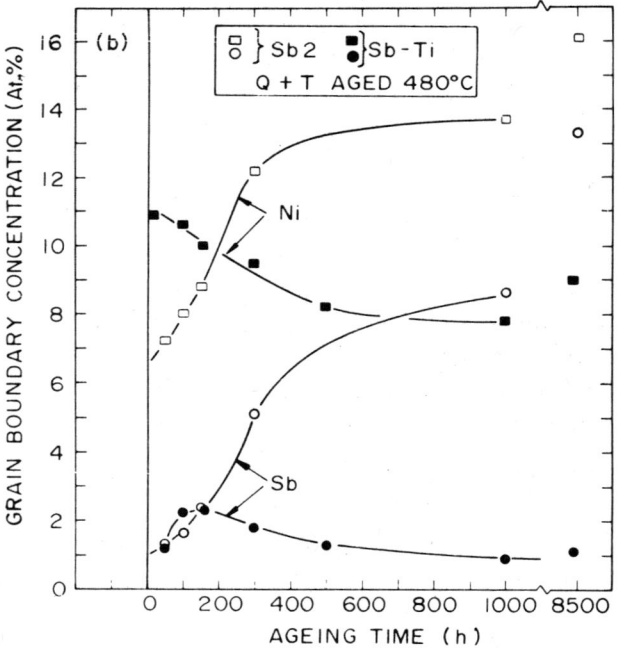

Fig. 12. Effect of the addition of 0.1 percent titanium on tin-induced embrittlement in a low-carbon Ni-Cr steel.[11]

weaker, as shown by Fig. 13. Some important insights can be gained from the solubility measurements of Kaneko et al.[12], as indicated by Fig. 14. They have shown that the attractive interaction between phosphorus and a number of alloy elements increases in strength in the order Ni, Mn, Si, Al, Cr, V, W, Mo, Nb, Ti, Zr. Thus, the tendencies for both cosegregation to grain boundaries and precipitation of a phosphide in the matrix increase in this order. (In the formalism of the Guttmann analysis[6] the interaction coefficient α increases in this order.) This means that we should see, first, an increase in segregation (and embrittlement) in ternary Fe-P-M alloys in this series and then a decrease due to the precipitation of phosphorus *in situ* in the matrix. This, in fact, is observed. Mulford et al[2] have shown that phosphorus segregation is enhanced more by chromium than by nickel, and Ohtani et al.[9] have found that phosphorus-induced embrittlement in a low-carbon Ni-Cr steel can be completely eliminated by titanium, as shown by Fig. 15. The effect of molybdenum is particularly interesting. Figure 16(a) contrasts the behavior of the Ni-Cr steel doped with 0.02 percent phosphorus with the same steel when 0.6 percent molybdenum is added. In the molybdenum-free steel there is substantial embrittlement ($\Delta TT \simeq 100$ C)

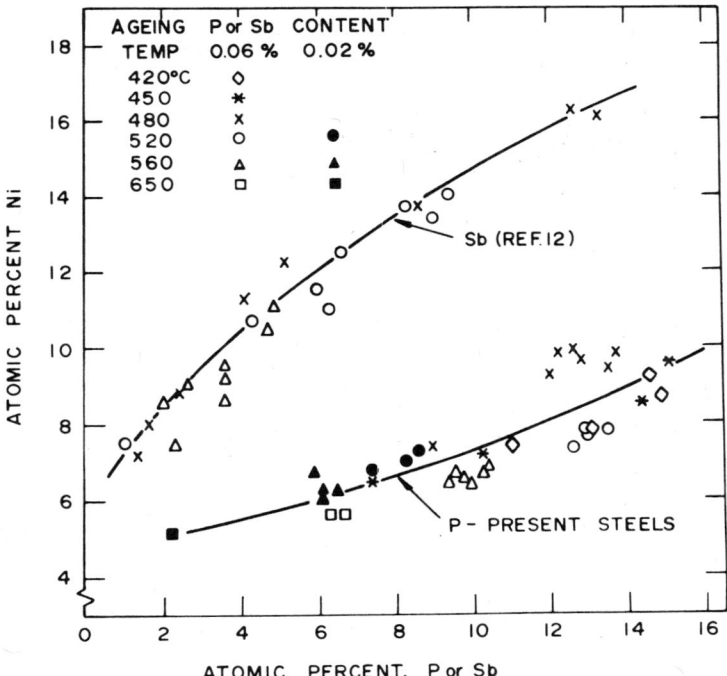

Fig. 13. Comparison of the relationships between grain-boundary concentrations of nickel and phosphorus and between nickel and antimony in embrittled Ni-Cr steel.[2]

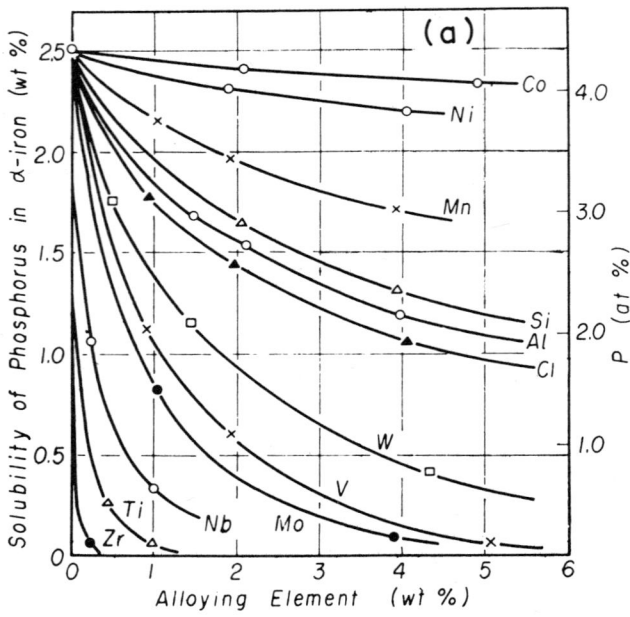

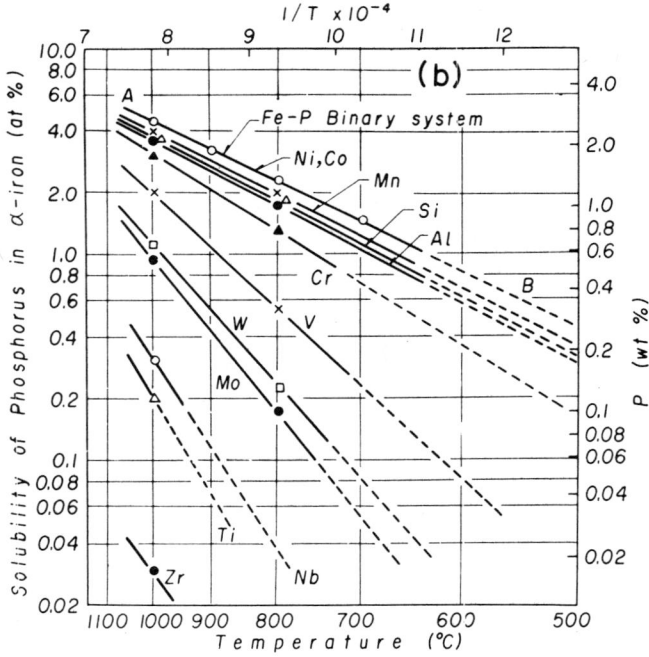

Fig. 14. Effects of various elements on the solubility of phosphorus in α-Fe: (a) solubility at 1000 C; (b) solubility for 1 percent alloying element as a function of temperature.[12]

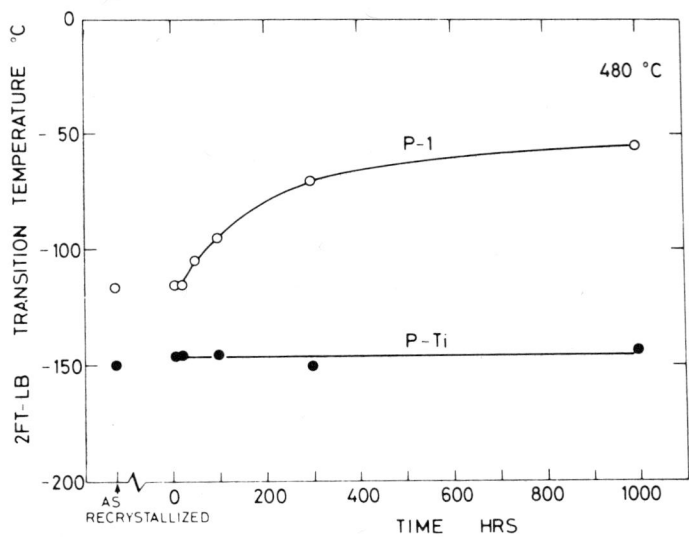

Fig. 15. Effect of the addition of 0.1 percent titanium on phosphorus-induced embrittlement in a low-carbon Ni-Cr steel.[9]

and a slight "overaging" effect due to the softening at 500 C. However, with molybdenum added, the embrittlement is virtually eliminated even though the hardness is significantly higher. Figure 16(b) shows the behavior of a similar molybdenum-treated steel, except the phosphorus level has been raised to 0.06 percent. During the first 100 hours at 500 C, embrittlement is slight and appears to have saturated. However, continued aging produces a large upward shift in transition temperature, which is accompanied by an increase in the amount of intergranular fracture and in the phosphorus and nickel concentrations on grain boundaries.[13] In view of the solubility data in Fig. 14, it would appear that 0.1 percent titanium can completely tie up 0.06 percent phosphorus in a low-carbon (<.01 percent) Ni-Cr steel. Similarly, 0.6 percent molybdenum in a 0.2 percent carbon Ni-Cr steel can tie up 0.02 percent phosphorus almost completely for at least 1000 hours at 500 C. However, at the 0.06 percent phosphorus level, the removal of molybdenum from solution by the formation of molybdenum-rich carbides is sufficiently great to break up the Mo-P interaction and to allow phosphorus segregation and concomitant embrittlement. Thus, the fact that the strong phosphide formers are also strong carbide formers means that a steel protected against short-time, phosphorus-induced embrittlement by molybdenum may nevertheless embrittle in long-time aging unless the molybdenum itself is somehow protected against precipitation by carbon. Hopefully, it will be possible to achieve the necessary balance of carbide-forming elements in a commercial steel such that embrittlement by phosphorus can be controlled.

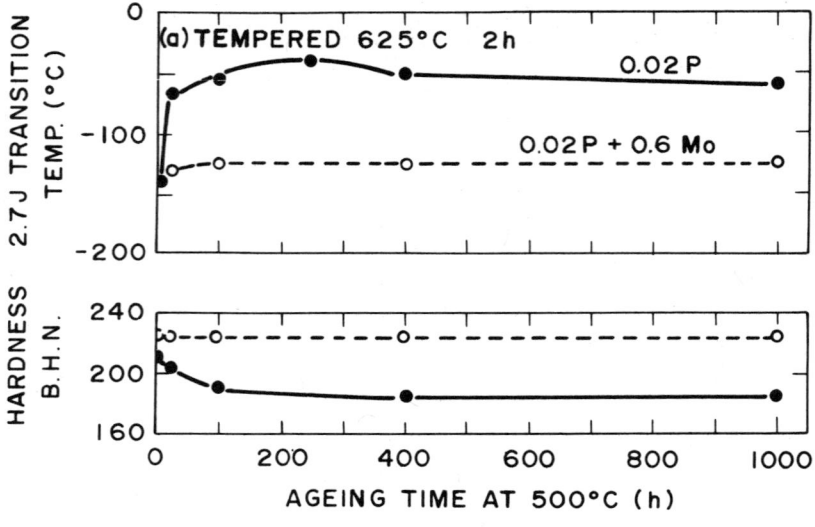

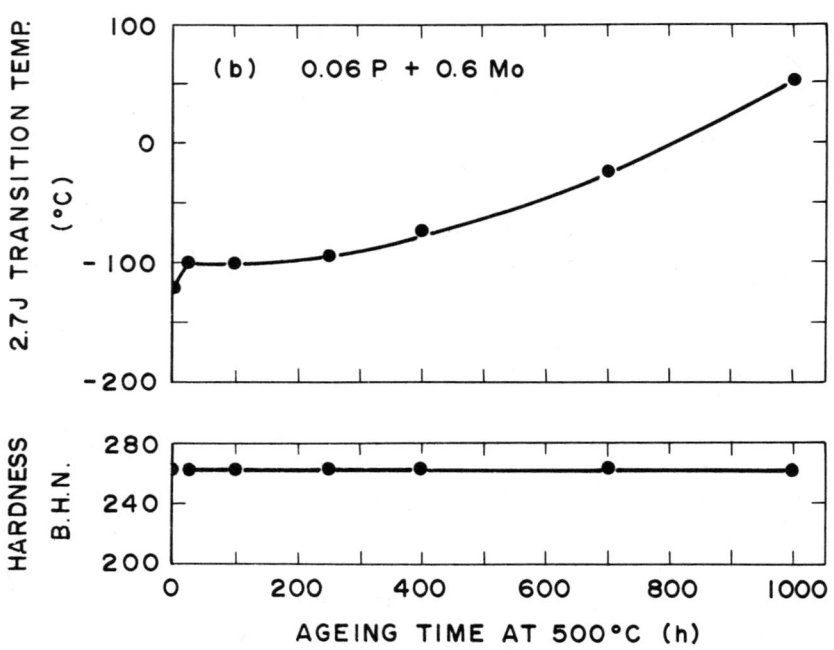

Fig. 16. (a) Embrittlement of a 0.2 percent Ni-Cr steel by 0.02 percent phosphorus with and without 0.6 percent molybdenum. (b) Embrittlement in a similar steel, except with 0.06 percent phosphorus.[13]

2.4 Hardness Effects

Returning to the effect of hardness, it appears to be a general principle that as the local hardness increases it becomes easier both to open an interface microcrack during plastic flow and to propagate that crack along a grain boundary. Physically, it is a question of the ease of plastic relaxation in a region of blocked shear, such as in Fig. 17(a), that determines the tendency for microcrack initiation, and the size of the plastic zone around a microcrack tip, as in Fig. 17(b), that governs the tendency for intergranular cracking (at a given grain-boundary impurity concentration). This, of course, is why the tendency for intergranular fracture increases with decreasing temperature in steels where the yield stress shows a similar increase. The hardness effect can also show up clearly in the overaging phenomenon, depicted by Fig. 16(a), and shown more dramatically in the work of Mulford et al.[2]; here, only a small amount of softening during aging can produce a substantial reduction in transition temperature. Alternatively, we can say that at a given test temperature it takes a lower and lower impurity concentration to produce intergranular fracture as the matrix hardness increases. An extreme example of this is the case of one-step temper embrittlement in an aircraft-quality 4340 steel[14] depicted in Fig. 18. Here is seen the classical embrittlement trough which occurs in impure steels when carbide formation commences in grain boundaries.[15] The fracture appearance at the minimum energy is largely intergranular, as shown, but the impurity concentration is so low that it does not produce resolvable peaks in the Auger spectrum determined by the normal defocused-beam method.

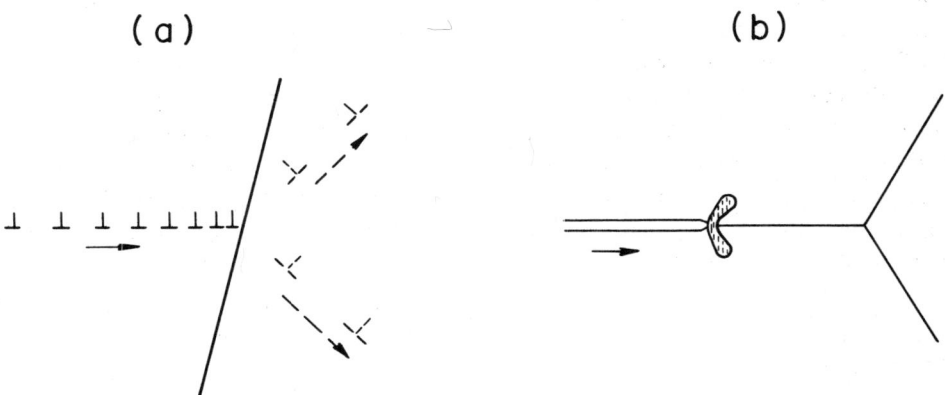

(a)   (b)

Fig. 17. Schematic showing the role of hardness in (a) crack nucleation at a blocked slip band, where plastic flow ahead of the dislocation pileup can inhibit crack nucleation, and (b) crack propagation, which can be inhibited by plastic flow at the crack tip.

314   C. J. McMAHON, JR.

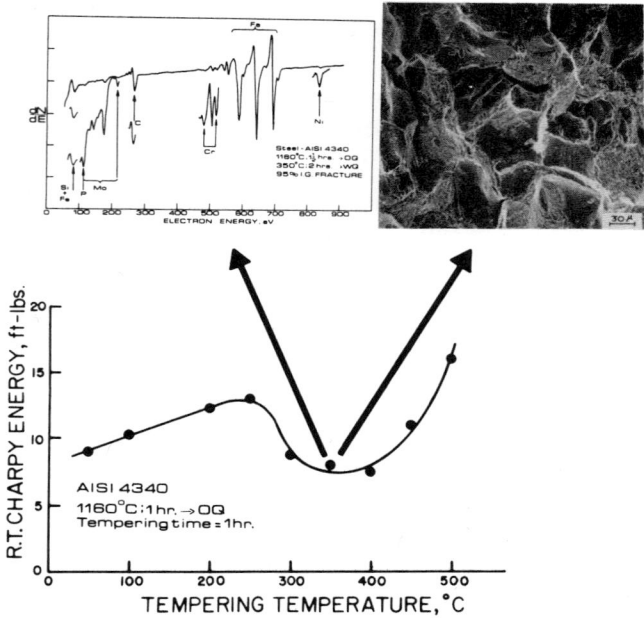

Fig. 18. One-step temper embrittlement of a 4340 steel showing the partly intergranular fracture mode and the Auger spectrum from such a fracture. The only possible segregated impurities are phosphorus and silicon; they are present, if at all, only in minute amounts.[14]

Another important manifestation of the hardness effect is found in the tendency for easy microcrack formation at grain-boundary carbides. This has been demonstrated in low-temperature tests by Rellick and McMahon[8], and it is important in the nucleation of cavities on sliding grain boundaries which contain carbides. This latter process has been studied recently at CERL, Leatherhead, by Caine[16], and Fig. 19 shows an example of intergranular void which has formed around a couple of carbide particles. The sliding grain boundary can be modeled by the dislocation array in Fig. 17(a), as shown by Smith and Barnby[17], whose analysis can be used to rationalize the effects of impurities on this process[1]. The demonstration that impurities can play a large role in void nucleation has been made by Tipler[18], and is illustrated by Fig. 20.

3 HIGH-TEMPERATURE CRACKING

It has come to be realized only recently that great improvements are possible in the high-temperature performance of alloy steels, providing that impurity effects can be brought under control. This was first shown clearly

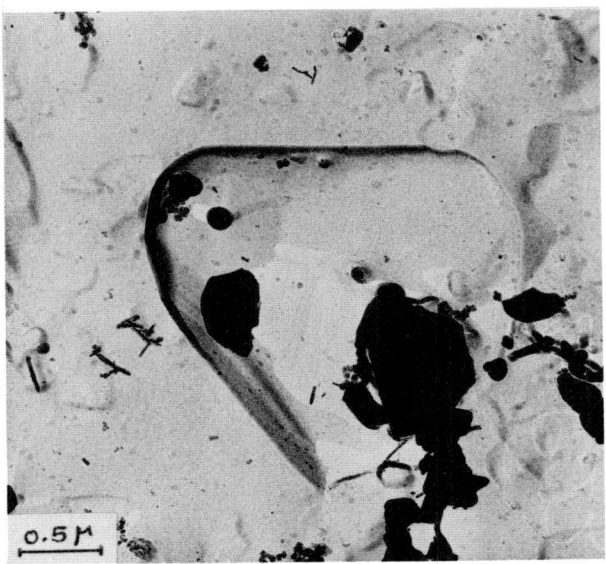

Fig. 19. Example of an intergranular void formed at carbides in a creep-tested 2-1/4Cr-1 Mo steel.[16]

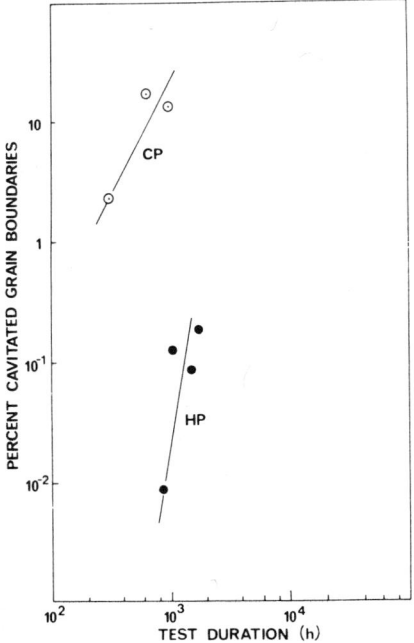

Fig. 20. Comparison of extent of grain-boundary cavitation during creep testing of 1/2Cr-1/2Mo-1/4V steel of commercial purity (CP) and high purity (HP).[18]

by Tipler[18]; the performance of the high-purity CrMoV steel depicted in Fig. 21 in terms of rupture strength and ductility is equivalent to that of a new ferrous "super" alloy developed for that temperature regime. The impurity effect becomes even more striking when it is noted that the high-purity specimens failed by the transgranular cup-and-cone mode associated with lower temperatures.

Impurity effects can be expected to be a major factor in the diffusive growth of intergranular cracks. This can be demonstrated quite simply[1] using the approach of Hull and Rimmer[19], Rice and Chuang[20], and Ashby and Fields[21]. Figure 22 shows the tip of an intergranular crack which is growing by diffusion under the influence of a normal stress. That is, atoms to the left of the crack tip diffuse along the crack surface and down the grain boundary (or, equivalently, vacancies enter the crack from the grain boundary). This process is driven by the work done by the stress normal to the boundary when material is inserted into the plane of the boundary (force x distance $= \sigma\Omega^{2/3}\cdot\Omega^{1/3} = \sigma\Omega$, where Ω is the atomic volume) minus the work done against the surface tension of the crack ($= 2\gamma_{sv}\Omega/r$, where γ_{sv} is the solid/vapor surface free energy and r is the radius of curvature of the crack surface at the tip). Hence, the net driving force is $(\sigma - 2\gamma_{sv}/r)\Omega$. Assuming

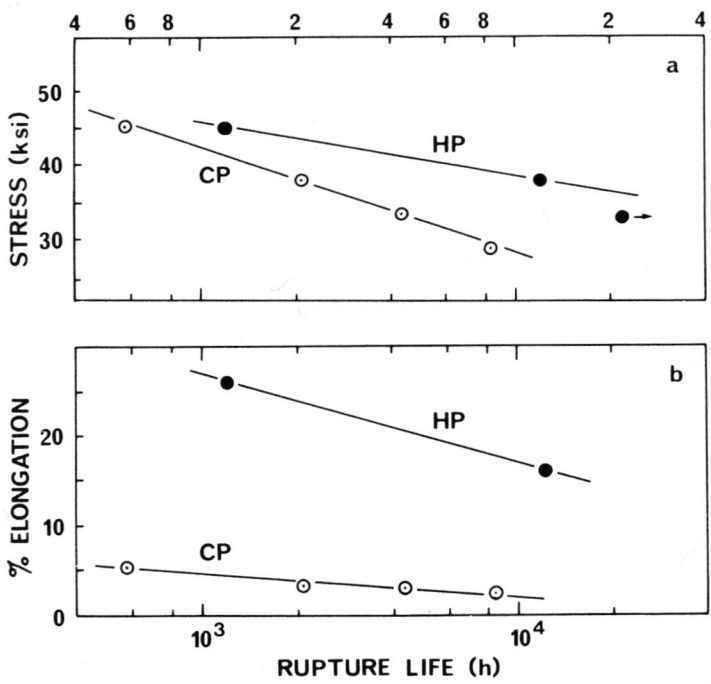

Fig. 21. Comparison of creep rupture strength and ductility in 1/2Cr-1/2Mo-1/4V steel of commercial purity and high purity.[18]

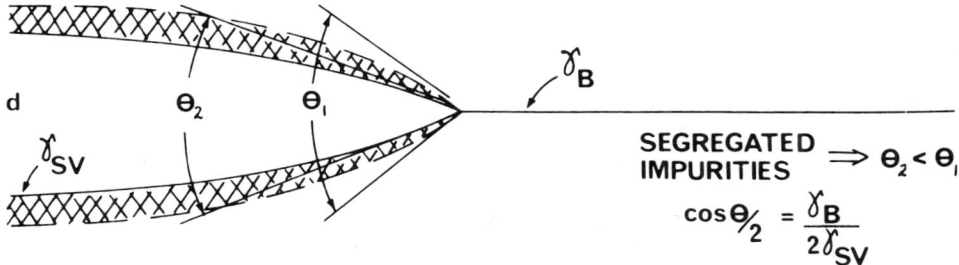

Fig. 22 Schematic showing the effect of surface-active impurities on the equilibrium shape of the tip of a grain-boundary crack at an elevated temperature, showing that such impurities result in thinner, sharper cracks which propagate faster by diffusion.

that local equilibrium is maintained by surface diffusion, the values of r and θ at the crack tip are those dictated by the local composition. Thus, $\cos \theta/2$ = $\gamma_{GB}/2\gamma_{SV}$. Studies of several systems by Hondros[22] have shown that $\gamma_{GB}/2\gamma_{SV}$ increases when a solute segregates to grain boundaries, as illustrated by Fig. 23. This means that θ is lowered by surface-active solutes. The crack therefore becomes sharper, as indicated by the dashed lines in Fig. 22, and this means that the crack advances farther for a given amount of grain-boundary diffusion. In addition, the fact that γ_{SV} decreases means that the driving force for the diffusive growth increases, as shown above. The behavior of the CrMoV steel depicted in Fig. 21 is presumably related to these factors.

Recent studies by King[23] at CERL, Leatherhead, have given further evidence of the impurity effects at elevated temperatures. In tensile tests at 550 C on CrMoV steel, the ductility was found to be strongly dependent on impurity content, as shown by Fig. 24. The commercial alloys showed the same trend as the laboratory alloys, except that some as-yet-undetermined component of the former provides an apparent scavenging effect which shifts the ductility drop to a somewhat higher impurity level. It is worth noting that the order of potency of the impurities, as denoted by the abscissa in Fig. 24, is the same as found in temper-embrittlement experiments.[24,25] It is important to note that impurity-induced cracking at elevated temperatures occurs over a wide range of strain rate, as indicated by the Tipler data (Fig. 21) and the results of the CERL hot tensile tests (Fig. 24). Apropos of the latter, the cracking which occurs in the heat-affected zone of welds* during a stress-relief heat treatment is effectively a hot tensile test. It is promoted by a coarse grain size and high hardness, in addition to segregated impurities.

---

*This has a variety of names, including stress-relief cracking and underbead cracking.

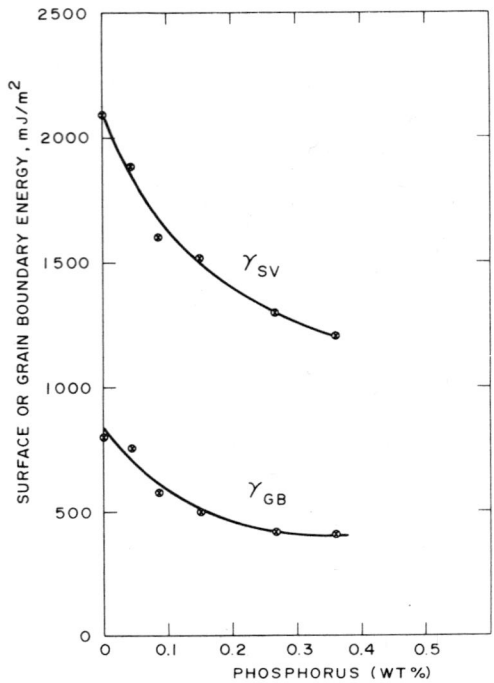

Fig. 23. Example of the apparently general observation that the reduction in γ_{SV} is greater than that of γ_{GB} when surface-active metalloids segregate to surface and grain boundaries.[22]

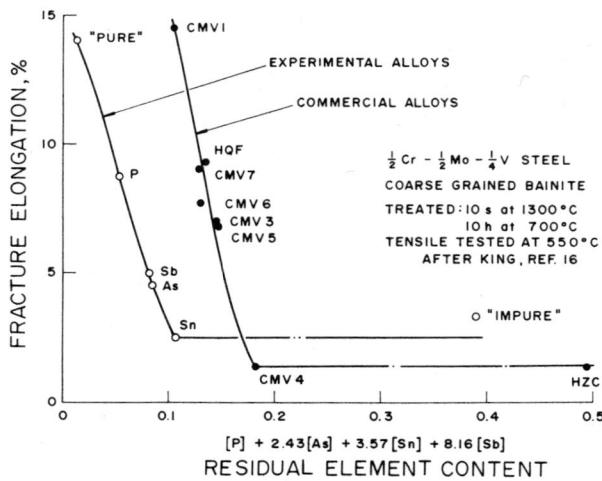

Fig. 24. Effect of various impurities on high-temperature tensile ductility in 1/2Cr-1/2Mo-1/4V steels, showing that the impurity effects are analogous to those observed in two-step temper embrittlement.[23]

4 ENVIRONMENTAL EFFECTS

There is an increasing body of evidence that environments which may promote slow transgranular crack growth at a fairly high stress-intensity level in a high-purity steel can cause intergranular cracking at a substantially lower stress intensity when the grain boundaries are contaminated by impurities. This has been demonstrated clearly in an HY 130 pressure-vessel steel[26], and some of these results are reproduced in Fig. 25. It is apparent that the stress-intensity threshold for crack growth in 0.1N H_2SO_4 is drastically reduced when the steel is temper embrittled. From the AES results in Fig. 25(b) it appears that the active impurities are phosphorus and silicon and that interactions with nickel and manganese are involved. This is presumably a case of hydrogen-assisted cracking, and the impurities seem to play a dual role in this case. They not only reduce intergranular cohesion, but also they somehow enhance the effect of hydrogen. That is, the sum of the hydrogen and impurity effects is greater than either of them acting alone. Our working hypothesis is that the hydrogen concentration at grain boundaries is enhanced by the presence of the impurities. Experiments on cracking in hydrogen gas are under way to study this further.

Slow crack growth in aqueous media is often considered to be due to hydrogen injected into the lattice as a result of the cathodic reaction in which hydrogen is reduced, especially where the recombination reaction $2H \rightarrow H_2$ is retarded. On the other hand, there are strong reasons to believe that in other cases the cracking mechanism involves true stress-corrosion cracking in which anodic dissolution occurs along the crack path, in conjunction with the rupture of a protective film by plastic flow. The analysis of the Hinckley Point turbine burst[27] indicates that this is a case in point. This phenomenon has not received sufficient study, but it has been suggested[28] that the enhancement of cracking by segregated impurities occurs by their retardation of repassivation after film rupture. This would serve to explain the intergranular crack mode.

5 SUMMARY

The design of alloy steels for pressure vessels has been approached from the standpoint of controlling the tendencies for intergranular fracture at ordinary and elevated temperatures and in aggressive environments. The impurity effects on toughness at ordinary temperature are becoming well understood and the principles of alloy design are almost at hand. Encouraging progress is being made in the remaining two areas, but there is still a great deal of work needed before a working understanding is achieved.

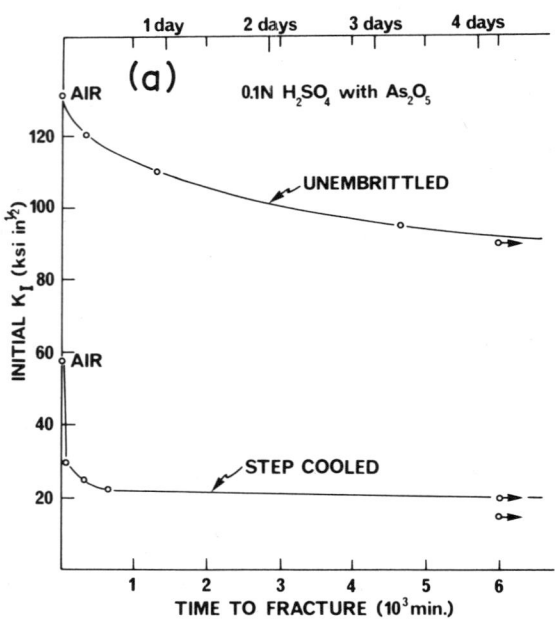

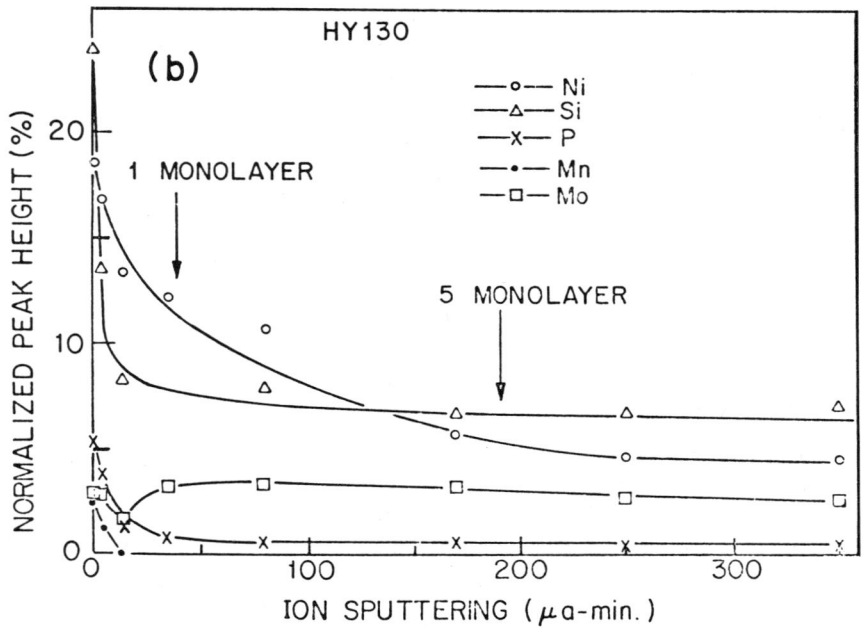

Fig. 25. Effect of two-step temper embrittlement on the resistance to H_2-induced cracking in HY-130 steel; (b) solute distribution in the near-grain-boundary region in this steel in the temper-embrittled condition.[26]

ACKNOWLEDGMENTS

This review was prepared under the joint auspices of the National Science Foundation, the American Iron and Steel Institute, and the Naval Air Systems Command. Discussions with Professors D. P. Pope, T. Egami, and W. D. Graham and Drs. H. C. Feng, S. K. Banerji, C. L. Briant, and A. H. Ücisik were most helpful. Use was made of experimental results of Dr. R. A. Mulford, A. K. Cianelli, and Dr. S. K. Banerji from the University of Pennsylvania and Drs. B. King, B. Caine, and F. P. Ford from CERL, Leatherhead, Surrey, for which the author is grateful.

REFERENCES

1. McMahon, C. J., Jr., in *Grain Boundaries in Engineering Materials* (Fourth Bolton Landing Conference), J. L. Walter et al. (Eds.), Claitor's Publishing Division, Baton Rouge, La. (1975), p. 525.
2. Mulford, R. A., McMahon, C. J., Jr., Pope, D. P., and Feng, H. C., *Met. Trans.*, **7A**, 1183 (1976).
3. Cianelli, A. K., Feng, H. C., Ucisik, A. H., and McMahon, C. J., Jr., submitted to *Met. Trans.*
4. Capus, J. M., *J. Iron and Steel Inst.*, **200**, 922 (1962).
5. Ohtani, H., Feng, H. C., McMahon, C. J., Jr., and Mulford, R. A., *Met. Trans.*, **7A**, 87 (1976).
6. Guttmann, M., These de Doctoral d'Etat, Orsay, March, 1974; *Surface Sci.*, **53**, 213 (1975).
7. Nageswararao, M., McMahon, C. J., Jr., and Herman, H., *Met. Trans.*, **5**, 1061 (1974).
8. Rellick, J. R., and McMahon, C. J., Jr., *Met. Trans.*, **5**, 2439 (1974).
9. Ohtani, H., Feng, H. C., and McMahon, C. J., Jr., *Met. Trans.*, **5**, 516 (1974).
10. McMahon, C. J., Jr., Furubayashi, E., Ontani, H., and Feng, H. C., *Acta Met.*, **24**, 695 (1976).
11. Ohtani, H., Feng, H. C., and McMahon, C. J., Jr., *Met. Trans.*, **7A** (1976), in press.
12. Kaneko, H., Nishizawa, T., Tamaki, K., and Tanifuji, A., *J. Japan Inst. Metals*, **29**, 166 (1965).
13. McMahon, C. J., Jr., Cianelli, A. K., and Feng, H. C., submitted to *Met. Trans.*
14. Banerji, S. K., Feng, H. C., and McMahon, C. J., Jr., unpublished research, University of Pennsylvania, 1975.
15. Klinger, L. J., Barnett, W. J., Fromberg, R. P., and Troiano, A. R., *Trans. ASM*, **46**, 1557 (1954).
16. Caine, B., unpublished research, CERL, Leatherhead, Surrey, England, 1974.
17. Smith, E., and Barnby, J. T., *Met. Sci. J.*, **1**, 1 (1967).
18. Tipler, H. R., International Conference on Properties of Heat Resistant Steels, Dusseldorf, May, 1972.
19. Hull, D., and Rimmer, D. E., *Phil. Mag.*, **4**, 673 (1959).
20. Chuang, T. J., and Rice, J. ., *Acta Met.*, **21**, 1625 (1973).
21. Ashby, M. F., and Fields, R. J., unpublished research, University of Cambridge, 1974.
22. Hondros, E. D., *Proc. Roy. Soc.*, **A286**, 479 (1965).
23. King, B., unpublished research, CERL, Leatherhead, Surrey, England, 1974-75.
24. Steven, W., and Balajiva, K., *J. Iron and Steel Inst.*, **193**, 141 (1959).

25. Low, J. R., Jr., Stein, D. F., Turkalo, A. M., and Laforce, R. P., *Trans. TMS-AIME*, **242**, 14 (1968).
26. Yoshino, K., and McMahon, C. J., Jr., *Met. Trans.*, **5**, 363 (1974).
27. Kalderon, D., *Proc. Inst. Mech. Engrs. (London)*, **186**, 341 (1972).
28. Ford, F. P., unpublished research, CERL, Leatherhead, Surrey, England, 1974-75.
29. Mulford, R. A., McMahon, C. J., Jr., Pope, D. P., and Feng, H. C., *Met. Trans.*, **7A** (1976), in press.

AGENDA DISCUSSION: NEEDS IN ALLOY DESIGN

G. S. Ansell*

*Rensselaer Polytechnic Institute
Troy, New York*

I. Levy**

*Battelle
Pacific Northwest Laboratories
Richland, Washington*

There is a wide diversity in the level of "needs" for alloy design. At one extreme are those applications for which the alloy requirements are readily defined by an existing design or application; e.g., the replacement of a structural part with an equivalent that costs less. At the opposite extreme are those applications where either changes in existing design or totally new directions make it difficult to define or anticipate alloy needs, e.g., the inner

*Chairman
**Secretary.

containment vessel for a fusion reactor. The former case generally is associated with a static or evolutionary systems-design process, while for the latter case, the systems-design process is fluid, in many cases synergistically changing with alloy design.

Chairman Ansell opened the discussion by posing the following questions:

a. To what extent and in what terms should the needs for alloy design be defined in terms of requisite alloy properties by the system designer?

b. Is there, or should there be, a difference between the process of alloy design for a static systems design and that for a fluid systems design?

c. Are there ways in which alloy design and systems design can be most effectively interrelated?

d. Can alloy-design theory presently in hand be utilized to guide systems design in advance of actual alloy development?

e. How should the needs for alloy design defined by these presentations be translated into alloy design parameters?

The discussion quickly focused upon the necessity for cooperative interaction between the systems designer and the materials specialist. Levy presented an example in the development of nuclear fuel elements which illustrated the problems that can result when this interaction fails. For these fuel elements, the alloy design was based upon the degradation expected under steady-state operating conditions. On the other hand, operating transients produce excursions in temperature and stress that accelerate degradation. The systems designer, aware of these transients, based his design on the maximum allowable strain that the fuel element could withstand over its life. The alloy developer never knew of the actual design requirement. With this gap in communications, there was no alloy development directed to insure the actual design requirement. Indeed, it is only during the past 2 years that testing has been started to determine the transient capabilities of these materials. Sims pointed out that the key element is communication—communication as insured by effective management. He cited the sequence of iterative interaction between mechanical design and alloy design in gas-turbine development. The materials designer must understand the mechanical designer and have some feeling for his problems (and vice versa). The two groups discuss the interactive effects systems and alloy design have on each other and the possible gains that can be achieved. Here, the mechanical design and the materials development are, in essence, parallel. Owen stressed that it is necessary to make a distinction between alloy development as applied to evolutionary design and that for applications in which there is little past experience. In evolutionary design, wherein one is seeking to improve performance, reduce cost, or slightly change function, there is not much difficulty. The real problem comes in those areas where the conceptual design is pulled "out of the air", such as in the case of nuclear reactors. Something that has never been made before. He suggested that we are not training metallurgists to participate in such a collaborative design process. It is essential to broaden a metallurgist's training in design so that he can become an effective member of the team. Frost observed that in his experience, in nuclear applications, the most effective alloy development came from materials specialists who had additional training in nuclear engineering. Bement pointed out that the thrust of the nuclear materials program at M.I.T. is to bring nuclear engineers and materials engineers together to do this type of interactive development.

Ashby suggested an alternative view to this process. In general we approach the relationship between the structure and properties by the use of modeling, while in fact, we want some models which work the other way— models which start with design and go back to the selection or specification for materials. Bement extended this argument to his experience in the use of performance modeling where one defines boundary conditions, establishes feedback in the system and interactions, analyzes sensitivities, and does history following. In this process one often ends up with a model which has so many adjustable parameters that it is worthless. Nevertheless, in the process of constructing the model, one does establish a systematic and formal communications link between the design engineer and the materials engineer. Ansell asked whether this concept was at a stage where one could now use systems analysis and operations research, coupling the system design and our understanding of materials fundamentals, to formalize this procedure in an effective manner. Bement responded that they had gone through this iteration enough times for rather specific problems to show that this approach can be useful.

McClintock pointed out the gap that exists between the type of property specification needed for effective mechanical design and the properties targeted in alloy design. A finite-element stress analyst would like to predict the rate of crack growth in a real system. From a fundamental basis we did not know how we could give it to him, as there is not yet a basis for this. Bush went further to state, "I submit that the alloy designer generally doesn't measure the right properties, and certainly the systems designer doesn't ask for the right properties." In pressure vessels, for example, one measures uniaxial tensile properties, but from a designer point of view, biaxial properties would be far better. Inspectability is the Achilles heel in many alloy systems. It is usually ignored in alloy design until too late.

Ansell then opened up the following issues. While the papers in this session have outlined needs for alloy design which are specific to particular applications, it is apparent that there are common requirements for improved alloy characteristics which would be useful for a range of applications, e.g., high-temperature stability, oxidation resistance, etc.

a. To what extent is it useful to design new alloys with improved properties irrespective of specific application?

b. Is it possible and desirable to use needs overviews of this type to define alloy design requirements in more general terms where alloy design could serve multiple applications?

c. Which is more effective, an alloy looking for a need or a need looking for an alloy?

One immediate response was the view that the alloy designer is actually a mythical person in most industries. What exist are component designers and alloy selectors. Most alloy development gets done just because somebody is trying to sell more of whatever they are making. They are not designing for some specific application. Decker countered this point. There is such a thing as an alloy designer. He goes about design in a systematic manner using enlightened empiricism. He argued that in the design process, it is necessary to set four or five quite precise property targets. These need to be quantitative targets that can be measured in the laboratory in order to optimize the design. Ashby continued the discussion, balancing enlightened empiricism and fundamental understanding in the design process. In his view, the development of totally new alloys results empirically, and not from a fundamental understanding. The empiricism may be slightly enlightened by our understanding of the fundamentals, but the contributions from our fundamental understanding will be small. For every alloy developed from fundamentals, there are a hundred that were produced purely by the empirical approach. On the other hand, how do we design better with the alloys we have already? Here, there is real hope that research in the fundamentals will improve that process. For example, there needs to be a better understanding of how to extrapolate creep and creep-rupture data. One wants to extrapolate not just in time but in stress state. Proper modeling and understanding of fundamentals should tell us how to do that. If one knows creep-rupture life in uniaxial tension, then what is the behavior under biaxial conditions? One needs to understand and formulate design criteria for the creep-fatigue interaction in materials. Current design practice against the combined effect of creep and fatigue is extremely crude and inadequate.

Jaffee introduced the problem of time in the alloy-development process. The mechanical designers are constrained by what alloy designers have contributed to the state of the art. He does not have time to wait for the new material. The alloy design process comes about when there is a major change in the capabilities of materials. Begley added that new alloy design is a very long-term proposition. No designer in his right mind would allow that opportunity. Jaffee suggested that progress is made in new alloy design when the ultimate user, with some knowledge of what is possible, sets targets and makes these targets generally known. These targets should be set not only for a given class of materials, but for several classes so that there is competition. For alloys with a large potential market, commercial incentive will respond. Where there is a small market, or the costs of design or code qualification are high, then there needs to be a different institutional response. An ERDA* or an EPRI* is needed. The institution, public or private, that stands to gain from the development has to pay the costs.

Ansell asked: How do the government sponsoring agencies approach their involvement in defining alloy requirements and the interrelationship between fundamental research and alloy design? Mayer responded that frequently they try to identify the kinds of systems that they will have 15 years from now. In other instances, they will look ahead to drastic improvement in materials properties. "The higher, faster, lighter, more powerful syndrome that everybody looks for. You can't go wrong in doing that. You know that you will always want a bigger payload. You won't always want it to go faster, but you will want better maneuverability, you will want the lighter system."

Stevens added: "The situation at ERDA is very, very spotty. Certainly, there is a good effort we've made in the reactor division to come up with specifications for materials for the second generation of the LMFBR. In the area of the fusion reactor, I think I would like to point out it took the materials people to get the fusion activity people to sit down and try to conceptualize what one of their machines might look like. It was necessary to do that so we could come up with some kind of estimate of what the environment of that first wall was going to be. And it has only been recently that that CTR group has developed a materials program. But there it still isn't really possible at this point to write down a good set of specifications for the first wall material. This is evolving."

As a result of this discussion, it was clear that the needs for alloy design are multifaceted. The immediate response focused on the identification of specific alloy requirements engendered by pressing future applications, e.g., the inner containment vessel for a fusion reactor.

While enticing, it became quickly evident that such identification of specific alloy requirements and applications was not particularly fruitful. When such requirements could be met by discrete or evolutionary modifications of existent materials, little difficulty was perceived in alloy design. However, discontinuous or major shifts in alloy requirements pose severe problems for alloy design. The long lead times required for both alloy and systems development, the technical communication gap between systems designers and alloy developers, and the fluid nature of the design process continuously change the validity of such lists of alloy requirements. The principal value in identifying future alloy requirements is in setting targets for alloy development to assure discipline in the design process, not

---

*ERDA is the Energy Research and Development Administration of the United States.
EPRI is the Electric Power Research Institute of the public and private utilities in the U.S.

necessarily in providing an alloy suitable for the purpose originally intended. As succinctly stated by Ashby, "things are built from existent materials".

On the other hand, there is a real need to strengthen the basis for effective alloy design. Two areas were identified where effort must be placed in order to enhance the design process. The first is to provide the translation between the usual basic materials properties tractable in alloy design and the engineering design parameters needed for application. The second lies in extending the basis for modeling the properties of materials: the structure-property and environment-property relationships. In these areas, future expectations were modest. Broadened educational and applications experience can assist in developing the linkage between physical properties and engineering applications. However, this gap will not close quickly. Alloy design will continue to be principally dependent on modeling, leading to an intuitive rather than a direct linkage between alloy development and application.

_____ *Part Three*

ELEMENTS OF ALLOY DESIGN—
ROLES AND LIMITATIONS

UTILIZATION AND LIMITATIONS OF PHASE TRANSFORMATIONS AND MICROSTRUCTURES IN ALLOY DESIGN FOR STRENGTH AND TOUGHNESS

G. Thomas

Department of Materials Science and Engineering
University of California
Berkeley, California

ABSTRACT

The development and characterization of microstructures resulting from phase transformations in martensitic steels and spinodal systems are outlined and discussed in relation to improvements in ambient temperature strength and toughness. Spinodals and martensites can both have dislocated, fine-grained microstructures when formed under appropriate conditions. In martensitic steels, the importance of retained-austenite morphology and stability on toughness at high strengths is emphasized. A method is described for improving the strength of low-carbon steels by developing duplex fine-grained structures of ferrite and martensite. It is suggested that coarsened spinodal alloys have good work-hardening strengthening potential if the discontinuous growth problem at grain boundaries can be overcome. Some ideas for the utilization of multiple phase transformations are also presented.

1 INTRODUCTION

From the viewpoint of structural alloy design the physical metallurgist is concerned with controlling microstructure and composition to produce desirable combinations of mechanical properties. The fundamental limitation of strengthening a given alloy system is that due to the elastic moduli. The theoretical tensile strength may be given approximately as $G/20$ where G is the shear modulus. If phase transformations are used to produce dislocation pinning by providing obstacles (particles or dislocations etc.), then the theoretical strength can be equated to the well-known Orowan pinning model, viz., $G/20 \approx G_b/l$ where l is the pinning distance between obstacles, and b is the Burgers vector of the mobile dislocations. Thus, at the theoretical limit, $l \sim 20|b| \sim 100$ A which requires high-resolution metallography for analysis. Of course there will be little or no plastic ductility at this strength level, so in design a compromise is attained to provide the best fracture resistance (usually large values of K_{Ic} toughness) and this usually means sacrificing strength. In any case, the major objective is to provide uniform obstacle distributions so that deformation by slip will occur uniformly and thus minimize local stress concentrations. Phase transformations leading to heterogeneous distributions of precipitates are thus usually undesirable.

In this paper, the utilization of two major transformations is discussed, viz., martensitic and spinodal, since these currently show great potential for manipulation and application of novel approaches to microstructure-property control. Ordered alloys (including ordered matrix solid solutions, interstitial ordering, and embrittlement) are not discussed. The properties considered are tensile strength and fracture toughness (as measured by K_{Ic} and Charpy) relevant at ambient temperatures, and the alloy systems described are experimental ones designed in an attempt to facilitate basic understanding of the problems involved.

A comprehensive review of this rapidly developing field of research is not intended but rather a summary of results and ideas of current work in my group. Emphasis is also placed on the need for detailed and quantitative electron metallographic analysis to properly characterize microstructure, even though this may not directly assist in elucidating the reaction mechanisms. An important aspect *not* covered in this paper (see paper by McMahon in this book) is the chemical analysis of small volumes to which new electron-optical techniques such as Auger analysis and micro-X-ray analysis using scanning-transmission instruments are being applied. Such work is essential for developing commercial alloys since impurities alone may be the chief limitation on performance.

2 SOME PROBLEMS OF UNIFORMITY
OF MICROSTRUCTURE

As mentioned above, a major concern for microstructure-property control is that of uniformity of microstructure and the need to identify desirable and undesirable features. The fundamentals of the mechanisms of phase transformations must therefore be understood and the microstructures properly and quantitatively characterized. A brief outline of some of these factors follows.

2.1 Nucleation and Growth Systems

For nucleation and growth (N&G) transformations, it is necessary to consider both homogeneous and heterogeneous reactions. In the absence of an applied stress, the free energy ΔF_n associated with nucleation has been expressed (e.g., Cahn[1]) by $\Delta F_n = -\Delta F_v + \Delta F_s \pm \Delta F_e$, where ΔF_v is the volume free energy, ΔF_s the surface free energy, and ΔF_e the strain energy (this is negative in heterogeneous nucleation, e.g., at dislocations).[2] In order to produce homogeneous nucleation, large supersaturations are needed, i.e., the ΔF_v term must dominate. In systems where quenching from the single-phase region precedes aging, inhomogeneities can be introduced e.g., large vacancy supersaturations and additional dislocations. Subsequent aging produces microstructures that may then not be uniform e.g., precipitation at grain boundaries (due to the influence of the surface and strain energy terms and the higher diffusion rates), precipitate-free-zones, and precipitation on defects in the matrix. In some cases different phases may appear, as is well known for Al-Cu base alloys where θ'' occurs in the matrix, θ' on dislocations, and θ' or θ at grain boundaries in the same alloy on aging to near maximum strength.[2] Thus, control of uniform microstructure is very difficult, especially if it is to be accomplished by only simple heat treatments. The Al-Zn-Mg system (commercial 7075) is a good example of one of the potentially highest strength aluminum alloys which is limited by the heterogeneity of precipitation, and a compromise between toughness and strength can be achieved by overaging. However, some experimental advances have been made with complex multiple treatments including thermal-mechanical treatments, especially for alleviating grain-boundary heterogeneities.[3,4]

2.2 Spinodal Systems

For perfectly homogeneous phase separations, such as in spinodals where the above equation does not apply, uniformity of structure is attained immediately after the transformation and can be attained by direct cooling from the solution temperature.[5-8] The problem that arises in the postnucleation period is that coarsening rates may differ at boundaries and within the matrix. Figures 1 and 2 compare light-optical and transmission-electron-optical micrographs of the structures after advanced aging in a Cu-Ni-Fe alloy. It might appear from Fig. 1 that a new type of grain-boundary process has occurred, but a recent detailed study of the morphology (Fig. 2), crystallography, and kinetics of aging[9] has demonstrated that this effect is due to accelerated coarsening, generally only at large-angle boundaries, and is accompanied by grain-boundary migration. Since for this alloy system the formation of the grain-boundary product is growth controlled, it cannot be due to a discontinuous or cellular reaction, both of which are nucleation controlled.[8] It is also interesting that the toughness of this CuNiFe spinodal alloy decreases[10] as a result of this grain-boundary coarsening. A detailed study of this coarsening problem is in progress so that attempts can be made to solve the boundary problem by a fundamental approach to design, with changes in alloying and/or heat treatment, e.g., to lower grain-boundary mobility.

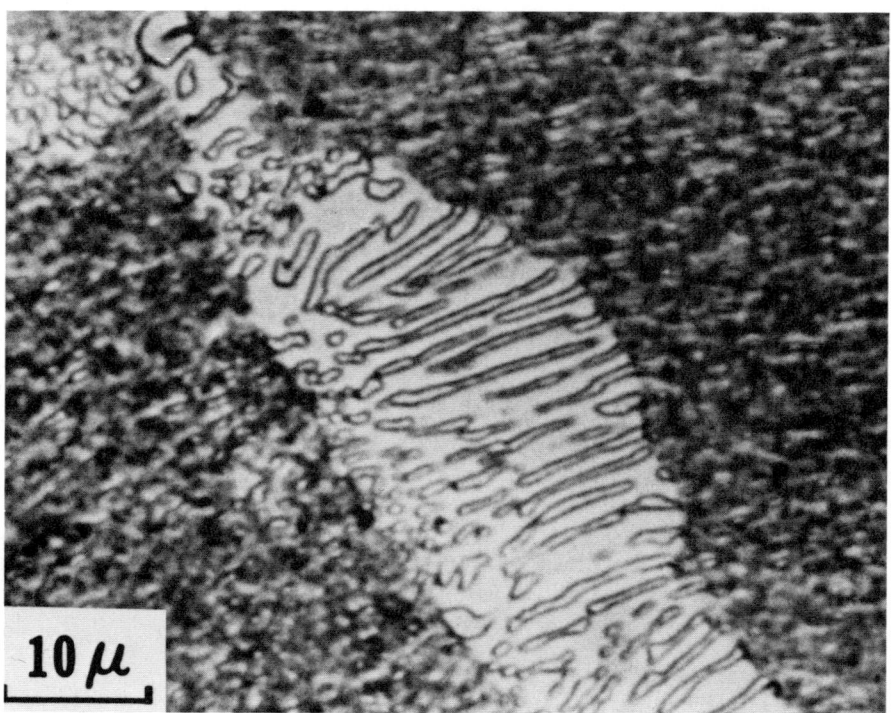

Fig. 1. Coarsened spinodal structure in 70Cu-19Ni-11Fe after aging for 500 hr at 748 C, showing the lamellar structure near the grain boundary. Courtesy R. Gronsky.

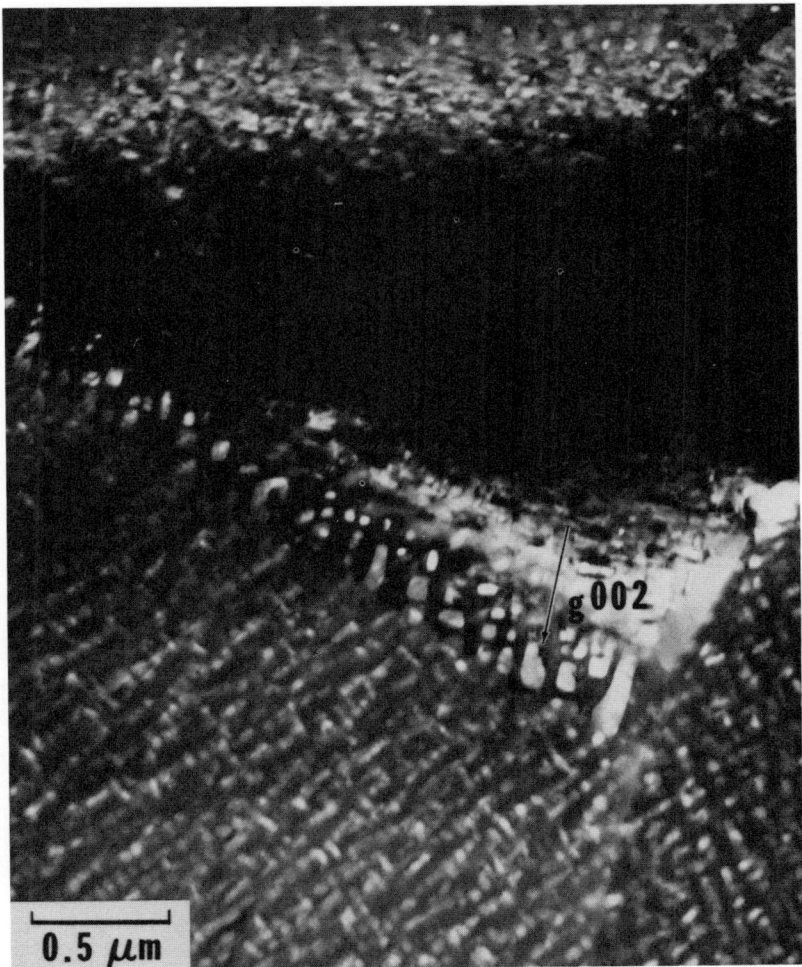

Fig. 2. Same alloy as Fig. 1 but aged 10 hours at 655 C, showing enhanced coarsening and boundary migration (in direction of $\vec{g} = 002$). The wavelength at the boundary region is almost 3 times that in the matrix. Note that the copper-rich phase is in lighter contrast due to preferential electropolishing. Courtesy R. Gronsky.

During spinodal coarsening, dislocation networks form at the interphase boundaries to relieve the coherency strains when coarsening proceeds and the phase compositions approach their equilibrium levels. An example for CuNiFe is shown in Fig. 3. In this way, walls of dislocations are generated periodically thus providing a regular "dispersion" of cell boundaries which should be good sources of strengthening (Section 3.3). However, the toughness will be limited if grain-boundary effects such as those described above cannot be overcome.

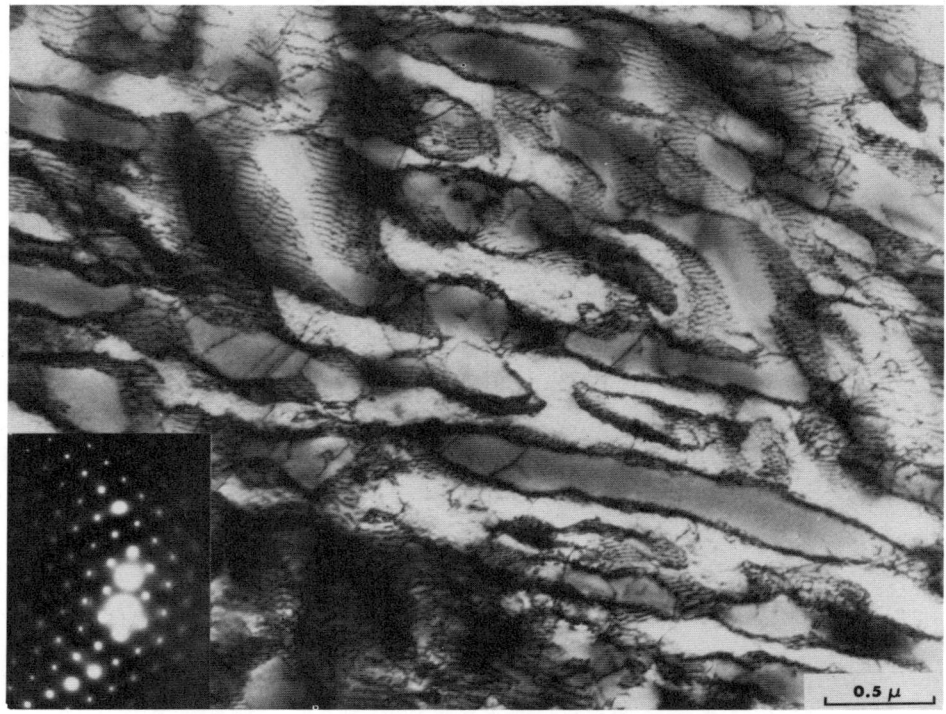

Fig. 3. High-voltage electron micrograph (650 kV) of overaged and 10 percent tensile-deformed CuNiFe alloy showing interphase dislocations. These have Burgers vectors of $a/2<110>$. Some slip dislocations are also visible, especially within the lighter contrast, copper-poor phase.

Despite the great potential for utilizing spinodal reactions to produce tough, strong alloys, the amount of work on spinodal systems has so far been extremely small, especially from the structure-property viewpoint (Section 3), in comparison with the efforts expended on examination of transformations of other kinds. The development of alloy theory to allow prediction of spinodal transformation in several component systems is clearly desirable.

2.3 Martensitic Systems in Iron-Base Alloys

The martensitic transformation can also produce a uniform microstructure provided that the internal substructure produced by the transformation shear and subsequent accommodation (slip or twinning) is unique.[11,12] Figure 4 shows electron micrographs typical of the dislocated and twinned martensites that are found in quenched iron-base alloys.

Fig. 4. Electron micrograph showing (a) part of a packet of dislocated lath martensite (tough) in Fe-5Ni-0.26C, and (b) twinned martensite (brittle) in Fe-6.8Mn-0.24C. Courtesy D. Huang.

It is now established that martensitic carbon steels have poor K_{Ic} toughness if the structure is twinned (say approximately 10 percent or more, i.e., when the carbon content exceeds ~0.3 percent). In our alloy design program, studies are being done primarily on high-purity, ternary Fe-C-X alloys, which are vacuum melted. The X refers to alloying elements, chromium, molybdenum, boron, etc. These alloys, while serving as model systems so as to provide basic information on the effect of alloying elements on microstructure, strength, and toughness by minimizing well-known problems of impurity embrittlement, have also turned out to provide excellent properties even in the untempered condition (see Fig. 5a), which is advantageous economically. Figure 5b shows results of current work on grain-refined dislocated martensites[14] in Fe-1Cr-1Mo-0.3C. These data can be compared with those for Fe/Mo/C alloys shown in Fig. 5a. It should be noted that these properties compare very favorably with commercial high-alloy steels, e.g., the 18 percent nickel maraging type.

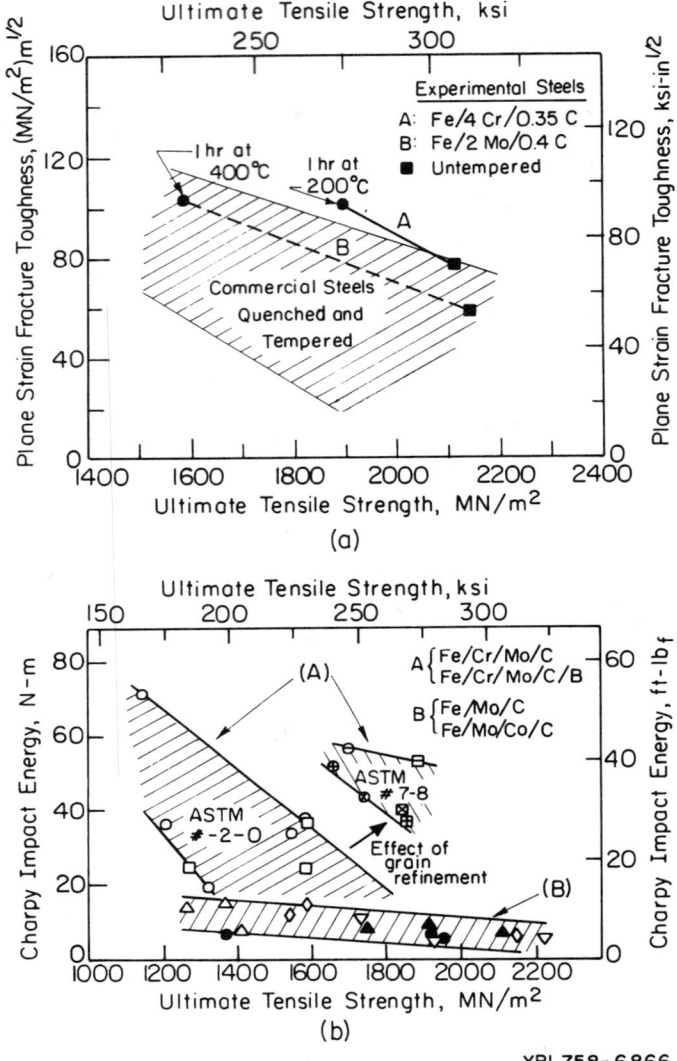

XBL758-6866

Fig. 5. (a) Relationship between plane-strain fracture toughness and ultimate tensile strength for two experimental ternary alloy steels developed recently. Note that the Fe-Cr-C steel has properties superior to those of the Fe-Mo-C steel and the other commercially available high-strength steels and 18Ni maraging alloys. Continuous films of stabilized interlath austenite could be identified in Fe-Cr-C steels, whereas, in Fe-Mo-C steels, the amount, if any, was too small to be identified (data from refs. 28 and 31). (b) Relation between Charpy-impact toughness and ultimate tensile strength for Fe-1Cr-1Mo-0.3C alloys and Fe-Mo-C alloys (from ref. 31). The Fe-Mo-C alloys have low impact toughness at all strength levels, while the Fe-1Cr-1Mo-0.3C alloys had good combinations of strength and toughness. Extensive in-terlath retained-austenite films (see Fig. 12) were observed by transmission electron microscopy, but none was detected in the Fe-Mo-C alloys. The effect of grain refinement for improving the properties is also shown.[14]

2.4 Microstructure Control for Martensitic Structural Steels

As a result of basic studies on the effects of alloying and microstructure on mechanical properties (reviewed in ref. 13), some guidelines can be given for designing strong, tough steels:

1. Obtain dislocated martensite with minimal twinning. This is largely determined by carbon content and total alloy content, both of which affect the M_s temperature. Thus, a lower limit of M_s of about 350 C is recommended (max carbon ~0.4 percent). This structure can be achieved by direct quenching, and sufficient hardenability is obtained by adding substitutional elements such as chromium, molybdenum, etc., i.e., the X component in the Fe-C-X series. Thus, the total amount of solute must be controlled to control M_s-M_f and hardenability. Hardenability is needed to prevent undesirable structures resulting from diffusional products (e.g., pearlite, upper bainite). Some autotempering may occur on quenching but this should not lead to embrittlement, provided the carbides nucleate only within martensite and not at boundaries. In this regard, lower bainite may also be a desirable structure.[13]

2. A composite structure wherein the martensite (or lower bainite) is surrounded by continuous thin films of austenite appears to greatly benefit K_{Ic} toughness at high strengths (Figs. 5 and 12). These films can again be obtained by control of composition and heat treatment through a knowledge of transformation kinetics and effects of alloying additions. The stability of this austenite is important; e.g., if on tempering (even autotempering) it decomposes to ferrite and carbide, the undesirable boundary carbide morphology occurs, resulting in embrittlement.[28] The addition of alloying elements such as aluminum and silicon are beneficial in this regard (see also the paper by Parker and Zackay in this book). If the austenite films are mechanically unstable, they may transform under stress. If they transform to martensite (so-called trip mechanism), this could be a useful toughening reaction during crack propagation.

It is the formation of grain-boundary carbides that probably limits the possibility of using continuous cooling to directly transform austenite to a mixture of dislocated ferrite and carbides (autotempered martensite or lower bainite) which can also be done to produce a dispersion-strengthened alloy steel (see, e.g., Honeycombe, ref. 15). Current work on bainite[16] confirms the view that upper and lower bainite form by different mechanisms and should be represented by separate curves in TTT diagrams. This means that careful evaluation of transformation kinetics and microstructure is needed if a particular alloy system is to be fully exploited; this is especially true for existing commercial alloys.[17] It cannot be emphasized too strongly that much of the enormous terminology that has developed to describe the microstructures in steels has arisen from inadequate resolution (Troostite, Sorbite, etc.) and has led to confusion. High-resolution electron microscopy is absolutely essential to characterize and thus clearly specify the microstructure, e.g., to distinguish upper and lower bainite, and to differentiate narrow bands of interlath austenite and cementite (Section 4.3). Furthermore, while grain/packet/lath/phase boundaries in martensitic steels provide strength by providing effective barriers to dislocation motion, the exploitation of this potential strengthening source without loss of toughness, requires careful design of heat treatments to avoid boundary decoration by brittle precipitates or other heterogeneities.

In summary, although both martensitic and spinodal reactions can produce uniform microstructures, and provide dislocations without external plastic deformation, grain-boundary effects are still probably the most important factors determining the mechanical properties of the alloys.

3 SPINODAL TRANSFORMATIONS AND ALLOY DESIGN

The attractiveness of the spinodal transformation stems from the uniform, homogeneous, usually periodic microstructures that are produced (see e.g., the matrix regions in Fig. 2). Since a spinodal microstructure can be characterized by the wavelength and amplitude of its periodic composition modulations (Fig. 6), these parameters must be accurately determined for correlation with desirable properties.

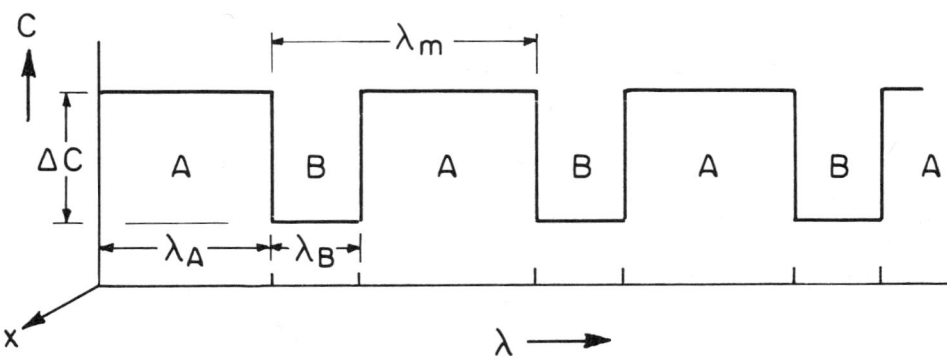

Fig. 6. Composition-wavelength variation for a spinodal with unequal volume fractions of the two phases. The change in composition Δc is assumed to vary linearly with change in lattice parameters Δa.

3.1 Coherent Spinodal Products

From a survey of the available experimental data for several alloy systems which appear to decompose spinodally[6,7,18-24], it is obvious that a universal theoretical explanation for strengthening in spinodal alloys has not been found. When correlations have been attempted, it is generally agreed that the yield strength σ_y is *not* dependent upon wavelength λ during the time the spinodal product remains coherent. The results obtained on the Cu-Ni-Fe spinodal alloys[6,7], which have served well as a model system, have shown that the yield strength of the coherent spinodal depends on the composition difference Δc between the two phases which is related to their difference in lattice parameters Δa (Fig. 6), independently of the wavelength and volume fraction. If the elastic constants of the two decomposed phases are nearly the same, then, on the basis of a theoretical treatment by Dahlgren[24], the maximum yield stress is given by $\sigma_{y\,max} \propto \Delta a_o$, where Δa_o is the difference in lattice parameters of the two spinodal products at their extreme tie-line compositions. For the same condition, the yield stress is predicted to be independent of wavelength and volume fractions of the spinodal product. When the elastic constants of the phases show appreciable differences, then some volume-fraction dependence is indicated in the yield stress.

Dahlgren[24] was careful to point out that while the decomposing phases are still coherent, the "extreme" compositions do not correspond with the limits of the miscibility gap, so that the phases are metastable.

3.2 Semicoherent Spinodal Product

During spinodal coarsening, the change in compositional differences Δc maximizes as the equilibrium tie-line conditions are attained, and so the yield stress $-\Delta a$ relationship no longer holds. At this stage of transformation, coherency is lost and interfacial misfit dislocations are generated (Figs. 3 and 7). The Burgers vectors of these dislocations in CuNiFe alloys are $a/p<110>$ (or $a/q<100>$ in other cubic alloys)[25], where the magnitudes of p and q depend on the misfit (normally p = 2, q = 1).

Thus, overaging has introduced a regular array of dislocation walls. Upon plastic deformation, the slip dislocations must overcome the interactions with the dislocations at the interface (Fig. 7) or even possibly push the interface dislocations out of the interface. The overall interaction distance is nevertheless λ, the compositional wavelength, and slip will occur preferentially in the softer phase. If the volume fractions of the two phases are unequal (Fig. 6), the yield stress is expected to be volume-fraction dependent.

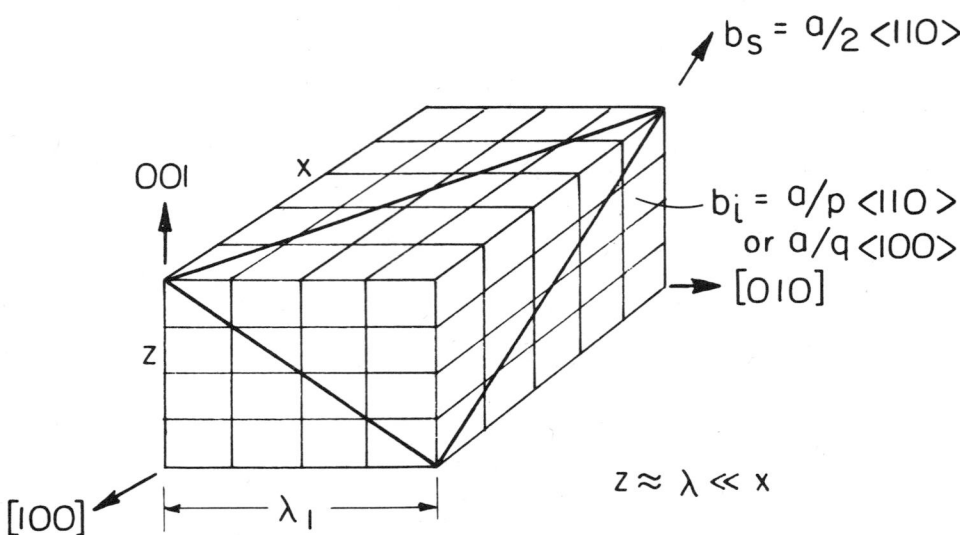

Fig. 7. The interfacial dislocations and a slip plane for a semicoherent spinodal (fcc system). Slip dislocations travel a maximum distance $\sqrt{2}\lambda$ before interacting again at the interface (assuming $z = \lambda$). The values of p and q are determined by the misfit.

3.3 Potential for Strengthening Semicoherent Spinodals

From the foregoing it is clear that the overaged, semicoherent, spinodal alloy systems are attractive from the dispersion and work-hardening viewpoint, especially if they can be produced by direct slow cooling from the solution temperature. On straining, a glide dislocation in such a material will experience two types of obstacles: (i) regions of elastic inhomogeneity, and (ii) periodic arrays of dislocations that delineate the regions of elastic inhomogeneity (Fig. 3).

Assume a dislocation source exists at the interface and a dislocation moves a distance $\sqrt{2}\lambda_A$ in phase A before coming to the next interface, and that the system, like CuNiFe, is fcc (Fig. 7). At this point, the slip dislocation must interact with the wall or forest of the two sets of interface dislocations. Thus, a strong interaction is expected at these junctions. Once this interaction is exceeded, the dislocation must either penetrate into the next phase B and travel a distance $\sqrt{2}\lambda_B$ before again becoming pinned due to interface interaction, or find a route through an adjoining "particle" of phase A so that slip essentially occurs entirely in phase A (assuming phase A is softer than phase B). A possible example of this behavior is visible in Fig. 3.

Work on dispersion-hardened materials, for which there are several contributions to the total hardening, suggests[26] that the independent contributions should be summed as:

$$\tau_{Total} = \sqrt{\tau^2_{el.\ inhom.} + \tau^2_{array}} \quad .$$

As the periodic arrays are stable because they consist of structural dislocations, the interfaces can be considered similarly to grain boundaries, in which case a Hall-Petch type of strengthening would be expected, i.e., yield strength is proportional to $(\lambda)^{-1/2}$. Basically, the question of dislocation motion is determined by the energy of the dislocation in the two phases viz., line energy $T \alpha G b^2$, and $T_A \sim G_A b_A^2$ in phase A and $T_B \sim G_B b_B^2$ in phase B. Since $|b_A| \cong |b|_B$, the important term is the shear modulus, and dislocation motion in phase B will be negligible if $G_B \gg G_A$.

It should be emphasized that the attractiveness of a spinodal system for this type of strengthening depends on the manner in which the dislocation substructure is produced and the way in which λ can be controlled. Thermal treatment alone, including perhaps merely furnace cooling, may induce the formation of regular high-density dislocation arrays at the interphase interfaces, without the need for mechanical processing of any kind. The limitation in exploiting these alloys rests with any grain-boundary problem, such as the accelerated coarsening in CuNiFe described in Section 2.2.

3.4 Summary

From an alloy-design point of view, the problem of identifying the strengthening mechanisms in spinodal alloys must be solved before they can be effectively exploited. It appears that the *only* way to circumvent the embrittling grain-boundary effects so that the potential strengthening due to semicoherent spinodals may be studied is by analysis of single crystals. Detailed electron-microscopy analysis can then be used to determine the nature of any slip-interfacial dislocation reactions, slip traces, etc., leading to an understanding of the strengthening process throughout the entire aging sequence, including any changes in strengthening mechanism. Finally, fracture-toughness measurement of polycrystals must necessarily be included in any study, even for systems in which heterogeneous grain-boundary precipitation may have been avoided, since lamellar structures, particularly spinodals, are well known to coarsen discontinuously.[9]

4 EXPLOITATION OF MARTENSITIC TRANSFORMATIONS

4.1 Strength of Martensite

The strength of martensite is determined principally by the carbon content in solution and the toughness by the microstructure, which in turn depends upon composition and heat treatment. Experimental steels have been designed to obtain improved properties merely by transformation to dislocated martensite, as outlined in Section 2.4. This transformation provides one of the most efficient means of obtaining uniformly dislocated microstructures, and tempering is not required to obtain desirable toughness at ambient temperatures, as shown in Fig. 5a.*

Although the strength can be increased merely by increasing the carbon content, beyond about 0.4 percent carbon, steels have low toughness. A simple linear hardening theory for polycrystals due to interstitial carbon is given by the well-known Nabarro theory $\tau_c = G\epsilon^2 C$, where τ_c = shear yield strength, G = shear modulus, ϵ = misfit, and C = atomic fraction of carbon. This equation predicts a strength of $\sim G/300$ at 0.35 percent carbon. The other important strengthening parameter is the dislocation density, and since the martensite is essentially work hardened as a result of the high dislocation content resulting from the transformation, the contribution from these dislocations can be written in terms of a flow stress $\tau_f \cong Gb\sqrt{\rho}$, where b = Burgers vector and ρ = dislocation density.

Thus, to double the strength due to this contribution, the dislocation density must be increased by at least a factor of 4. Such high dislocation densities may be achieved by deformation, e.g., ausforming.

4.2 Limitations of Carbon Content: Design to Eliminate
Quench Cracking (Experimental Fe-4Cr-0.4C Steel)

Since one of the most economical ways of increasing the strength of a steel is to raise its carbon content, it is worth considering what might be done to overcome the limitation of this procedure, namely, the onset of embrittlement due to quench cracking at the prior austenite grain boundaries, which occurs even in relatively high-purity alloys. We have recently investigated this problem in an experimental Fe-4Cr-0.4C steel[27], since a similar steel but with a lower carbon content, viz., 0.35 percent, has excellent combinations of high strength and toughness even in the untempered

---

*However, although K_{Ic} is fairly high, in these experimental steels (Fig. 5) the critical plastic-zone size is very small if the strength is also high, e.g., at a K_{Ic} of 80 ksi $\sqrt{\text{in.}}$ and a strength of 300,000 psi, the critical value of plastic-zone size is only ~ 0.01 inch on the basis of the McClintock criterion (loc. cit).

condition (see Fig. 5a)[28]. In order to eliminate undesirable alloy carbide morphologies from the microstructure of these Fe-Cr-C alloys (for example, at prior austenite grain boundaries), it was necessary to austenitize the Fe-4Cr-0.4C steel at high temperature ($>$1000 C), but this practice led to intergranular cracking during the quench in either oil or water (Fig. 8a). After such high-temperature treatments, the grain size was large ($>$300 μ, i.e., ASTM 0.5, Fig. 8a) and the microstructure contained extensive twinning but little evidence of autotempering (Fig. 8b), while scanning electron microscopy showed largely intergranular fracture (Fig. 9a). On the basis of the study of the effect of the heat-treatment variables on the intergranular cracking tendency[27], it was concluded that the amount of carbon in solution (γ) and martensite packet size* are the two most important factors influencing cracking tendency, viz., an increase in either causes an increase in cracking. In order to achieve desirable fracture-toughness properties so that the steels can be utilized in engineering applications, it is essential to austenitize at a high enough temperature to dissolve all carbides. At the same time, in order to avoid quench cracking at high levels of carbon in solution, it is necessary to obtain a small grain size. Clearly, for martensitic structures, these requirements cannot be achieved in a single heat treatment.

Multiple heat treatments were then designed[27] and the results are summarized in Fig. 9. The corresponding microstructures and mechanical properties are summarized in Figs. 8 and 10, respectively.

These results show that by ideally combining the benefits of high-temperature austenitization and small grain size in the double treatments, the resulting mechanical properties of quenched structures are optimized and are superior to those of conventionally produced lower bainite in the same steel as shown in Fig. 10. Since it is proposed that intergranular quench cracking results from the accommodation of strain generated by impingement of two or more growing martensitic packets at the grain boundaries, a small martensite packet size (i.e., small austenite grain size), autotempering, and the establishment of good intercrystalline cohesion (by elimination of undissolved carbides at the boundaries) are some of the factors which contribute favorably to a solution of the intergranular-quench-cracking problem.

---

*A martensite packet is defined as a group of dislocated martensite laths usually slightly misoriented or twin related.

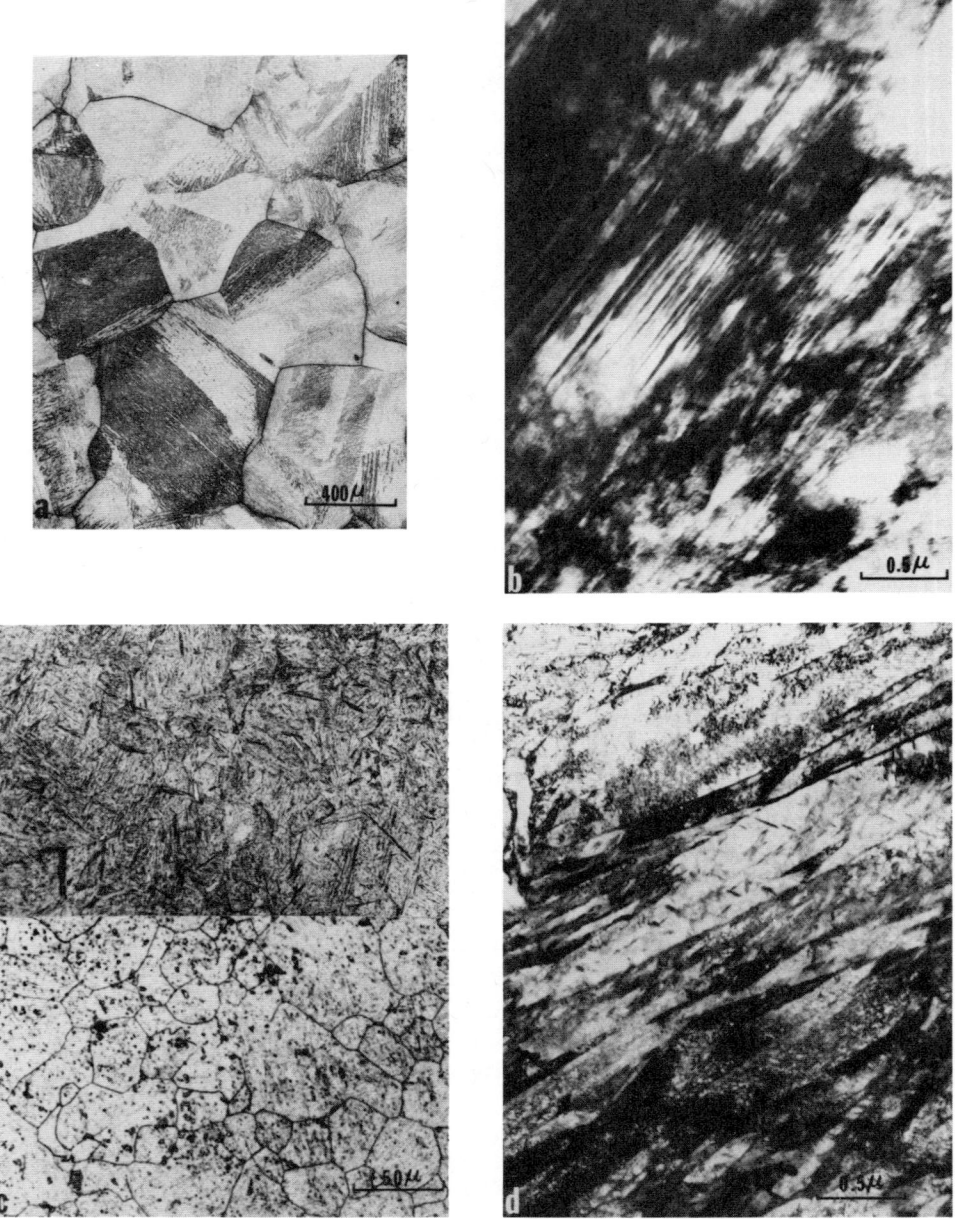

Fig. 8. Fe-4Cr-0.4C steel quenched from 1200 C; note intergranular cracks (a) and twinned plates (b). (c) and (d) show grain refinement after the double treatment shown in Fig. 9d. Two etchants are needed in (c) to distinguish the grain structure. In (d), dislocated, autotempered martensite and interlath retained austenites are resolved (compare with Fig. 8b). This structure is beneficial for good toughness—see Figs. 9 and 10. Courtesy B.V.N. Rao.

SEM FRACTOGRAPH

HEAT-TREATMENT

MECHANICAL PROPERTIES

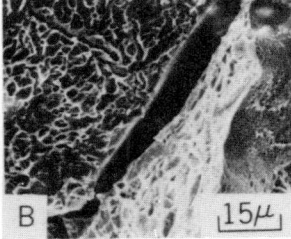

A |100μ|

Intergranular fracture in quench-cracked specimens.

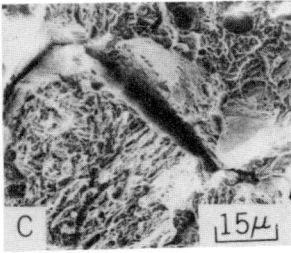

B |15μ|

Unhealed intergranular cracks produced after the first quenching.
A mixture of intergranular and quasi-cleavage fracture modes
are present.

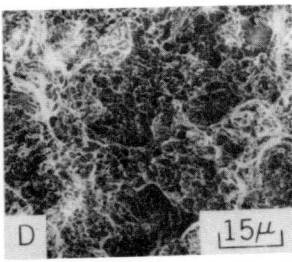

C |15μ|

The poor intercrystalline cohesion is evident. A mixed
fracture mode consisting of quasicleavage and ductile fracture
is obtained in areas where the crack path is transgranular.

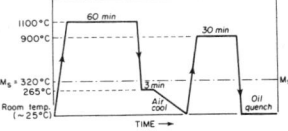

D |15μ|

Quench-crack-free specimen showing high energy transgranular fracture.

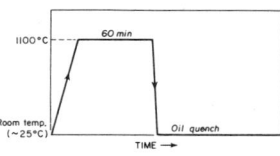

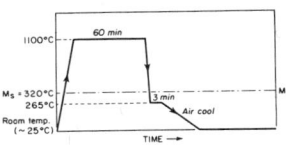

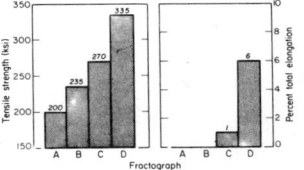

Fig. 9. Elimination of intergranular quench cracking in a 0.4 percent carbon martensitic steel. Carbon in solution and martensite packet size are the two most important parameters influencing intergranular quench cracking in Fe-4Cr-0.4C steel, viz., an increase in either causes an increase in the transformation strains and the resultant impingement stresses of two growing martensitic packets. Conventional high-temperature austenitization followed by quenching (A) invariably leads to intercrystalline cracking because of the large martensite packet size and the carbon being entirely in solution. After this treatment the steel failed prematurely at 200 ksi and at zero elongation. Conventional grain refinement, involving repeated austenitization and quenching (B) does not result in any significant improvement in mechanical properties since intergranular cracks produced during first quench do not heal on subsequent heating. Although the transformation strains are reduced in the interrupted-quenched specimen (C), the intercrystalline cohesion remains poor because of the large martensite packet size. The specimen after this treatment does, however, yield plastically, followed by failure at 270 ksi. When this specimen was reaustenitized at 900 C to refine the grain size (D), a substantial improvement in mechanical properties was achieved. A tensile strength of 335 ksi and a 6 percent elongation have been obtained. In this case, although the specimen retained most of the carbon in solution, the much-reduced martensite packet size resulted in elimination of intergranular cracks. Courtesy of B.V.N. Rao.

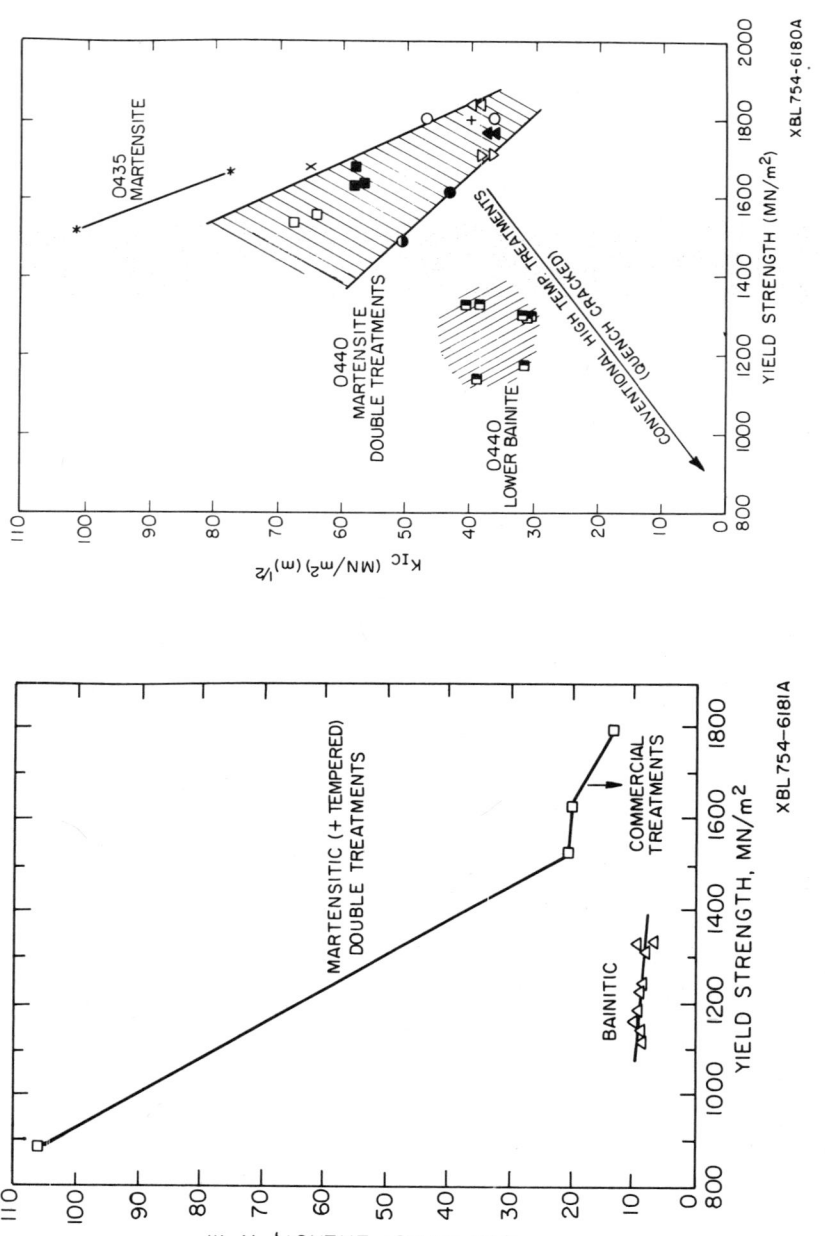

Fig. 10. (a) Relationship between: (a) yield strength and Charpy-V-notch energy, and (b) yield strength and plane-strain fracture toughness for the martensitic double treatments and single conventional lower bainitic treatments of Fe-4Cr-0.4C steel (see Fig. 9). Courtesy B.V.N. Rao.

4.3 Retained Austenite Identification and Significance in Quenched (and Tempered) Steels

As discussed earlier, the results on martensitic steels have shown that retained austenite has an extremely important effect on fracture toughness (measured by K_{Ic} and Charpy) at a particular strength level. For example at 250,000-psi quenched-strength levels, the K_{Ic} value for Fe-Cr-C was 70 ksi $\sqrt{in}$. yet only 54 ksi $\sqrt{in}$. for Fe-Mo-C (Fig. 5a), i.e., the Fe-Cr-C steel is much tougher by the K_{Ic} criterion. Detailed electron metallography and diffraction have shown that in the Fe-Cr-C steel[28], continuous sheets or fibers of interlath austenite are present; whereas, if present in the Fe-Mo-C steel, they were not resolved.[31]

Since such thin layers of austenite are not detected by X-ray analysis and since the analysis is not a trivial matter, it is worth emphasizing the need for proper characterization, using electron microscopy and diffraction. First, in most cases, the volume fraction of retained austenite in structural steels is small, and hence the austenite diffraction reflections are often weak. Careful tilting in order to bring the austenite films into a strong diffracting condition and the matrix into strong transmitting condition is necessary to enable the austenite diffraction spots to be recognized. Second, in most structural steels, autotempering occurs during quenching and so the diffraction pattern is further complicated by the presence of extra reflections from the carbide particles. A clear distinction of austenite reflections from carbide reflections is thus not straightforward because of the similar lattice spacings occurring in the patterns, and because spherical aberration limits the spatial resolution in selected area diffraction to ~2 μ at 100 kv, although this is greatly improved at higher voltages. A choice of appropriate orientation for unambiguous characterization of structure is thus necessary. Analysis[32] has shown that a <111> martensite orientation is suitable and a constructed pattern is shown in Fig. 11. This pattern is to be compared with an actual analysis, shown in Fig. 12, for which the (200)γ reflection spot was used to obtain the dark-field image of retained austenite films. While the amount of retained austenite can vary with treatment[27,33], detection now seems to be quite general[32,35] even in low carbon steels (0.1 percent), as revealed by the electron-microscopy analysis just described which is strongly recommended for all such studies.

4.4 Control of Untransformed Austenite: Stability

Current work shows that isolated particles of retained austenite are not as beneficial as continuous films (Figs. 12 and 13), suggesting that austenite may act as a crack stopper at the lath boundaries, or that it promotes crack branching along the boundaries, perhaps in a similar manner to fracture in

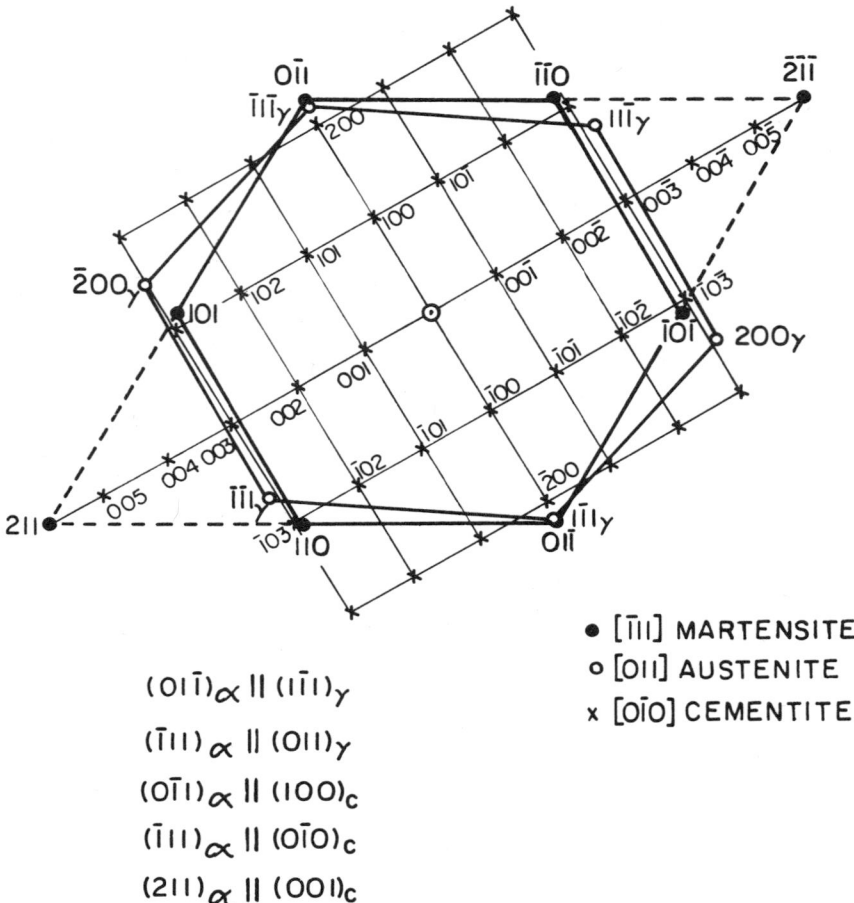

$(01\bar{1})_\alpha \parallel (1\bar{1}1)_\gamma$

$(\bar{1}11)_\alpha \parallel (011)_\gamma$

$(0\bar{1}1)_\alpha \parallel (100)_c$

$(\bar{1}11)_\alpha \parallel (0\bar{1}0)_c$

$(211)_\alpha \parallel (001)_c$

Fig. 11. Calculated electron-diffraction pattern for martensite in $[\bar{1}11]$ with retained austenite and Widmanstätten cementite.

certain fiber composites.[34] Such a mechanism could explain the fractographs often observed in our steels (Fig. 13) as was discussed earlier (Section 2.4). It has been observed that increasing the austenitizing temperature can increase the toughness.[17] Current work on Fe-4Cr-0.3C steels shows that increasing the austenitizing temperature from 870 C to 1200 C increases the value of K_{Ic} from 64 ksi $\sqrt{\text{in.}}$ to 82 ksi $\sqrt{\text{in.}}$, although the properties seem to be optimized by 1000 to 1100 C treatment. The criterion for the critical value of the plastic-zone size discussed by McClintock in his paper (this Colloquium) is $r_c = 1/2\pi \, [K_{Ic}/Y]^2$, where Y has some value between the yield and ultimate tensile strength. Thus the corresponding values for r_c are

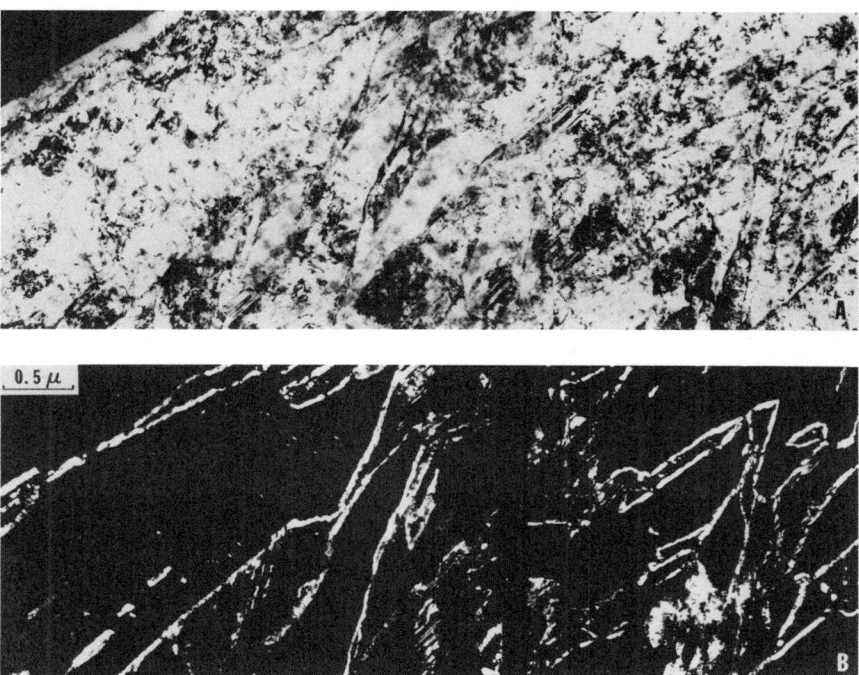

Fig. 12. Bright-field image (a) and dark-field image (b) of an Fe-1Cr-1Mo-0.3C steel quenched from 870 C into ice water. Note that the interlath retained-austenite films do not show good contrast in the bright-field image but are very clear in the (200)γ dark-field image. This is not always the case, but it serves to illustrate the importance of the proper dark-field imaging. Courtesy Y.-L. Chen.

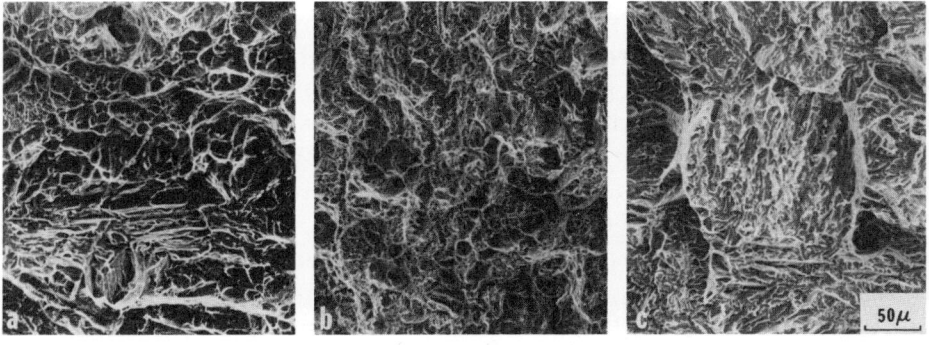

Fig. 13. Fractographs (SEM) showing improvement in Charpy toughness for Fe-1Cr-1Mo-0.3C steel (a) quenched from 1200 C (σy = 194 ksi, Cv = 28 ft-lb) and (b) double treated (grain refined) and quenched (σy = 196 ksi, Cv = 42 ft-lb). (c) shows loss of toughness due to tempered martensite embrittlement resulting from transformation of austenite producing carbide particles at interlath boundaries; (σy = 172 ksi, Cv = 14 ft-lb). Treatment same as for Fig. 13a, but after 1 hr, 350 C tempering (see Fig. 5b). Courtesy Y.-L. Chen.

250 μ after the 870 C austenitizing and ~430 μ after the 1200 C austenitizing, taking Y as the ultimate tensile strength (250 ksi and 200 ksi, respectively). Since the grain size increases from 30 μ to 260 μ with this increase in austenitizing temperature, it may be that the grain size is the most important factor determining the characteristic distance over which the critical fracture stress must be exceeded for crack propagation (as measured by K_{Ic}). Since increasing grain size also increases hardenability, there may also be an effect due to retained austenite (see Parker and Zackay, *loc. cit*).* Carbon segregates to the austenite and could stabilize it against transformation because of a lower M_s locally. Additional experiments are now in progress to obtain more quantitative data on these variables and to obtain Charpy data. The influence of austenitizing temperature on austenite carbides and martensite substructure, which already has been shown to be important for the Fe-4Cr-0.4C steels discussed above[27], will receive special emphasis.

With reference to the observed differences in toughness between the Fe-Cr-C and Fe-Mo-C steels (Fig. 5a), the greater amount of retained austenite in the former alloy is to be expected since molybdenum has a very strong effect in limiting the austenite range of stability (~2 percent molybdenum compared to ~13 percent chromium), so that each alloying element may have a different effect on the amount and possible distribution of retained austenite. Alloy design programs to study this problem are also in progress at Berkeley.

4.5 Summary

In this section it has been shown that simple control of composition and heat treatment to produce dislocated martensite and continuous films of retained austenite leads to excellent combinations of ambient strength and toughness, whereas interlath carbides lead to embrittlement (Fig. 13). Also, the limitation due to quench cracking on increasing carbon concentration to increase strength of martensite steels can be alleviated by multiple, though probably expensive, heat treatments. Further work is necessary to study in more detail the effects of alloying to control retained austenite and processing to achieve grain refinement.

The martensitic transformation can also be exploited as a possible means of strengthening softer phases by controlling the transformation (either thermally by control of M_s temperature or mechanically by control of M_d) so as to form a dispersion of martensite with austenite. It is known that duplex austenitic-martensitic alloys, e.g., 304 stainless steels, can be

---

*Recent work suggests that undissolved carbides and inclusions are more important in governing K_{Ic}, whereas grain size is more important in governing Charpy values (Carlsen, Rao, Thomas).

strengthened this way. Furthermore, the flow stress has been found to vary linearly with volume fraction of martensite[35] irrespective of the way that martensite was produced. In the following section, an approach is described in which martensite is used as a strengthening dispersion with ferrite, in low-carbon steels.

5 DUPLEX STRUCTURES AND STRENGTHENING OF LOW-CARBON STEELS

While a considerable effort in research and development has gone into improving the strength and toughness of medium- and high-carbon steels, much less effort has been directed towards understanding more completely the structure-property relationship in low-carbon steels. With the increasing problem of materials and fuel shortages, however, there is now an urgent need for design engineers to effect weight reductions in transportation systems such as automobiles by achieving economical increases in strength of steels or new alloys. Interest is developing in high-strength, low-alloy steels, but room exists for improvements in plain carbon steels.

For plain low-carbon steels, several approaches have been used in the past, e.g., rapid heating and cooling cycling through the austenite transition temperature for grain refinement.[36] However, if the final transformation product is a mixture of ferrite and pearlite, the strength level is not as great as is required. On the other hand, if the cycling is done so as to attempt to produce 100 percent martensite, the ductility is poor and undesirable microstructures (e.g., upper bainite) can often result because of the low hardenability.

In considering commercial automobile steels, e.g., INNA and 1010 (basically Fe-0.5Mn-0.1C with INNA having 0.01 percent nitrogen also), a novel way of thermal cycling which involves annealing in the two phase ($\alpha+\gamma$) field has been used. This heat treatment and subsequent cycling treatment is compared with conventional cycling for grain refinement in Fig. 14. The initial austenitizing treatment consisted of annealing in argon at 1100 C (30 minutes) so as to dissolve all carbides, followed by ice-brine quenching to obtain 100 percent martensite as the starting microstructure.[36] Details of the experimental conditions and range of cycling treatments investigated will be published elsewhere[39], but the initial grain size was ASTM #2 ($\sim$80 μ) which is refined by the cycling process, e.g., after the second cycle, the grain size is $\sim$18 μ.

By holding in the $\alpha+\gamma$ temperature range, the α and γ phases will attain the compositions specified by the tie line corresponding to the holding temperature. The alloy will then consist of low-carbon ferrite and higher carbon austenite. Upon quenching, the austenite transforms to martensite which electron microscopy shows to be dislocated (see Fig. 15). The ferrite

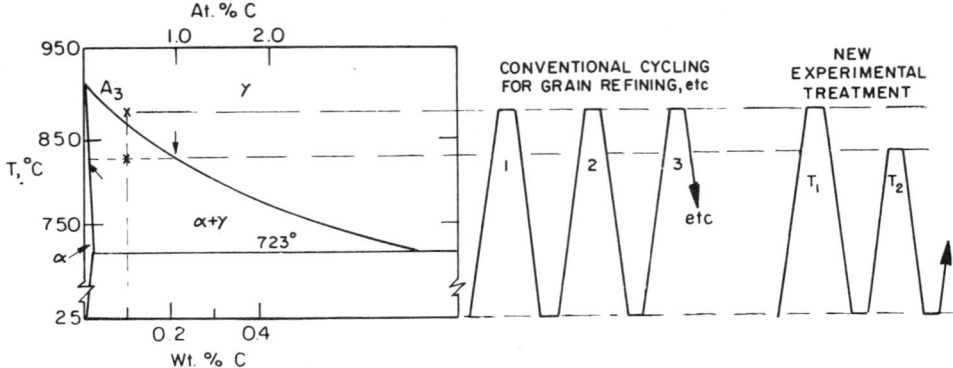

Fig. 14. Principle of heat treatment to produce martensite dispersions in ferrite. The conventional grain-refining cycling is also shown in comparison with that following two-phase annealing. Courtesy J.-Y. Koo.

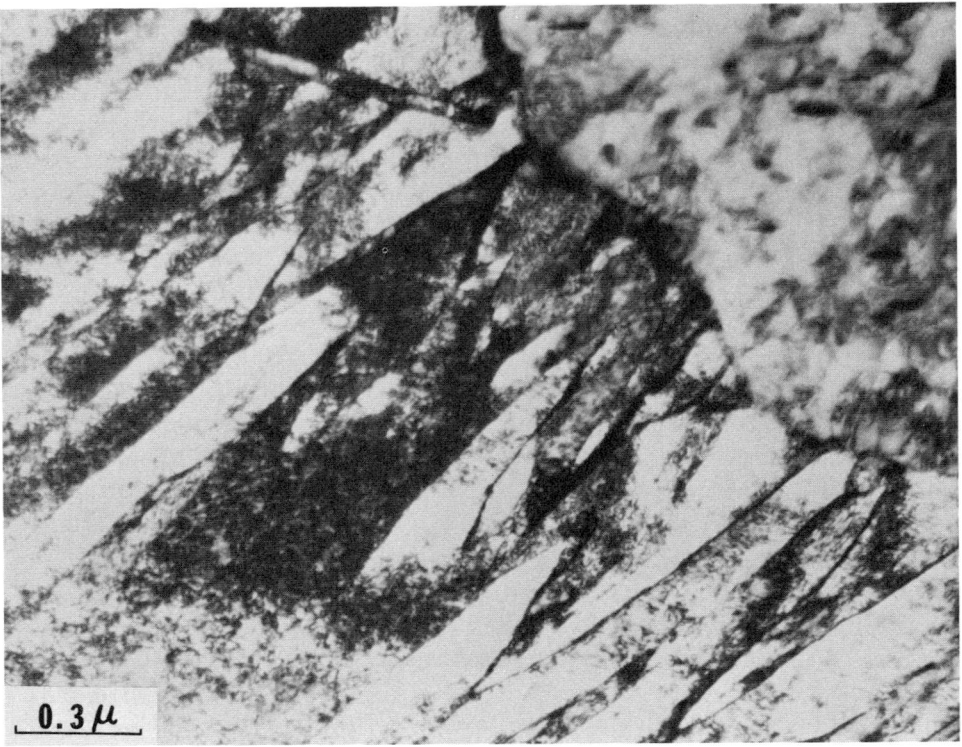

Fig. 15. Dislocated martensite and ferrite in doubly treated "INNA" steel. Some heterogeneous precipitation occurs on dislocations in the ferrite, especially after tempering. Courtesy J.-Y. Koo.

which does not transform back to austenite during the reheating to the $\alpha+\gamma$ field contains subboundaries which are formed as a result of dislocation generation and rearrangement during cycling. These walls (~ 1 μ apart) become sites for carbide (or nitride) precipitation on tempering, which can be a further source of strengthening.

This method of using the two-phase ($\alpha+\gamma$) field has also been utilized by Jin et al.[37] for iron-nickel and maraging steels and Snape and Church[38] for low-alloy steels. However, the transformation behavior and beneficial results obtained for the 0.5Mn-0.1C steels differ from those in the higher alloy steels.

The effects of the ferrite-martensite mixtures on the yield strength and ductility (percent elongation) are shown in Fig. 16. This graph also contains the commercial specifications for the INNA steel. It can be seen that the new heat treatments described here can double the yield strength and also the tensile strength (due to the greater work-hardening capability) at an acceptable uniform elongation level of 10 percent. The method could be useful as a finishing treatment for improving strength (i.e., after forming). The improved mechanical properties can be interpreted in a manner used for duplex structures such as fiber composites, for the condition that the microstructural constituents (ferrite and martensite) are equally strained. The flow stress can then be expressed empirically as $\sigma_f = \sigma_\alpha f(\alpha) + [1-f(\alpha)]\sigma_m$, where σ_α = flow stress for ferrite, σ_m = flow stress for martensite, and $f(\alpha)$ = volume fraction ferrite. Such a relationship is in agreement with the data in Fig. 16. It should be noted that σ_α is determined principally by the ferrite grain size and whether or not it is also precipitation hardened. As discussed earlier, σ_m is determined by the carbon content, which can be varied by the annealing temperature in the $\alpha-\gamma$ range (Fig. 14).

These results indicate the improvements that can be made in these low-carbon steels. Since several million tons are produced each year, some savings in material are clearly possible if these strengthening treatments can be adopted economically by industry.

6 CONCLUDING REMARKS: USE OF MULTIPLE TRANSFORMATIONS

Steels are systems where several transformations occur during production; for example, during the martensite transformation, precipitation of carbides can occur simultaneously or consecutively. However, much more potential exists for combining more than one type of phase transformation in the same alloy system, especially as a means for eliminating mechanical stages in processing to produce microstructures with useful combinations of properties. For example, consider the ausforming treatment (see Fig. 17).

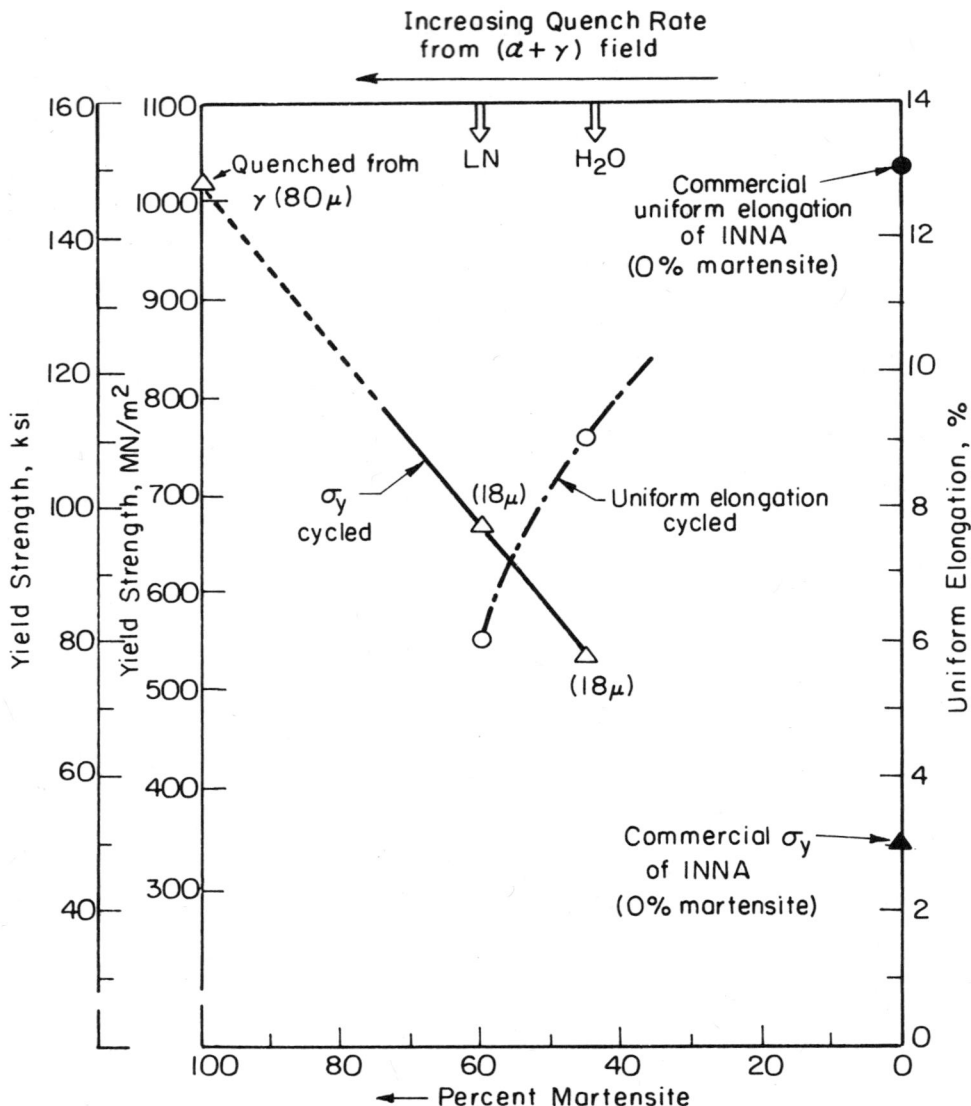

Fig. 16. Summary of mechanical properties of "INNA" steel showing influence of volume fraction of martensite and grain size. Courtesy J.-Y. Koo.

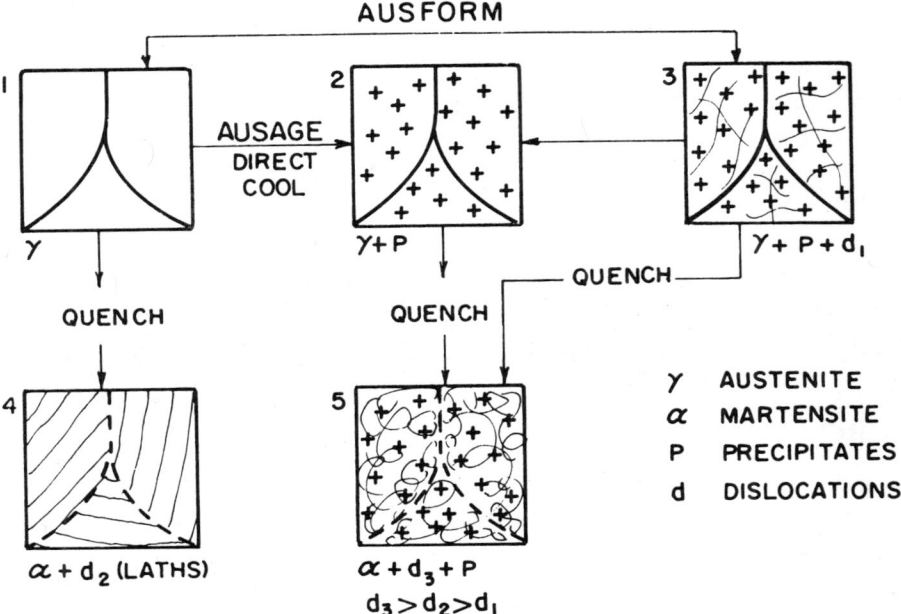

Fig. 17. Suggestions as to how multiple phase transformations may be useful in providing maximum dispersion strengthening through dislocation generation (martensite) and precipitation: 1 → 2 conventional martensite; 1 → 3 → 5 conventional ausforming; and 1 → 2 → 5 — first ausage, then transform to martensite or first spinodally decompose to disperse particles in austenite, and then quench so as to transform one of the phases to dislocated martensite and produce dislocation multiplication.

This process involves deformation of metastable austenite during which heterogeneous nucleation of carbides on dislocations occurs.[40] On quenching, the carbon-depleted austenite transforms to martensite, and provided that the martensite is dislocated (C $\lesssim$ 0.3 percent), the steel has greater toughness and strength than are obtained by conventional martensitic transformation. This approach is limited by the austenite transformation kinetics in that the pearlite and bainite transformations need to be well separated. However, similar results can be obtained *without* the deformation if precipitation can be induced in austenite prior to transformation. Examples of this process include ausaging FeNiTi-type alloys to precipitate Ni_3Ti[41] or spinodally decomposing CuNiFe prior to quenching to form martensite[42]. For these processes to be effective, it is essential to produce dislocated martensite so that dislocation multiplication can occur at the precipitates during the martensitic transformation (Fig. 17). The composition of the alloy limits the usefulness of multiple treatments since it determines whether or not martensite can be formed (athermally below M_s or by deformation below M_d), but, nevertheless, much can be done to explore these ideas further.

ACKNOWLEDGMENTS

This work was done under the auspices of the United States Energy Research and Development Administration through the Inorganic Materials Research Division of the Lawrence Berkeley Laboratory. I want to thank members of my research group for their invaluable contributions and for providing me with the data and illustrations. I also wish to thank various companies, especially INCO, Republic Steel, Climax Molybdenum, and Daido Steel who have generously donated alloys made to our specifications. I am grateful to Dr. Noel Kennon for his critical, helpful review of the manuscript.

REFERENCES

1. Cahn, J. W., *Acta Met.*, **5**, 169 (1957).
2. Nicholson, R. B., *Phase Transformations in Metals,* American Society for Metals, Cleveland, Ohio (1970), p. 269.
3. Thomas, G., *J. Inst. Metals (London),* **89**, 253 (1960–61).
4. Lorimer, G., and Nicholson, R. B., in *Mechanisms of Phase Transformations,* Institute of Metals, London, England (1969), p. 36; Lorimer, G. W., Nasir, M. J., Nicholson, R. B., Nuttall, K., Ward, D. E., and Webb, J. R., in *Electron Microscopy and Structure of Materials,* G. Thomas (Ed.), University of California Press, Berkeley, California (1971), p. 222. See also Doig, P., Edington, J., and Jacobs, M. H., *Phil. Mag.,* **31**, 285 (1975).
5. de Fontaine, D., *Ultrafine-Grain Metals,* J. J. Burke and V. Weiss (Eds.), Syracuse University Press, Syracuse, New York (1970), p. 93.
6. Butler, P., and Thomas, G., *Acta Met.,* **18**, 347 (1970).
7. Livak, R., and Thomas, G., *Acta Met.,* **19**, 497 (1971).
8. *Phase Transformations*, American Society for Metals (1972), e.g., p. 487.
9. Gronsky, R., and Thomas, G., Lawrence Berkeley Laboratory Report No. 3523, January, 1975; *Acta Met.,* in press.
10. Livak, R. J., and Gerberich, W. W., in *Electron Microscopy and Structure of Materials,* G. Thomas (Ed.), University of California Press, Berkeley, California (1971), p. 647.
11. Wayman, M., in *Modern Diffraction and Imaging Techniques in Materials Science,* S. Amelincx et al. (Eds.), North Holland Press (1970), p. 187.
12. Thomas, G., *Met. Trans.,* **2**, 2372 (1971).
13. Thomas, G., *Iron and Steel Intern.,* **46**, 451 (1973).
14. Chen, Y.-L., Ph.D. thesis, University of California, Berkeley; Lawrence Berkeley Laboratory Report No. 5179 (1976).
15. Honeycombe, R.W.K., *Structure and Strength of Alloy Steels,* Climax Molybdenum Co. publication (1975).
16. Huang, D. H., Ph.D. Thesis, University of California, Berkeley; Lawrence Berkeley Laboratory Report No. 3713 (1975).
17. Parker, E. R., and Zackay, V. F., in *Alloy Design,* J. K. Tien and G. S. Ansell (Eds.), Academic Press (in press).
18. Douglass, D. L., and Barbee, T. W., *J. Mater. Sci.,* **4**, 121 (1969).
19. Carpenter, R. W., *Acta Met.,* **15**, 1297 (1967).
20. Richman, R. H., and Davies, R. G., *Trans. AIME,* **236**, 1551 (1966).
21. Saito, K., and Watanabe, R., *J. Phys. Soc., Japan,* **22**, 681 (1967).

22. Badia, F. A., Kirby, G. N., and Mihalison, J. R., *Trans. ASM*, **60**, 395 (1967).
23. Schwartz, L. H., Mahajan, S., and Plewes, J. T., *Acta Met.*, **22**, 601 (1974).
24. Dahlgren, S. D., Ph.D. Thesis, University of California, Berkeley; Lawrence Berkeley Laboratory Report No. 16846 (1966).
25. Bouchard, M., Livak, R. J., and Thomas, G., *Surface Sci.*, **31**, 275 (1972).
26. Brown, L. M., and Ham, R. K., in *Strengthening Methods in Crystals*, A. Kelly and R. B. Nicholson (Eds.), Elsevier Publishing Co. (1971), p. 9.
27. Rao, B.V.N., and Thomas, G., *Mat. Sci. and Eng.*, **20**, 195 (1975).
28. McMahon, J. M., and Thomas, G., in *Microstructure and Design of Alloys*, Institute of Metals, London, England (1973), Vol. I, p. 180.
29. Speich, G. R., and Szirmae, A., *Trans. AIME*, **245**, 1063 (1969).
30. Brobst, R. P., and Krauss, G., *Met. Trans.*, **5**, 457 (1974).
31. Clark, R. A., and Thomas, G., *Met. Trans.*, **6A**, 969 (1975).
32. Rao, B.V.N., Koo, J.-Y., and Thomas, G., *Proc. 33rd EMSA Conference*, Claitors Publishers (1975), p. 30.
33. Lai, G. Y., Wood, W. E., Clark, R. A., Zackay, V. F., and Parker, E. R., *Met. Trans.*, **5**, 1663 (1974).
34. Cook, J., and Gordon, J. E., *Proc. Roy. Soc.*, **A282**, 508 (1964).
35. Manganon, P. R., and Thomas, G., *Met. Trans.*, **1**, 1577, 1587 (1970).
36. Grange, R. A., *Met. Trans.*, **2**, 65 (1971).
37. Jin, S., Morris, J. W., and Zackay, V. F., *Met. Trans.*, **6A**, 141 (1975).
38. Snape, E., and Church, N. L., *J. Metals*, **23** (1972).
39. Koo, J.-Y., and Thomas, G., to be published, Lawrence Berkeley Laboratory Report No. 3980 (1975); *Mater. Sci. and Eng.*, in press.
40. Johari, O., and Thomas, G., *Trans. ASM*, **58**, 563 (1965).
41. Cheng, I.-L., and Thomas, G., *Met. Trans.*, **3**, 503 (1972).
42. Vercaemer, C., and Thomas, G., *ibid.*, **3**, 2501 (1972).

DISCUSSION on Paper by G. Thomas

HAASEN: Is anything known on the fatigue behavior of your spinodal alloy as a function of the microstructure?

THOMAS: Not as far as I know; we have done no fatigue experiments. Creep is planned especially for the semicoherent spinodals.

ANSELL: One clear structural distinction between plate and lath martensite is the discontinuous nature of plate formation resulting in the separation of the plates by austenite as contrasted to the continuous, or packet, nature of lath martensite. It would thus appear that your criterion for toughness being retained austenite between martensite plates is simply the equivalent of stating that higher toughness is expected in plate martensites which are not completely transformed as compared to partially transformed lath martensites. Indeed, lath martensites, by definition, can never have the separating distribution of retained austenite which you have described, while all incompletely transformed plate martensites must always have such a retained-austenite distribution.

THOMAS: You recall that in my slide showing the tensile versus K_{Ic} data for our Fe-4Cr-0.35C and Fe-2Mo-0.4C steels, the Fe-Cr-C one has better toughness than the latter at the same strength level. Both show dislocated martensite but only the Fe-Cr-C alloy has the retained austenite even though M_s-M_f are similar. Thus the point I'm making is that to improve toughness you want to produce dislocated martensite surrounded by retained austenite films (as a composite). I certainly do not want to confuse the terminology pertinent to martensite. The other point I'm emphasizing is that to identify this austenite requires painstaking electron microscopy and diffraction—it is not found easily by X-rays (if at all). I discuss this in more detail in my paper and in a recent one (Rao, Koo, Thomas, EMSA, 1975, p. 30).

SHERBY: I was wondering what you would predict in terms of the type and morphology of martensite that might be created if the grain size of the prior austenite is in the submicron range. I fully appreciate that this might be very difficult to do with the steels you have been studying, but does your work permit you to extrapolate to such grain-size ranges and allow prediction of the type of martensite developed and the mechanical characteristics of the end product? We have recently developed submicron ferrite grains in plain-carbon ultrahigh-carbon steels (1.3 to 1.9 percent carbon) by special thermal-mechanical processing procedures; it is likely that submicron austenite grains (containing spheroidized cementite) can be obtained upon transformation heating. One may thus have the opportunity to look into martensite formation from submicron austenite.

THOMAS: In fully martensitic high-strength steels, if the grain size is very small there might be a change from dislocated to twin morphology due to impingement stresses being high at boundaries which would lead to embrittlement, but this is a guess on my part. We do not specifically grain refine much below 20 μ in our studies—even in the duplex ferritic-martensitic 1010-type steels I discussed. In the latter, purely thermal cycling limits us to about 15 μ. We have not tried thermal-mechanical grain refining, but already more work needs to be done on very fine-grained strong steels, especially to raise fracture toughness.

BEMENT: In your analysis of the shear stress for your incoherent spinodal alloy you used a root-mean-square superposition of the shear moduli and boundary barrier terms. Why do you use this instead of a linear superposition?

THOMAS: This is only a suggestion based on Brown and Ham's analysis (ref. 26 of my paper). My own feeling is that since the interface dislocations are very stable, the interfaces can be considered to be like grain boundaries, in which case I would expect a Hall-Petch ($\lambda^{-1/2}$) strengthening where λ is the spinodal wavelength (as discussed in my paper).

BEMENT: The development of homogeneity in martensitic steels should also take into account thorough thickness variations, especially where kinetics are involved in the formation and distribution of some of the transformation products. To what extent would you limit the thickness of your model systems to satisfy this homogeneity requirement?

THOMAS: In our experimental vacuum steels I described (Fe-Cr-C and Fe-Mo-C), our hardenability data (based on Jominy tests) show they are at least as good as commercial 4340.

KEAR: In the Cu-Ni-Fe spinodal system you showed an interesting TEM photomicrograph of discontinuous coarsening by grain-boundary migration. Have you any thoughts on how this phenomenon can be controlled through alloy design?

THOMAS: Now that we have identified the grain-boundary problem to be one of discontinuous coarsening, rather than a cellular or competing precipitation process, we must clearly retard boundary migration. One possibility would be to use inclusions at grain boundaries. In the Cu-Ni-Fe alloys, perhaps carbides will do the job. However, we have not yet tried this but the problem you raise is one we want to tackle next. A fourth element may be effective in boundary pinning, but this will need careful evaluation from the chemical viewpoint. Texture control needs also to be explored in efforts to limit these mobile boundaries.

ALLOY DESIGN WITH OXIDE
DISPERSOIDS AND PRECIPITATES

J. K. Tien

Henry Krumb School of Mines
Columbia University
New York, New York

ABSTRACT

The mechanical behavior of particle strengthened alloys is reviewed and at times extended. On the basis of this knowledge, we discuss and conclude that second phase particles, coherent or incoherent, can certainly enhance the flow strength, creep resistance, and stress-rupture life of alloys. Unfortunately, we also conclude that, at this time, particles are usually not beneficial alloy-design elements if enhanced uniaxial ductility, plane-strain ductility, stress-rupture ductility, and toughness are called for. Such properties as fatigue-crack propagation resistance appear to require, for example, both high strength and high ductility, a situation which can come to pass only when the perennial conflict between strength and ductility is resolved in particle strengthened systems in particular and in any other material system in general. Wherever possible, we distinguish between the role of coherent versus incoherent particles in alloy design.

1 INTRODUCTION

There are high expectations that particles, be they precipitates or inert dispersoids, can be dependable elements in structural-alloy design. Are these expectations fulfilled? Certainly, the reliance of the turbine industry on the precipitation-strengthened superalloys and the expanding sales of precipitation-strengthened aluminum alloys call for an affirmative answer. However, there are many other age-, marage-, and dispersion-hardened systems that have not gained commercial acceptance in spite of many years of costly optimization. These systems no doubt could be victims of the whims of economics and demand. However, one is still tempted to question whether particle-strengthened systems possess the versatility of properties normally called for in structural applications. For example, the turbine industry often pays a high penalty in the form of expensive hafnium additions and directional solidification in order to achieve the required minimum ductility in the strong turbine superalloys.

In what follows, we attempt to assess through existing knowledge the status of particles as structural-alloy design elements. Unfortunately, such indisputably positive attributes of particle-strengthened alloys such as their good yield strength and creep resistance are the only properties that are fairly well understood. In contrast, the potentially negative attributes of these alloys, such as their poor ductility, fracture toughness, and perhaps inferior fatigue-crack propagation resistance, are not well understood. Worse yet, important properties such as plane-strain ductility and crack propagation under creep-loading conditions have received almost no attention. Hence, along with the appraisal of the state of the art in our knowledge of the particle-strengthened systems from the structural-alloy design point of view, we also indicate areas where further research needs to be done in order to improve the understanding.

2 PARTICLES AND FLOW STRESS

Concentrated work has gone into developing the particle-strengthening mechanisms and exhaustive and often elegant reviews are now available.[1-3] The compacted knowledge can be summarized as follows.

The increase in yield strength achievable by any strengthening mechanism depends on how effectively the strengthening microstructure can obstruct the motion of the mobile dislocations. In dispersion-strengthened or incoherent precipitation-strengthened alloys, the dislocations encounter the particles and try to bypass them by the Orowan bowing mechanism. The resulting yield strength depends therefore on the critical Orowan stress required to achieve the bypass process and is given by

$$\tau = \tau_o + \alpha \frac{Gb}{\ell} \quad , \tag{1}$$

where τ is the flow (yield) stress of the alloy, τ_o is the flow stress of the unstrengthened matrix which itself may have been increased by solid-solution strengthening, G is the shear modulus of the alloy, b is the Burgers vector of the glide dislocations in the alloy, ℓ is the average interparticle spacing, and α is a geometrical constant depending on the particle shape and distribution and on the method of space averaging used to convert volume fraction to interparticle spacing. Reference 1 provides a sound review on these geometric aspects of particle strengthening. In addition to the Orowan contribution, the inert dispersion-strengthened alloys derive additional strength from the dislocations substructures introduced into these alloys during their thermomechanical processing, which in conjunction with directional recrystallization can even result in directionally aligned grains with high aspect ratios where it will be possible to orient the weak grain boundaries parallel to the load axis. Directional solidification performs this same grain-orientation function for precipitation-strengthened alloys.

In coherent precipitation-strengthened systems, the dislocations can bow between the particles when the interparticle spacing is large, say, when it is ten times the particle size. In nickel-base superalloys, for example, the $\gamma'(Ni_3(AlTi))$ precipitates are about 0.3 micron in size. The flow stress in this case can again be estimated by the Orowan model, Eq. (1), with slight modifications.[1] However, as the interparticle spacing decreases, which can occur if either the volume fraction is increased or the particle size is decreased, or both, the dislocations will start to cut through the coherent precipitates on the slip systems that are common to matrix and precipitate. In so doing, the mobile dislocations or superlattice dislocation pairs[4] begin to test the resistance of the ordered states in the precipitates. In the extreme case when the coherent particles are very close together, which is the case in the well-endowed, 40 to 60 volume percent γ' superalloys, the resulting flow stress of the system is fairly accurately predicted by[1,2,4]

$$\tau = \gamma_{APB}/2b + (\tau_o + \tau_p)/2 - T/br \quad , \tag{2}$$

where τ_o and τ_p are the flow stresses in the matrix and the precipitate, respectively, γ_{APB} is the antiphase-boundary energy of the ordered precipitate, T is the line tension of the dislocations, and r is the equiaxed particle size. It should be noted that the major contribution ($\sim$80 percent) to the flow stress in Eq. (2) comes from the first term which is the contribution of the degree of order in the strengthening precipitates.

If the strengthening precipitates are coherent or semicoherent with the matrix, the coherency strain between the matrix and precipitate lattices can also contribute to strengthening, essentially by increasing the interaction distance between dislocation and particles. This leads to the additional contributions of $\Delta\tau$ to the flow stress given by[5]

$$\Delta\tau = CG|\epsilon|^{3/2}(r/b)^{1/2}f^{1/2} \quad , \tag{3}$$

where f is the volume fraction of the precipitate and ϵ is the coherency strain due to the misfit between the lattice parameters of the precipitate and the matrix and is usually controlled by minor alloying additions.

Orowan strengthening, say, by ~3 volume percent thorium oxide dispersoids of ~0.1-micron size in nickel (TD-nickel) results in a strengthening factor* of about three, i.e., the shear flow strength of TD-nickel[3] is about 200 MN/m^2, compared with 70 MN/m^2 for nickel[6]. On the other hand, the strengthening factor for high-volume-fraction (~50 percent) coherent precipitates, Eq. (2), is about eight to nine—the typical shear yield strength of a cast, nickel-base superalloy is 600 MN/m^2.[4] Substructure strengthening in oxide-strengthened alloys can also be quite significant. After thermomechanical processing, the shear yield strength of TD-nickel can be as high as 400 MN/m^2.[3] Coherency strain strengthening is usually not very significant.

Aside from differences in strengthening factors, a major distinction between coherent precipitation and inert dispersion strengthening systems is the temperature dependence of these factors. As a consequence of the relatively nonreacting nature of the oxide dispersoids, dispersion strengthening persists to near the solidus temperature. Coherent precipitation strengthening, on the other hand, is temperature limited by thermally activated processes. For example, all γ' strengthened nickel-base superalloys experience a sharp fall off in yield strength to solid-solution levels at about 760 to 860 C, which is hundreds of degrees below the γ' solvus temperature, and at least 100 degrees below the temperature where significant precipitate coarsening or dissolution can take place.[7,8] This fall off is not well understood. It is believed to be due to slip-caused disordering of the otherwise ordered precipitates, resulting in a dynamic decrease in the antiphase-boundary energy.[9] It is worthwhile to note that the most recent alloy-development achievement, namely, the yttria dispersion-strengthened superalloys, indeed combines the high-strength qualities of precipitation strengthening at intermediate and ambient temperatures with very high temperature strength capability of inert dispersion strengthening.[10]

---

*Throughout this paper we refer to the term "factor" in the following sense: "factor" is defined as the property of the alloy with particles minus the property of the matrix alloy in the absence of particles, divided by the matrix property.

Currently, another difference between systems with precipitates and inert dispersoids rests with particle distribution and volume fraction. Because of the high strengthening factors associated with particle strengthening, a uniform distribution of the particles is a must or else severe strain gradients can easily be produced. As a consequence of nucleation, growth, and coarsening factors too detailed to be discussed here, uniform distributions of even high volume fractions of fine coherent precipitates can be produced. In fact, in the case of superalloys, the γ' precipitates are so uniformly distributed that they appear in periodic cube arrays, Fig. 1. Unfortunately, this is not the case in inert-oxide dispersion-strengthened alloys. The powder metallurgical processes by which these systems are being currently produced can uniformly distribute only about 2 to 4 volume percent of fine oxide dispersoids.

In summary, it appears that from an alloy-design viewpoint, the strengthening factor is increased if

- Volume fraction is high
- Particles are fine
- Particle distribution is uniform
- Particles are inert for high-temperature applications.

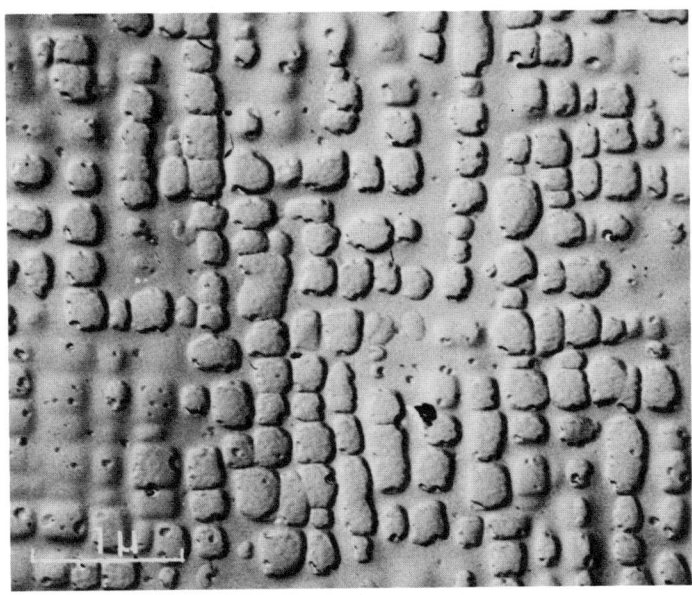

Fig. 1. Typical replica transmission electromicrograph of a high-volume-fraction (~50 percent) γ'-strengthened nickel-base superalloy, showing homogeneity of precipitate distribution in the matrix.

3 PARTICLES AND DUCTILITY

While the strengthening factors associated with particles may be positive, the ductility factor is not. The ductility factor between typical nickel solid-solution alloy and nickel-base superalloy is roughly -0.8[11], and that between nickel-20 percent chromium solid-solution alloy and TD-NiCr is roughly -0.75[12]. This factor is also negative in age-hardened or maraging steels, as in aluminum-base alloys.[13] Further, the ductility factor appears to become even more negative at elevated temperatures.[3,11]

Although more work needs to be done, sufficient understanding exists to allow the conclusion that the negative ductility factor may, in most cases, be an unavoidable consequence of particle strengthening. Hence, again the perennial conflict between strength and ductility.

Particles, especially hard oxides and intermetallic compounds, deform less than the softer matrix. As a consequence, accommodation dislocations are produced near the particles during the initial stages of deformation.[14] This certainly provides for plenty of dislocations, but their ease of movement is also constrained accordingly, thus resulting in early and fairly extreme work-hardening rates which do suppress mechanical instabilities. However, as deformation progresses, this early positive work-hardening factor becomes a detriment since it also tends to prevent further accommodation between particle and matrix. As a consequence the particles now act as centers of void initiation, and the subsequent coalescence of these voids can cause premature fracture of the system.

Void or crack initiation at the particle can occur either by cracking of the particle, if it is brittle, or by decohesion of the particle-matrix interface. Certainly, both these processes are less likely when the particle is coherent and ductile. Interface decohesion is more likely in the case of the brittle dispersoids and incoherent precipitates encountered in the strong alloys, while the brittle fracture mode is more popular with the unintentional inclusions in the commercial steels. Gurland and Plateau[15] have treated the idealized case where all particles behave identically, and derived the total strain at fracture ϵ_T for the interface decohesion mode as

$$\epsilon_T = \epsilon_o + 1/2 \ln[1/\lambda^4 - 1/k\lambda^2 + 4(a/s)^4/(9kf^2\lambda^2)] \qquad (4)$$

and for the particle cracking mode as

$$\epsilon_T = \epsilon_o + 1/2 \ln[1 + 4(a/s)^4/(9kf^2\lambda^2)] \quad , \qquad (5)$$

where a is the half width of the void, s is the critical near-neighbor void spacing for rapid ductile fracture, λ is the length-to-width ratio of the particles, ϵ_o is the strain for void nucleation, k is the strain concentration factor around a particle, and again f is volume fraction of particles. It is clear from these equations that the strain to fracture or total ductility of an alloy strengthened by hard particles decreases with the volume fraction of the particles. Also a high aspect ratio of the particle shape in a direction normal to the tensile stress would impair the ductility, since it affects the λ value unfavorably.

More recently, Argon and co-workers[16-18] have reported analytical and experimental results that not only confirm that particles can be detrimental to ductility, but, in addition, show that interaction between plastic zones around particles can lead to even greater reductions in ductility. They conclude that when the volume fraction is around 0.01 or less, the particles behave as if they are isolated from each other, i.e., the Gurland and Plateau treatment[15] is sufficient to predict alloy ductility. However, when the volume fraction of particles approaches, say, 0.1, interaction effects come into play and the interfacial stresses are enhanced due to the overlap of the plastic zones around particles. This results in interface separation at very low strains, i.e., plastic zone overlap is found to reduce ϵ_0 in Eq. (4) or Eq. (5). It should be noted that even if the average volume fraction is low, any local inhomogeneities of particle distribution can result in localized lowering of ϵ_0, and hence, cause premature void nucleation and reductions in ductility. We believe this to be the cause of the wide scatter one usually finds in the ductility data of dispersion-strengthened alloys produced by powder metallurgy.

We certainly do not mean to imply that one can never alter the ductility factor of particles from negative to positive. Certainly, manufacturing difficulties aside, one can imagine many schemes of duplex or even triplex distributions of brittle, ductile, coherent, and incoherent particles that, when cleverly spaced in the alloy, will not only result in strengthening but also lead to increased ductility and toughness. Until these schemes come to pass, minimization of tensile-ductility loss in particle-strengthened alloys can be achieved when

- Volume fraction of particles is kept minimum and homogeneous
- The particles are fine and equiaxed, and coherent with high particle-matrix interfacial strength.

4 PARTICLES AND PLANE-STRAIN DUCTILITY

Although tensile ductility is a good indicator of the reliability of an alloy under uniaxial loading conditions, there are many instances in which the ductility of an alloy needs to be evaluated under more complex states of stress. A typical example and a common situation is the state of plane-strain loading encountered in regions of stress concentrations in large structures such as in the vicinity of small cracks, flaws, or inclusions in large structural components. Such considerations have recently prompted studies in the evaluation of the ductility of some steels and aluminum alloys under plane strain constraints imposed through proper choice of specimen geometries.[19-21] These studies show that the plane-strain ductility of materials in general is less than their uniaxial ductility (see Table I). Further, it appears that plane-strain ductility decreases even more sharply with yield-strength level than does uniaxial ductility. Apart from these interesting experimental observations, there is very little understanding available at this time to explain why this is the case. In what follows we try to provide a qualitative rationalization based on the premise that local interfacial stresses and interface decohesion are important in deciding alloy ductility.

It is known that triaxial stresses add on to local stresses around the particle and enhance the particle-matrix interfacial stresses.[15-17] Under such conditions, the interfacial stress σ_{int} is given by

$$\sigma_{int} = \sigma_T + \sigma_y(\bar{\epsilon}_p) \quad , \tag{6}$$

where σ_T is the local triaxial stress and $\sigma_y(\bar{\epsilon}_p)$ is the flow stress of the particle free matrix corresponding to the local plastic strain around the particle. The value of $\sigma_y(\bar{\epsilon}_p)$ is dependent on the matrix yield strength and work hardening: the higher either of these properties are, the higher $\sigma_y(\bar{\epsilon}_p)$ will be and, hence, the higher the chances are of interface decohesion and premature fracture. The local triaxial stress σ_T depends on the state of stress in the vicinity of the particle and will be higher in a plane-strain situation than in a plane-stress situation. Hence, we conclude that the higher interfacial stresses at the particle-matrix interface under plane-strain test conditions as against the plane stress, or uniaxial, test conditions would result in a higher chance of interface separation and hence poorer ductility under plane-strain conditions. Further, the apparent decreases in plane-strain ductility as yield

Table I. Plane-Strain Ductility of Some Materials Compared with Their Uniaxial Ductility Values

Alloy	σ_y, MN/m^2	ϵ_f—Tensile	ϵ_f—Plane Strain	$\dfrac{\epsilon_f\text{—Plane Strain}}{\epsilon_f\text{—Tensile}}$
A517-F steel[19]	770	0.99	0.6	0.61
HY-130 steel[19]	995	1.05	0.5	0.48
18-Ni marage[19]	1750	0.73	0.15	0.21
7075 overaged[21]	175	0.42	0.22	0.52
7075 peakaged[21]	515	0.67	0.15	0.22

strength is increased, Table I, can also be explained by Eq. (6). According to this equation, for the same particle content, the particle-matrix interface stresses will be higher in an alloy with higher flow stress. This, superimposed on the already high triaxial stresses, leads to a larger probability of void initiation by particle-matrix decohesion. Accordingly, the attendant reduction in ductility can therefore be larger for the stronger alloys under plane-strain conditions.

It is interesting to note that high local interfacial stresses can come about near particles, even in the absence of obvious plane-strain conditions. For example, we believe that particles appear to be the culprits in causing plane-strain-type-ductility reduction when mild steels are exposed to hydrogen, even during the initial stages of deformation when the specimen as a whole is in the uniaxial stress state. A recent mechanism shows that this may be the case because particle or inclusion interfaces can steal hydrogen from the Cottrell atmospheres on mobile dislocations as these dislocations fly by the particles.[22-23] The stolen hydrogen in turn can pressurize any voids associated with the particles and cause even greater increases in the interfacial stresses in Eq. (6) and, hence, a localized decrease in ductility.

Even though the above rationalizations are capable of explaining the phenomenological observations on plane-strain ductility, they do not provide us with any detailed understanding as to how the loss of plane-strain ductility due to addition of particles depends on such details as particle volume fraction; particle size, shape, and distribution; and particle coherency with the matrix. For example, although coherent particles may not cause a high uniaxial-ductility loss, they do cause a severe drop in plane-strain ductility.[21] Hence, apart from noting that the increase of particle-matrix interfacial strength can improve plane-strain ductility, we are not in a position to set any other alloy design guidelines for improving plane-strain ductility of particle strengthened alloys. Further study is therefore needed to clarify the role of precipitates and dispersoids on plane-strain ductility. Until such a clarification is available, the particle-strengthened systems should be used with great caution for applications under plane-strain conditions.

5 PARTICLES AND FRACTURE TOUGHNESS

An excellent review of the state of understanding of fracture toughness of alloys is included in the paper by Hahn in the present Colloquium.[24] Hence, we summarize only some of the major aspects of fracture toughness as they may apply to particle-strengthened systems and refer the interested reader to Hahn's paper for further details.

Fracture toughness can be defined as the ability of the material to resist the propagation of a crack under monotonic (noncyclic) loading and is represented by K_{Ic}, the critical-stress-intensity factor which the material can sustain under plane-strain conditions before catastrophic failure occurs. Most of the theoretical treatments aimed at relating fracture toughness to other material properties are based on continuum mechanics[25-27] and, hence, do not really distinguish between particle strengthening and any other modes of strengthening. Thus, one has to surmise the effect of second-phase particles on fracture toughness through experimental observations.

In a recent review of the experimental results obtained in aluminum-base alloys, Hahn and Rosenfield[21] came to the conclusion that increasing volume fraction of large (1 to 10 μ) incoherent inclusions reduces fracture toughness, as these inclusions crack easily on matrix deformation. They also observed that the intermediate-size (roughly 0.1 μ) incoherent inclusions are more resistant to cracking and, hence, do not affect toughness as much. However, the fine coherent precipitates in aluminum alloys which, unlike the inclusions, are cut by the dislocations tend to reduce toughness drastically, particularly in the stronger alloys. This is probably due to loss in work-hardening capacity as dislocations sweep through the coherent precipitates, resulting in such mechanical instabilities as large-shear-band formation. Although details of this coherent-particle behavior are apparently not clear, it is interesting that, as mentioned, coherent particles have very similar effects on plane-strain ductility of these alloys.[21] This certainly implies the not unexpected close relationship between plane-strain ductility and fracture toughness.

It should be noted here that in the evaluation of toughness of particle-strengthened alloys, one should also assess the effects of particles on the critical plastic zone size, r_{Ic}, at fracture, defined as $(K_{Ic}/\sigma y)^2$. McClintock has pointed out in this colloquium that the toughness one can expect from an alloy depends on relative values of r_{Ic} with respect to typical microstructural distances such as interparticle or inclusion spacing and grain size. This approach certainly deserves further study.

In summary, we conclude that, like plane-strain ductility, the understanding of the effects of particles on fracture toughness is at best qualitative and phenomenological, and even the latter in the limited sense. There appear to be sufficient data to show that the toughness factor of particles, be they coherent or incoherent, is negative, much like the plane-strain (and uniaxial) ductility factors.

6 PARTICLES AND FATIGUE-CRACK PROPAGATION

The current level of proficiency in either materials processing or non-destructive quality control is certainly not sufficient to preclude flaws or cracks in structural components. Fracture toughness is only one factor reflecting the resistance of the material to the growth of these flaws. Under cyclic loading, the type of resistance is measured through the magnitude of the per-cycle crack-propagation rate, da/dN, at known levels of the cyclic stress intensity factor ΔK. The phenomenological aspects of fatigue-crack propagation behavior have been the subject of many recent reviews; see refs. 28 to 30, for example.

With respect to structural-alloy design, we present for discussion Fig. 2, which is a graphical glimpse at fatigue-crack-propagation rates of various metals, steels, aluminum-base alloys, titanium-base alloys, and metallic glasses collected from recent publications.[31,32] It should be clear from Fig. 2 that fatigue-crack propagation rates are apparently insensitive to the many microstructural parameters manipulated during alloy design, including the presence of inclusions or particles. This could be a result of a true insensitivity of fatigue-crack-propagation rates to the microstructural details of material systems. More likely, we conclude that the microstructural parameters affect fatigue-crack propagation rates in such a manner that they more or less counterbalance each other, leading to apparently nonvarying fatigue-crack-propagation rates. The rationale behind this conclusion rests with a recently developed fatigue-crack-propagation theory.[33,34]

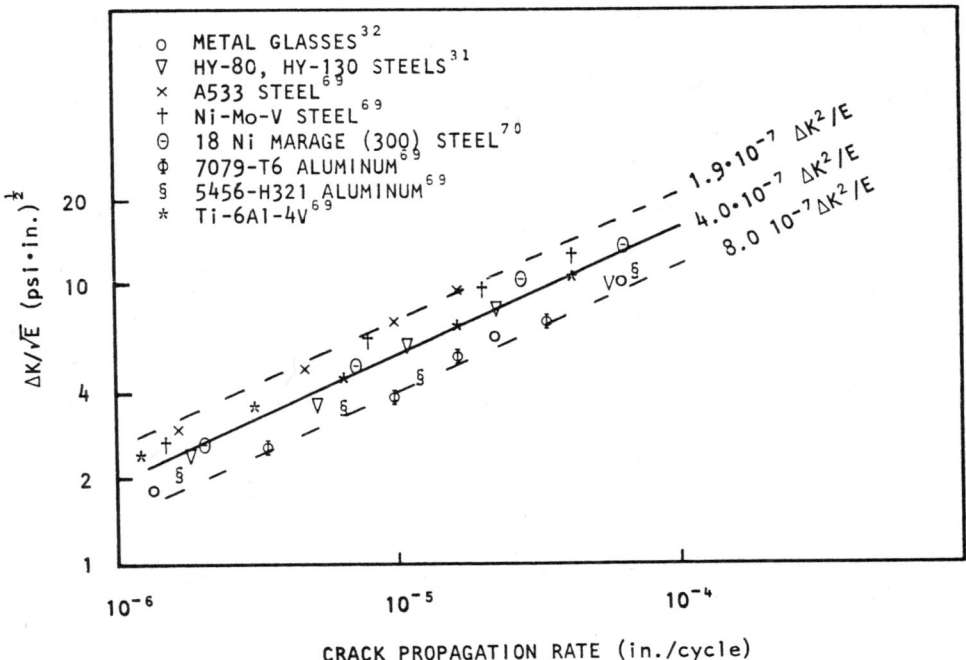

Fig. 2. Fatigue-crack propagation rates in various metals and alloys showing the apparent insensitivity of da/dN to microstructure.

This theory is based on the failure criterion that an existing crack will propagate into the element of material ahead of the crack when the strain level in this element exceeds that particular fracture strain of the material corresponding to the state of stress in the element. Accordingly, this fracture strain can be approximated by the plane-strain ductility in the case of a small crack propagating in a thick structural component, while it will be closer to the tensile ductility when the crack is large. The theory yields

$$da/dN = \frac{1}{\beta \pi \sigma_{y.c}^2 (n' + 1)} (\sigma_{y.c}/E\epsilon_F)^{n'+1} \Delta K^2 \quad , \tag{7}$$

where β is an almost constant factor that is slightly dependent on crack tip blunting and fatigue stress (R) ratio, $\sigma_{y.c}$ is the cyclic yield stress, n' is the cyclic work-hardening exponent, and ϵ_F is the afore-discussed fracture strain under the state of stress prevalent near the crack tip.

Experimental data on fatigue-crack propagation indicate that da/dN is related to ΔK by a power law with the exponent varying from 2 to 6. However, as reflected by the data in Fig. 2, most of the results show a ΔK dependence closer to the second, as predicted by Eq. (7). Therefore, it appears that the above model is in reasonably good functional agreement with experimental data. The interesting aspect of Eq. (7) rests, however, with the predicted inverse correlation between da/dN and the other more mundane and such already discussed properties as strength and ductility.

Accordingly, Eq. (7) allows us to estimate the effects of second-phase particles on fatigue-crack growth rates through the particle effects on these material properties. Here again we run into ignorance problems in that there is not enough information on some of these properties, including the cyclic material properties and plane-strain ductility, to generally evaluate the da/dN factor of particles. However, on the basis of available data, some quick calculations of the ratio of the da/dN in the particle-strengthened alloy to da/dN in the unstrengthened matrix, both at the same ΔK level, do provide some interesting conclusions. Table II indicates for aluminum-base alloys and steels the effect of precipitation strengthening on the fatigue-crack-propagation rates under plane-stress conditions. In the case of the aluminum alloys chosen, the overall effect of the decrease in ductility and work hardening with the increase in cyclic yield strength due to precipitation hardening is, unfortunately, a slight increase in the crack propagation rate in the precipitation-strengthened alloy as compared with that in the unstrengthened alloy, i.e., a negative da/dN resistance factor. In the case of the steels chosen, precipitation hardening results in a positive da/dN resistance factor, but only slightly. These examples and Eq. (7) in general support our original contention that fatigue-crack propagation rates are insensitive to microstructural details because of cancelling factors. In fact, it appears that

Table II. Calculated Fatigue-Crack-Propagation-Rate Ratios of
Strengthened Alloy to Unstrengthened Matrix[a]

Alloy	σ_y, MN/m²	$\sigma_{y.c}$, MN/m²	n	n'	ϵ_F	E, 10³ MN/m²	$\frac{(da/dN)_{Alloy}}{(da/dN)_{Matrix}}$
1100 Al	100	60	0.2	0.15	2.1	70	
2024 T351 Al	380	325	0.032	0.065	0.28	70	2.9
1005 steel	200	285	0.16	0.12	1.6	210	
18-Ni maraging steel	1900	1500	0.015	0.08	0.81	180	0.62

(a) All mechanical properties are quoted from ref. 68.

the perennial conflict between strength and ductility is loudly advertised when it comes to manipulating da/dN resistance through alloy or microstructural design.

For what it is worth at this time, the alloy-design guidelines for fatigue-crack growth resistance are

- The same particle considerations already discussed leading to high flow strength—assuming that these same considerations will affect cyclic flow strength in the same manner
- Whatever particle considerations will increase cyclic-work-hardening exponents
- The same particle considerations already discussed that will enhance ductility—plane stress or plane strain as the case may be.

Again, this last item is, as already discussed aplenty, microstructurally conflicting with the first two items.

7 PARTICLES AND CREEP AND STRESS RUPTURE

Many of the successful particle-strengthened systems are those which are currently used in elevated-temperature applications, say, at temperatures at or above 0.5 of the absolute solidus of the respective system. For example, nickel-base and cobalt-base superalloys are used in applications where the temperature ranges between 600 and 1000 C. The temperature limit for TD-nickel and thoria-dispersion-strengthened and corrosion-resistant nickel-chromium-aluminum, iron-chromium-aluminum, or cobalt-chromium-aluminum alloys is 1150 C. Similarly, the contemplated temperature limit of the yttria-dispersion-strengthened superalloys is also about 1150 C, or even slightly higher. Accordingly, the effects of particles, be they incoherent oxides or coherent precipitates, on creep and stress-rupture properties are important considerations in structural alloy design. As we will see, these high-temperature alloys are "super" mainly because creep and certain stress-rupture factors of particles are very positive.

As reflected by structural engineering design usage, the creep and stress-rupture properties of interest are the minimum or steady-state creep rate $\dot{\epsilon}_s$, the stress-rupture time t_r, and the final stress-rupture ductility ϵ_r. Often a quantity known as "creep strength" is referred to in structural design. However, this particular quantity is explicitly governed by $\dot{\epsilon}_s$ since it is defined as the stress level necessary to result in a given steady-state strain value (usually about 2 percent) after either 100 or 1000 hours of elapsed loading time.

The overall steady-state creep rate of an alloy can be viewed for alloy-design purposes as a superposition of the creep rates due to diffusional creep, grain-boundary-sliding creep and dislocation or intergranular creep processes. That is

$$\dot{\epsilon}_s = \dot{\epsilon}_{diff} + \dot{\epsilon}_{gbs} + \dot{\epsilon}_{disl} \quad . \tag{8}$$

Diffusional creep is a result of the biased diffusional flux of atoms in the direction of the tensile stress resulting in a time-dependent elongation of the specimen in this direction. The expressions for creep rate due to such a diffusional flow have been derived by Herring[35], Nabarro[36], and later by Coble[37] and these can be written in a general form as

$$\dot{\epsilon}_{diff} = CD_{B,G}b^3\sigma/kTd^n \quad , \tag{9}$$

where C is a geometric constant of the alloy, σ is the applied stress, d is the grain size of the alloy, and n is an exponent which is dependent on the diffusion path, being 2 for bulk diffusion and 3 for diffusion along grain boundary, $D_{B,G}$ is the bulk or grain-boundary self-diffusion coefficient, and kT has the usual meaning. This mode of creep is usually significant only at very high temperatures ($>0.8T_m$). However, as we will see in the next section, although this mode may not produce much strain at lower temperatures, it can cause grain-boundary chemistry changes leading to property degradation.

Grain-boundary-sliding creep on the other hand is a result of the accommodation of high shear-strain gradients across grain boundaries in polycrystalline alloys through sliding of adjacent grains along the grain boundary. The creep rate due to grain-boundary sliding has been derived to be[38,39]

$$\epsilon_{gbs} = \beta b^3 D_B \sigma^2/kTGd \quad , \tag{10}$$

where β is a geometrical constant.

It is clear from Eqs. (9) and (10) that creep rates due to diffusional and grain-boundary-sliding processes can be reduced, if not eliminated, by minimizing the grain-boundary area per unit volume, either by going to monocrystals or by designing very coarse-grained or aligned polycrystalline materials.

Dislocation or intragranular creep, as the name implies, is the result of the thermally activated motion of dislocations within the grains, and it cannot be eliminated. On the basis of current analyses of the phenomenological as well as the analytical aspects of the dislocation creep process[40-43], the steady-state creep rate can be written as

$$\dot{\varepsilon}_{disl} = A\left[(\sigma-\sigma_i)^m/E\right]\exp{(-Q/RT)}\quad, \tag{11}$$

where Q is the true activation energy for creep and m is the stress exponent of the steady-state dislocation creep process, and σ_i is the internal stress which must be overcome in the dislocation multiplication process and which can be small compared to σ the applied stress.

The steady-state creep rate in directionally recrystallized TD-nickel is seven or eight orders of magnitude smaller than the creep rates in nickel in spite of their nearly equal modulus-modified activation energies which is that for vacancy self-diffusion. This is mainly because the creep-stress exponent m in Eq. (11) is much larger in TD-nickel as compared to nickel, and $(\sigma-\sigma_i)/E$ is much less than unity under creep-loading conditions. Even though the creep-resistance factor for TD-nickel is thus very large and positive, the sharp stress dependence is not entirely desirable in structural design since the design safety factors have to be extremely high.

In the case of directionally solidified or monocrystalline nickel-base superalloys, the creep-resistance factor is also very positive and is of the order of 10^7 or 10^8. The reason for this low creep rate is the slightly large stress exponent in these alloys and an activation energy which is nearly three times higher than that for vacancy self-diffusion and which is believed to reflect the activation energy of interstitial self-diffusion. Further, this transfer from interstitial to vacancy diffusion control of dislocation motion in superalloys, and the attendant lowering of creep rate, is believed to be due to the development during creep of very fine-mesh accommodation dislocation networks that cage the fine γ' precipitates.[40-42]

A theory for stress-rupture life or ductility has yet to be developed. However, empirically there appears to be an inverse correlation between $\dot{\varepsilon}_s$ and t_r[44,45] where

$$(\dot{\varepsilon}_s)^\alpha t_r = \text{Constant} \tag{12}$$

The exponent α which is a function of the alloy system and microstructure is found to be close to 1 in many cases, indicating that Eq. (12) is a reflection of some strain-controlled criterion for creep failure, in which case the "constant" on the right-hand side is then at least a function of the stress-rupture ductility of the system. On this empirical basis, one can speculate that, since particles increase creep resistance, i.e., decrease $\dot{\epsilon}_s$, they will increase t_r, which happens to be the case; but, they also will decrease the stress-rupture ductility, which unfortunately also happens to be true. Typical experimental value of stress-rupture ductility for a Ni-20% Cr solid solution alloy is about 45 percent[46]; for the precipitation strengthened nickel-base superalloy Udimet-700, it is about 8 to 12 percent[47]; for the dispersion-strengthened alloy TD-nickel, it is 4 to 9 percent; and for the dispersion-strengthened superalloy MA-753, it is 3 to 20 percent[48]. And so, again we witness the trade-off between strength (in this case lowering of $\dot{\epsilon}_s$ and enhanced t_r) and ductility.

Accordingly, we conclude that creep rates can be greatly decreased and stress-rupture life increased if

- The system is monocrystalline, coarse grained, or aligned polycrystalline
- The volume fraction of particles is high and uniform
- The particles are fine and inert
- The particles are fine and coherent.

However, the alloy design guidelines for increased stress-rupture ductility, which is a property that is not at all understood are highly likely to be the inverse of the above.

8 PARTICLES AND THEIR THERMAL STABILITY

Thermally activated processes, i.e., diffusion, can help dislocations to move and grain boundaries to slide; they can also give rise to time-dependent morphological and chemical instabilities of the particles themselves. These instabilities include the familiar precipitate coarsening or Ostwald ripening; stress coarsening; grain-boundary chemistry changes leading to strengthening-precipitate dissolution; the nucleation and growth during service of brittle and acicular phases in multicomponent alloys; and the ravages of surface-strengthening as well as substrate-strengthening microstructure by oxidation and hot corrosion, which is discussed by Pettit in another paper in this Colloquium.[49] Unplanned for instabilities can be quite insidious, since they can result in significant changes in properties during service. They also severely restrict the application of such otherwise elegant phenomenological approaches to mechanical behavior characterization as the deformation maps of Weertman[50] and Ashby[51]. In what follows, we discuss morphological and chemical instabilities through selected examples, and, whenever possible, we attempt to postulate alloy-design steps that may minimize these instabilities.

In multicomponent precipitation alloys, stress-induced diffusion of various solute atoms from grain boundaries under compression to those under tension can cause preferential dissolution of the strengthening precipitates at grain boundaries transverse to the stress axis. This has been observed in many superalloys at temperatures about or above 900 C, e.g., see Fig. 3.[8] Such dissolution can and does enhance grain-boundary sliding, resulting in shorter stress-rupture lives. The kinetics of such preferential dissolution processes have been worked out and these can be summarized as[52]

$$d\xi/dt = - \frac{BD\Omega\sigma\delta}{kT\Delta c_e d^2} \quad , \tag{13}$$

where $d\xi/dt$ is the dissolution rate, D is the appropriate diffusion coefficient depending on the diffusion path, δ is a parameter which equals the grain size for bulk-diffusion-controlled dissolution and equals the grain-boundary thickness when grain-boundary diffusion is controlling the dissolution, and Δc_e is the difference in the equilibrium solute content of the precipitate

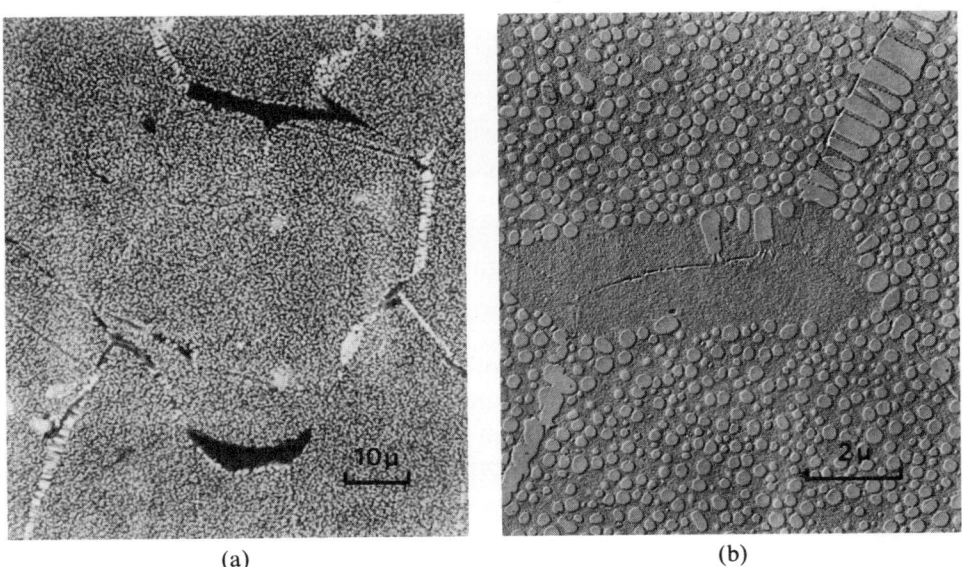

(a)   (b)

Fig. 3. Denudation and enrichment of γ' precipitates in a nickel-base superalloy at grain boundaries transverse and parallel to the applied tensile-stress direction, respectively, after 100 hours of creep at 1000 C and 50 MN/m². The tensile stress axis is vertical in both the lower magnification (a) and higher magnification (b) views. During compressive creep, denudation occurs along grain boundaries parallel to the compressive-stress axis and enrichment occurs along boundaries transverse to the compressive-stress axis.[8]

phase and the matrix phase. From this equation it is clear that coarse grain size, small diffusion coefficients through alloying, large Δc_e values, and the use of directionally solidified materials can result in the minimization of the stress-induced dissolution of the grain-boundary precipitates.

Another example of particle instability at elevated temperatures is stress coarsening. For example, Fig. (4) shows the effects of applied-stress sense on the morphological changes of the γ' precipitates in a nickel-base superalloy monocrystal.[7] Stress orientation can also affect the final morphology of the precipitates.[53] The driving force for such stress coarsening of coherent precipitates has been identified to be proportional to the product of the lattice mismatch and the difference in elastic moduli between the precipitate and the matrix phases.[53] Hence, stress coarsening could be minimized by the use of strengthening precipitates having matching lattice parameters with the matrix, which may be experimentally feasible, or precipitates having matching elastic modulus with the matrix, which may not be practical.

Conventional particle coarsening can be minimized, of course, by minimizing solubility, diffusion coefficient of solutes, and particle surface energy.[54,55] Indeed, the very coherent γ' precipitates already possess low surface energy, and thoria or any other oxide dispersoids are insoluble in most metals (except at very high temperatures).

Long-time loading at high temperatures can also result in the nucleation and growth of undesirable phases with long incubation times. For example, stress aging can cause the formation of acicular carbides at transverse grain boundaries which act as stress concentrators and cause sudden intergranular failure and curtailed stress-rupture ductility in the alloy.[56] Such topologically close-packed phases as sigma, mu, etc., have also plagued the superalloys.[57]

The lack of long-time stability of multicomponent alloys is still a major problem for the structural-alloy designer. One obvious solution is to make the alloy with fewer components and depend on inert particles for strengthening. This indeed is the attractiveness of the dispersion-strengthened metals and binaries. However, the major thrust in designing stable alloys should be in the direction of theoretical studies to predict the stability of phases in multicomponent alloys, such as the PHACOMP studies[58-59] and the more recent pair-potential model studies[60-63].

9 PARTICLES AND CREEP CRACK PROPAGATION

Preexisting flaws can also propagate under creep-loading conditions. The importance of creep-crack propagation was not realized until a few years ago, and we are not aware of any theoretical model for this mode of crack growth. In what follows, we will present an approximate and as yet

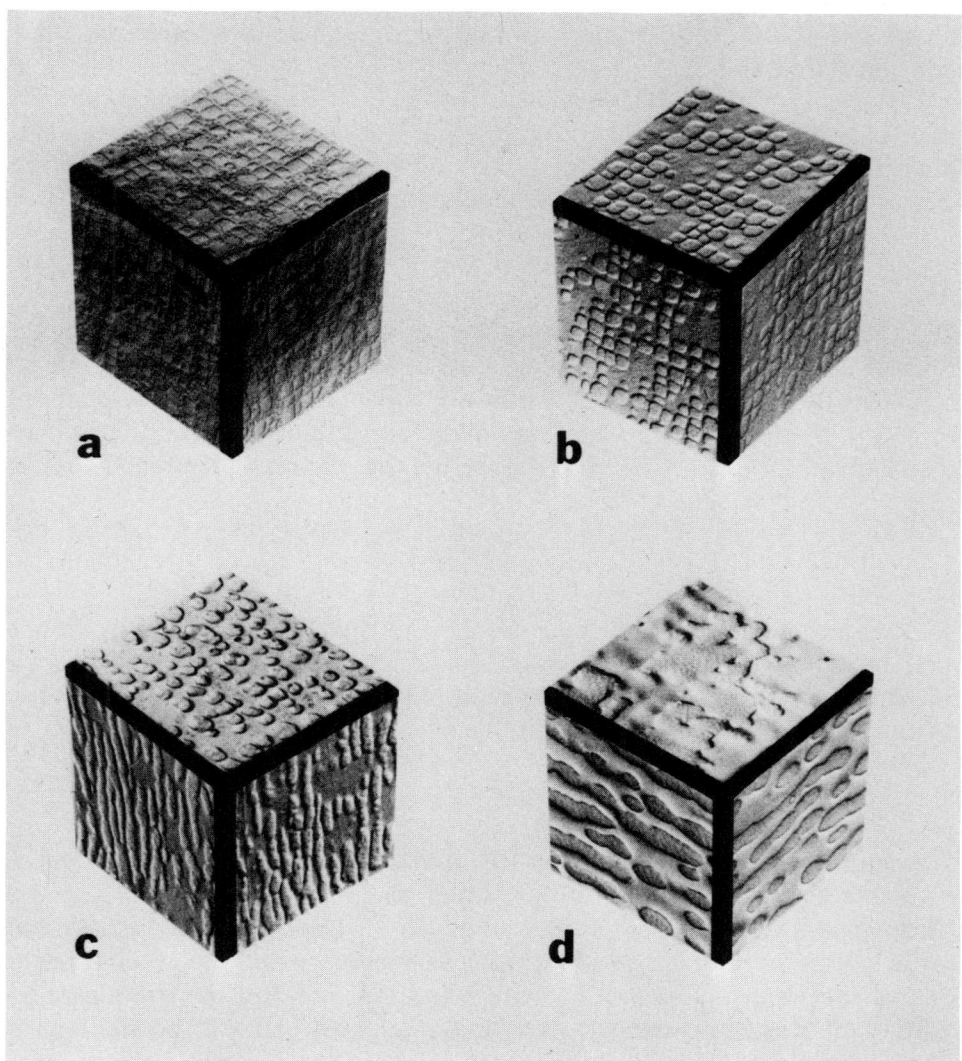

Fig. 4. <100> oriented cube representation of coherent γ' precipitate morphology in nickel-base-superalloy single crystals (a) before stress coarsening, and after 100 hours annealing at 955 C under (b) zero applied stress, (c) 155 MN/m^2 compressive stress, and (d) 155 MN/m^2 tensile stress. Stress axes were normal to the top of (c) and (d) cubes.

unpublished model[64] which is fairly consistent with the scant experimental results and which can provide some alloy-design guidelines for creep-crack-propagation resistance.

Physically, this model assumes that the high local stresses ahead of a crack in a component under creep loading lead to an enhanced creep rate which results in stress rupture and crack advance. We estimate the crack tip stress σ using the Orowan-Irwin elastic approach as

$$\sigma = \sigma_a \sqrt{a/\rho} \; , \tag{14}$$

where σ_a is the applied or nominal stress, a is the crack length, and ρ is the effective crack tip radius under creep loading conditions which depends on the extent of crack tip blunting. ρ can also be looked upon as the length of the region ahead of the crack tip where these high local stresses prevail. The creep rate at this local stress level can be obtained using a phenomenological creep equation, say Eq. (11), i.e.,

$$\dot{\epsilon} = A[(\sigma_a\sqrt{a/\rho} - \sigma_i)/E]^m \; e^{-Q/RT} \tag{15}$$

The time it takes for this crack tip element to undergo stress rupture can be obtained from the pseudoempirical equation relating minimum creep rate and stress rupture life at a particular applied stress level and temperature, for example, Eq. (12), i.e.,

$$t_r = C/\dot{\epsilon}^\alpha \tag{16}$$

This equation is pseudoempirical in the sense that when α is unity it represents a strain-controlled failure criterion for stress rupture, where the critical strain needed for rupture is equal to C. Experimentally, it is found that α is usually slightly less than unity. The significance of C when α is not equal to one is not clear, but it could be an indication of the stress-rupture ductility of the alloy. Be this as it may, Eq. (16) can still be used on a phenomenological basis to predict stress-rupture life for a given steady-state creep rate. Accordingly, the stress-rupture time for the crack tip element of effective size ρ can be written as

$$t_r = CA^{-\alpha} \; e^{\alpha Q/RT}[(\sigma_a\sqrt{a/\rho} - \sigma_i)/E]^{-m\alpha} \tag{17}$$

The crack-propagation rate $\dot{a}$ can then be obtained as

$$\dot{a} = \rho/t_r = [(A^\alpha e^{-\alpha Q/RT})\rho/C][(\sigma\sqrt{a/\rho} - \sigma_i)/E]^{m\alpha} \tag{18}$$

Noting further that σ_i, the internal stress, is much less than the local stress near the crack tip and the stress intensity factor K is equal to $\sigma\sqrt{a}$, we can rewrite Eq. (18) as

$$a = [A^\alpha e^{-\alpha Q/RT}/C\rho^{(1/2m\alpha-1)}](K/E)^{m\alpha} \qquad (19)$$

This equation predicts a power-law dependence of creep crack-growth rate on the stress-intensity factor, with the exponent being a product of the stress exponent of steady-state creep m and stress-rupture life exponent α. Similar functional behavior is indeed observed experimentally in the few studies that have so far been reported.[65-67]

On the basis of the above equation for creep crack-growth rates, one can set forth some guidelines for alloy design. It appears that all the particle considerations that will reduce steady-state creep rate and increase stress-rupture ductility (higher C) would reduce creep crack-growth rates. Here again, we are confronted with the perennial conflict between "creep strength" and stress-rupture ductility. However, on the basis of the observation that the creep-resistance factor of particles is very large and positive, we can expect this to offset the negative stress-rupture-ductility factor of particles, leading to a positive creep-crack-propagation resistance factor of particles.

It is difficult to identify the material properties or microstructural characteristics that go into the parameter ρ in Eq. (19). It is possible that ρ, the characteristic distance over which the material undergoes very high creep rates, can be the spacing between voids or triple point cracks on a grain boundary, etc. In most cases, ρ will then be proportional to grain size. On the basis of this hypothesis, one would then expect the creep-crack-growth rate to decrease with increasing grain size, which appears consistent with recent experimental evidence.[67]

10 CONCLUSION

The foregoing appraisal of the effect of particles on the many important material properties relevant to alloy design for structural applications is summarized in Tables III and IV.

The conflicting alloy-design requirements on the various particle parameters are readily apparent. It is clear that we metallurgists must study not only such positive attributes of particles as strength and creep resistance, but also such potentially negative attributes of particles in alloys as ductility, toughness, and flaw-growth resistance under creep and fatigue loading conditions. Finally, it is also clear that we no longer have the luxury of confining our interests to uniaxial mechanical properties, but must begin to understand such properties as plane-strain ductility and toughness.

Table III. Summary of the Positive and Negative Attributes of Particles

Property	Inert Dispersoids	Coherent Precipitates
Yield Strength	Very positive	Very positive
Uniaxial Ductility	Negative	Negative
Plane-Strain Ductility	Negative	Negative
Fracture Toughness	Negative	Negative
Fatigue-Crack-Propagation Resistance	Not known	Negative— Marginally positive
Creep Strength or Creep Resistance	Very positive	Very positive
Stress-Rupture Resistance (Rupture Life)	Very positive	Very positive
Stress-Rupture Ductility	Negative	Negative
Resistance to Thermal Instability	Positive	Negative
Resistance to Creep Crack Growth	Positive	Positive

Table IV. Summary of Guidelines for Alloy Microstructural Design with Particles

Property to be Maximized	Particle or Microstructural Parameter:	Coherent Precipitates or Inert Dispersoids?	Particle Volume Fraction	Particle Size	Grain Size
Yield Strength		Either	High	Fine	Fine
Uniaxial Ductility		?	Low	Fine	?
Plane-Strain Ductility		?	Low	Fine	?
Fracture Toughness		?	Low	Fine	Fine
Fatigue-Crack-Propagation Resistance		?	?	?	Fine
Creep and Stress-Rupture Resistance		Either	High	Fine	Coarse
Stress-Rupture Ductility		?	?	?	?
Resistance to Thermal Instabilities		Inert dispersoids	—	—	Coarse
Resistance to Creep Crack Propagation		Either	High	Fine	Coarse

ACKNOWLEDGMENTS

I wish to thank Sampath Purushothaman for his help in gathering information used in this manuscript. I am also grateful to the National Aeronautics and Space Administration and to the National Science Foundation for partially supporting some of the experimental and analytical work included in this paper through grants NASA-NSG-3050 and NSF-DMR-75-09878.

REFERENCES

1. Brown, L. M., and Ham, R. K., *Strengthening Methods in Crystals*, A. Kelly and R. B. Nicholson (Eds.), John Wiley and Sons, New York (1971), p. 9.
2. Stoloff, N. S., *The Superalloys*, C. T. Sims and W. C. Hagel (Eds.), John Wiley and Sons, New York (1972), p. 79.
3. Wilcaux, B. A., and Clauer, A. H., *ibid.*, p. 197.
4. Copley, S. M., and Kear, B. H., *Trans. TMS-AIME*, **16**, 675 (1966).
5. Fleisher, R. L., *The Strengthening of Metals*, Reinhold, New York (1964), p. 93.
6. *Properties of Metals and Alloys*, International Nickel Company, New York, 1972.
7. Tien, J. K., and Copley, S. M., *Met. Trans.*, **2**, 215 (1971).
8. Tien, J. K., and Gamble, R. P., *Met. Trans.*, **2**, 1663 (1971).
9. Tien, J. K., *Proceedings of the Second International Conference on Superalloys—Processing*, P. W-1, MCIS, Battelle, Columbus Laboratories, Columbus, Ohio (1972).
10. Benjamin, J.S., *Met. Trans.*, **1**, 2943 (1970).
11. *Alloy Data Book*, Special Metals Corporation, New Hartford, New York (1970).
12. Thompson, A. W., *Met. Trans.*, **5**, 1855 (1974).
13. *Metals Handbook*, Vol. 8, American Society for Metals, Metals Park, Ohio (1974).
14. Ashby, M. F., *Strengthening Methods in Crystals*, A. Kelly and R. B. Nicholson (Eds.), John Wiley and Sons, New York (1971), p. 137.
15. Gurland, J., and Plateau, J., *Trans. ASM*, **56**, 442 (1963).
16. Argon, A. S., Im, J., and Needleman, A., *Met. Trans.*, **6A**, 815 (1975).
17. Argon, A. S., Im, J., and Safoghe, R., *Met. Trans.*, **6A**, 825 (1975).
18. Argon, A. S., and Im, J., *Met. Trans.*, **6A**, 839 (1975).
19. Clausing, D. P., *Int. J. Frac.*, **6**, 71 (1970).
20. Barsom, J. M., and Pellegrino, J. V., *Eng. Frac. Mech.*, **5**, 209 (1973).
21. Hahn, G. T., and Rosenfield, A. R., *Met. Trans.*, **6A**, 653 (1975).
22. Tien, J. K., *Proceedings of the Conference on the Effect of Hydrogen on Behavior of Metals*, Jackson Hole, Wyoming, September, 1975, to be published by TMS-AIME.
23. Tien, J. K., Thompson, A. W., Bernstein, I. M., and Richards, R. J., unpublished research, submitted to *Met. Trans.*
24. Hahn, G. T., Paper IV-2, this Colloquium Proceedings.
25. Hahn, G. T., and Rosenfield, A. R., *ASTM-STP, No. 432*, 5 (1968).
26. Krafft, J. M., *Appl. Mat. Res.*, **3**, 88 (1964).
27. Rice, J. R., *Proceedings of Third International Conference on Fracture*, Munich (1973), Vol. 2, p. I-441.
28. Hahn, G. T., and Simon, R., *Eng. Frac. Mech.*, **5**, 523 (1973).
29. Schwalbe, K., *Eng. Frac. Mech.*, **6**, 325 (1974).
30. Walton, D., and Ellison, E. G., *Int. Met. Rev.*, **17**, 100 (1972).

31. Barsom, J. M., *ASTM-STP*, No. 486, 1 (1971).
32. Davis, L. A., Paper III-5, this Colloquium Proceedings.
33. Purushothaman, S., and Tien, J. K., *Scripta Met.*, **9**, 923 (1975).
34. Purushothaman, S., and Tien, J. K., unpublished research, Columbia University, New York (1975), submitted to *Met. Trans.*
35. Herring, C., *J. Appl. Phys.*, **21**, 437 (1950).
36. Nabarro, F.R.M., *Report of a Conference on the Strength of Solids*, London (1948), p. 75.
37. Coble, R. L., *J. Appl. Phys.*, **34**, 1679 (1963).
38. Raj, R., and Ashby, M. F., *Met. Trans.*, **2**, 1113 (1971), *ibid.*, **3**, 1937 (1972).
39. Langdon, T. G., *Phil. Mag.*, **22**, 689 (1970).
40. Malu, M., and Tien, J. K., *Acta. Met.*, **22**, 145 (1974).
41. Malu, M., and Tien, J. K., *Scripta. Met.*, **9** (October 1975), in press.
42. Malu, M., and Tien, J. K., unpublished research, Columbia University, New York (1975).
43. Nix, W. D., Paper presented at the Seventh Annual TMS-AIME Spring Meeting at University of Toronto (May, 1975).
44. Larson, F. R., and Miller, J., *Report NACA-TN-2890* (1953).
45. Monksman, F. C., and Grant, N. J., *Proc. ASTM*, **56**, 834 (1956).
46. Shahinian, P., and Achter, M. R., *Trans. ASM*, **51**, 244 (1959).
47. Aning, M., and Tien, J. K., unpublished research, Columbia University, New York (1975).
48. Benjamin, J. S., and Cairns, R. L., *Modern Developments in Powder Metallurgy*, **5**, 47 (1970).
49. Pettit, F. S., Paper V-1, this Colloquium Proceedings.
50. Weertman, J., *Trans ASM*, **61**, 681 (1968).
51. Frost, H. J., and Ashby, M. F., *Rate Processes in Plastic Deformation*, American Society for Metals, Metals Park, Ohio (1973), p. 70.
52. Aning, K., and Tien, J. K., unpublished research, Columbia University, New York (1975).
53. Tien, J. K., and Copley, S. M., *Met. Trans.*, **2**, 543 (1971).
54. Lifshitz, I. M., and Slyozov, V. V., *J. Phys. Chem. Sol.*, **19**, 35 (1961).
55. Wagner, C., *Z. Elektrochim.*, **65**, 243 (1961).
56. Decker, R. F., *Proceedings of the Symposium on Steel Strengthening Mechanisms*, Climax Molybdenum Company, Greenwich, Connecticut (1969), p. 147.
57. Sims, C. T., *The Superalloys*, C. T. Sims and W. C. Hagel (Eds.), John Wiley and Sons, New York (1972), p. 259.
58. Woodyatt, L. R., Sims, C. T., and Beattie, H. J., *Trans. AIME*, **236**, 519 (1966).
59. Boesch, W. J., and Slaney, J. J., *Met. Prog.*, **86**, 109 (1964).
60. Machlin, E. S., *Acta Met.*, **22**, 95 (1974).
61. Machlin, E. S., *op. cit.*, p. 109.
62. Machlin, E. S., *op. cit.*, p. 367.
63. Machlin, E. S., *op. cit.*, p. 1433.
64. Purushothaman, S., and Tien, J. K., unpublished research, Columbia University, New York (1974).
65. Siverns, M. J., and Price, A. T., *Int. J. Frac.*, **9**, 199 (1973).
66. Neate, G. J., and Siverns, M. J., *Proceedings of International Conference on Creep and Fatigue in Elevated Temperature Applications*, Publication No. 13, Institution of Mechanical Engineers, Philadelphia (1973).
67. Floreen, S., *Met. Trans.*, **6A**, 1741 (1975).

68. *Cyclic and Monotonic Properties of Materials*, Scientific Research Staff, Ford Motor Company, Dearborn, Michigan (1973).
69. Bates, R. C., and Clark, W. G., *Trans. ASM*, **62**, 380 (1969).
70. Van Swam, L. F., Pelloux, R. M., and Grant, N. J., *Met. Trans.*, **6A**, 45 (1975).

DISCUSSION on Paper by J. K. Tien

ANSELL: I don't understand the basis for your conclusion that in all two-phase alloys, increasing the dislocation substructural density improves the creep properties. For oxide-dispersion-strengthened alloys, just the opposite effect has been observed in both TD-Nickel and Al-Al$_2$O$_3$ SAP-Type alloys. Indeed, for the latter alloys, the steady-state creep rate is approximately 6 orders of magnitude lower for the alloys after full recrystallization as compared to that observed for the as-extended, heavily dislocated alloys.

WILCOX: I would like to raise three points regarding alloy design with inert oxide dispersions: (a) for ambient-temperature strengthening, (b) for high-temperature strengthening, and (c) the role of substructure in high-temperature strengthening of a dispersion-containing alloy.

(a) Several years ago Clauer and I performed some experiments on TD-Nickel (Ni-2 vol % ThO$_2$) aimed at comparing the relative strengthening behavior of dispersoids and a fine dislocation subgrain structure, as measured by room-temperature yield strength. We found that for a given particle or subgrain spacing, the subgrain strengthening was 6 or 7 times more potent. This suggests that for ambient temperature applications, dispersion hardening may not be a terribly exciting proposition, particularly in view of the relatively high cost of producing a uniform oxide dispersion versus the relatively lower cost of mechanical working to produce a fine dislocation substructure.

(b) The second point deals with the effect of dispersions on helping to develop highly elongated grains during thermomechanical processing— which in turn improves high-temperature strength. Your point about minimizing transverse boundaries by having highly elongated grains (the grain aspect ratio, or GAR effect) is absolutely true. The GAR effect has been shown by us and researchers at International Nickel to be a very important high-temperature-strengthening consideration.

(c) The third point relates to the influence of a dislocation substructure on high-temperature creep strength of a dispersion-containing alloy. This situation is not completely clear. Previous workers (Webster, and more recently Grant) claim that the dispersion stabilizes a dislocation substructure produced by TMP, and this results in high-temperature strengthening. However, during their TMP, other microstructural features, such as the GAR, change, so that their comparisons may not be valid. Clauer and I recently did some experiments on TDNiCr, which is free of a dislocation substructure, as received from the manufacturer. We prestrained the alloy 2 to 4 percent at room temperature and annealed specimens at 1000 C. This resulted in a random dislocation array, pinned on the ThO_2 particles with no change in GAR. Subsequent creep tests at 600 to 800 C revealed no significant creep strengthening compared with the substructure-free starting material. This observation tends to refute the arguments of Webster and Grant.

HAASEN: I don't understand the last term in your equation for the flow stress of an alloy with coherent precipitates ("intermediate case"). How could a proportionality-to-particle radius to the 5/2 power arise?

The high activation energy in creep of directionally recrystallized TD-nickel appears to be more likely caused by dislocation-solute dragging effects than by self-interstitial diffusion. (See the related phenomenon in grain boundary movement.)

TIEN: There is little doubt that dislocation substructures contribute significantly to the ambient temperature strength of the TD-Nickel type of systems. This is demonstrated by typical numbers that I have quoted in the paper. I do agree with Prof. Ansell and Dr. Wilcox on the point that dislocation substructure effects on creep rates in dispersion-strengthened systems are not as well understood as they are in coherent-precipitation-strengthened systems. However, it is my feeling that there is a substructure effect—not so much the substructures produced by prior mechanical work but those produced during the creep process itself.

Finally, I thank Prof. Haasen for pointing out a typographical error in the flow-stress equation of coherent precipitates, which has been corrected in the paper. His comment regarding the possibility of solute drag of moving dislocations being the cause of the very high activation energies observed during creep of monocrystals of γ'-strengthened superalloys is interesting. However, I don't believe that it applies, since solid solutions of nickel do not possess high creep activation energies. These high energies are observed when both precipitates and fine-mesh dislocation substructures are present. We believe that this results in a very high density of interstitial-type jogs on moving screw dislocations—so high a density that the nonconservative motion of the jogs is rate controlled by the emission of interstitials.

DESIGN OF HETEROGENEOUS MICROSTRUCTURES BY RECRYSTALLIZATION

*E. Hornbogen**

Ruhr-Universität Bochum
Institut für Werkstoffe
Bochum, West Germany

ABSTRACT

The theory for the control of grain size and shape of heterogeneous alloys by thermal and mechanical treatment is described. The amount, size, and distribution of particles, coherency with the matrix phase, and interaction with dislocations determine the defect structure of the alloy. Recrystallization of supersaturated solid solutions will produce a variety of microstructures depending on the order in which recrystallization or precipitation occur. Thermomechanical treatments are described for producing the microduplex structure in heterogeneous alloys.

*Prepared during a sabbatical leave at Battelle's Columbus Laboratories 1974/75.

1 MICROSTRUCTURAL ELEMENTS

The structure of metals can contain six microstructural elements: the zero- to three-dimensional defects and two types of anisotropy (Table I).[1,2] Useful metallic materials usually require several of these elements. The 21 combinations of two are shown in Table II. An important aspect of alloy design is the optimization of these parameters, so that the material shows the desired macroscopic properties.

This paper is mainly concerned with the production of microstructures that contain the element 3 (grain and phase boundaries). However, both types of anisotropy (elements 5 and 6) are related to our theme. Particles (element 4) are sometimes needed as tools to control the grain structures. Dislocations (element 2) have to be considered as one source of the driving force for recrystallization and in connection with the formation of sub-boundaries and new recrystallized grains.

Table I. Six Elementary Microstructural Features of Metallic Materials

1	S	0-dimensional	Vacancies, solute atoms
2	D	1-dimensional	Dislocations
3	B	2-dimensional	Grain and phase boundaries
4	P	3-dimensional	Particles
5	X	Crystal anisotropy	Crystal texture, single crystal
6	M	Microstructural anisotropy	Aligned grains or particles

Table II. Combinations of Two Microstructural Elements

	S	D	B	P	X	M	Examples	
M	MS	MB	MB	MP	MX	MM	MB	aligned grain structure
X	XS	XD	XB	XP	XX		XB	polycrystal with texture
P	PS	PD	PB	PP			PB	polycrystal with particles
B	BS	BD	BB				BB	grain + phase boundaries, for example duplex structure
D	DS	DD					DB	cold-worked polycrystal
S	SS						SB	polycrystalline solid solution

2 DEFORMED HETEROGENEOUS METALS

The occurrence of recrystallization requires that energy be stored in the material in the form of defects.[3] These can be obtained by thermal or radiation treatments. The most common way, however, is the introduction of defects by plastic deformation. If they provide the driving force for recrystallization, the average energy density (i.e., dislocation density) is of primary importance.[4,5] For the formation of new crystallites, the local defect structure becomes very significant. The following sites are listed in the sequence of their effectiveness in aiding the formation of recrystallized grains: (1) original grain boundaries, (2) transformation bands, (3) intersections of twins and bands of concentrated slip, (4) coalescing subgrains, and (5) continuously growing subgrains.

If a second phase is dispersed in the alloy, the defect structure is modified, depending on whether: (a) coherent particles (or holes) are deforming with the matrix, (b) small particles are by-passed by dislocations, and (c) large particles act as dislocation sources.

Characteristic for case (a) is that the plastic strain becomes confined to small planar portions of the volume, because of a local decrease in critical shear stress due to cutting of particles.[6] Work hardening of particular slip planes in case (b) favors an even distribution of dislocations across the volume.[7] If the particles of the second phase become large and widely spaced, the dislocation distribution again becomes inhomogeneous, but differently from case (a). The interfaces of the particles act as dislocation sources[8], and flow around the particles leads to a high amount of lattice curvature in their close environment. This leads to the establishment of an additional nucleation site for recrystallized grains. The incoherent interface of large particles and a high dislocation density and lattice curvature favor the formation of recrystallization fronts. An additional requirement for this to occur is a particle spacing above a critical value—as a rule, 2 to 3 times the dislocation cell size—so that a sufficient number of dislocations can rearrange.[3,9] An equivalent effect on defect distribution is found at the α-β interfaces of duplex structures.

3 RECRYSTALLIZATION OF SUPERSATURATED SOLID SOLUTIONS

This problem has attracted much recent attention and was included in several reviews.[9-12] The situation is characterized by the fact that in addition to the defects, the alloy contains atoms that have a tendency to precipitate. Therefore, sequential and simultaneous precipitation and recrystallization reactions occur, by which a wide variety of different microstructures can be obtained.

From the equations for the start of recrystallization and precipitation, a survey can be obtained for the temperature dependence of the different reactions (Fig. 1):

$$t_R = t_{RO} \exp \frac{Q_R(\rho)}{RT} \tag{1}$$

$$t_A = t_{AO} \exp \frac{[\Delta G_N(T,\rho) + Q_D]}{RT} \tag{2}$$

$$t_A > t_R \quad ,$$

where t_R and t_A are the times at which recrystallization or precipitation start at a temperature T. The activation energy Q_R is a function of the dislocation density ρ, ΔG_N mainly of the undercooling for the precipitation process. The activation energy for diffusion Q_D is regarded as temperature indepen-

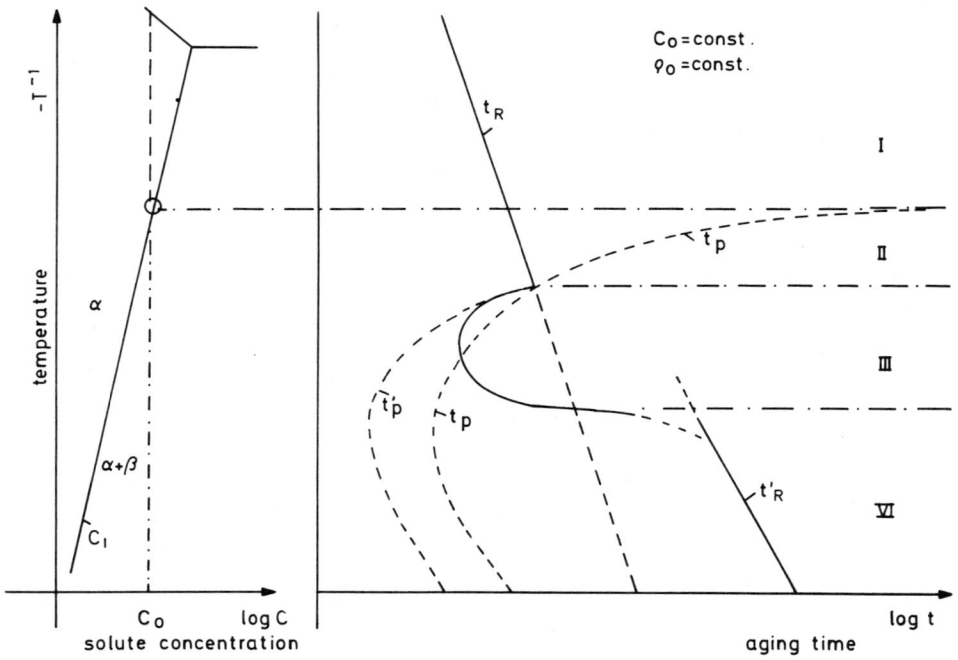

Fig. 1. Diagram of temperature-time-start of reaction for an alloy of the composition C_o: t_R—start of recrystallization (particle free), t_P—start of precipitation (dislocation free), t_R'—start of recrystallization after preceding precipitation, t_P'—start of precipitation at a dislocation density ρ_o.

dent. This leads to the different shape of the two functions $t_R = f(T)$ and $t_A = f(T)$ from which four temperature ranges can be defined (Fig. 1):

$T > T_E = T_I$ recrystallization in the solid solution

$T_I > T > T_{II}$ recrystallization precedes precipitation

$T_{II} > T > T_{III}$ recrystallization and precipitation occur simultaneously

$T_{III} > T$ precipitation retards recrystallization.

$T_I = T_E$ is the equilibrium temperature of the particular alloy composition. The temperatures T_{II} and T_{III} depend on defect density, on composition, and especially on the nucleation behavior of the defect alloy.[13] The significance of temperature T_{II} is demonstrated by measurements of reaction rates of a deformed Al-Fe alloy (Fig. 2). At the higher temperature ($T_I > T > T_{II}$), there are two maxima indicating the sequence: recrystallization → precipitation, while at a lower temperature ($T_{II} > T > T_{III}$) only one reaction can be recognized.

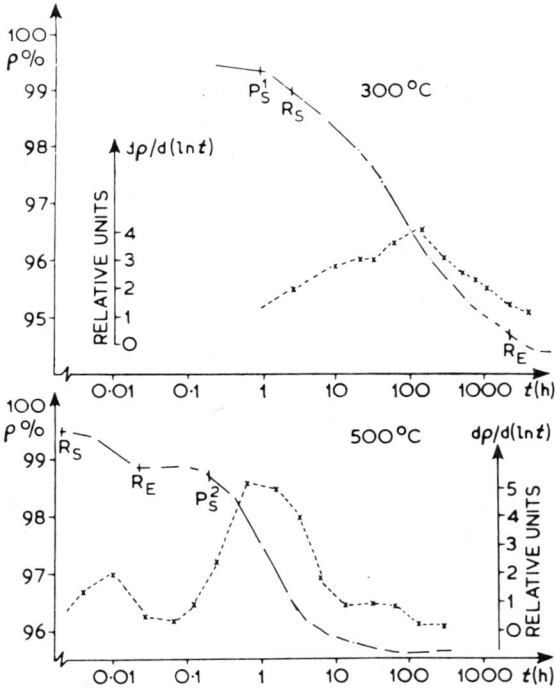

Fig. 2. Sequence of recrystallization and precipitation in an Fe-0.042 wt % Al alloy after 50 percent cold work. Aging temperature 500 C: stage II, recrystallization occurs before precipitation. Aging temperature 300 C: stage III or IV, recrystallization and precipitation occur simultaneously (measurements of electrical resistivity ρ).

Increasing the defect density or decreasing the solute concentration leads to a decrease in T_{II}, so that for such conditions, recrystallization may always precede precipitation (range II).

Range III is characterized by the fact that with increased supersaturation, precipitation and recrystallization occur simultaneously by a new reaction which is designated as "combined discontinuous reaction" (Fig. 3a). In range IV, precipitation occurs by individual heterogeneous nucleation and so rapidly that the pinning force of the particles effectively inhibits the motion of any recrystallization front. The only reaction that can occur is particle controlled subgrain growth, which is termed "combined continuous reaction" (Fig. 3b). The rate of the discontinuous combined reaction and the transition to the continuous mechanism can be derived by quantitatively evaluating and summing up the forces p_i, that act at the reactive front:

$$\text{range III} \quad \sum_{i}^{n=3} p_i = p_R + p_C - p_P > 0 \tag{3}$$

$$\text{range IV} \quad \sum_{i}^{n=3} p_i = p_R + p_C - p_P < 0 \tag{4}$$

The forces are defined as the change in free energy per unit area A across the reaction front that moves in x-direction: $p = (1/A)(dG/dx)$. The driving forces are caused by the decrease in defect density $\rho_0 - \rho_1$, $p_R \approx \mu b^2 (\rho_0 - \rho_1)$ or the change in concentration $c_0 - c_1$, $p_C \approx (RT/V_m) c_0 \ln (c_0/c_1)$. μ = shear modulus, b = atomic spacing, R = gas constant, V_m = molar volume. Opposed to these driving forces is the pinning force, which exists if particles have formed ahead of the reaction front: $-p_P = (3f\gamma_{\alpha\alpha}/2r_P)$, where f is the volume portion, r_P is the radius, and $\gamma_{\alpha\alpha}$ is the energy of the grain boundary. This force has been calculated for evenly distributed particles that are blanking off discs with an area of $r_P^2\pi$ from the grain boundary. The energy $r^2\pi\gamma_{\alpha\alpha}$ has to be provided for each particle to move the reaction front.[17] There are several other possibilities for grain boundary—particle interactions, including dissolution and reprecipitation, transformation metastable → stable, and dragging.[18,19]

The discontinuous combined reaction can lead to accelerated recrystallization and to the highest possible rate of approach to equilibrium at a given temperature. It usually produces a relatively coarse lamellar aggregate of phases (Fig. 3a).

A mixed microstructure of dislocations arranged to subboundaries and a fine dispersion of particles is formed by the continuous reaction. It approaches equilibrium at the slowest possible rate. The properties that follow from this microstructure can be characterized as "hard" either mechanically or in respect to hysteresis in ferromagnets or superconductors.

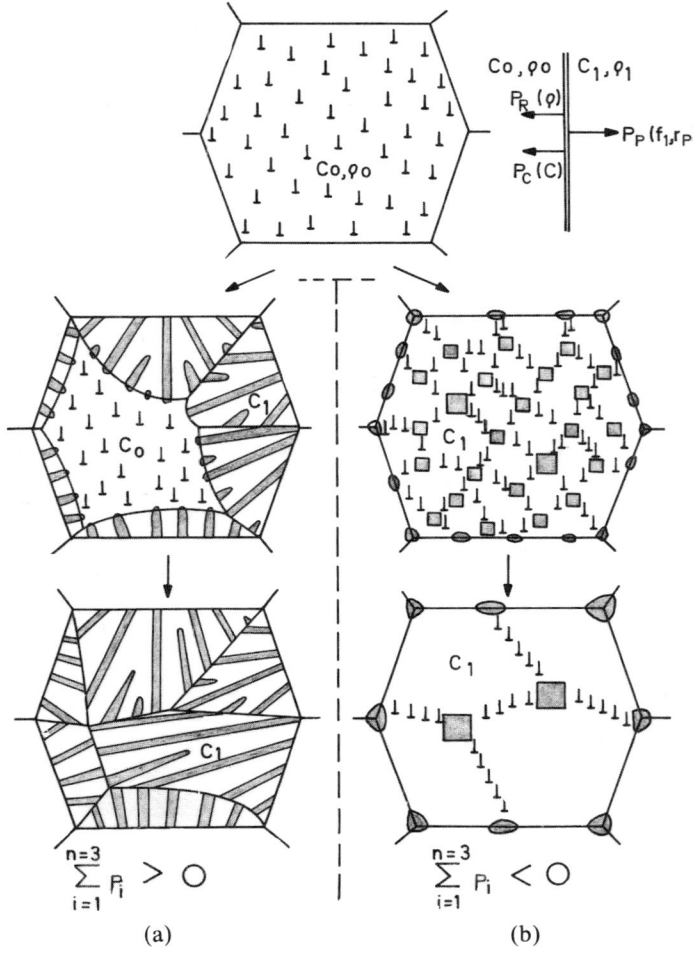

Fig. 3. Schematic representation of combined recrystallization and precipitation reactions: (a) discontinuous and (b) continuous.

It follows from Eqs. (3) and (4) that the occurrence of the continuous or the discontinuous reaction is a function not only of the temperature but also of the defect density. If, for example, the dislocation density is increased by increasing amounts of cold work ϵ, not only the driving force p_R is increased but also the force $-p_P$, if these dislocations provide sites of easy heterogeneous nucleation of the equilibrium phase. This in turn reduces the concentration c_1 and therefore the available chemical driving force p_C. The portions of the types of microstructure which have formed at a given temperature in three different alloys after increasing amounts of cold work are shown in Fig. 4. The transition to the continuous process at high defect density always indicates that the individual formation of a more stable

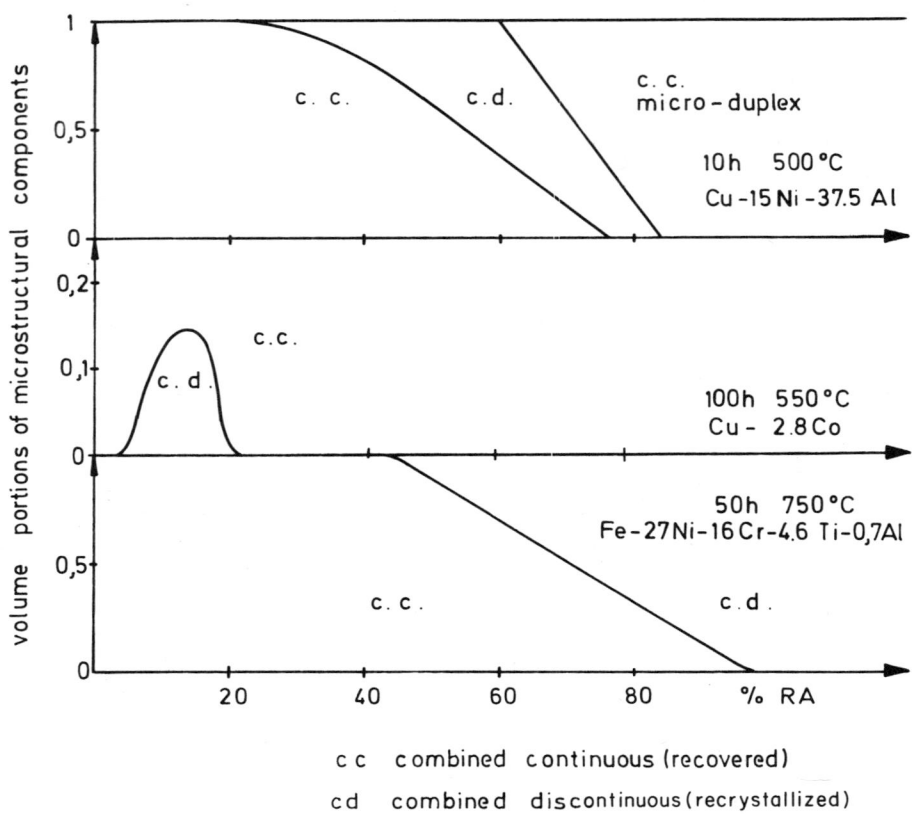

c c   combined continuous (recovered)

c d   combined discontinuous (recrystallized)

Fig. 4. Microstructures obtained as a function of the amount of cold work (R.A.) for three alloys and heat treatments as indicated.

precipitate is facilitated by dislocations or the defect structures formed by their rearrangement. Figure 5 shows the microstructure and an analysis of the driving forces for a precipitation-hardening austenitic steel after thermomechanical treatment. No continuous process occurs after high amounts of deformation in this alloy (Fig. 4) because no copious individual heterogeneous nucleation is induced by the defects.

4 RECRYSTALLIZATION INVOLVING A SECOND UNDEFORMED PHASE

It has been discussed in the preceding sections that dispersed particles can provide sites of easy nucleation of recrystallized grains, and that they can impede the motion of reaction fronts. Therefore, whether particles

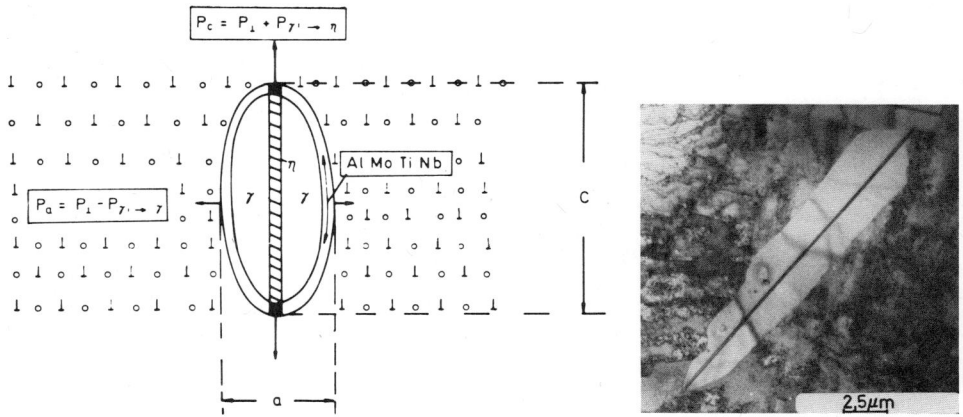

Fig. 5. Combined discontinuous reaction in a precipitation-hardening austenitic steel (alloy 3, Fig. 4)—treatment: 50% rolled, 50 hours, 790 C—and schematic representation of the reaction. $P_c = P\perp + P\gamma' \rightarrow \eta$, driving force in the elongated direction c; $P_a = P\perp - P\gamma' \rightarrow \gamma$, driving force in the a-direction; $\gamma' \rightarrow \eta$, transformation of the metastable phase γ' into stable η; $\gamma' \rightarrow \gamma$, resolution of metastable γ'.

accelerate or retard recrystallization depends on the special conditions (Fig. 6). It also can occur that recrystallization starts rapidly at many particles and stops as soon as the reaction fronts have reached the neighboring particles. Subsequent grain growth can then be extremely slow because of the impeding effect of the particles.

The effect that particles have on the initiation can be quantitatively interpreted by a model in which it is assumed that the minimum radius of curvature R of the recrystallization front is equal to the particle radius r_P. For the motion of a front with this curvature, a critical driving force or dislocation density ρ_C is necessary:

$$\mu b^2 (\rho_C - \rho_0) \geq \gamma_{\alpha\alpha} r_P^{-1} \tag{5a}$$

The radius above which a particle can initiate recrystallization at a dislocation density ρ_C is (for $\rho_0 \approx 0$):

$$r_P = \frac{\gamma_{\alpha\alpha}}{\mu b^2 \rho_C} \tag{5b}$$

The critical dislocation density for a particle size r_P is:

$$\rho_C = \frac{\gamma_{\alpha\alpha}}{\mu b^2 r_P}. \tag{5c}$$

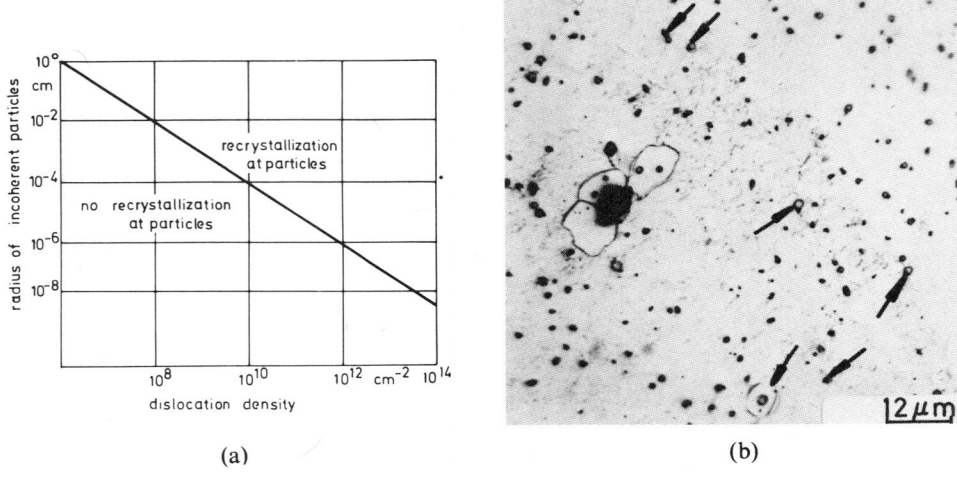

(a)   (b)

Fig. 6. Nucleation of recrystallization at particles: (a) calculated minimum particle radius for the formation of one nucleus in α-iron; (b) Fe + 0.23 wt % C steel with Fe_3C particles ($d_p \approx 1$ μm) and slag inclusion ($d_p \approx 10$ μm). The smaller particles are of the minimum size to initiate nucleation at 50 percent cold work (corresponding to $\rho_o \approx 10^{11}$ cm^{-2}).

This indicates that smaller particles can initiate recrystallization with increasing defect density (see Fig. 7). In Eq. (5), the inhomogeneity of the dislocation density around a particle is neglected as well as the fact that a recrystallization front has not only to be moved but also to be created. Because both corrections will affect our result with opposite sign, the assumptions made in Eq. (5) can be regarded as a reasonable estimate.

Several investigations have shown that not the particle radius r_P but also the spacing $\triangle_P$ can be the important factor that determines whether recrystallization is accelerated or retarded ($r_P^3/\triangle_P^3 = 0.554\ 4\pi/3f$). Prerequisite for the application of Eq. (5) is: $\triangle_P \gg r_P$. If this condition is fulfilled and the dislocation density is much higher than that required by Eq. (5), a large number of nuclei can form at a particle.

The condition for retardation of recrystallization follows from the impeding force of particles defined in connection with Eq. (4). If $p_R - p_P < 0$, no recrystallization can take place; for $p_R - p_P > 0$, the velocity of the reaction front is slowed down, depending on the value of p_P caused by the particle dispersion.

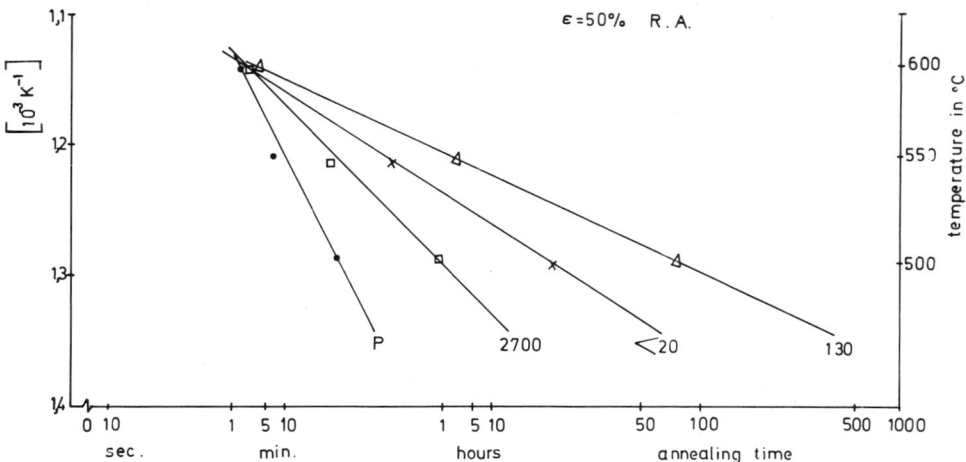

Fig. 7. Temperature dependence of the start of recrystallization in Fe + 0.23 wt % C steel with different carbide particle diameters (in nm) before deformation (P = ferrite-pearlite microstructure).

5 THE ORIGIN OF DUPLEX STRUCTURES

These structures are of interest because they can provide superplastic behavior or a favorable combination of yield strength and toughness. Their typical topological features are given in Table III. The occurrence of a duplex structure is favored by volume portions close to $f_\alpha = f_\beta = 0.5$, which determine the position of most duplex alloys in the phase diagram. For example $\alpha+\beta$-brass-, $\alpha+\gamma$-iron-, and $\alpha+\beta$-titanium-type alloys are suitable for the production of duplex structures—usually by a combined recrystallization and precipitation reaction (Section 3). In most cases, the additional objective is the formation of a very fine grained structure—a so-called microduplex structure.

Figure 8 indicates three important thermomechanical treatments that can be applied. Treatment I leads to precipitation of the second phase from a deformed supersaturated solution. To obtain a duplex structure, this phase must form individually at evenly distributed nucleation sites (i.e., not only at the original grain boundaries[24,25]). Also, the occurrence of a discontinuous combined reaction has to be avoided. Usually, a high amount of cold work is necessary to obtain the desired structure (Figs. 4b, 9).

In treatment II, the alloy is deformed at the temperature of homogenization and then quickly cooled into the two-phase range. This treatment is essential for Fe-C alloys, because precipitation of Fe_3C can never be avoided during heating from room temperature to the $\alpha+\gamma$-range.[26]

Table III. Topological Features of One-Phase Polycrystals, Dispersions, and
and Duplex Structures of the Phases α and β

		α-Polycrystal	α+β Dispersed	α+β Duplex	β+α Dispersed	β-Polycrystal
Boundaries	$\alpha\alpha$	+	+	+	−	−
	$\alpha\beta$	−	+	+	+	−
	$\beta\beta$	−	−	+	+	+
Junctions	$\alpha\alpha\alpha$	+	+	−	−	−
	$\alpha\beta\beta$	−	+	+	−	−
	$\alpha\alpha\beta$	−	−	+	+	−
	$\beta\beta\beta$	−	−	−	+	+

Type of Microstructure (spanning header over middle columns)

(a) + occurring; − not occurring.

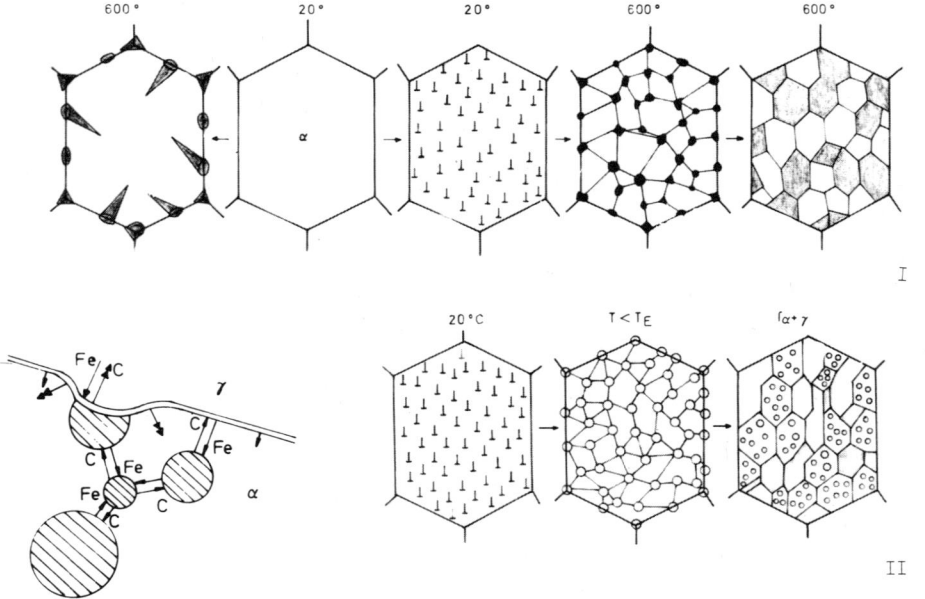

Fig. 8. Schematic representation of the formation of microduplex structures. I, example Fe-Ni: net and Widmannstätten structure in the undeformed alloy (cf. Fig. 9c); after deformation, there is even distribution of α-nuclei that grow to form a duplex structure (cf. Fig. 9a,b); II, example Fe-C: carbides precipitate during heating into the γ-field and stay inside the α-crystallites in a metastable equilibrium (cf. Fig. 9d).

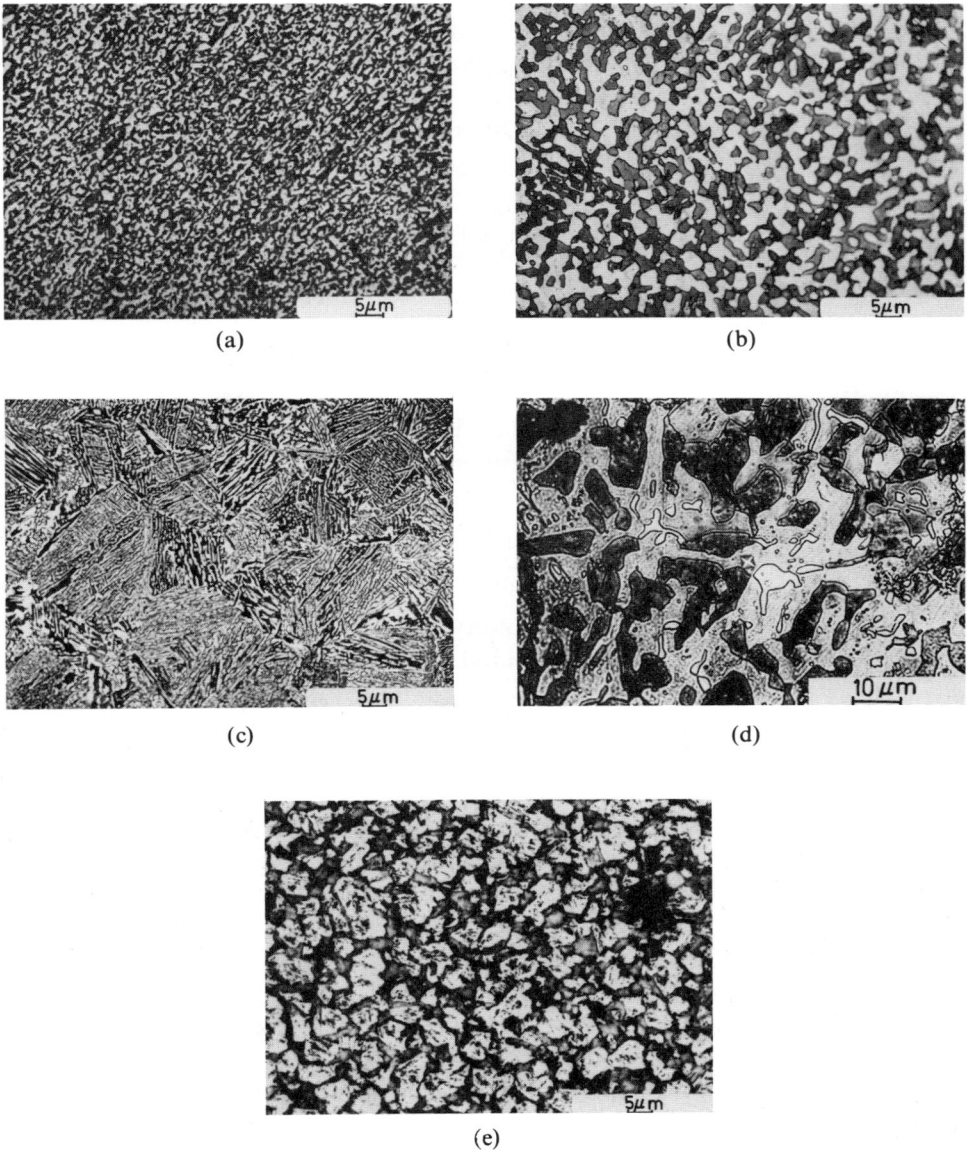

Fig. 9. Duplex microstructures in iron alloys: (a) microduplex structure: Fe + 9.8 wt % Ni, 90 percent deformation, 100 hours, 630 C; (b) growth of a duplex structure, cf. Fig. 13, 40 percent deformation, 1000 hours, 630 C; (c) nucleation of ferrite at austenite and martensite boundaries: 0 percent deformation, 100 hours, 630 C; (d) ferrite component containing growing Fe₃C particles in Fe-0.5 wt % C steel, 10 percent deformation, 10 hours, 745 C; (e) real duplex structure obtained by 40 percent deformation and direct cooling into the $\alpha+\gamma$ field, ~45 percent deformation at 900 C, 80 minutes, 775 C.

Further grain refinement is obtained by a sequence of deformation and tempering at different temperatures (treatment III). The annealing at a temperature $T_2 \neq T_1$ in the $\alpha+\beta$-range requires the formation of α in β and β in α and thus leads to a further breakup of the primary duplex structure.[27] Martensitic transformation of one of the phases during plastic deformation or cooling occurs in brasses and steels and can be used in addition to modify the duplex microstructure.

6 TEXTURE AND MICROSTRUCTURAL ANISOTROPY

Only those recrystallization textures directly correlated to the mechanisms discussed in the preceding sections are considered here. The four temperature ranges indicated in Fig. 1 are of direct relevance for the control of textures. The texture obtained in range II is the same as in the homogeneous solid solution range I, because recrystallization is unaffected by the subsequent precipitation. In range IV, recrystallization and therefore reorientation of large-volume portions is inhibited. Therefore the original (rolling) texture is preserved (Fig. 10).

For range III and for alloys containing a coarse dispersion of hard particles, the situation is more complicated. Nucleation of many new crystallites at large particles leads to random orientation. Growth selectivity can, however, produce a sharp texture if the particle spacing Δ_P is large enough to allow a sufficient path of travel for the grain boundaries of the favorable orientation.[28] If particles have precipitated from a concentrated solid solution before the recrystallization front has annihilated the defects, a transformation from the texture of the concentrated solution to that of the pure metal is observed, for example, in Al-Cu alloys. This can be explained by the purifying effect of precipitation so that the higher growth selectivity of pure metals can become effective in the alloy. A random orientation can be expected if the combined discontinuous reaction occurs and the chemical force p_C dominates, just as it has been found after discontinuous precipitation.

Out of the many cases of microstructural anisotropy, only the alignment of grain and phase boundaries by solid-state reactions is discussed briefly.

The simplest way to obtain an uniaxial grain structure is by the deformation and annealing of a polycrystalline alloy that contains a dispersion of particles at the grain boundaries. If the conditions are such that the recrystallization front is held up by the particles, the grain shape, determined by mode and amount of deformation, is preserved to a large extent. Other methods to obtain a uniaxial grain structure are directional solidification, recrystallization[29,30], phase transformation[31-34], and solid-state reactions under uniaxial external stress or in a magnetic field.

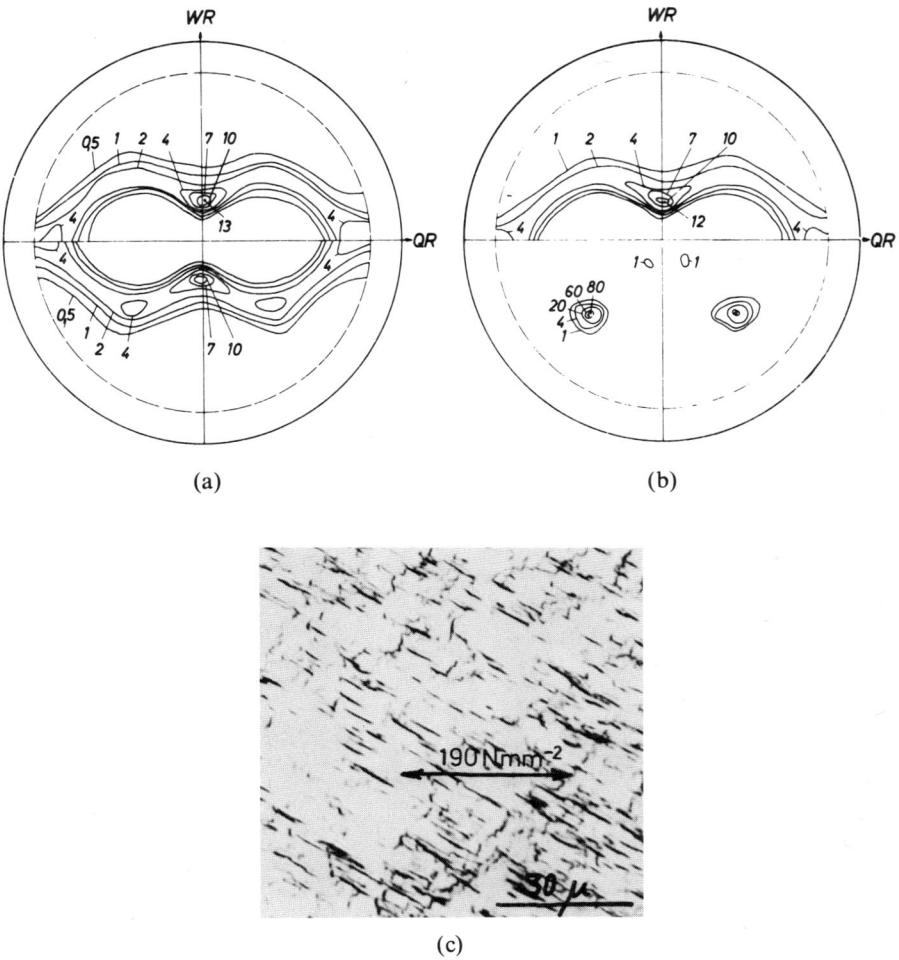

(a)   (b)

(c)

Fig. 10. Anisotropy: (a,b) copper rolling texture after 95 percent cold work (upper half of pole figures) and annealing texture after 4 hours 200 C (lower half of pole figures). (a) Pure copper; (b) Cu + 0.04 vol % B₄C, particle diameter $d_p \approx 2$ μm, sharp cube texture; and (c) alignment of bainitic particles in β-brass Cu + 60 wt % Zn, quenched and aged under stress 10 min at 250 C.

Recently, directional recrystallization has attracted attention as a method to obtain very high grain aspect ratios in high-temperature, nickel-base superalloys. An ultrafine-grained, hot-extruded structure is used as the starting structure. A high-temperature zone is moved parallel to the axial directions of a bar, as indicated schematically in Fig. 11. The velocity is dictated by the minimum time necessary to heat the alloy to $T > T_R$ to complete recrystallization and the requirement that equiaxial grain or subgrain

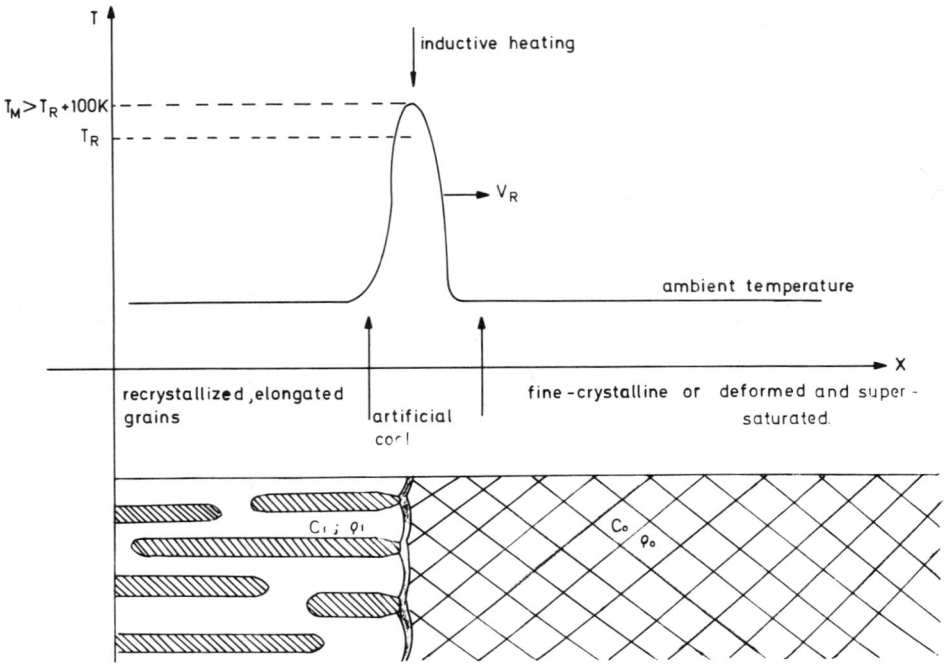

Fig. 11. Schematic representation of directional recrystallization of a supersaturated and defect solid solution. Recrystallization and directional precipitation occur in a steep temperature gradient.

growth be minimized. The optimum is attained if the temperature gradient is as steep as possible.

A similar process will lead to an aligned two-phase structure if eutectoid decomposition or discontinuous precipitation takes place. Particle spacings obtained by these solid-state reactions are one order of magnitude smaller than those for directional solidification. Different from directional recrystallization, a negative temperature gradient has to be applied for directional eutectoid decomposition. The velocity is dictated by the requirement that no new nucleation occurs in the zone of growth. The most perfectly aligned structures are obtained if the eutectoid grows from a single crystal. Therefore a two-stage directional process yields the best results: growing a single crystal by directional solidification and subsequently the eutectoid at lower temperatures. There is a chance that aligned two-phase microstructures may be obtained by a directional discontinuous combined (recrystallization plus precipitation) reaction. This possibility has not yet been explored.

7 GRAIN GROWTH OF HETEROGENEOUS ALLOYS

If the alloys are shaped or used at elevated temperatures, the stability of the microstructure which was produced by recrystallization becomes important. The driving force for all growth processes is the density of the interfacial energy which can be expressed as function of the grain size $r_{\alpha\alpha}$ for grain structures as $p_g = 3\gamma_{\alpha\alpha}/2r_{\alpha\alpha}$. The rate of the reactions depends on whether the atoms have to jump only a distance of the order of magnitude of the thickness of the boundary or much larger distances and on whether boundary diffusion, $D_{\alpha\alpha}$ or $D_{\alpha\beta}$, pipe diffusion along dislocations cores D_P, or volume diffusion D_V is effective. The time laws for growth which are obtained on this basis are shown below in systematic order (Fig. 12)[35]:

$$r_{\alpha\alpha} \propto (D_{\alpha\alpha}t)^1 \quad \text{recrystallization, secondary recrystallization} \qquad (6a)$$

$$r_{\alpha\alpha} \propto (D_{\alpha\alpha}t)^{1/2} \quad \text{grain growth}^{[36]} \qquad (6b)$$

$$r_P \propto (D_V t)^{1/3} \quad \text{growth of dispersed particles}^{[37]} \qquad (6c)$$

$$r_{\alpha\beta} \propto (D_{\alpha\beta}t)^{1/4} \quad \text{growth in duplex structures}^{[27,38]} \; (D_{\alpha\beta} \gg D_V) \qquad (6d)$$

$$r_P \propto (D_P t)^{1/5} \quad \text{growth of particles connected by dislocations for}$$
$$D_P \gg D_V{}^{[39]}. \qquad (6e)$$

The $t^{1/3}$ law has been confirmed for growth of a very large number of metallic dispersions, while the $t^{1/2}$ law is rarely found for grain growth. This is explained by accumulation of impurities during growth, even in high-purity metals and by changes in the grain-boundary structures towards low-energy configurations. The $t^{1/4}$ law for the growth of duplex structures is caused by diffusion paths corresponding to the grain size. The low power is in turn the reason why microduplex structures are preferred to one-phase polycrystalline structures, if long-time stability of a fine-grained structure is required for superplastic deformation (Fig. 13a).

In order to obtain thermal stability, the two grain-growth mechanisms [Eqs. (6b) and (6d)] are often modified by a dispersion of particles. The condition at which grain growth is controlled by particle growth is obtained by equating the driving and the pinning forces, $p_g = p_P$[17]:

$$r_{\alpha\alpha}c \geqslant c\,\frac{r_P}{f} \quad, \qquad (7a)$$

where c is a geometrical factor with the order of magnitude of one which considers shape, size distribution, and local distribution of the particles.

Correspondingly, a critical volume fraction f_c of particles of a size r_P can be defined that stabilizes a given grain size $r_{\alpha\alpha}$. This volume fraction

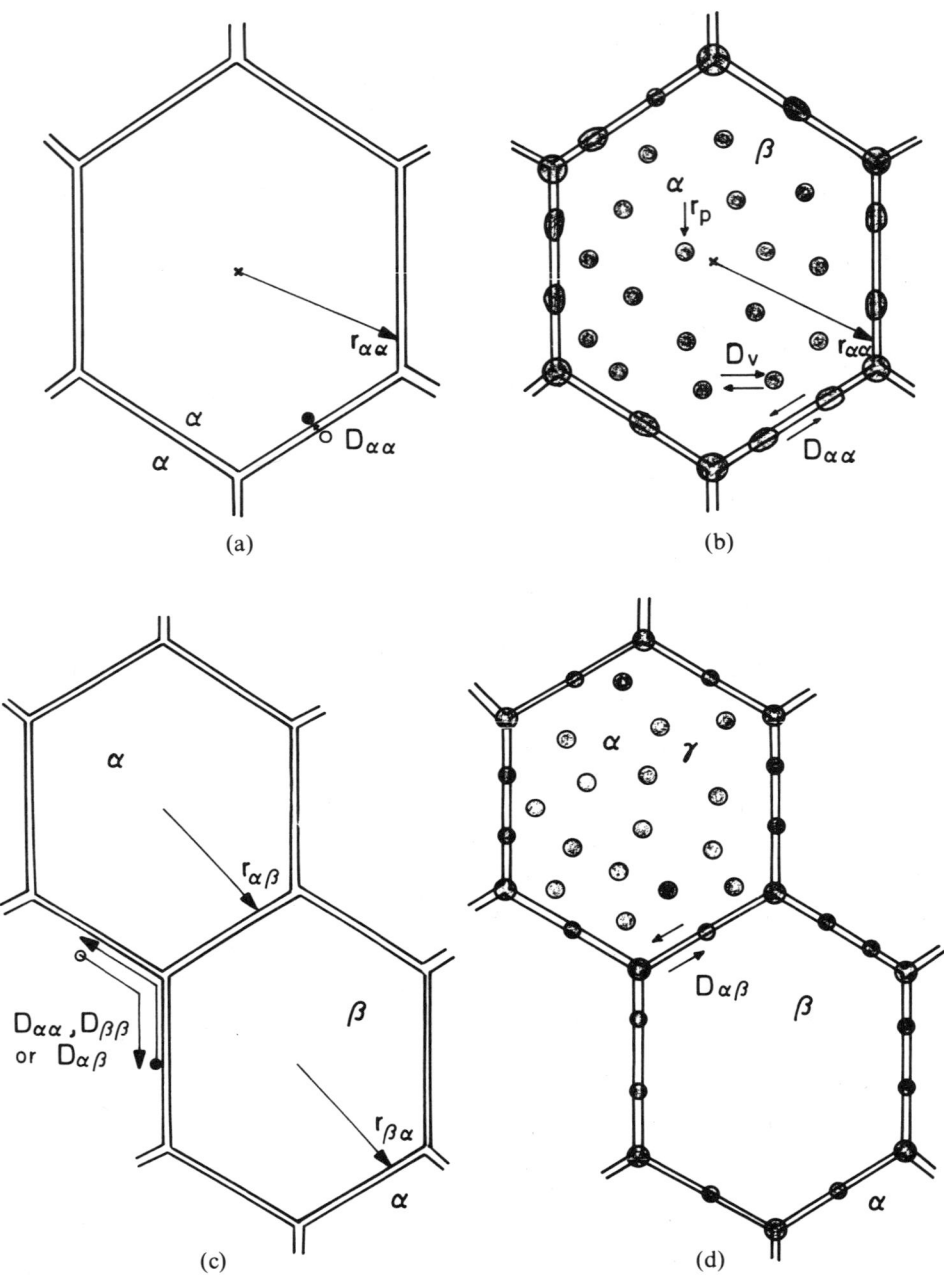

Fig. 12. Schematic representation of growth reactions: (a) grain growth, controlled by grain-boundary diffusion across the boundary; (b) grain growth, controlled by particle growth; (c) growth of duplex structures (cf. Figs. 9a and 9b); and (d) growth of duplex structures controlled by particles.

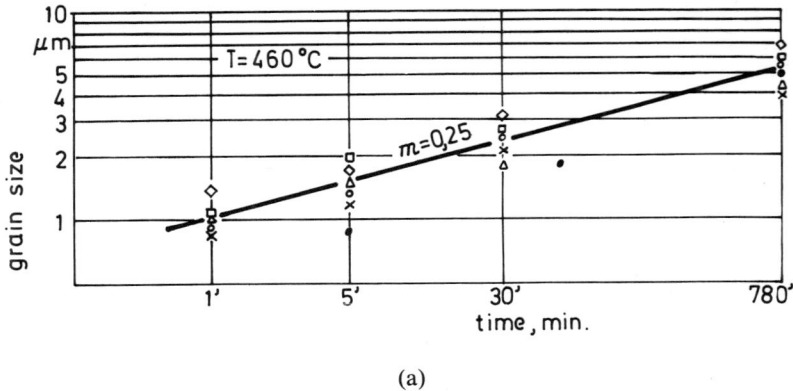

(a)

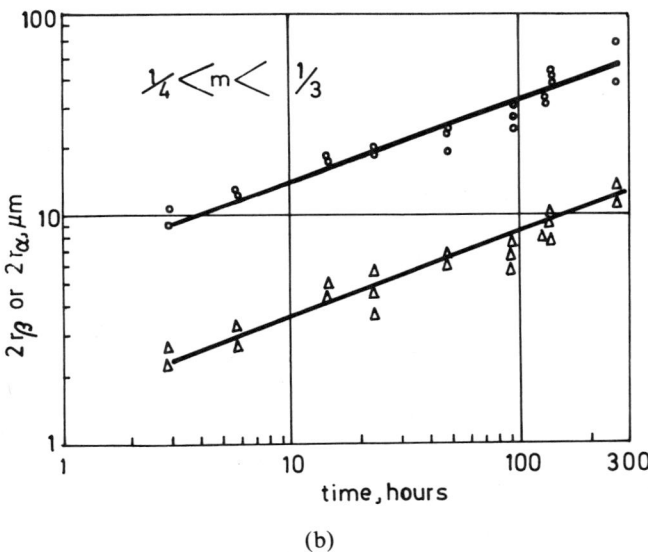

(b)

Fig. 13. Examples of growth of heterogeneous microstructures: (a) Cu + 42 wt % Zn, growth of a duplex structure; (b) Ti + 5.25 wt % Al + 5.5 wt % V, growth of β-grains controlled by α-particles at grain boundaries.[40]

can be very much reduced to f'_c if it is sufficient to locate the particles at the grain boundary:

$$f_c \geqslant c \, \frac{r_P}{r_{\alpha\alpha}} \quad , \quad f'_c \geqslant c \, \frac{r_P^2}{r_{\alpha\alpha}^2} \tag{8a,b}$$

Finally, it has to be mentioned that grain growth is not zero if the conditions given in Eqs. (7) and (8) are fulfilled; it is however controlled by growth of the pinning particles [Eq. (6c)]:

$$r_{\alpha\alpha c}(t) = c \, \frac{r_P(t)}{f} \tag{7b}$$

If the function $r_P(t)$ is known, the controlled grain growth can be computed. As an example, the growth of α-Ti crystallites controlled by β-Ti particles is shown in Fig. 13b. The exponent of the time dependence is $1/4 < 0.3 < 1/3$, indicating that boundary and volume diffusion are responsible for the growth of the β-particles.

REFERENCES

1. Hornbogen, E., *Prakt. Metallog.*, **5**, 51 (1968).
2. Hornbogen, E., *Z. Metallk.*, **64**, 867 (1973).
3. Doherty, R. D., *Metal Sci.*, **8**, 132 (1974).
4. Higgins, G. T., *Metal Sci.*,**8**, 143 (1974).
5. Cahn, R. W., *Can. Met. Quarterly*, **13**, 253 (1974).
6. Hornbogen, E., and Zum Gahr, K. H., *Metallography*, **8**, 181 (1975).
7. Klein, H. P., *Z. Metallk.*, **61**, 564 (1970).
8. Staniek, G., and Hornbogen, E., *J. Mater. Sci.*, **9**, 879 (1974).
9. Köster, U., *Metal Sci.*, **8**, 151 (1974).
10. Hornbogen, E., and Kreye, H., *Textures in Research and Practice*, J. Grewen and G. Wassermann (Eds.), Springer, Berlin (1969), p. 274.
11. Hornbogen, E., *Prakt. Metallog.*, **9**, 349 (1970).
12. Köster, U., *Recrystallisation of Metallic Materials*, F. Haessner (Ed.), Riederer, Stuttgart (1971), p. 215.
13. Hornbogen, E., in *Nucleation*, A. C. Zettlemoyer (Ed.), Dekker, New York (1969), p. 309.
14. Kreye, H., and Hornbogen, E., *J. Mater. Sci.*, **5**, 89 (1970).
15. Kreye, H., and Brenner, U., *J. Mater. Sci.*, **9**, 1775 (1974).
16. Minuth, E., and Hornbogen, E., *Prakt. Metallog.*, **11**, 650 (1974).
17. Zener, C., and Smith, C. S., *Trans. AIME*, **175**, 15 (1949).
18. Ashby, M., and Centamore, R.M.A., *Acta Met.*, **16**, 1081 (1968).
19. Hornbogen, E., *Metall*, **27**, 270 (1973).
20. Leslie, W. C., Michalak, J. T., and Aul, F. W., *Iron and Its Dilute Solid Solutions*, Interscience (1963), p. 116.
21. Doherty, R. D., and Martin, J. W., *Trans. ASM Quart.*, **57**, 874 (1964).
22. Corti, C. W., and Cotterill, P., *Scripta Met.*, **6**, 1047 (1972).
23. Kamma, C., and Hornbogen, E., *J. Mater. Sci.*,**11** (1976), to be published.
24. Ansuini, F. J., and Badia, F. A., *Met. Trans.*, **4**, 15 (1973).
25. Minuth, E., and Hornbogen, E., Proceedings of International Conference on Metallography, Leoben, Austria, 1974, *Arch. Eisenhüttenw.*, **41**, 883 (1970).
26. Brock, C., Hornbogen, E., and Stratmann, P., *Arch. Eisenhüttenw.*, **47**, 513 (1976).
27. Mäder, K., and Hornbogen, E., *Scripta Met.*, **8**, 979 (1974).

28. Haessner, F., Hornbogen, E., and Mukherjee, M., *Z. Metallk.,* **57,** 171 (1966).
29. Cairus, R. L., Curwick, L. R., and Benjamin, J. S., *Met. Trans.,* **6,** 179 (1975).
30. Bailey, P. G., and Kutchera, R. E., Technical Report AFML-TR-73-294 (1973).
31. Livingston, J. D., *J. Crystal Growth,* **24/25,** 94 (1974).
32. Chadwick, G. A., and Edmonds, D. V., *Chemical Metallurgy of Iron and Steel,* Iron and Steel Institute, London (1973), p. 264.
33. Livingston, J. D., and Cahn, J. W., *Acta Met.,* **22,** 495 (1974).
34. Gurevich, Ya. B., et al., *Soviet Phys. Doklady,* **16,** 992 (1972).
35. Hornbogen, E., *Metall,* **29,** 248 (1975).
36. Frois, C., and Dimitrov, O., *Mem. Sci. Rev. Met.,* **59,** 643 (1962).
37. Wagner, C., *Z. Elektrochem.,* **95,** 581 (1961).
38. Ardell, A. J., *Acta Met.,* **20,** 601 (1972).
39. Kreye, H., *Z. Metallk.,* **61,** 108 (1970).
40. Levine, E., Greenhut, I., and Margolin, H., *Met. Trans.,* **4,** 2519 (1973).

DISCUSSION on Paper by E. Hornbogen

THOMAS: With reference to your method of producing duplex structures in Fe-C alloys similar to ours, in our method we produce the dislocations by fast quenching to martensite (100 percent) to give dislocated martensite, and reheat to austenite or directly into the $(\alpha+\gamma)$ two-phase field to produce our uniform duplex $(\alpha + \text{martensite})$ grains. Thus it is not really necessary to hot roll the austenite in order to introduce dislocations there (at heat in 1010 type steel). Perhaps you need heavy deformation if you want to do this in higher carbon steels—is this a limitation?

HORNBOGEN: Independent of the carbon content we find that we need the dislocations of at least 40 percent of plastic deformation (by rolling) to produce a real duplex structure. At smaller amounts of deformation, the distribution of phases is still uneven because the nucleating phase forms at austenite or martensite grain boundaries. The amount of dislocations produced by the martensitic transformation alone was never sufficient to form a duplex structure which corresponds with our definition (see Table III).

TIEN: Not that many years ago I was taught that the "nucleation" stage of recrystallization involved the diffusion-controlled rotation and coalescence of small neighboring grains or subgrains to an orientation such that one side of the now auger grain is founded by a high-mobility, incoherent grain boundary which can then migrate and initiates the grain-growth process. I believe this model was attributed to J.C.M. Li and was published as an Appendix to a recrystallization paper of Hsun Hu's in the circa 1962 ASM Symposium publication on *Recovery and Recrystallization.* Do you feel that this model is still physically correct?

HORNBOGEN: There are several sites at which high-angle boundaries which will act as recrystallization fronts can form in a deformed homogeneous alloy: they are listed in Section 2 of my paper. Two models for subgrain growth exist. One is assuming continuous growth which implies disappearance of small subgrains; the second is subgrain rotation (Li, Hu). Both models agree with the fact that the angle of tilt and twist increases as the subgrains grow, so that a high-angle boundary structure can develop which induces recrystallization. This stage of the reaction can terminate the continuous combined reaction in heterogeneous alloys. However, nucleation at noncoherent particles can become the most effective mechanism in heterogeneous alloys which overrides all the other mechanisms mentioned above.

SHEWMON: Would you relate the cooling and working cycles you discuss to those used in commercial practice for the low-alloy, high-strength steels that Embury discussed?

HORNBOGEN: Yes, as one example may serve the explanation of delamination fracture in HSLA-steels along recrystallized ferritic grains; the ferrite is being highly deformed by rolling while it is cooled and its solubility decreases. The conditions for range II (see Fig. 1 of my paper) are to be expected, i.e., recrystallization followed by precipitation. A filmlike precipitate will form because of the imperfect (noncoincidence) structure of the grain boundaries which in turn will act as cleavage plane for delaminations as described in Embury's paper.

ALDEN: Do you anticipate a large excess vacancy concentration in the heavily worked alloys and, if so, how do you think it affects the subsequent phase decomposition?

HORNBOGEN: Vacancy concentrations as they can be expected in metals will not have much effect on the driving force p, but rather on the mobility of a discontinuous reaction front. A vacancy current across the reaction front can be derived from the net volume range connected with a precipitation reaction. Vacancies produced by high amounts of cold work will have annihilated at dislocations until the reaction front arrives and therefore will have no effect on the velocity of the reaction front of a discontinuous reaction. The effect of vacancies on individual formation of particles in continuous reactions will be restricted to the enhancement of diffusion, since the formation of dislocation rings which act as nucleation sites is not to be expected because of the high dislocation density of heavily worked alloys.

PROCESSING AND PROPERTIES
OF SUPERPLASTIC ALLOYS

T. H. Alden

Metallurgy Department, U.B.C.
Vancouver, B.C., Canada

ABSTRACT

The special utility of superplastic alloys depends basically on the remarkable variation with temperature and grain size of the mechanical properties of ultrafine-grained materials. At the first level, the property may be the flow stress, which is extremely small at high temperature but increases rapidly as the temperature falls. This permits a variety of hot-working processes at small forces, yet provides the unique service toughness which is conferred by small grain size. In addition, superplastic alloys processed by free stretching at high temperature show unique ductility which allows single-die sheet forming into complex shapes. The engineering work on these processes is now in a fairly advanced state, especially for the Zn-Al eutectoid alloy and some titanium alloys.

Ultrafine grain size is achieved and maintained most easily (though not uniquely) in concentrated two-phase alloys. This restriction is a fairly severe one, especially since many such alloys contain intermediate phases which, while ductile at high temperature, are brittle at service temperatures. An additional requirement, if a homogeneous microstructure is to be achieved, is that the alloy be hot worked in an initially single-phase condition or at least that the phases be intimately mixed. The latter criterion is usually satisfied in as-cast eutectics, which are the most common type of superplastic alloy.

1 INTRODUCTION

While the refinement of grain size for improvement of the service properties of alloys is an old practice of physical metallurgy, a qualitative change and advance in this practice has accompanied the recent development and use of *superplastic alloys*. This change has several components:

1. The structure of superplastic alloys is properly called *ultrafine-grained*, with grain sizes near or below 1 μm, or reduced by roughly an order of magnitude from prior values.

2. The production of small grain size is the initial rather than the final processing step. Further, this art has achieved a new level of sophistication in terms of precision control of alloy composition, heat treatment, temperatures of deformation, and degree of deformation.

3. Grain boundaries have a major role in the control of properties during *deformation processing* as well as (in many cases) during service.

4. Superplastic alloys usually have a concentrated *two-phase* microstructure, appropriately named microduplex. A discussion of these topics is the basis of this paper.

2 FLOW STRESS

The flow stress of superplastic alloys varies strongly with temperature and, at high temperature, with grain size. Data taken from the literature are shown in Figs. 1 and 2. At high temperature and small grain size, the flow stress is exceptionally low. Low flow stress permits forming at low pressure and in lightweight dies, both in expansion- and compression-forming processes. At least as much as high strain-rate sensitivity, this softness is responsible for the extraordinary ductility and formability of superplastic alloys.

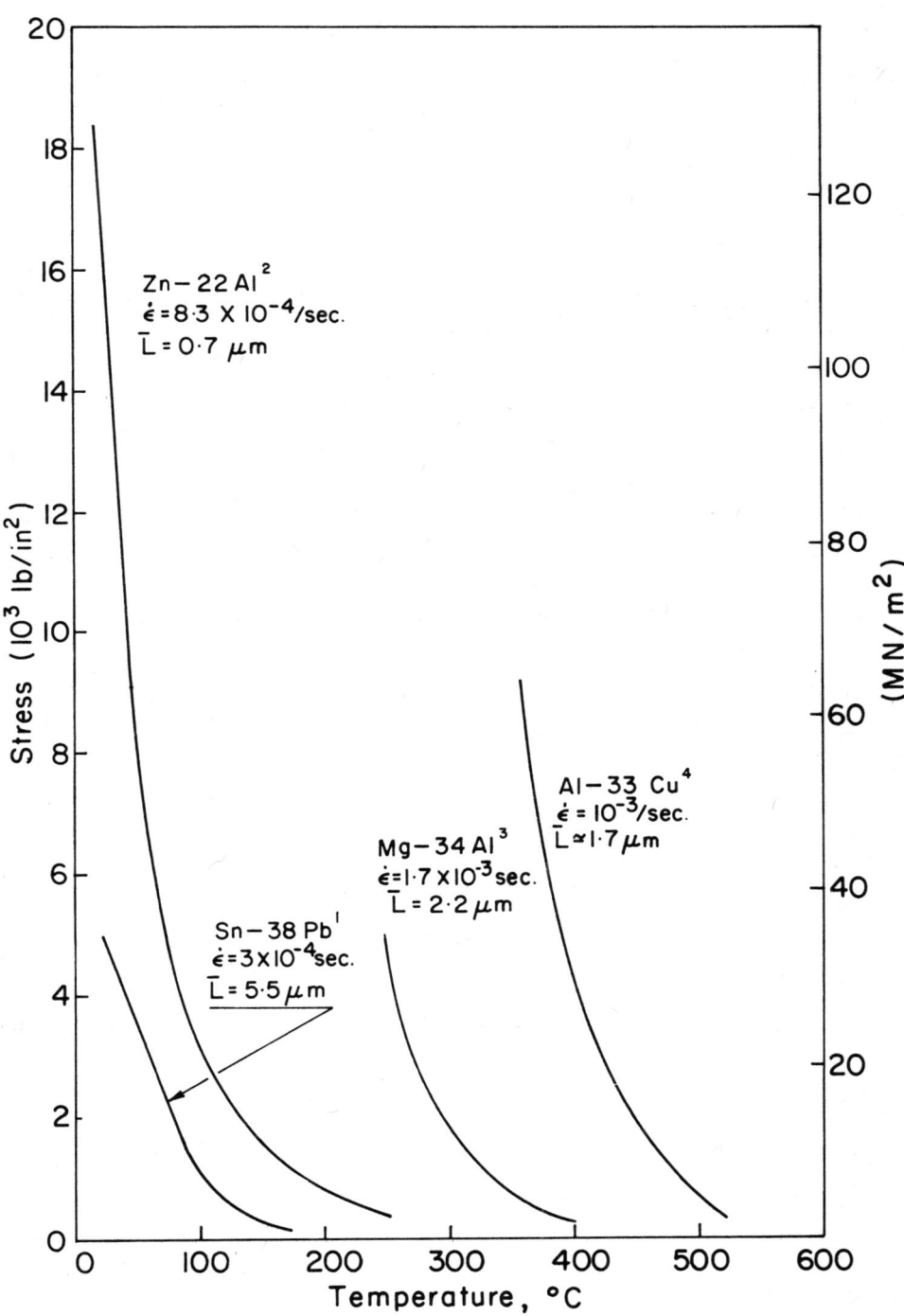

Fig. 1. Flow stress versus temperature for various superplastic alloys.

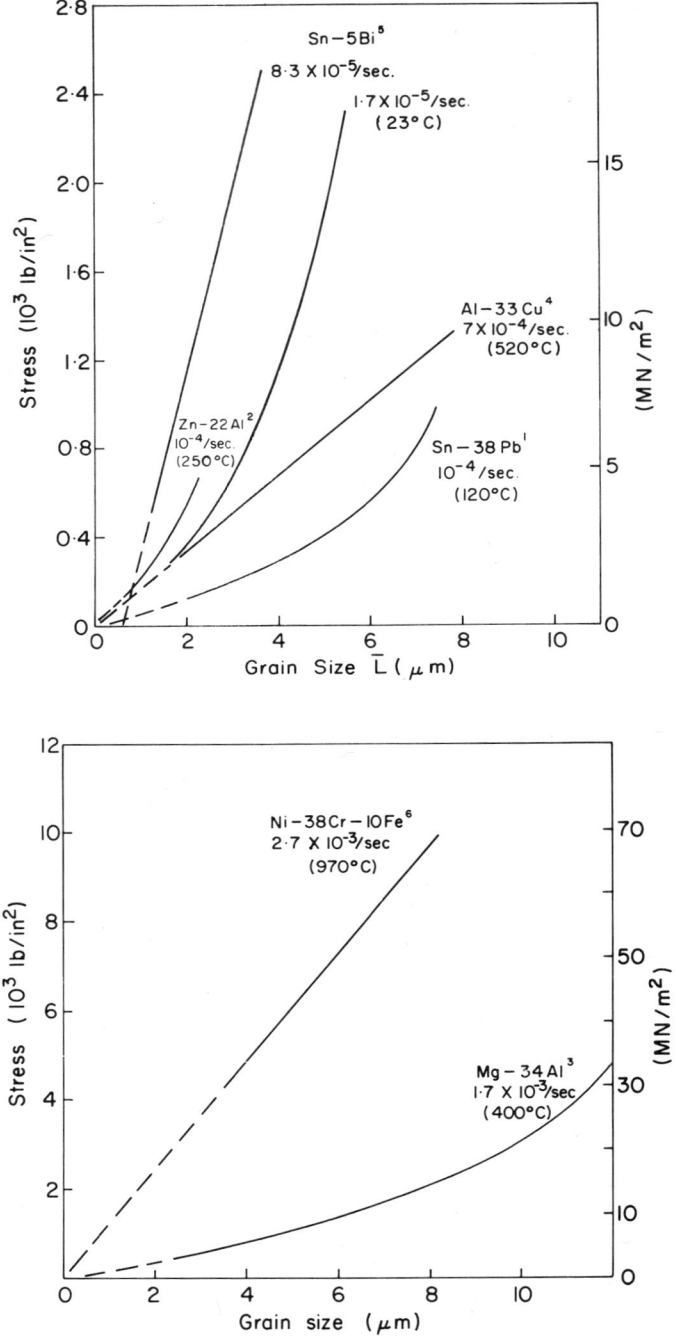

Fig. 2. Stress versus grain size for various alloys at temperatures at which they exhibit superplasticity.

3 STRAIN-RATE SENSITIVITY

The many unique features of the deformation of superplastic alloys were identified within a few years following its rediscovery and have been discussed in recent reviews.[7-9] These features include three-stage strain-rate hardening, retention of equiaxed grains, absence of strain hardening and extensive grain-boundary sliding, in addition to the well known high strain-rate sensitivity. In a sense, the culmination of these studies has been the beautiful theory of Ashby and Verral[10], in which change in *shape* of the sample is caused by change in *place* of the grains. This place change is accomplished by a combination of grain-boundary sliding and diffusional flow of the grains (principally by grain-boundary diffusion), Fig. 3. The process is basically Newtonian viscous. According to the authors[10], "The strongest evidence that the present model is a good description of superplasticity is . . . that the microstructural and topological predictions of the model for large strains match those observed in superplastic alloys".

The equation given by Ashby and Verrall describes stage 1 (as a threshold stress) and stage 2 of the three-stage log stress–log strain rate curve, while stage 3 is described by the semiempirical equations of dislocation creep.[11] The result for lead is in poor agreement with the experimental curve for Pb-5Cd.[12] The theoretical curve shows a pronounced S-shape with an $m_{max} \simeq 0.95$, while experimentally $m_{max} \simeq 0.35$ and stage 1 is not well defined. It may be possible to understand this low m_{max} by recalling that the Newtonian mechanism occurs as a *transition process,* i.e., it is "trapped" between mechanisms of low rate sensitivity. Thereby the *stress range of the transition,* defined as the difference in stress level of the stage 1 and stage 3 curves at an appropriate transition strain rate, will be the major factor which determines m_{max}.[13] This concept is illustrated in Fig. 4, for which the curves were calculated assuming a threshold process for stage 1. As the threshold stress σ_o increases, m_{max} decreases, in this case to $m_{max} = 0.38$ at $\sigma_o = 2 \times 10^7$ dyne/cm^2. A similar result should hold if stage 1 is an extended region of low (roughly constant) slope, rather than a threshold process.

A related idea may explain why zinc- and tin-base alloys have superior superplastic properties.[13] Because of their noncubic crystal structures, they are difficult to deform by slip in fine-grained polycrystalline form. The result is a high stress level in stage 3 and an expanded stress range (at high strain rate!) for rate-sensitive flow.

These ideas may form a basis for development of superplastic alloys. Compositional and microstructural factors which harden an alloy against slip deformation should not, in general, impede grain-boundary sliding and diffusional accommodation. (The special case is grain refinement which has a beneficial effect on both processes.) Attractive possibilities are fine-grained intermediate phases or solid-solution hardened alloys.

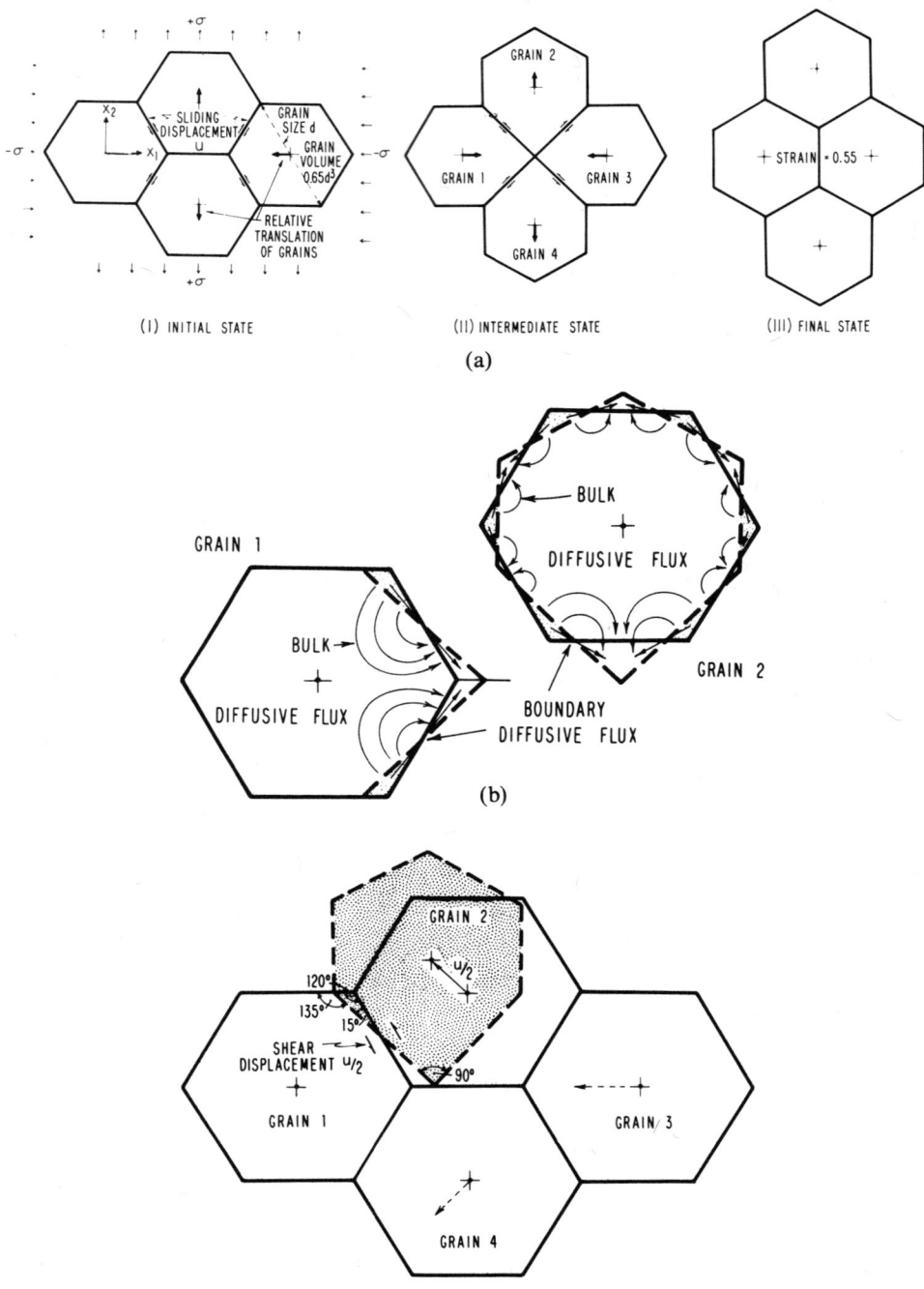

Fig. 3. (a) Geometry of deformation by diffusion-accommodated grain-boundary sliding; (b) diffusive shape change of the two-dimensional grains; (c) sliding displacement.[10]

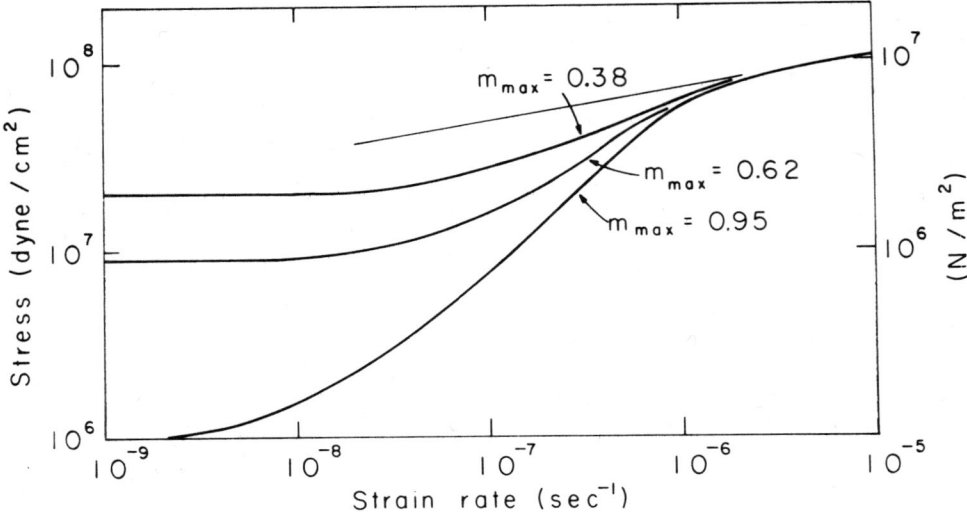

Fig. 4. Variation of theoretical logarithmic stress/strain-rate curves and m_{max} with threshold stress σ_o; as σ_o increases, m_{max} decreases.

4 GRAIN-SIZE REFINEMENT

Grain-size refinement may be achieved by thermal processing or mechanical processing (at elevated temperature) or by various combinations of heat treatment and hot or cold working.

4.1 Thermal Processing

As is well known, the Zn-Al eutectoid alloy decomposes to a microduplex structure when quenched from above 300 C and transformed near room temperature.[2] The phases are nearly pure zinc and an aluminum-zinc solid solution in roughly equal volume fractions. The mechanism of this transformation is only partly known. Initial phase separation occurs by the growth of nuclei formed primarily at the α' grain boundaries.[14] The product is elastically strained, hard, and of nonequilibrium composition, but its micromorphology is unknown. Subsequently, it seems to recrystallize to the microduplex product (Fig. 5), accompanied by a dramatic softening and sharpening of the X-ray diffraction lines.[14]

It is tempting to speculate that the initial decomposition product remains single-phase, face-centered cubic but is of widely varying composition. Indeed, the X-ray evidence is consistent with this idea, since in this product the prominent "lines" are in fact "halos", with the fringe positions

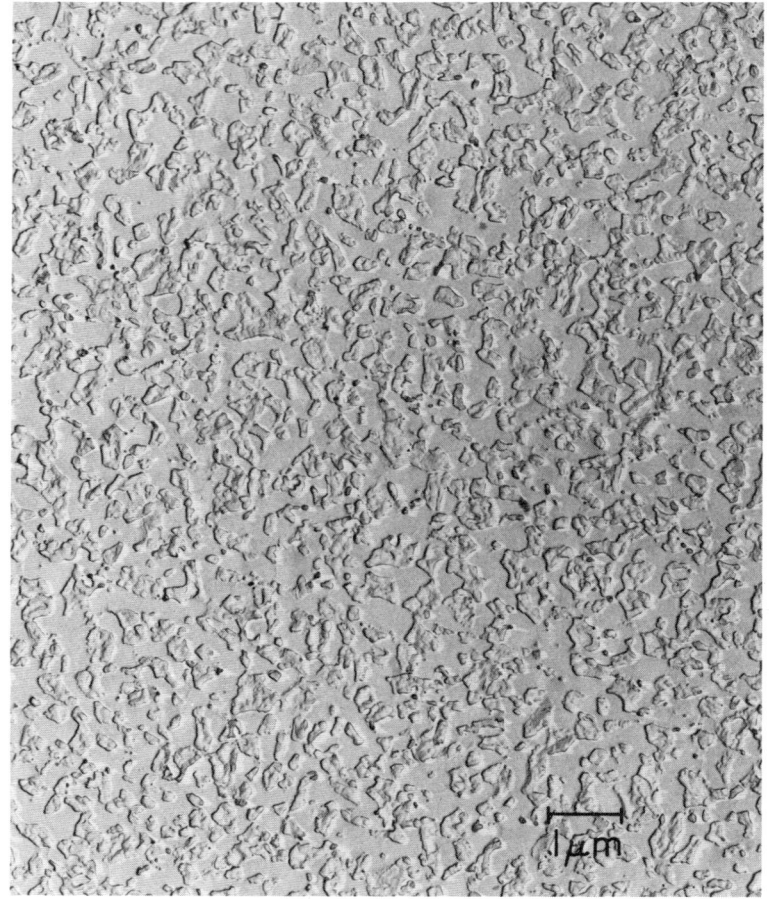

Fig. 5. Fine-grained Zn-22 Al, transformed $\alpha' \rightarrow \alpha + Zn$ at 0 C; the roughened phase is Zn[2].

corresponding to the lattice parameters of α' and α, respectively. Subsequently, recrystallization and precipitation of zinc may occur "simultaneously" in the manner postulated for extruded Sn-5Bi[5] and various steels and stainless steels[15] (cf. section 4.3). The grain size of the recrystallized material is less than 1 μm.

Repeated phase change by means of several heating-cooling cycles can also produce grain refinement. Schadler used multiple quenches between ice water and a salt bath at 825 C to achieve a 1 to 2-μm grain size in a commercial 1340 steel (1.87 percent manganese, 0.42 percent carbon).[16] This steel was rate sensitive near 725 C but at fairly low strain rates of 10^{-3} to 10^{-2}/min. The grains are $\alpha + \gamma$ at a final annealing temperature 733 C, and α + pearlite after air cooling. Similar results were reported recently for an Fe-Ni-Ti alloy, although this alloy was not shown to be superplastic.[17]

4.2 Mechanical Processing

Isothermal plastic deformation, usually hot working, can produce fine-grained structures in certain alloys. The best-known example is *eutectic alloys,* which have a fine structure in the cast state without the necessity of special casting procedures like quenching from the melt.[18] The structure is "fine" in the sense that the phases are small in scale and intimately mixed. In the cast condition, however, the structure is not fine grained and, as a result, not superplastic. For example, the tin matrix in the cast Pb-Sn eutectic contains very few grain boundaries.[18] Hot working converts this fine mixture to a (recrystallized) microduplex structure. The many interphase boundaries, which are quite immobile, stabilize the structure against grain growth.

Hot working is also useful in precipitation alloys. A good example is Sn-5Bi.[5] Extruded to a large area reduction in a single operation, the alloy emerges from the die at a temperature near the solvus; subsequently, recrystallization and precipitation occur "simultaneously" so that grain growth is inhibited by the bismuth particles.

4.3 Thermomechanical Treatment

The most advanced studies of thermomechanical processing to produce microduplex structures have been done on various nickel alloys, stainless steels, and low-alloy steels by workers at the P.D. Merica Laboratory.[15,19,20] The method depends generally on simultaneous recrystallization and precipitation, as in the above example, but may involve combinations of hot working at ever-decreasing temperature, cold working, and annealing. The basic process, hot working across a solvus line, may be altered in the case where precipitation is slow by holding for various times in the two-phase region prior to final working. A variant requires annealing only in the single phase region, followed by quenching, cold work, and finally aging above the recrystallization temperature (Fig. 6).

In order to obtain microduplex structures in low-alloy steels, Snape and Church[20] added aluminum up to about 2 percent in order to raise the temperature of the $\gamma \rightarrow \alpha + \gamma$ transition into the commercial hot-working range, 820 to 1200 C. It was found that the smallest uniform grain size was achieved when two conditions were fulfilled: (1) the temperature at the start of rolling was near the temperature for the start of ferrite precipitation and (2) the finish temperature was high enough that adequate austenite remained for grain-size stabilization. The latter condition was more easily satisfied in a steel containing 1 percent nickel, which lowers the A_{r1} temperature.

A very recent development is microduplex processing of inexpensive ultrahigh-carbon steels, 1.3 to 1.9 percent carbon.[21] These steels are solution

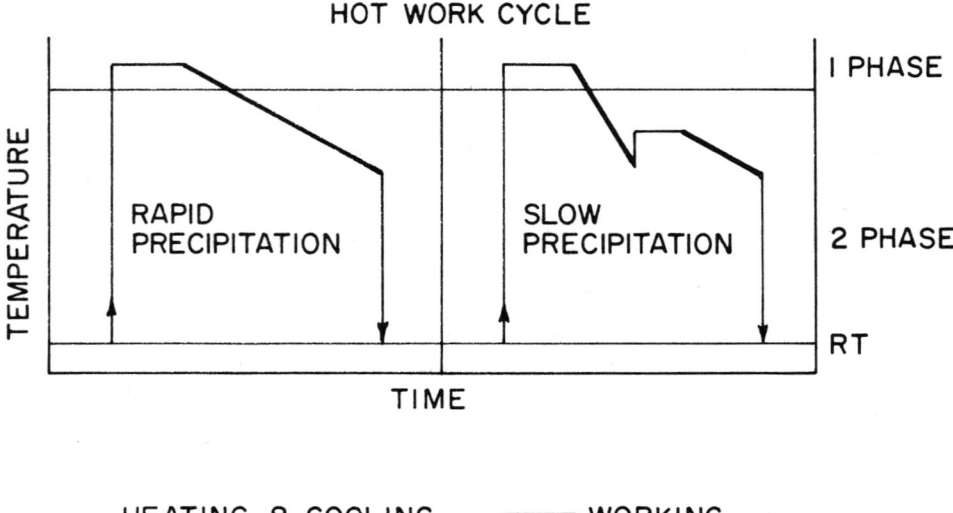

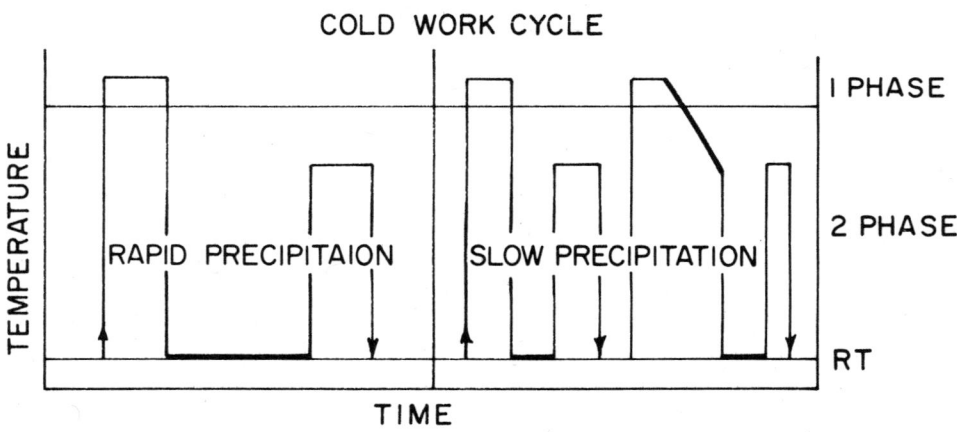

Fig. 6. Thermomechanical treatment to achieve microduplex structures in alloys which are single phase at high temperature and double phase at low temperature; optimum processing may involve several steps including hot and cold working and annealing.[15]

treated at 1150 C to form single-phase austenite, and then worked continuously by rolling at ever-decreasing temperature, concurrently with the precipitation of cementite. The final working temperature is near 550 C, which is sufficiently high to achieve spheroidization of the pearlite. The result is a very fine microstructure of spheroidized cementite in micrograin ferrite (Fig. 7).

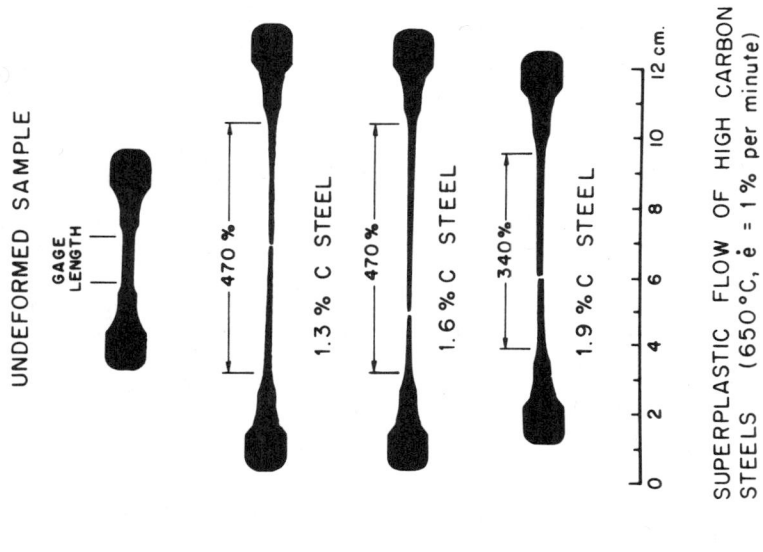

UNDEFORMED SAMPLE

GAGE
LENGTH

470 %

1.3 % C STEEL

470 %

1.6 % C STEEL

340%

1.9 % C STEEL

0 2 4 6 8 10 12 cm.

SUPERPLASTIC FLOW OF HIGH CARBON
STEELS (650°C, ė = 1% per minute)

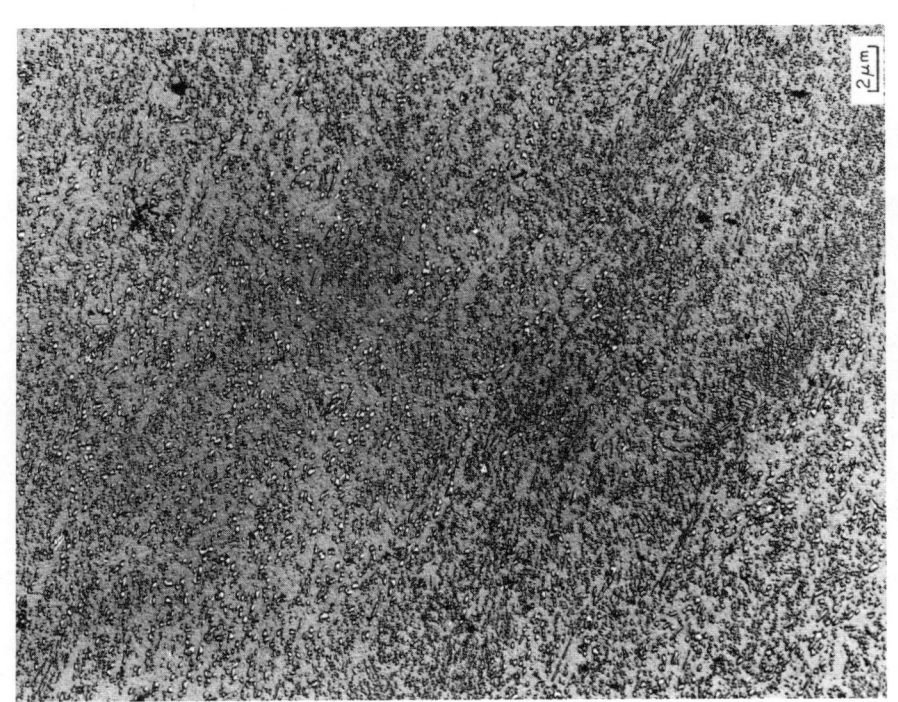

2μm

Fig. 7. Microstructure of fine-grained, 1.3 percent ultrahigh-carbon steel.[21] This structure is moderately superplastic at 650 C.

5 DEFORMATION PROCESSING

As early as 1934, Pearson showed the potential for nonconventional forming in superplastic alloys by free expansion of Sn-Bi tubes with internal pressure.[22] This work was lost, however, and modern interest follows the demonstration by Backofen et al. of the high strain-rate sensitivity of Zn-22Al.[23] Other essential material properties, subsequently established and common to some other superplastic alloys include low flow stress and resistance to fracture, all at quite high strain rate.[24]

Extensive development work on sheet thermoforming of Zn-22Al has been done by Fields and co-workers at I.B.M.[25] The various characteristics of the alloy which make it economic for this process include convenient working temperature (250 to 275 C), useful service properties similar to zinc die casting alloys, heat treatability, and low cost. Various techniques were adapted from the plastics industry to produce a variety of shapes and to help control the final sheet-thickness distribution.

From a metallurgical viewpoint, there are a number of interesting features of these techniques.[24] It was found, for example, that 50 percent cold reduction as a final step in the preparation of sheet lowers the flow stress substantially and reduces the forming time by the factor $1/4$. In addition, the final thickness is somewhat more uniform, apparently as the result of additional flow stabilization by moderate strain hardening. Material not cold rolled softens rather than hardens during superplastic extension (Fig. 8).

This procedure does not seem to have been applied to other superplastic alloys (a recent exception is low-alloy steels[26]), but a similar effect of cold work was noted in Pb-5% Cd[12]. The mechanism of easier flow may involve enhanced diffusion (pipe diffusion) along dislocations and cell walls introduced by this cold work. The mechanism should be operative in any alloy.

In certain operations, for example the drawing of a simple cup, it is useful to control the initial temperature of the sheet. (The dies are always hot.) If initially cold, the sheet will flow more readily near its rim, since this region will heat more quickly and flow at lower stress. By this means, excessive thinning of the cup at its center is prevented. Variations of the technique in terms of preheat temperature and holding time prior to forming depend on design objectives and on sheet thickness.[24]

A remarkable benefit is obtained if the dies are at a temperature *above* 275 C. During forming, the die is in contact with the sheet for an insufficient time for its temperature to exceed 275 C. Subsequently, however, the fully formed part transforms to coarse-grained α' and is *hardened* thereby, thus facilitating its removal from the die. Air cooled from this temperature, it has superior strength and creep resistance in comparison with the original (superplastic) microstructure (Fig. 9).

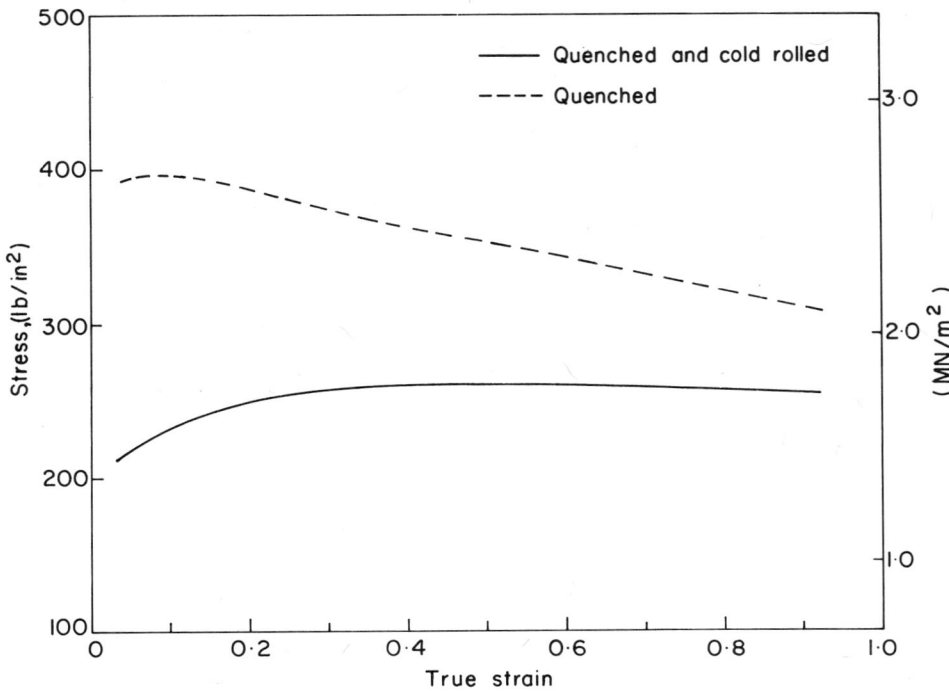

Fig. 8. Stress versus strain at constant extension rate 0.2 in./min; the cold-rolled sheet (QCR) is softer and hardens at strains up to 0.5 while the quenched sheet softens above 0.1.[24]

6 COMPRESSION FORMING

The combination of low flow stress and high ductility in superplastic alloys makes them exceptionally useful in compression forming. At the first level, such forming may be simply hot rolling, which can be accomplished readily in, for example, normally difficult two-phase austenite plus ferrite alloys, and at roll forces smaller even than those with single-phase materials at similar temperatures.[27] Such behavior, however, only hints at the remarkable formability and resistance to fracture of these alloys. In upset forging tests of Zn-22Al, Stewart[28] was able to obtain 96 percent reduction in thickness without any edge cracking (Fig. 10). Even cracks existing in the preform were reported not to grow. When forged in a closed die, a complex part may be formed in a single step and with outstanding precision of shape (Fig. 11).

The very low flow stress of this alloy at low strain rate means that such forgings can be produced in light presses, though a heavier press is, of course, required for rapid production.

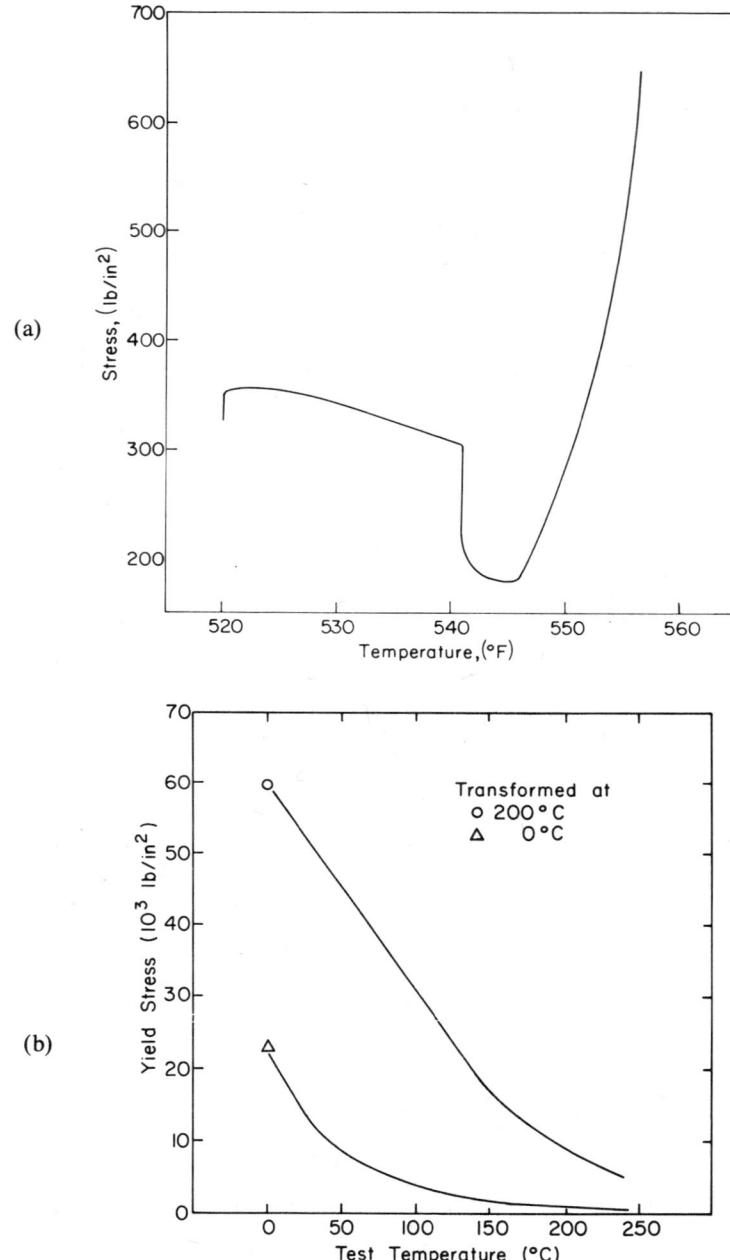

Fig. 9. (a) Yield stress on continuous heating for near-equilibrium microstructures in Zn-22 Al; a rapid hardening occurs above 290 C as a result of transformation and grain coarsening in α'.[24] (b) Yield stress versus temperature for Zn-22 Al, isothermally transformed at 200 C and 0 C; in air cooling from above 275 C, transformation occurs near 200 C (the C-curve nose) and the material is hard.[2]

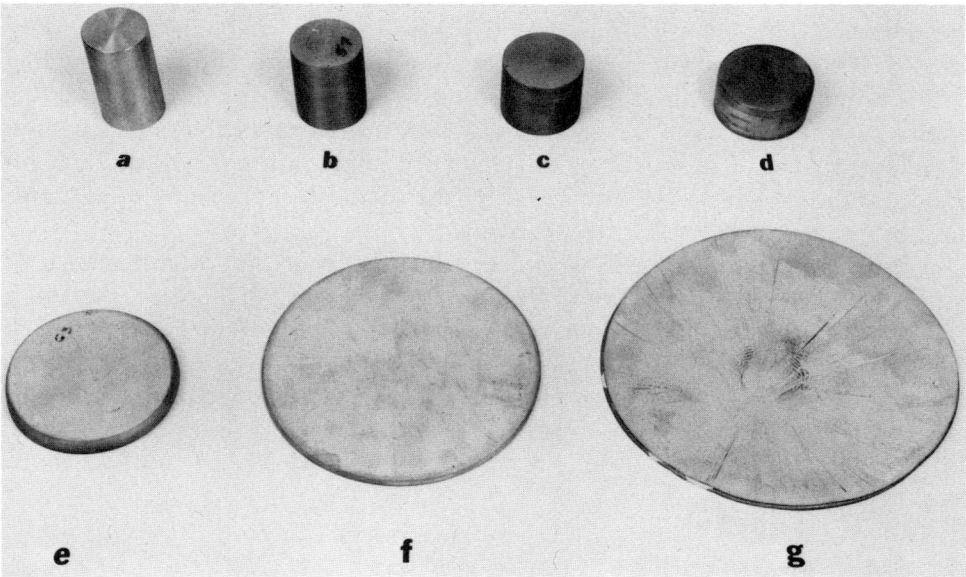

Fig. 10. Forged specimens of Zn-22Al after various height reductions at 260 C: (a) 0 percent, (b) 22 percent, (c) 38 percent, (d) 57 percent, (e) 82 percent, (f) 93 percent, and (g) 96 percent. Initial sample is 1.4 inches in diameter and 2.5 inches in height.[26]

Fig. 11. Single-step closed die forging of complex part from Zn-22Al at 260 C; forging rate was 0.5 in./min, maximum load 100 tons, and base diameter of the finished part about 2.5 inches.[26]

The resistance of superplastic alloys to cracking is primarily the result of their extremely low flow stress and consequent difficulty of crack propagation.[29] Mechanistically, high-temperature fracture is principally intergranular, with cracks forming at triple points and voids along the boundary surface. Triple-point cracks relieve high stresses at the ends of sliding grain boundaries. Grain-boundary sliding is of course the dominant deformation process in superplastic alloys[5,13], but in this case, high stress cannot develop because of rapid slip or diffusion accommodation[10,30]. These accommodation mechanisms are enhanced by the softness of the individual grains and by the small grain size, respectively.

A final point concerns the general cleanliness and homogeneity of structure in superplastic alloys. Because they are usually hot worked in a single-phase regime, hard impurity phases, if present, will tend to dissolve or be dispersed and be rendered relatively harmless in initiation of fracture. In cases where this is not accomplished, purification by "gettering" may be used to remove the harmful impurity.[27]

7 SERVICE PROPERTIES

The mechanical properties of ultrafine-grained alloys have been discussed extensively in a recent Symposium.[31] For high-melting-temperature alloys, flow, fracture, and fatigue strength increase in accordance with an inverse square-root law to the finest grain size. Toughness is exceptional. Low-melting-point materials must be considered separately, as previously noted. In this case, creep resistance is poor even at ambient temperatures, unless the microstructure is altered to a coarser grain. This can usually be done by means of a final-step heat treatment.

REFERENCES

1. Zehr, S. W., and Backofen, W. A., *ASM Trans. Quart.,* **61**, 300 (1968).
2. Alden, T. H., and Schadler, H. W., *Trans. Met. Soc. AIME,* **242**, 825 (1968).
3. Lee, D., *Acta Met.,* **17**, 1057 (1969).
4. Holt, D. L., and Backofen, W. A., *ASM Trans. Quart.,* **59**, 755 (1966).
5. Alden, T. H., *Acta. Met.,* **15**, 469 (1967).
6. Hayden, H. W., and Brophy, J. H., *ASM Trans. Quart.,* **61**, 542 (1968).
7. Johnson, R. H., *Design Eng.* (March, 1969).
8. Davies, G. J., Edington, J. W., Cutler, C. P., and Padmanabhan, K. A., *J. Mater. Sci.,* **5**, 1091 (1970).
9. Baudelet, B., *Mem. Sci. Rev. Met.,* **68**, 479 (1971).
10. Ashby, M. F., and Verrall, R. A., *Acta Met.,* **21**, 149 (1973).
11. Sherby, O. D., and Burke, P. M., *Prog. Mater. Sci.,* **13**, 325 (1967).
12. Alden, T. H., *ASM Trans. Quart.,* **61**, 559 (1968).

13. Alden, T. H., in *Plastic Deformation of Materials,* Arsenault (Ed.), Academic Press, New York (1975), p. 225.
14. Garwood, R. D., and Hopkins, A. D., *J. Inst. Met.,* **81,** 407 (1952–53).
15. Gibson, R. C., and Brophy, J. H., in *Ultrafine Grain Metals,* Syracuse University Press, Syracuse, New York (1970), p 377.
16. Schadler, H. W., *Trans. Met. Soc. AIME,* **242,** 1281 (1968).
17. Jin, S., Morris, J. W., Jr., and Zackay, V. F., *Met. Trans. A.,* **6A,** 141 (1975).
18. Cline, H. E., and Alden, T. H., *Trans. Met. Soc. AIME,* **239,** 710 (1967).
19. Hayden, H. W., Gibson, R. C., Merrick, H. F., and Brophy, J. H., *ASM Trans. Quart.,* **60,** 3 (1967).
20. Snape, E., and Church, N. L., *J. Metals,* **24,** 23 (January, 1972).
21. Sherby, O. D., Walser, B., Young, C. M., and Cady, E. M., "Superplastic Ultrahigh Carbon Steels", *Scripta Met.,* **9** (5), 569–573 (May, 1975).
22. Pearson, C. E., *J. Inst. Metals,* **54,** 111 (1934).
23. Backofen, W. A., Turner, I. R., and Avery, D. H., *ASM Trans. Quart.,* **57,** 981 (1964).
24. Fields, D. S., Jr., and Stewart, T. J., *Int. J. Mech. Sci.,* **13,** 63 (1971).
25. Fields, D. S., Jr., and Hubert, J. F., *Superplastic Metal Forming,* IBM Corporation, Endicott, New York (1975). See also *Met. Eng. Quart.,* **13,** 1–20 (November, 1973).
26. Stewart, M. J., *Met. Trans.,* **6A** (8), 1672–1674 (August, 1975).
27. Gibson, R. C., Hayden, H. W., and Brophy, J. H., *ASM Trans. Quart.,* **61,** 85 (1968).
28. Stewart, M. J., *Can. Met. Quart.,* **12,** 159 (1973).
29. Hart, E., in *Ultrafine Grain Metals,* Syracuse University Press, Syracuse, New York (1970), p. 247.
30. Alden, T. H., *J. Australian Inst. Met.,* **14,** 207 (1969).
31. *Ultrafine Grain Metals,* Syracuse University Press, Syracuse, New York (1970).

DISCUSSION on Paper by T. H. Alden

SHERBY: Bement's discussion on materials exhibiting high strain-rate sensitivity where m values of unity are observed is interesting. I would like to suggest that this type of potential superplastic behavior should be classified as internal-stress superplasticity. This type of superplastic behavior includes such examples as creep flow (1) under cyclic phase transformation in allotropic metals and (2) under thermal-cycling conditions for polycrystalline metals that exhibit high anisotropy in their thermal-expansion coefficients. In these cases, it can be shown that the presence of an internal stress can lead to very high values of m, even exceeding unity. Alden, in his paper, considered only fine-structure superplasticity, where one apparently relies on grain-boundary shearing as a principal mode of deformation to obtain high strain-rate sensitivity. In this latter case, the strain-rate sensitivity exponent seems to center around m $\simeq$ 0.5. Experimental results have revealed that such a value persists over three orders of magnitude of strain rate (e.g., Ni-Fe-Cr microduplex alloys and Al-Zn). On a related subject, I would like to mention that Rai and Grant recently published a paper where they proposed that Region I (of the logarithm stress versus logarithm strain-rate curve) does not represent a region of constant structure. Specifically, they suggest and show evidence that the low value of m observed here was not meaningful and was simply due to grain growth occurring at low stresses (long-time tests), leading to higher strengths than the strength predicted from an extrapolation of the line associated with m ≃ 0.5. I believe their new results make analyses of the behavior of fine-grained superplastic materials a bit easier than thought earlier.

Alden's thoughts on how to enhance the superplastic effect are very interesting. He presented the argument that if we can inhibit slip-creep processes and enhance grain-boundary sliding, then the range of strain rates where superplastic flow can be observed is extended. This consideration is important if superplastic materials are to be used extensively in technological applications. Typically, the maximum rate of superplastic flow, before ordinary slip creep takes over, is 1%/min. If we can extend the range of superplastic behavior to strain rates of 100 to 1000%/min, then the economics of superplastic forming will be enhanced manyfold. One aspect that does not appear to have been considered in the past to enhance superplasticity is the nature of the second phase in two-phase superplastic materials. Thus, if one can change the composition and properties of the second phase in such a way that slip creep is inhibited and grain-boundary shearing is not adversely affected, then the transition strain rate can be increased.

ALDEN: There is no question but that grain growth occurs during the deformation of superplastic alloys, particularly if they have ultrafine grain sizes (near 1 μm). This effect can produce a pseudo stage 1 and increase the stress in stages 2 and 3, leading to an erroneously large m value, as has been shown by Rai and Grant.[1] However, in the Pb-Sn eutectic alloy, which has been extensively studied, grain growth is slow and barely affects the shape of the log stress—log strain-rate curve, yet this curve clearly exhibits three-stage strain-rate hardening.[2] The best data are taken from constant-stress creep tests.[3,4] In stage 1, the creep rate decreases very slowly with time (~ x 2 in 25 to 50 hours), probably as a result of grain growth. However, the data points[3] are the *initial* creep rates. In stage 2, the creep rate is constant, independent of prior creep history; the stress-strain rate points are thus unambiguously determined. In stage 3, primary creep is observed, indicative of ordinary strain hardening (by accumulation of dislocations). Here, steady-state strain rates are plotted, and these rates are somewhat below those determined in Instron tests.[2] In this latter case only is the strain variation of properties a significant issue.

1. Rai, G., and Grant, N. J., *Met. Trans.,* **6A**, 385 (1975).
2. Cline, H. E., and Alden, T. H., *Trans. Met. Soc. AIME,* **239**, 710 (1967).
3. Alden, T. H., in *Plastic Deformation of Materials,* Arsenault (Ed.), Academic Press, New York (1975), p. 225.
4. Surges, K., M.A.Sc. Thesis, University of British Columbia (1969).

This plot and other log stress—log strain-rate plots determined for coarser-grained samples in which grain growth should be slow[1], show S-shaped curves and m_{max} values exceeding 0.7. In my opinion, the idea that $m = 0.5$ is a "typical" rate sensitivity for superplastic alloys remains uncertain[2] and that it is a maximum value is wrong. For the purpose of further evaluation of the fundamental plasticity properties of superplastic alloys, constant-stress creep tests are clearly superior to constant crosshead speed tests and should be employed.

BEMENT: You did not discuss in your presentation the influence of anisotropy on superplastic behavior. This may be an important consideration in the behavior of Zircaloy fuel cladding in nuclear service and may account for both observed plastic strains greater than the predicted strain limit and lower than predicted failure rates. In this system, "soft" grains would be surrounded by "hard" grains from the standpoint of texture constraints on dislocation creep. In addition, irradiation hardening would further impede dislocation creep but would provide high point-defect supersaturations for grain-boundary creep, but at service temperatures less than $0.5T_m$. Experiments have shown irradiation creep to be Newtonian in nature and to exhibit an enhanced strain-rate sensitivity. Would you please comment on the likelihood of a superplastic response in anisotropic materials under such conditions?

ALDEN: Please see the comments by Sherby.

HIRTH: With regard to the equation describing superplastic deformation, there is the difficulty with the stress exponent in the Verrall-Ashby model as noted by Sherby. In addition, it would appear that coalescence would tend to be promoted by enhanced α-α and β-β contact, so perhaps a reseparation aspect could be considered in the model. It is worth noting that there is a model by Chaudhari involving pileup and diffusional relaxation of grain-boundary dislocations which gives the strain rate proportional to σ^2 and L^{-3}, which is of the empirical form noted by Sherby.

ALDEN: The Ashby-Verrall model was formulated for a pure metal or single-phase alloy. Although new grain junctions will be formed during strain, these will not lead to coalescence unless the orientations happen to coincide. This coincidence may occur occasionally and lead to slow grain

---

1. Alden, T. H., *Acta Met.*, **15**, 469 (1967).
2. Alden, T. H., in *Plastic Deformation of Materials*, Arsenault (Ed.), Academic Press, New York (1975), p. 225.

growth, which does occur. However, a study of growth in Sn-1% Bi indicated that enhanced diffusion was a more likely cause.[1]

Chaudhari's model is one of a group of recovery creep models which
have been applied to superplasticity. These models suffer from a variety
of faults, in that they predict deformation characteristics which are found
usually in coarse-grained alloys but not in superplastic alloys.[2] In an ad
hoc fashion, such models can predict rate-sensitive deformation, but
strong inverse grain-size dependence of the strain rate can be achieved
only by extreme and artificial assumptions.[3]

CAMPBELL: The relatively low range of strain rates over which the rate
sensitivity is high enough to give superplastic behavior is a serious limitation with regard to possible production processes. The rate sensitivity of
many alloys increases at very high rates, when dislocation damping
mechanisms are brought to operate, so that superplasticity might be attainable at such rates. However, a recent investigation[4] of fine-grained
lead-tin eutectic showed that at shear strain rates of order 10^3 sec^{-1}, the
behavior was nonsuperplastic.

1. Clark, M. A., and Alden, T. H., *Acta Met.,* **21**, 1195 (1973).
2. Alden, T. H., in *Plastic Deformation of Materials*, Arsenault (Ed.), Academic Press, New York (1975), p. 225.
3. Alden, T. H., *Acta Met.,* **17**, 1435 (1969).
4. S. Cydens and J. D. Campbell, *Inst. Phys. Conf. Series*, No. 21 (1974), p. 62.

METALLIC GLASSES

L. A. Davis

Materials Research Center
Allied Chemical Corporation
Morristown, New Jersey

ABSTRACT

A new area of materials technology was inaugurated with the recent discovery that one may produce filaments of iron-, nickel-, and/or cobalt-base metallic glasses rapidly and continuously from the melt. These materials offer great promise for technological utilization because of their combination of cost effectiveness and, for example, unique magnetic or mechanical proprties. The compositions, formation, and structure of metallic glasses are briefly reviewed. Major emphasis is on their distinctive mechanical behavior.

1 INTRODUCTION

Metallic glasses are formed by quenching from the liquid (or vapor) so rapidly that crystallization is bypassed. Synthesis of such materials from the melt was first reported by Klement and co-workers[1,2] in 1960 and 1963 for a $Au_{75}Si_{25}$ alloy. This pioneering work led to major efforts over the next decade to synthesize metallic glasses and to characterize their structures and properties. The advent of technological interest occurred in the last few years, coincident with the demonstration that one can produce continuous, uniform cross section ribbons and wires of these materials at high rates (up to ~1800 m/min) by rapid quenching directly from the melt.[3] In addition, alloys amenable to this process based on relatively inexpensive elements such as iron, nickel, and cobalt were identified.[4]

It has been demonstrated that glassy metallic alloys exhibit unique electrical[5], magnetic[6], chemical[7], and/or mechanical properties. With the emphasis of the present colloquium on structural alloys, this paper focuses on the mechanical properties of metallic glasses. A brief review is given of the compositions, formation conditions, and structure of these alloys. The reader is referred to Jones[8], Giessen and Wagner[9], and Cargill[10] for further background information.

2 KINETICS, ALLOYS, AND STRUCTURES

The formation of a metallic glass is largely governed by kinetics; in principle, any liquid may be cooled to the metastable glassy state if the rate of cooling is high enough.[11] Assuming that no crystal nuclei exist in a melt, the required cooling rate is minimized by high solid/liquid interfacial energy γ, high entropy of fusion β, and high viscosity τ (see, e.g., Turnbull[12]). According to Jones[8] it is uncertain how γ varies among various materials, but the tendency for glass formation certainly increases in going from metallic to molecular materials coincident with large increases in β and τ. Glass formation is also favored by a high ratio of glass transition temperature to melting point (T_g/T_m).[8] Hence, any relative decrease of T_m as at a eutectic composition is favorable.

Elemental metallic glasses have been synthesized by vapor deposition on substrates cooled to 4.2 K. However, such glasses exist only at low temperatures, e.g., the crystallization temperature for cobalt is ~40 K.[13] Metallic glasses metastable to temperatures well above ambient (typically of the order of 400 C) always contain major portions of at least two elements, either two metals, e.g., $Cu_{50}Zr_{50}$[14] (subscripts in atomic percent) or a metal and a metalloid, e.g., $Pd_{80}Si_{20}$[15]. These systems, particularly the latter type, have unusually stable liquid phases, i.e., deep eutectics, owing to a strong interaction between the constituent elements.[16] Alloying may also increase T_g.[16]

For practical purposes, it is the near-eutectic alloys which are accessible to glass formation from the melt with current rapid quenching ($\simeq 10^6$ K/sec) technology. While $TM_{\sim 80}X_{\sim 20}$ eutectics based on iron, nickel, and cobalt exist (TM = transition metal and X = metalloid), the corresponding binary glasses have not been synthesized by melt quenching; glasses of technical interest are ternaries, quaternaries (e.g., $Ni_{40}Fe_{40}P_{14}B_6$: METGLAS[R] 2826*), or higher.

Although glass formation is primarily a function of cooling kinetics, one can present structural criteria for the observed glassy alloys. For example, Nowick and Mader[17] attributed the stability of concentrated metal-metal alloys ($\sim A_{50}B_{50}$) to a significant atomic size difference of the constituent atoms, which may allow more efficient filling of space[18,19]. It may also be argued (see, e.g., Polk[20,21]) that $TM_{80}X_{20}$ alloys are stabilized by high atomic packing density and by compositional short-range order. Cargill[19] has demonstrated that the Bernal[22,23] model [dense random packing (DRP) of hard spheres] produces a good fit to the radial distribution functions of several glassy alloys dominated by a single element. However, DRP is at least 14 percent less dense than crystalline close packing, while metallic glasses are typically only 1 to 2 percent less dense than their corresponding crystalline phases.[10] Polk suggested, therefore, that a model, having the correct density, appropriate for $TM_{80}X_{20}$ could be generated by placing the metalloid atoms in the larger polyhedral holes of a DRP metal structure. The short-range order of such a structure is then similar to that in the equilibrium intermetallic phases, e.g., Pd_3Si, which leads to a glassy alloy with a high degree of composition short-range order. As noted below, this probably plays an important role in the highly heterogeneous plastic flow in these materials. It is interesting to note also that the strong TM-X interaction which favors short-range order in the glass also favors comigration, e.g., of Ni-P or Mn-P, to grain boundaries in steels, leading to temper embrittlement.[24]

3 MECHANICS

The requirement for rapid heat extraction places a constraint on the form of glassy material produced, i.e., at least one dimension must be small. Hence thin ribbons and small-diameter wires are a natural product; in the following we consider in detail the mechanics of these filaments.

3.1 Ductility

Glassy-alloy filaments or splats can be severely deformed by cold rolling without cracking. This was apparently first noted by Duwez.[25] Pampillo

---

*Registered trademark, Allied Chemical Corporation, Morristown, New Jersey.

and Chen[26] reported that 2-mm-diam. $Pd_{77.5}Cu_6Si_{16.5}$ rods may be cold rolled to strip $\sim$0.1 mm thick. More recently, Takayama and Maddin[27] achieved up to 40 percent reduction on rolling Ni-Pd-P strips.

Ductility in bending is also evidenced by glassy-alloy specimens.[27-29] One may, for example, place a small loop of strip in the gap between the platens of a micrometer and close the gap so that the strip is pressed back flat against itself. Gilman[30] has noted that a strip of steel shim stock of similar strength and dimensions will not sustain such a sharp bend. It is well known, of course, that silica-glass filaments cannot be sharply bent. In the first case, the glassy alloy is presumably superior because it is a homogeneous elastic continuum. In principle, it contains no hard particles, voids, or phase boundaries where local triaxial stresses can arise and promote crack initiation. In the later case, the strongly directional bonding of SiO_2 precludes ductile behavior.

On loading in uniaxial compression, gross plastic deformation may also be achieved[26,31,32]; Pampillo and Chen[26] effected height reductions on the order of 40 percent in small cylinders of $Pd_{77.5}Cu_6Si_{16.5}$. In this case, the alloy is so highly stable that rods $\sim$2 mm in diameter can be quenched to the glassy state.

In each case noted—rolling, bending or compression—the mode of deformation is inhomogeneous, occurring in the form of highly localized shear deformation bands. These bands occur on planes of maximum shear stress, i.e., on planes whose normals lie at $\pm$45 degrees to the compression or rolling axis or on planes parallel to the bending axis. Figures 1 and 2 show the tension surface shear steps and the shear offset profile, respectively, of bent strips. The width of the shear bands is not resolvable on such micrographs. Using electron-microscope replicas, Masumoto and Maddin[33] found them to be $\sim$200 A thick in $Pd_{80}Si_{20}$. As corresponding shear offsets of $\sim$2000 A were found, these bands contain enormous strain.

Macroscopic ductility is achieved in uniaxial compression, for example, because multiple glide bands are generated. On loading in uniaxial tension, deformation is concentrated in a singularly intense shear band and failure occurs at low elongation by shear rupture. The true strain to failure remains large, however.

3.2 Strength

3.2.1 Tension

Metallic glasses are distinguished by exceptional strength. For alloys based on palladium, tensile strengths of the order of 150 kg/mm^2 are observed[32], while for those based on nickel and/or iron, tensile strengths of

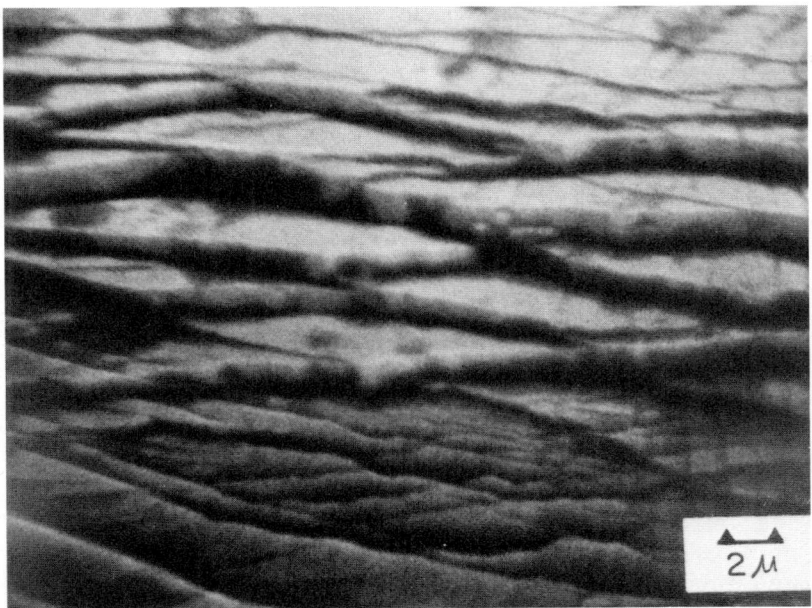

Fig. 1. Shear offsets on the tension side of a bent Ni-Fe base metallic glass strip. Scanning electron micrograph (SEM) courtesy of C. A. Pampillo.

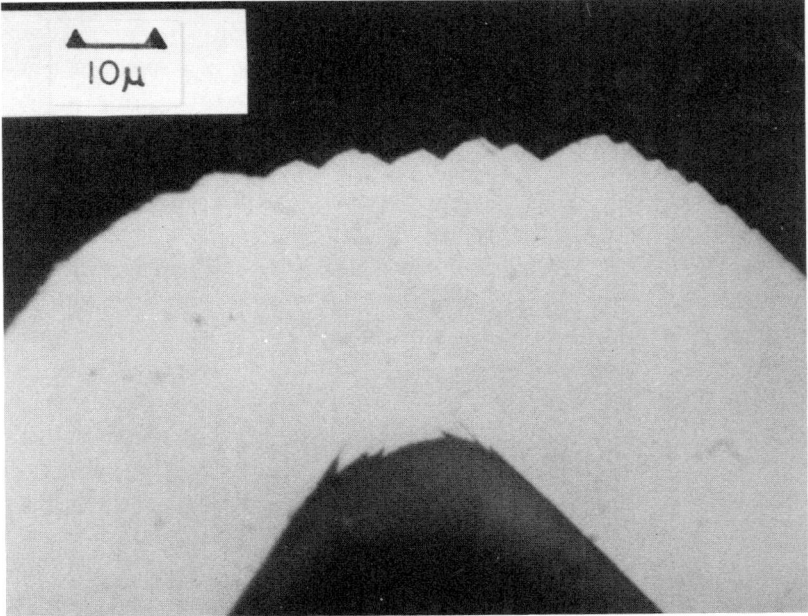

Fig. 2. Shear offset profile of a bent metallic glass strip.[34]

the order of 200 to 300 kg/mm^2 are observed[34-38]. There appears to be some confusion in the literature, however, concerning the nature of the tensile deformation/failure behavior. Yield stresses as low as one-half the ultimate tensile (fracture) strength have been reported.[28,33,38,39] Such stresses are estimated from a departure from linearity of tensile load-elongation (L-e) curves. It has been the author's experience, however, that the L-e curve may be linear or nonlinear (monotonic load increase in the nonlinear case), apparently depending on the peculiarities of the specimen-grip interaction and configuration.[32] Moreover, for all data reported, save for those noted below, the tensile stress was determined for specimens which failed in the antiplane-strain shear mode (Mode III; see below). This failure mode, also classified as "low energy tear"[40], is typically observed in relatively thin sheets of polycrystalline metals. It invariably occurs at a stress below the general yield stress of the material, where general yielding is defined, by Knott[41], as the condition where "it is no longer possible to trace a path between opposite loading points through elastically deformed material only". At the risk of belaboring this point, it may also be noted that such low yield stresses would imply strong work hardening in these materials. The occurrence of highly inhomogeneous shear flow argues against this; Cottrell[42] notes that even a trace of work hardening will preclude intensely localized deformation. Argon[43] cites negative work hardening as a condition for shear localization. Clearly work hardening must be absent in these materials. We conclude that the tensile failure stresses reported in the literature are typically less (on the order of 10 to 20 percent) than the yield stress of the material.

Tensile yielding may be observed in glassy alloy specimens when the tearing mode of failure is suppressed, as in the case of wires[32] or low-aspect-ratio strips.[37] Failure of a glassy wire occurs through a zone of maximum shear stress at ~45 degrees to the tensile axis.[32] In the case of a sheet, plasticity theory[44] predicts that necking will occur in a zone whose normal is perpendicular to the thickness vector and at $\theta = 35.3$ degrees to the tensile axis. According to Argon[45], θ will increase slightly if the ratio σ_y/E is large, where σ_y and E are the tensile yield stress and Young's modulus, respectively. For metallic glasses, σ_y/E is exceptionally large (~0.02) so that $\theta \simeq 37$ degrees, which corresponds to the failure angle observed for specimens with a hand-polished, low-aspect-ratio gage section.[37] The mode of failure is shear rupture (see below). Because of the propensity of these materials for shear deformation, it is presumed that coincident with imperceptible necking, unstable shear and shear rupture occur. Hence yielding and failure are coincident and σ_y is equal to the tensile failure stress. Table I lists σ_y for those alloys for which it is known.

Table I. Mechanical Properties of Metallic Glasses

Alloy	H, kg/mm^2	σ_y, kg/mm^2	H/σ_y	E, 10^3 kg/mm^2	E/σ_Y
$Ni_{36}Fe_{32}Cr_{14}P_{12}B_6$ (METGLAS^R 2826A)	880[46]*	278 tension[47]	3.16	14.5[46]	52
$Ni_{49}Fe_{29}P_{14}B_6Si_2$ (METGLAS^R 2826B)	792[48]	243 tension[48]	3.26	13.2[46]	54
$Pd_{77.5}Cu_6Si_{16.5}$	498[32]	157 comp.[26] 147 tension[32]	3.17	8.97[46]	57
$Pd_{64}Ni_{16}P_{20}$		158 comp.[31]			

*References listed at end of paper.

3.2.2 Compression

Pampillo and Chen[26] have shown that shear bands activated by compressive loading of $Pd_{77.5}Cu_6Si_{16.5}$ remain active, independent of any geometrical instability associated with the shear displacement. This was demonstrated on a specimen which was first compressed to initiate deformation and then repolished to produce smooth lateral surfaces. By photographing the specimen before and after polishing it was evident that the same deformation bands were active on reloading. This is direct evidence of the lack of work hardening in these materials. Hence in the curve of Fig. 3 the compressive yield stress is the plateau stress. The roundness of the curve at low strain no doubt derives from extrinsic factors such as slight misalignment of the specimen ends. We note, for example, that the most intense shear bands observed by Pampillo and Chen (ref. 26, Figs. 5 and 6), presumably the first nucleated, appear to occur at an edge of the specimen in contact with a loading platen.

Values of the compressive yield stress (σ_{yc}) for two alloys are shown in Table I. It will be noted that σ_{yc} is slightly greater than σ_{yt} for $Pd_{77.5}Cu_6Si_{16.5}$, indicating a small hydrostatic pressure or normal stress effect on yielding. The existence of this effect was confirmed[32] by interrupted compression tests at ambient pressure and 6.2 kb; $d\ln\sigma/dP \simeq 5 \times 10^{-3}/kb$.

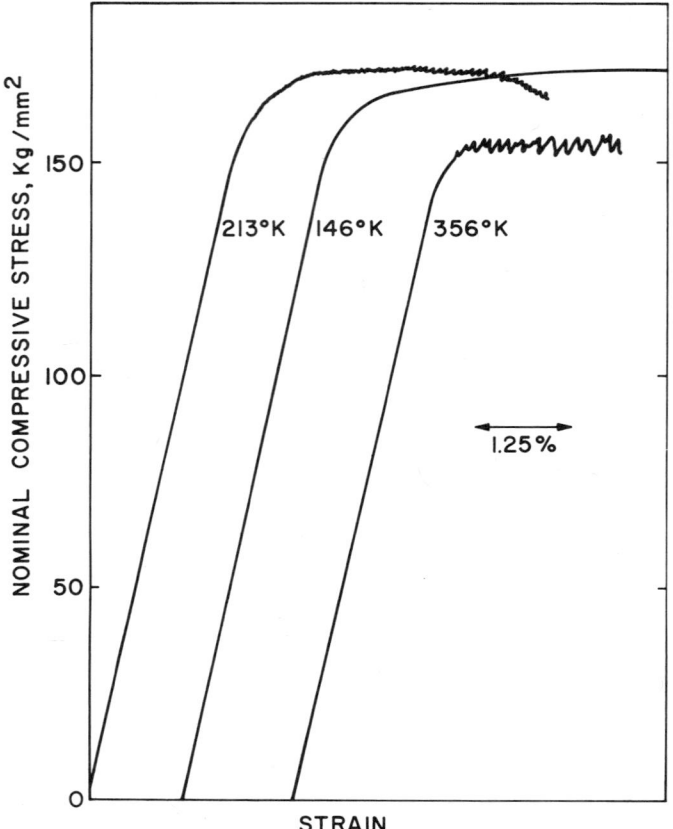

Fig. 3. Compression stress-strain curves at different temperatures for Pd$_{77.5}$Cu$_6$Si$_{16.5}$.[26]

3.2.3 Hardness

Metallic glasses possess extreme hardness, consistent with their high strength.[48] Table I contains hardness values (H) for three of the glasses listed, from whence it is evident that the ratio H/$\sigma_y \simeq 3.2$. This value is about equal to that typically observed for polycrystalline metals[49] and is predicted approximately (for the two-dimensional case) by slip-line field analysis.[50] The ratio predicted for covalently bonded glasses[51,52] is less (decreasing with increasing σ_y/E) because of the relative propensity of such glasses for elastic behavior as opposed to plastic behavior.

3.2.4 Flow Mechanism

The observation of deformation-induced surface slip bands has been cited as evidence for the existence of dislocations in crystalline materials.[53] Similarly, the observation of such bands infers the existence of dislocations in metallic glasses. More precisely, evidence is provided by the observation that shear bands terminate within glassy-alloy specimens.[26] The discontinuity between sheared and unsheared material, first considered as an elastic distortion by Volterra[54], is, by definition, a dislocation. Gilman has proposed[30,55-57] that atomic size dislocations exist in glassy materials; hence one can imagine that a shear band is composed of a pileup of microscopic dislocations. The slip vector of such a dislocation will fluctuate in direction and magnitude along the dislocation line[30,55-57] but its mean value will be dictated by some structural parameter, such as the average nearest-neighbor distance in the glass. Such dislocations are invisible in the electron microscope because the structural periodicity required for diffraction contrast is absent. However, if they exist, one expects that their long-range elastic stress fields would produce an internal stress. The flow properties of metallic glasses under ambient conditions appear dominated by such a stress field, i.e., in a range of ~200 C about ambient; the stress for plastic flow (or failure stress) varies in proportion to the elastic stiffness.[26,36,58]

The stress required for dislocation motion (yield stress) in a metallic glass has been estimated by Li[59] for the case of a dislocation in a dislocation lattice. The estimated shear stress $\tau = \mu/80$, where μ is the shear modulus, is within a factor of 2 of the observed value. Gilman[30] has suggested that the yield stress is determined by dilatation at the dislocation core generated when the dense, random-packed structure is disturbed as atoms move past one another. He derives the expression

$$\sigma_y = [8\pi\epsilon^2/(1+\alpha)]B \quad ,$$

where ϵ is the strain perpendicular to the direction of motion, B is the bulk modulus, and $\alpha = 3B/4\mu$. The value of ϵ may be estimated[30] to be ~0.04. Using representative values[60] of B and μ for $Pd_{77.5}Cu_6Si_{16.5}$, one then finds $\sigma_y \simeq 140$ kg/mm^2, in reasonable agreement with the data given in Table I.

Correlated shear events may be expected in an amorphous material if the structural relaxation time is long, i.e., when $T \ll T_g$. When some structure exists, the effect will be enhanced, e.g., when a local shear event occurs which tends to destroy compositional short-range order in a metallic glass, the probability of an additional event occurring in the same vicinity is increased.[56] A series of local events occurring at a flow front constitutes a dislocation. Polk and Turnbull[61] were the first to suggest that the destruction of short-range order would favor inhomogeneous plastic flow. Pampillo and

Chen[26] later showed that shear bands in $Pd_{77.5}Cu_6Si_{16.5}$ can be preferentially etched. This indicates a change in local structure. The etching susceptibility is eliminated by an elevated-temperature anneal ($\sim$50 C below T_g). In these materials, such a treatment is likely to increase order, supporting the suggestion that order is destroyed in regions where plastic flow is concentrated.

3.3 Fracture

3.3.1 Axial Symmetry or Plane-Stress Conditions

Tensile failure of glassy-metal wires and strips (plane stress) occurs by shear rupture through a zone of intense shear deformation. In the case of general yielding, this zone develops over the entire cross section of the specimen prior to rupture; in the case of tearing of a strip (antiplane-strain failure), the shear deformation and shear rupture events occur at a crack front which propagates across the specimen in the manner of a screw dislocation (shown schematically in Fig. 4), i.e., the crack front propagates on a plane of maximum shear stress in a direction perpendicular to the tensile axis.

In both cases, yielding and tearing, the microscopic mode of failure is ductile. It is not surprising, therefore, that the nature of the observed fracture-surface features is the same. An example is shown in Fig. 5 for a $Pd_{77.5}Cu_6Si_{16.5}$ wire. A smooth, featureless zone, corresponding to a shear offset, lies in an arc to the right of the fracture surface. The remainder of the surface is marked by a ridge or vein pattern. The observation of this pattern was first noted by Leamy et al.[28], who used stereo pair-scanning electron micrographs to show that the veins are hills, not steps, on a smooth background. Pampillo and Reimschuessel[62] confirmed the finding of Leamy et al. that the vein patterns on the opposing fracture surfaces are essentially mirror images of one another, allowing, of course, that a relative shear displacement of the two surfaces has occurred.

One may recognize, therefore, that the vein pattern is created by the interaction of locally generated shear disk cracks. Such cracks presumably nucleate at isolated imperfections such as microscopic voids or hard particles and propagate in a manner analogous to expansion of a dislocation loop. A vein or ridge is left where two such cracks collide.

3.3.2 Plane Strain

By means of electric-discharge machining one may place a small hole ($\sim$0.1-mm in diameter) in the center of a metallic glass "panel". If such a panel is then subjected to cyclic loading, fatigue cracks will nucleate at opposite sides of the hole and propagate until the critical crack length is

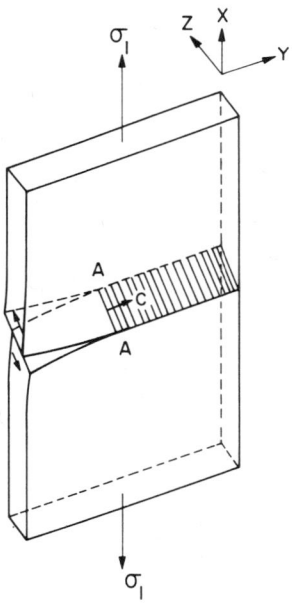

Fig. 4. Tearing mode of failure (antiplane strain, mode III) in a thin sheet; the crack front (marked AA) propagates in the direction of the arrow C.

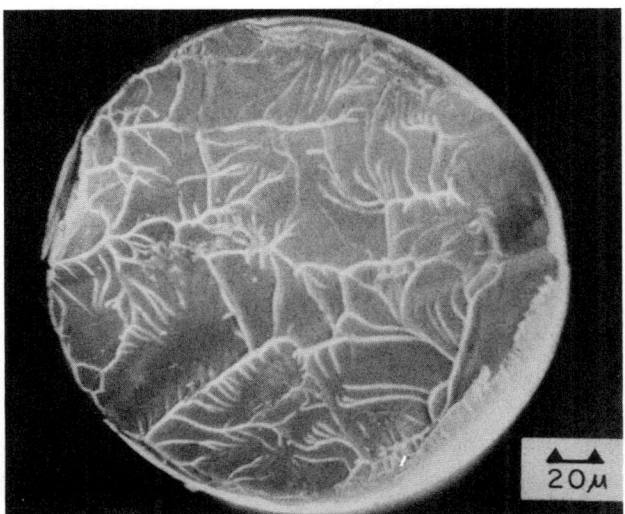

Fig. 5. Fracture surface of a $Pd_{77.5}Cu_6Si_{16.5}$ filament tested at 1 atm (SEM).[32] The direction of view is parallel to the specimen axis. The smooth shear offset region is an arc on the right of the micrograph (the "brush" markings in the shear region are an artifact caused by handling) and hence the fracture surface slopes down from left to right at an angle to the specimen axis of ~45 degrees.

reached for catastrophic failure.[63] In this case, square fracture is initially observed, i.e., the fracture surface is macroscopically at 90 degrees to the tensile axis except for shear lips at its edges (Fig. 6). Depending on the thickness of the strip, a square-to-slant (45-degree tear) transition may occur. For Ni-Fe base metallic glasses where $\sigma_y \simeq 243$ kg/mm^2 (Table I), fully plane strain conditions obtain for strips ~70 μ thick and no transition (0.8-mm-wide panel) is observed.[63] For thinner specimens such as that of Fig. 6 (~43 μ thick), a slant transition does occur (outside the field of view).

In detail, the plane-strain fracture surface is distinguished by classical chevron markings[64,65] (shown in Fig. 6 and for a different specimen in Fig. 7), the apices of which point toward the origin of the crack. These chevrons exhibit a sawtooth morphology with surfaces inclined to the tensile axis and marked by a fine-scale, equiaxed vein pattern. From this we conclude that local failure occurs by shear rupture. Under conditions of plane-strain constraint, the local yield stress is raised to three times its normal value. We expect that this accounts for the reduced scale of the observed vein pattern, as compared, for example, with that of Fig. 5.

3.4 Fracture Toughness

The fracture toughness of glassy metallic alloy specimens has been examined by Kimura and Masumoto[66] and by the author[68]. Kimura and Masumoto employed tear tests to determine the mode III fracture toughnesses of three metallic glasses. The present author examined the thickness dependence of fracture toughness in Ni-Fe base metallic glasses. Glassy alloy strips 25 μ (Ni$_{48}$Fe$_{29}$P$_{14}$B$_6$Al$_3$), 43 μ (Ni$_{39}$Fe$_{38}$P$_{14}$B$_6$Al$_3$) and 72 μ thick (Ni$_{49}$Fe$_{29}$P$_{14}$B$_6$Si$_2$) were tested in single-edge-notch (Ni$_{48}$) and center-cracked-panel (Ni$_{39}$ and Ni$_{49}$) configurations. Fracture toughnesses were calculated using the formula given by Irwin et al.[67]:

$$K_Q = \sigma\sqrt{\pi a}$$

(with a suitable multiplicative correction factor for the case of single-edge-notch geometry), where a is the crack half-length.

Fracture-toughness values for the Ni-Fe base glasses are plotted as a function of specimen thickness in Fig. 8. As the hardnesses of these glasses, and hence their yield stresses, are equivalent within experimental error, we conclude that the variation in toughness is essentially a thickness effect. Toughness is expected to increase with decreasing thickness as the state of stress at the crack tip changes from plane strain to plane stress. The data of Kimura and Masumoto for Fe$_{80}$P$_{13}$C$_7$ ($K_{IIIc} \simeq 320$ kg/mm$^{3/2}$) are plotted in Fig. 8 at ~6 μ, which approximates two times the shear lip thickness observed (see, e.g., Fig. 6) for the Ni-Fe glasses.

(a)

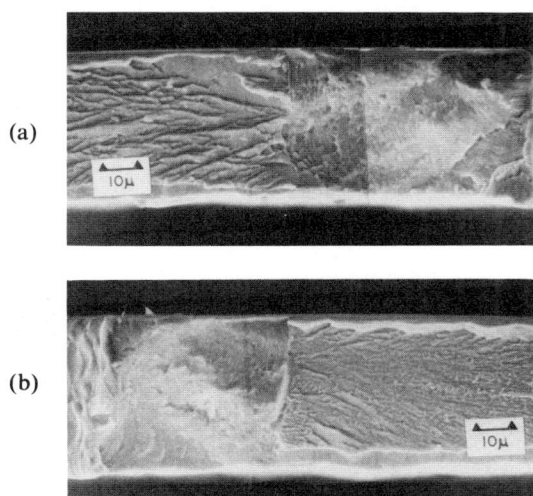

(b)

Fig. 6. Parts a and b show sections of a continuous fracture surface, located on either side of an EDM hole, of a $Ni_{39}Fe_{38}P_{14}B_6Al_3$ metallic glass specimen examined in the scanning electron microscope.[63] The edges of the hole lie at the middle of this partial montage. Adjacent to either side of the hole is the fatigue crack. A sharp boundary separates the fatigue crack and the rapid fracture surface, which lies at the outside extreme of each micrograph. A classical chevron pattern is evident on the brittle fracture surface. Shear lips lie on either side of the specimen. Except for the shear lips, the fracture surface is macroscopically at 90 degrees to the tensile axis. Outside the view of these micrographs, the fracture surface assumed a slant configuration.

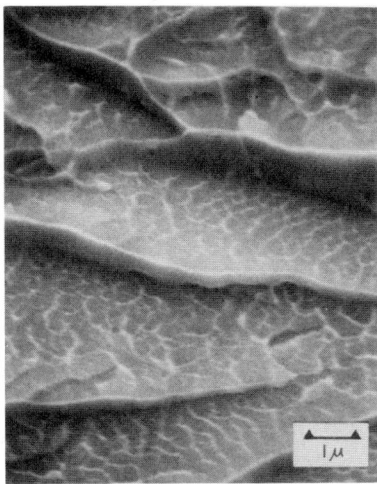

Fig. 7. A view of the chevron pattern surface for a $Ni_{48}Fe_{29}P_{14}B_6Al_3$ specimen. A fine-scale, equiaxed vein pattern is evident on the chevron surfaces. The crack propagation direction is to the left.

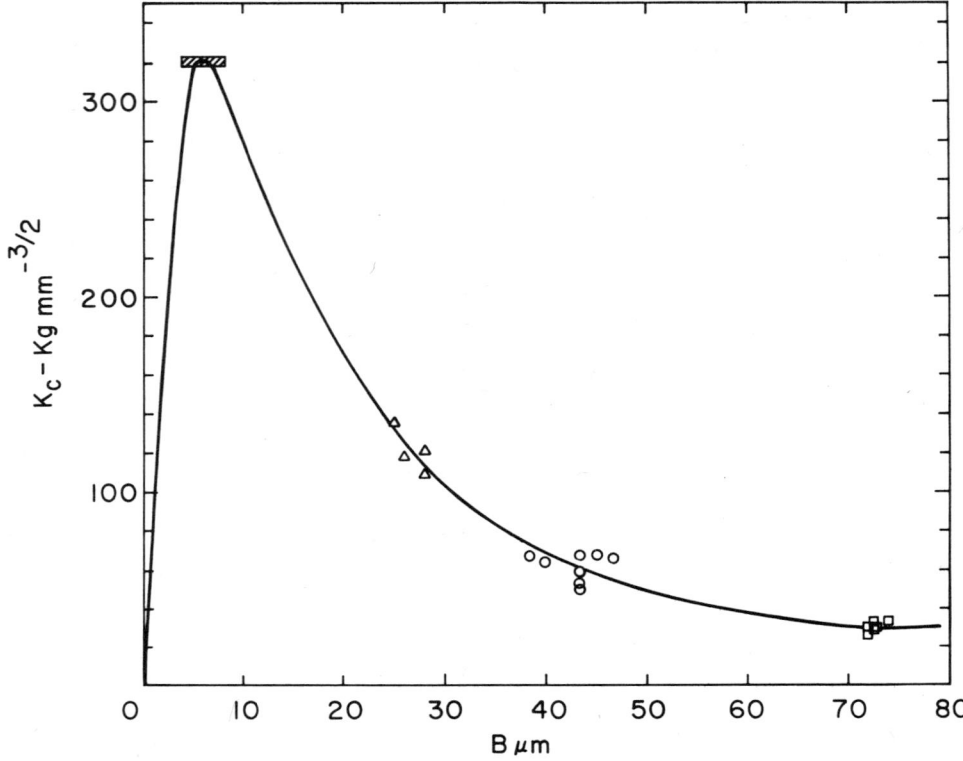

Fig. 8. Fracture toughness, K_c, as a function of thickness for various metallic glasses.[63] Triangles, circles, and squares are for the Ni_{48}, Ni_{39}, and Ni_{49} alloys (see text), respectively. The shaded bar represents data of Kimura and Masumoto[66] for $Fe_{80}P_{13}C_7$ obtained from tear tests.

One is accustomed to the use of massive plates for determination of valid plane-strain fracture-toughness (K_{Ic}) values. As a rule of thumb, Knott[68] cites the condition

$$B \geqslant 2.5 \, (K_{Ic}/\sigma_y)^2 \quad , \tag{1}$$

where B is the specimen thickness. Accordingly, low-strength steels require thick sheets. For the Ni-Fe base metallic glasses, σ_y is exceptionally large ($\sim$243 kg/mm^2) and the measured toughnesses for the 72-μ sheets are modest ($\sim$30 kg/$mm^{3/2}$). Inserting these values into Eq. (1), one finds $B \geqslant 38$ μ. Hence we conclude that $K_{Ic} \simeq 30$ kg/$mm^{3/2}$ for the Ni-Fe glasses.

To place these toughness values in perspective, one can compare them with values observed for high-strength steels. Using the yield stress as a comparative parameter, Fig. 9 shows a summary of K_{Ic} data for steels presented by Zackay et al.[69] The K_{Ic} value for the ferrous metallic glasses is as good or better than one might expect, given their high yield stresses; the comparison is even more favorable if one allows for their lower moduli (approximately two-thirds that of steel). It is a factor of potential importance for composite applications that K_{Ic} for metallic glasses is about two orders of magnitude greater than that for typical brittle-reinforcement materials.

A point of some interest is the relationship between shear ductility and fracture toughness. We noted above the superior bend ductility of metallic glasses relative to that, for example, of steel sheets. Toughness values for steels are, however, higher, which appears to be inconsistent. We note, then, that bending produces biaxial stresses; crack initiation occurs at sites where local triaxial stresses develop, i.e., at inclusions, voids, phase boundaries, etc.[70] Such sites are largely absent in glassy-alloy strips. Triaxial stresses do exist, however, near the crack tip in a plane-strain toughness test. For metallic glasses the local stress required for plastic flow is then $\sim 3Y \simeq 730$ kg/mm^2 (1.04 x 10^6 psi), which is at least one-third larger than that required for high-strength steels. At these stress levels one probaby needs little in the way of a defect to nucleate shear disc cracks ahead of the main crack which then coalesce to produce cumulative catastrophic failure.

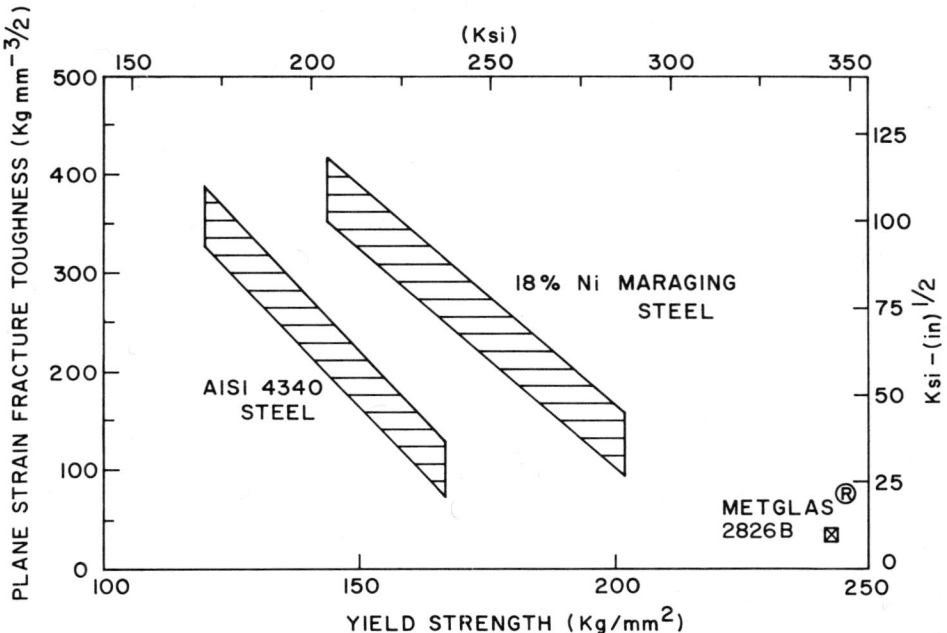

Fig. 9. Plane-strain fracture toughness versus yield stress for ferrous materials (adapted from Zackay et al.[69]).

446   L. A. DAVIS

4. CONCLUSION

Metallic glasses metastable to temperatures well above ambient always contain major portions of at least two elements, e.g., in the ratio $TM_{80}X_{20}$ (TM = transition metal, X = metalloid). Such glass compositions lie near deep eutectics where in a given system the cooling rate required for glass formation is minimized. It is these alloys which are accessible to glass formation from the melt with current rapid-quenching ($\simeq 10^6$ C/sec) technology. Alloys well removed from the eutectic require excessive critical cooling rates to suppress crystallization.

Metallic glasses are readily prepared in the form of thin strips. When such strips are bent, rolled, or indented, numerous highly localized intersecting shear bands are observed. It is this ductility which is most notably in contrast to the mechanical behaviors of silicate glasses. On compression of small cylinders, intersecting shear bands are also active and no work hardening is observed, i.e., these materials appear to approximate elastic—perfectly plastic behavior. On tensile loading, deformation occurs in a singularly intense shear band and failure occurs at low elongation by shear rupture through this band. The true shear strain at failure is of the order of 10 to 100. Whether failure occurs coincident with general yielding or by a lower stress tearing mode is a function of the aspect ratio of the sample.

Metallic glasses exhibit exceptional strength, e.g., the yield strength σ_y of $Ni_{49}Fe_{29}P_{14}B_6Si_2$ is ~243 kg/mm^2 (345 ksi). As Young's modulus for this alloy is ~13.5 x 10^3 kg/mm^2 (19 Mpsi), $\sigma_y/E \simeq 0.02$, which is typical for metallic glasses. As metallic glasses readily undergo highly localized shear deformation, they follow the classical relationship between hardness and yield strength observed for crystalline metals ($H \simeq 3\sigma_y$).

The plane-strain fracture toughness K_{Ic} observed for Ni-Fe glasses is $\simeq$ 30 kg/mm$^{3/2}$. This value appears to be consistent with those for high-strength steels if one allows for the higher yield strength and lower modulus of the metallic glass.

Because of their high strengths, high moduli (as compared with ordinary glasses), and toughness (as compared with brittle-reinforcement fibers, e.g., of graphite), it is anticipated that metallic glass filaments will find usage in matrix-reinforcement applications.

ACKNOWLEDGMENTS

The author is indebted to Drs. Donald Polk and John J. Gilman and Professor James C. M. Li for comments on the manuscript.

REFERENCES

1. Klement, W., Willens, R. H., and Duwez, P., *Nature*, **187**, 869 (1960).
2. Duwez, P., and Willens, R. H., *Trans. AIME*, **227**, 362 (1963).
3. *Chem. Eng. News*, **19**, 24 (1973).
4. Polk, D. E., and Chen, H. S., *J. Noncryst. Solids*, **15**, 165 (1974).
5. Tsuei, C. C., and Hasegawa, R., *Solid State Comm.*, **7**, 1581 (1969).
6. Egami, T., Flanders, P. J., and Graham, C. D., Jr., *Appl. Phys. Lett.*, **26**, 128 (1975).
7. Naka, M., Hashimoto, K., and Masumoto, T., *J. Japan Inst. Metals,* **38**, 835 (1974).
8. Jones, H., *Rep. Prog. Phys.*, **36**, 1425 (1973).
9. Giessen, B. C., and Wagner, C.N.J., in *Liquid Metals*, S. Z. Beer (Ed.), Marcel Dekker, New York (1972), p. 633.
10. Cargill, G. S., III, in *Solid State Physics*, F. Seitz, D. Turnbull, and H. Ehrenreich, Eds., Academic Press, New York, in press.
11. Cohen, M. H., and Turnbull, D., *Nature*, **203**, 964 (1964).
12. Turnbull, D., *Contemp. Phys.*, **10**, 473 (1969).
13. Bennett, M. R., and Wright, J. G., *Phys. Stat. Sol. (a)*, **13**, 135 (1972).
14. Ray, R., Giessen, B. C., and Grant, N. J., *Scripta Met.*, **2**, 357 (1968).
15. Crewdson, R. C., Ph.D. Thesis, California Institute of Technology, Pasadena, California (1966).
16. Turnbull, D., *Jour. Phys.*, **35**, Colloque 4, p. 1 (1974).
17. Nowick, A. S., and Mader, S., *I.B.M. J. Res. & Dev.*, **9**, 358 (1965).
18. Cargill, G. S., III, *J. Appl. Phys.*, **41**, 12 (1970).
19. Cargill, G. S., III, *J. Appl. Phys.*, **41**, 2248 (1970).
20. Polk, D. E., *Scripta Met.*, **4**, 117 (1970).
21. Polk, D. E., *Acta Met.*, **20**, 485 (1972).
22. Bernal, J. D., *Nature*, **185**, 68 (1960).
23. Bernal, J. D., *Proc. Roy. Soc.*, **280A**, 299 (1964).
24. Guttmann, M., *Surface Sci.*, in press.
25. Duwez, P., unpublished data.
26. Pampillo, C. A., and Chen, H. S., *Mater. Sci. Eng.*, **13**, 181 (1974).
27. Takayama, S., and Maddin, R., *Acta Met.,* **23**, 943 (1975).
28. Leamy, H. J., Chen, H. S., and Wang, T. T., *Met. Trans.*, **3**, 699 (1972).
29. Chen, H. S., Leamy, H. J., and O'Brien, M. J., *Scripta Met.*, **7**, 415 (1973).
30. Gilman, J. J., *J. Appl. Phys.*, **46**, 1625 (1975).
31. Chen, H. S., *Scripta Met.*, **7**, 931 (1973).
32. Davis, L. A., and Kavesh, S., *J. Mater. Sci.*, **10**, 453 (1975).
33. Masumoto, T., and Maddin, R., *Acta Met.*, **19**, 725 (1971).
34. Chen, H. S., and Polk, D. E., *J. Noncryst. Solids*, **15**, 174 (1974).
35. Polk, D. E., and Pampillo, C. A., *Scripta Met.*, **7**, 1161 (1973).
36. Pampillo, C. A., and Polk, D. E., *Acta Met.*, **22**, 741 (1974).
37. Davis, L. A., *Scripta Met.*, **9**, 339 (1975).
38. Masumoto, T., and Maddin, R., *Mat. Sci. Eng.*, **19**, 1 (1975).
39. Chen, H. S., and Wang, T. T., *J. Appl. Phys.*, **41**, 5338 (1970).
40. Tetelman, A. S., and McEvily, A. J., Jr., *Fracture of Structural Materials*, John Wiley, New York (1966), p. 94.
41. Knott, J. F., *Mat. Sci. Eng.*, **7**, 1 (1971).
42. Cottrell, A. H., *The Mechanical Properties of Matter*, John Wiley, New York (1964), p. 322.
43. Argon, A. S., in *Polymeric Materials*, ASM, Metals Park, Ohio (1975), p. 411.

44. Hill, R., *The Mathematical Theory of Plasticity*, Oxford University Press, London, England (1967), p. 300.
45. Argon, A. S., in *The Inhomogeneity of Plastic Deformation*, ASM, Metals Park, Ohio (1973), p. 161.
46. Davis, L. A., unpublished data.
47. Takayama, S., and Davis, L. A., to be published.
48. Davis, L. A., *Scripta Met.*, **9**, 431 (1975).
49. Tabor, D., *The Hardness of Metals*, Oxford University Press, London, England (1951).
50. Hill, R., *The Mathematical Theory of Plasticity*, Oxford University Press, London, England (1967), p. 213.
51. Marsh, D. M., *Proc. Roy. Soc.*, **A279**, 420 (1964).
52. Marsh, D. M., *Proc. Roy. Soc.*, **A282**, 33 (1964).
53. Hirth, J. P., and Lothe, J., *Theory of Dislocations*, McGraw-Hill, New York (1968), p. 3.
54. See A.E.H. Love, *The Mathematical Theory of Elasticity*, Dover (1944), p. 221.
55. Gilman, J. J., in *Dislocation Dynamics*, A. R. Rosenfield, G. T. Hahn, A. L. Bement, Jr., and R. I. Jaffee (Eds.), McGraw-Hill, New York (1968), p. 3.
56. Gilman, J. J., in *Physics of Strength and Plasticity*, A. S. Argon (Ed.), M.I.T., Cambridge, Mass. (1969), p. 3.
57. Gilman, J. J., *J. Appl. Phys.*, **44**, 675 (1973).
58. Pampillo, C. A., *J. Mater. Sci.*, **10**, 1194 (1975).
59. Li, J.C.M., in *Frontiers in Materials Science*, L. E. Murr and C. Stein (Eds.), Marcel-Dekker, New York (1976), p. 527.
60. Dutoit, M., and Chen, H. S., *Appl. Phys. Letters*, **23**, 357 (1973).
61. Polk, D. E., and Turnbull, D., *Acta Met.*, **20**, 493 (1972).
62. Pampillo, C. A., and Reimschuessel, A. C., *J. Mat. Sci.*, **9**, 718 (1974).
63. Davis, L. A., *J. Mater. Sci.*, in press.
64. Cottrell, A. H., *The Mechanical Properties of Matter*, John Wiley, New York (1964), p. 366.
65. Tetelman, A. S., and McEvily, A. J., Jr., *Fracture of Structural Materials*, John Wiley, New York (1966), p. 102.
66. Kimura, H., and Masumoto, T., *Scripta Met.*, **9**, 211 (1975).
67. Irwin, G. R., Kies, J. A., and Smith, H. L., *Proc. Am. Soc. Testing Materials*, **58**, 640 (1958).
68. Knott, J. F., *Fundamentals of Fracture Mechanics*, Butterworths, London (1973).
69. Zackay, V. F., Parker, E. R., Morris, J. W., Jr., and Thomas, G., *Mat. Sci. Eng.*, **16**, 201 (1974).
70. Hahn, G. T., and Rosenfield, A. R., *Met. Trans.*, **6A**, 653 (1975).

DISCUSSION on Paper by L. A. Davis

WILCOX: I have two questions: (a) Could you comment about the general aqueous corrosion behavior of metallic glasses? Do they behave as one big grain boundary or as a grain-boundary-free material? (b) Would you discuss the general characteristics of the glassy transition temperature, e.g., some typical T_g values for various alloys, the ratio of T_g to T_m (melting point), etc.?

DAVIS: With appropriate addition of chromium to Fe-Ni base metallic glasses, exceptional corrosion resistance is achieved. For example, such alloys appear to exhibit no active region in their anodic-polarization behavior in dilute H_2SO_4. The critical chromium concentration is ~8 at.%[1], as compared with 12 at.% for crystalline alloys. The observed corrosion resistance has been attributed to rapid formation of a particularly thick, chromium-containing oxide film. The homogeneity of these materials, including the lack of boundaries, is also considered to contribute to their corrosion resistance.[7]*

The ratio of the glass transition and liquidus temperatures (T_g/T_l) is of the order of 0.5 for a variety of alloys. For a low-melting alloy such as $Au_{81}Si_{19}$, $T_g \simeq 290$ K; for $Pd_{80}Si_{20}$, T_g is ~655 K. For the Ni-Fe base glasses T_g is about 700 K.[4]* The glass transition temperature is conveniently identified by calorimetry; the specific heat rises abruptly near T_g (by about 5 cal/cm K). The glass transition temperature is also related to the structural relaxation time (τ) of the material. Several degrees above T_g, $\tau \simeq 1$ second; several degrees below T_g, it is of the order of 10^3 to 10^4 seconds.

ALDEN: What is the effect of temperature on yield stress and do these alloys exhibit creep?

DAVIS: The temperature dependence of the yield stress (σ_y) of a metallic glass has been determined only for $Pd_{77.5}Cu_6Si_{16.5}$.[26]* Over a temperature range of ~200 to 500 K, σ_y appears to scale with Young's modulus, E. Below 200 K, σ_y rises more rapidly than E (or the shear modulus, μ), and above 500 K, it falls more rapidly than E. For nickel- and/or iron-base glasses, only the temperature dependence of the tearing stress $(\sigma_f$, mode III failure) has been determined.[36]* As experience indicates that σ_f scales roughly with σ_y, we can suggest that the dependence of σ_y on T for the Ni-Fe glasses is similar to that of the palladium alloy, i.e., athermal behavior obtains in the intermediate temperature region (~300 to 500 K) and thermally activated deformation plays a role at very high or low temperature.

At temperatures above the glass transition temperature (T_g) in the absence of crystallization, a glassy (liquid) alloy exhibits Newtonian viscous flow [H. S. Chen and D. Turnbull, J. Chem. Phys., **48**, 2560 (1968)]. At temperatures slightly less than T_g a regime of transient creep, followed by steady state creep, is observed. The steady state creep rate is proportional to stress (loc. cit.).

---

*See list of references given above.

TIEN: Many potential applications will open up, e.g., as first wall of the CTR, gas turbine hardware, etc., if only the crystallization temperatures of the metallic glasses can be raised to, say, about 1000 C. Do you feel that this may be possible? And if so, when will such a product be available for evaluation?

DAVIS: Known glassy alloys exhibit crystallization temperatures (T_c) ranging up to ~700 C, e.g., for binary alloys rich in niobium or tantalum (with nickel). One hopes that higher T_c glassy alloys may be synthesized, but the time scale is a matter of conjecture.

AGENDA DISCUSSION: ELEMENTS OF ALLOY DESIGN— ROLES AND LIMITATIONS

R. F. Decker*

Vice President, R&D
INCO, Inc.
New York, New York

B. A. Wilcox**

Battelle-Columbus
Columbus, Ohio

1 INTRODUCTION

A key ingredient in the design of alloys is the manipulation of two critical elements, chemical composition and microstructure, to produce alloys of desired properties for a specific application—or to produce one alloy for a multitude of applications. For example, chemical composition is varied to increase strength via solid solution or precipitation hardening, to

---

*Chairman.
**Secretary.

improve corrosion resistance, and to improve the resistance to fracture.

Metallurgists have demonstrated in countless cases that microstructure is often a major factor in alloy design. One classic example is the Hall-Petch equation which relates the flow stress, σ_y, of a metal or alloy to the grain size, d, by the relation:

$$\sigma_y = \sigma_o + kd^{-1/2} \tag{1}$$

This relation is generally applicable at ambient temperatures. Recently, a high-temperature analogue of the Hall-Petch equation has related the creep and yield strength to the degree of elongation of grains, or the grain aspect ratio. This has been demonstrated for dispersion-strengthened nickel-base alloys.[1,2] At a high temperature, say 1100 C, the stress for a given rupture life, the stress to produce a given minimum creep rate, or the slow-strain-rate tensile yield strength increases with increasing grain aspect ratio, L/ℓ, as illustrated schematically in Fig. 1. In the linear (solid line) portion of the graph, the high-temperature strength is given by

$$\sigma = \sigma_e + K(L/\ell - 1) \quad , \tag{2}$$

where σ_e and K are as defined in Fig. 1.

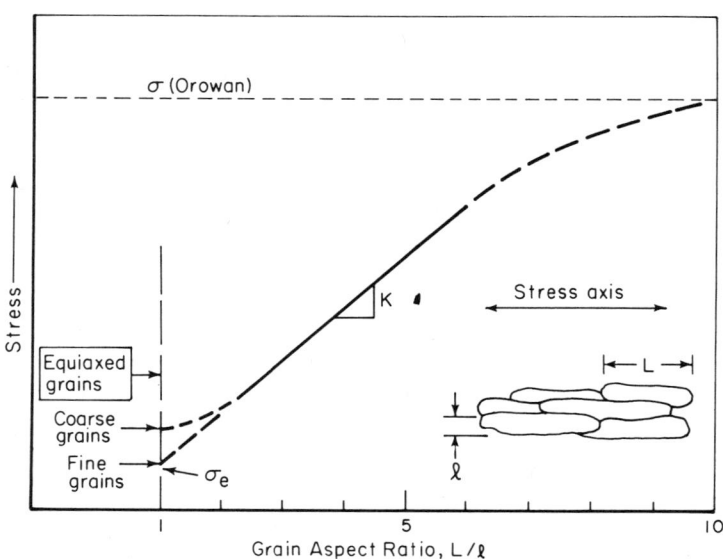

Fig. 1. The grain aspect ratio effect. At elevated temperatures the creep and yield strength increase with increasing degree of grain elongation. (Example is for dispersion-strengthened nickel-base alloys at a high temperature, e.g., 1100 C).

2 DISCUSSION

In addition to grain size and shape, Chairman Decker introduced for discussion other key elements in alloy design: cell structure and dislocation content, second phases at grain boundaries, homogeneously distributed second phases, and nonequilibrium structures (e.g., metallic glasses)—and he asked how mathematical modeling could help alloy designers in relating microstructure to properties.

In discussing grain size and shape, the Chairman asked whether it is necessary to have a certain number of grains across a component cross section—particularly in directionally solidified or dispersion hardened alloys with high grain aspect ratios. This appears to be the case for conventional equiaxed superalloys, as shown in Fig. 2. Wilcox responded that the grain-aspect-ratio effect holds even for very thin sheets (0.001 inch thick) where, in places, there may be only one grain across a cross section. This, however, is the case for uniaxial tension. Where multiaxial stresses exist, as in cooled turbine vanes, there may be a problem if only one grain is in a load-bearing cross section.

Kear discussed experience with monocrystals and directionally solidified (D.S.) alloys. There is really no substitute for monocrystals if one is thinking of very thin sections, particularly if hoop stresses are superimposed on normal tensile stresses. Hahn asked if the orientation also played a significant role. Kear replied that it did, very definitely, particularly for creep strength and thermal fatigue. The (100) orientation is preferred for D.S. superalloys or monocrystals because it gives a combination of high creep strength and a low modulus [20×10^6 psi compared with 45×10^6 psi for the (111) orientation]. The low modulus improves thermal-fatigue resistance..

Levy commented that it may be desirable to have a number of grains in a cross section when corrosion is a factor. There should be enough transgranular grain boundaries so that there is not a direct corrosion path across a part.

Ashby pointed out that Fig. 2 could be explained by Hutchinson's analysis of Taylor factors for polycrystalline slip versus single-crystal slip in creep.

In dispersion-hardened alloys there still can be Orowan strengthening at high temperatures, provided the grains are sufficiently elongated. Clauer made the point that here plastic flow by grain-boundary sliding is minimized and deformation occurs by dislocation processes. For very high grain aspect ratios, the strength approximates the Orowan stress (see Fig. 1). Also, for creep of single crystal TDNiCr, Nix has found a threshold stress below which measurable creep does not occur. This stress is approximately the Orowan stress. The threshold stress gives rise to the very high stress exponents often observed during creep of dispersion-hardened metals.

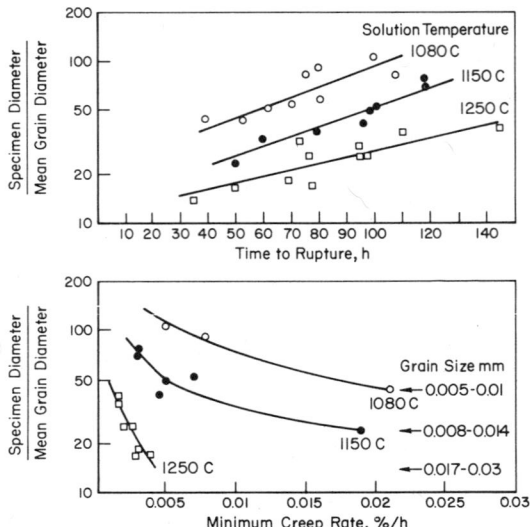

Fig. 2. Influence of ratio of specimen diameter to mean grain diameter and solution-treatment temperature on the creep properties of wrought superalloy tested at 14.2 kgf/mm² at 870 C[3].

Hornbogen discussed the design of microduplex alloys which can have a very fine grain size, as fine as 0.1 μm, if careful double thermal treatments are done. The properties are described in his paper. Sherby discussed some progress in ultrahigh-carbon steel grain refinement, and expanded the discussion to superplastic behavior of these fine-grained steels. They have been able to hot press atomized powders of these steels at 600 C by superplastic forming. Sherby, however, was not too optimistic, in general, about the practical applications of superplastic alloys. VerSnyder countered that in Pratt & Whitney's F-100 engines, the discs are made by a superplastic forming process. Alden found the ability to produce 0.1-μm grain sizes very extraordinary, and questioned how stable such small grains would be. If stable at elevated temperatures, such structures could help superplastic forming by a factor of 10³. Hornbogen suggested adding third phases to stabilize 0.1-μm grain sizes in microduplex alloys.

Chairman Decker opened discussion on the role of dislocation content and distribution in strengthening by recounting three mechanisms described in Session III by which high dislocation contents and fine cell structures (<1 μm) could be generated: spinodal decomposition (Thomas), creep of nickel-base alloys which generated a dislocation substructure around γ' (Tien), and microduplex processing of ferrous alloys (Hornbogen). Thomas emphasized the superior stability of the first array. McMahon mentioned another way to introduce such dislocation distributions in tool steels—by deforming martensite and then having the carbides dissolve and migrate to the dislocation networks. This is called "white structure", and is often detrimental; but if it could be produced in large quantities, there might be some applications.

Wilcox discussed the role of a dislocation substructure on the creep of alloys containing inert dispersions. There is now evidence that a substructure does little or nothing to improve high-temperature creep strength in such alloys. This has been demonstrated for TD NiCr, SAP-type alloys, 7075 Al, and 2024 Al. Ansell confirmed this for SAP-type alloys, where he found that the presence of a substructure increased the creep rate by a very large amount. Ashby also found for single crystals of Cu-BeO (produced by internal oxidation) that a substructure increased the creep rate rather than reducing it. It may be, however, as suggested by Ansell, that when second-phase particles are shearable (e.g., γ' in nickel-base alloys) substructure creep strengthening can be realized.

The presence of second phases at grain boundaries can influence both room-temperature and elevated-temperature mechanical behavior. Retained austenite at martensite grain boundaries can improve the fracture toughness of steels, as discussed by V. F. Zackay, by changing the stress distribution around a propagating crack. This can result in crack blunting, and very tough materials can be produced. High toughness can be achieved with about 15 percent retained austenite of the right morphology in a 200-ksi steel. This situation is analogous to tungsten carbide—cobalt matrix cutting tools. The virtue of "tripping" the second phase remained questionable, after considerable debate. McClintock discussed the effect of second phases in terms of crack propagation. By introducing a discontinuous bimodal distribution of second phases in the right order, crack propagation can be greatly hindered. The austenite thickness will, of course, be important, but the efficacy in crack blunting will depend on the plastic zone size. Hirth stated that this general concept had been applied to beta-titanium alloys where crack meandering was achieved by having the right distribution and morphology of retained alpha. Owen commented that the properties of the martensite, as well as the distribution and morphology of retained austenite, very much affected the toughness of steels. For example, dislocated martensite is more ductile than twinned martensite. Tien asked whether interfaces should be coherent or noncoherent for good fracture resistance. McClintock replied that incoherent interfaces are preferred for crack blunting, but the opposite is true for crack initiation.

Chairman Decker asked: When there is grain boundary sliding in creep, is there a possibility of inhibiting sliding by dispersing particles in the boundary? Levy commented that blocky carbides in superalloys did help high-temperature creep by inhibiting grain-boundary sliding. Also, the formation of such carbides resulted in zones denuded of carbon at grain boundaries, and this improved tensile ductility as well as the creep-rupture life. Kear pointed out that in some cases, microcracks can be initiated at grain boundaries in superalloys by slip bands intersecting grain-boundary carbides, which can enhance creep cavitation. However, this can be controlled to some extent by heat treatment to develop suitable γ' morphologies.

Metallic glasses are an exciting development which point the way toward much future research and development of alloys with highly metastable structures. Decker asked whether dispersed particles could be used to minimize or eliminate the highly localized shear bands which form on deformation. Davis replied that at this point not much really was known about the influence of second-phase particles on deformation of metallic glasses. Ashby pointed out that second-phase particles would probably not help promote homogeneous slip. The difficulty in metallic glasses is the nucleation of slip, and once it is nucleated, deformation occurs catastrophically. When a particle with a different modulus is present, then flow (probably catastrophic) will occur at a stress lower than in the absence of the particles. Hornbogen commented that such particles might, however, improve the fracture toughness. Wilcox related that the presence of second-phase particles (ThO_2) in chromium lowered the ductile-to-brittle transition temperature, and that the probable mechanism was slip dispersal by the particles once deformation had been initiated. This, however, might not be relevant to metallic glasses, where plastic flow is extremely localized. Jaffee suggested that metallic glass, which is not as brittle as graphite fibers, might have an application in fiber-reinforced composites. Davis mentioned that it is possible to draw wire and strip of metallic glasses, and this does not affect the subsequent ductility or strength.

The design of alloys by the synthesis approach involves mathematical modeling. An example is Phacomp, which correlates chemical composition with phase stability. Intuition also plays a great role, as was demonstrated by Clarence Bieber in the development of many of today's commercial superalloys. Bieber carried around a pack of yellowed sheets which contained a running log of properties, such as stress-rupture life and creep ductility, versus composition. He made approximately 20 different heats a week in the melt shop and kept a continuous record of the progress of a given alloy. This chemical approach, call it intuition or enlightened empiricism, proved highly successful. Today Bieber's type of operation would be very expensive, and our present sophisticated metallurgists ought to be able to optimize properties more cost effectively through the use of models relating structure to properties.

Modern examples of mathematical modeling in alloy design are the Ashby creep-mapping method, the Hall-Petch analysis, and Phacomp. Owen mentioned that the British had been very successful in using a modified Hall-Petch approach to design high-strength, low-alloy steels.

They were successful in predicting yield strength, but prediction of fracture properties proved very difficult.

Mayer brought up Phacomp as an alloy design tool. It seemed to him that General Electric used the method extensively, whereas Pratt & Whitney did not. Sims explained that Phacomp was derived from the terms "phase computation", and the methodology was developed to try to predict when brittle phases (such as sigma) would form. The technique, in simple terms, uses electron-hole density to assist in predicting compound formation. The method worked for U-700, and has been extended to other alloys. In a sense, it is like a regression analysis where composition is fed into a computer program and the output predicts whether or not sigma phase will form. Alloy melters are given Phacomp specifications, which must be met in production. This means that all customers of the vendors use (if only indirectly) the concept of Phacomp. Mayer concluded that this was a good example where some fairly fundamental thinking resulted in a practical alloy design tool. Jaffee pointed out that Phacomp was an adaptation of work by Hume-Rothery and Beck. Their method employed group number to predict phase stability, and was described in the first Battelle Colloquium.

Tien briefly described the work of Machlin, who uses a pair-potential calculation to predict phase formation. Although relatively new, the method has been extremely successful to date. Within about 2 to 3 years, the technique should be computerized, and ultimately will be another tool which will complement Phacomp.

Sherby commented on deformation mapping as applied to superplasticity. If dislocation creep could be made more difficult by alloy design, e.g., control of second phases, then substantial grain-boundary sliding could occur at higher strain rates. This would delight the technologist, who could then employ higher rate forming processes.

Alloy design for fracture resistance was broached by McClintock. Features such as plastic-zone size and crack-tip-opening displacement can be mathematically modeled, and together with microstructural considerations can be applied to predict fracture toughness. In recent years, there has been a rapid increase in our knowledge of what controls fracture toughness, and major advances have been made by a combined mechanics-metallurgical approach.

Bement brought up another modeling tool—computer calculation of multicomponent phase diagrams, which are based on the Brewer-Engel theory (electronic characteristics), or Kauffman's modeling, which is based on thermodynamic considerations. Rudy has successfully utilized the thermodynamic approach to study refractory compounds (W-Ti-C, W-Hf-C) which are used as cutting tools.

Sims added some other important aspects of alloy design, which encompass both the designers and the metallurgists. He included the important contributions of alloy vendors, who have helped design and develop the complex, but relatively well understood, superalloys. The whole of superalloy alloy development can really be considered a success—it is a good model for future alloy designers. Some of the "superalloy concepts" could be readily applied to other areas, such as design of alloys for a variety of nuclear applications. Successful alloy development for the future will be undertaken by both enlightened empiricists and "metallurgical fundamentalists". Shewmon pointed out that the situation in design of alloys for nuclear applications, particularly for the breeder reactor, is not so straightforward, and will require a great number of very difficult experiments before the fundamentals are known. The problems of swelling, irradiation-induced defects, and fission-product attack will have to be solved.

Chairman Decker concluded the discussion on a bullish note. Alloy design has come a long way in the last decade, and the degree of enlightment has markedly improved. This has been very helpful to the alloy designer, and the future holds even greater expectations.

REFERENCES

1. Benjamin, J.S., and Bomford, M.J., *Met. Trans.*, **5**, 615 (1974).
2. Wilcox, B. A., and Clauer, A. H., *Acta Met.*, **20**, 743 (1972).
3. Richards, E. G., *J. Inst. Met.*, **96**, 365 (1968).

STRUCTURE/PROCESSING/PROPERTY/ PROOF RELATIONSHIPS

MANIPULATION OF SUPERALLOY MICROSTRUCTURES AND PROPERTIES BY ADVANCED PROCESSING TECHNIQUES

B. H. Kear and A. F. Giamei

Materials Engineering and Research Laboratory
Pratt & Whitney Aircraft
Middletown, Conn.

ABSTRACT

The degree of microstructural and property control now attainable in superalloys is quite remarkable. The principal developments in the coupled areas of alloy design and advanced processing are reviewed. In the area of directional-solidification processing, the relative advantages of oriented single-crystal growth, as opposed to textured columnar grain growth, are discussed. Recent developments in superalloy-powder processing, including problems and solutions in the areas of consolidation and forming are described. Some of the metallurgy associated with the fabrication of a totally disappearing diffusion-bonded joint is discussed.

1 INTRODUCTION

During the past two decades, continuing developments in the coupled areas of superalloy design and processing have led to some striking advances in the performance capabilities of the superalloys used in high-temperature structural applications in the gas-turbine engine. Although the period of the 1950's was marked by an almost explosive growth in alloy development, since about 1965 the emphasis has shifted to process development. In this productive period, several important processing innovations were made, including directional solidification, powder processing and consolidation, hot isostatic pressing, superplastic forging, and transient-liquid-phase diffusion bonding. In retrospect, it can be seen that the successful introduction of these new processes has depended in large measure on the acquisition of a fairly detailed understanding of the basic metallurgy involved.

The various processing developments that have occurred in the areas of solidification, forming, and joining are described in the following sections. Moreover, an attempt is made to highlight the areas where microstructural manipulations have been critical to the eventual successful execution of the various processing-development activities.

2 SOLIDIFICATION

A frequent mode of failure in creep of nickel-base alloys is by grain-boundary cavitation and cracking, particularly along boundaries oriented transverse to the applied stress. In 1960, VerSnyder and Guard[1] demonstrated that this problem could be overcome by directional solidification, so as to develop a columnar grain structure with all the grain boundaries aligned parallel to the applied stress. Following this lead, in the mid-1960's, Pratt & Whitney Aircraft developed a variety of commercial processes for the directional solidification of bars, ingots, slabs, and more complex shapes. The most notable achievement was the successful fabrication of cast-to-size, columnar-grained, gas-turbine airfoils.[2]

In its original form, the directional-solidification (DS) process involved pouring the melt into a preheated ceramic shell mold, supported on a water-cooled copper chill plate, and controlling the temperature gradient in the mold by an arrangement of zone heaters to promote solidification in a direction normal to the chill plate. Later, refinements in the process were introduced to permit faster growth rates under conditions of steeper thermal gradients. In one scheme[2], provision was made for the gradual withdrawal of the mold from the hot zone into a chamber where radiation cooling could occur freely (Fig. 1). In another scheme[3], the mold was lowered from the hot

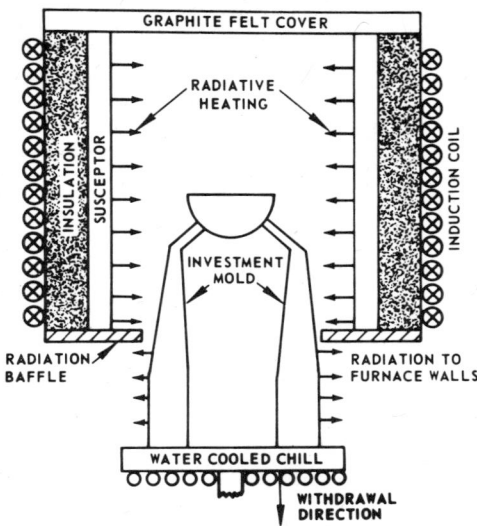

Fig. 1. Apparatus used for the directional solidification of castings by the withdrawal technique.

zone, through an arrangement of heat shields, into a cooling medium consisting of rapidly circulating liquid tin (Fig. 2). The liquid-metal-cooling (LMC) technique ensures intimate contact between mold and "liquid chill" during solidification, so that thermal gradients are not only steep but also remain fairly uniform throughout the entire solidification process.

In practice, the maximum allowable growth rate that can be tolerated for processing of columnar-grained material is approximately 15 in./hr, since at higher rates the pronounced solid/liquid interfacial curvature that develops causes grain divergence, i.e., there is a loss in directional uniformity of the columnar-grain structure. Moreover, under high growth rates, the characteristic <100> preferred orientation of the colmnar-grain structure becomes less than ideal. Under favorable low-growth-rate conditions, the spread in <100> orientations about the longitudinal axis of the columnar-grained material is only about 7 degrees at a distance of approximately 1 inch from the chill plate.[4]

Procedures for processing monocrystalline material emerged quite naturally from the procedures developed for processing columnar-grained material. The simplest method devised for obtaining near <100> orientation monocrystals was to use the chill zone of columnar-grained material for "self-seeding" the alloy monocrystal. This was done merely by inserting a tubular constriction in the mold at the optimum distance of approximately 1 inch from the chill, so as to exclude all but one grain.[5] To obtain monocrystals in a precise <100> orientation, or for that matter in any other

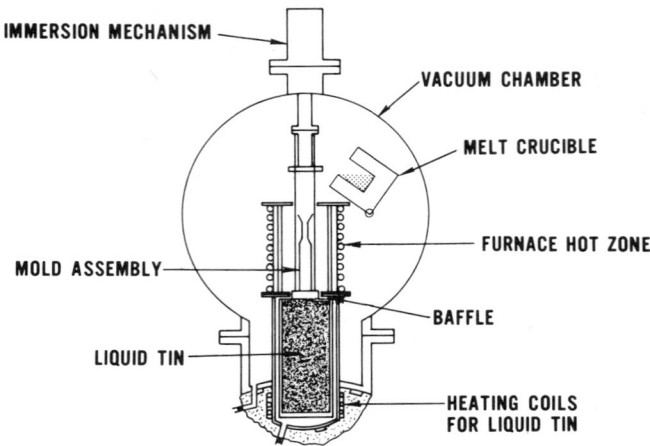

Fig. 2. Apparatus used for the directional solidification of castings by the liquid-metal-cooling technique.

orientation, conventional seeding techniques have to be used in conjunction with the DS process. Using these various procedures, monocrystalline material has now been produced in as many different sizes and shapes as is the case for columnar-grained material. For example, monocrystals can now be obtained in the form of wires, sheets, ingots, airfoils, and even springs.[6] Using the LMC technique of solidification, monocrystals have been grown from the melt in any orientation at rates up to 200 in./hr, which contrasts with the maximum permissible growth rate, approximately 15 in./hr, for columnar-grained material. Much higher growth rates can be tolerated in monocrystal solidification because curvature of the solid/liquid interface merely changes the morphology of the dendrites and not their orientation.

Concurrently with this process-development activity, a continuing effort has been made to define the important microstructure/property relationships that are characteristic of directionally solidified nickel-base alloys, particularly alloys of the high γ' volume fraction type. Directional solidification results in a general improvement in the mechanical properties of conventional γ' precipitation-hardened nickel-base alloys, particularly in the creep-rupture and thermal-fatigue properties at intermediate termperatures.[4] These improvements in properties derive as much from the <100> preferred orientation of the directionally solidified material as from the elimination of transverse grain boundaries[7], except in materials where the boundaries are particularly weak. The scale of the microstructure also exerts some influence on properties. As a general rule, it can be stated that the overall properties of the material improve with refinement of the microstructure, i.e., with decreasing size of dendrites, γ/γ' eutectic pools, MC carbides, and other microconstituents. The major dimensions of the

MC carbides strongly influence fatigue properties, since they determine effectively the average size of internal flaws.[8] This is because large platelike MC carbides are either precracked in the as-cast material, or crack easily on loading below the yield strength of the material. Again, large γ/γ' eutectic pools, which cannot be eliminated by heat treatment, also impair the overall creep properties, because the coarse γ' in the γ/γ' eutectic is not as creep resistant as the normal fine distribution of cooling γ' in matrix γ. Suitably refined microstructures can be produced by high-rate solidification under conditions of steep thermal gradients, as in the LMC technique of solidification (Fig. 3). Such refined structures can be rendered completely homogeneous by a normal high-temperature solutionizing heat treatment. High-rate solidification is beneficial also in that it reduces the size of shrinkage pores and minimizes effects due to melt convection. Solute convection develops in the partially solidified, "mushy" zones of castings whenever a sufficiently large difference in density develops due to solute segregation, and when the resistance to fluid flow is low. Fluid flow by solute or concentration-driven convection can give rise to extraneous equiaxed surface grains, commonly known as "freckles" [9,10](Fig. 4), and can also cause significant compositional shifts in the bulk melt[11]. Freckling is the result of the disruption of the normally aligned dendritic structure by the motion of solute-rich interdendritic melt in well-defined channels in the mushy zone.[12] Sharpening the gradient inhibits freckle formation because of the reduced width of the mushy zone, shorter local solidification time, and greater resistance to fluid flow in the narrower interdendritic passages of the refined structure. The alloy Mar-M200 is particularly freckle prone because inverse segregation of the heavy element tungsten and normal segregation of lighter elements, such as titanium and aluminum, makes for an unstable mushy zone. An alloy design approach that can be used to solve this problem is to replace one half of the tungsten with the normally segregating heavy element tantalum (distribution coefficient <1) so as to compensate for the inverse segregation of tungsten (distribution coefficient >1) and thereby stabilize the mushy zone.

The single most significant event in alloy design associated with DS occurred following preliminary castability tests on gas-turbine airfoils having internal cooling passages. In certain strong alloys of the high γ' volume fraction type, such as Mar-M200, it was found that the circumferential hoop stresses that developed during shrinkage of the hollow casting about its ceramic core were sufficient to cause extensive cracking of the longitudinal grain boundaries. This hot-cracking problem has been overcome in several alloys by making controlled additions of hafnium, typically in the range of 1.0 to 3.0 weight percent.[13,14] In Mar-M200, such additions of hafnium are sufficient to generate, by normal dendritic segregation, a relatively high volume fraction of hafnium-rich γ/γ' eutectic in the interdendritic/grain

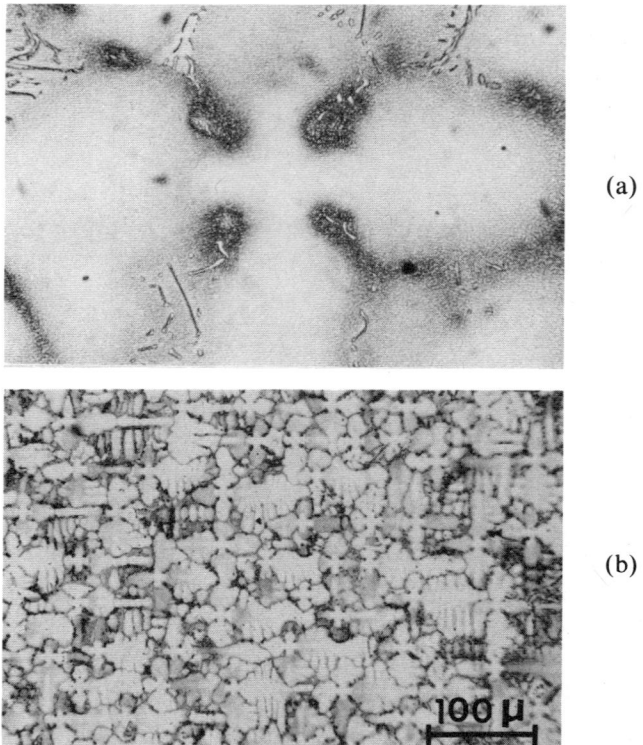

(a)

(b)

Fig. 3. Dependence of the dendrite size on growth rate in [100] single crystals of Mar-M200 grown by the liquid-metal-cooling technique; (a) 2.8×10^{-3} cm/sec, (b) 8.4×10^{-2} cm/sec.

Fig. 4. Chains of equiaxed surface grains, or "freckles" in a 10-cm-diameter [100] single crystal of Mar-M200 grown at a slow rate.

boundary regions of the material (Fig. 5). The improved castability may be attributed to the decoration of the grain boundaries with hafnium-rich eutectic, since tests have shown that hafnium-rich γ' is stronger than the normal $\gamma+\gamma'$ structure (Fig. 6), at least in the intermediate temperature range of interest with respect to hot cracking. Another tangible benefit derived from hafnium-doping of Mar-M200 has been a general improvement in the transverse properties of directionally-solidified columnar-grained material, without impairment of the excellent longitudinal properties.[15] The improvement in creep-rupture properties at intermediate temperatures has been ascribed to the presence of strong, ductile, hafnium-rich γ' along the grain boundaries, which reduces stress concentrations by slip dispersal, thereby inhibiting crack initiation and possibly crack growth in creep.[16] Another beneficial effect is the formation of a more favorable distribution of chromium-rich $M_{23}C_6$ carbides along grain boundaries (Fig. 7), which should inhibit grain-boundary sliding in creep without sacrificing ductility. Some evidence has been obtained that hafnium doping also improves the intrinsic cohesive strength of grain boundaries in nickel-base alloys, apparently because of the ability of this highly reactive element to combine with deleterious tramp elements such as sulfur (Fig. 8), which normally segregate to grain boundaries and cause serious embrittlement.[17].

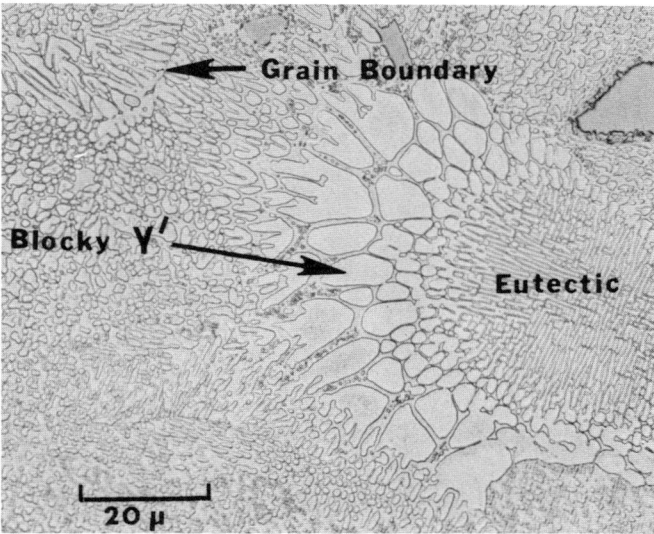

Fig. 5. Microstructure of hafnium-doped Mar-M200 showing relatively large amounts of interdendritic/grain boundary γ/γ' eutectic; the blocky γ' in the eutectic is rich in hafnium.

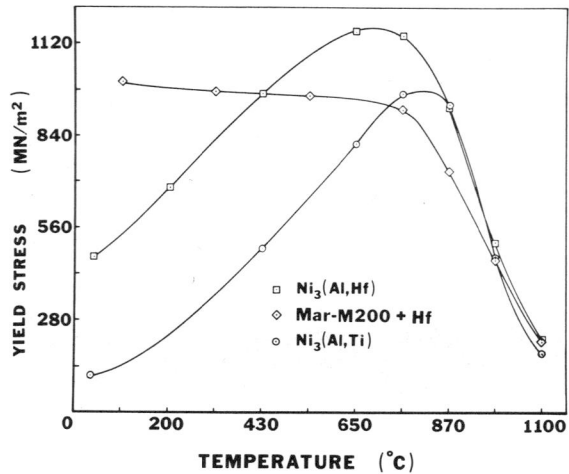

Fig. 6. Yield stress-versus-temperature curves for hafnium-doped Mar-M200 and alloyed γ' containing 2.73 atomic percent hafnium or titanium.

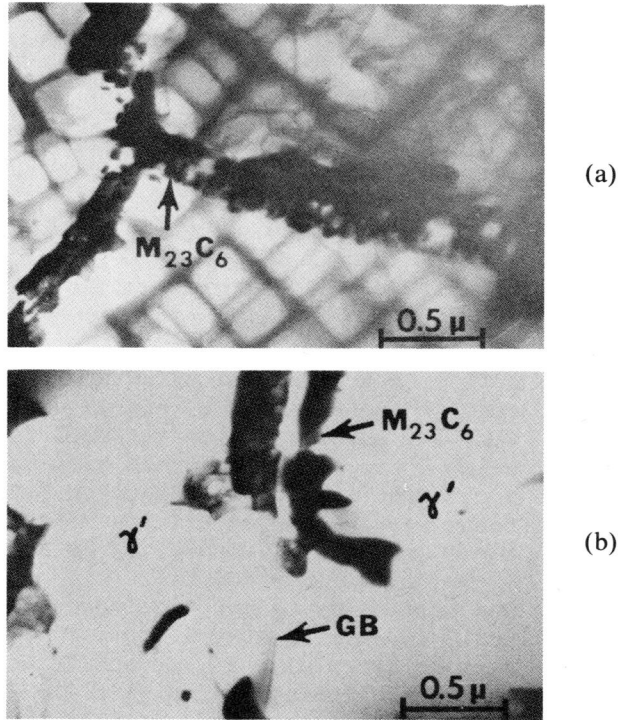

Fig. 7 Transmission electron micrographs of γ' and $M_{23}C_6$ carbide distribution at the grain boundaries of heat-treated (a) Mar-M200 and (b) Mar-M200 plus 2 weight percent hafnium.

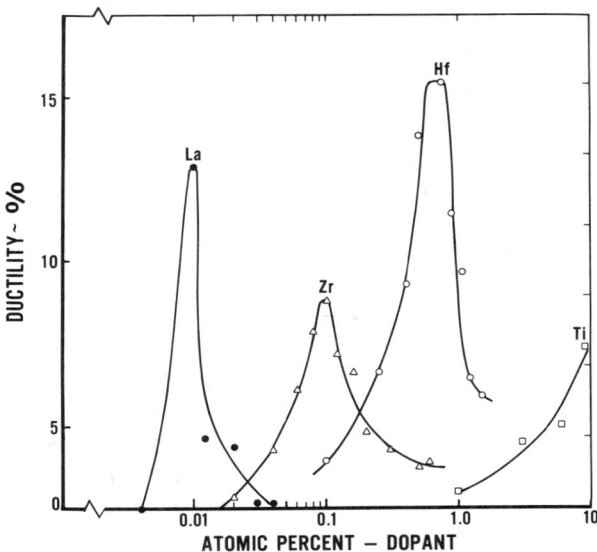

Fig. 8. Ductility enhancement in a sulfur-contaminated Ni-Al-Ta superalloy by the addition of various reactive-element dopants.

A more recent innovation in alloy design has been the introduction of the new generation of eutectic superalloys, e.g., the $\gamma/\gamma'+\delta$ eutectic, which is essentially a γ' precipitation-strengthened nickel-base alloy containing reinforcing lamellae of the Ni_3Nb (δ) phase.[18] A eutectic superalloy of this type can be processed at reasonable rates with an optimally aligned microstructure only under conditions where the thermal gradients are steep, as in the LMC process (Fig. 9). The dependence of the interlamellar spacing, λ, on growth rate, R, for the $\gamma/\gamma'+\delta$ eutectic follows the familiar relationship $\lambda^2 R = $ constant, at least for the case of fully plane front solidification using the LMC process. A similar functional relationship has been found for the dependence of the high-temperature stress-rupture life on interlamellar spacing.[19] In other words, the rupture life in $\gamma/\gamma'+\delta$ increases with decreasing interlamellar spacing according to a Hall-Petch type of relationship. Currently, much of the work in this area is concerned with defining the optimum conditions for growth of suitably refined and aligned microstructures in complex shapes, such as gas-turbine airfoils, and with alloy modifications to improve corrosion resistance, ductility, and thermal-fatigue strength, without sacrificing creep strength and castability. Unfortunately, experience has demonstrated that alloying to improve the general properties of an eutectic superalloy invariably increases its melting range, which makes the material more difficult to grow with an optimally aligned microstructure, at least at reasonable rates. Extensive alloying also frequently results in an unacceptable loss in creep strength. For these reasons, some effort is now being directed to eutectic-alloy design primarily for creep strength and castability, with less concern for other materials properties of interest in gas-turbine applications. Such an approach necessarily will entail making better use of protective coatings.

It is anticipated that alloy monocrystals will eventually replace columnar-grained material in certain high-temperature structural applications in the gas-turbine engine. For example, in the casting of a hollow airfoil, the complete absence of grain boundaries in the monocrystal part would eliminate the grain-boundary hot-cracking problem prevalent in columnar-grained material. The anisotropy in mechanical properties characteristic of alloy monocrystals can also be exploited in this application simply by selecting an orientation that best matches the material properties with performance requirements.[7] Again, because the properties of monocrystals are extraordinarily reproducible in a given orientation, it should be possible to obtain substantial weight savings by designing the air-foil to tighter specifications. Lighter turbine blades would require lighter disks, shafts, bearings, etc., resulting in a generally more compact and efficient engine, with attendant large economic benefits. As a final point, it should perhaps be emphasized that the growth of alloy monocrystals by the LMC process appears to be a viable economic process since high growth rates have already been achieved. As mentioned earlier, the potential exists for growth of monocrystals by the LMC process at rates up to approximately 200 in./hr. The prospects for the growth of eutectic-superalloy monocrystals also appear to be good, but much remains to be done to demonstrate feasibility on a practical scale in a commercial foundry.

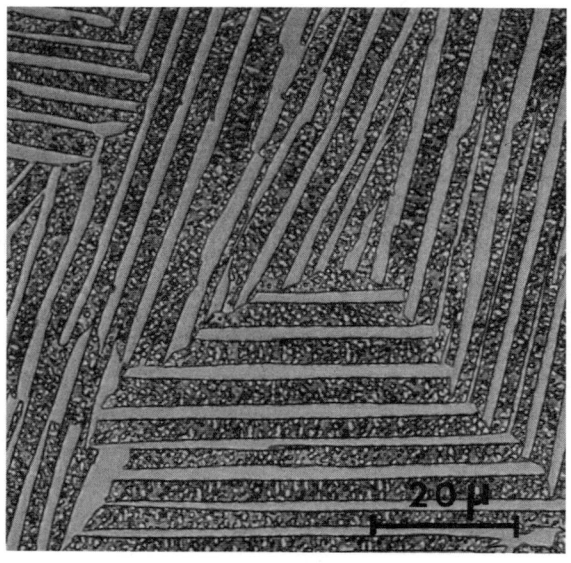

Fig. 9. Microstructure of the $\gamma/\gamma'+\delta$ lamellar eutectic alloy.

3 FORMING

The response of nickel-base alloys to conventional hot working seems to be related to the volume fraction of the γ' hardening phase. Low γ' volume fraction alloys, such as Waspaloy, can be worked easily above the γ'-solvus temperature, i.e., at temperatures where the alloy reverts to single-phase γ. Intermediate γ' volume fraction alloys, such as Astroloy, experience failure by grain-boundary cracking when worked above the γ'-solvus temperature, but can be worked in steps with intermediate anneals at temperatures just below the γ' solvus. Apparently, in this alloy[20], precipitation of γ' at grain boundaries is sufficient to prevent grain-boundary sliding and triple-point cracking (Fig. 10). High γ' volume fraction alloys, such as Mar-M200, are considered to be unworkable by conventional hot-working operations. However, successful fabrication of such alloys has been accomplished by controlled-strain-rate, isothermal forging, and superplastic forging.

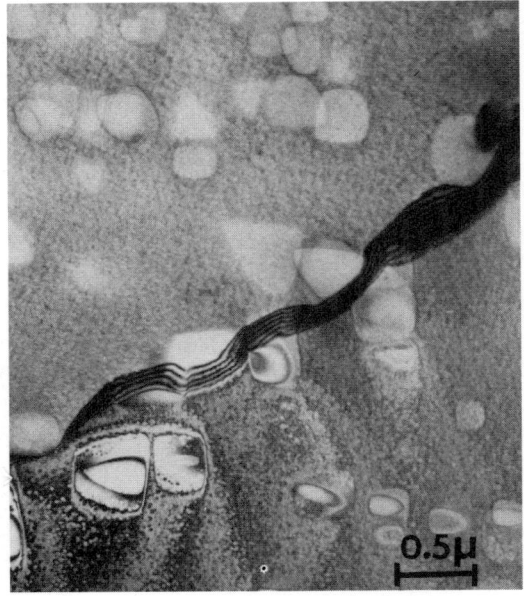

Fig. 10. Transmission micrograph of Astroloy after aging at a temperature just below the γ'-solvus temperature. Note perturbations in the grain boundary due to precipitation of fairly massive γ' particles along the boundary.

The difficulties encountered in the hot working of ingots of high γ' volume fraction alloys may be attributed to their lack of constitutional uniformity in the as-cast condition. This became clear following tests on perfectly homogeneous material, where no difficulties were experienced in forging, provided that the deformation was carried out isothermally with controlled deformation rates. At temperatures above the γ' solvus [Fig. 11(a)], the material invariably undergoes recrystallization. At temperatures below the γ' solvus [Fig. 11(b)], recrystallization can be suppressed under moderate deformation rates, in which case the material develops a uniform dislocation substructure. Since the yield strength of superalloys is strain-rate sensitive at high temperatures[21], the deformation below the γ'-solvus temperature should be conducted preferably at low deformation rates, at least initially, so as to avoid the buildup of dangerously high stresses in the early stages of ingot deformation. After sufficient deformation has been sustained by the material, the deformation rate may be increased, because diffusive slip in γ' characteristically occurs under a diminishing stress as the dislocation density increases. Very high deformation rates, however, should be avoided because this causes recrystallization.

The presence of a refined dislocation substructure in the forged material improves the overall mechanical properties in the low to intermediate temperature range.[22] Typically, significant improvements are obtained in yield strength, ultimate strength, and low- and high-cycle fatigue strength, without serious losses in ductility and in creep strength at least up to 1400 F. Applications for such substructure-strengthened materials include disks and shafts in the gas-turbine engine.

Since recrystallization of superalloys can be avoided by controlled deformation below the γ'-solvus temperature, it follows that alloy monocrystals can be forged while preserving their essentially monocrystalline character. This has already been done on a laboratory scale for several alloy systems [Fig. 11(c)]. Best results in terms of microstructural control and property improvements have been obtained using perfectly homogeneous alloy monocrystals, i.e., diffusion-annealed LMC-processed monocrystals. Currently, there is a limit on the size of bars or ingots that can be fabricated with suitably refined microstructures by the LMC process. Further development in the forging of alloy monocrystals, therefore, must await advances in the state of the art of DS processing. An important alloy design concept pertinent to the hot-deformation processing of alloy monocrystals is the use of trace elements for pinning the dislocations in the substructure. According to recent ideas[23], high values of ΔH and low values of the steady-state creep rate characteristic of creep in nickel-base alloys are a consequence of strong interactions between dislocations and certain large misfitting atoms, especially when they occur in combination with one another, e.g., boron and zirconium. Advances in our understanding of such

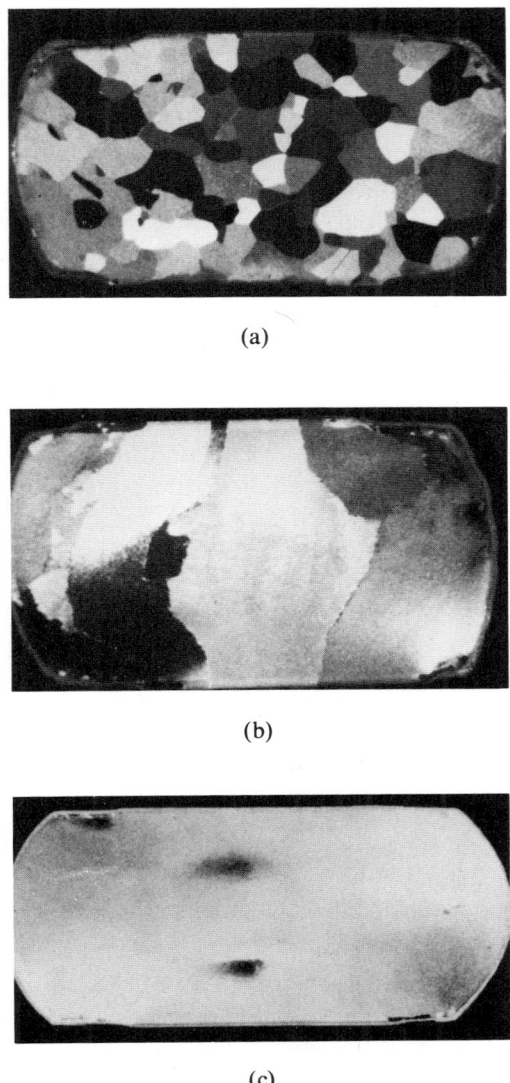

(a)

(b)

(c)

Fig. 11. Samples of a homogeneous Ni-Al-Ta superalloy after hot deformation processing in compression (ϵ=0.92): (a) columnar-grain material after deformation above the γ'-solvus temperature, showing evidence for recrystallization; (b) columnar-grain material after deformation below the γ'-solvus temperature, showing no evidence for recrystallization; (c) single-crystal material after deformation below the γ'-solvus temperature, showing no evidence for recrystallization. Magnification - 7.5$\times$.

interactions, particularly the synergistic effects obtained with combinations of two or more trace-element additions, almost certainly will lead to an increased temperature capability for substructure-strengthened material.

An important development in the hot deformation processing of superalloys followed from the discovery that high-rate deformation by hot extrusion below the γ'-solvus temperature causes extensive recrystallization. When the extrusion parameters are properly adjusted, the scale of the grain structure is comparable with the γ' particle size, and the material exhibits superplastic properties at high temperatures.[24,25] This is the basis for the Gatorizing[TM] forging process, where a billet is first made superplastic by hot extrusion and is then worked into shape by closed-die superplastic forging, using a carefully controlled deformation rate. Following deformation processing, the material is normally heat treated to obtain the optimum grain structure and γ' particle distribution.

Although Gatorizing was performed originally on vacuum-induction-melted ingots, this approach was soon abandoned in favor of processing of powder compacts, since experience demonstrated that the gross heterogeneities characteristic of as-cast material tended to persist throughout processing, with adverse effects on mechanical properties, particularly in the advanced superalloys. Gatorized products formed from powders exhibit remarkably uniform microstructures and mechanical properties. Currently, Gatorizing is finding useful application in the manufacture of gas-turbine disks and even complete rotors. A particular advantage of Gatorizing in such applications is that since extremely close tolerances can be guaranteed in closed-die superplastic forging, the need for costly finishing operations, say, by machining, is minimized. Die life is also extended.

Consolidation of nickel-base alloy powders by hot extrusion gives a satisfactory superplastic product, irrespective of the method of manufacture or source of alloy powder. On the other hand, direct consolidation of powders by hot isostatic pressing (HIP) may or may not give a superplastic product, depending on the initial grain size of the powder particles. A sufficiently fine grain size in the particles is clearly a prerequisite for developing a fine-grained superplastic microstructure in the HIP-consolidated material. This requirement clearly calls for very high cooling rates in powder processing. Some success has already been obtained with both the rotating-electrode and argon-atomization processes, using auxiliary cooling schemes. An alternative approach is to subject the powder particles to cold rolling. The deformation introduced into the particles will cause intraparticle recrystallization during HIP consolidation, so that the resulting product will exhibit superplastic properties. When such techniques have been fully developed, it seems likely that HIP will replace hot extrusion as the preferred method for making powder preforms for superplastic forging. Powder consolidation by HIP is also being seriously considered for making near net shape disks, and even blades. In the latter case, temperature and/or stress gradient annealing can be used to convert the fine-grained superplastic structure into a more favorable, creep resistant, directionally aligned, columnar grain structure.

Early in the development of superalloy powder technology, it became apparent that the available high-strength, castable alloys were not necessarily suitable for powder processing and consolidation. A particular problem was the tendency in certain alloys for the powder particles to be heavily decorated with MC carbides, which persisted even after consolidation and heat treatment.[26,27] Such networks of carbides seriously weaken the consolidated product, since they provide easy paths for crack propagation, even when grain-boundary migration has occurred across the prior-particle boundaries. Final grain size is also limited. A particularly undesirable thin-platelet carbide morphology is found in the larger particles, which cool at the slowest rates[28] (Fig. 12). The evidence pointed to growth of these carbide platelets in the partially solidified melt, just prior to completion of solidification. Experimentally, it has been established that platelet formation is a function of the thermodynamic stability of the MC carbide—the greater the carbide stability, say, as indicated by its melting point, the lesser is the tendency for platelet formation. Using this information for guidance[29], some progress has now been made in resolving the carbide problem through alloy modifications. In IN-100, for example, a substantial improvement in MC carbide distribution and properties was obtained by replacing vanadium with a more stable carbide-forming element, e.g., niobium (Fig. 13). The addition of hafnium also increases the volume fraction of γ/γ' eutectic, which occurs in the HIP-consolidated material in the form of a uniformly fine distribution of relatively coarse hafnium-rich γ' particles. Such a bimodal distribution of γ', i.e., coarse hafnium-rich eutectic γ' interspersed with fine cooling γ', is particularly beneficial since it improves the low-cycle fatigue strength without sacrificing creep properties[30], at least in the intermediate temperature range of interest for gas-turbine disk applications.

Extensive development work has also been carried out on powder processing of inert-particle dispersion-strengthened superalloys using so-called "mechanical alloying".[31] This involves the use of a high-energy ball mill to fuse together particles of superalloy and dispersoid prior to consolidation by hot extrusion or HIP. Following consolidation, the optimum microstructure, including grain size, morphology, and orientation, is obtained by appropriate thermomechanical treatments.[32]

Fig. 12. Extraction replica of an IN-100 powder particle. Note the thin MC carbide platelets with triangular or sawtoothed morphologies.

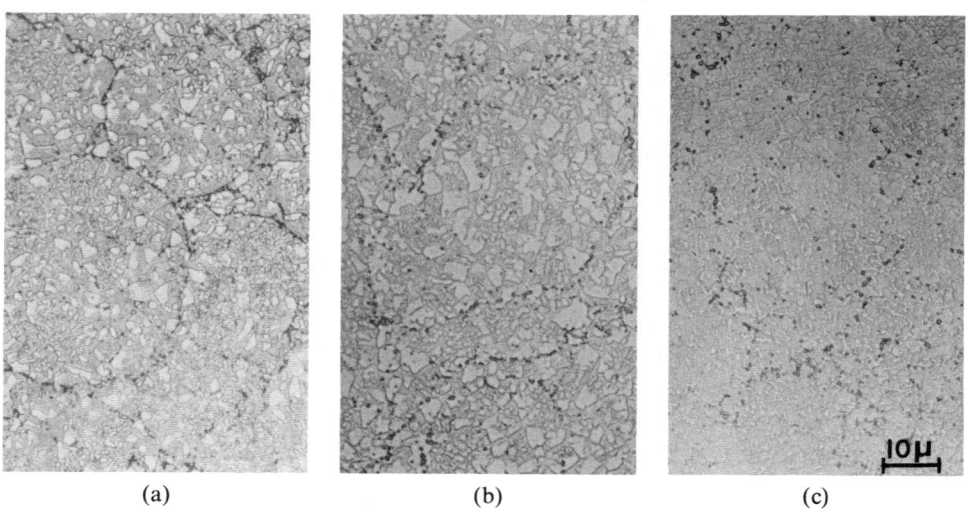

(a)   (b)   (c)

Fig. 13. Effect of alloying on the MC carbide distribution in hot isostatically pressed (2100 F/3 hr/15 ksi) powder of IN-100: (a) normal alloy, (b) modified alloy (1.5Nb, 0.0V), and (c) modified alloy. Note decoration of prior particle boundaries with MC carbides in (a) and absence of such decoration in (c).

4 JOINING

Metal-joining processes are used extensively in the manufacture of individual superalloy gas-turbine components, or parts, and even in the assembly of clusters of parts. The usefulness of metal joining derives largely from the convenience and flexibility it offers in design and manufacture, which generally makes for significant cost benefits and performance advantages.

Fusion welding is of limited utility in the joining of the high-strength γ' precipitation-hardened alloys because of their generally poor weldability[33], i.e., sensitivity to fusion-weld or heat-affected-zone cracking. A few specially designed weldable superalloys are available, but these are restricted to applications in the low to intermediate temperature range. Notable among these alloys is Alloy 718, which depends for its strength on the precipitation of a metastable DO_{22} γ'' phase based on Ni_3Nb[34] [Fig. 14(a)]. In this case, it is the rather sluggish precipitation of the γ'' phase that is responsible for the good weldability, because sufficient time is normally available during cooling of the weldment to permit the necessary plastic accommodation to occur in the γ matrix phase prior to precipitation of the γ'' phase. Attempts have been made to increase the temperature capability of this type of alloy, which, in effect, means increasing the thermal stability or resistance to coarsening of the γ'' phase. The most ingenious approach to solving the γ'' coarsening problem has been to use a small volume fraction of fine cuboidal-shaped γ' particles as preferred nucleation sites for the three variants of the γ'' phase.[35] The resulting "compact" γ'/γ'' precipitate morphology [Fig. 14(b)] is more resistant to coarsening and appreciably extends the useful temperature range of the alloy.

High-temperature brazing is widely used in the manufacture of gas-turbine components and parts[36], but not generally for joining critical superalloy components that are exposed to high temperatures and corrosive/erosive environments. This is because it is difficult to design a filler material that meets all the property requirements, both physical and mechanical, including complete compatibility with the substrate, or workpiece. This limitation in conventional brazing stimulated interest in the possible use of filler materials that could be subsequently diffused away, so as to obtain base-metal properties in the joint. Such a diffusion-brazing operation can, in principle, be carried out in a number of different ways. Currently, the technique that has met with the widest acceptance is referred to as transient-liquid-phase (TLP^R) diffusion bonding[37], which is essentially a hybrid of brazing and diffusion bonding.

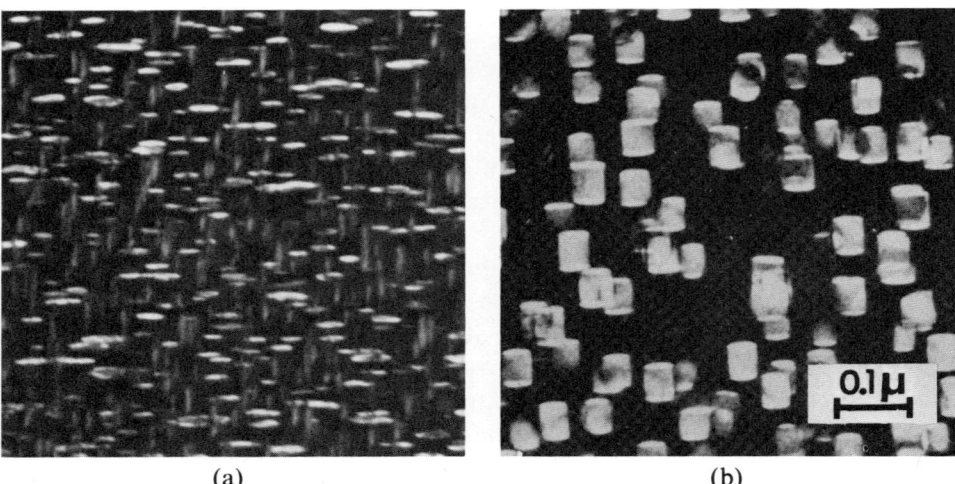

(a)   (b)

Fig. 14. Dark-field electron micrographs of precipitates in (a) Alloy 718 and (b) modified Alloy 718. In (a), only two of the three variants of the DO_{22} γ'' phase are visible. In (b), the cuboidal-shape γ' particles are completely enveloped in a thin shell of the three variants of γ'', giving a so-called "compact" γ'/γ'' morphology.

In its most advanced form, TLP bonding relies on the wetting of the contacting surfaces of the workpieces by a thin interlayer of a low-melting-point, boron-rich alloy. Under isothermal conditions and a slight normal pressure, preferably in vacuum or in argon atmosphere, the molten interlayer gradually alloys with the matching workpieces, and its melting point rises as the boron diffuses away from the interface. The bonding is judged to be complete when no melt remains at the interface. Subsequently, heat treatment is employed to erase all traces of the original interface. With this bonding technique, good joints have been made in complex-shaped parts, using simple tooling and mating-surface preparation, and in a variety of nickel-base superalloys and even dissimilar alloy combinations (Fig. 15). Furthermore, bond strengths comparable with base-metal strengths have been achieved, even in high-temperature stress-rupture tests.

Success with TLP bonding depends in large measure on careful interlayer design. The interlayer should have a composition that closely matches the composition of the superalloy substrate. At the same time, it should contain melt depressants in the minimum concentrations required to reduce the melting point to a level below the incipient melting of the substrate. Furthermore, the overall composition of the interlayer must be adjusted to avoid the formation of extraneous phases in the bonded region. A cursory examination of relevant phase diagrams provides some guidance with respect to interlayer design. One thing that becomes immediately clear

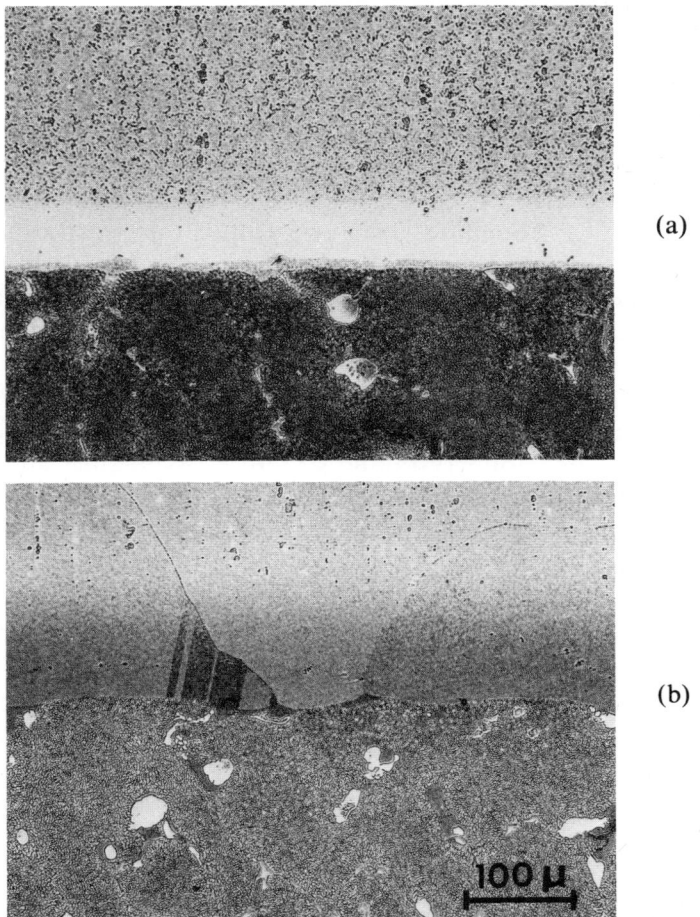

(a)

(b)

100 μ

Fig. 15. Diffusion-bonded joint between Udimet 700 (top) and IN-100 (bottom): (a) 4 hr at 2000 F, (b) 4 hr at 2100 F + 24 hr at 2125 F.

is that for a Ni-Cr, or Ni-Cr-Co base interlayer composition, the range of possible melt depressants is not large—the most promising being B, Si, Mn, Cb and Ti. However, tests have shown that all these melt depressants, with the exception of B, give rise to unwanted, stable compounds at the bonded interface. Another feature in favor of B as an additive is its rapid diffusivity at the high temperature of bonding. Accordingly, B has assumed the position as preferred melt depressant in most, if not all TLP interlayers. Interestingly enough, even when certain desirable elements, such as Al, Ti and C are excluded from the interlayer composition, so as to avoid compound formation, these elements are easily replenished in the bonded region by diffusion from the superalloy substrate.

A typical example of one of the family of interlayers designed for joining Udimet 700 is TLP-21, with nominal composition Ni-15Cr-15Co-5Mo-2.5B. In this interlayer, the Cr, Co and Mo match their concentrations in the superalloy. The Al and Ti are deliberately left out of the interlayer, but these elements are subsequently acquired from the substrate by diffusion. This particular interlayer has a melting range which permits its use at bonding temperatures $\geqslant$2050F.

Although TLP bonding was developed originally for joining polycrystalline material, it has been shown recently that it works equally well for monocrystalline material. In particular, it has been demonstrated that two or more segments of superalloy monocrystal with identical orientations can be diffusion bonded together to form a single larger monocrystal. Moreover, it appears that there is no restriction on the types of superalloy monocrystal that can be joined by this process in order to produce a composite monocrystalline structure. A possible application for such a composite monocrystal already exists. Thus, the ideal structure for a gas-turbine blade is considered to be a composite of fatigue resistant root section, creep and thermal fatigue resistant airfoil section, and erosion resistant tip section. In principle, the three component monocrystals comprising the composite blade can be joined together so that there are no grain boundaries.

REFERENCES

1. VerSnyder, F. L., and Guard, R. W., *Trans. Am. Soc. Metals,* **52,** 485 (1960).
2. VerSnyder, F. L., and Shank, M.E., *Mater. Sci. & Eng.,* **6,** 213 (1970).
3. Tschinkel, J. G., Giamei, A. F., and Kear, B. H., Patent Application No. 180,597.
4. Piearcey, B. J., Kear, B. H., and Smashey, R. W., *Trans. Quart. ASM,* **60,** 634 (1967).
5. Kear, B. H., Copley, S. M., Hornbecker, M. F., and Sink, L. W., *Trans. AIME,* **245,** 1361 (1969).
6. Copley, S. M., Kear, B. H., Hornbecker, M. F., and Duhl, D. N., *Mater. Sci. & Eng.,* **10,** 119 (1972).
7. Kear, B. H., and Piearcey, B. J., *Trans. AIME,* **238,** 1209 (1967).
8. Gell, M., and Leverant, G. R., *Trans. AIME,* **242,** 1869 (1968).
9. Preston, J., *Transactions of Tenth Vacuum Metallurgy Conference,* E. L. Foster (Ed.), New York (1968), p. 569.
10. Giamei, A. F., and Kear, B. H., *Met. Trans.* **1,** 2185 (1970).
11. Flemings, M. C., and Nereo, G. E., *Trans. AIME,* **242,** 50 (1968).
12. Copley, S. M., Giamei, A. F., Johnson, A. F., and Hornbecker, M. F., *Met. Trans.,* **1,** 2193 (1970).
13. Cochardt, A. W., U.S. Patent No. 3,005,705.
14. Hockin, J., and Taylor, W., Second paper presented at Second World Conference on Investment Castings, Dusseldorf, Germany, 1969.
15. Duhl, D. N., and Sullivan, C. P., *J. Metals,* **23,** 38 (1971).
16. Doherty, J. E., Kear, B. H., and Giamei, A. F., *J. Metals,* **23,** 59 (1971).
17. Kear, B. H., Giamei, A. F., and Doherty, J. E., *Proceedings of the Third International Conference on the Strength of Metals and Alloys,* Cambridge, England (1973), p. 134.

18. Thompson, E. R., and Lemkey, F. D., *Metallic Matrix Composites,* K. G. Kreider (Ed.), Academic Press, New York (1974).
19. Thompson, E. R., George, F. D., and Breinan, E. M., *Proceedings of the Conference on In-Situ Composites,* NMAB 308-2 (1972).
20. Oblak, J. M., Owczarski, W. A., and Duvall, D. S., *Met. Trans.,* **2,** 1499 (1971).
21. Leverant, G. R., Gell, M., and Hopkins, S. W., *Mater. Sci. & Eng.,* **8,** 123 (1971).
22. Oblak, J. M., and Owczarski, W. A., *Met. Trans.,* **3,** 617 (1972).
23. Strutt, P. R., Khobaib, M., and Kear, B. H., to be published.
24. Moore, J. B., and Athey, R. L., U.S. Patent No. 3,519,503.
25. Kear, B. H., Oblak, J. M., and Owczarski, W. A., *J. Metals,* **24,** 25 (1972).
26. Moskowitz, L. N., Pelloux, R. M., and Grant, N. J., *Proceedings of the Second International Conference on Superalloys,* Metals and Ceramics Information Center, No. 72-10 (1972).
27. Larson, J. M., *Modern Developments in Powder Metallurgy,* Hausner (Ed.), Plenum Press, New York (1974), Vol. 8, p. 537.
28. Grey, D. A., Runkle, J. C., and Genereux, P. D., Seventh Annual Spring Meeting TMS-AIME, University of Toronto, May, 1975.
29. Grey, D. A., to be published.
30. Bartos, J. L., Seventh Annual Spring Meeting TMS-AIME, University of Toronto, May, 1975.
31. Benjamin, J. S., *Met. Trans.,* **1,** 2943 (1970).
32. Benjamin, J. S., and Bomford, M. J., *Met. Trans.* **5,** 615 (1974).
33. Prager, M., and Shira, C. S., *Welding Research Council Bulletin,* No. 128 (1968).
34. Paulonis, D. F., Oblak, J. M., and Duvall, D. S., *Trans. ASM Quart.,* **62,** 611 (1969).
35. Cozar, R., and Pineau, A., *Met. Trans.,* **4,** 47 (1973).
36. Pattee, H. E., *Welding Research Council Bulletin,* No. 187 (1973).
37. Duvall, D. S., Owczarski, W. A., and Paulonis, D. F., *Welding J.,* **54,** 203 (1974).

DISCUSSION on Paper by B. H. Kear

TIEN: 1. You spoke of the "toughness" of Ni₃(Al, Hf)—that it is strong as well as tough. Would you give me an idea of its K_{Ic}, its plane-strain ductility, and its A and B parameters in Paris' per cycle crack growth law at whatever temperature that you have data for?

2. Have you come across any problems of long-term property degradation because of hafnium?

KEAR: My comments on the properties of Ni₃(Al, Hf) γ′ phase were made in connection with the effect of hafnium doping on the properties of Mar-M200. A marked improvement in grain-boundary ductility is obtained by doping Mar-M200 with ~2 weight percent hafnium, apparently because the grain boundaries become decorated with hafnium-rich γ′ due to segregation. Moreover, as I mentioned, independent tests have shown that hafnium-rich γ′ is stronger at intermediate temperatures than Mar-M200. In this connection, therefore, it is reasonable to state that hafnium-rich γ′ is stronger and more ductile, i.e., tougher, than Mar-M200, at least in the intermediate temperature range of interest. No data have been obtained for K_{Ic} in $Ni_3(Al, Hf)$, nor have any tests been performed to determine its fatigue-crack growth characteristics. However, some data have been obtained on the relative fatigue properties of a hafnium-rich alloyed γ' and Mar-M200 (details will appear in the December issue of *Met. Trans.*) In brief, it has been found, somewhat surprisingly, that the fatigue behavior of alloyed γ' (fine slip material) is actually inferior to that of Mar-M200 (coarse slip material) at any temperature of interest. For example, the 10^7 cycle endurance limit for the superalloy exceeds that of γ'. Moreover, the relative fatigue strength (endurance limit/yield strength) of the superalloy is also superior, since the yield strength of Mar-M200 changes only slightly between room temperature and 800 C, while that of alloyed γ' increases markedly. This is of interest in connection with current speculation on the influence of a bimodal distribution of γ' on the low-cycle fatigue properties of superalloys. One might suppose that massive γ' particles distributed throughout a normal fine distribution of γ' in γ would be beneficial, since γ' exhibits a very homogeneous slip mode, and, therefore, should promote slip dispersal. The fatigue data on single-phase γ', however, strongly suggest that this will not be the case, and that the fatigue-crack propagation behavior will be impaired, since it takes far lower stresses to move "fatigue" dislocations in γ' than it does in the superalloy. On the other hand, it still seems reasonable that slip dispersal would be beneficial in retarding crack initiation. Some recent, new low-cycle fatigue data on HIP consolidated material seems to be in accord with this general picture of the fatigue properties of alloys containing a bimodal distribution of γ'.

On your second question, long-term property degradation is a common problem in all superalloys exposed to high temperatures, but usually this is due to γ' particle coarsening and various phase changes that occur along grain boundaries. These effects do not appear to be any different in hafnium-doped alloys, provided that appropriate adjustments in alloy compositions are made to compensate for the increase in volume fraction of γ/γ' eutectic, e.g., the chromium and molybdenium concentrations in the alloy preferably should be reduced so as to avoid their precipitation, say, in the form of carbides, around the γ/γ' eutectic.

WRIGHT: Frank VerSnyder stated the other day that a long-term goal in superalloy design is to get rid of hafnium additions, presumably for reasons of practical processing problems, whereas some of the effects of hafnium you have shown seem indispensable. Would you care to comment?

KEAR: As I explained, hafnium doping of certain advanced nickel-base superalloys has two important benefits. First, it increases the strength of the grain-boundary regions of the cast alloy so that they more strongly resist grain-boundary cracking during cooling after solidification, particularly in hollow parts containing ceramic cores. Second, it improves the creep-rupture properties of the material, particularly the rupture ductility. This is true of polycrystalline material as well as columnar-grained directionally solidified material tested in the transverse direction. However, as emphasized by VerSnyder, the use of hafnium is not without its own special problems. In conventionally cast polycrystalline material, a new phenomenon, known as "patchy porosity", i.e., shrinkage porosity highly localized in certain regions of the cast part, makes its appearance. This has been attributed to the increase in the volume fraction of γ/γ' eutectic in hafnium-doped alloys, and the effect this has on "feeding" of solidification shrinkage. This problem is specific to conventionally cast polycrystalline material, and is not encountered in directionally solidified material. Another problem that is common to all types of solidification processing is the increase in the amount of entrapped "dross" (mainly clusters of hafnium-rich oxide particles) in the cast parts. This is due to the fact that the density of the hafnium-rich oxide particles is comparable with that of the superalloy melt, so that there is little tendency for normal phase separation to occur under the influence of buoyancy forces. Several approaches are now being actively pursued to resolve this problem, and, as pointed out by VerSnyder, the main emphasis has been towards finding a substitute for hafnium. Another promising approach has been to minimize possibilities for contamination of the melt by making use of more sophisticated vacuum-melting technology. It should be mentioned also that the use of monocrystals will eliminate the need for hafnium additions, and, in fact, additions of several other ingredients normally found in superalloys. Clearly, this opens up some interesting new possibilities for alloy design for monocrystal applications.

With respect to the processing of parts by powder-metallurgy techniques, additions of hafnium appear to be so generally beneficial that this is likely to become standard practice in the near future. Two important benefits of hafnium-doping are (i) a reduction in susceptibility to decoration of prior-particle boundaries with MC carbides and (ii) a marked refinement in the scale of hafnium-rich γ/γ' eutectic. The latter is believed to be a contributing factor to the striking improvement in low-cycle fatigue properties of the consolidated material.

McCLINTOCK: In the powder process for polycrystalline disks, what are the prospects for particles of various sizes (a few times the typical fatigue-crack growth per cycle) that will tend to deflect the crack and retard the fatigue-crack propagation rate? These would, of course, probably cause some decrease in the stress for fatigue-crack nucleation, but perhaps a net improvement in design for some applications would be attained.

KEAR: The various methods that have been devised for the processing of powders certainly could be used to obtain the types of microstructures that you have in mind. For example, it should be a relatively easy task to obtain a uniform distribution of soft (or hard) particles in a conventional superalloy matrix by the technique of "mechanical alloying". What we need at this point in time is a much better definition of what constitutes an optimum microstructure from the fracture-mechanics viewpoint. Perhaps you can help us in this regard.

Leverant has obtained some new data on the fatigue-crack propagation behavior in eutectic superalloys that has some bearing on your question. He has found that for cases where the cyclic plastic zone size at the crack tip is larger than the interlamellar, or interfiber, spacing in superalloy eutectics, then there is no effect of spacing on the fatigue-crack propagation rate. Again, in powder-consolidated superalloys, it has been found that a bimodal distribution of γ' improves resistance to fatigue-crack initiation, rather than propagation.

LEVY: In LMFBR fuel elements, grain boundaries are the source of failures in austenitic alloys: they are embrittled by irradiation (helium, precipitates, etc.); they are weakened by boron leaching by the sodium coolant; they are notched by fission-product attack; and they are highly stressed during transient overpower conditions. It would, thus, be very desirable to have monocrystal fuel elements. Is it technologically possible to produce 6-foot-long, 0.1230-inch OD x .015-inch wall-thickness monocrystal tubing of austenitic steels (for example)?

KEAR: Ceramic shell molds with leachable cores have been used to fabricate tubular monocrystalline superalloys with the dimensions 4 inches long, $^1/_2$-inch OD $\times$ $^1/_{16}$-inch wall thickness. The material has been used to investigate the influence of combined stresses on the creep properties of superalloy monocrystals, e.g., a uniaxial tensile stress combined with a circumferential hoop stress derived by internal gas pressurization. We believe, on the basis of this experience, that it should not be too difficult to fabricate short pieces (up to approximately 12 inches maximum in length) of tubular material with the much thinner wall thickness that you have specified. However, there is little prospect of obtaining *uniform* cross-section tubular material in longer pieces. This is because of difficulties associated with maintaining dimensional stability in shell molds, particularly the proper alignment of core and shell.

A more practical approach to meeting your requirements would be to process the material by the edge-defined film-fed growth technique of solidification, developed by Tyco Laboratories. This process makes use of a die to control the cross section of a continuously pulled monocrystal, and has been used successfully in the fabrication of tubular crystals of sapphire and other materials. An optimum degree of wetting between melt and die seems to be an essential requirement of this process. Assuming that a suitable die material can be found for a stainless steel melt, it seems to us that the EFG process offers the best prospects for successful fabrication of thin-wall tubular monocrystals in a continuous or semicontinuous manner.

SIMS: Since the manufacture of monocrystals is deeply dependent upon the thermal properties of the compounds involved and the efficiency of heat transfer, would you care to comment on the size limits to which the process may be applied?

KEAR: We have encountered no difficulties in the manufacture of monocrystals of nickel and its alloys up to approximately 4 inches in diameter. Should the need arise, we feel that we could produce monocrystals up to approximately 8 inches in diameter using current practice. However, growth of monocrystals with dimensions much larger than this probably will require some modifications in processing, e.g., the use of a continuously fed thin melt zone to maintain sharper gradients and tolerable growth rates.

Using existing technology, in order to achieve monocrystal growth in bars with large diameter/length ratios it is necessary to reduce the growth rate as the monocrystal diameter is increased. This is because the direction of heat flow is into the melt in the hot region, down through the mushy zone where the heat baffles are located, and out of the solid monocrystal in the cold region. Clearly, under these conditions the efficiency of heat transfer must decrease with increasing monocrystal diameter. In a practical system, the ceramic shell mold, i.e., the container for the melt, is a serious obstacle to efficient heat transfer because of its relatively poor thermal conductivity. The abstraction of heat from the ceramic shell mold itself occurs most efficiently using the liquid-metal-cooling technique of solidification, i.e., where a bath of rapidly circulating liquid tin maintains intimate contact with the shell mold throughout solidification.

Another point that should perhaps be emphasized is that the actual shape of a part, rather than its net cross-sectional area, is more important in setting an upper size limit for monocrystal solidification. For example, a thin sheet having the same overall cross-sectional area as a bar would be much easier to grow at high rates, because its geometry is naturally more conducive to efficient heat transfer. Fortunately, many of the parts of interest, e.g., gas-turbine airfoils, are relatively thin sections, which are suitable for high-rate solidification.

STRAIN CONTROL FATIGUE AS A TOOL TO INTERPRET FATIGUE INITIATION OF ALUMINUM ALLOYS

T. H. Sanders, Jr.
D. A. Mauney
J. T. Staley

Alcoa Laboratories
Alcoa Center, Pennsylvania

ABSTRACT

A wide variety of aluminum alloys in standard tempers were investigated using strain control fatigue and transmission electron microscopy. The log of the number of reversals to failure increased linearly with decreasing log of the plastic-strain amplitude down to a critical level which was alloy and temper dependent. Below the critical level of plastic strain, failure times were shorter than would be predicted by extrapolating the high-plastic-strain-amplitude data. Transmission electron microscopy revealed that specimens tested above the critical level deformed homogeneously, while those tested below the critical level deformed heterogeneously; consequently, microstructures which would maintain homogeneous slip at low plastic deformation should provide better resistance to fatigue initiation.

1 INTRODUCTION

Much of the past structural-aluminum-alloy development has been directed toward manipulating the microstructure to produce high static strengths. This was followed by extensive efforts to increase the fracture toughness and resistance to stress-corrosion cracking. After these properties were optimized, fatigue design data were generated to characterize the material under various load environments. However, recent programs reflect the importance of fatigue as a critical design parameter rather than a secondary one.

Plastic strain is thought to be the controlling parameter in fatigue.[1] The interaction of the plastic deformation with the microstructure is believed to be important whether fatigue initiation or fatigue-crack growth is considered.

Strain control fatigue (SCF) has shwn particular promise as a tool in understanding the fundamental relationship between the microstructure and the dislocations formed during cyclic straining.[2-10] By controlling the total strain per cycle and monitoring the stress, the cyclic plastic response per cycle can be readily determined. This type of fatigue experiment has generally been referred to as low-cycle fatigue since failure at measurable, macroscopic plastic strain occurs in less than 5×10^4 cycles for most ductile structural materials.[11]

The goal of this work was to establish a correlation between SCF parameters and the microdeformation characteristics of various aluminum-alloy microstructures to provide a foundation for future development of microstructures more resistant to fatigue initiation.

2 MATERIAL SELECTION

The alloys chosen for this investigation represent a complete survey of the strengthening mechanisms typical in commercial aluminum products. The strengthening microstructures ranged from subgrain and dislocation strengthening in 1350 alloy, solid-solution strengthening by magnesium in 5083, finely dispersed, needlelike transition precipitates of Mg_2Si in 6061, and high density of coherent G.P. zones in 2024, to the high density of G.P. zones and η' (transition $MgZn_2$-type precipitate) and incoherent η in differing amounts in X7046 and 7075.

The alloys studied were plant-fabricated material and were in commercial tempers. The alloys, tempers, products, and nominal compositions are listed in Table I.

Table I. Material Description

Alloy	Temper	Product	Nominal Composition[a]							
			Si	Cu	Mn	Mg	Cr	Zn	Zr	Fe
1350	H17[b]	Rod	99.50 percent minimum aluminum							
5083	O[c]	Plate	<0.4	<0.4	0.7	4.45	·0.15	<0.25	—	<0.4
6061	T651[d,e]	Rod	0.6	0.27	<0.15	1.0	0.2	<0.25	—	<0.7
2024	T351[f,g]	Rod	<0.5	4.4	0.6	1.5	<0.1	<0.25	—	<0.5
X7046	T63[h]	Plate	<0.2	<0.25	0.17	1.3	0.13	7.2	0.12	<0.4
7075	T651[d,e]	Rod	<0.4	1.6	<0.3	2.5	0.3	5.6	—	<0.5
7075	T7351[c,h]	Rod	<0.4	1.6	<0.3	2.5	0.3	5.6	—	<0.5

(a) Composition and temper information from Aluminum Association via *Aluminum Standards and Data, 1974-75*, Fourth Edition, Aluminum Association, New York (January, 1974).
(b) Indicates a 65 percent increase in tensile strength by cold working.
(c) Full annealed to lowest strength condition.
(d) T6 indicates solution heat treated and then artificially aged.
(e) T-51 indicates stress relieved by stretching to 1-1 2 to 3 percent permanent set between solution heat treatment and aging.
(f) T3 indicates solution heat treated and then cold worked and naturally aged.
(g) T63 indicates solution heat treat, period of natural age, and then artificially aged.
(h) Solution heat treated and then a two-step low to high temperature artificial age.

3 EXPERIMENTAL PROCEDURE

The alloys were characterized prior to fatigue testing. The phases present were determined by the Guinier-deWolfe X-ray technique. Foils were prepared by conventional dimpling and polishing techniques and observed in a Hitachi transmission electron microscope equipped with a metallurgical stage. The degree of recrystallization and approximate ASTM grain size were determined by a transmission X-ray technique using unfiltered copper radiation. Monotonic and elastic-modulus tensile properties were determined in accordance with ASTM Standard Recommended Practices E-8 and E-111, respectively.

Multiple-specimen SCF tests were conducted according to testing practices currently being discussed in the ASTM Subcommittee E-9.08 on fatigue under cyclic strain. SCF specimens were machined to a constant-diameter reduced section of 0.360 in. (0.914 cm) by 0.720 in. (1.83 cm) long with a surface finish better than 30 rms μin. (0.76 μm). The specimens were stored and tested in a laboratory environment with the relative humidity controlled to 50 ±10 percent.

Total strain SCF testing was carried out on an MTS closed-loop electrohydraulic testing system using an 0.5-in. (1.3 cm)-gage-length axial extensometer attached to the reduced section of the specimen. A 10,000-pound (44.5 kN) dynamic-capacity load cell in series with the specimen was used to monitor load. The system was calibrated to a maximum strain error of 1.2 percent of reading or ±0.0001-in. (0.00025 cm), whichever was larger, and a maximum load error of 0.5 percent of reading or ±10 pounds (44.5 N), whichever was larger. The load-strain hysteresis loops were monitored at logarithmic increments of specimen life on an X-Y plotter. Throughout each test, the strain and load were continuously recorded on an oscillograph.

The SCF tests were started in the tension direction and a constant strain rate of 0.01 sec^{-1} was used throughout the tests. Specimen failure was defined throughout this work as the number of cycles at fracture. Details of SCF tests and analysis of the data are given in the Appendix.

Electron microscopy foils were prepared from fractured specimens taken from salient points along the fatigue-life plot. To document the fatigue deformation, disks 0.25 in. (0.635 cm) thick were spark machined from an area approximately 0.25 in. (0.635 cm) below the fracture surface and foils located approximately 0.06 in. (0.15 cm) from the machined test specimen surface. The foil location was chosen to insure that the observed deformation structure was related to the fatigue process rather than to the deformation induced by the plastic zone preceding the propagating fatigue crack.

4 RESULTS

4.1 Structure Analysis, Pretest

The results of the Guinier-deWolfe phase analyses, the grain sizes, and the degrees of recrystallization are shown in Table II. The microstructures of the alloys prior to straining are shown in Fig. 1.

4.2 Phenomenological

The results of the strain-controlled tests are shown in Figs. 2 through 8, and the monotonic tensile and cyclic parameters are summarized in Table III.

The data presented in Figs. 2a and 2b through 8b indicate that the cyclic-dependent change in stress for constant total-strain amplitude correlates with differences in microstructure characteristic of the several alloys. The strain-hardened 1350-H17 cyclically softened and the annealed 5083-0 cyclically hardened. Precipitation-hardened room-temperature-aged 2024-T351 containing finely dispersed, coherent G.P. zones hardened during cyclic deformation to a lesser degree than did 5083-0. The alloys which contained the transition precipitates β' (Mg_2Si) in 6061-T651 and η' ($MgZn_2$) in

Table II. Material Characterization by X-Ray

Alloy	Constituent Phases	Dispersoid Phases	Strengthening Mechanism	Grain Size (ASTM)
1350-H17	$Al_{12}Fe_3Si$ $FeAl_6$	—	Dislocations	Unrecrystallized
5083	Mg_2Si $Al_{12}(Fe,Mn)_3Si$ $CrAl_7$	Cr_2Al_{11} $(Fe,Mn)Al_6$	Magnesium Solid solution	5.0
2024-T351	Al_7Cu_2Fe Al_2CuMg $Al_{12}(Fe,Mn)_3Si$	$Al_{20}Cu_2Mn_3$	G.P. zones	6.0
6061-T651	Mg_2Si $Al_{12}(Fe,Mn)_3Si$	—	$\beta'(Mg_2Si)$	6.0
X7046-T63	Mg_2Si $Al_{12}(Fe,Mn)_3Si$	$ZrAl_3$	G.P. zones and $\eta'(MgZn_2)$	Unrecrystallized
7075-T651	Al_7Cu_2Fe Mg_2Si $(Fe,Mn)Al_6$	$Al_{12}Mg_2Cr$	G.P. zones and $\eta'(MgZnCuAl)$	7.0
7075-T7351	Al_7Cu_2Fe Mg_2Si $(Fe,Mn)Al_6$	$Al_{12}Mg_2Cr$	$\eta'(MgZnCuAl)$ and incoherent η	7.0

X7046-T63 and 7075-T651 slightly hardened only at the higher plastic-strain amplitudes and appeared to be cyclically stable at low-plastic-strain cycling. The overaged 7075-T7351 which contained the incoherent η ($MgZn_2$) appeared to be cyclically stable.

The plots of plastic-strain amplitude versus cyclic life (Figs. 2c through 8c) for each alloy plainly reflect the linearity of the Coffin-Manson relationship[1] down to a critical level of plastic strain which was alloy dependent. Below this critical level there was a definite departure from single-slope behavior except for 1350-H17 which showed no departure in the regime tested. The slope of the line in the high-plastic-strain regime was analytically determined as described below.

The monotonic true fracture ductility, ϵ_f, was considered to be a data point at one strain reversal based on the assumption that cyclic plastic deformation follows the same mechanism as monotonic plastic deformation. The only exception was EC-H17, where the monotonic ductility was of such

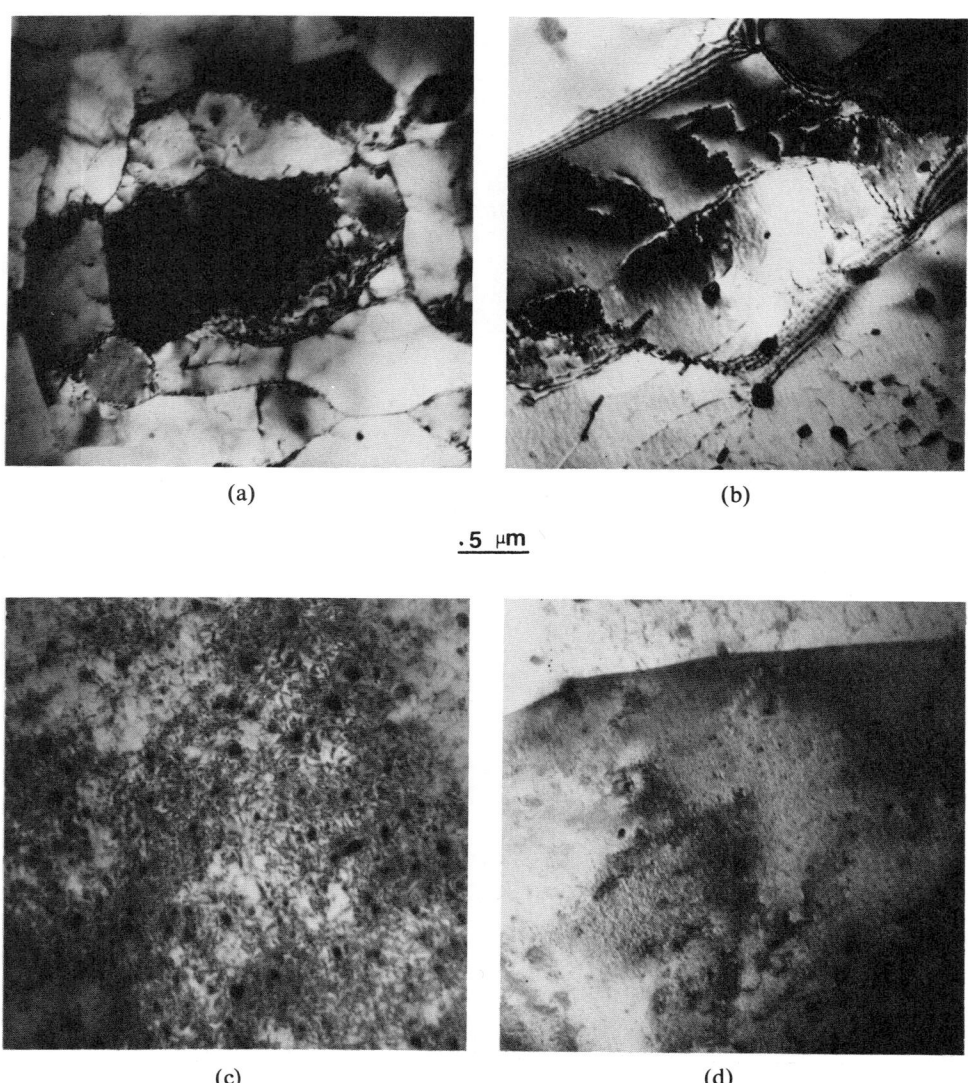

.5 μm

Fig. 1. (a) 1350-H17, unrecrystallized structure with a cold-work structure. (b) 5083-O, recrystallized structure with manganese and chromium dispersed phases. (c) 2024-T351, recrystallized structure with a large volume fraction of heterogeneously and homogeneously precipitated G.P. zones. (d) 6061-T651, recrystallized structure with a high volume fraction of partially coherent β' (Mg_2Si) precipitates.

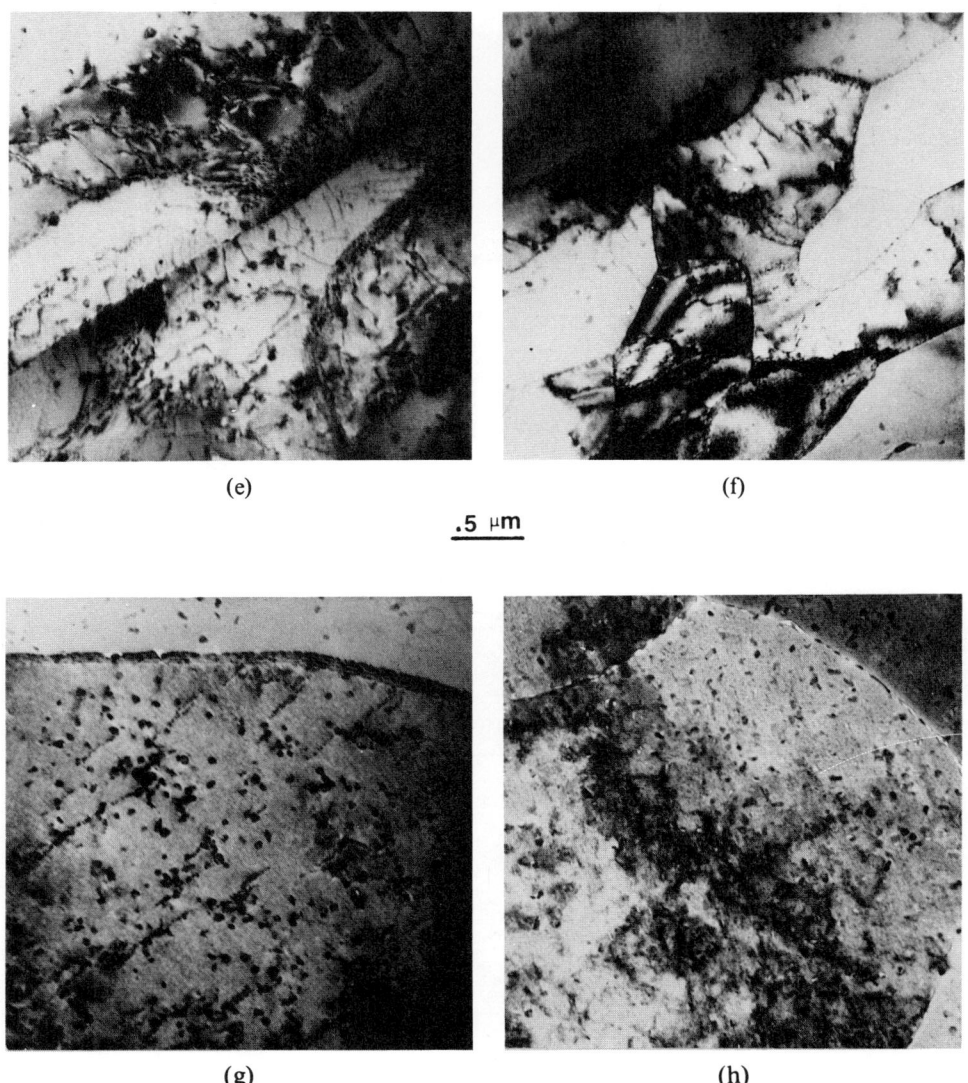

(e)   (f)

.5 μm

(g)   (h)

Fig. 1 (continued). (e) and (f) X7046-T63, unrecrystallized structure with a high volume fraction of G.P. zones and η' (MgZn$_2$) and finely distributed, partially coherent ZrAl$_3$ dispersoids.
(g) 7075-T651, recrystallized structure with a high volume fraction of G.P. zones and η'
(MgZnCuAl). (h) Overaged 7075 recrystallized structure with η' and η precipitates.

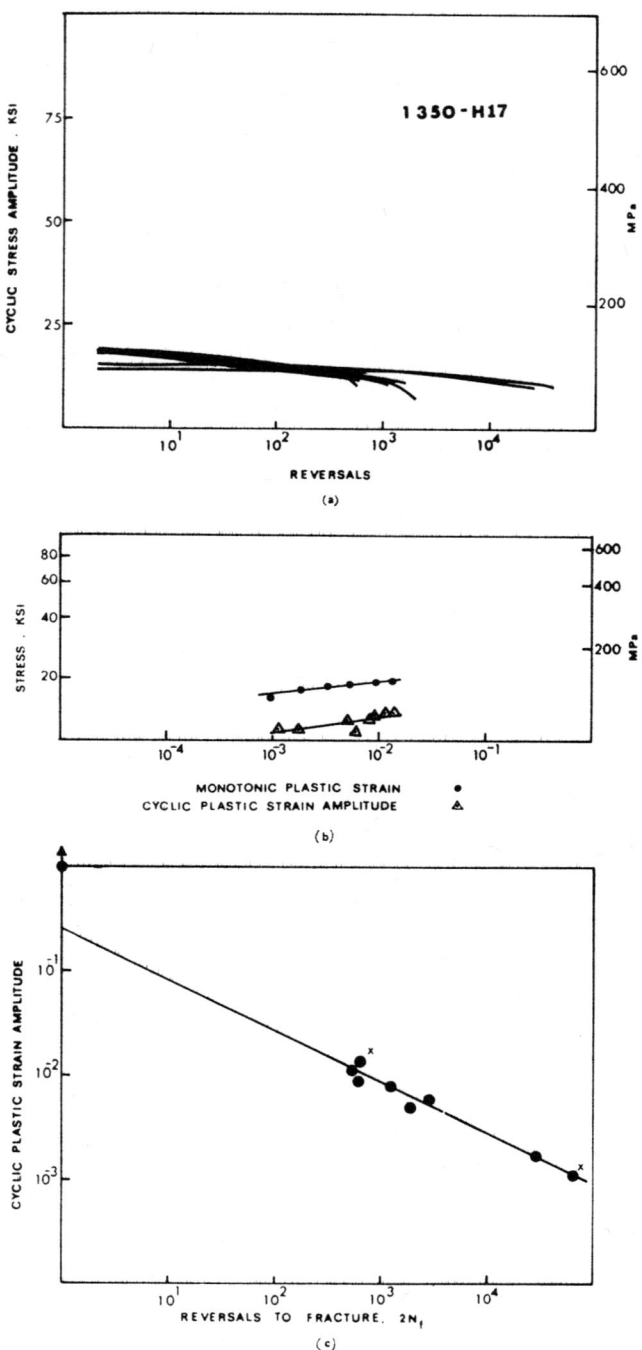

Fig. 2. SCF behavior of 1350-H17: (a) stress amplitude versus cycles, (b) cyclic and monotonic stress-strain relations, and (c) Coffin-Manson plot.

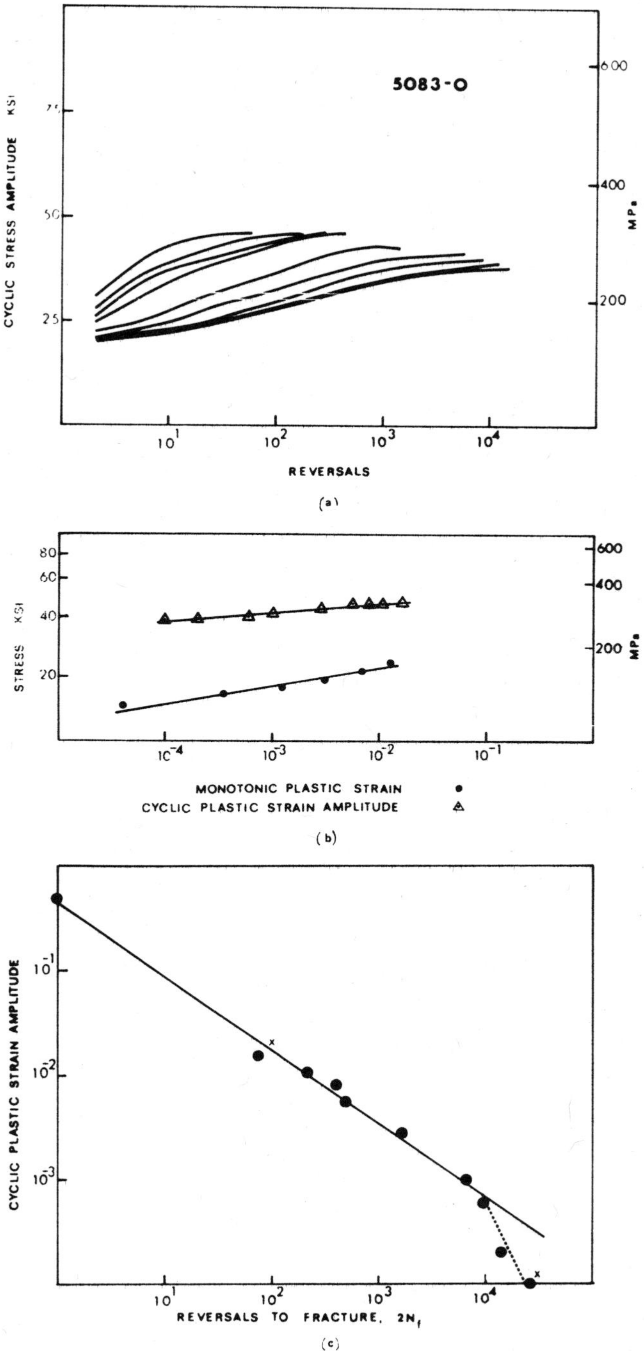

Fig. 3. SCF behavior of 5083-0: (a) stress amplitude versus cycles, (b) cyclic and monotonic stress-strain relations, and (c) Coffin-Manson plot.

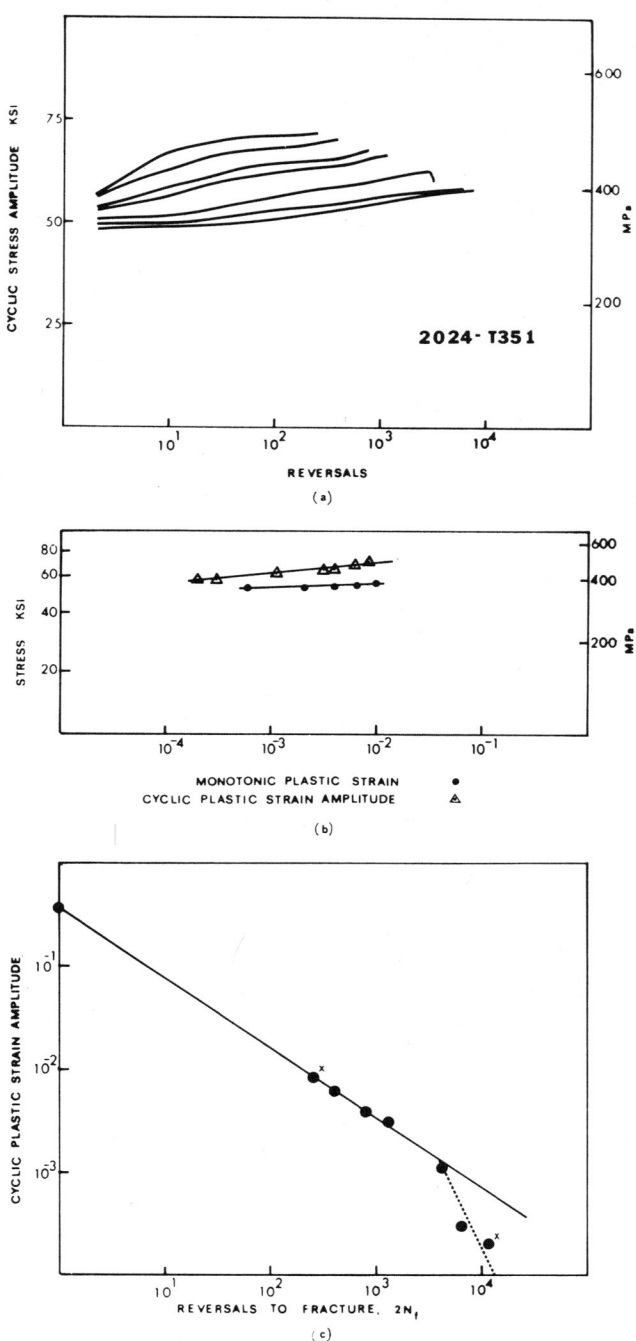

Fig. 4. SCF behavior of 2024-T351: (a) stress amplitude versus cycles, (b) cyclic and monotonic stress-strain relations, and (c) Coffin-Manson plot.

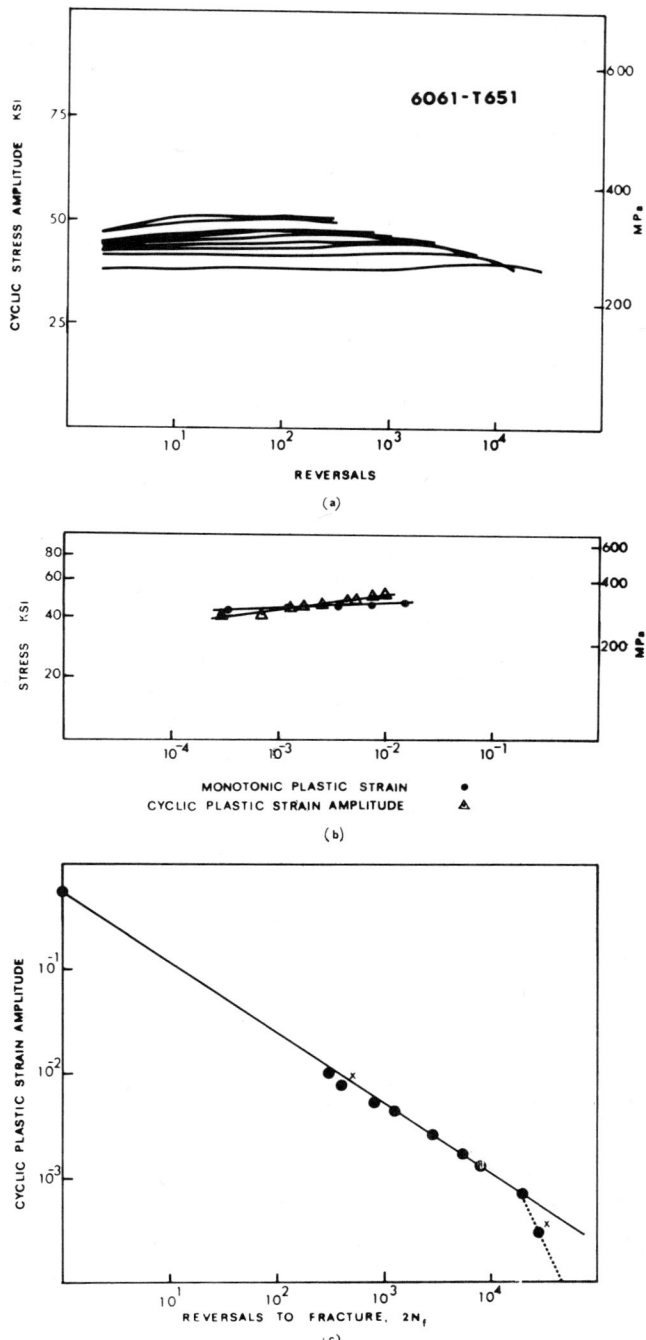

Fig. 5. SCF behavior of 6061-T651: (a) stress amplitude versus cycles, (b) cyclic and monotonic stress-strain relations, and (c) Coffin-Manson plot.

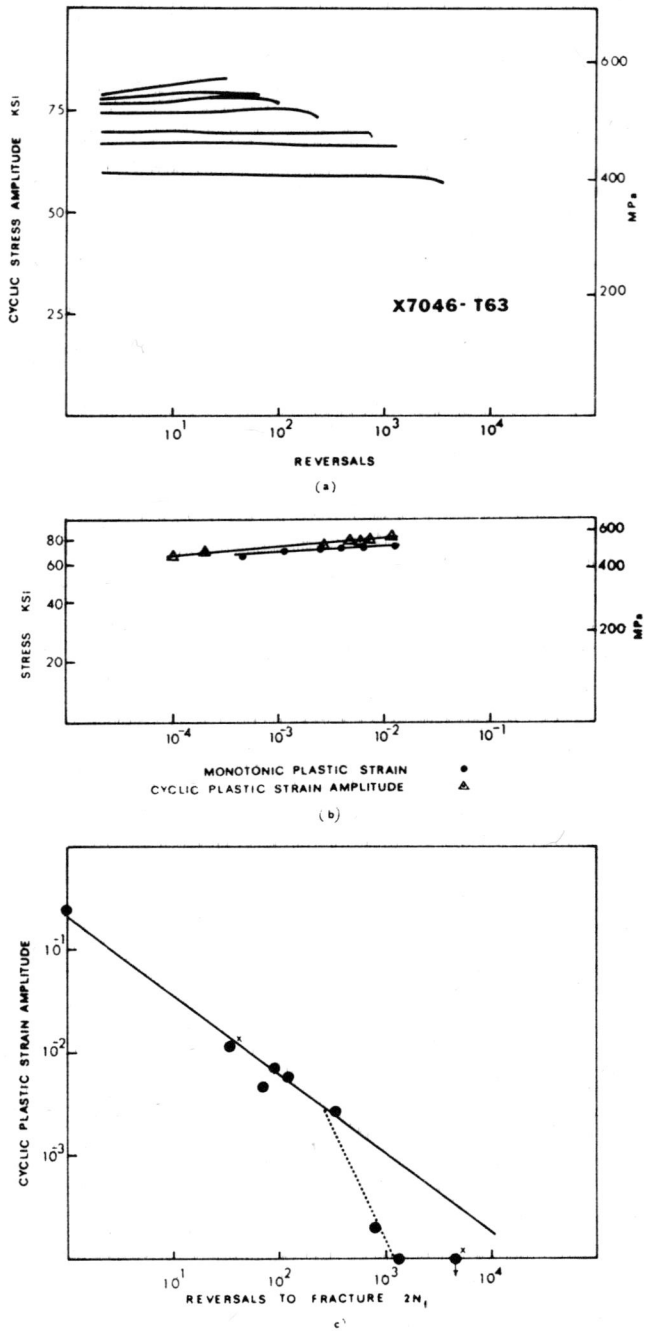

Fig. 6. SCF behavior of X7046-T63: (a) stress amplitude versus cycles, (b) cyclic and monotonic stress-strain relations, and (c) Coffin-Manson plot.

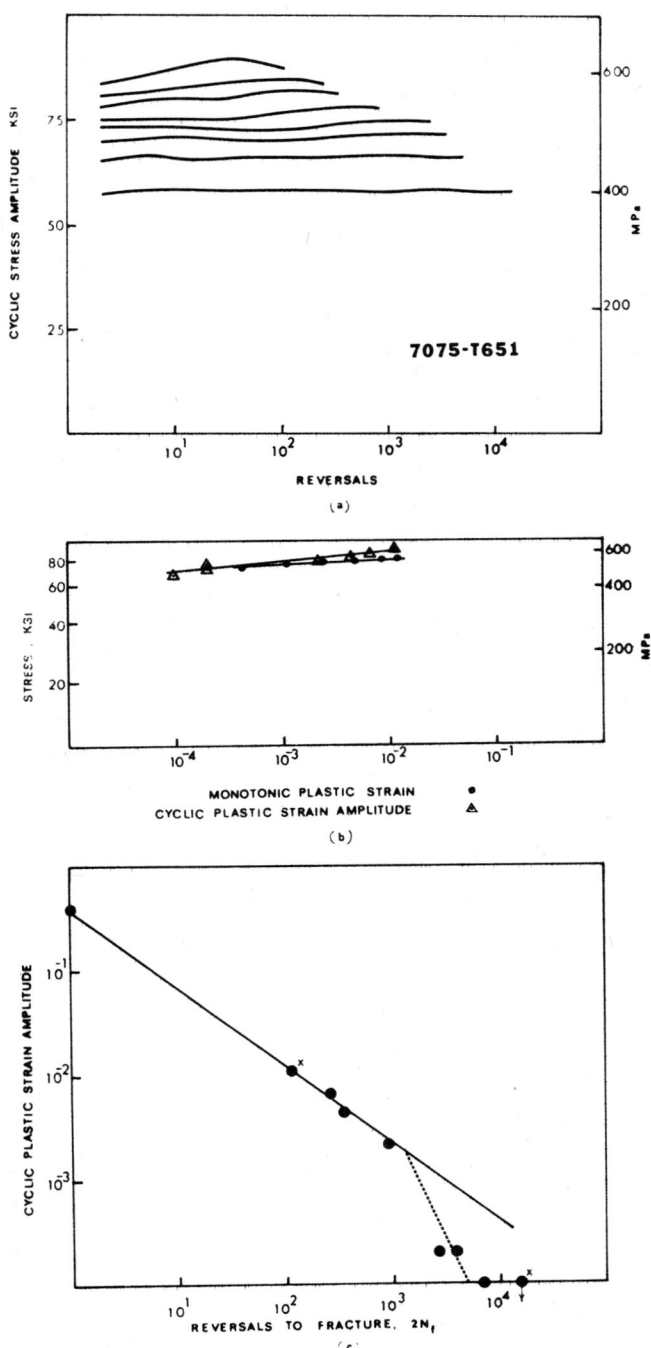

Fig. 7. SCF behavior of 7075-T651: (a) stress amplitude versus cycles, (b) cyclic and monotonic stress-strain relations, and (c) Coffin-Manson plot.

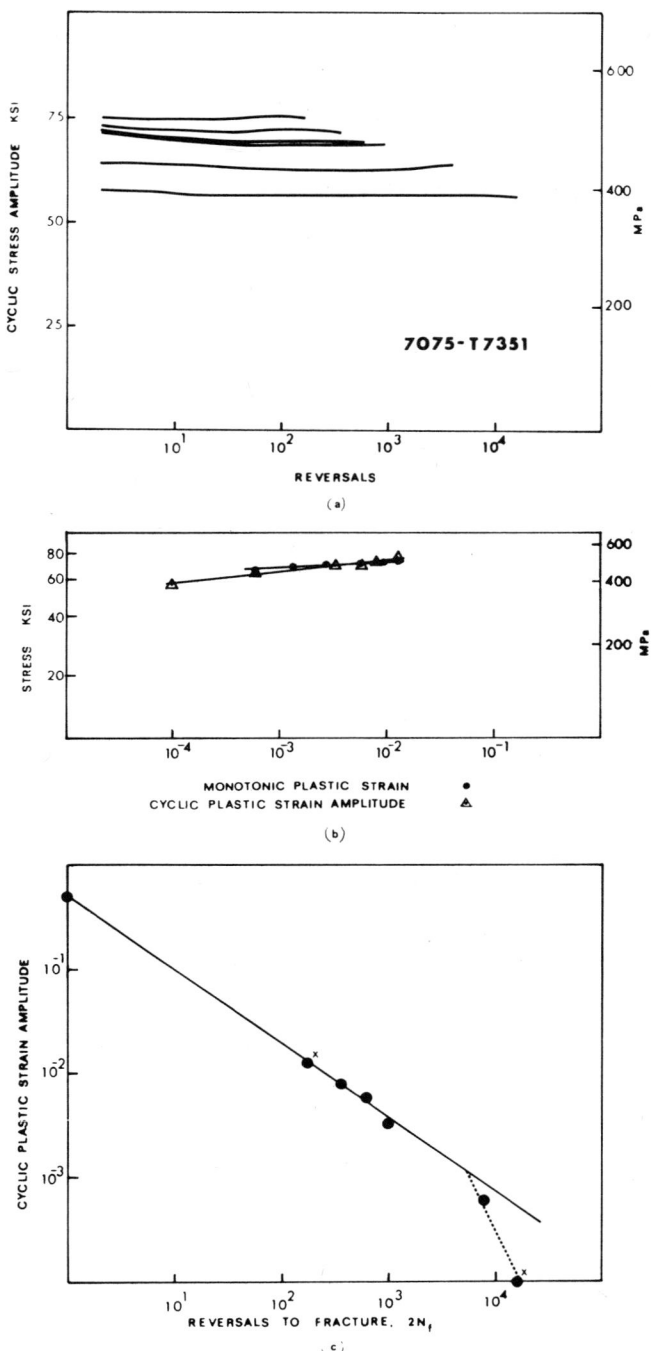

Fig. 8. SCF behavior of 7075-T7351: (a) stress amplitude versus cycles, (b) cyclic and monotonic stress-strain relations, and (c) Coffin-Manson plot.

Table III. Monotonic Tensile and Cyclic Parameters

	1350-H17	5083-O	2024-T351	6061-T651	X7046-T63	7075-T651	7075-T7351
Monotonic Properties(a)							
S_y, ksi (MPa)	17.7 (122)	19.0 (131)	52.9 (365)	43.3 (299)	70.7 (487)	77.6 (535)	69.9 (482)
S_u, ksi (MPa)	18.9 (130)	40.0 (276)	71.7 (494)	48.5 (334)	76.2 (525)	85.3 (588)	78.6 (542)
RA, %	84	36	30	42	21	32	39
ϵ_f	—	0.45	0.36	0.54	0.24	0.39	0.49
n	0.061	0.089	0.020	0.019	0.030	0.025	0.026
E, ksi (MPa)	10140 (69913)	10180 (70189)	10800 (74463)	10270 (70810)	10250 (70672)	10350 (71361)	10400 (71706)
Cyclic Properties(b)							
$S_y{}'$, ksi (MPa)	11.5 (79.3)	42.2 (291)	64.3 (443)	44.3 (305)	76.1 (525)	80.2 (553)	67.1 (463)
$\epsilon_f{}'$	0.24	0.43	0.36	0.50	0.21	0.39	0.50
n'	0.074	0.042	0.052	0.072	0.040	0.050	0.057
c*	-0.49	-0.70	-0.68	-0.68	-0.80	-0.75	-0.71

(a) S_y, 0.2% offset yield strength
S_u, ultimate strength
RA, reduction in area
ϵ_f, true fracture ductility
n, strain–hardening exponent
E, modulus of elasticity.

(b) $S_y{}'$, 0.2% offset cyclic yield strength
n, cyclic–strain–hardening exponent
c, fatigue–ductility exponent
$\epsilon_f{}'$, fatigue–ductility coefficient.

*Indicates slope above the critical level of plastic strain amplitude.

a large magnitude that the constant-volume assumption used in the true-fracture-ductility calculation was not considered valid. A reiterated, linear, "least-squares" regression-analysis procedure was employed by adding lower plastic-strain-amplitude data points after the first iteration. This initial iteration contained the true fracture ductility and the first three additional data points. Subsequent data points were added until the standard deviation about the regression line significantly increased. This lowest data point and all subsequent lower plastic strain amplitudes were not considered to obey the single-slope behavior as described by the Coffin-Manson relationship. In the high-plastic-strain region of the strain-life curve, a slope of approximately −0.7 described the behavior for all the alloys investigated except 1350-H17, which was approximately −0.5.

The critical strain level at which the data departed from this relationship was estimated by superimposing the plots of plastic strain versus number of reversals to failure for all the alloys. This superpositioning was accomplished by maintaining each alloy's plot at coincident strain levels and sliding the plots along the reversals-to-failure axis until a "by eye" judgment was made that all the lower plastic-strain data points fell near and along the same straight line. A best-fit "by eye" straight line was then drawn through the data and transferred, in the form of a dashed line, to each of the alloy plots with the plots still superimposed. The point where this dashed line intersects the high-plastic-strain line is defined as the critical plastic-strain level for each alloy.

4.3 Structure Analysis, Posttest

Figures 9 through 16 show the postfracture microstructures of specimens representing test results as indicated by the "X" in Figs. 2c through 8c. Over the range of data taken, 1350-H17 exhibited single-slope strain-life behavior and developed a similar deformation microstructure at both high and low plastic-strain amplitudes. However, for the other alloys investigated, where there was a significant departure from single-slope behavior, the deformation microstructures of specimens tested above and below the break in the Coffin-Manson plot differed. The microstructures of specimens tested above this critical level of plastic strain had a uniform, high density of dislocations (Figs. 10, 12a through 16a). In contrast, microstructures of specimens tested below the critical level typically contained regions of nonuniform dislocation density (Figs. 11, 12b through 16b). These observed relative differences in uniformity of the dislocation structure were insensitive to foil orientation and, therefore, not a contrast effect.

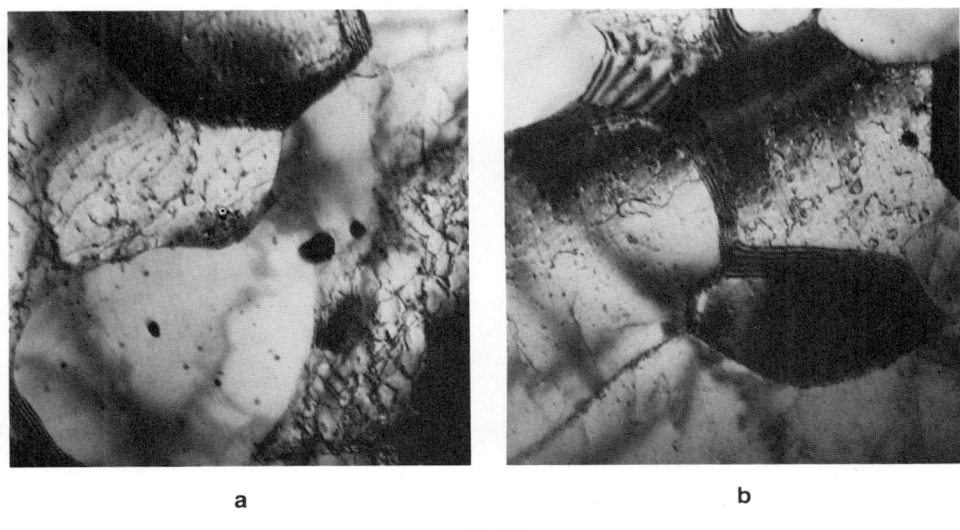

a   b

.5 μm

Fig. 9. Deformation structure of 1350-H17 cycled to failure at (a) high plastic strain amplitude (PSA) and (b) low PSA cycling.

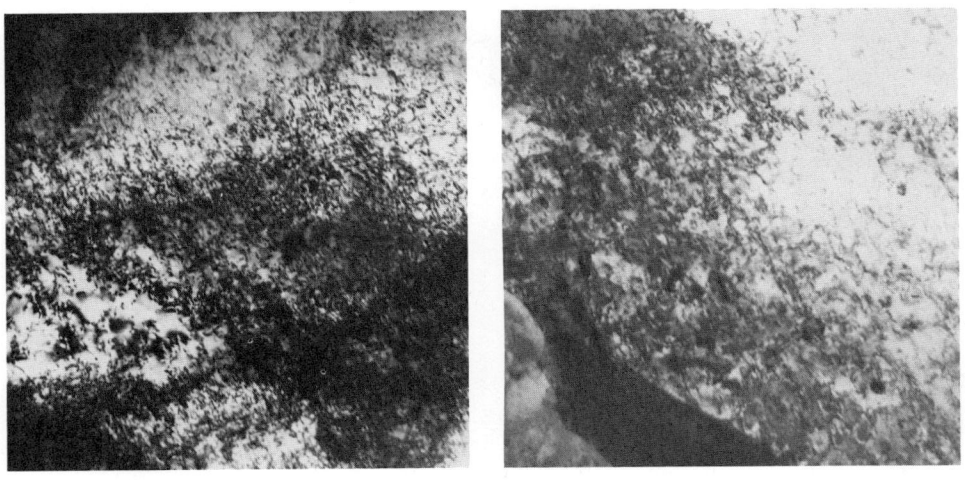

.5 μm

Fig. 10. Deformation of structure of 5083-O cycled to failure at high PSA.

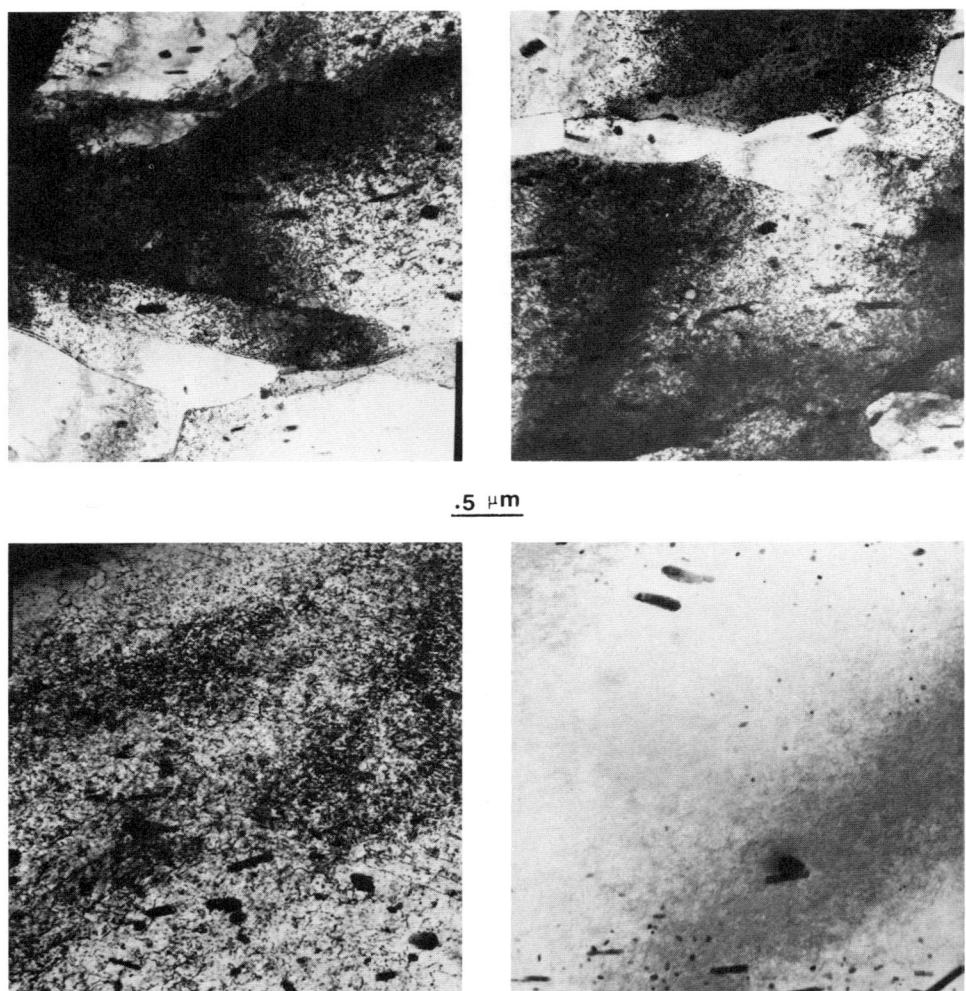

.5 μm

Fig. 11. Deformation structure of 5083-O cycled to failure at low PSA showing variations in structure within the same foil.

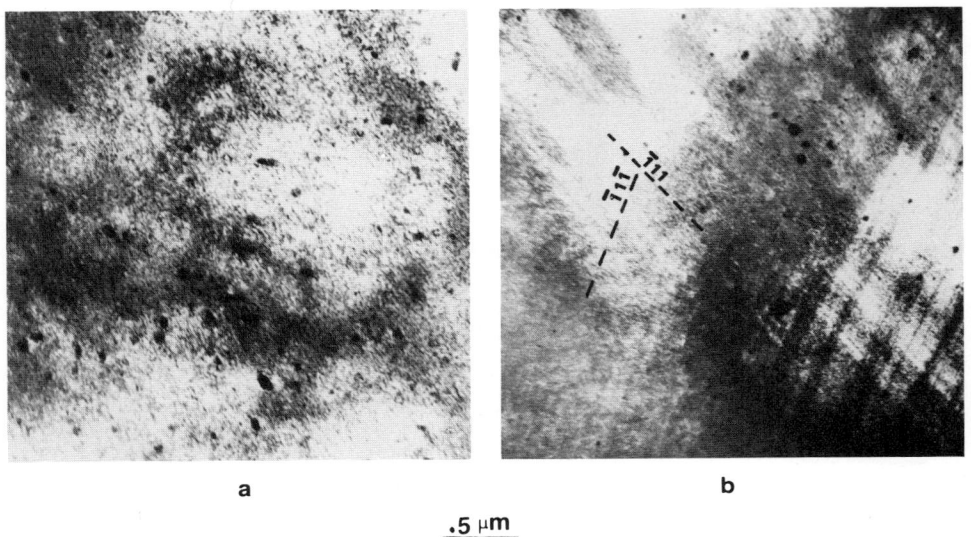

Fig. 12. Deformation structure of 2024-T351 cycled to failure at (a) high PSA and (b) low PSA.

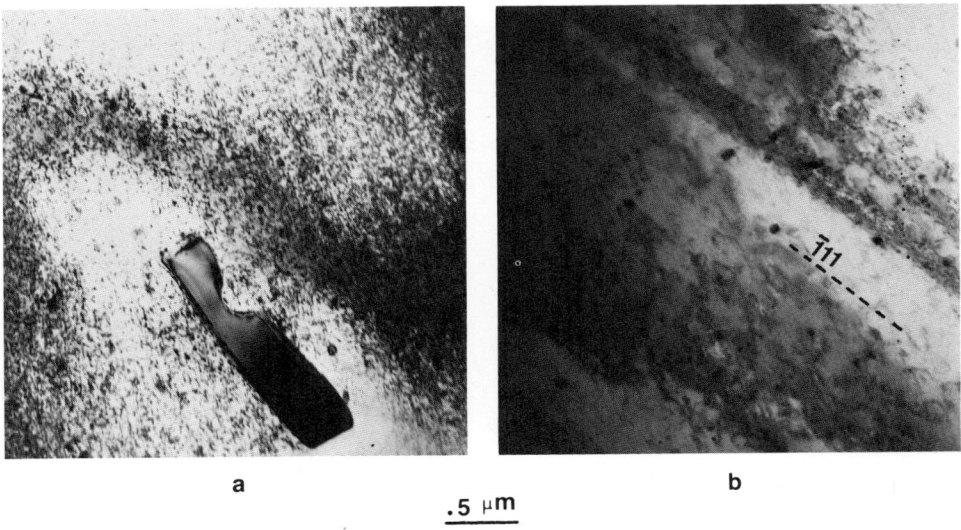

Fig. 13. Deformation structure of 6061-T651 cycled to failure at (a) high PSA and (b) low PSA.

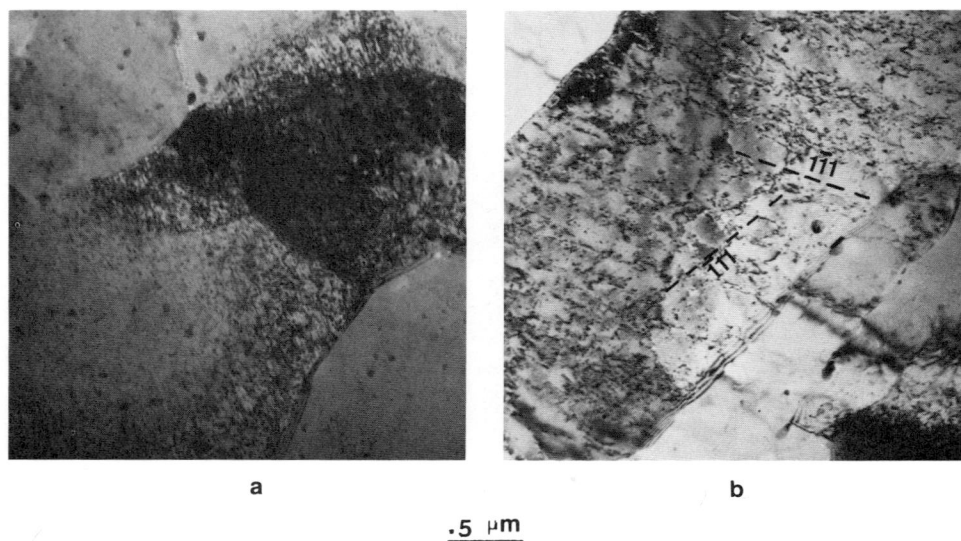

a   b

.5 μm

Fig. 14. Deformation structure of X7046-T63 cycled to failure at (a) high PSA and (b) low PSA.

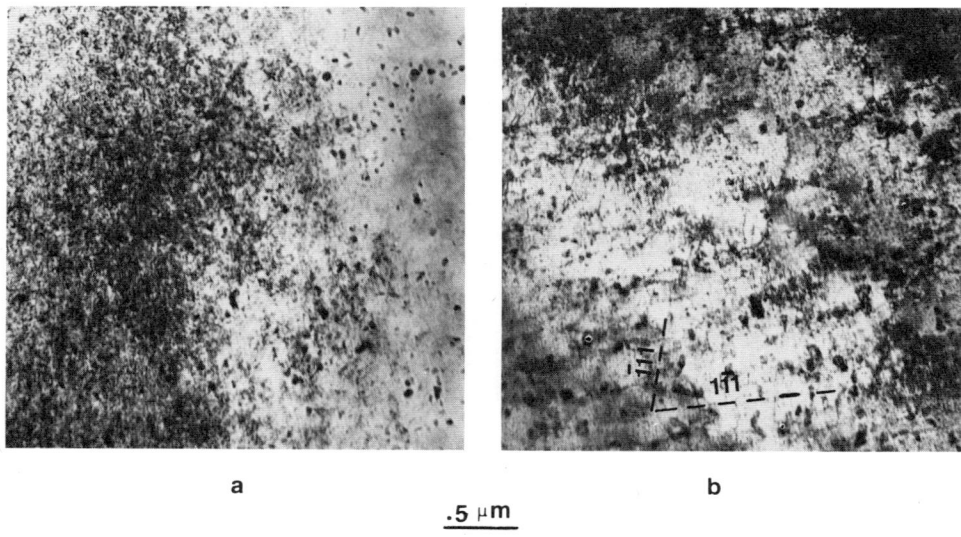

a   b

.5 μm

Fig. 15. Deformation structure of 7075-T651 cycled to failure at (a) high PSA and (b) low PSA.

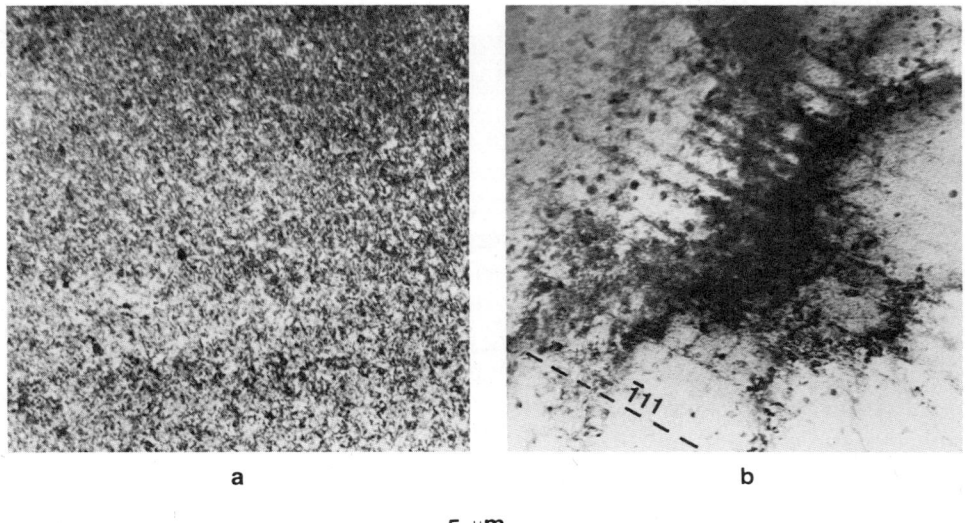

a   b

.5 μm

Fig. 16. Deformation structure of 7075-T7351 cycled to failure at (a) high PSA and (b) low PSA.

5 DISCUSSION

5.1 Failure Criterion

Using the plastic-strain amplitude versus cyclic-life data, Coffin[13] and Manson[14] proposed that fatigue life could be related explicitly to the plastic strain amplitude. Mathematically, the expression can be written as:

$$\frac{\Delta\epsilon_p}{2} = \epsilon'_f \, (2N_f)^c \quad,$$

where

$\dfrac{\Delta\epsilon_p}{2}$ = the plastic-strain amplitude

ϵ'_f = the fatigue-ductility coefficient

$2N_f$ = the number of reversals to failure

c = the fatigue-ductility exponent.

One objective of research on fatigue mechanisms is to provide a foundation for the development of microstructures that are more resistant to fatigue initiation. Consequently, possible criteria of fatigue-crack initiation were sought so that a suitable criterion would be established to define the number of reversals for use in a Coffin-Manson analysis. Like the definition of tensile yield stress, the definition of fatigue-crack initiation is arbitrary and depends on the sensitivity of the measuring device. Unlike the clearly defined tensile yield stress, however, the definition of fatigue-crack initiation has not been standardized. Suggested definitions include the smallest crack detectable either unaided or with magnification, a crack long enough to cause a significant decrease in the load-carrying capability of a structural component, or a crack sufficient in size to cause fracture of a cylindrical fatigue specimen. In this investigation, crack detection by observation of the specimen surface was impractical because the extensometer prevented direct observation of the complete specimen surface. An alternative failure criterion for crack initiation based on a 2-ksi (13.8-MPa) load drop gave relationships similar to those obtained using fracture as the failure criterion.* Moreover, use of specimen fracture permits extrapolation to monotonic fracture ductility, ϵ_f. Consequently, the less ambiguous and easier to determine number of cycles to specimen fracture was used to define N_f for this investigation.

5.2 Related Work

Other investigators have used Coffin-Manson plots to relate microstructure to fatigue resistance. They also found that, for some microstructures, the data could not be described by a single-slope relationship.

Saxena and Antolovich[7] recently studied the effect of stacking-fault energy, SFE, on the fatigue deformation process in the Cu-Al system. They observed that for the high-SFE alloys, the fatigue-life plot was linear over the range in plastic-strain amplitude tested. However, for the low-SFE alloys the fatigue life showed a distinct break. They associated the break with the inability of the dislocations to cross slip at low plastic strain amplitudes when the SFE was low.

Sanders and Starke[9] recently investigated an Al-Zn-Mg-Cu alloy and related results of SCF to microstructure. They concluded that variations in precipitate size, distribution, and spacing could alter the dislocation-precipitate interactions and that these interactions were dependent upon the magnitude of the plastic-strain amplitude. Consequently, multiple slope

---

*The ratio of the life defined by load drop to the life defined by fracture was not dependent on alloy or strain amplitude and had a mean value of 0.87. Therefore, the incremental differences of the log failure lines were constant, and the slopes and critical values for departure from single-slope behavior were the same for these two failure criteria.

behavior in Coffin-Manson plots would be expected when the homogeneity of the deformation process is affected by the magnitude of the cyclic plastic strain.

5.3 Consequences of Heterogeneous Deformation

The presence or absence of localized deformation bands is significant for two reasons: (1) the bands can act as initiation sites for fracture[23] and (2) the magnitude of microplastic strain cannot be accurately measured by a conventional extensometer. Because the extensometer measures the average rather than the localized strain in the deformation bands, the Coffin-Manson plots for microstructures which deform homogeneously at high plastic strains and heterogeneously at low plastic strains must exhibit multiple-slope behavior.

The graphical explanation for the correlation between the change in slope in the plot of plastic-strain amplitude versus cyclic life and the development of heterogeneous deformation follows. Figure 17 depicts two

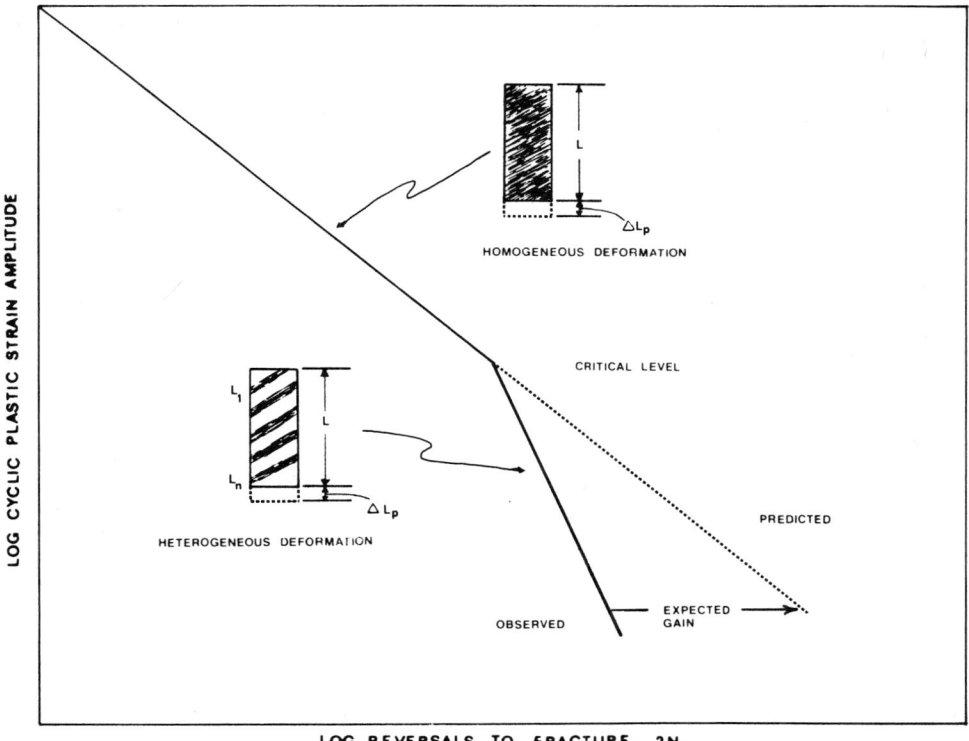

Fig. 17. Departure from single-slope behavior of a Coffin-Manson plot.

cylinders of material representing the specimen test section between extensometer knife edges. The top cylinder represents the homogeneous deformation that takes place above the critical level of plastic-strain amplitude. Here
the reversed plastic deflection, ΔL_p, per cycle is uniformly distributed over
the length, L, of the cylinder of material, resulting in a macroscopic plastic
strain that agrees in magnitude with the microscopic plastic strain.
Therefore, this portion of the plot can be represented by the empirical
Coffin-Manson relationship which assumes a continuum. The lower cylinder
represents the heterogeneous deformation below the critical level. Here the
reversed plastic deflection per cycle is concentrated in only a portion of the
cylinder, $L_1 + L_2 + L_3$, and results in a measured macroscopic plastic strain
that is lower in magnitude than the microscopic plastic strain. Assuming
that cyclic life is determined by the microscopic plastic strain and follows
the Coffin-Manson relationship, the lower value of the measured
macroscopic plastic strain will cause a change in slope toward shorter cyclic
life on a log-log plot.

    To compare the alloys on an equivalent basis, the plastic-strain
amplitudes for each alloy were normalized by dividing by the true fracture
ductility, ϵ_f. A plot of the log of the normalized plastic-strain amplitude versus the log of the number of reversals to failure for all the alloys is shown in
Fig. 18. In this composite plot, all the alloys have a similar behavior in the

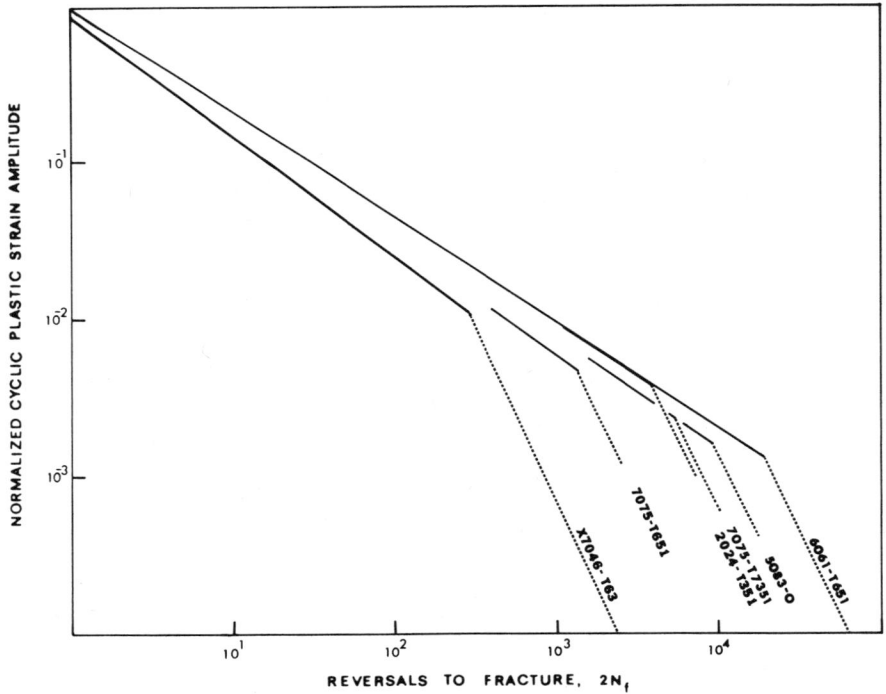

Fig. 18. Normalized Coffin-Manson plots.

high-plastic-strain-amplitude regime. However, in the low-plastic-strain regime, below the departure from single-slope behavior, the variations between the alloys become significant.

For the precipitation-hardening alloys, the position of the break in decreasing order of normalized plastic-strain amplitude was X7046-T63, 7075-T651, 2024-T351, 7075-T7351, and 6061-T651. We believe that the relative location of the break relates to the ability of each of the microstructures to maintain homogeneous deformation at low levels of plastic-strain cycling.

5.4 Deformation in Age-Hardening Alloys

In precipitation-hardening alloys, the increase in flow stress during aging is due to the interactions of dislocations with preprecipitates and precipitates. Coherent and partially coherent precipitates may be penetrated by dislocations since the slip systems of the precipitates and the matrix are generally coincident. Precipitates are more or less effective in retarding dislocation movement because of differences in stacking-fault energy, lattice parameter, Peierls force, or specific volume between the matrix and the second-phase particles.[15] The strength and microdeformation characteristics of a precipitation-hardening alloy will thus depend upon the degree of coherency, size, spacing, and uniformity in the distribution of precipitates. The 7XXX alloys are effective in producing high strengths because they develop a high density of closely spaced coherent and semicoherent precipitates. Microdeformation mode of 7XXX alloys depends on the degree of aging and on the stress.[16-18] Dense slip bands develop in materials aged in the vicinity of peak strength when they are strained a few percent. With additional stress, the bands may widen and intersect or, alternatively, cross slip may initiate between the bands. In either case, microdeformation at high strains becomes homogeneous. The tendency to form bands at low plastic strains is altered by either underaging or overaging, but the actual mechanism associated with their formation is not thoroughly understood. The presence of the deformation bands has been attributed to different conditions: (1) a lower than average precipitate density caused by inherent difficulties in obtaining a rapid, uniform quench[19], or (2) localized weakening which occurs by interaction of dislocations with the precipitates, thus reducing the effectiveness of the precipitates in a narrow region[20-37].

The precipitation-hardenable alloys in this investigation developed dense bands of dislocations when cycled at low-plastic-strain amplitudes. When cycled at high-plastic-strain amplitudes, however, they developed a uniform distribution of dislocations. This difference in the uniformity of deformation suggests that a process such as cross slip to avoid particles was possible at high plastic strains, but was not favorable at low plastic strains. Therefore, the proportion of the structure containing concentrated slip bands increased as the plastic-strain amplitude decreased.

Differences in SCF test behavior among X7046 and the two tempers of 7075 are consistent with the Mg:Zn ratios and the morphology and coherency of the precipitates. In X7046 the magnesium and zinc are in the ratio of $MgZn_2$. In the T63 temper, the microstructure consists of small, coherent precipitates having the approximate composition of $MgZn_2$ in a solute-depleted aluminum matrix. This combination of very hard precipitates in a soft matrix would accentuate the tendency toward concentration of slip; dislocations would find less resistance in a region where previous dislocations had cut precipitates than in virgin areas. The composition of 7075, however, is such that there is at least 1 percent atomic weight of magnesium in excess of the amount necessary to form either the $MgZn_2$ or $MgZnCuAl$ precipitates. The excess magnesium will be in solid solution and will contribute to the overall strength of the matrix. Consequently, slip in 7075-T6 would be more homogeneous than in X7046-T63 because the difference between the strength of the matrix and the precipitate is smaller. Continued aging of 7075 beyond peak strength to the T73 temper produces a large number of precipitates which are large, incoherent, and spaced further apart. Dislocations do not cut these particles but bypass them by an Orowan mechanism. As a consequence, the slip is less localized. The homogenization of slip shifts the departure from linearity to lower plastic strain amplitudes.

The precipitate morphology of the naturally aged 2024-T351 is different from the precipitate morphology of 7XXX alloys in naturally aged or T6-type tempers. Therefore, the precipitate-dislocation interactions are in some ways different. Because of the small difference in volume between equal numbers of aluminum atoms and of clusters of aluminum, zinc, and magnesium atoms, aluminum alloys containing zinc and magnesium develop zones which maintain an approximately spherical morphology during natural aging and in the initial stages of artificial aging. The strength after natural aging is a consequence of small variations of alloy chemistry in the vicinity of the zones and can be explained in terms of solute or cluster drag on the dislocation. In contrast, clusters of aluminum, copper, and magnesium atoms occupy a smaller volume than equal numbers of aluminum atoms. Consequently, 2024 develops platelike zones surrounded by an elastic strain field. Thus, in addition to solute or cluster drag on the dislocations, the elastic strain field adds to the drag on a moving dislocation. Therefore, significant strength can be achieved by natural aging.

During natural aging, the zones are small and closely spaced. Therefore, the structure of 2024-T351 is more homogeneous than the structure of artificially aged 7XXX alloys and will deform more homogeneously.

The tendency for more homogeneous deformation in 6061 is attributed to a volume fraction of precipitate-interparticle spacing effect. Owing to the relatively low amount of magnesium and silicon in 6061, only a small amount of Mg_2Si is available to form zones and precipitates and the inter-particle spacing in the T6 temper is correspondingly large. Consequently, the precipitates provide minimum interference to the motion of the dislocations at low-plastic-strain amplitudes, and deformation is homogeneous.

5.5 Deformation in Al-Mg Alloys

Another feature of microstructure that may cause the deformation process to be dependent upon the plastic-strain amplitude is represented by 5083-O, a magnesium—solid-solution-hardened aluminum alloy. Manganese and chromium are added to this alloy to form high temperature precipitates which retard recrystallization and inhibit grain growth if recrystallization does occur. These dispersoids characteristically have a nonuniform distribution which is a consequence of segregation during solidification of the ingot. Typically, they are 1 to 5 μm in size, are incoherent with the aluminum matrix, and as such act as dislocation sources during plastic deformation (Fig. 19).[38,39] Nonuniform distribution of dislocation sources would produce a nonuniform distribution of dislocations at low-plastic-strain amplitudes because only favorable sources can operate. With high-plastic-strain cycling, however, more sources can operate and thus produce homogeneous microscopic deformation.

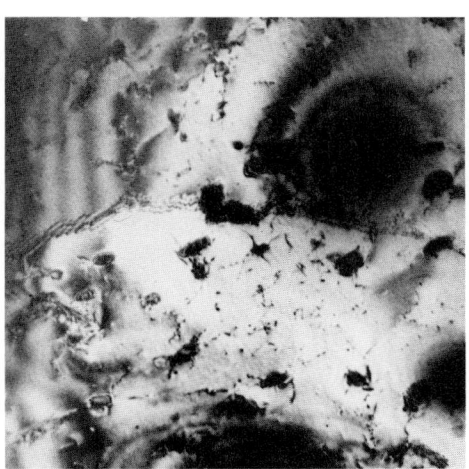 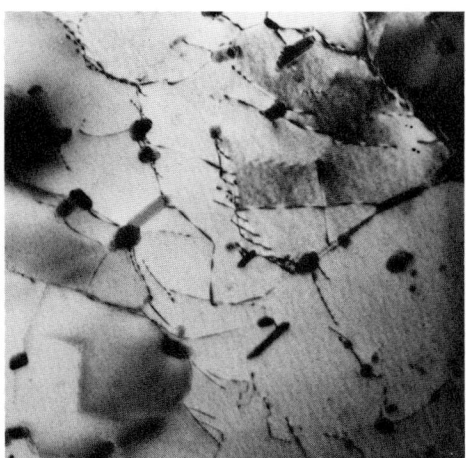

.5 μm

Fig. 19. 5083-O deformed approximately 1 percent showing that the incoherent interfaces between dispersoids and the matrix have acted as dislocation sources.

6 SUMMARY AND CONCLUSIONS

A variety of commercial aluminum alloys in standard tempers was investigated using strain control fatigue and transmission electron microscopy. The log of the number of reversals to failure increased linearly as the log of the plastic-strain amplitude decreased down to a critical level of plastic-strain-amplitude cycling. Below the critical level of plastic strain, the number of cycles to failure was lower than would be predicted by extrapolating the high-plastic-strain-amplitude data. The deviations from single-slope behavior of a Coffin-Manson plot related to the relative inability of the microstructure to develop homogeneous slip during low-plastic-strain cycling. Cycling above the critical level of plastic-strain amplitude produced deformation structures characterized by uniform distributions of dislocations. The uniform microdeformation was essentially the same as the macroscopic deformation measured by the extensometer. Cycling below the critical level of plastic-strain amplitude, however, produced deformation structure characterized by nonuniform distributions of dislocations. Consequently, the highly localized microdeformation was much higher than the macroscopic deformation measured by the extensometer. As a result, the fatigue lives were dictated by these areas of highly localized deformation and thus were shorter than predicted by extrapolating macroscopic deformation from the high-plastic-strain-amplitude regime.

This work indicates that strain-control-fatigue measurements in conjunction with transmission electron microscopy can be an effective tool in an alloy development program to evaluate microstructures that would promote homogeneous slip at low-plastic-strain cycling and thus produce greater resistance to fatigue-crack initiation.

ACKNOWLEDGMENTS

The authors would like to thank Mr. R. A. Kelsey for initiating the investigation. We would also like to express our sincere appreciation to Mr. B. A. Walker for his assistance in completing the strain-control-fatigue tests. We are indebted to Dr. W. G. Fricke, Jr., for the many helpful discussions.

REFERENCES

1. Sandor, B. I., *Fundamentals of Cyclic Stress and Strain,* University of Wisconsin Press, Madison, Wisconsin (1972).
2. Feltner, C. E., and Laird, C., *Acta Met.,* **15**, 1621 (1967).
3. Feltner, C. E., and Laird, C., *ibid.,* **15**, 1633 (1967).
4. Calabrese, C., and Laird, C., *Mater. Sci. Eng.,* **13**, 141 (1974).

5. Calabrese, C., and Laird, C., *ibid.,* **13**, 159 (1974).
6. Calabrese, C., and Laird, C., *Met. Trans.,* **5**, 1785 (1974).
7. Saxena, A., and Antolovich, S. D., *Met. Trans.,* **6A**, 1809 (1975).
8. Sanders, T. H., Jr., Ph.D. Thesis, Georgia Institute of Technology, Atlanta, Georgia (1974).
9. Sanders, T. H., Jr., and Starke, E. A., Jr., *Met Trans.* (to be published).
10. Chien, K. H., and Starke, E. A., Jr., *Acta Met.,* **23**, 1173 (1975).
11. *1975 Annual Book of ASTM Standards,* Part 10, American Society for Testing and Materials, Philadelphia, November, 1975, p. 611.
12. *Aluminum,* Vol. I, K. R. Van Horn (Ed.), American Society for Metals, Metals Park, Ohio (1967).
13. Coffin, L. F., *Trans. ASME,* **76**, 931–950 (1954).
14. Manson, S. S., *NACA Tech. Note,* No. 2933 (July, 1953).
15. Gleiter, H., and Hornbogen, E., *Mater. Sci. Eng.,* **2**, 285–302 (1967–68).
16. Unwin, P.N.T., and Smith, G. C., *J. Inst. Metals,* **97**, 183 (1960–61).
17. Stubbington, C. A., *Acta Met.,* **12**, 931 (1964).
18. Speidel, M. O., *Proceedings of Conference on Fundamental Aspects of Stress-Corrosion Cracking,* The Ohio State University, Columbus, Ohio (1967), p. 561.
19. Lynch, S. P., *Metal Science Journal,* **7**, 93 (1973).
20. Polmear, I. J., and Bainbridge, I. F., *Phil. Mag.,* **4**, 1293 (1959).
21. McEvily, A. J., Clark, J. B., Utley, E. C., and Herrstein, W. H., *Trans. Met. Soc., AIME,* **227**, 1093 (1963).
22. Forsyth, P.J.E., *J. Australian Inst. Met.,* **8**, 52 (1963).
23. Stubbington, C. A., *Acta Met.,* **12**, 931 (1964).
24. Stubbington, C. A., and Forsyth, P.J.E., *ibid.,* **14**, 5 (1966).
25. Abel, A., and Ham, R. K., *ibid.,* **14**, 1495 (1966).
26. Gleiter, H., *ibid.,* **16**, 455 (1968).
27. Lynch, S. P., and Ryder, D. A., *Aluminum,* **49**, 748 (1973).
28. Calabrese, C., and Laird, C., *Mater. Sci. Eng.,* **13**, 141 (1974).
29. Hanstock, R. F., *J. Inst. Metals,* **83** (1954–55).
30. Broom, T., Molineux, J. H., and Whittaker, V. N., *ibid.,* **84**, 357 (1955–56).
31. Broom, T., Mazza, J. A., and Whittaker, V. N., *ibid.,* **86**, 17 (1957–58).
32. Broom, T., and Whittaker, V. N., *Nature,* **177**, 486 (1956).
33. Seitz, F., *Advances in Physics,* **1**, 43 (1952).
34. Broom, T., *ibid.,* **3**, 26 (1954).
35. Turnbull, D., in Reb. Conference on Defects in Crystalline Solids, Physical Society (1955), p. 203.
36. Clark, J. B., and McEvily, A. J., *Acta Met.,* **12**, 1359 (1964).
37. Averback, B. L., *Trans. Amer. Soc. Metals,* **41**, 262 (1949).
38. Hirth, J. P., and Lothe, J., *Theory of Dislocations,* McGraw-Hill, New York (1968).
39. Swann, P. R., in *Electron Microscopy and Strength of Crystals,* Interscience Publishers, New York (1963), p. 167.

APPENDIX

Strain Control Fatigue Testing

The cyclic stress-strain response from SCF testing can be expressed similarly to the monotonic stress-strain response with the total strain equal to the sum of the elastic and plastic portions of strain:

$$\frac{\Delta \epsilon_t}{2} = \frac{\Delta \epsilon_e}{2} + \frac{\Delta \epsilon_p}{2} = \frac{\sigma_a}{E} + \epsilon'_f (2N_f)^c \quad ,$$

where

$\Delta \epsilon_t$ = total strain range

$\Delta \epsilon_e$ = elastic strain range

$\Delta \epsilon_p$ = plastic strain range

ϵ'_f = fatigue ductility coefficient

σ_a = stress amplitude

c   = fatigue ductility exponent

E   = elastic modulus

$2N_f$ = reversals to failure.

A schematic of the quantities associated with a generalized hysteresis loop is presented in Fig. 20.

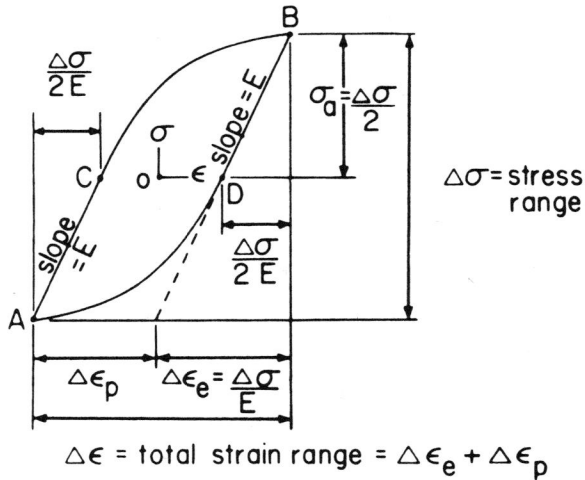

$$\Delta \epsilon = \text{total strain range} = \Delta \epsilon_e + \Delta \epsilon_p$$

Fig. 20. Quantities associated with the hysteresis loop.[1]

In SCF testing, the independent variable, stress, is monitored and plotted as a function of the number of cycles. The cyclic-dependent material responses under constant strain control are shown in Fig. 21. An increase in stress amplitude indicates cyclic hardening; constant stress amplitude, saturation; and a reduction in stress amplitude, cyclic softening. By cycling several specimens to failure at constant-strain amplitudes, the empirical Coffin-Manson relationship can be plotted. Mathematically, this relationship is of the form

$$\frac{\Delta \epsilon_p}{2} = \epsilon'_f \, (2N_f)^c \quad ,$$

and when plotted as

$$\log \frac{\Delta \epsilon_p}{2} = \log (\epsilon'_f) + c \log (2N_f) \quad ,$$

results in a linear plot. Two relationships are plotted in Fig. 22. Considering two materials having equal values of ϵ'_f, the material having the smallest absolute value of c has a greater tolerance to cyclic plastic straining.

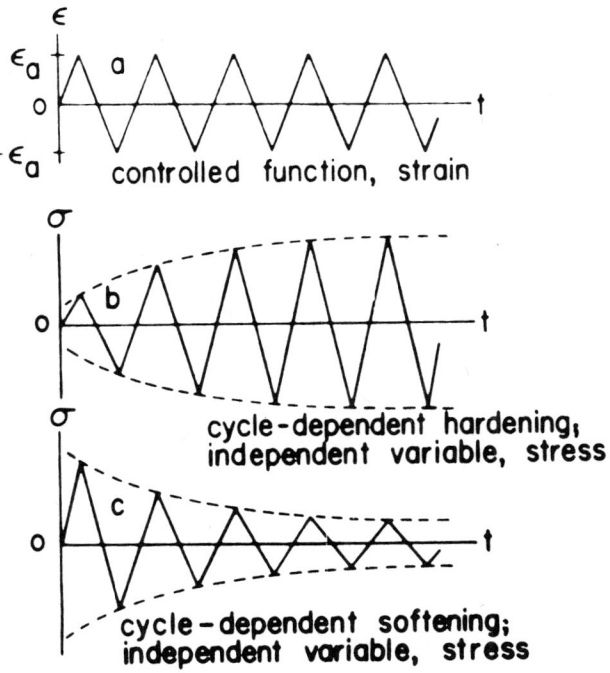

Fig. 21. Cycle-dependent material responses under strain control.[1]

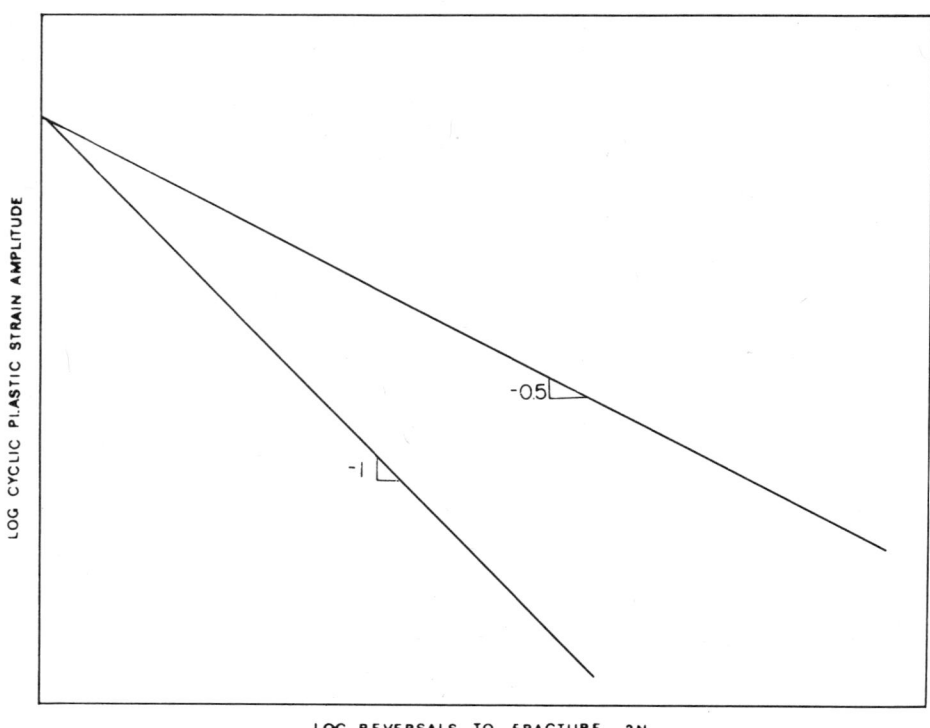

Fig. 22. Relationship between the plastic strain amplitude, $\triangle\epsilon_p/2$, and the number of reversals to failure, $2N_f$, at two different values of fatigue ductility exponent, c.

A cyclic stress-strain curve can be obtained by plotting the values of stress amplitude versus the corresponding values of strain amplitude. The stress amplitude is determined at saturation; however, if saturation does not occur, the stress amplitude at half-life is generally used. If the log (σ_a) versus log ($\triangle\epsilon_p/2$) is plotted, the slope n', may be determined. The slope is the cyclic-strain-hardening exponent and is similar to that obtained by plotting the monotonic stress-strain values.

DISCUSSION on Paper by T. H. Sanders, Jr.

McCLINTOCK: Wouldn't it be easier for the designer, in fitting to high-cycle-fatigue data, to express the results in terms of total strain?

SANDERS: Yes, when we communicate data of this type to designers we will need to use total strain amplitude. However, our purpose here was to use plastic-strain amplitude as an interpretative tool in studies of deformation processes.

BEMENT: Coffin in recent years has investigated the effects of cycle frequency and hold times on the relative effects of creep relaxation and environmental effects on departures from linearity at low-plastic-strain amplitudes on a Coffin-Manson plot. Those effects are best represented by plotting the log of the plastic-strain range against the log of the "frequency-modified" fatigue life. This approach would appear to be especially advantageous in your studies to discriminate the effects of non-uniform deformation and stress-corrosion cracking.

SANDERS: We agree that the use of the "frequency-modified life" approach may be a way to separate nonhomogeneous deformation effects from corrosion effects. For this reason we expect to address this approach in the future.

TIEN: I believe that fatigue frequency could be an interesting variable for your plastic-strain-controlled LCF tests on aluminum alloys, especially with respect to using this test as a sorting tool for alloy design. I would expect the slip mode to change over the normal laboratory frequency range—a fact that may change the relative response of your alloys to LCF.

SANDERS: We agree that frequency may have an effect on the deformation process and therefore relative alloy response. We will address this variable in future studies through the effect of strain rate.

THOMAS: When I studied slip distribution, microstructure, and fracture in aluminum alloys many years ago, I found a strong temperature dependence of slip mode and particle distribution. At low temperatures one finds confined planar slip (high slip steps); often the particles, specifically Mg_3Al_2, cracked. At ambient temperatures, the slip steps were finer and no particle fracture was observed; presumably, cross slip is easier. I wonder if fatigue will show similar trends. Certainly the effect of temperature on slip and fracture in fatigue should be investigated. The effect of magnesium as a solid-solution strengthener was very potent on the slip-fracture-temperature dependence mentioned above. Also, particle fracture itself may not be too deleterious to the toughness of the alloy.

SANDERS: We anticipate the effect of temperature on dispersion of slip to be the same in fatigue as during monotonic deformation on the assumption that the mechanism of cyclic plastic and monotonic plastic deformation are the same. We hope to consider this question in the future. The question of the effect of particle fracture on toughness is not in the scope of this study.

RADIATION-DAMAGE CONSIDERATIONS*

B.R.T. Frost

Materials Science Division
Argonne National Laboratory
Argonne, Illinois

ABSTRACT

The designs of nuclear-fission and fusion power plants do not, in general, appear to make unusual demands on materials in terms of mechanical-property requirements. Superficially, the needs of the designer can be met by commercially available alloys. However, the radiation environment produces unique effects on the composition, microstructure, and defect population of these alloys, resulting in time-dependent and time-independent changes in mechanical properties. These changes must be quantified for the designer.

To illustrate these problems, the materials needs of the core of a Liquid-Metal Fast-Breeder Reactor (LMFBR) and of the first wall of a fusion reactor are discussed. In the case of the LMFBR core, the phenomenon

*Work supported by the U.S. Energy Research and Development Administration.

of void swelling causes serious design problems, and a search is being made for a low-swelling alloy that has adequate mechanical properties. The fusion reactor poses different problems because the neutron energy is high (14 MeV) and is accompanied by a high flux of charged particles. The long-term choices for a wall material have been narrowed to vanadium and niobium alloys.

In the search for low-swelling alloys, it has become clear that minor elements play an important role in determining the nature of the radiation effects. The segregation of minor elements to void surfaces and the dispersion and reformation of second-phase precipitates are two important radiation-induced phenomena that require additional study in view of their influence on void swelling and high-temperature properties.

1 INTRODUCTION

The mechanical properties required of alloys for use in nuclear plants are basically similar to those required in any power plant. In water-moderated fission reactors operating at temperatures below 800 F (430 C), the materials generally deform elastically, and design emphasis is placed on mechanisms of crack initiation and propagation. In gas-cooled and sodium-cooled fission reactors and in fusion reactors, materials are generally operating above 400 C. In this case, the design emphasis is on high creep strength and ductility and on creep-fatigue interactions (cumulative damage), i.e., on time-dependent properties. However, the temperatures and stresses involved are within the ranges encountered in modern steam and gas turbine systems.

To meet these requirements and to have the capability of withstanding the generally corrosive environments of these reactors, conventional austenitic and ferritic steels have been used, together with zirconium alloys that were specially developed for use in water reactors. For the future, increasing emphasis will be given to precipitation-hardened alloys and to the bcc metals and alloys. This will permit increased operating temperatures and thermal efficiencies.

The unique feature of fission and fusion reactors is the radiation environment. Energetic neutrons, gamma rays, and charged particles produce radiation damage and localized heating within the bulk of the alloys. Radiation damage, for the purposes of this discussion, involves the production of lattice defects and transmutation products. The production of defects may lead to enhanced diffusion (including enhanced creep), defect cluster formation, and void formation. The dispersion of precipitates into the lattice under energetic particle bombardment is a special case; a dynamic, flux-dependent equilibrium is established between dispersion and reprecipitation, which is enhanced by radiation. Transmutations can produce hydrogen and helium gases and other metallic species, e.g., niobium becomes transmuted to zirconium, so that the pure niobium wall of a fusion power reactor would become a Nb-10% Zr alloy after approximately 20 years. Thus, although conventional alloys have often been selected for nuclear applications, their long-term behavior in service is quite different from that in a nonnuclear plant.

2 EFFECTS OF RADIATION ENVIRONMENT ON ALLOY STRUCTURES AND PROPERTIES

A basic, and as yet only partially solved, problem is to understand the special effects of the radiation environment on alloy structures and properties. The problem is complex because two types of processes occur simultaneously: (1) flux-dependent processes, which include diffusion (atomic transport), diffusion creep, and stress relaxation; and (2) dose- or fluence-dependent processes, which produce changes in chemical composition and microstructure. These two types of processes are interactive, i.e., the radiation-enhanced creep rate will change with time, even when all other factors are held constant, because the microstructure changes. Hence, it is difficult to select or design alloys for nuclear applications because we do not adequately understand these time-dependent and time-independent processes and their interrelationships. In the current research approach, an attempt is made to isolate the important radiation-induced phenomena and to test the effects of different initial microstructures and compositions on these in a systematic manner.

2.1 Radiation Environment in Liquid-Metal Fast-Breeder Reactor

In an effort to give more physical meaning to the present discussion, two examples of the importance of considering radiation effects in the selection of alloys will be cited from nuclear systems currently under development. The first is the Liquid-Metal Fast-Breeder Reactor (LMFBR). The core of this reactor is a compact heat source, which is assembled from a number of closely packed hexagonal subassemblies, as illustrated in Fig. 1, each of which contains hundreds of small-diameter, metal-clad fuel elements (Fig. 2). The temperature range to which the fuel assemblies and rods are exposed is approximately 400 to 650 C.

The neutron flux is energetic and of high intensity, typically in the range of 10^{15} to 10^{16} n/cm^2/sec. The fuel-element tubes are required to resist the internal pressure that arises from fuel swelling and fission-gas release. A

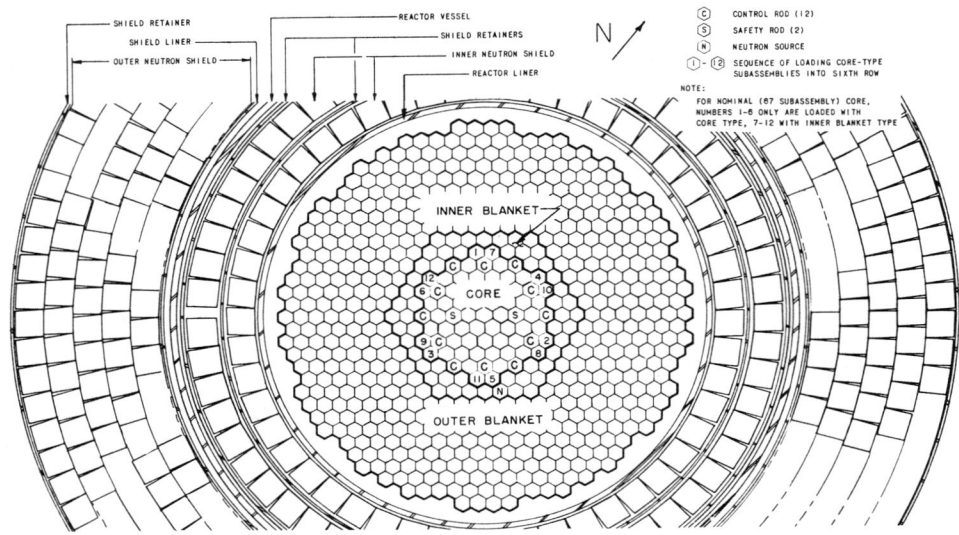

Fig. 1. Plan view of a fast-breeder reactor (EBR-II) showing the packing of hexagonal assemblies.

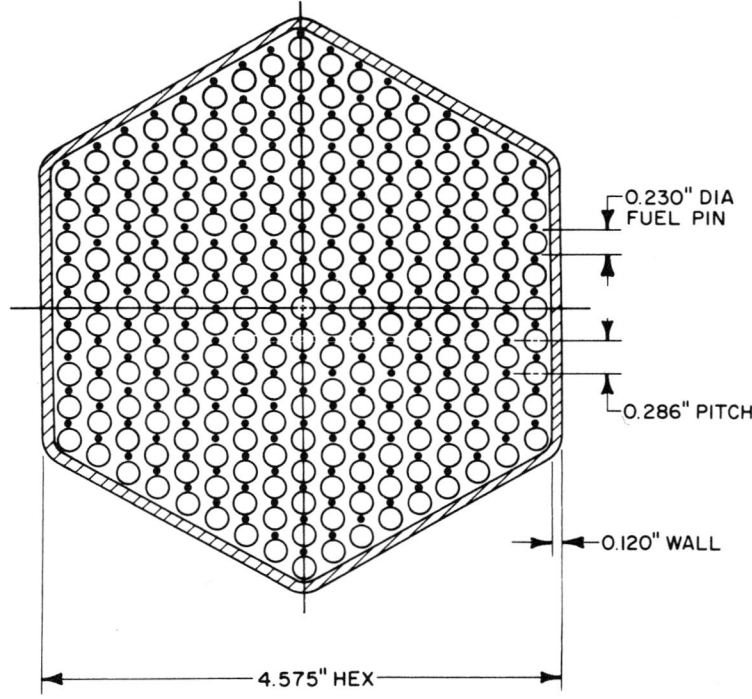

Fig. 2. Cross section of a typical LMFBR fuel subassembly showing the regular close packing of many small fuel elements.

universal choice for these tubes has been 20 percent cold-worked type 316 stainless steel, because of its good creep strength, reasonable ductility, ease of fabrication and manipulation, and resistance to hot sodium. During its lifetime exposure to the neutron flux, the stainless steel swells, because of the formation of voids, in the manner shown graphically in Fig. 3. It also loses nonuniform ductility as a result of combination of interior hardening of the grains and grain-boundary weakening. The neutron flux is non-uniform across the reactor core; hence, nonuniform swelling of fuel assemblies will occur, leading to distortion.[1] Figure 4 shows the manner in which the subassemblies are held in the core, whereas Fig. 5 shows the direction of distortion due to swelling. The distortion could make the unloading of the core impossible. This problem cannot be solved by spacing the hexagonal cans farther apart, since that will lead to an unacceptably high loss of neutrons that are needed to breed new fuel. Some relief is afforded by radiation-induced stress relaxation or creep; the magnitude of radiation-induced creep is illustrated by results of a uniaxial creep test conducted on type 316 stainless steel in the core of the EBR-II fast reactor, shown in Fig. 6. The creep may be sufficient to relax the interference stresses caused by bowing, but it is difficult to strike the correct balance between the two processes. Hence, a search is being made for a swelling-resistant alloy that also has adequate strength and residual ductility.

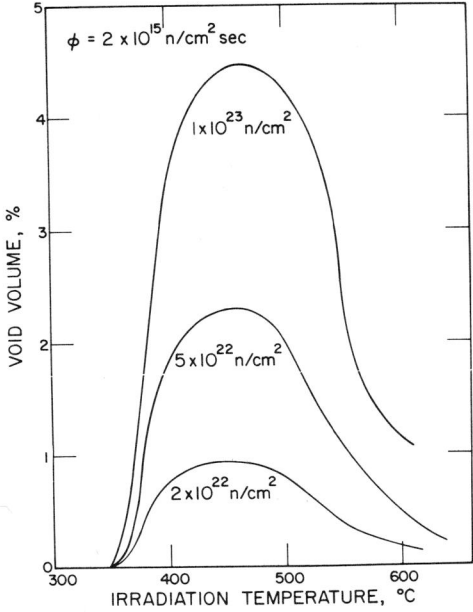

Fig. 3. Approximate curves of void swelling versus irradiation temperature in stainless steel at three fluence levels.

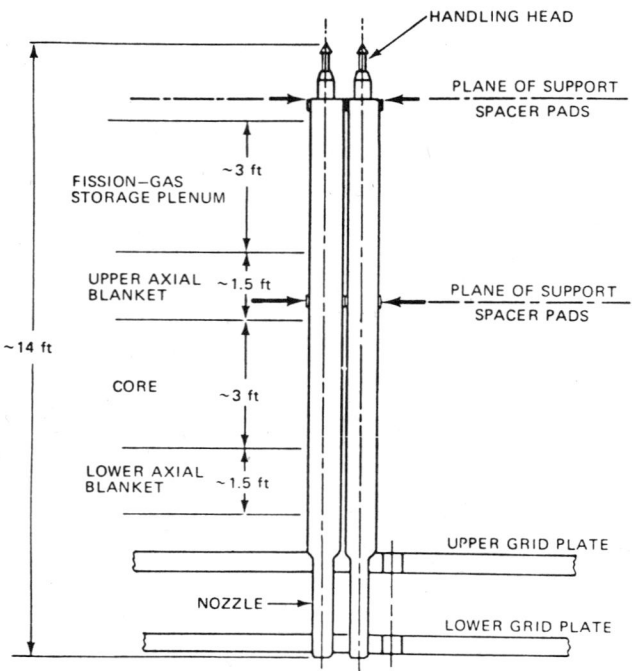

Typical LMFBR core-support concept.

Fig. 4. Vertical section of an LMFBR core showing the points of support for the sub-assemblies.

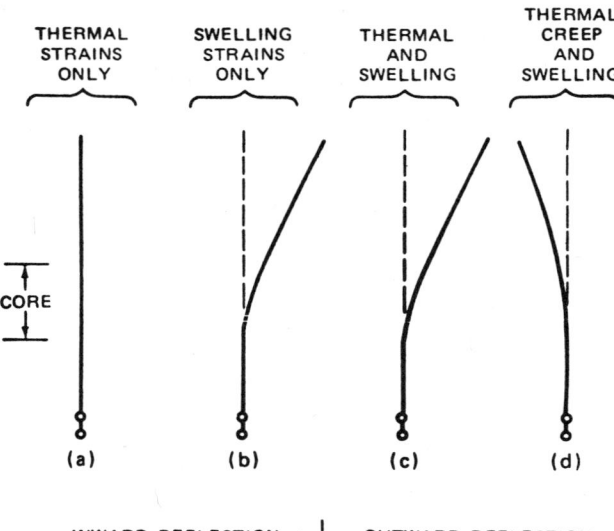

Fig. 5. Deflection of subassemblies due to differential swelling and compensation by creep.

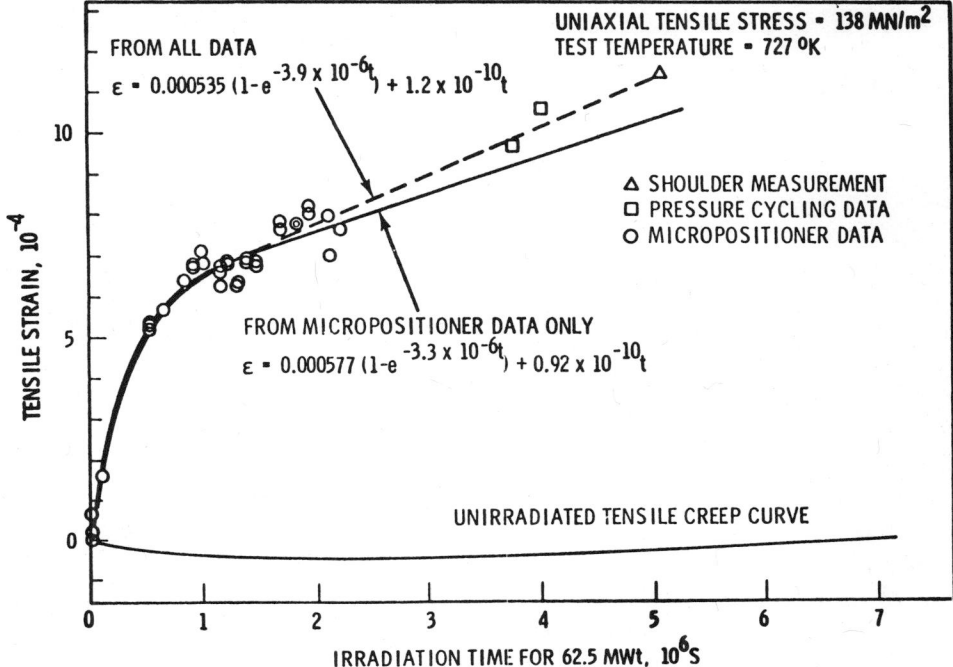

Fig. 6. Radiation-creep curve for type 316 stainless steel in the core of the EBR-II reactor.

Void swelling, the microstructural form of which is illustrated in Fig. 7, is caused by the biased flow of interstitials to dislocations and other sinks, leaving a net excess of vacancies that cluster in association with helium produced by (n,α) transmutations in, for example, iron, chromium, and nickel.[2,3] An increase in the number of sinks for the vacancies should reduce swelling; this has been sought through the use of two-phase alloys such as γ' in a nickel matrix (e.g., Nimonic, PE16), by a change in stacking-fault energy, and by changes in minor alloying elements that alter the void-nucleation process.

This is essentially a "biological" approach in which many alloys are being tested. Hence a rapid screening method is required. In current fast reactions, several years are required to attain an adequate neutron dose, which is unacceptable. To overcome this problem, energetic charged particles (typically 4 Mev) are used to produce intense but localized damage in alloy samples; void formation is analyzed by transmission-electron microscopy, which also permits interrogation of other microstructural changes. This screening process has yielded valuable indications of the compositional factors that produce alloys with low swelling rates. Figures 8a and 8b show the results of tests on a range of commercial and synthetic Fe-Cr-Ni alloys,

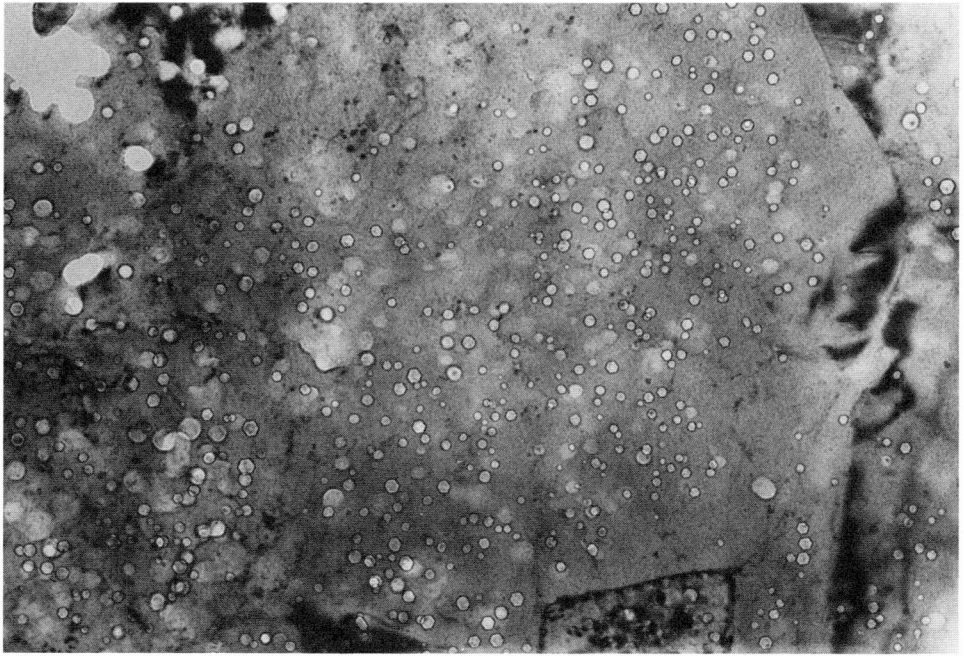

Fig. 7. Transmission-electron micrograph of irradiated stainless steel showing voids.

some of which contain other elements such as aluminum and molybdenum. It should be noted that the swelling rate becomes low at nickel contents in excess of 40 wt %. In Fig. 9, the swelling rate of an Fe-15Cr-20Ni alloy is plotted as a function of minor alloy additions. The effectiveness of titanium and silicon additions is significant; a combination of these two elements at approximately the 1 wt % level of each has been shown to reduce drastically the swelling of type 316 stainless steel. These results show that several low-swelling alloys are available which fall within the allowed composition limits for existing commercial alloys. This is important because of the long time required for code acceptance of new alloys.

2.2 Radiation Environment in Fusion Reactors

To date, fusion reactors have not demonstrated scientific feasibility, and they probably will not be in commercial service before the year 2000 A.D. However, because of the optimism generated by recent plasma physics experiments, considerable thought has been given to the probable materials requirements of such reactors and to the peculiar environment to which the materials will be subjected. The most popular reactor concept at this time is

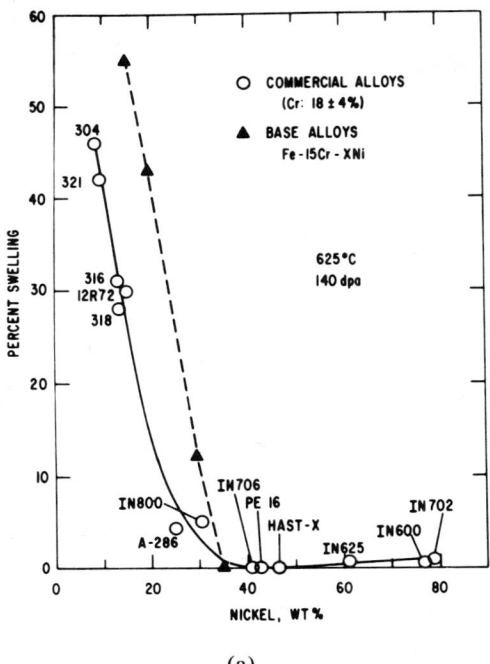

(a)

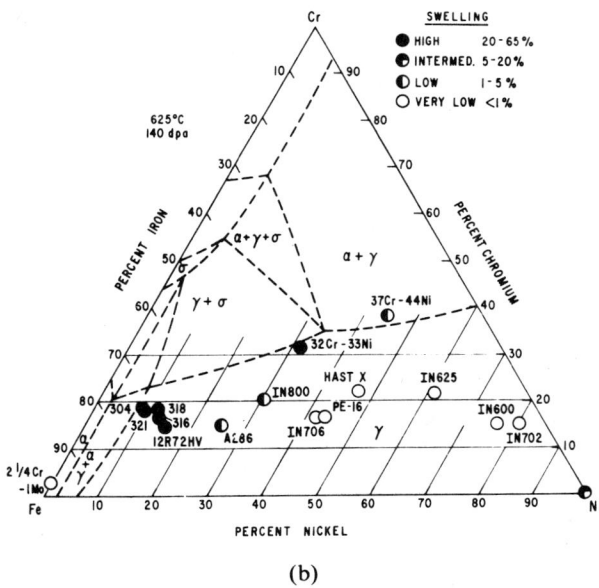

(b)

Fig. 8. (a) Fe-Cr-Ni phase diagram at the 625 C isotherm showing swelling behavior as a function of composition. (b) Same data plotted in the form of swelling versus nickel content of commercial synthetic or pure alloys.

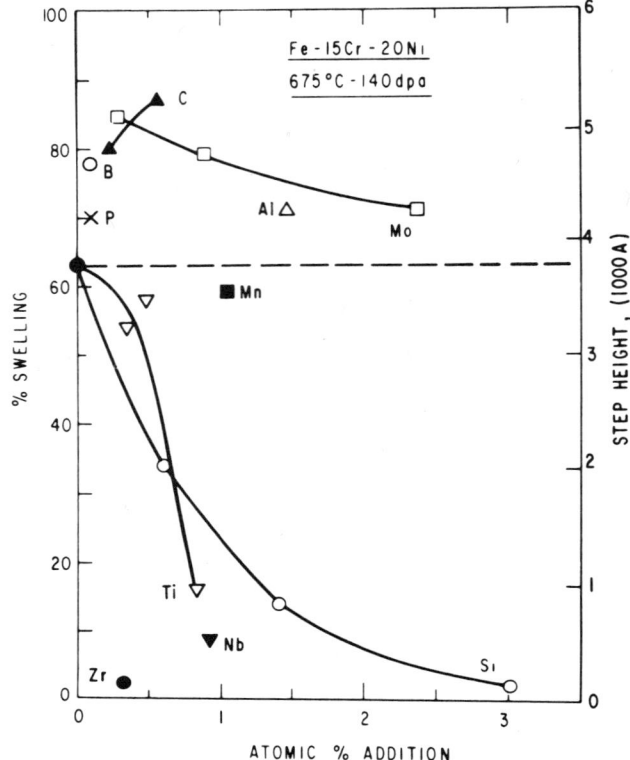

Fig. 9. Swelling of an Fe-15Cr-20Ni alloy at 675 C and 140 dpa as a function of various minor alloy additions.

the Tokamak, shown schematically in Figs. 10 and 11. The most probable reaction for early fusion reactors is

$$D + T \rightarrow {}^4He + n + 17.6 \text{ Mev} \cdot$$

Deuterium and tritium are ionized in the vacuum chamber of the Tokamak, compressed magnetically, and heated ohmically to an energy in excess of ~10 kev. The break-even point (or scientific feasibility) will be demonstrated when $n\tau \geqslant 10^{14}$, where n is the number of ions/cm^3, and τ is the confinement time in seconds. As shown in Fig. 12, the first wall will be subjected to bombardment by 14-Mev neutrons, charged and neutron particles, and short-wavelength radiation (bremsstrahlung). The neutrons will cause damage, including surface sputtering, displacements, voids, gas bubbles, and metal-atom transmutations.[4] They will pass into the lithium blanket where most will be captured to produce tritium, but some will be slowed down and will diffuse back into the metal wall and cause some damage and

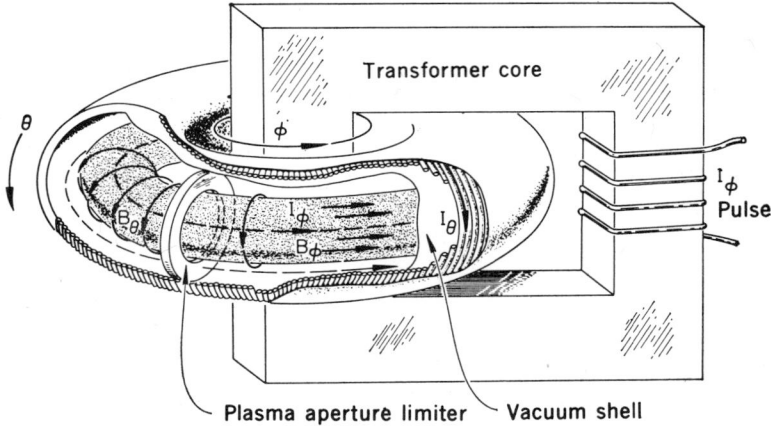

Fig. 10. Schematic of a Tokamak fusion device.

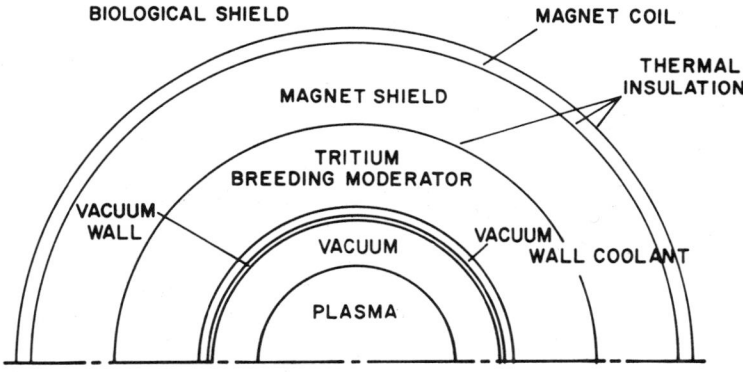

Fig. 11. Cross section of a fusion reactor.

transmutations. The charged and neutral particles and energetic neutrons will cause sputtering from the inner wall.[5] Since plasma energy losses vary with Z^4, where Z is the atomic number of the sputtered wall atom, the wall must be made of (or coated with) a low-Z material.

The radiation-damage processes in the fusion reactor first wall will include void swelling, radiation creep, and helium-induced loss of ductility. Thus, the selection of a suitable alloy for this application has some of the same elements of choice as the LMFBR fuel tube. However, other factors are different, e.g., the displacement damage produced by 14-Mev neutrons versus fission neutrons, the transmutation rates, and the surface effects. The magnitude of these differences is indicated in tabular form in Fig. 13, where the damage and transmutation rates in the niobium wall of a fusion reactor

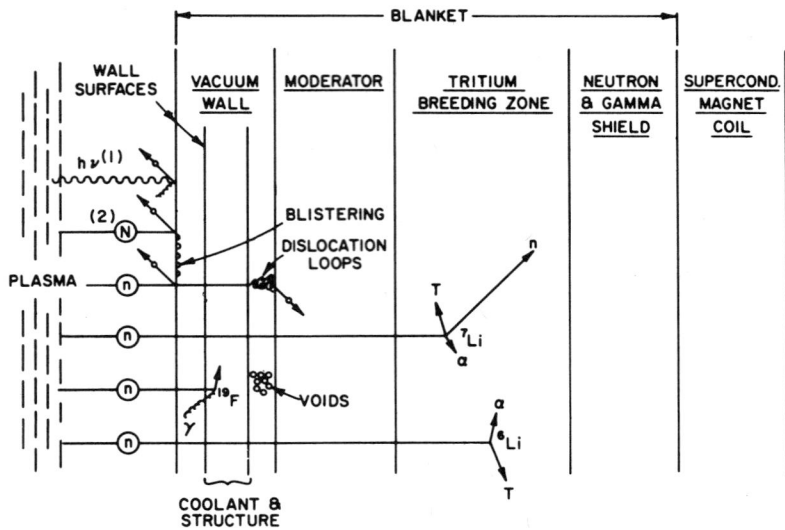

Fig. 12. Possible radiation processes that occur in the first wall of a fusion reactor.

NEUTRON-RADIATION DAMAGE

	FUSION REACTOR NIOBIUM WALL	EBR II FISSION REACTOR STAINLESS STEEL CLADDING
NEUTRON FLUX	4×10^{15} n/cm^2 sec	4×10^{15} n/cm^2 sec
NEUTRON ENERGY	14.1 mev	1 mev
DISPLACEMENT RATE	200 atoms/year	100 atoms/year
HYDROGEN PRODUCTION	900 ppm/year	300 pp/year
HELIUM PRODUCTION	300 ppm/year	20 ppm/year
ZIRCONIUM PRODUCTION	10^4 ppm/year	—
VOLUME SWELLING	(1%/year??)	3%
OPERATING TEMPERATURE	600°- 1000°C	470°C

Fig. 13. Neutron effects in fusion- and fission-reactor materials.

are compared with these rates for stainless steel in a fast fission reactor. Although the damage or displacement rates differ by only a factor of two, the transmutation rates differ by much greater factors. The latter will cause large changes in mechanical and physical properties.

From a design viewpoint, the fusion reactor first wall will be subjected to high thermal stresses but not high loading stresses. Design tolerances can be fairly loose, but differential swelling and excessive radiation-creep rates are to be avoided. The outer surface of the wall may be in contact with liquid lithium, which is used to generate tritium fuel.

In early experimental fusion reactors, wall temperatures may be as low as 400 C, so that ferritic or austenitic steels could be used (although data on 14-Mev neutron effects in these materials will be needed). More efficient reactors will have higher wall temperatures, possibly as high as 1000 C and certainly in excess of 600 C. Since lithium is known to corrode nickel-base alloys and even stainless steels at these temperatures, the only materials with an adequate combination of properties are the refractory metals niobium, molybdenum, vanadium, and their alloys. Nuclear considerations, such as the level of reduced radioactivity (which is important from the standpoint of maintenance and ultimate disposal), favor vanadium and its alloys as a long-term choice for first-wall material, with niobium as a second choice. Fission reactor and ion bombardment irradiations have been conducted on vanadium[6], niobium[7], and their alloys. The pure metals have high swelling rates, but small alloying additions (e.g., 1 percent titanium or zirconium) reduce these considerably. No measurements of irradiation creep have been made, although it has been shown that realistic levels of helium reduce the ductility markedly at temperatures above 700 C (Fig. 14).

3 ADDITIONAL FACTORS COMPLICATING ALLOY SELECTION FOR NUCLEAR REACTORS

These two examples of design problems illustrate the ways in which the radiation environment increases the difficulty of the task of alloy selection for nuclear reactors. In the course of the work outlined above, additional complicating factors have become apparent. One has been mentioned, i.e., the development of a different size and spacing of precipitates[8] in two-phase alloys, which will obviously influence the high-temperature mechanical properties. This is evident from the microstructural changes shown in a Ni-Al alloy during irradiation (Fig. 15). Closely related to this effect is the promotion of diffusion-controlled phase transformations by irradiation. An alloy that normally is thermally stable may become unstable in a fast-neutron flux.

Minor alloy elements (or impurities) have been shown to exert a major influence on radiation processes. First, it has been shown that the size and

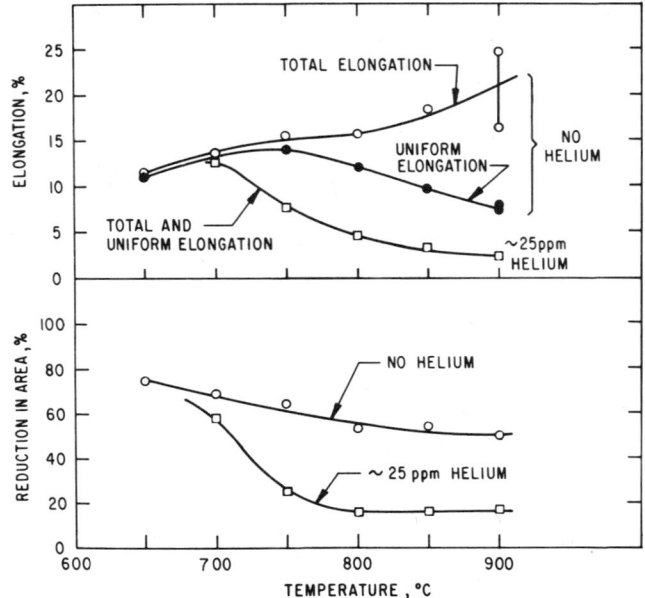

Fig. 14. Ductility as a function of temperature in vanadium containing 0 and 25 ppm of injected helium.

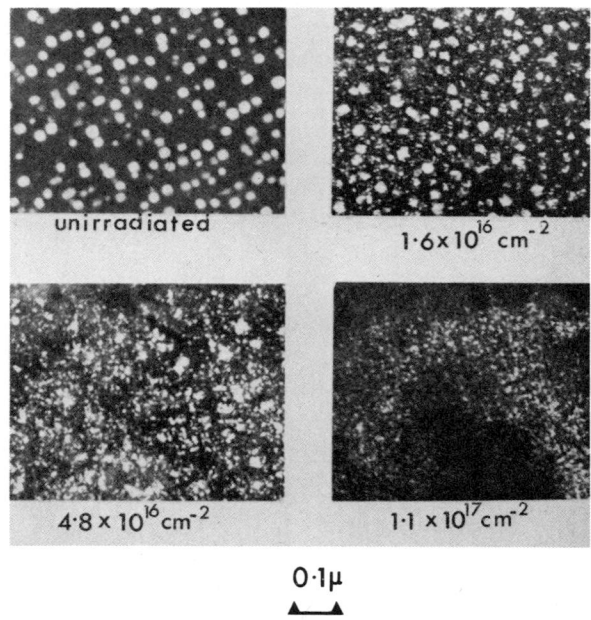

Fig. 15. The change in Ni$_3$Al precipitate size and density in a nickel-aluminum alloy as a function of charged-particle irradiation dose at 550 C.

number density of dislocation loops in the bcc metals is strongly influenced by the level of the interstitial elements in solution.[9] Second, the void-swelling behavior of niobium and vanadium is strongly influenced by the oxygen content of the metal.[10] For example, niobium forms an ordered void lattice at approximately 780 C when it contains oxygen in excess of approximately 250 ppm, but it does not form an ordered lattice at 10 ppm; larger, randomly arranged voids are formed that produce a considerably higher swelling rate (Fig. 16). It is speculated that a very fine ordered precipitate of NbO may be formed at the higher oxygen levels, which provides nucleation sites for voids. As shown in Fig. 16, an ordered lattice of voids is not observed at 828 C but the influence of oxygen content on void size is evident.

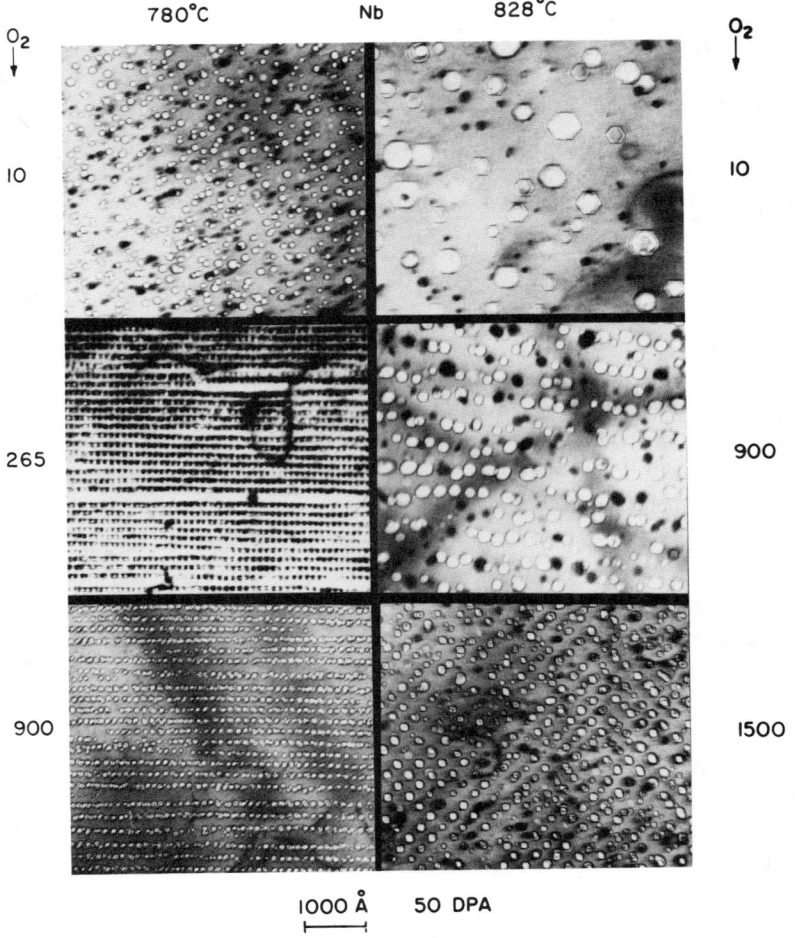

Fig. 16. Void size and number density in niobium containing various levels of oxygen after nickel ion bombardment.

Careful transmission electron microscopy has shown that solutes with negative size factors segregate toward void surfaces.[11] A particular case is an Fe-18Cr-8Ni-1Si alloy in which the silicon clearly segregated at the void surfaces during electron irradiation (Figs. 17a and 17b). It has been shown theoretically that this can lead to a reduction in void growth[12], and a kinetic model of the segregation process has been developed[13]. Possibly related to this effect is the empirical finding (mentioned earlier) that the addition of about 1 wt % titanium and 1 wt % silicon to type 316 stainless steel markedly reduces the swelling rate.[14]

A general conclusion from the findings on minor-element effects is that the specifications normally used for alloys, such as ASME type 316, have limits that are too broad for nuclear applications. To optimize between adequate mechanical properties and minimum radiation effects, it is generally necessary to specify narrow composition ranges (especially of the minor elements) and to use special processing methods such as double vacuum melting. Some indications have been found that special thermomechanical working processes may be beneficial, e.g., a process that produces fairly large carbide precipitates crossing grain boundaries results in less deterioration in ductility during service.[15] This area has not been well explored, and caution must be exercised in proceeding without taking into account the possible effects of radiation in modifying thermally stable microstructures.

4 SUMMARY

In summary, the designer of nuclear plants generally works within stress and temperature ranges that appear to be capable of being met by commercially available alloys. However, the radiation environment produces unusual effects on these alloys; one is the change of microstructure by void and cluster formation together with precipitate modification, and the other is the change in diffusion rates or the "effective temperature". The former is time dependent, the latter is time independent and both affect mechanical properties. Hence, the nuclear designer must be provided with data such as creep rates and ductility limits for given stress levels as functions of neutron dose, dose rate, and spectrum as well as temperature. This makes alloy improvement and optimization a slow and expensive process. Not surprisingly, the designer often attempts to bypass this process by using safety factors of large magnitude or by attempting to design around the problems. However, the economic incentives to develop new alloys are sufficiently strong to provide convincing arguments for following this lengthy and difficult process.[1] Consequently, sizable alloy development programs in support of advanced nuclear reactors are under way in most of the major industrialized countries.

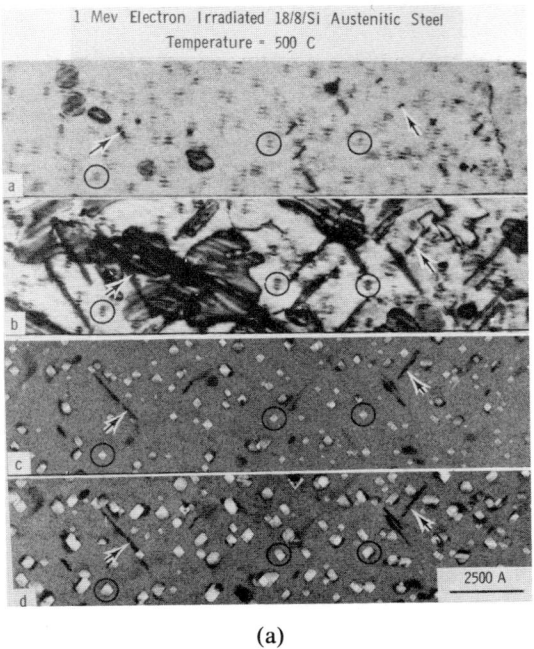

(a)

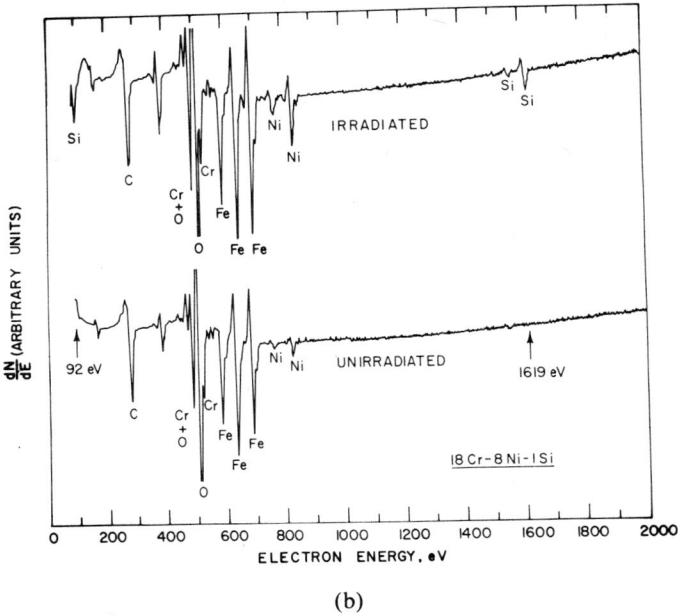

(b)

Fig. 17. (a) Transmission-electron micrographs of electron-irradiated 18/8/Si stainless steel. Silicon has segregated to the void surfaces—note the circled regions. (b) Auger spectroscopy of surfaces in 18/8/Si stainless steel showing silicon segregation after irradiation.

REFERENCES

1. Huebotter, P. R., and Bump, T. R.; J. W. Corbett and L. C. Ianniello (Eds.), USAEC Report CONF-710601 (1972), pp. 84–124.
2. Bullough, R., and Perrin, R. C., *ibid.*, p. 768.
3. Katz, J. L., and Wiedersich, H., *ibid.*, p. 825.
4. Draley, J. E., Frost, B.R.T., Gruen, D. M., Kaminsky, M. S., and Maroni, V. A., *Interscience Energy Conversion Conference, Proceedings*, published by Society of Automotive Engineers, Inc., New York (1971), p. 38.
5. Wiedersich, H., Kaminsky, M. S., and Zwilsky, K. M., *Surface Effects in Controlled Fusion*, North-Holland Publishing Co. (1974).
6. Santhanam, A. T., Taylor, A., and Harkness, S. D., *Nucl. Met.*, **18**, 302 (1973).
7. Loomis, B. A., Taylor, A., Klippert, T. E., and Gerber, S. B., *Nucl. Met.*, **18**, 332 (1973).
8. Nelson, R. S., Hudson, J. A., and Mazey, D. J., *J. Nucl. Mater.*, **44**, 318 (1972).
9. Downey, M. E., and Eyre, B. L., *Phil. Mag.*, **11**, 53 (1965).
10. Loomis, B. A., Taylor, A., and Gerber, S. A., IEEE Publication No. CH0843-NPS, pp. 46–48 (1974).
11. Okamoto, P. R., and Wiedersich, H., *J. Nucl. Mater.*, **53**, 336 (1974).
12. Brailsford, A. D., *J. Nucl. Mater.*, **56**, 7 (1975).
13. Lam, N. Q., and Johnson, R. A., Paper presented at Conference on Computer Simulation for Materials Applications, Gaithersburg, Maryland, April 19–21, 1976.
14. Bloom, E. E., and Stiegler, J. S., private communication.
15. Harries, D. R., and Roberts, A. C., *ASTM Special Tech. Pub.*, No. 426, p. 21 (1967).

DISCUSSION on Paper by B.R.T. Frost

McMAHON: The work that you discussed with regard to the search for low-swelling alloys illustrates two points that have some generality. The first is that the process of alloy design must sometimes wait for the right experimental technique to come along. In the present case, the use of perforated masks by Johnston et al. enabled them to measure swelling by simple profilometer traces on free surfaces, since the surface of a Ni^+-irradiated sample then rises only in the regions exposed by the holes in the mask. This eliminated the necessity for laborious examinations by TEM. The other point is related to the first; it is often helpful in alloy design to be able to have a critical test of relatively short duration so that large numbers of different compositions can be examined quickly. This empirical approach can then get you up to the first plateau from which basic scientific knowledge can point the ways for further advance. Until that first plateau is reached, however, the abstract approach is likely to be unproductive.

FROST: I am pleased that you mentioned the work of Johnson et al. since they not only refined the heavy-ion screening test, but they also showed that it may not be necessary to use two-phase alloys. Simply moving to another region of the Fe-Cr-Ni γ field has beneficial effects. On your second point, I would comment only that some rough guidelines for the selection of alloys have to be developed for the screening process to keep the number to be tested within reasonable bounds.

ASHBY: How have the French, British, and Russians dealt with swelling of core materials in their prototype fast reactors?

FROST: At the present time the countries that you mentioned are operating demonstration reactors of about 300 MW(e) output. These are not intended to be economic so that core swelling can be dealt with by early unloading of fuel elements. Both the British and the French have strong development programs for low-swelling alloys, and the British are using PE16 and type 321 steel in the P.F.R. to get early experience with these lower swelling alloys. It does not seem that the Russians are as far advanced in the development of low-swelling alloys, although we are aware that they are working on the problem.

LEVY: I am greatly concerned over the fact that the well-documented (for at least 12 years) grain-boundary embrittlement of austenitic steel and nickel alloys by irradiation damage has not prevented an expensive LMF-BR cladding-material development program from centering its efforts almost solely upon austenitic steels and nickel alloys in an effort to solve the swelling problem. Swelling is important, but it will not be the source of fuel-pin failure; it will be grain-boundary embrittlement (now aggravated by corrosion effects). Fuel-pin failure has had not only economic, but also safety implications. The recognition of the implications of fuel failure does not seem to be factored into this alloy development program.

FROST: I believe that the reactor designers have resigned themselves to working with low residual ductilities, e.g., 0.1 percent creep strain in 10^4 hours. However, I agree that more attention needs to be given to the fracture processes in cladding alloys. Although low swelling is the primary aim of the LMFBR alloy development program, many other parameters and phenomena have to be taken into account (such as compatibility with fuel and coolant) in the alloy optimization process—and are being taken into account in the ERDA program.

TIEN: We have been told by the ERDA-CTR group to concentrate on the "low swelling" PE16 (a nickel-base γ-γ' alloy) in our deuterium plasma shock studies. It is my understanding that PE16 has been selected to be the first wall material in Princeton's break-even Tokamak. Would the use of PE16 prevent the swelling-induced bowing or other mechanical-instability problems in the fast breeder without introducing other problems? Are there plans to actually use PE16?

FROST: PE16 is one of the alloys being tested in the U.S. alloy development program, and, to the best of my knowledge, is being used in the British P.F.R. Since it contains γ', it is probable that the dispersoid size and distribution will be modified by irradiation, but that does not seem to produce any marked deterioration of properties.

TIEN: You mentioned that the coherent γ' particles, or more to the point, the $\gamma \gamma'$ interfaces can be sinks for excess vacancies. I would think that everything else being equal, the highly coherent γ' interfaces are not very efficient sinks. Conversely, are there plans to evaluate incoherent particles (such as fine carbide particles) as vacancy sinks to minimize swelling? I do think more work is necessary in this area.

FROST: I agree that more efficient second phase sinks are possible. As I mentioned elsewhere in this discussion, W. Johnston's work indicates that a second phase may not be needed—a change in composition (perhaps affecting stacking-fault energy or short-range order) has equally beneficial effects.

DECKER: The data on void formation in austenite versus nickel, silicon, titanium, zirconium, and niobium, reminds one of alloying effects in alloy design of heat treating hardware to resist carburizing and nitriding corrosion. Some of these elements lower the solubility and activity of carbon and nitrogen in austenite. Perhaps removal of carbon and nitrogen from solid solution in austenite by the beneficial reactive elements is the mechanism of reducing void formation.

On γ' instability under irradiation, Ostwald-Wagner theory predicts that alloying of the γ' with elements of low γ solubility and diffusivity would help. Candidate additives would be niobium and tantalum.

FROST: Removal of carbon and nitrogen from solution might inhibit the clustering of these elements that could form nucleation sites for voids. I am not sure that we have the tools to demonstrate this at present. The suggestion for stabilizing γ' is interesting and should be pursued.

SHEWMON: Doesn't the sodium corrosion rate rise continuously with nickel content? If so, is there any effective upper limit on the nickel content of the alloy you can use? Would this cause any other problems?

FROST: The solubility of nickel in sodium rises rapidly at temperatures above about 650 C, and the solution rate rises with nickel content in the alloy. At present there is no strong intent to raise cladding temperatures above 650 C, so this is probably not a serious restraint. If it becomes a serious problem, a compromise can be made since the intermediate nickel alloys, like PE16, seem to have nickel solution rates close to that of type 316 stainless steel. This would put the upper limit of nickel around 40 percent.

BEGLEY: Could you comment on any trends in either composition or structure with respect to suppression of swelling as opposed to shifting the temperature range for maximum swelling as illustrated below in Case I versus Case II.

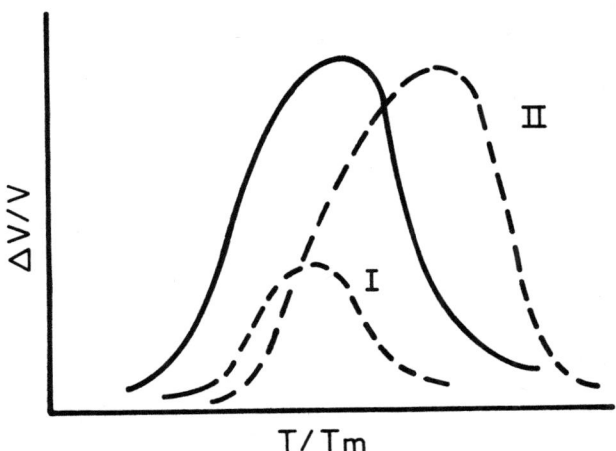

FROST: The nature of swelling is such that it generally occurs in the range 0.3 to 0.5 T_m (where T_m is the absolute melting temperature). This is the temperature range where vacancies are fairly mobile, but not so mobile that they easily reach sinks. This range is a function of crystal structure and binding energy. So it is not likely that alloying can only cause a shift in this range if it produces a marked change in crystal structure or binding energy. However, materials with a high T_m, such as molybdenum, have their swelling range above the operating temperature of the LMFBR core. There are reactor-physics reasons for not using the high-melting-point bcc metals, with the exception of niobium and vanadium. These metals dissolve oxygen and carbon from solution in sodium and become embrittled. Hence, the general approach at present is to reduce the peak height, as shown in your Case I.

BUSH: Much of your discussion was devoted to relatively short-time behavior. Another problem relevant to long-term properties of irradiated structural alloys covers thermally or radiation induced diffusion and mass transport. Two possible cases are: (1) the total removal of all interstitials from alloys and removal through the sodium sink leading to pronounced changes in mechanical properties and (2) a similar action where some substitutional components may redistribute across a cross section with change in chemical and mechanical properties.

Do you visualize these as significant problems? Is there evidence that a high neutron flux might influence these diffusion processes?

FROST: These are potential problems. Decarburization of steels in sodium has been observed under certain circumstances, and this could be accelerated by irradiation. The most likely place for it to occur is in the secondary sodium-steam circuit, which is made up of ferritic and austenitic steels, but the neutron flux there is negligible. There is no evidence yet of a composition change across the wall of fuel-element cladding that could affect swelling and mechanical properties. I regard the mass transport of oxygen through the fuel to the cladding, with subsequent corrosive attack, as a more serious problem. We are trying to solve that by the use of getters and by the adjustment of the fuel oxygen:metal ratio.

THOMAS: I would again like to emphasize the necessity for detailed characterization of materials so that we understand their structures even *before* complications arise from radiation, etc. For example, you showed a sharp drop in swelling in Fe/15Cr/xNi with increasing nickel. Two strong effects occur as you increase nickel: (a) a sharp increase in stacking-fault energy and (b) short-range order sets in around 25 percent nickel. These effects seem to coincide with reduced swelling. The S.R.O. is difficult to detect by X-rays or electron diffraction; we found it by studying superdislocation behavior, and Brookhaven checked our specimens by neutron diffraction. We know nothing about effects of minor elements on these properties.

Another point is that you showed the dissolution of ordered particles upon irradiation, but no one has given data on what happens in the "solid solution"; perhaps S.R.O. Thus a careful analysis of the "solid solution" is necessary in addition to the more obvious effects on particles. Finally, it may be interesting to note that we are presently studying effects of electron radiation (HVEM) in S.R.O. alloys.

FROST: This is an interesting point that should be looked at more carefully. Generally, radiation reduces the degree of order, although it is possible that rearrangements in the locality of thermal spikes or the process of vacancy drag of solutes *might* produce short-range ordering effects, which in turn *might* influence void nucleation.

PAXTON: In designing alloys as variants of type 316 stainless steel, for example, there are some classical problems which show up after simple aging, and which are sensitive to small amounts of alloying elements (e.g., silicon on σ phase). In the accelerated tests, are you guarding against the possibility of long-term disasters for reasons entirely different from irradiation?

FROST: We hope that we are guarding against long-term thermal-aging effects by looking at phase diagrams so that we can avoid sigma-phase embrittlement, etc. However, we cannot perform 30-year aging tests, nor can we accelerate them, so we have to hope that past experience can guide us. We have to bear in mind that enhanced diffusion in the radiation environment may accelerate phase separation, and the re-solution of precipitates may modify their distribution and morphology.

STRENGTH OF METALS AND ALLOYS AT HIGH STRAINS AND STRAIN RATES

J. D. Campbell, A. M. Eleiche* and M.C.C. Tsao**

University of Oxford
Oxford, England

ABSTRACT

The effects of strain rate and strain-rate history on the strength of metals at high strains have been investigated in two series of tests at room temperature using the split Hopkinson-bar method to obtain shear strain rates up to 3000 s^{-1}. In the first series, various materials were subjected to nearly constant low and high strain rates in torsion, and apparent rate sensitivities were determined by comparison of the flow stresses at a given strain. The flow stresses were found to increase with rate even at large shear strains (~0.5), the effect of adiabatic heating being relatively small. In the second series, strain-rate changes of about six orders of magnitude were rapidly imposed at shear strains up to 0.6. The results indicate that the flow

*Now at Massachusetts Institute of Technology, Cambridge, Massachusetts.
**Now at Westinghouse Electric Corporation, Pittsburgh, Pennsylvania.

stress depends on the strain-rate history, and may be greater or less than that obtained at the same strain and strain rate in a constant-rate test. The differences in the observed behavior of copper, titanium, and mild steel are discussed, and it is shown that the data for copper are consistent with a proposed form of constitutive relation involving a functional of the strain rate.

1 INTRODUCTION

The flow behavior of materials at high strains and strain rates is important in many metal-processing operations and in mechanics problems involving impact and fracture. In most of these applications the strain rate imposed on the material varies greatly during the deformation. Experimental data for high-rate deformation, on the other hand, are usually obtained at approximately constant strain rates, and such data are often interpreted in terms of a mechanical equation of state relating stress, strain, strain rate, and temperature.

It is known, however, that in general the flow stress immediately following a sudden change of strain rate or temperature differs from that corresponding to such an equation, i.e., that the stress is history dependent.[1-12] This is interpreted on the microscale in terms of a dependence of the dislocation structure of the material on the previous deformation history, while on the macroscale it implies that an equation of state does not exist and suggests that the stress should be represented by a functional of one or more of the state variables. Various approaches to the problem of formulating constitutive relations of this type for history-dependent materials have been recently discussed by Valanis[13], Dillon and Perzyna[14], Krempl[15], and others.

Most of the investigations involving sudden changes of strain rate have been carried out at low rates ($<10^{-2}$ s^{-1}); few have employed large jumps to high rates, and apparently none has been performed at large strains. The present paper describes two series of tests undertaken at room temperature to determine for several metals and alloys the effects of strain rate and strain-rate history at such strains.

From the results obtained, a direct comparison is made between the apparent rate dependence of the flow stress, observed in constant-rate tests, and the intrinsic rate dependence at a given dislocation structure. The observed macroscopic behavior of three materials is analyzed and possible constitutive relations for variable-rate deformation are discussed.

2 APPARATUS AND TEST METHODS

At low rates of strain, conventional screw-driven or servo-hydraulic test machines may be used for both constant-rate and rate-change tests. At medium rates (up to about 10^2 s^{-1}), open-loop hydraulic machines have been used for both types of test.[16-20] At high rates (10^2 to 10^4 s^{-1}), the only experimental method which permits accurate measurement is that introduced by Kolsky[21], in which a short specimen is loaded by stress waves propagated along uniform elastic bars. This method, originally developed for compression loading, has been adapted for torsional loading of cylindrical specimens[22-25]; investigations carried out using two machines of this type are described below.

In each machine, two collinear elastic bars are supported horizontally on guides, the specimen being sandwiched between the bars and attached to them with an epoxy cement. A large-amplitude torque is initially stored in a part of the input bar remote from the specimen, this torque being reacted by a friction clamp gripping the bar. This clamp is suddenly released by fracture of a high-strength steel bolt, thus initiating equal positive and negative torsional waves at the position of the clamp. The positive wave, of magnitude equal to half that of the initially stored torque, propagates towards the specimen. The incident and reflected waves in the input bar and the transmitted wave in the output bar are measured by means of resistance strain gauges and an oscilloscope; the specimen behavior is deduced from these measurements by the standard method for split Hopkinson-bar tests. The measurement of mean specimen strain has also been checked using an optical method to record the relative rotation of the specimen flanges during a dynamic test; this check indicated satisfactory agreement with values derived from the analysis of the waves in the bars.[26] The use of the torsional mode eliminates uncertainties caused by geometrical dispersion of the waves in the bars and by radial inertia effects in the specimen. It also makes possible the use of very short gauge lengths, so that higher strain rates can be achieved. The effect of varying specimen gauge length has been investigated by Nicholas and Lawson.[25]

In the first apparatus, aluminum-alloy bars 22.2 mm in diameter were used, their lengths limiting the effective test duration to about 200 μs. Shear strain rates up to about 3000 s^{-1} were attainable, thus giving a maximum shear strain of 0.6. The rise time of the input wave was about 25 μs; the strain rate increased from zero to its nominal value in this time, falling slightly during the remainder of the test. Specimens with a mean diameter of 10.2 mm, a wall thickness of 0.51 mm, and a gauge length of 1.27 to 2.54 mm were used. Full details of the apparatus have been given elsewhere.[27-29]

The second apparatus, based on the same principles, is of improved design and incorporates titanium-alloy bars 25.4 mm in diameter and with a length such that the maximum test duration is about 1000 μs. Strain rates over 4000 s^{-1} have been attained, corresponding to a maximum strain of more than 4; in the work reported here, however, the strain rate was limited to 1000 s^{-1}. The specimens used in this apparatus had a mean diameter of 16.26 mm, a wall thickness of 0.38 mm, and a gauge length of 1.27 to 2.54 mm. The machine has been described in detail in earlier reports.[29-31]

In order to investigate the effects of strain-rate history, the second apparatus has been fitted with a low-speed drive attached to the output bar. By this means, the specimen can be prestrained at a low rate (of order 10^{-3} s^{-1}) before the dynamic test is initiated. The specimen is not unloaded before dynamic straining occurs; thus a direct measurement may be made of the response to a strain-rate jump of about six orders of magnitude. The apparatus is instrumented to allow recording of the preloading torque and twist, so that the complete test history is measured.

Figure 1 shows a schematic layout of the second apparatus with its associated instrumentation, and Fig. 2 shows typical oscillograms obtained in constant-rate and rate-jump tests on copper, mild-steel, and titanium specimens, using this apparatus.

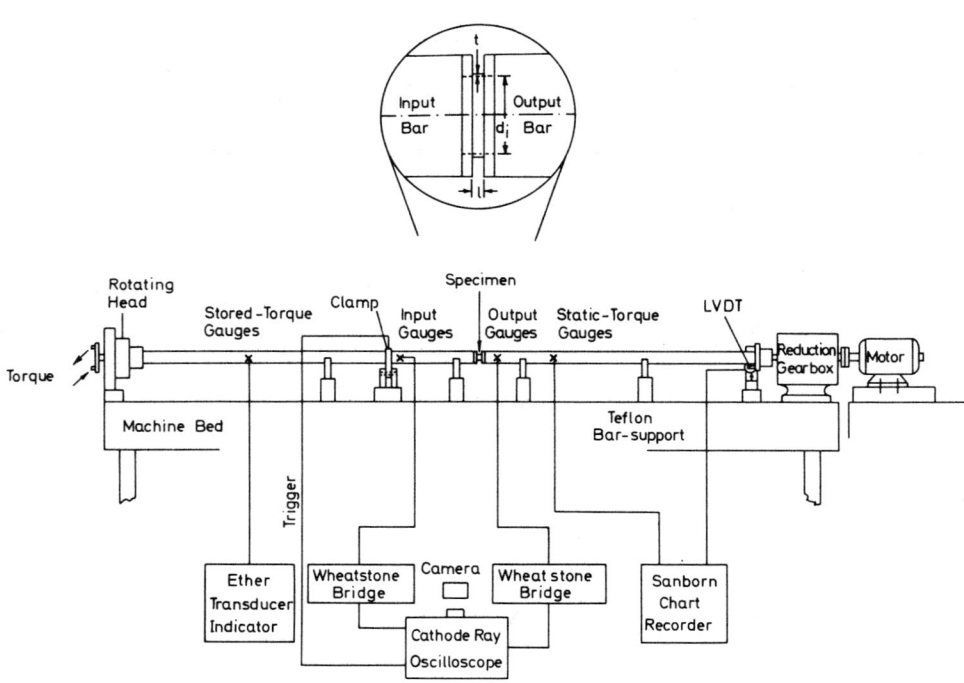

Fig. 1. Schematic layout of torsional Hopkinson-bar apparatus.

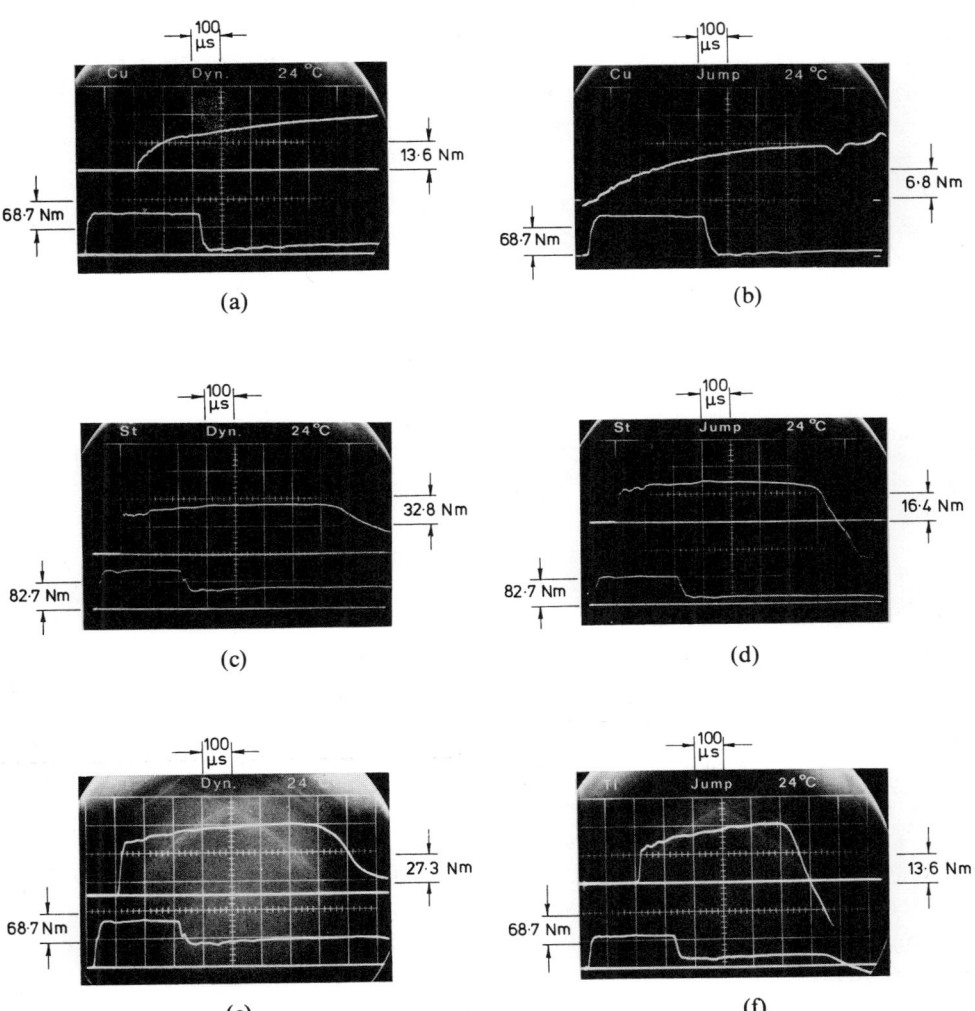

Fig. 2. Oscillograms obtained in second series of tests at strain rate $\dot{\gamma} \simeq 900$ s^{-1} and prestrains α: (a) copper, $\alpha = 0$; (b) copper, $\alpha = 0.268$; (c) mild steel, $\alpha = 0$; (d) mild steel, $\alpha = 0.200$; (e) titanium, $\alpha = 0$; (f) titanium, $\alpha = 0.150$. Time delay on upper trace: (a), (c), (d), (e), 100 μs; (b), 250 μs; (f), no delay.

3 RESULTS

3.1 Constant-Rate Tests[28]

Six materials (see Table I) were tested at a quasi-static rate of 0.004 s^{-1} and at various dynamic rates in the range 600 to 3000 s^{-1}, using the first apparatus.

Table I. Materials Tested at Constant Rates

Material	Composition, wt %	Condition	Grain Density, mm^{-2}
Mild steel (En 2A)	0.075 C, 0.40 Mn	Annealed at 750 C	690
Stainless steel (En 58B)	18Cr, 8 Ni	As received	—
Titanium	Commercial purity	Annealed at 700 C	340
Copper	99.96 Cu	Annealed at 550 C	1320
Brass	64 Cu, 35 Zn	As received (α-β)	—
Aluminum	99.74 Al	Annealed at 350 C	—

All these materials showed positive mean strain-rate sensitivities μ_{12}, defined as

$$\mu_{12} = (\tau_2 - \tau_1)/\ln (\dot{\gamma}_2/\dot{\gamma}_1) \quad , \tag{1}$$

where τ_1, τ_2 are flow stresses measured at an arbitrary strain γ (<0.5) in quasi-static and dynamic tests at rates $\dot{\gamma}_1$, $\dot{\gamma}_2$, respectively.

For copper, brass, and aluminum, μ_{12} was found to increase with γ; for the steels and titanium it was essentially independent of γ.

Since tests were not carried out at intermediate strain rates, the continuous variation of the rate sensitivity, defined by $\mu = (\partial\tau/\partial\ln \dot{\gamma})\gamma$, could not be determined; however, the trend of the results indicated that μ increased considerably with $\dot{\gamma}$ or τ for mild steel and titanium, within the range covered.

If it is assumed that a single thermally activated process is responsible for the observed rate dependence, and that the dislocation structure is a function of γ only, mean activation volumes V_{12}^* may be calculated from the relation

$$V_{12}^* = kT/\mu_{12} \quad , \tag{2}$$

where k is Boltzmann's constant, T is the absolute temperature and μ_{12} is defined by Eq. (1). Table II shows values of μ_{12} and $V_{12}*$ for the various materials tested, together with values of the flow stress ratio τ_2/τ_1, for $\gamma = 0.1$, $\dot{\gamma}_1 = 10^{-3}$, and $\dot{\gamma}_2 = 10^3$ s^{-1}. For mild steel and titanium, values of μ_2 and $V_2* = kT/\mu_2$ are also given; these are determined from the dynamic test results only, and relate to strain rates in the neighborhood of $\dot{\gamma}_2$.

The values given in Table II are not corrected for adiabatic heating, since this is very small at a strain of 0.1. For mild steel, for example, the temperature rise is less than 10 C, which corresponds to a change in τ_2 of less than 2 percent.

The results for mild steel, titanium, and copper are comparable to those found in previous dynamic torsion tests on similar materials.[12,27,29,32] However, studies involving sudden large changes in rate[9,10,12,20] have shown that at moderate strains the corresponding stress increments may not be equal to differences obtained by comparing the results of constant-rate tests. The second Hopkinson-bar apparatus was therefore used to investigate this effect in detail at large strains.

Table II. Strain-Rate Sensitivities Obtained From Constant-Rate Tests

Material	μ, MN/m^2		V*, nm^3		
	μ_{12}	μ_2	$V_{12}*$	V_2*	τ_2/τ_1
Mild steel	6.6	31	0.61	0.13	1.51
Stainless steel	1.5		2.7		1.05
Titanium	9.9	28	0.41	0.14	1.52
Copper	1.5		2.7		1.34
Brass	3.8		1.1		1.30
Aluminum	1.0		4.0		1.43

3.2 Rate-Change Tests

Three materials were chosen for investigation: mild steel, commercially pure titanium, and copper. The steel was En 1B free-machining mild steel (0.125% C, 1.15% Mn, 0.37% S) supplied and tested in the hot-rolled condition; its ferrite grain density was 690 mm^{-2}. The titanium was commercially pure (IMI 130), hot-rolled and annealed at 700 C to give a grain density of 75 mm^{-2}. The copper was high-conductivity copper, nominally oxygen-free; analysis showed 31.5 ppm of oxygen, 70 ppm of silver, and smaller amounts of other elements; the material was annealed at 400 C giving a grain density of 360 mm^{-2}.

Since the purity and condition of these materials differed appreciably from similar materials tested in the earlier study, constant-rate tests were first carried out at strain rates of order 10^{-3} and 10^{3} s^{-1}. The results of such tests are plotted in Figs. 3, 4, and 5; they show the same general features as those described by Table II, though the steel has yield and flow stresses considerably higher than those of the En 2A steel.

In rate-change tests, the output-bar oscillograms for titanium showed a well-defined yield point similar to that observed in dynamic tests without prestraining (see Fig. 2, e and f). This yield point was reached in a time of 5 to 10 μs, i.e., within the rise time of the loading wave. For mild steel, a fairly definite yield point was also reached in 5 to 10 μs. For copper, however, the output-bar traces showed no discernible yield point in room-temperature tests; in this respect they are similar to those obtained in earlier work[23] using smaller strain-rate changes.

Incremental stress-strain curves were derived for each rate-change test, and plotted with origin at the values (τ_{0}, γ_{0}) at the instant of initiation of the dynamic test. Typical curves obtained in this way are given in Figs. 3, 4, and 5. In each figure, the adiabatic temperature rise, computed from the plotted curves on the assumption that all the plastic work is converted into heat, is shown as a function of strain for the dynamic tests. For copper and titanium, the temperature difference between the dynamic tests is nearly constant (about 8 and 23 C, respectively, for the examples given), while for mild steel it decreases with increasing strain; the maximum difference for the curves shown is about 17 C. The actual temperature differences are smaller than these because of heat transfer and because some of the plastic work is converted into stored energy.

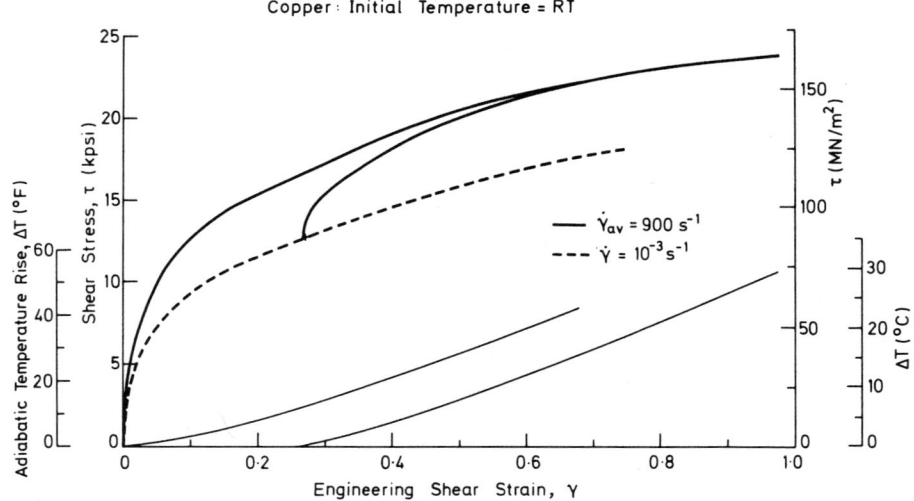

Fig. 3. Stress-strain curves for constant-rate and rate-jump tests on copper.

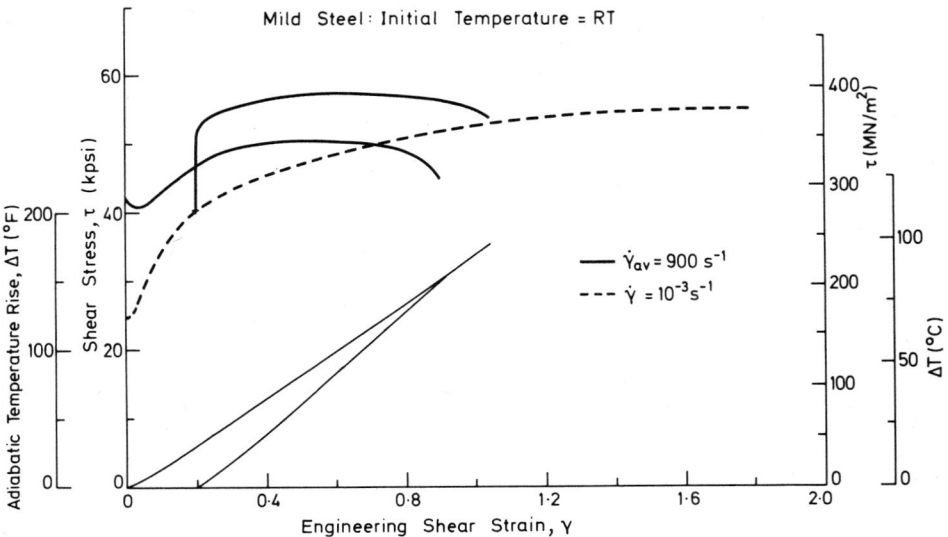

Fig. 4. Stress-strain curves for constant-rate and rate-jump tests on mild steel.

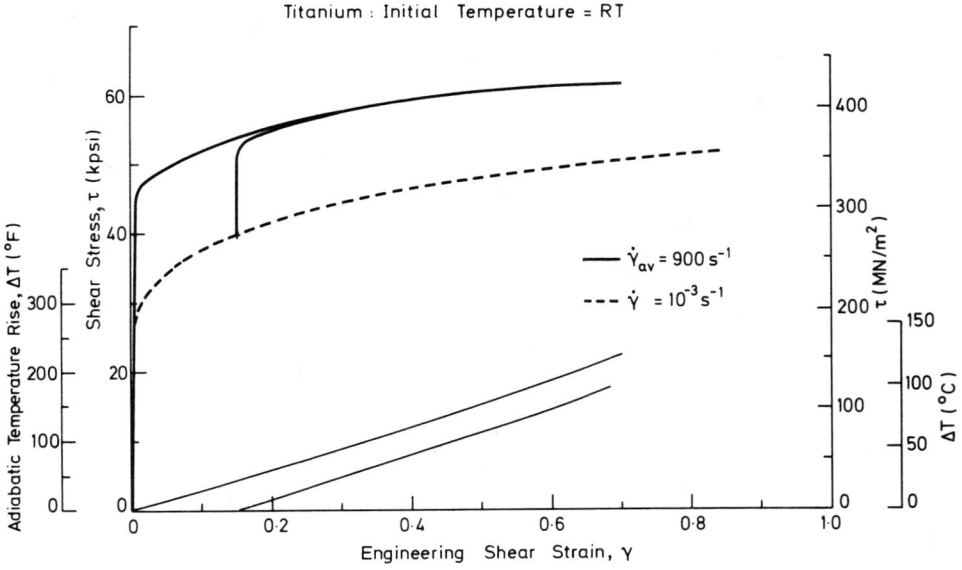

Fig. 5. Stress-strain curves for constant-rate and rate-jump tests on titanium.

4 DISCUSSION

4.1 Effect of Dislocation Structure

The flow stress of a given alloy, at a specified temperature, depends on the dislocation structure as well as on the instantaneous rate of strain. Since the dislocation structure may be a function of the strain-rate history, rather than of the current plastic strain alone, it is to be expected that the flow stress is not uniquely defined by the current strain and strain rate. Thus, as has recently been pointed out[33], a comparison of flow stresses obtained at constant strain rates cannot in general be used to establish the kinetic law controlling flow at a given dislocation structure. The values of μ in Table II, and the values of V* derived from them, therefore do not necessarily relate to the intrinsic rate sensitivity of the materials tested.

Although it is usually assumed that in a rate-change test the dislocation structure remains constant during the jump of strain rate, the possibility cannot be ruled out that a significant change in the mobile dislocation density occurs during such a jump. The increment in flow stress measured during a rate-change test, which is taken to specify the intrinsic rate sensitivity of the material, therefore relates to the mean mobile dislocation density during the jump rather than to the value just before the jump.

The present results show that for copper this intrinsic rate sensitivity is very much lower than the apparent rate sensitivity μ_{12}; for titanium it is nearly equal to μ_{12}, while for mild steel it is about $2\mu_{12}$ at a prestrain of ~0.2.

From the theory of the split Hopkinson-bar experiment, the rate of change of stress in the specimen may be written

$$\dot{\tau} = G[K(T_i - T_t) - \dot{\gamma}^p] \quad , \tag{3}$$

where G is the specimen shear modulus, T_t and T_i are the transmitted and incident torques respectively, $\dot{\gamma}^p$ is the plastic shear strain rate in the specimen, and K is a constant depending on the bars and specimen dimensions. From the oscillograms, $T_t/T_i \ll 1$, so Eq. (3) becomes

$$\dot{\tau} \simeq G(KT_i - \dot{\gamma}^p) \quad , \tag{4}$$

where, for the conditions used in the rate-change tests, $KT_i \simeq 10^3 \text{ s}^{-1}$. Thus, a large initial stress rate $\dot{\tau}$ indicates a plastic strain rate $\ll 10^3 \text{ s}^{-1}$; at "yield", $\dot{\gamma}^p$ increases rapidly, reducing $\dot{\tau}$ to a small value. The oscillograms show that for titanium and mild steel a significant increment of stress is required to increase the plastic strain rate from 10^{-3} to 10^3 s^{-1}, while for copper very little change of stress is needed. It appears therefore that most of the observed rate dependence of the flow stress of copper is caused by changes in the dislocation structure with strain-rate history. These changes take place quite gradually with increasing strain, so that after a large jump in strain rate, the flow-stress curve only slowly approaches the curve corresponding to high-rate tests (see Fig. 3). For titanium, on the other hand, most of the observed rate dependence is accounted for by the intrinsic rate sensitivity of the material. For mild steel the observed rate dependence of the flow stress is less than that corresponding to the intrinsic rate sensitivity, and it therefore appears that the dislocation structure developed at high rates is such as to soften the material; this may be related to the fact that in mild steel slip becomes finer at high rates.

The results obtained in the rate-jump tests may be compared with those of Frantz and Duffy[10] for aluminum and Klepaczko[11] for copper. By means of explosive loading, these workers obtained a torsional loading wave of rise time ∼10 μs, and observed an initial very rapid rise in the transmitted pulse; for aluminum, this rise was followed by a small yield drop. The time to reach yield was 3 or 4 μs for aluminum and about 10 μs for copper. In the present tests, as has been noted, no such yield point was detected for copper at room temperature; further tests at low temperatures have, however, shown fairly well defined yield points, though no yield drops.

Nicholas[9] has described rate-jump tests in shear carried out on aluminum, mild steel, and titanium. In these tests the dynamic strain rate was limited to 25 s^{-1}, and no significant stress increment was detected for aluminum. For mild steel, the initial stress increment, at a prestrain $\gamma \simeq 0.017$, was about half that expected from constant-rate tests, though the flow stress subsequently rose above the value obtained at 25 s^{-1}. For titanium, the initial stress increment (at $\gamma \simeq 0.034$) was also about half that expected from constant-rate tests, the subsequent flow stress remaining below that obtained at 25 s^{-1}. The differences between these results and those of the present investigation may result from the use of a lower dynamic rate applied more slowly.

4.2 Macroscopic Flow Relations

The development of constitutive equations capable of describing the mechanical behavior of real materials for arbitrary thermal and strain-rate histories is clearly a difficult matter. However, for the room-temperature behavior described above, an attempt can be made. Thus, for titanium, it appears that a mechanical equation of state relating stress, strain, and strain rate is valid, to a first approximation. Also, after correcting for temperature rise in the dynamic tests, the stress difference due to a change of strain rate is nearly independent of strain, so that the appropriate relation is of the form

$$\tau = f(\gamma) + g(\dot{\gamma}) \quad , \tag{5}$$

where the function $f(\gamma)$ corresponds to a rate-independent stress due to long-range stress fields in the material, while the function $g(\dot{\gamma})$ relates to a thermally activated deformation process characterized by the activation volume V^*.

For copper and mild steel, the situation is more complex, as the results do not correspond to a mechanical equation of state. For these materials, the flow stress may be expressed in terms of a hereditary integral or functional of strain rate. For copper, the stress difference between a jump-test curve and the low-rate curve is approximately independent of prestrain up to a value of 0.6. Thus, for a strain-rate jump from $\dot{\gamma}_1$ to $\dot{\gamma}_2$ at a prestrain α, the flow stress at a strain $\gamma > \alpha$ may be expressed as

$$\tau = f_1(\gamma) = f_2(\gamma,\dot{\gamma}_1) + f_2(\gamma-\alpha, \dot{\gamma}_2) - f_2(\gamma-\alpha, \dot{\gamma}_1) \quad , \tag{6}$$

where $f_1(\gamma)$ is the flow stress at vanishingly small rate and $f_2(\gamma,\dot{\gamma})$ represents the overstress, i.e., the additional stress required at a constant finite rate $\dot{\gamma}$. For $\alpha = 0$ and $\dot{\gamma}_1 = \dot{\gamma}_2 = \dot{\gamma}$, this equation takes the same form as the mechanical equation of state involving overstress proposed by Malvern.[34]

Assuming linear superposition of the overstress in Eq. (6), the general expression for a test in which the strain rate varies with strain according to $\dot{\gamma} = \eta(\gamma)$ is

$$\tau = f_1(\gamma) + f_2(\gamma,\eta_0) + \int_0^\gamma f_2' [\gamma-\alpha, \eta(\alpha)] \, \eta'(\alpha) d\alpha \quad , \tag{7}$$

where $\eta' = d\eta/d\gamma$, $f_2' = \partial f_2/\partial \dot{\gamma}$ and $\eta_0 = \eta(0)$.

For mild steel, the overstress varies appreciably with α so that a more general formulation than Eqs. (6) and (7) is required; however, to a first approximation, Eq. (6) describes the observed behavior.

4.3 Application to Flow Behavior of Copper

The constant-rate tests of copper at low and high rates both conform to the power law $\tau \propto \gamma^n$ within an accuracy of a few percent, the value of n being essentially the same at both rates. It is also known[35] that for copper the apparent rate sensitivity μ is, over a wide range of rates, essentially independent of $\dot{\gamma}$ but increases with γ. Thus it appears that an appropriate constitutive relation for this material, for tests at constant rate $\dot{\gamma}$, is

$$\tau = A\gamma^n [1 + m \ln (1 + \dot{\gamma}/B)] \quad , \tag{8}$$

where A, B, m, and n are constants. Equation (8) is a particular form of Malvern's equation, elastic strains being neglected. Differentiation gives the apparent rate sensitivity as

$$\mu = \left(\frac{\partial \tau}{\partial \ln \dot{\gamma}}\right)_{\gamma} = \frac{\dot{\gamma}}{\dot{\gamma}+B} \, m \, A\gamma^{n} \qquad (9)$$

Thus, for rates above 10B, $\mu \simeq mA\gamma^{n}$, while μ becomes small at rates less than B/10. The constant m therefore governs the rate sensitivity at a given strain, while B represents a characteristic strain rate at which the behavior changes from rate insensitive to rate sensitive. The constants A, B, and m are easily related to a $\tau - \ln \dot{\gamma}$ plot, as shown in the inset of Fig. 6.

From Eqs. (6) and (8), the flow stress in a rate-jump test is given by

$$\tau = A \left[1 + m \, \ln(1+\dot{\gamma}_1/B)\right]\gamma^{n} + \left[m \, A \, \ln \frac{B+\dot{\gamma}_2}{B+\dot{\gamma}_1}\right] (\gamma - \alpha)^{n} \quad , \qquad (10)$$

the first term relating to the quasi-static prestraining and the second to the dynamic straining for $\gamma > \alpha$.

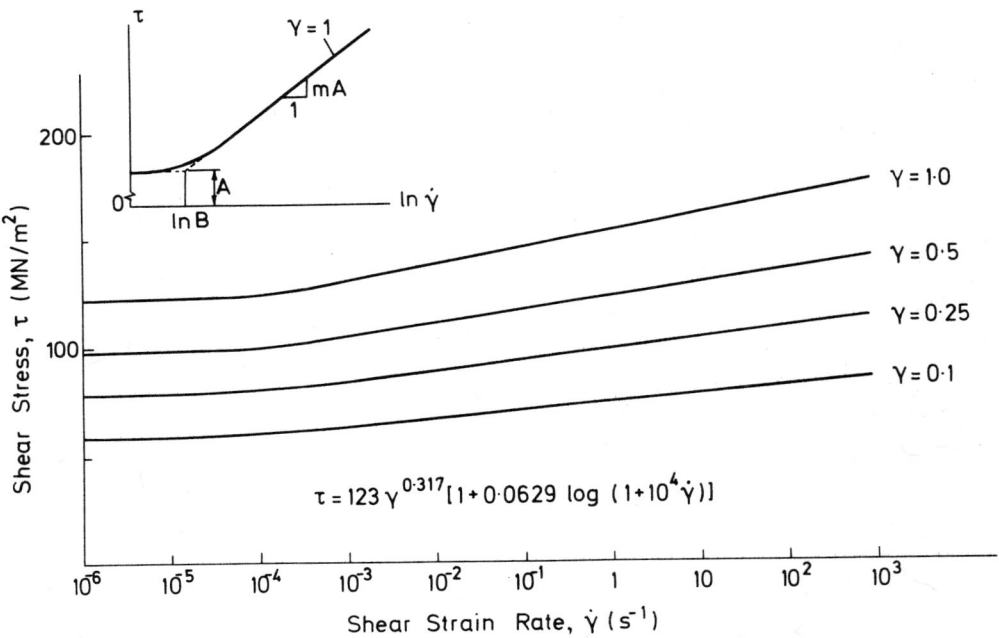

Fig. 6. Variation of flow stress of copper with logarithm of strain rate $\dot{\gamma}$, at different strains γ, according to Eqs. (8) and (11).

We choose $B = 10^{-4}$ s^{-1}, so that $\mu \simeq$ constant for $\dot{\gamma} > 10^{-3}$ s^{-1}, and from the constant-rate results for copper we find the values of the other constants to be

$$A = 123 \text{ MN/m}^2, \ m = 0.0273, \ n = 0.317 \ . \tag{11}$$

Thus, taking $\dot{\gamma}_1 = 3 \times 10^{-3}$ s^{-1} and $\dot{\gamma}_2 = 900$ s^{-1}, Eq. (10) becomes

$$\tau = 135 \ \gamma^{0.317} + 42 \ (\gamma - \alpha)^{0.317} \text{ MN/m}^2 \ . \tag{12}$$

Figure 7 shows the experimental curves for $\alpha = 0, 0.085, 0.268$, and 0.580, $\dot{\gamma}_1 = 3 \times 10^{-3}$ s^{-1}, $\dot{\gamma}_2 \simeq 900$ s^{-1}, together with points computed from Eq. (12). The agreement is within the experimental accuracy, except at large strains where the strength is reduced significantly by adiabatic heating.

5 CONCLUSIONS

Constant strain-rate tests in pure shear at quasi-static rates of order 10^{-3} s^{-1} and at dynamic rates of order 10^{3} s^{-1} have shown that for various

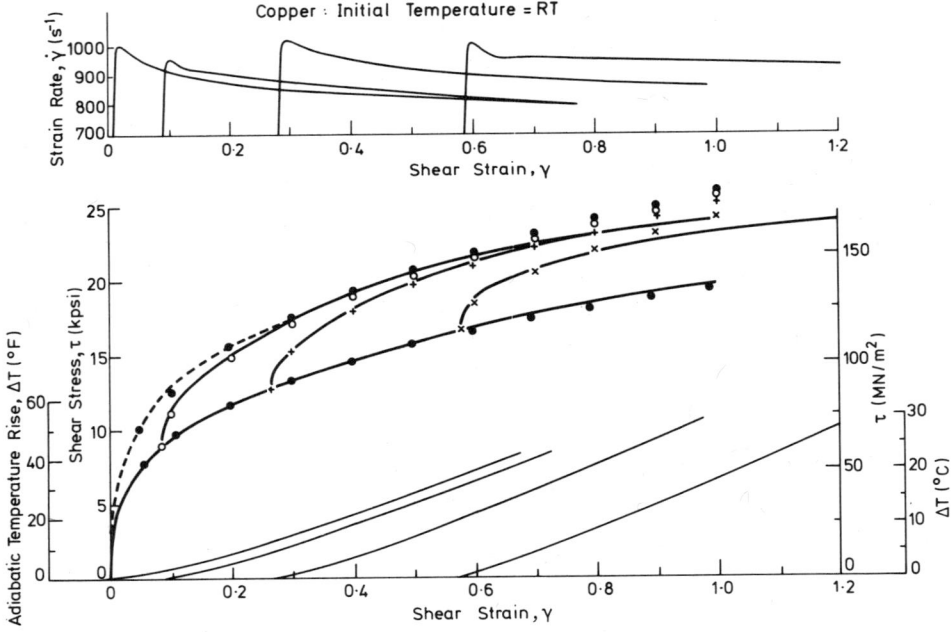

Fig. 7. Experimental stress-strain curves for copper at prestrains $\alpha = 0, 0.085, 0.268, 0.580$; $\dot{\gamma}_1 = 3 \times 10^{-3}$ s^{-1}, $\dot{\gamma}_2 \simeq 900$ s^{-1}. The plotted points are computed from Eq. (12).

pure metals and alloys at room temperature, the flow stress at strains up to 0.5 increases with rate over this range. At a strain of 0.1 the increases varied from 5 percent for a stainless steel to 52 percent for commercially pure titanium. For mild steel and titanium, the apparent rate sensitivity $\mu = (\partial\tau/\partial\ln\dot{\gamma})\gamma$ increased with strain rate, but not with strain, while for aluminum, copper, and brass, μ increased with strain.

Further room-temperature tests, in which the strain rate was increased within a few microseconds from $\sim 10^{-3}$ s^{-1} to $\sim 10^{3}$ s^{-1} at various strains have shown that for copper, titanium, and mild steel, the response to the rate increase depends little on the prestrain α, for values of α up to 0.6. The nature of the response is closely related to the difference between the flow-stress curves obtained from the constant-rate tests at low and high rates. Thus, for copper the stress increases quite gradually after the change in rate, the high-rate flow stress being approached only after a large additional strain has occurred. For titanium and mild steel, however, the stress increases very rapidly after the change in rate; the work-hardening rate then drops suddenly to a value similar to those obtained in constant-rate tests.

For all three materials, the total flow stress at a given strain is significantly affected by strain-rate history, though less so for titanium than for copper or mild steel. For titanium and copper the flow stress following the change in rate is in general less than that obtained in a constant-rate test at the high rate; for mild steel, the converse is true.

The material behavior at constant rates can be represented by equations of the type

$$\tau = f_1(\gamma) + f_2(\gamma,\dot{\gamma})   ,$$

where the functions f_1 and f_2 represent the rate-independent and rate-dependent components, respectively. For the copper tested, an appropriate form of this equation is

$$\tau = A\gamma^n + mA \gamma^n \ln (1 + \dot{\gamma}/B)   ,$$

where $A = 123$ MN/m^2, $B = 10^{-4}$ s^{-1}, $m = 0.0273$, and $n = 0.317$.

Generalizations of these equations have been proposed for application to variable-rate tests, and they have been shown to describe the results of the rate-change tests on copper within the experimental accuracy.

It is concluded that, in general, the flow stress of metals and alloys at high strains and strain rates is a function of the strain-rate history, and that, depending on the material, it may be above or below that predicted for the same strain by assuming a mechanical equation of state. Correspondingly, the plastic strain rate at a given stress and strain may differ greatly from that given by such an equation.

ACKNOWLEDGMENT

This research has been sponsored in part by the Air Force Materials Laboratory (LLN), United States Air Force, under Grant No. AFOSR 71-2056.

REFERENCES

1. Orowan, E., *J. West Scot. Iron Steel Inst.*, **54**, 45 (1946-47).
2. Dorn, J. E., Goldberg, A., and Tietz, T. E., *Metals Tech. A.I.M.M.E.,* T.P. 2445 (1948).
3. Tietz, T. E., and Dorn, J. E., *Cold Working of Metals*, American Society for Metals, Cleveland, Ohio (1948), p. 163.
4. Vasiliev, L. I., Bulina, A. S., and Zagrebenikova, M. P., *Dokl. Akad. Nauk S.S.S.R.*, **90**, 767 (1953).
5. Vasiliev, L. I., and Eremina, L. I., *Ibid.*, **93**, 1019 (1953).
6. Trozera, T. A., Sherby, O. D., and Dorn, J. E., *Trans. A.S.M.E.*, **49**, 173 (1957).
7. Sylwestrowicz, W. D., *Trans. Met. Soc. A.I.M.E.*, **212**, 617 (1958).
8. Klepaczko, J., *Archw. Mech. Stossow.*, **19**, 211 (1967).
9. Nicholas, T., *Exp. Mech.*, **11**, 370 (1971).
10. Frantz, R. A., Jr., and Duffy, J., *J. Appl. Mech.*, **39**, 939 (1972).
11. Klepaczko, J., and Duffy, J., *Mechanical Properties at High Rates of Strain*, J. Harding (Ed.), Inst. of Physics, London (1974), p. 91.
12. Klepaczko, J., *Mat. Sci. Eng.*, **18**, 121 (1975).
13. Valanis, K. C., *Arch. Mech.*, **23**, 517 (1971).
14. Dillon, O. R., and Perzyna, P., *Arch. Mech.*, **24**, 727 (1972).
15. Krempl, E., *Acta Mech.*, **22**, 53 (1975).
16. Lindholm, U. S., and Yeakley, L. M., *Exp. Mech.*, **7**, 1 (1967).
17. Cooper, R. H., and Campbell, J. D., *J. Mech. Engng. Sci.*, **9**, 278 (1967).
18. Lindholm, U. S., *Mechanical Behavior of Materials Under Dynamic Loads*, U. S. Lindholm (Ed.), Springer, New York (1968), p. 77.
19. Briggs, T. L., and Campbell, J. D., *Acta Met.*, **20**, 711 (1972).
20. Campbell, J. D., and Briggs, T. L., *J. Less-Common Metals*, **40**, 235 (1975).
21. Kolsky, H., *Proc. Phys. Soc.*, **62B**, 676 (1949).
22. Baker, W. E., and Yew, C. H., *J. Appl. Mech.*, **33**, 917 (1966).
23. Campbell, J. D., and Dowling, A. R., *J. Mech. Phys. Solids*, **18**, 43 (1970).
24. Duffy, J., Campbell, J. D., and Hawley, R. H., *J. Appl. Mech.*, **38**, 83 (1971).
25. Nicholas, T., and Lawson, J. E., *J. Mech. Phys. Solids*, **20**, 57 (1972).
26. Stevenson, M. G., and Campbell, J. D., *J. Strain Analysis*, **10**, 172 (1975).
27. Lewis, J. L., and Campbell, J. D., *Exp. Mech.*, **12**, 520 (1970).
28. Tsao, M.C.C., and Campbell, J. D., Report No. 1055/73, Department of Engineering Science, University of Oxford (1973); also AFML-TR-73-177, Air Force Materials Laboratory, Wright-Patterson Air Force Base, Ohio (1973).
29. Clyens, S., M.Sc. Thesis, Department of Engineering Science, University of Oxford (1973).
30. Eleiche, A. M., and Campbell, J. D., Report No. 1106/74, Department of Engineering Science, University of Oxford (1974).
31. Stevenson, M. G., and Campbell, J. D., Report No. 1098/74, Department of Engineering Science, University of Oxford (1974).
32. Lawson, J. E., and Nicholas, T., *J. Mech. Phys. Solids*, **20**, 65 (1972).

33. Kocks, U. F., Argon, A. S., and Ashby, M. F., *Thermodynamics and Kinetics of Slip*, Pergamon Press, Oxford (1975), p. 257.
34. Malvern, L. E., *J. Appl. Mech.*, **18**, 203 (1951).
35. Lindholm, U. S., *J. Mech. Phys. Solids*, **12**, 317 (1964).

DISCUSSION on Paper by J. D. Campbell

BUSH: How sensitive do you consider the data to be to the test conditions? Certainly your data appear to be internally consistent; however, if one were to conduct punching shear with a tensile component would your pure shear data check to a reasonable degree or are we examining relative rather than absolute properties?

CAMPBELL: In an earlier investigation [A. R. Dowling, et al., *J. Inst. Met.*, **98**, 215 (1970)], quasi-static and dynamic punching tests were performed on various metals and alloys. A comparison was made between the quasi-static behavior of copper in this type of test and in pure torsion of thin-wall tubes; the correlation was fairly good, although there are difficulties in relating the two because of the changing geometry in the punching test. The presence of a moderate tensile component in the punching test should not seriously affect the results; for example, according to von Mises' yield criterion, a tensile stress of half the shear stress would cause a reduction of only about 4 percent in the shear yield stress. The present data should therefore be applicable to punching shear with reasonable accuracy.

THOMAS: In high-strain-rate deformation such as shock loading, the mode of deformation changes from slip to twinning, such as in copper and nickel at room temperature. Do you find twinning in your experiments, and if so, what effects will this have on your analysis of the strain rates, etc.?

CAMPBELL: Because of the large hydrostatic component, very high stress levels are developed under flat-plate shock loading, although the strains remain relatively small. The level of stress required to produce macrotwinning in copper appears to be about 20 GN/m^2, although microtwins have been observed at stresses an order of magnitude smaller [*Inelastic Behavior of Solids*, M. F. Kanninen et al. (Eds.), McGraw-Hill, New York, p. 531]; in the high-rate tests described in the present paper, the hydrostatic component is zero and the stress levels are much smaller still. It is therefore almost certain that at the relatively low stresses and very large strains obtaining in the present tests, slip remains the dominant deformation mode. This conclusion is supported by metallographic examination made in an earlier investigation [*Mechanical Properties at High Rates of Strain,* J. Harding (Ed.), Institute of Physics, London, p. 85], which showed that the twin density in the gauge length of copper specimens strained at room temperature in pure shear at rates up to 10^3 sec^{-1} did not differ significantly from that in the undeformed flanges.

BEMENT: Wouldn't you agree that the uniformity of strain is as important as the total strain in examining equation-of-state deviations at high strain rates? Fragmentation studies have illustrated the importance of structural heterogeneities and the resulting nonuniform strain on adiabatic shear. Furthermore, the development and instability of cellular subwalls can result from intrinsic shear nonhomogeneities and differences in screw and edge dislocation velocities.

CAMPBELL: Yes, I agree that uniformity of strain is important. While we are not at present able to measure the instantaneous strain distribution along the gauge length, we have verified that in high-rate tests to moderately large strains, the final distribution is macroscopically uniform; we have found no indication of localized adiabatic shear. Changes in the dislocation distribution such as those you mention very probably contribute to the observed dependence of the flow stress on strain-rate history.

ALDEN: I have written a microstructural theory of plastic flow in strain-hardened metals in which the slip is athermal and the obstacle dislocation microstructure cellular. Two predictions of this theory seem to be in agreement with your results for copper. These are: (1) on increase of strain rate there is no abrupt elastic increase of stress and (2) the subsequent stress-strain curve is initially strongly curved and joins (roughly) with the curve obtained at the higher rate. During this transient, the free slip area for dislocation loops increases, and ultimately the cellularity of the microstructure changes—in this case, on increase of strain rate it will decrease.

CAMPBELL: I think that your theory is an interesting approach to the problem of including the effects of changing microstructure in the analysis of strain-rate-dependent flow in metals and alloys. The results we have obtained in jump tests on copper at room temperature are qualitatively similar to those predicted by your theory; for titanium and mild steel, however, our results indicate that the "isostructural" rate sensitivity is relatively large. Further results we have obtained for copper over a wide range of temperatures show that at low temperatures there is a finite jump of stress when the strain rate is suddenly increased, and that in general the incremental stress for a given incremental strain increases with increasing temperature. It seems probable that it will be necessary to include the effects of both thermal activation and changing dislocation structure in a general theory of rate effects in metals.

Microstructural Effects in Wear of Metals

N. P. Suh

Department of Mechanical Engineering
Massachusetts Institute of Technology
Cambridge, Massachusetts

ABSTRACT

Various metallurgical factors which affect the wear of metals are considered in terms of the delamination theory of wear. In particular, those factors which control the surface traction, the deformation of the subsurface, and the subsequent crack nucleation and propagation process are discussed. The evidence for the existence of a "soft, nonworkhardening" layer due to the dislocation instability at the surface and the presence of a flow-stress gradient is reviewed. The microstructure is shown to have significant effects on the subsurface crack nucleation and propagation process, thus on the wear rate. The role of second phase particles in crack nucleation is the same both in wear and in bulk deformation. However, the effect of the particle morphology in wear differs from that on the fracture behavior under uniaxial and torsional loading conditions. The depth of the subsurface crack, thus the wear-sheet thickness, is a strong function of the frictional coefficient and the applied load. The condition for crack propagation is discussed in terms of the state of stress existing in the subsurface over which an asperity has just moved.

1 INTRODUCTION

Until very recently, the wear of metals, like other surface-related phenomena, had not received much attention from materials scientists and engineers. This was in part due to the critical problems posed by such bulk phenomena as fracture, fatigue, and creep. However, the relative importance of these bulk phenomena, *vis-à-vis* surface phenomena, has diminished with the advent of fracture mechanics. Fracture and fatigue can now at least be handled from a practical point of view. However, the mechanisms of friction and wear are yet to be fully understood, in spite of the fact that the durability of machines and structures is often controlled by wear and corrosion.

A theory has recently been proposed, the delamination theory of wear[1], which unlike previous theories emphasizes the importance of the microstructure of metals since they affect the deformation of the surface and subsurface layer, and subsequent crack nucleation and propagation below the surface. The purpose of this paper is to review the sliding wear of metals in terms of the delamination theory and to show that a similar mechanism is operative in such wear processes as fretting, abrasive, and erosive wear.

Metals wear by different mechanisms, depending on the environmental and loading conditions. Wear can be controlled by a diffusion mechanism, by a cutting action of abrasives, and by chemical reactions, as well as by the delamination mechanism. In a given situation there can be more than one wear mechanism operating simultaneously, but often the wear rate is determined by a single dominant mechanism. In this paper, only those wear processes which are controlled by the mechanism of subsurface deformation and crack nucleation and propagation are examined. Following the detailed discussion of the validity of the delamination theory of wear for the sliding case, other wear processes such as fretting, abrasive, and erosive wear are discussed in order to show the importance of microstructural effects. The applicability of the delamination theory to fretting wear is examined. It should be noted that in the case of sliding wear and fretting wear, the delamination process is affected by all those factors, such as oxidation and adhesion, which influence surface traction.

2 DESCRIPTION OF THE DELAMINATION THEORY OF WEAR

2.1 Background

The important features of the delamination theory of wear can best be understood by contrasting with the adhesion theory of wear which has dominated the thinking of tribologists until recently.

Adhesion theories are based on the postulate that wear under sliding conditions is a consequence of asperity-asperity interactions between two opposing surfaces.[2-5] In particular, it is commonly assumed that the asperity junctions *adhere* to each other and the weaker asperities are subsequently sheared off by the action of the adhering junctions. Consequently, much attention has been devoted in the past to the mechanism of adhesion between different materials and to the determination of the actual area of contact at interfaces. Some of the adhesion theories often cited are those of Archard for wear[2] and Green for friction[3]. Existing experimental evidence is sufficient to support the idea that the adhesion does indeed occur, though perhaps not to the extent assumed by the proponents of the adhesion theories. The disputable point is the role of adhesion in sliding wear. The delamination theory of wear has postulated that the primary role of adhesion is to increase the surface traction which in turn causes greater deformation of the surface which causes crack nucleation and propagation in the subsurface. These postulates were later confirmed by experimental results. There are some gross friction and wear phenomena that have been known to deviate from the postulates of the adhesion theories, for example:

1. The work required to create a given volume of wear particles by the adhesion mechanism, as given by the upper bound solutions of plasticity, is much less than the external work done. The difference is about two to three orders of magnitude.

2. The aspect ratio of wear particles is much different from one, although the adhesion models imply an equiaxial wear particle.

3. Microstructural effects on wear cannot be accounted for by the adhesion theory.

4. The coefficient of friction tends to be much larger than that predicted by adhesion theories.

2.2 The Wear Process According to
the Delamination Theory of Wear

The above known deviations of the experimental results from the predictions of the adhesion theory motivated the examination of the basic hypothesis of the adhesion theory. As a consequence, the delamination theory of wear was conceived and formulated which predicted previously unknown phenomena such as the existence of subsurface cracks and a soft, nonworkhardening surface layer, the role of hard particles, the mechanism of the early stages of fretting wear, the shape of wear particles, and the primary role of hardness in wear of metals. These predictions have subsequently been verified by experiments. These experiments together with theoretical analyses are presented in this paper.

The delamination theory of wear refers to the following particular set of mechanisms which cause wear of metals under sliding conditions:

1. When two sliding surfaces come in contact, normal and tangential loads are transmitted through contact points. Small asperities of the softer surface are easily deformed and fractured by the repeated loading action, forming small wear particles of various shapes and eventually creating a relatively smooth surface. At the same time, bulk deformation of the surface layer takes place because of the surface traction.

2. *The plastic deformation is due to both the motion and the generation of dislocations. However, the dislocations *very* near the surface are subjected to a sufficient image force to be pulled out of the surface as the coherent oxide layer is broken by the action of moving hard asperities. The dislocations below a critical depth, where the image force is not strong enough, are stable, and as they accumulate they workharden the subsurface. Consequently, the deformed material exhibits a flow-stress gradient near the surface: as the distance from the surface increases, the hardness initially increases, reaches a maximum value, and then decreases.

3. A consequence of the flow-stress gradient is that the hard asperities more easily "penetrate" the surface of the softer metal, displacing and spreading the softer metal at the surface. Therefore, the surface traction is caused by the plowing of the softer surface by hard asperities as well as by the adhesive force. Furthermore, once the major asperities are removed, the contact is not just an asperity-to-asperity contact but, rather, the same loading history is repeated all along the softer surface as the asperity of the harder surface plows the softer surface.

4. As the subsurface deformation continues, cracks are nucleated away from the surface. Crack nucleation very near the surface is not favored for two reasons: a triaxial state of compressive stress existing just below the contact points and the existence of the "soft" metal near the surface, which does not workharden appreciably owing to the instability of dislocations and thus prevents stress concentration. Crack nucleation is preferentially initiated at a hard particle-matrix interface.

5. Upon further deformation, these cracks extend and propagate, joining the neighboring ones. The cracks tend to run parallel to the surface since the loading condition is similar all along the surface. The location of a long crack is likely to be controlled by the tensile state of loading existing near the surface over which the asperity has just passed.

---

*The existence of a flow-stress gradient is not necessary for the existence of the subsequent mechanisms, but is useful in understanding the mechanics of surface traction.

6. When these cracks finally shear to the surface, thin, long, *wear sheets* delaminate. The thickness of the wear sheet is controlled by the magnitude of the normal and the tangential load at the surface.

In the next few sections, the work my associates and I have done at M.I.T. to check the validity of the above physical reasoning is presented together with those results reported by others.

3 FLOW-STRESS GRADIENT AT THE SURFACE
(i.e., THE EXISTENCE OF A "SOFT" LAYER)

Adhesion theorists had assumed that the flow stress at the *very* surface is either the same as or larger than the bulk stress. The delamination theory postulated that the very surface layer of metals is softer and non-workhardening because of the instability of dislocations very near the surface, and that there is a flow-stress gradient near the surface because of its proximity to the low-dislocation-density zone and the resulting low rate of dislocation multiplication.

Two different kinds of experiments have been done to verify these postulates. Direct evidence for the flow-stress gradient is obtained by microhardness measurements of the surface: by varying the indentation load, the hardness can be plotted as a function of depth. Less-explicit evidence, which shows that the very surface layer does not workharden appreciably, is obtained through the application of this idea to the minimization of the wear rate of metals.

Direct microhardness tests have been made on worn surfaces of metals by Kirk et al.[6] and Koba[7] to verify the delamination theory. Later it was found that Savitskii[8] had published similar results earlier*. Figure 1 shows the results obtained by Savitskii and Kirk. Savitskii obtained the hardness variation in aluminum by varying the indentation load, while Kirk et al. obtained their results by indenting the subsurface below the wear track after sectioning a copper specimen. The flow stress is indeed low at the very surface and increases to a maximum value before it decreases again. This is not the case with well-annealed metals. Whitehouse[9] measured the hardness variation using a very fine indenter with a very light load ($\sim$100 μN) on well-annealed copper, which indicated that the hardness increases continuously all the way to the surface. The experimental results of Savitskii show that the peak hardness in aluminum occurs at about 10 to 15 μm, while Kirk et al. found the peak hardness in copper at about 80 μm. Similar hardness variations in steel cannot be measured using this type of experimental technique, probably because the flow stress varies over an extremely shallow depth. The normal hardness variation in steel reported in

---

*A similar flow-stress gradient near the surface of copper single crystals which have undergone *bulk* deformation was reported by Fourie.[41]

the literature is that due to phase transformation associated with thermal softening or quenching during the wear test.

According to dislocation theory, the image force on a dislocation varies inversely as a function of depth from the surface. The depth at which the image stress is equal to the friction stress acting on dislocations is inversely proportional to the dislocation friction stress. This depth is of the order of $1/20$ μm for silicon-iron and 10 to 20 μm for well-annealed single crystal copper with a dislocation density of only 50 cm^{-2}.[1] For most of the commercial-grade metals, this thickness is of the order of $1/10$ μm. This depth seems to be smaller than the depth at which the subsurface hardness reaches a peak value. It is difficult to draw any definitive conclusions on the thickness of the "soft" layer, since the friction stress acting on dislocations and the dislocation dynamics during wear tests are not known for these particular specimens. It is reasonable to expect that other factors will also influence the flow stress variation, such as recovery and the formation of microcracks in the highly deformed surface layer.

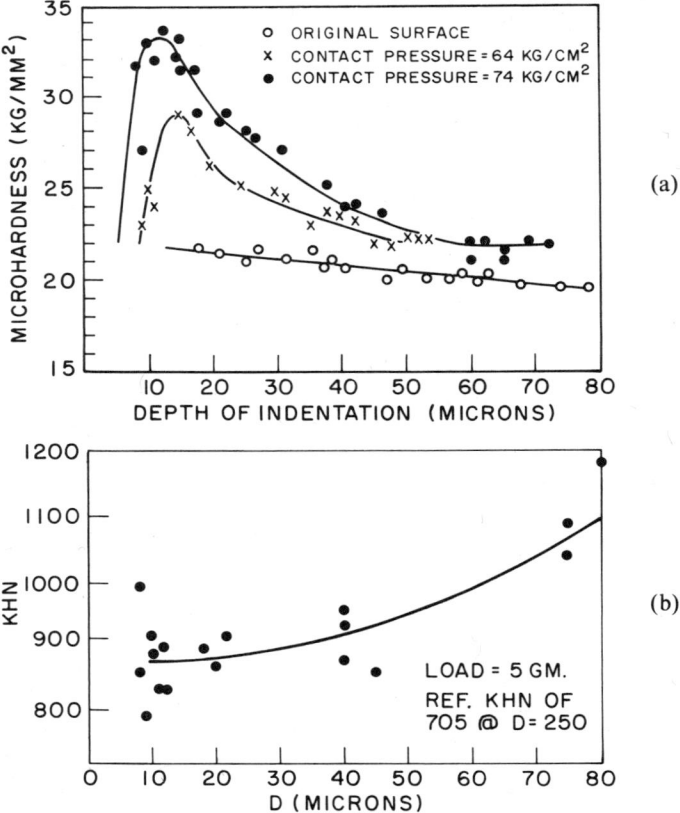

Fig. 1. Microhardness variation near the surface: (a) aluminum after sliding at different pressures[8]; (b) copper after sliding under a 682-g normal load. The bulk hardness is 705 KHN.[6]

Starting from the postulate that there is a soft, nonworkhardening layer near the surface due to the instability of dislocations, a method of minimizing sliding wear has been devised by plating a hard substrate with a very thin layer of a soft metal.[10] It was reasoned that when the thickness of the plated layer is *less* than its critical value, the accumulation of dislocations within the plated layer can be prevented and the resulting soft plated layer will deform continuously without workhardening. Thus this minimizes the plastic deformation of the substrate and forestalls wear by delamination. If the thickness of the plated layer is thicker than its critical value, the delamination mechanism will function within the plated layer, creating wear sheets of the plated metal. This reasoning was subsequently verified by experiments.[10]

Figure 2 is a photograph of a nickel-plated AISI 4140 specimen and an unplated 4140 steel specimen.[11] The nickel-plated specimen was wear tested by a cylinder-cylinder test for 2 hours in an argon atmosphere at a normal load of 2.25 kg (a Hertzian stress of 175 kg/mm^2), whereas the unplated specimen was tested for 1/2 hour under the same normal load and atmosphere. The wear coefficients are less than 2×10^{-5} for the plated specimens and 4×10^{-2} for the unplated specimens. It should be noted that the average stress in the latter specimen was much less than that of the plated specimen, since the area of contact increased during the wear test. The difference between the wear rates of the specimens is astonishing. The test on the plated specimen was stopped since no wear could be observed after 2 hours of testing. The thickness of the nickel was about 0.5 μm. As the thickness of the nickel layer is increased, the wear rate increases because of the delamination occurring in the plated layer. It should be noted that for the plated layer to be effective, the plated metal has to be compatible with the substrate.

A more critical test of the existence of a critical depth for the soft layer is illustrated using gold-plated specimens. A thick layer of gold (25 μm) was plated on an AISI 1020 steel substrate and wear tested.[11] Figure 3 shows the early stage of crack development in the plated layer and the deformation of the asperities of the substrate. In all experimental cases with plated surfaces, it was observed that the plated layer wears down to a thinner stable layer when the thickness of the plated metal layer exceeds a critical value. The resultant wear particles can accelerate the subsequent wear process by acting as abrasive particles. The existence of a critical thickness could not have been predicted by the adhesion theory.

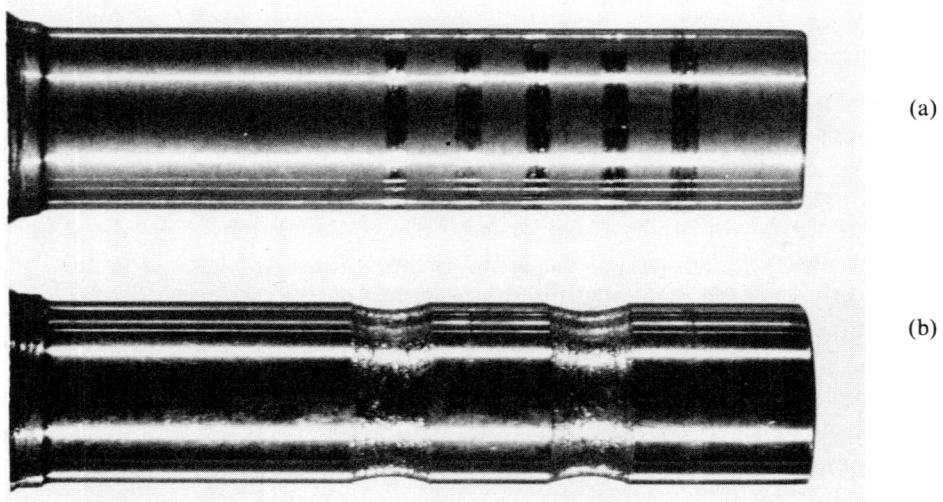

Fig. 2. Comparative wear of (a) nickel-plated (~0.5 μm thick) AISI 4140 specimen and (b) unplated specimen.[11] The plated specimen was tested for more than 2 hours and unplated specimen was tested for 1/2 hour, both in argon. The normal load was 2.25 Kg (Hertzian stress = 175 kg/mm^2).

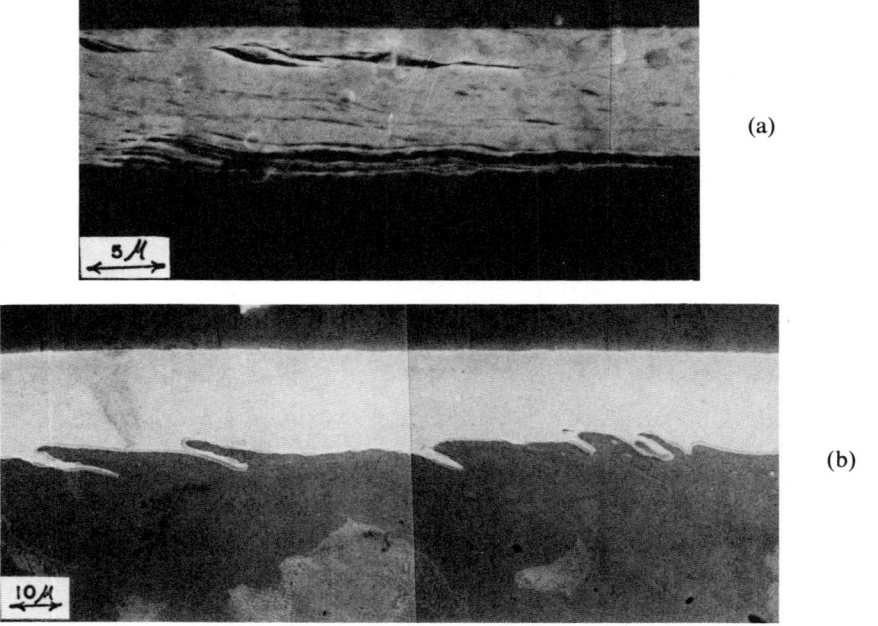

Fig. 3. Cross-sectional view of the gold-plated AISI 1020 specimen. Note the crack in the plated layer in (a) and the deformation of the surface asperity of the substrate in (b).

4 SURFACE TRACTION

The delamination theory and the adhesion theory differ as to the role of surface traction. The adhesion model of Archard implies that wear is affected by the magnitude of the normal load but not by the magnitude of the tangential load. The delamination theory predicts that the magnitudes of both normal and tangential loads affect wear particle size and wear rate. Experimental evidence for the importance of both components is presented in a later section. Here, the mechanisms by which a tangential load arises are discussed.

In the past, three mechanisms for the generation of frictional force between two sliding surfaces have been considered: adhesion between asperities, deformation of the softer asperities by harder asperities, and the plowing of the softer surface by hard asperities. Much attention has been given to the first two mechanisms in determining the tangential load using the adhesion theory.[3,4,12-14] While it is generally accepted that adhesion plays an important role in friction and wear, there is serious doubt that the contribution of plowing to the frictional force, especially in the case of fcc metals, can be neglected.

Plowing is a possible mechanism since the flow-stress gradient existing near a worn surface favors the penetration of the softer surfaces by hard asperities, spreading the softer material at the surface. Experimental results support the existence of this mechanism as illustrated in Fig. 4, which shows the topography of a plowed surface. In order to check this possibility

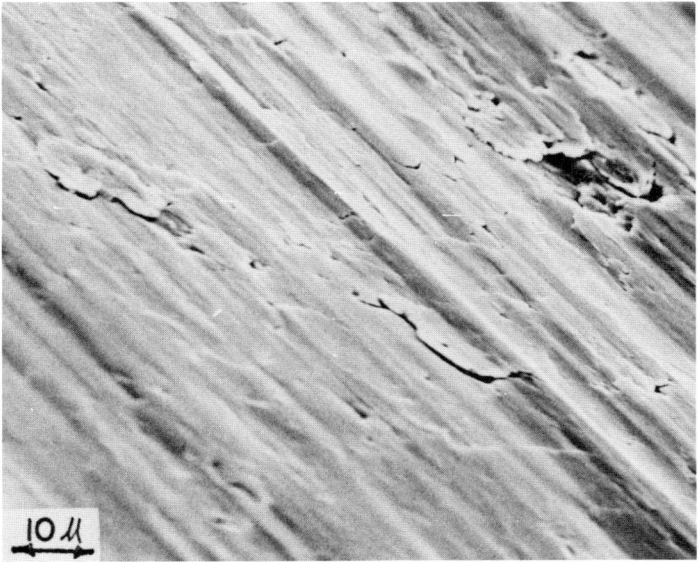

Fig. 4. Topography of a plowed surface. Material: AISI 1020; normal load: 2.25 kg; sliding speed: 180 cm/min.

theoretically, the displacement field of an elasto-plastic solid in plane strain under the action of a much harder asperity is being investigated by a finite element method using the computer program ADINA developed by Bathe.[15] The numerical results obtained for the displacements of the surface are shown in Fig. 5.[16] The results show that the hard asperities can penetrate into and spread the softer metal. Since the asperities have a gradual slope of the order of 10^{-3}, the area of contact should increase as the asperities penetrate. When there is motion of the penetrated asperities tangential to the surface, the softer metal must flow around the asperities. This plowing action increases the frictional force over that arising from adhesion alone.

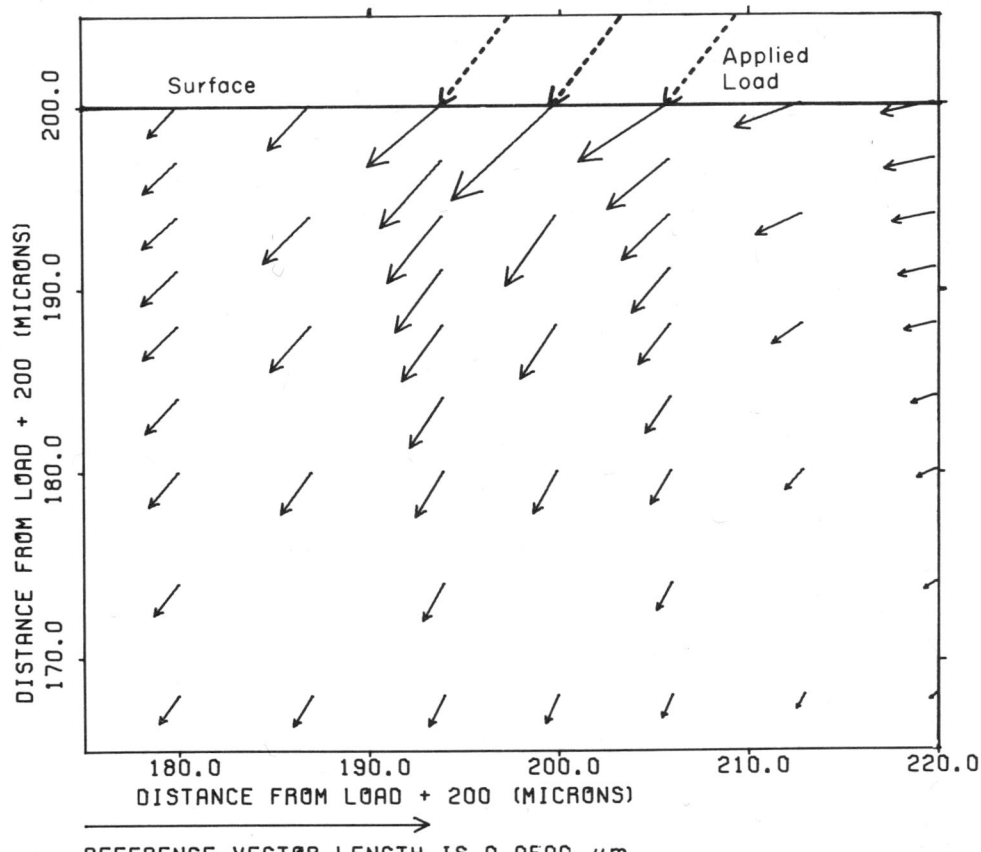

Fig. 5. Theoretically determined displacement of the surface, with tangential stress equal to yield and $\mu = 3/4$. Plane-strain solution for an elasto-plastic solid.

5 DEFORMATION OF THE SUBSURFACE
BY SURFACE TRACTION

An important consequence of the surface traction all along the surface is "uniform" plastic deformation of the surface layer. This is shown to be the case experimentally, as seen in Fig. 6, which presents a cross-sectional view underneath the wear tracks of AISI 1020 and 1095 specimens. It should be noted that even the carbides in the pearlite region of the 1095 specimen deformed with the matrix. The equivalent strain as a function of the distance away from the surface is plotted in Fig. 7.[17] It should be noted that the maximum equivalent (effective) strain shown in the figure is several orders of magnitude greater than that observed in bulk fracture of tensile specimens. Such a large strain is due to the combined effect of the high compressive state of triaxial loading under asperities, the soft nature of the deformed metal surface, and the strain accumulation due to cyclic softening of the metal under constant-stress cyclic loading.

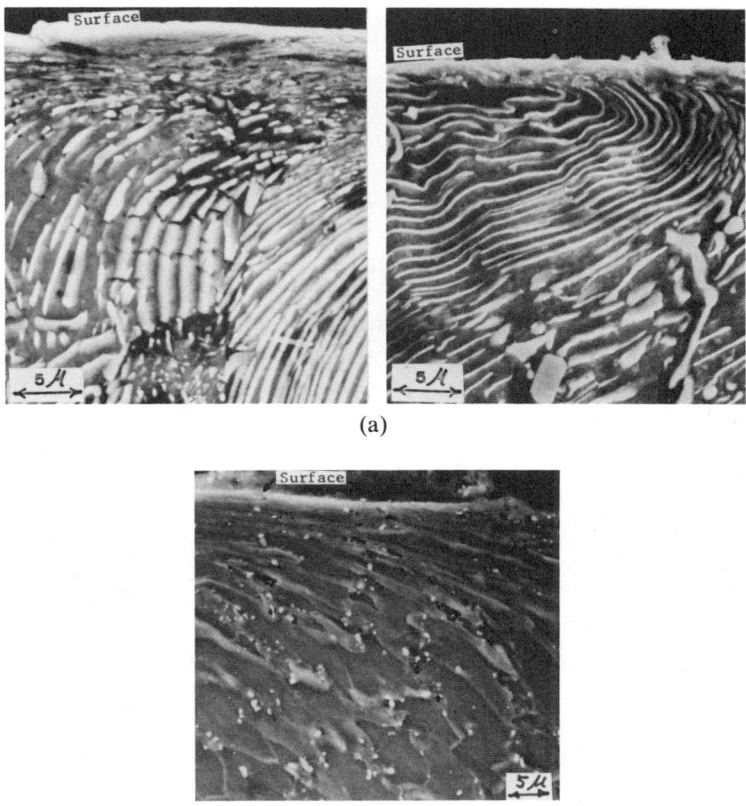

(a)

(b)

Fig. 6. Cross-sectional view underneath the wear track parallel to the sliding direction (note the large plastic deformation): (a) AISI 1095 specimen; (b) AISI 1020 specimen.[10,28]

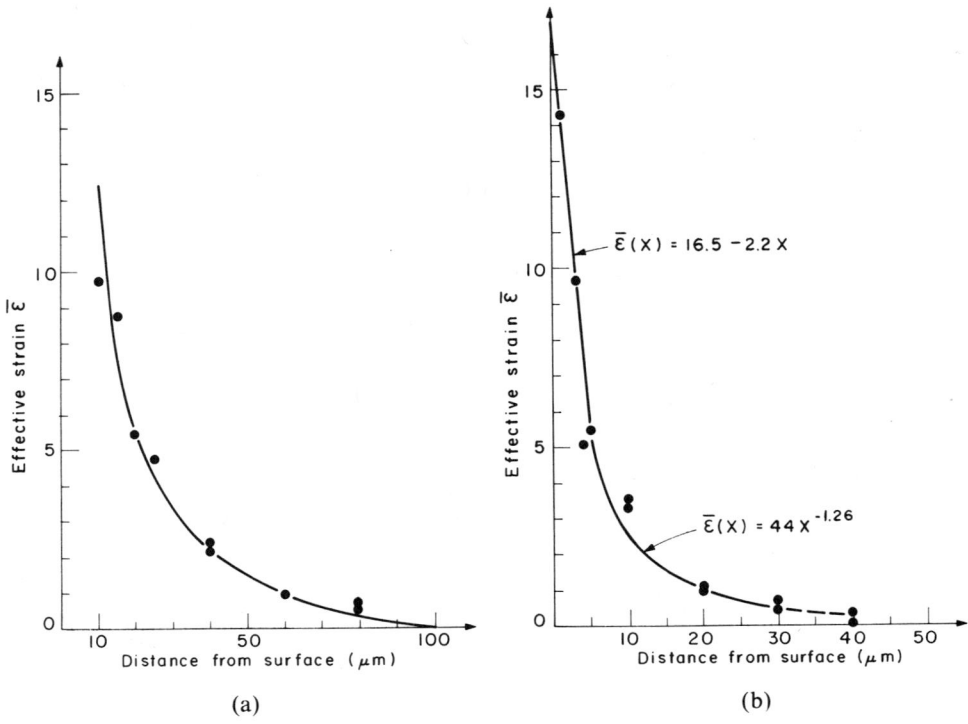

Fig. 7. Plot of the equivalent strain as a function of the distance away from the surface: (a) OFHC copper (normal load 2.1 kg, i.e., Hertzian stress of 163 kg/mm^2); (b) AISI 1020 steel.[17]

The state of stress during loading is illustrated in Fig. 8. Note that the maximum tensile stress along the direction normal to the surface occurs at a finite distance below the surface. As the asperity moves along the surface, the material in this region experiences a compression-tension type of cyclic loading. Below this region, the cyclic loading is only compressive. It should be noted that there is also a tensile stress in the direction parallel to the surface over a very shallow depth very near the surface. The actual magnitude of this tensile stress is likely to have been smaller had the analysis included the flow-stress gradient near the surface.

Each time the surface is subjected to cyclic loading of constant stress amplitude, plastic strain will accumulate until the surface is ready to delaminate.* Because of cyclic softening of metals and the presence of microcracks, the stress-strain behavior is expected to differ from that exhibited in the monotonic-loading case. In monotonic loading, the plastic-strain accumulation is normally assumed to be impossible when the metal is reloaded by the same set of stresses because of workhardening and elastic

---

*Accumulation of plastic strain at rolling contacts was investigated by Merwin and Johnson.[42]

unloading of the material. Even unloading may not be purely elastic, since the metal may recover part of the plastic strain inelastically. However, the exact constitutive relationship under this type of loading is not yet available, nor is the answer of whether or not the cyclic behavior of a surface differs from that of the bulk.

The analysis given in this section is very approximate at best, since the material was assumed to be isotropic. It is known that metals acquire texture during the wear process[18-21], making the metal highly anisotropic and affecting the adhesion and surface traction. In the case of fcc metals, the plane of the surface after deformation is determined to be a (111) plane, whereas in bcc metals it is a (011) plane. These are slip planes. In the case of hcp metals, some claim that the (0001) plane is the surface plane after deformation[20], while others disagree[18]. This is the slip plane for all hcp metals, although other slip planes are activated in metals with a c/a ratio less than ideal.

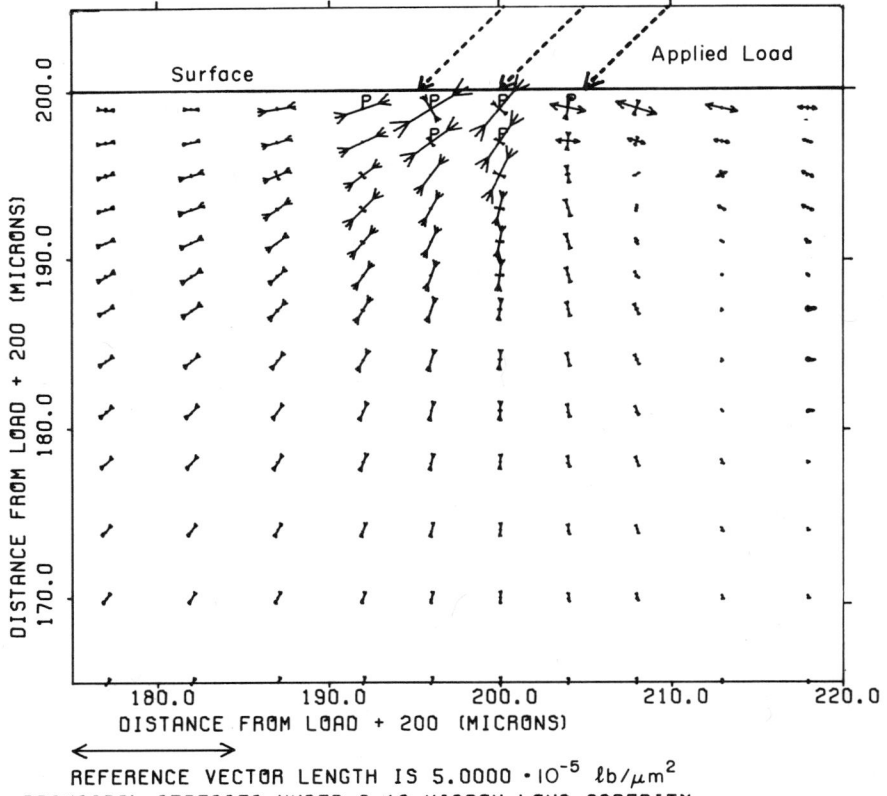

REFERENCE VECTOR LENGTH IS 5.0000 · 10^{-5} $lb/\mu m^2$
PRINCIPAL STRESSES UNDER A 10 MICRON LONG ASPERITY,
MU=1 (P INDICATES PLASTIC)

Fig. 8. State of stress during loading, with tangential stress equal to yield and $\mu = 1$, arrows pointing outward indicate tensile stress and arrows pointing inward compressive stress. Plane-strain solution for an elasto-plastic solid.

6 WEAR SHEET FORMATION BY CRACK
NUCLEATION AND PROPAGATION

6.1 General Observations

In order to check whether subsurface cracks are present as predicted by the delamination theory, worn specimens were sectioned perpendicular to the surface and parallel to the wear track. Micrographs of crack nucleation near the surface and crack propagation between two adjacent inclusions are presented in Fig. 9. Extended cracks in OFHC copper and AISI 1020 steel are shown in Fig. 10. These are typical micrographs. A wear sheet lifting off a worn surface after the subsurface crack shears to the surface is shown in Fig. 11. Wear particles generated without lubricants under different loads using different materials are shown in Fig. 12. These results show that the mechanism postulated by the delamination theory of wear is, in general, correct.

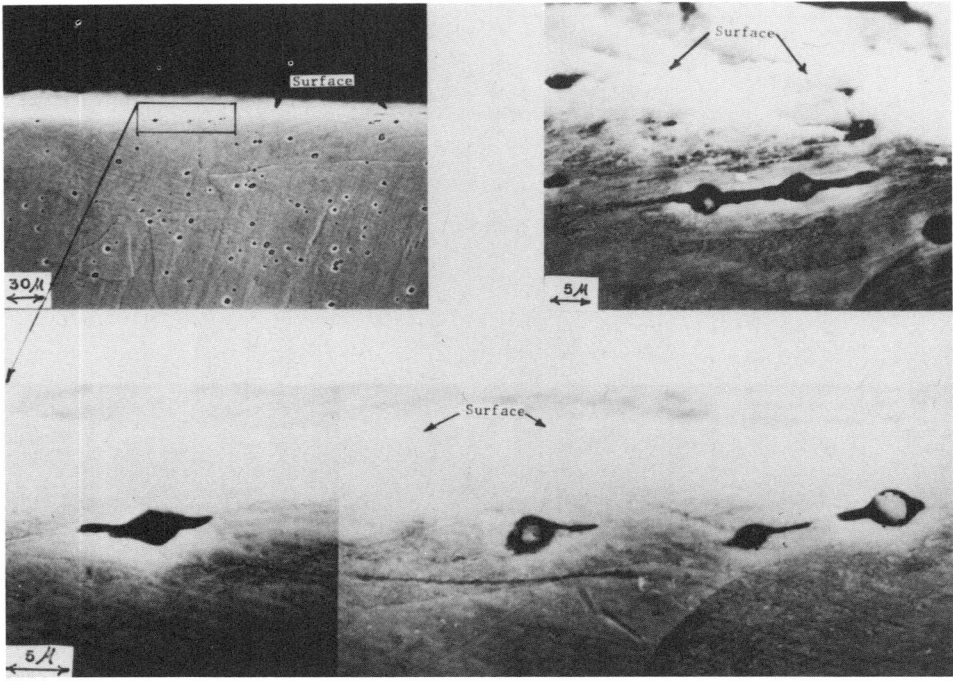

Fig. 9. Crack nucleation near the surface and crack propagation between two adjacent inclusions (material: Fe-1.3 Mo).[22]

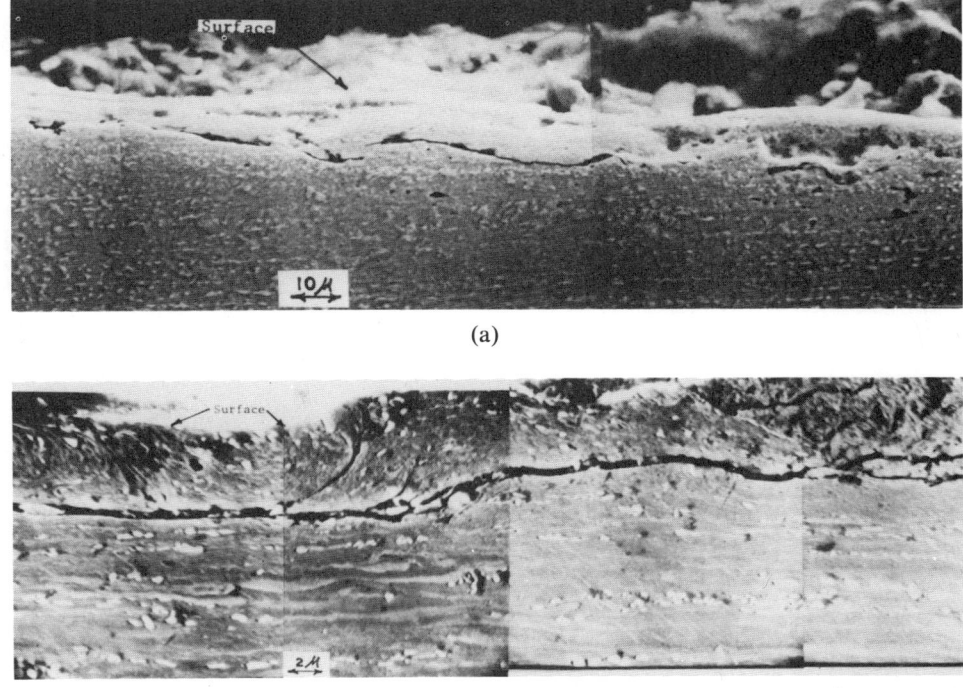

(a)

(b)

Fig. 10. Subsurface crack in (a) OFHC copper and (b) AISI 1020 steel.[22]

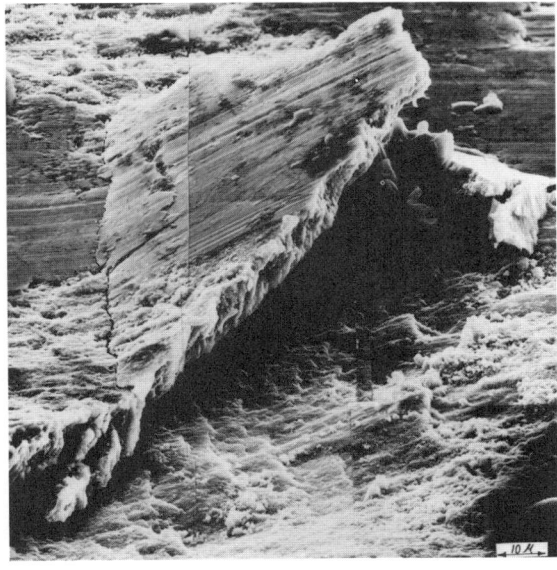

Fig. 11. Micrograph of a wear sheet lifting off; note that the sliding direction is from right to left.[23]

AISI 4140 Steel Wear Particles for a
Normal Load of 0.4 Kg (100 Kg/mm^2)

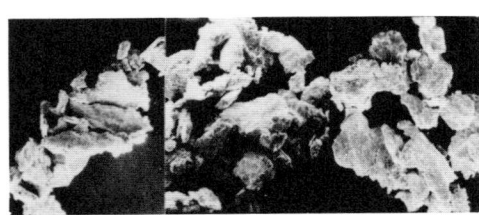

AISI 1020 Steel Wear Particles for a
Normal Load of 0.4 Kg (100 Kg/mm^2)

400 μ

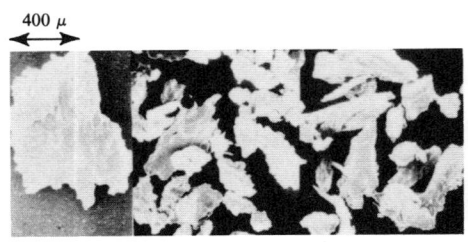

AISI 4140 Steel Wear Particles for a
Normal Load of 2.25 Kg (170 Kg/mm^2)

400 μ

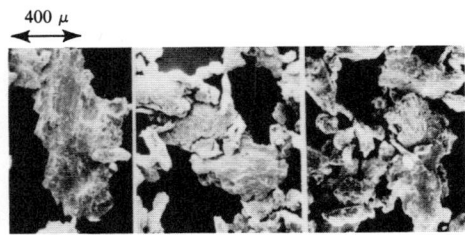

AISI 1020 Steel Wear Particles for a
Normal Load of 2.25 Kg (170 Kg/mm^2)

20 μ

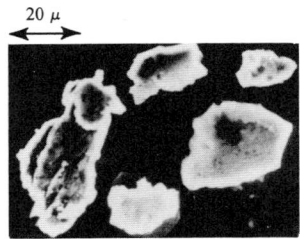

AISI 52100 Steel Wear Particles for a
Normal Load of 4.5 Kg (310 Kg/mm^2)

Fig. 12. Wear-particle shapes for various steels tested in dry argon at speed of 1.8 m/min.

6.2 Crack Nucleation

In bulk deformation of metals, crack nucleation is favored at the hard inclusion-matrix interface.[24,25] This is also true during delamination wear. However, the wear process is unique in that there are two opposing mechanisms affecting crack nucleation: the high compressive state of triaxial loading very near the asperity contact suppresses the crack-nucleation process, while the large plastic strain field promotes it. The hydrostatic pressure decreases with distance from the surface, as shown in Fig. 13. Therefore, cracks will nucleate where the condition for crack nucleation is satisfied.

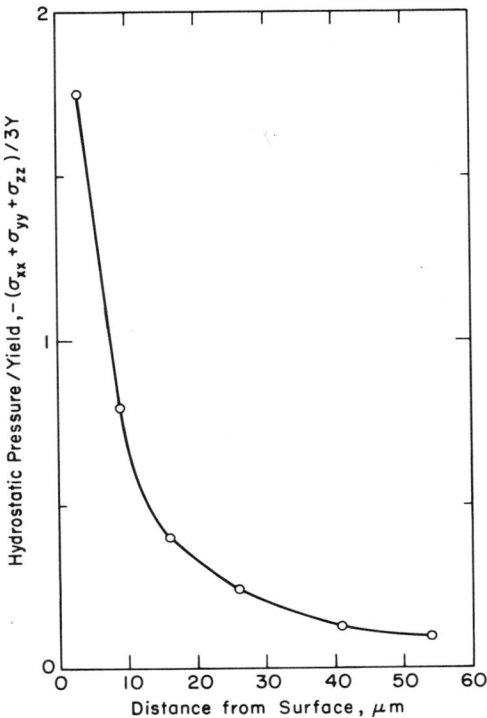

Fig. 13. Plot of hydrostatic pressure underneath the slider as a function of the depth.

The exact condition for crack nucleation is not known for the case considered here. However, there is evidence that the crack-nucleation process does not differ, at least qualitatively, from those discussed for uniaxial bulk deformation by Argon et al.[25,26], although the total strain and the strain gradient are more severe in delamination wear than in bulk deformation. Cracks are most readily nucleated between adjacent large particles, and more cracks are detected around large second-phase particles than around small inclusions. Crack nucleation is usually by the decohesion of the matrix-particle interface rather than by the fracture of particles in metals with equiaxial particles. In metals with elongated particles of large aspect ratios, the fracture of particles is more prevalent. Crack nucleation is detected over the major portion of the plastically deformed region, well below the depth at which crack propagation occurs.

For crack nucleation around hard particles, two conditions must be satisfied: (1) the local stress must be equal to the strength of the particle-matrix interface or the cohesive strength of the particle and (2) the elastic-strain energy released upon decohesion or fracture must be sufficient to supply the surface energy of the crack or the void. According to Argon et al.[25], the energy condition is always satisfied by particles larger than 100 A in the case of normal bulk deformation; or, conversely, voids cannot form around particles smaller than 100 A because the elastic energy released upon crack nucleation is less than the surface energy requirement. The interface stress between the particle and the matrix is developed because of the interfacial displacement incompatibility upon plastic yielding of the matrix.[27] The interface stress is *greatly* intensified when two inclusions are in such a close geometric proximity that the strain field of the matrix material between the particles is nearly homogeneous. In the case of delamination wear, these interface tensile stresses are counterbalanced by the triaxial state of compressive stress existing underneath the slider. Because of this high hydrostatic pressure, large plastic deformation is required for crack nucleation in delamination wear.

In zone-refined, zone-leveled iron with 500 ppm tungsten, it was difficult to identify the region of crack and/or void nucleation. However, a few large cracks were detected very near the surface, indicating that cracks nucleated and that a very large strain is necessary for crack nucleation and/or propagation in metals with no inclusions.[23] At this time, the exact mechanics of crack nucleation in clean, single-phase metals are not known. Several plausible possibilities have been considered: crack nucleation at impurity inclusions, vacancy formation during the large plastic deformation and subsequent coalescence to form a void, crack nucleation due to dislocation pileups at dislocation cells, and void formation at triple points due to the large local stress developed during deformation and texture development.

6.3 Crack Propagation

Crack nucleation does not guarantee crack propagation. Crack propagation may occur only where the hydrostatic pressure is negative, such as the region over which the slider has just moved. When there is a positive hydrostatic pressure, the cracks will close and support all of the normal load and part or all of the tangential load. Therefore, any portion of the crack in the compressively loaded region will not contribute much to the stress concentration at the crack tip in the tensile region. Furthermore, a maximum stress concentration at the crack tip is always reached when the crack tip is located at some fixed distance away from the asperity contact point, regardless of the total crack length. That is, the maximum stress-intensity factor for a crack at a given depth is the same regardless of its total length (provided that the crack is longer than a certain minimum length). The minimum stress-intensity factor is zero when the crack is entirely in the compressive region. The fatigue-crack growth rate is generally proportional to the change in stress intensity to some power; since the change in stress intensity remains constant, the crack growth rate should become constant after a short transient period of initial crack growth. Thus, when the wear rate is controlled by crack propagation, the wear rate should be proportional to the distance slid (proportional to the number of asperity passages or load cycles).

This argument is supported by many phenomenological observations of wear rate versus sliding distance and indirectly supported by the observation that the curve of wear rate versus coefficient of friction has the same general shape as the curve of the depth of the subsurface crack nearest the surface versus coefficient of friction shown in Fig. 14.

The tensile stress parallel to the surface (see Fig. 8) may also propagate cracks perpendicular to the surface when the material is incapable of undergoing plastic deformation or when there are no preexisting subsurface cracks and voids. However, in ductile materials the soft layer can easily blunt stress concentration and the high hydrostatic pressure at the surface may close up any surface cracks as the slider moves over the surface. Furthermore, since the tensile state of stress is confined to a region very close to the surface, the cracks perpendicular to the surface cannot propagate extensively.

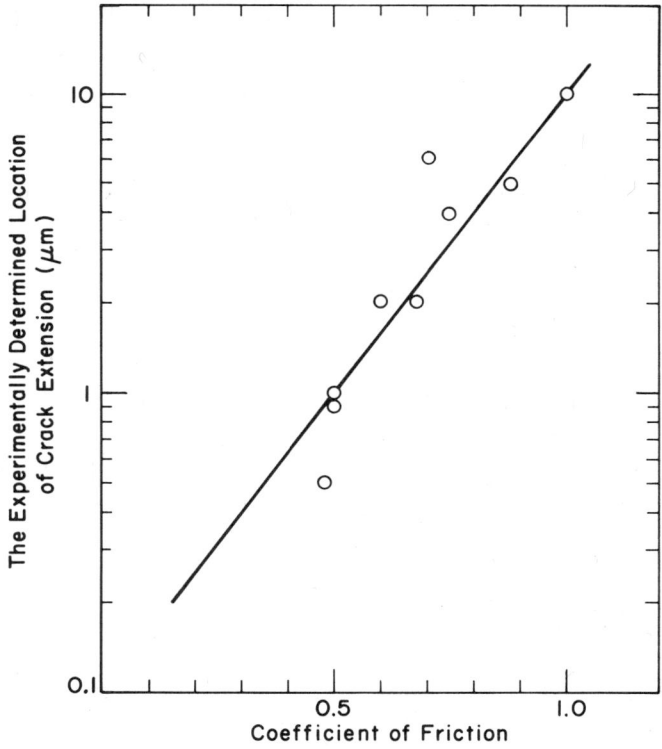

Fig. 14. Experimentally determined crack depth as a function of the coefficient of friction. All experimental points were obtained using a normal load of 2.25 kg (Hertzian stress 175 kg/mm^2).

6.4 Microstructural Effects

According to the delamination theory of wear, inclusions and second-phase particles can influence the wear process in two opposing ways. The increase in the density of small particles (i.e. $<$ 100 A) increases the hardness and minimizes the deformation of the surface, reducing wear. On the other hand, large particles increase the number of crack-nucleation sites, thus increasing the wear rate, unless the volume fraction of the particles is so high that they again begin to increase the flow stress. In view of the two opposing effects the particles could have on the wear process, one can expect to have the maximum wear rate when the increase in hardness cannot compensate for the increase in the number of crack-nucleation sites when the particle size varies over a wide range.

The above implication of the delamination theory was investigated by Jahanmir, Abrahamson, and Suh[28] using zone-refined pure iron, iron-ruthenium, and iron-molybdenum solid solutions, and AISI 1018, 1045, and 1095 steels. The pure iron was practically inclusion-free, while the iron solid solutions had a number of inclusions 1 to 2.5 μm in diameter. The steels were heat treated to have three types of carbide structure. AISI 1018 steel was tested in the spheroidized and pearlitic conditions, AISI 1045 was tested after spheroidization, and AISI 1095 was tested in the spheroidized, pearlitic, and bainitic conditions. The heat treatment was also varied to obtain fine and coarse structures. The volume fractions of the cementite in steels were determined from the stoichiometric relation.

The results of the wear test for pure iron, solid solutions, and spheroidized steel are shown in Fig. 15.[28] The tests were done using an AISI 52100 slider at 180 cm/min and a normal load of 0.45 kg (i.e., a Hertzian stress of 100 kg/mm^2). The wear resistance shown in the figure is a dimensionless quantity defined as the inverse of the wear coefficient, i.e., $K = (W/S)(3H/L)$, where W is the volume lost, S is the sliding distance, H is the hardness in kg/mm^2, and L is the normal load. It should be noted that the wear resistance is normalized with respect to hardness. The results shown in Fig. 15 indicate that the wear resistance decreases initially with increasing volume fraction of particles down to a minimum value of 0.07 volume fraction. Beyond this volume fraction, the wear resistance increases again, as was postulated by the delamination theory of wear.

The wear behavior depends on the morphology of the second-phase particles. Among the various structures studied for AISI 1095 steel, coarse pearlite and bainite have the lowest and the highest wear coefficients, respectively, and the spheroidized steels have lower wear resistance than pearlitic steels. When the structure consists of spheroidized particles, cracks propagate between these particles nearly parallel to the surface. Some of the long iron carbides in the pearlite structure undergo typical cohesive fracture observed in inclusions with large aspect ratios during bulk deformation.

However, what is impressive in this case is the large number of elongated carbide grains undergoing plastic deformation without fracturing, presumably due to the large hydrostatic pressure involved in the wear process.

It is interesting to note that the manner in which the particle morphology influences the wear rate differs from the fracture behavior under uniaxial and torsional loading conditions. Whereas the pearlitic structure has higher wear resistance than the spheroidized structures, Rosenfield et al.[29] and Lemmon and Sherby[30] have reported better ductility for spheroidized steels than pearlitic steels in bulk deformation. They found that the pearlite plates fracture preferentially at lower strains than that required for crack formation in the spheroidized steels. The difference may be caused by the fact that in sliding wear the long axis of the carbides in pearlites becomes parallel to the surface after deformation. Then the cracks created in the carbide by fracture would be oriented perpendicular to the surface, making it less favorable for crack propagation.

Microstructural effects in wear differ from those in bulk behavior mainly because the loading condition existing at the surface is unique. The "ductility" of the surface is higher than that in bulk deformation because of the

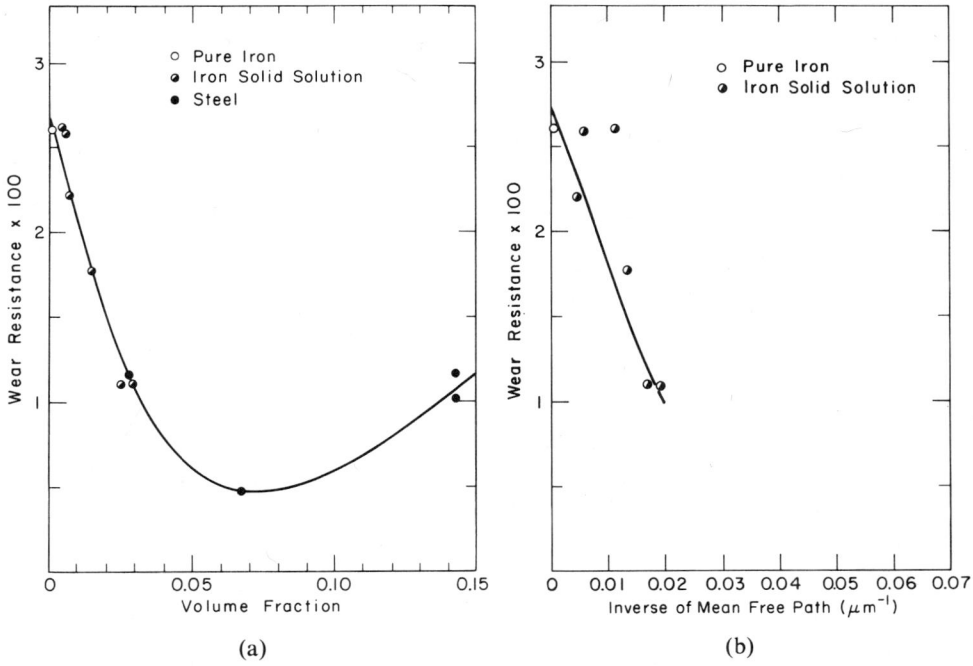

Fig. 15. Effect of volume fraction and mean free path on the wear resistance, $1/K -$ $SL/3VH$, of iron and steel.[28]

high hydrostatic pressure and low workhardening rate, and the mechanical energy consumed is high because of the cyclic nature of loading. Whether or not the concepts used in fracture mechanics are applicable in determining the crack propagation rate is yet to be investigated.*

7 THERMODYNAMIC CONSIDERATIONS

One of the requirements of a correct wear theory is that it satisfy the conservation laws of nature. The fact that the external work done during sliding cannot be accounted for by the adhesion theory is the major shortcoming of the theory. The difference between the actual energy consumed and that predicted by the adhesion theory is two to three orders of magnitude.

The work done by the slider is consumed to plastically deform and plow the surface and the subsurface layer, to create new surface, and to supply elastic strain energy associated with dislocations. The plastic deformation of the subsurface is cyclic and the energy consumed is dissipated in the form of heat. It has been well established that the strain energy stored is only a small fraction (ca. 5 to 10 percent) of the work done in plastic deformation. It was also shown that the surface energy is much smaller than the plastic work done in fracture of metals. Therefore, it is clear that the energy consumed during cyclic plastic-strain accumulation is responsible for most of the energy consumed in a wear process.

It was stated earlier that plastic-strain accumulation under cyclic loading at constant stress amplitudes is possible because of the cyclic softening of metals. An approximate analysis has been done by Suh and Sridharan[31] to determine the order of magnitude of the energy consumed during wear. It was done on the basis of the delamination theory of wear, an empirically determined plastic-strain field near the surface at the onset of delamination of a wear sheet, and a constitutive relation which takes into account the effect of cyclic softening.[30] The result of analysis states that the coefficient of friction μ and wear factor κ [which is defined as $\kappa = (W/SL)$] are related to each other by material constants when the work done is assumed to equal the energy consumed. Under the assumptions made in the analysis, which include the magnitude of net strain accumulated under cyclic loading, the analysis predicts the same order of magnitude for the energy consumed and for the external work done. A more careful investigation is being done to check the validity of the assumptions made in the analysis.

---

*Although the delamination process is failure under cyclic loading, it differs from a classical fatigue process in that the soft layer and ductile fracture play an important role and that the magnitude of plastic strain is extremely large.

8 MICROSTRUCTURAL EFFECTS IN OTHER WEAR PROCESSES

8.1 Fretting Wear

Fretting wear occurs between two tightly fitting surfaces when there is a small-amplitude, oscillating slip between them. The early stage of fretting wear is caused by the wear mechanisms postulated by the delamination theory of wear.[1,32,33] Once the wear particles are generated, they may undergo oxidation and act as abrasive particles between the surfaces, accelerating the wear process. However, the critical stage of fretting wear is the generation of wear particles by delamination process. Figure 16 shows a fretted Ti-6Al-4V surface taken by Waterhouse[33], which was tested in air at 107 cycles per second under a mean stress of 247 MN/m^2 and alternating stress of 83.5 MN/m^2. The wear particles are in the form of thin sheets.

One of the unique and controversial phenomena observed in fretting wear is the dependence of the wear rate on the amplitude of the oscillatory motion, as shown in Fig. 17.[34] In terms of the delamination theory of wear, the decrease in the wear rate at low amplitudes can be explained by noting that cracks cannot link together between two inclusions (i.e., crack-nucleation site) when the plastic strain is small. Fretting-wear rate depends on the microstructure since the distance between crack-nucleation sites *vis-à-vis* the amplitude of oscillating slip is the rate-determining process. Once the amplitude of oscillating slip is sufficient to link the cracks together, the early stage of fretting wear is not any different from sliding wear.

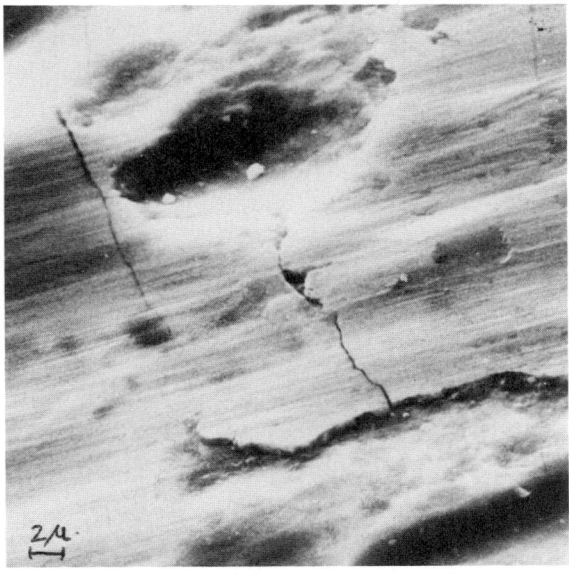

Fig. 16. Micrograph of a fretted Ti-6Al-4V surface.[34]

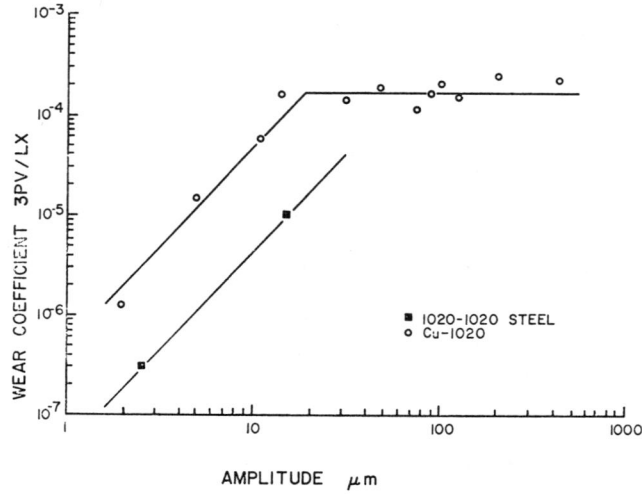

Fig.17. Fretting-wear coefficient versus the amplitude of oscillating slip.[24]

One of the major problems associated with fretting wear is fretting fatigue. When fretting fatigue is initiated at cracks created by fretting wear, fretting fatigue can be prevented if fretting wear can be reduced. Although lubrication of the interface is a possible solution, in some applications lubricants cannot be applied. A better solution is to plate a hard substrate with a *thin* layer of soft metal to minimize the delamination wear, as discussed earlier for sliding wear.[10,11] Then, fretting fatigue can also be prevented. Such experimental results were recently reported by Ohmae et al.[35], although the importance of the thickness of the plated layer was not recognized by them.

The environmental effect on fretting was recently discussed by Van Leeuwen in terms of the delamination theory of wear.[36] He postulated that since water may oxidize bare metal according to the reactions

$$Fe \rightarrow Fe^{2+} + 2e$$

$$Fe^{2+} + 2\ H_2O \rightarrow Fe\ (OH)_2 + 2H^+ \quad ,$$

the resulting hydrogen ions or protons could be transported by the dislocation as Cottrell clouds, as they move from a region near the surface into the subsurface region to pile up at hard particles. The dislocations would then deposit the hydrogen at the interface between the matrix and the obstacle and cause hydrogen embrittlement cracking at that location. This transport of hydrogen by dislocation motion, rather than by thermally activated diffusion, would then explain the accelerated fretting corrosion in a water-containing environment.

8.2 Abrasive Wear

Abrasive wear can be caused either by hard abrasive particles entrapped between two sliding surfaces or by the asperities of a hard surface sliding on a soft surface. The abrasive-wear volume is, to a first approximation, proportional to the normal load and the distance slid, and inversely proportional to hardness. Many wear processes change into abrasive wear when hard loose wear particles are generated.

The mechanism of abrasive wear consists of two modes of material removal: cutting of the surfaces by hard particles or asperities and delamination by subsurface deformation. The surface traction associated with the cutting action deforms the surface layer which causes subsurface crack nucleation and propagation. Depending on the sharpness and hardness of abrasive particles, either the delamination or the cutting action may dominate the mechanism. The harder the surface, the lesser will be the cutting action. When the delamination mechanism dominates, the microstructural effects discussed for sliding wear should be equally applicable. Figure 18 shows the subsurface deformation of an annealed AISI 1018 steel pin (0.64 cm in diameter) which was abraded against 240-grit emery paper at a speed of 6.3 cm/sec and a normal load of 0.4 kg. The subsurface cracks should be noted.

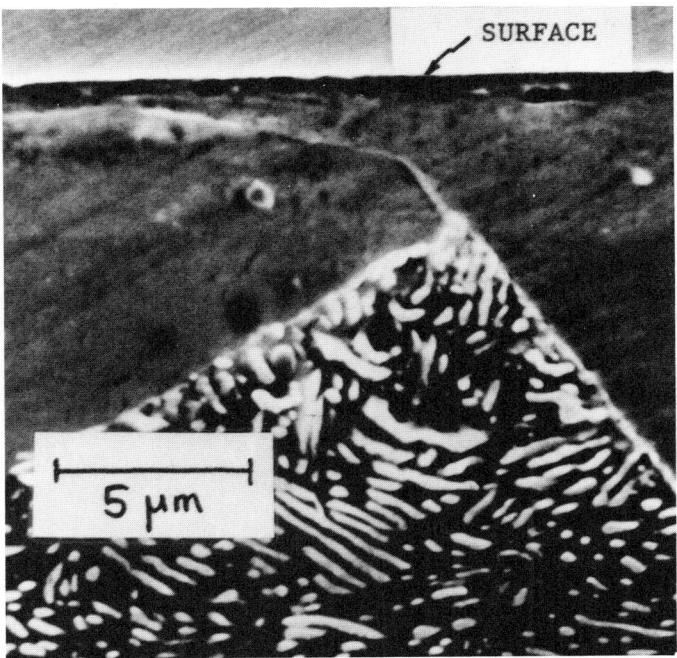

Fig.18. Cross-sectional view of an abraded AISI 1020 steel specimen.

8.3 Erosion by Solid-Particle Impingement

When metal surface is gradually worn away by the action of solid particles impinging on a surface, it is called erosion by solid-particle impingement. This is a problem in pipes which transport solid-particle slurries, in gas turbines, and in aircraft propellers. This problem was investigated by Finnie[37] and Bitter[38] who proposed wear mechanisms for this type of erosive wear. Finnie's model assumes that each of the solid particles acts as a cutting tool when it strikes the surface, removing some of the surface material. Bitter[38] proposed that the normal velocity component of the impinging solid particle causes wear by a fatigue process, while the tangential components cause wear by deformation.

The surfaces of 2024-T4 aluminum impacted at various speeds at an impingement angle of 45 degrees and room temperature are shown in Fig. 19. It should be noted that the material is displaced a great deal, but the amount of material removed by actual cutting action is very small. Figure 20 shows the aluminum specimens impacted at 280 m/sec at room temperature and at 150 m/sec at 200 F.[39] The specimen impacted at the higher speed shows a sign of melting, but the evidence is not conclusive. The specimen impacted at 200 F shows typical mud cracks, indicating that the surface has undergone oxidation at a high temperature or has melted. Figure 21 shows a cross-sectional view of the subsurface cracks in copper and steel. It should be noted that the cracks tend to be perpendicular to the direction of impact and that subsurface cracks exist at a finite distance away from the surface. The subsurface cracks extend deeper into the specimen impacted by particles travelling perpendicular to the surface. The mechanics of the surface deformation are such that we have a maximum deformation at the surface. Since crack nucleation is not favored wherever there is a high

Direction of Impact

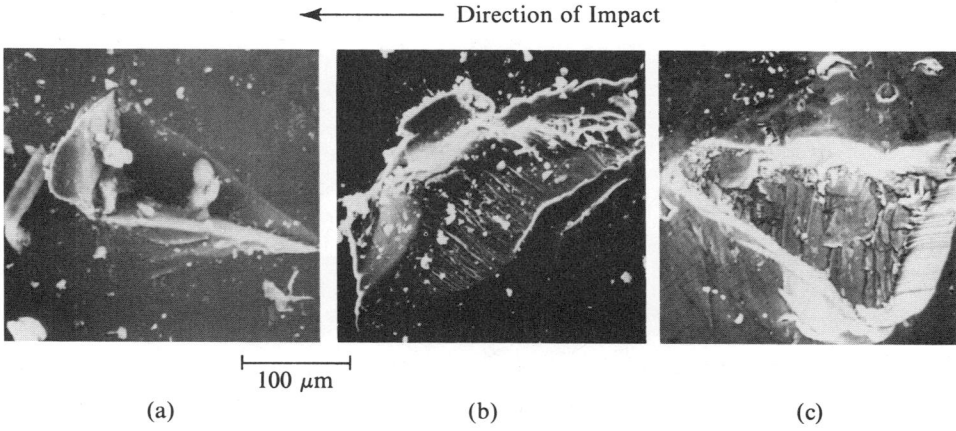

100 μm

(a)   (b)   (c)

Fig. 19. Surface of 2024-T4 aluminum impacted by silicon carbide particles at an impingement angle of 45 degrees. Impact velocity: (a) 32 m/sec, (b) 156 m/sec, (c) 310 m/sec.

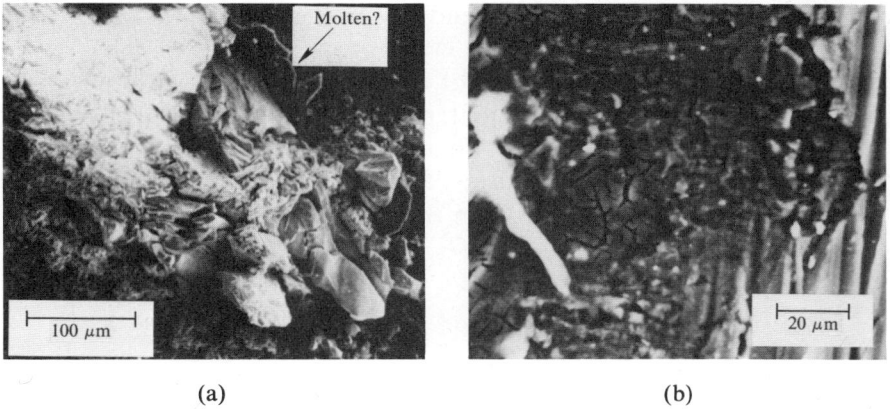

(a)   (b)

Fig. 20. Surfaces of 2024 aluminum impacted by silicon carbide particles at an impingement angle of 90 degrees: (a) temp 17 F, velocity 280 m/sec; (b) temp 200 F, velocity 150 m/sec.[39]

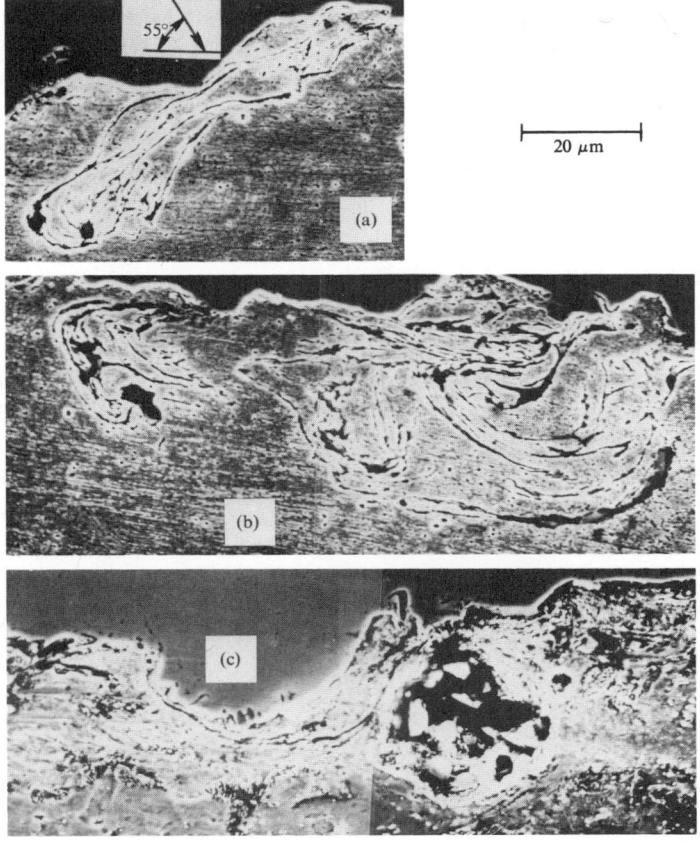

Fig. 21. Subsurface microcracks. (a) Copper (cold rolled and annealed at 375 C for 12 hours); abrasive 254 μm SiC; velocity 105 m/sec. (b) The same as (a) except the direction of impact was 90 degrees to the surface. (c) Steel (cold rolled AISI 1020 steel annealed at 70 C for $53^{1}/_{2}$ hours). Otherwise the same as (b).

hydrostatic pressure, cracks will tend to nucleate away from right underneath the impacted surface. Since crack propagation is similarly affected by hydrostatic pressure, the depth of stable crack propagation will vary, depending on the position of the material relative to the interface between the particle and the surface and on the direction of particle motion. It should be noted that the mechanism of erosion by liquid-particle impingement seems to be quite different from the case considered here.[40]

9 CONCLUSIONS

In such wear processes as sliding, fretting, abrasion, and erosion, deformation of the surface layer and the crack nucleation and propagation process in the subsurface control the wear process. Consequently the microstructure of the surface affects the wear resistance of the surface, as it controls hardness and ductility of the material. Although there are deviations, the crack-nucleation process in wear is similar to that in bulk deformation. The effect of particle morphology on crack propagation in wear differs from that in fracture of uniaxially and torsionally loaded specimens. The delamination theory of wear provides a reasonable theoretical basis on which to explain sliding- and fretting-wear phenomena and to design sliding surfaces metallurgically to minimize wear.

ACKNOWLEDGMENT

The author has been fortunate to have had the association of many intellectually stimulating and technically competent colleagues and students in carrying out the work on the delamination theory of wear. Said Jahanmir, Dr. Ernest P. Abrahamson, II, Jonathan Fleming, and Chung L. Leung were particularly instrumental in obtaining many of the results reported in this paper. The author also wishes to thank Prof. Frank A. McClintock and Dr. Arthur P. L. Turner who made many helpful comments on this paper.

The work was supported by the Advanced Research Projects Agency of the Department of Defense and was monitored by ONR under contract number N00014-67-A-0204-0080. The author wishes to express his deep appreciation to Dr. Edward van Reuth and Lt. Richard Miller for their personal interest in the work.

REFERENCES

1. Suh, N. P., *Wear*, **25**, 111 (1973).
2. Archard, J. F., *J. Appl. Phys.*, **24**, 981 (1953).
3. Green, A. P., *Proc. Roy. Soc. (London)*, **A228**, 181 (1955).
4. Rabinowicz, E., *J. Appl. Phys.*, **32**, 1440 (1961).
5. Bowden, F. P., and Tabor, D., *Friction and Lubrication of Solids*, Clarendon Press, Oxford, England (1954).
6. Kirk, J. A., and Swanson, T. D., to appear in *Wear*.
7. Koba, H., S. M. Thesis, Department of Mechanical Engineering, M.I.T. (1973).
8. Savitskii, K. B., cited in Kragelskii, V. I., *Friction and Wear*, Butterworth, Washington (1965).
9. Whitehouse, D. J., *Proceedings of International Conference on Product Engineering*, Tokyo, Japan (1974), Part II, p. 39.
10. Jahanmir, S., Suh, N. P., and Abrahamson, E. P., II, *Wear*, **32**, 33 (1975).
11. Jahanmir, S., Abrahamson, E. P., II, and Suh, N. P., *Wear* (in print).
12. Buckley, D. H., *Proceedings of the 39th Meeting of the Structures and Materials Panel* (1974); NATO Report AGARD-CP-161, p. 13-1.
13. Green, A. P., *J. Mech. Phys. Solids*, **2**, 197 (1954).
14. Gupta, P. K., and Cook, N. H., *Wear*, **20** 73 (1972).
15. Bathe, K. J., Acoustics and Vibration Laboratory, Department of Mechanical Engineering, M.I.T., Report No. 82448-1 (1975).
16. Fleming, J., S. M. Thesis, Department of Mechanical Engineering, M.I.T. (in progress).
17. Agustsson, G., S. M. Thesis, Department of Mechanical Engineering, M.I.T. (1974).
18. Wheeler, D. R., and Buckley, D. H., *Wear*, **33**, 65 (1975).
19. Scott, V. D., and Wilman, H., *Proc. Roy. Soc. (London)*, **A247**, 353 (1958).
20. Goddard, J., Harker, H. J., and Wilman, H., *Proc. Phys. Soc.*, **80** (3), 771 (1962).
21. Huppman, W. J., and Clegg, M. A., Paper presented at the ASLE-ASME International Lubrication Conference, New York, October, 1972.
22. Suh, N. P., Jahanmir, S., Abrahamson, E. P., II, and Turner, A.P.L., *J. Lub. Tech., Trans. ASME*, **96**, 631 (1974).
23. Jahanmir, S., Suh, N. P., and Abrahamson, E. P., II, *Wear*, **28**, 235 (1974).
24. Rosenfield, A. R., *Met. Rev.*, **13**, 29 (1968).
25. Argon, A. S., Im, J., and Safoglu, R., *Met. Trans. A*, **6A**, 825 (1975).
26. Argon, A. S., and Im, J., *Met. Trans. A*, **6A**, 839 (1975).
27. Ashby, M. F., *Phil. Mag.*, **14**, 1157 (1966).
28. Jahanmir, S., Abrahamson, E. P., II, and Suh, N. P., *Proceedings of 3rd North American Metalworking Research Conference*, Carnegie Press (1975), p. 854.
29. Rosenfield, A. R., Hahn, G. T., and Embury, J. D., *Met. Trans.*, **3**, 2797 (1972).
30. Lemmon, D. C., and Sherby, O. D., *J. Mater.*, **4**, 444 (1969).
31. Suh, N. P., and Sridharan, P., Paper presented at the 3rd International Tribology Conference, Paisley College of Technology, September 1975. *Wear*, **34**, 291 (1975).
32. Waterhouse, R. B., and Taylor, D. E., *Wear*, **29**, 337 (1974).
33. Waterhouse, R. B., *Proceedings of 39th Meeting of the Structures and Materials Panel*, Munich, Germany, October, 1974, NATO Report AGARD-CP-161, p. 8-1.
34. Stowers, I. F., Ph.D. Thesis, Department of Mechanical Engineering, M.I.T. (1974).
35. Ohmae, N., Nakai, T., and Tsukizoe, T., *Wear*, **30**, 299 (1974).
36. van Leeuwen, H. P., *Proceedings of the 39th Meeting of the Structures and Materials Panel*, Munich, Germany, 1974, NATO Report AGARD-CP-161, p. 13-19.
37. Finnie, I., *Proceedings 3rd U.S. National Congress on Applied Mechanics* (1958), p. 527.
38. Bitter, J.G.A., *Wear*, **6**, 5 and 169 (1963).
39. Sherman, C. J., S. M. Thesis, Department of Mechanical Engineering, M.I.T., 1971.

40. Speidel, M. O., and Keser, G., Brown Boveri, Research Report, KLR-74-10, Presented at the 4th International Conference on Rain Erosion and Associated Phenomena, Meersburg, Germany, 1974.
41. Fourie, J. T., *Can. J. Phys.*, **45**, 777 (1967).
42. Merwin, J. E., and Johnson, K. L., *Proc. Inst. Mech. Engrs.*, **177**, 676 (1963).

DISCUSSION on Paper by N. P. Suh

WILCOX: I was a little bit confused by your observation that a soft layer improved wear resistance, since we normally think the wear resistance is improved by making a material harder. Is the soft layer essentially a solid lubricant?

SUH: In general, harder metals wear less than softer metals. This is due to the fact that harder metals deform less under a given set of surface traction, thus delaying the subsurface crack nucleation and propagation process. Hardness can also affect the wear process by influencing the magnitude of the frictional force acting at the surface. The role of the *thin*, soft metal layer plated on a hard substrate is to reduce the frictional-force component which controls the deformation and crack-propagation process that occurs in the substrate under a given loading condition. All lubricants lower the frictional force. What is unique about the soft metal layer is that it deforms continuously without wearing if the thickness of the layer is less than a critical value. Perhaps it should be reemphasized that wear of metals under sliding conditions is affected by "ductility" as well as by hardness when delamination wear occurs.

HORNBOGEN: Is it not so that delamination wear will require rather ductile materials in which a compressive stress in the surface can be produced by high amounts of plastic deformation? If not much plastic deformation can occur, it may be more likely that the wear rate is determined by critical crack growth or fatigue crack growth that starts at the surface.

SUH: The delamination theory of wear applies to wear of materials under sliding conditions. Metals must deform for the delamination process to take place or have preexisting subsurface cracks. In the case of ball bearings and other brittle solids, cracks are often initiated at the surface under the influence of the tensile stress parallel to the surface (see Fig. 8). I believe that this type of failure of ball bearings is due to small indentations made on the surface by wear (or impurity) particles during rolling and spinning of ball bearings. Fatigue cracks can propagate from the roots of these indentations without involving gross plastic deformation. One obvious way of eliminating this type of wear is to filter out solid particles from the lubricating oil. Cracks are sometimes initiated at the subsurface of bearings when inclusions are present.

BUSH: Could one alter the delamination characteristics by a strong preferred orientation of hard fibers normal to the wear surface imbedded in a soft matrix?

SUH: Yes, we have performed experiments with fiber-reinforced composites which seem to indicate that the wear rate is a strong function of the fiber orientation.

WRIGHT: Would you care to comment briefly on differences you have observed in sliding and erosive wear behavior at elevated temperatures compared to room temperature. Do the mechanisms change, for instance.

SUH: Dr. W. Ruff of NBS told me that subsurface cracks are not present at elevated temperatures. Our tests were done at low temperatures. There are also indications that metals may be melting at high impingement velocities and also when the specimen is heated to elevated temperatures (see Fig. 20). Strains involved at high-speed particle-impingement conditions are very large.

DESIGN OF STRUCTURAL ALLOYS WITH HIGH-TEMPERATURE CORROSION RESISTANCE

F. S. Pettit

Pratt & Whitney Aircraft
Division of United Technologies Corp.
Middletown, Connecticut

ABSTRACT

In order for an alloy to possess high-temperature corrosion resistance, the reaction of the alloy with the environment must result in a reaction product that effectively deters all subsequent reaction. It is shown that oxide reaction product barriers, in particular Al_2O_3, Cr_2O_3, and SiO_2, are the most effective barriers to develop corrosion resistance in alloys. The conditions that must be satisfied in order for such oxides to be formed as continuous barriers on the surfaces of alloys are discussed and described. The stability of these continuous oxide barriers to corrosion induced by ash deposits is considered and methods to improve the adhesion of oxide barriers are examined. Techniques and procedures to be followed in designing alloys for corrosion resistance are proposed.

1 INTRODUCTION

In designing alloys for use at elevated temperatures, the alloys must not only be as resistant as possible to the effects produced by reaction with oxygen, but resistance to attack by other oxidants in the environment is also necessary. In addition, the environment is not always only a gas since, in practice, the deposition of ash on the alloys is not uncommon. It therefore is more realistic to speak of the high-temperature corrosion resistance of materials rather than their oxidation resistance.

The development of high-temperature corrosion resistance in structural alloys requires the formation of reaction-product barriers on the surfaces of alloys which inhibit reactions between the alloys and the environments. The most resistant alloys are those for which the barriers are most effective in deterring all subsequent reactions. In some cases, the barriers necessary to develop the desired resistance to corrosion can be formed on structural alloys by appropriate compositional modifications. In many practical applications for structural alloys, however, the required compositional changes are not compatible with the required physical properties of the alloys. In such cases, the necessary compositional modifications are developed through the use of coatings on the surfaces of the structural alloys and the desired reaction-product barriers are developed on the surfaces of the coatings.

In this paper, the barriers that have been used to develop high-temperature corrosion resistance in alloys are examined. The critical properties of such barriers are then discussed. In particular, the factors which affect the formation, growth, and adhesion of protective reaction-product barriers are examined, and the influence of alloy composition, alloy microstructure, and processing on these properties is considered.

2. PROTECTIVE BARRIERS SUITABLE FOR DEVELOPING HIGH-TEMPERATURE CORROSION RESISTANCE IN ALLOYS

In principle, reaction products formed on the surfaces of alloys affect the corrosion rates of alloys when transport of the reactants through these products is the rate-controlling step. In order for a reaction-product barrier to be effective in developing corrosion resistance, it must therefore be formed as a continuous layer over the alloy and this layer must be relatively free from cracking and spalling. The type of reaction-product barrier that is developed on an alloy is dependent upon the composition of the alloy and the composition of the environment in which the alloy is used.

In cases where alloys are exposed to complex environments, the reaction-product barriers may be composed of discrete layers or zones of different phases or the different phases may be mixed with one another. The development of discrete zones of the different reaction products is a necessity for corrosion resistance since this condition most effectively affords the use of certain reaction products to inhibit subsequent corrosion. In the event that the reaction-product barrier does contain mixtures of phases rather than discrete zones, it is necessary to modify the composition of the alloy to remove this condition. For example, when iron is corroded in argon-SO_2 mixtures at about 800 C, iron oxide and iron sulfide form cooperatively to produce a lamellar structure and the presence of the sulfide network causes more severe corrosion than what is observed in pure oxygen.[1] This condition does not occur when 10 percent* chromium is added to the iron, since discrete zones of reaction products are formed.

In many environments encountered in practice, oxygen is usually present. Furthermore, oxide reaction products are usually thermodynamically more favorable because of the greater affinity of most metallic elements for oxygen compared with the other oxidants that may be present in different environments. As a result of this condition, it is common for the reaction-product zone which is adjacent to the environment to be an oxide, as shown in Fig. 1.[2] Oxide-reaction-product barriers generally are also more effective in inhibiting reactions between alloys and various environments than other types of barriers, as is evident upon comparing the parabolic rate constants for growth of sulfides and oxides on different alloys, Table I.[3] As a result of these conditions, the development of corrosion resistance in alloys generally involves the use of oxide-reaction-product barriers. This paper therefore considers barriers of this type only. In those rather few applications where sufficient oxygen is not present to permit the formation of even the very thermodynamically stable oxides, it will be necessary to use another type of barrier. The general factors to be considered in selecting and effectively utilizing such a barrier, however, will be similar to those discussed in this paper for oxides.

3 GROWTH RATES OF OXIDE BARRIERS

Parabolic rate constants for the growth of different oxide barriers as a function of temperature are compared in Fig. 2. It can be seen that Al_2O_3, SiO_2, and Cr_2O_3 are the more effective barriers to oxidation. This condition is also true for hot corrosion induced by ash deposition since the added problems incurred by such a condition do not involve transport through protective barriers but, rather, the avoidance of physical breakdown of such barriers.

---

*All compositions are given in weight percent.

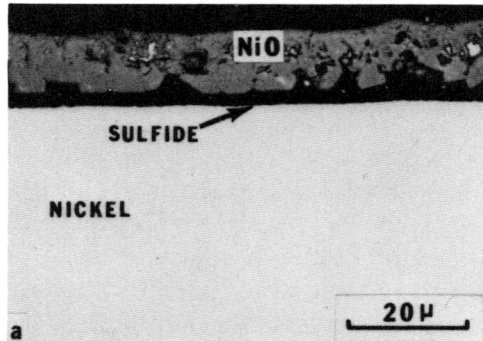

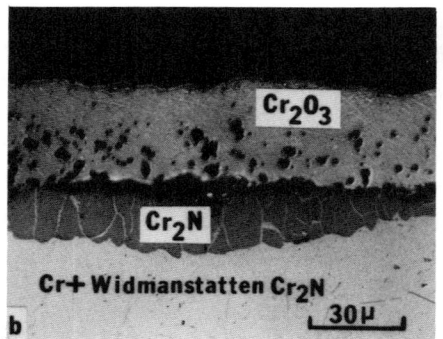
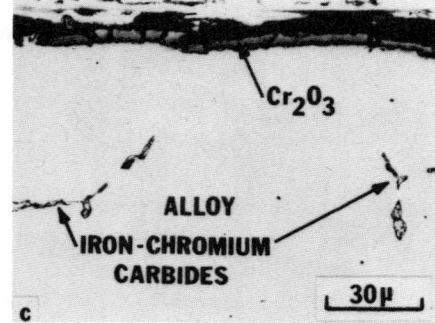

Fig. 1. Formation of sulfide, nitride, and carbide phases beneath external oxide scales: (a) nickel heated in flowing SO_2 for 8 hours at 1000 C; (b) chromium oxidized for 17 hours in air at 1200 C; (c) Fe-25Cr alloy heated in a flowing CO-CO_2 mixture with $CO/CO_2 = 5$ at 900 C for 20 hours.

Table I. Parabolic Rate Constants of Sulfidation and Oxidation of Some Binary and Ternary Alloys

Alloy	Temperature, C	k_p, g^2 cm^{-4} min^{-1} Sulfidation	Oxidation
Fe-20Cr	1000	5.9×10^{-4}	1.2×10^{-9}
Fe-20Cr	900	1.1×10^{-4}	1.0×10^{-10}
Ni-20Cr	1000	5.3×10^{-4}	3.0×10^{-9}
Co-20Cr	800	1.6×10^{-5}	5.0×10^{-11}
Fe-25Cr-5Al	1000	6.5×10^{-5}	1.6×10^{-11}
Ni-20Cr	800	6.0×10^{-5}	1.2×10^{-10}

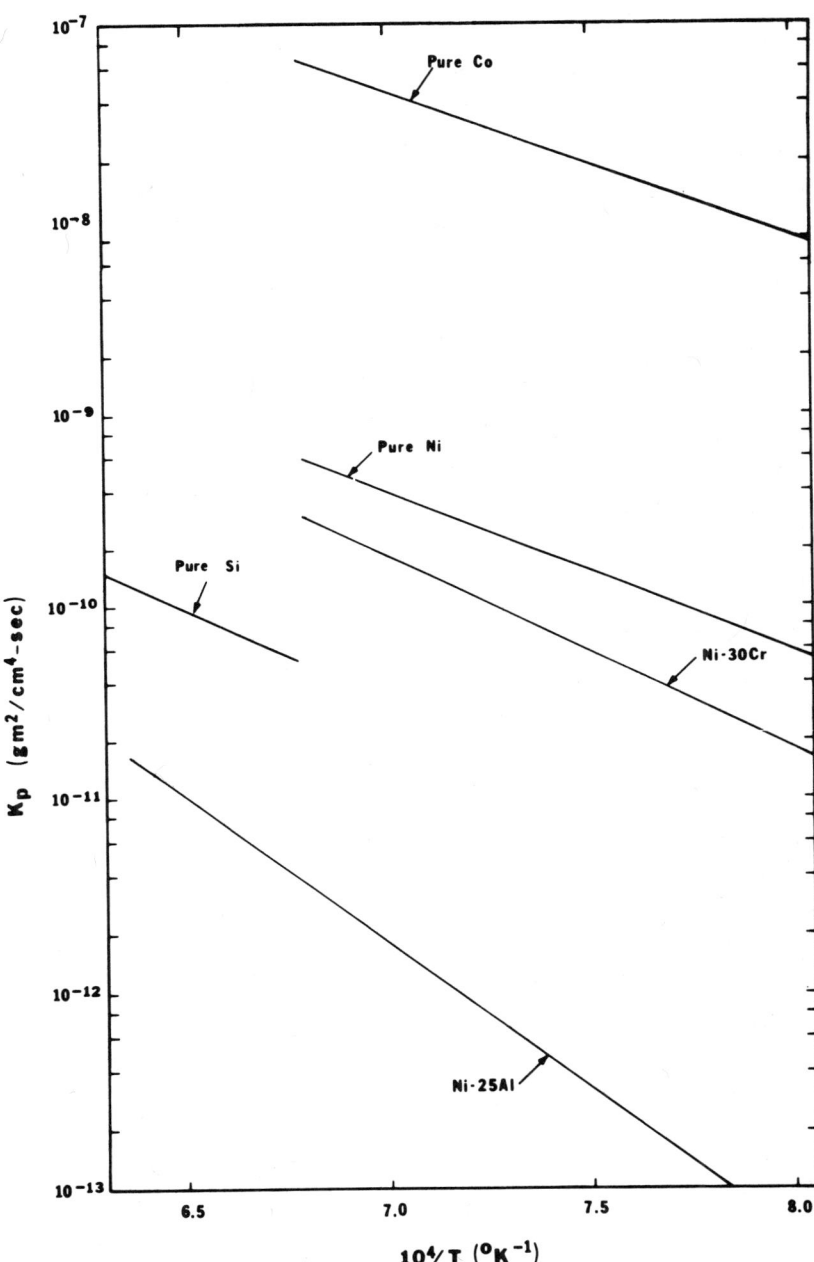

Fig. 2. Temperature dependence of the parabolic rate constants obtained for the growth of various oxide barriers on pure metals (CoO on cobalt, NiO on nickel, SiO$_2$ on silicon) and alloys (Cr$_2$O$_3$ on Ni-30Cr, Al$_2$O$_3$ on Ni-25Al). The oxidizing environments for the data presented were 0.1 atm of oxygen for nickel, Ni-30Cr, and Ni-25Al; 0.125 atm of oxygen for cobalt; and 1 atm of oxygen for silicon.

Some consideration has been given to using oxide barriers other than Al_2O_3, Cr_2O_3, or SiO_2 to develop high-temperature corrosion resistance in alloys. Some single oxides such as BeO or MgO as well as certain complex double oxides (e.g., $NiAl_2O_4$) may be worth evaluating. The factors which are currently limiting the corrosion resistance of alloys, however, are the development and adhesion of the currently available barriers. It therefore is believed that the need to obtain additional oxide barriers is not great.

Possibilities also exist to decrease the growth rates of Al_2O_3 and Cr_2O_3 barriers. For example, results recently obtained from studies with Al_2O_3 scales show that the growth of such barriers is controlled by diffusion of oxygen through grain boundaries in the barriers.[4] In Fig. 3 it can be seen that Al_2O_3 scales of different thicknesses are developed on the surface of alloys as a result of the different sizes of grains in the oxide scale. Caplan and Sproule[5] have also shown that the scales of Cr_2O_3 developed on chromium are of nonuniform thickness since the scale grows much faster at grain boundaries than at other locations in the scale. Such results indicate that substantial reductions in the growth rates of Al_2O_3 and Cr_2O_3 scales may be achieved if the density of grain boundaries in these scales can be reduced. It may also be possible to reduce growth rates of scales by using impurities to modify the concentration of defects.

4 DEVELOPMENT OF PROTECTIVE OXIDE BARRIERS ON ALLOYS

4.1 Oxide-Barrier Development in Oxygen

In considering the factors that affect the development of protective oxide barriers on alloys, it is convenient to examine first the formation of the barrier without the complication of other oxidants in the environment and then subsequently treat this more complicated condition. As reaction between an alloy and the oxygen in the environment takes place, most practical alloys and coatings do not contain enough of that element whose oxide is the most thermodynamically stable to permit the alloy surface to become totally covered with this oxide. Other elements in the alloy, whose oxides are less stable, are therefore oxidized. As oxidation is continued, conditions can be established whereby the amount of the most stable oxide is increased owing to interdiffusion in the alloy, and a continuous barrier of this oxide can be formed. The theory to describe this selective oxidation process has been developed by Wagner[6-9] and by Rapp[10,11]. More recent analyses of this process with emphasis on the diffusional aspects have further refined this theory.[12] From the viewpoint of the alloy designer, the point to be emphasized is that protective oxide barriers are developed on alloys during

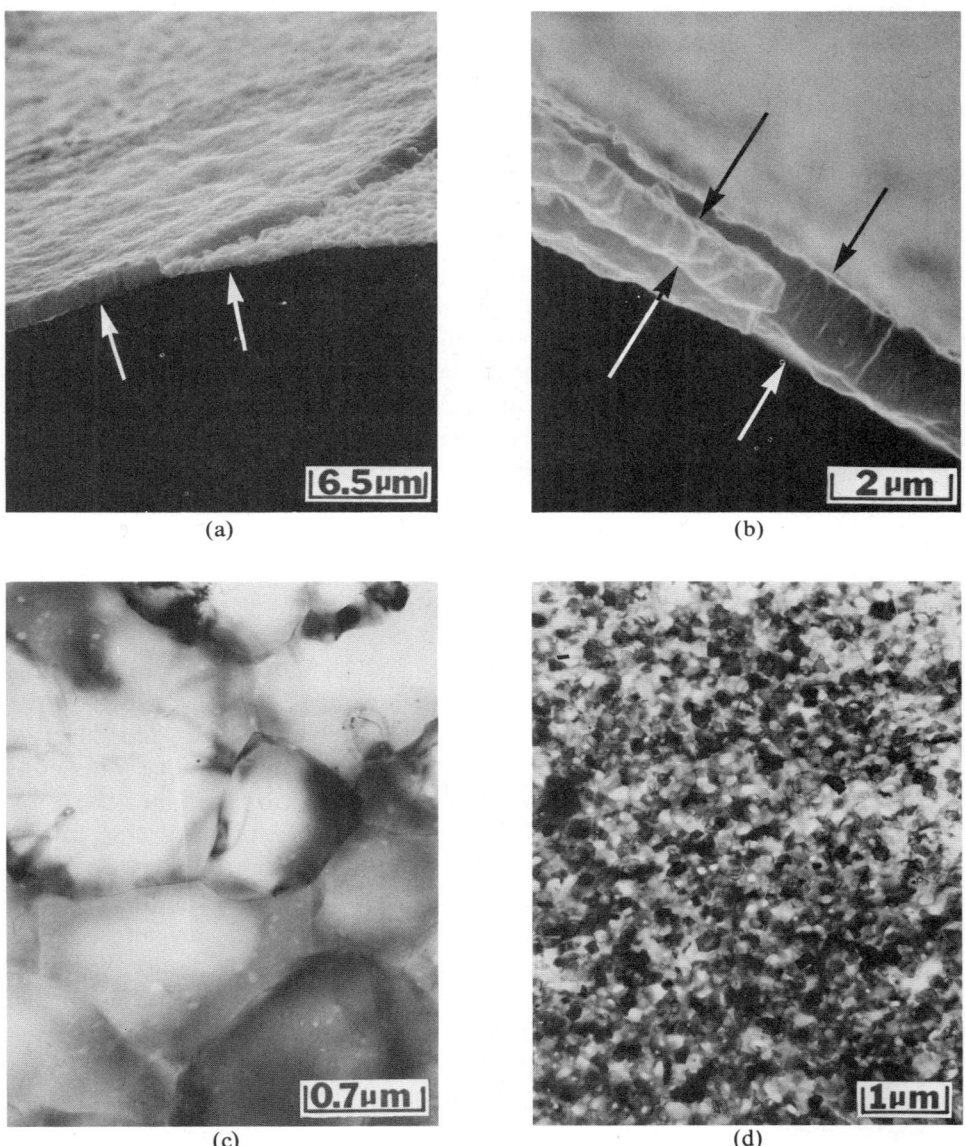

Fig. 3. Photographs showing that the thicknesses of Al_2O_3 scales which are formed during oxidation of alloys are dependent upon the size of grains in these scales: (a) the surface of an Al_2O_3 flake that has been removed from an alloy; (b) thickness of flake (black arrows) along fracture indicated by white arrows in (a); (c) and (d) oxide grains in thin and thick portions of oxide scale, respectively.

a transient oxidation period in which diffusional processes take place and the volume fraction of the desired oxide is increased. With such a process in mind, the intent is therefore to modify the alloy so as to permit selective oxidation to occur at the lowest possible concentration of the selected element.

One of the most effective techniques to obtain corrosion resistance in alloys is through appropriate control of the composition of the alloy. The desired oxide barrier can be obtained simply by increasing the concentration in the alloy of the metallic component of the oxide. In many alloy systems, requirements other than corrosion resistance do not permit using this procedure. It is therefore necessary to use other elements in the alloy to aid in the selective-oxidation process. The high-temperature corrosion theory is not at a stage where one can choose the type and amount of these other elements without a certain amount of trial and error. The level of knowledge is sufficient, however, to indicate the general approach to be followed in order to meet with success; namely, conditions should be established whereby the flux of oxygen into the alloy must be decreased relative to the flux from the alloy of the element to be selectively oxidized.

A variety of elements have been found to aid in the selective oxidation of other elements in alloys. The compositional limits at which protective layers of Al_2O_3 and Cr_2O_3 are formed on NiCrAl alloys are presented in Fig. 4. Examination of this diagram shows that chromium aids in the selective oxidation of aluminum and that aluminum has a similar effect on the selective oxidation of chromium.[13] The effect of chromium on the selective oxidation of aluminum may be a secondary-oxygen-getter effect. This concept, developed by Wagner[14], proposes that the flux of oxygen into the alloy is reduced as a result of the development of a layer of the oxide of the secondary getter. The addition of zinc to silver-aluminum or copper-aluminum alloys permits the selective oxidation of aluminum at lower aluminum concentrations because zinc acts as a secondary oxygen getter.[15]

The observed beneficial effect of aluminum on the selective oxidation of chromium is not a secondary-oxygen-getter effect because aluminum has a greater affinity for oxygen than has chromium. Highly oxygen active elements such as scandium[16] and hafnium[17] have been observed to promote the selective oxidation of chromium in alloys. It appears as though such elements, including aluminum, may help in the development of Cr_2O_3 barriers by causing the flux of chromium from the alloy to be increased. The mechanism by which such elements increase the transport of chromium is not clear but may involve oxidation of these elements to form oxide particles. The effects produced by oxide particles in alloys are discussed subsequently.

Some refractory metals have been observed to favorably affect the selective oxidation of chromium in alloys. For example, as can be seen in Fig. 5, a protective barrier of Cr_2O_3 is not developed on a Co-20Cr alloy

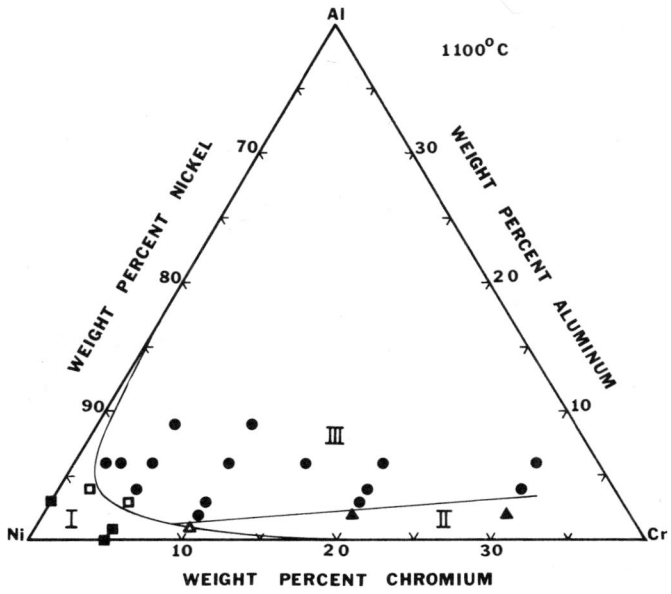

Fig. 4. Compositional limits at which continuous oxide barriers of NiO(I), Cr₂O₃(II), and Al₂O₃(III) are developed during oxidation of Ni-Cr-Al alloys. The data points in this diagram correspond to the compositions of the alloys studied. Solid squares, triangles, and circles indicate compositions for which the oxide barriers were NiO, Cr₂O₃, and Al₂O₃, respectively. Open symbols represent compositions where evidence for two types of barriers was observed (i.e., nonuniform oxidation).

during oxidation in oxygen at 1000 C but such a barrier is developed on a Co-20Cr-12W alloy. The mechanism by which the tungsten causes the chromium to be selectively oxidized is not available, but the oxidation resistance of X-40 and Mar M 509 is probably better than that of binary cobalt-base alloys with the same chromium concentrations because of this effect.

The microstructure of alloys can have important effects on the selective oxidation of elements. The results presented in Fig. 6 show that fine-grained microstructures are more favorable for the development of protective Cr₂O₃ barriers than coarse-grained structures. Oxide dispersions in alloys are also effective in decreasing the amount of chromium required for protective-barrier formation.[18,19] Models to account for the observed influence of alloy grain boundaries or oxide particles have been developed. Some features of the models still require elucidation. It appears that grain boundaries and inert oxide particles may affect the selective-oxidation process through nucleation effects.[19]

In developing corrosion resistance in alloys, it is useful to attempt to have the alloy as homogeneous as possible because compositional variations

(a)

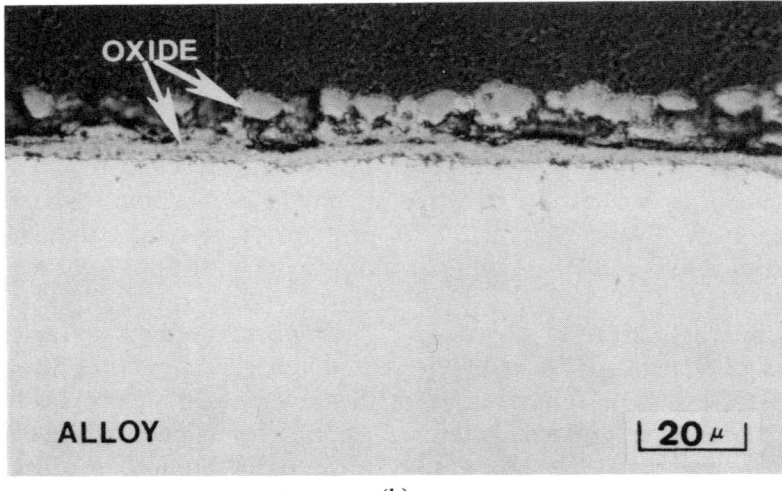

(b)

Fig. 5. Oxide scales formed on Co-20Cr after 2 hours of oxidation (a) and on Co-20Cr-12W after 20 hours of oxidation at 1000 C in 1 atm of oxygen. After 20 hours of oxidation, weight gains of 30 and 1.8 mg/cm^2 were observed for the Co-20Cr and Co-20Cr-12W specimens, respectively. A continuous barrier of Cr$_2$O$_3$ had been developed over the Co-20Cr-12W specimen but not over the Co-20Cr specimen.

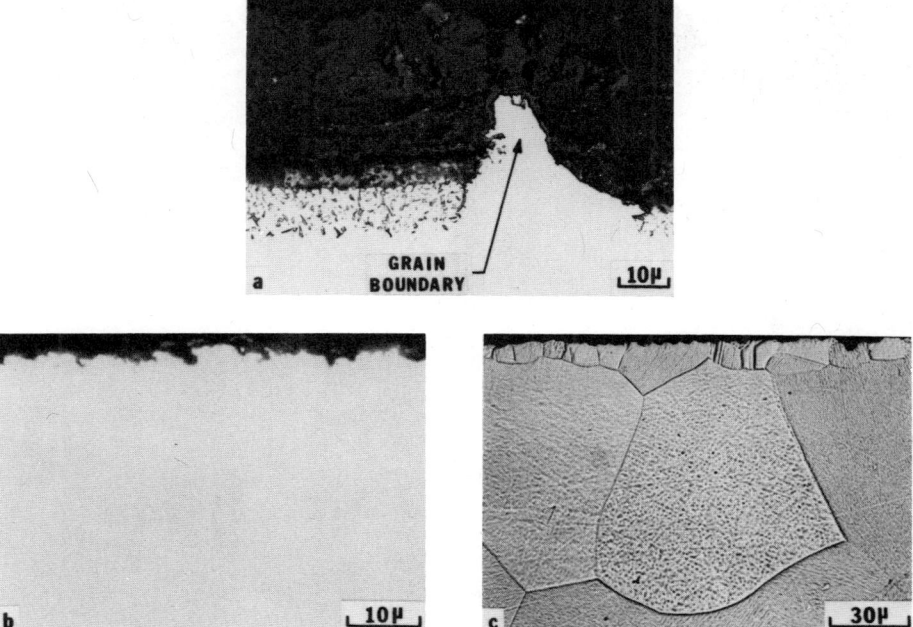

Fig. 6. Effect of surface condition on a Ni-10Cr alloy oxidized for 20 hours in 0.1 atm of oxygen at 900 C; (a) electropolished specimen, (b) grit-blasted specimen, (c) specimen described in (b) in etched condition. The small grains formed upon recrystallization of the deformed surface layer caused a continuous barrier of Cr_2O_3 to be formed.

can seriously disrupt the selective-oxidation process. Most structural alloys are multiphased and, consequently, compositional inhomogeneities are unavoidable. When wide compositional differences exist among the phases, very severe preferential oxidation can prevail, as shown in Fig. 7, and for such conditions, coatings must be used to obtain adequate corrosion resistance.

4.2 Oxide-Barrier Development in Complex Environments

In environments which contain other components in addition to oxygen, it is necessary to consider the effects of such components on the selective-oxidation process. It is important to emphasize that the complex or real-life environments are not always a gas but frequently consist of liquid coatings that are formed on alloys as a result of ash deposition.

The effects that other components in the gas or liquid may have on selective oxide barrier formation have not been studied in great detail. It

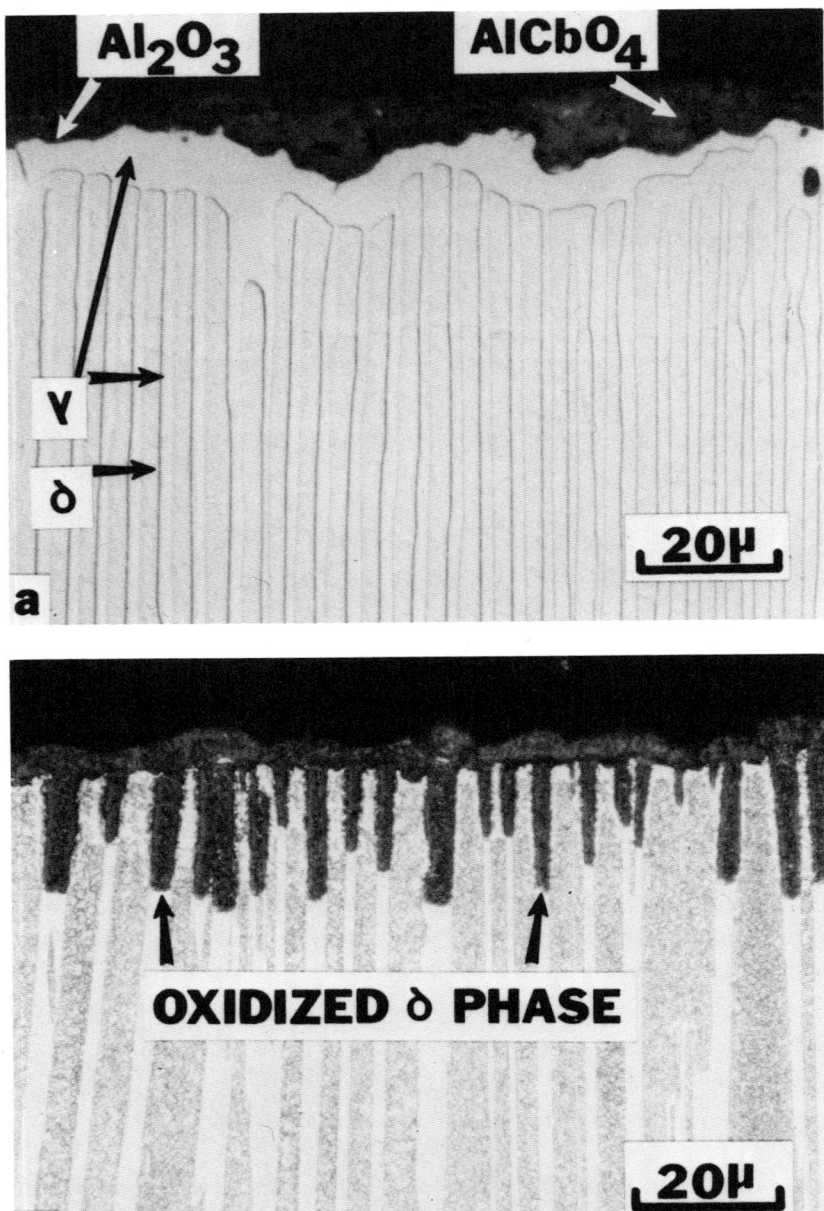

Fig. 7. Structures developed during the oxidation of a directionally solidified Ni-Cb-Cr-Al alloy. (a) Preferential oxidation of the strengthening δ (Ni₃Cb) phase is not excessive because a diffusion zone denuded of this phase is developed during oxidation at 1100 C. (b) During oxidation at 700 C such a denuded zone cannot be formed and severe preferential oxidation of the δ-phase occurs.

appears as though many of the additional components in practical environments (e.g., sulfur, carbon, nitrogen) will have adverse effects on the selective-oxidation process. Graham et al.[20] have observed that the selective oxidation of aluminum in a Ni-8Cr-6Al alloy at 1000 C takes place in oxygen (150 torr) but not in a CO-CO_2 gas mixture with CO/CO_2 of 220. In the latter case, the aluminum is oxidized internally and the Al_2O_3 is formed as a discontinuous subscale. Carbon from the gas reacts with chromium in the alloy and chromium is therefore not available to participate in the selective-oxidation process. It is reasonable to suppose that elements such as sulfur and nitrogen may be capable of producing similar effects. To offset such adverse effects, the only recourse, short of removing these undesirable components from the environment, is to utilize as many as possible of those techniques described previously as favoring the selective-oxidation process.

When the environment also consists of a liquid deposit, the effects of different components in the liquid are the same as those described previously for the gas environment. An additional undesirable feature, however, is that the oxide barrier may react or dissolve in the liquid. A comprehensive discussion of the mechanisms of ash-deposition-induced corrosion, or hot corrosion as it is commonly called, is beyond the scope of this paper and material is available in the literature on this subject.[21-26] It is sufficient here to state that the hot-corrosion degradation of alloys can be divided into two successive stages: an initiation stage and a propagation stage. The time required to initiate the propagation modes is dependent upon the following factors: alloy composition, alloy microstructure, gas composition, gas velocity, ash composition, ash deposition rate, temperature, thermal fluctuations, erosion, and specimen geometry. Some examples showing the effects of composition on the initiation of hot corrosion are presented in Fig. 8. Examples presented in Fig. 9 illustrate the effects of alloy microstructure, cyclic oxidation conditions, and the amount of salt deposition on the initiation of hot corrosion.

It appears that there are two different types of modes by which hot-corrosion degradation is propagated. One propagation mode has been described as an acidic fluxing process since the oxide barrier is prevented from being formed, or is destroyed, by reactions which involve the donation of oxide ions from the oxide barrier to the liquid-ash deposition. The other propagation mode consists of a combined basic fluxing-sulfidation process. This latter mode involves destruction of protective oxide barriers due to reaction with oxide ions in the ash deposit. Oxide-ion production in the ash is dependent upon removal of sulfur from the ash by the alloy. Severe degradation occurs when the alloy rapidly removes sulfur from the ash. Such conditions are conducive to rapid attack not only because the oxide ions dissolve the protective barrier, but also because the sulfur in the alloy prevents the protective oxide barrier from being reestablished.

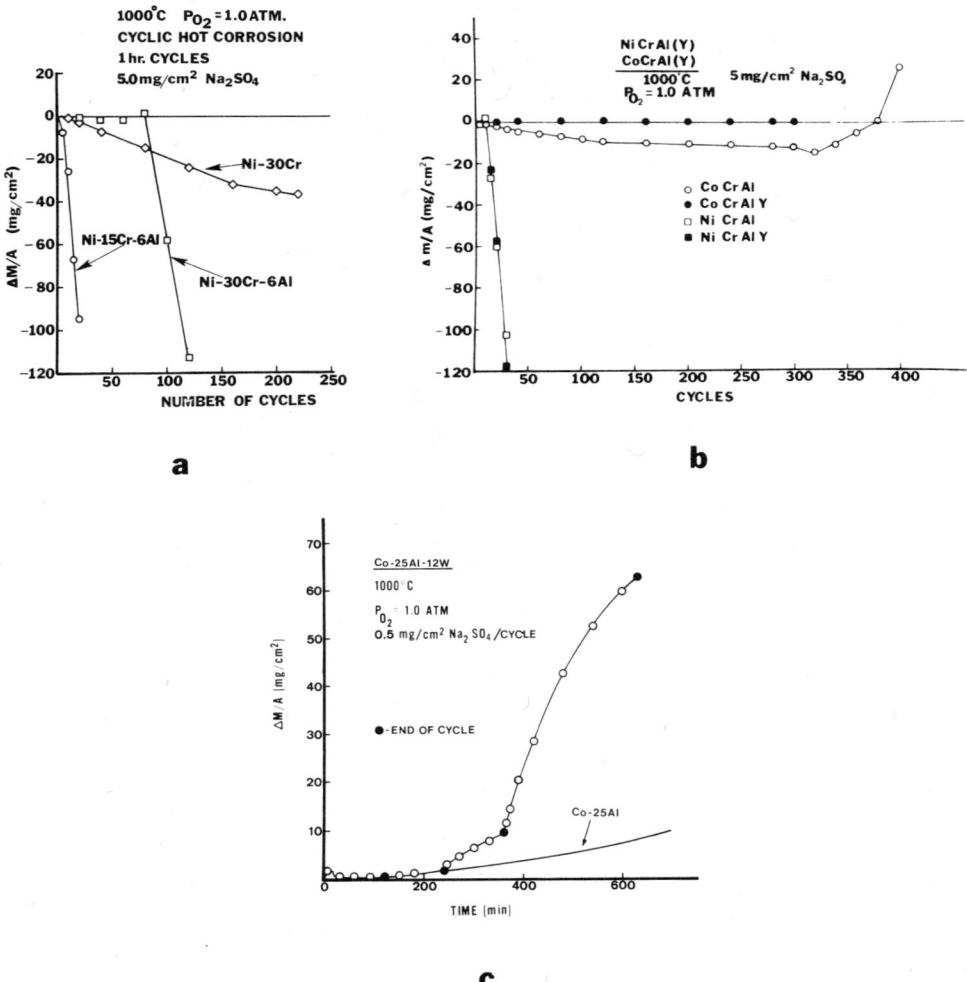

Fig. 8. Data obtained from the cyclic oxidation of Na₂SO₄-coated specimens. (a) The time to initiate severe hot-corrosion attack, as indicated by the specimen weight losses, is dependent on the chromium and aluminum concentration of the NiCrAl alloys. (b) CoCrAl(Y) alloys are much more resistant to the initiation of hot-corrosion attack than NiCrAl(Y) alloys with the same concentrations of chromium and aluminum. (c) The hot-corrosion attack of Co-25Al is much more severe when tungsten is added to this alloy.

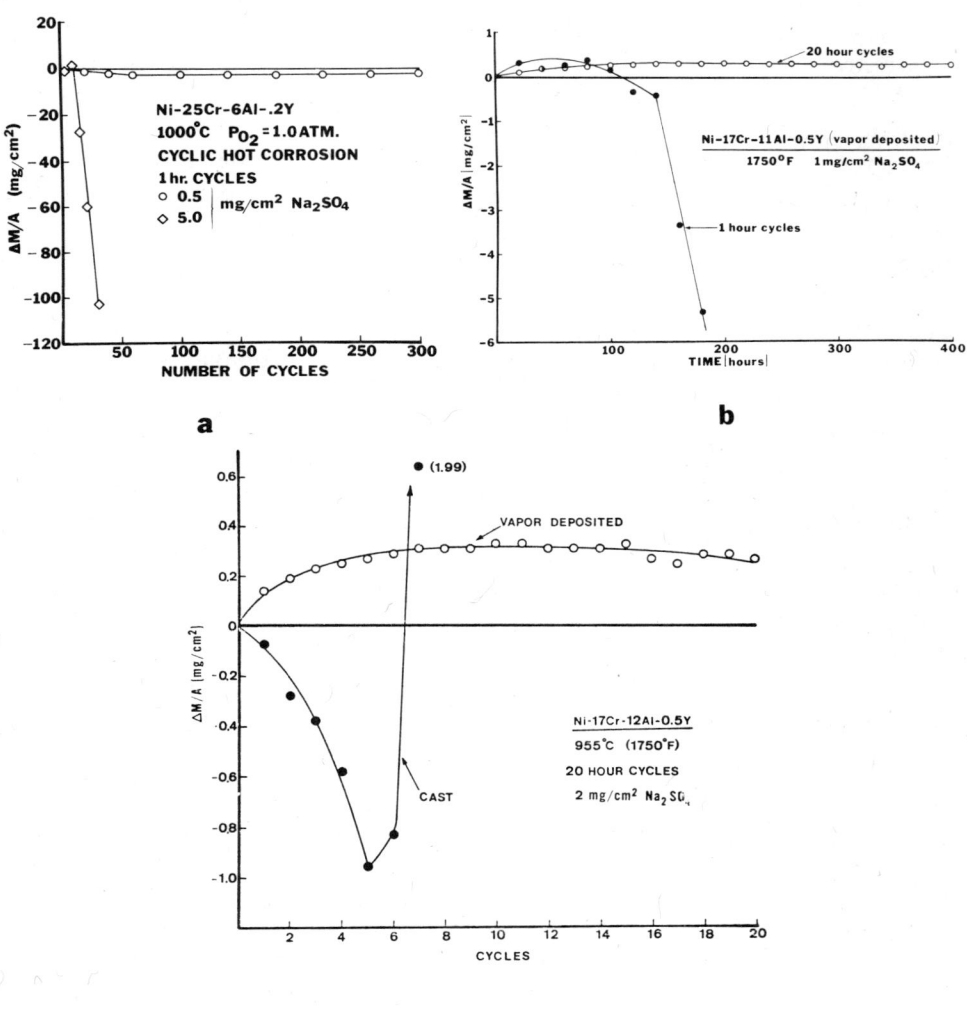

Fig. 9. Experimental data that show that the Na₂SO₄-induced hot corrosion of alloys is dependent upon (a) the amount of the deposit, (b) the number of thermal cycles imposed upon the specimens, and (c) the fabrication condition of the specimens.

The microstructural features of the two propagation modes are compared with each other as well as with those of an alloy with a protective Al$_2$O$_3$ barrier (see Fig. 10). The severe degradation via the acidic-fluxing propagation mode (Fig. 10b) occurs because certain elements in the alloy make the ash deposit an acidic flux. This type of propagation is self-sustaining in that, once initiated, subsequent ash deposition is not required for the degradation to continue. The features of the basic fluxing-sulfidation mode are presented in Figs. 10c and 10d. In this mode, sulfur is removed from the ash, the protective barrier is not stable in the oxide-ion-enriched deposit, and the sulfur in the alloy also prevents the protective barrier from being formed. Also in this mode, ash must continually be supplied in order that oxide ions can be produced and sulfur can be added to the alloy. This mode is therefore not self-sustaining. While the features of the propagation modes are important in developing the total hot-corrosion theory, the important point for the alloy designer concerned with corrosion resistance is to avoid the initiation of the propagation modes since an unacceptable corrosion rate will result upon initiation of either of the propagation modes.

To control hot corrosion it is necessary for oxide barriers to develop on the alloys which are as stable as possible in the liquid-ash deposit. The reaction of Al$_2$O$_3$, Cr$_2$O$_3$, and SiO$_2$ with ash deposits having different compositions has not been studied in detail. The data available indicates that Al$_2$O$_3$ and Cr$_2$O$_3$ are as suitable as any of the available oxide barriers for use under conditions involving ash deposition.

Refractory-metal oxides such as MoO$_3$ must be added to the ash deposit in order to initiate the acidic-fluxing propagation mode. Resistance to this type of corrosion can therefore be developed in alloys by keeping the concentration of tungsten, molybdenum, and vanadium as low as possible. It also appears that as the chromium content of the alloy is increased, larger amounts of tungsten, molybdenum, or vanadium can be tolerated without the development of a severe sensitivity to hot-corrosion attack.

Chromium can also be used to make alloys more resistant to the initiation of the basic fluxing-sulfidation propagation mode. The aluminum concentration must be carefully controlled. Increasing the aluminum content will usually result in increased resistance to this type of degradation for alloys which rely on Al$_2$O$_3$ barriers for corrosion resistance. In the case of alloys where the oxide barrier is Cr$_2$O$_3$, however, aluminum causes the time to initiate severe attack to be decreased. Since the oxides of the refractory metals cause attack via the acidic-fluxing mode, elements such as tungsten and molybdenum can be used to inhibit the onset of degradation via basic fluxing-sulfidation, but the danger exists of going over into the acidic-fluxing mode of degradation.

Cobalt can be used in place of or as a partial replacement for nickel to improve the resistance of alloys to degradation by the basic fluxing-sulfidation degradation mode (Fig. 8b). Cobalt-base alloys are especially

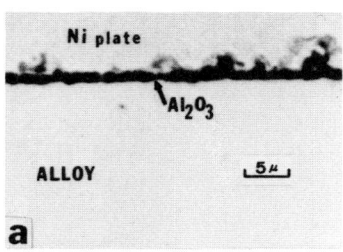

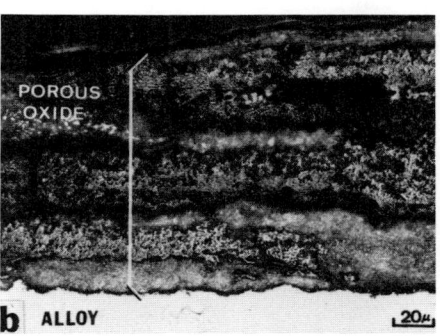

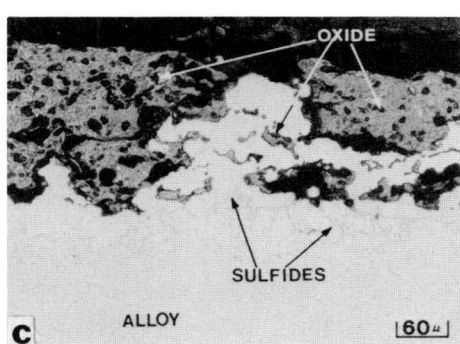

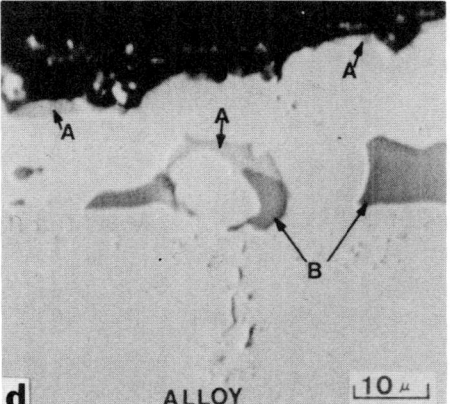

Fig. 10. Microstructures of (a) protective barrier of Al_2O_3, (b) nonprotective reaction product formed during acidic fluxing propagation mode, and (c) and (d) nonprotective reaction product and sulfide morphology of basic fluxing-sulfidation propagation mode, respectively.

more resistant than nickel-base alloys to this type of attack when aluminum is a component of the alloy systems. A substantial difference does not exist between the resistances of cobalt- and nickel-base alloys to the initiation of the acidic-fluxing propagation mode.

5 ADHERENCE OF PROTECTIVE OXIDE BARRIERS

The effectiveness of continuous reaction product barriers in protecting alloys from corrosion environments is greatly reduced when such barriers crack and spall under the action of growth and thermally induced stresses.

The adhesion of oxide barriers to alloys is therefore of critical importance to the design of alloys for corrosion resistance.

Oxygen-active elements such as yttrium and cerium can be used to significantly improve the adhesion of Al_2O_3 and Cr_2O_3 barriers.[4,27] Dispersed oxide particles (e.g., Al_2O_3, ThO_2, Y_2O_3) in alloys can also be used to very effectively improve the adhesion of Al_2O_3 and Cr_2O_3 barriers.[28] Typical cyclic-oxidation data obtained for alloys with oxygen-active elements or oxide particles are presented in Fig. 11. It can be seen that both of these techniques are equally effective in improving the adhesion of oxide barriers. It appears that the important consideration is not whether the metal or metal oxide is added to the alloy, but rather the distribution of the metal or its oxide in the alloy. Photographs of sections through cyclic-oxidation specimens with about the same composition but with different fabrication conditions are presented in Fig. 12. The specimen with the cast fabrication condition has undergone much more degradation than the one with the vapor-deposited fabrication condition. The yttrium is very uniformly distributed in the vapor-deposited alloy and this condition results in more effective utilization of yttrium.

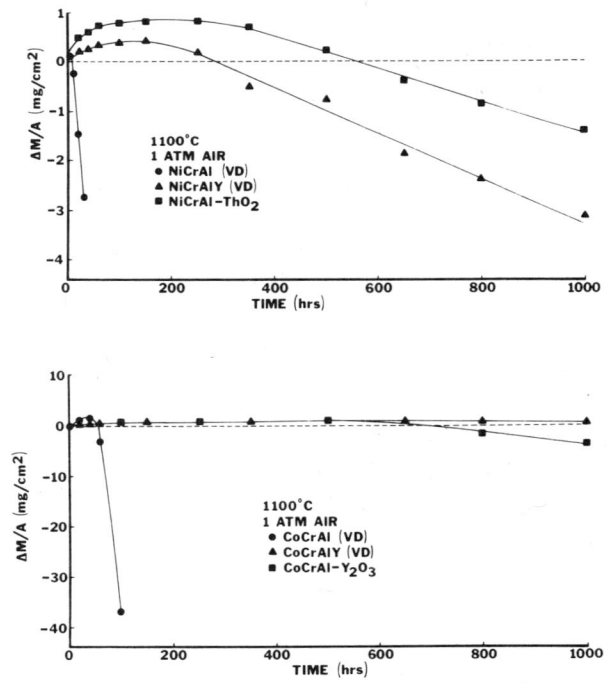

Fig. 11. Data on weight change versus time obtained for the cyclic oxidation of alloys with yttrium and with oxide particles. These data show that yttrium or oxide particles can be used to improve the adherence of Al_2O_3 scales on alloys.

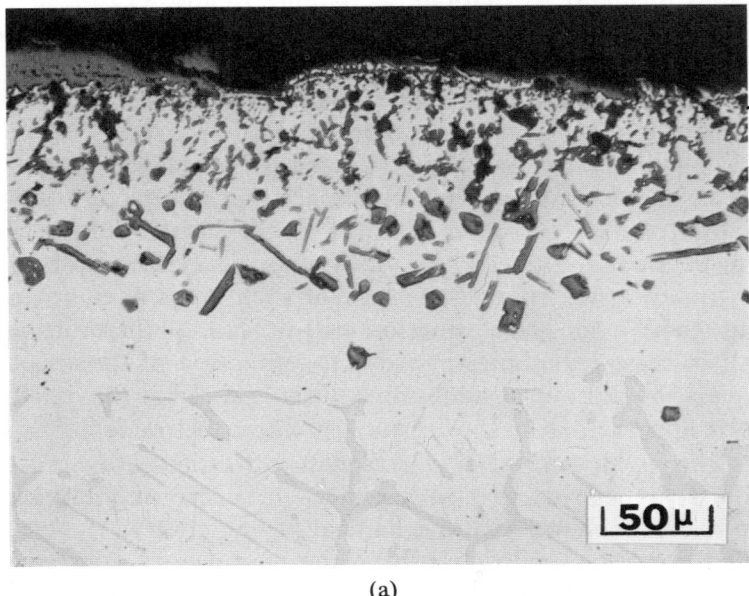

(a)

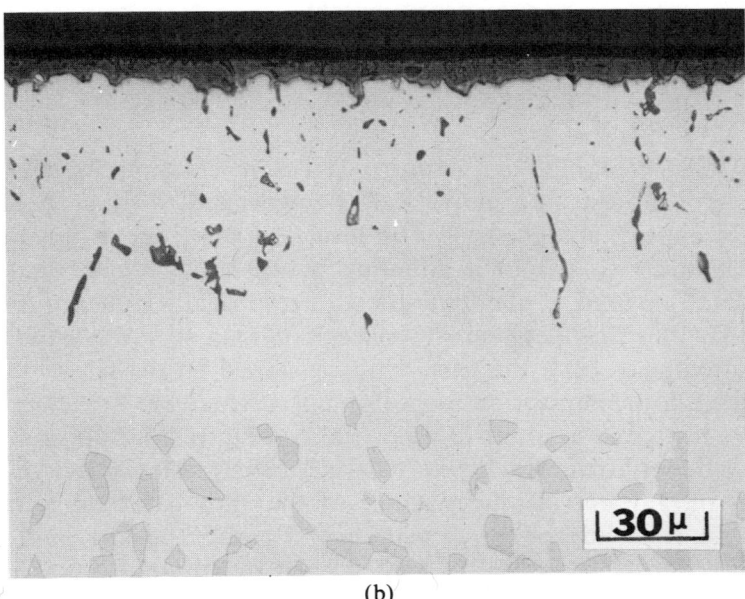

(b)

Fig. 12. Microstructures of CoCrAlY specimens having different fabrication conditions after cyclic oxidation testing at 1000 C in air: (a) as-cast alloy after 500 hours of testing; (b) vapor-deposited alloy after 1000 hours of testing.

Precious metals (e.g., platinum, rhodium) in alloys can also be used to improve the adhesion of Al_2O_3 scales. A minimum of about 10 percent of the precious metal is required to obtain improved adhesion.

The mechanism by which oxygen-active elements, oxide particles, and precious metals improve the adhesion of oxide barriers requires further definition. When any of these techniques are used, the development of voids at the oxide barrier-substrate interface is either completely prevented or greatly inhibited. The absence of voids at the barrier-substrate interface appears to be a condition which results in improved oxide-scale adhesion. It has been found, however, that the lack of such voids does not necessarily mean that good oxide-scale adhesion will prevail. It therefore appears as though all of these techniques improve the adhesion of the scales by some other process in addition to inhibiting void formation at the oxide-substrate interface. It appears that in most instances where the oxide barriers are very adherent, the oxide-substrate is irregular on either a macroscale or a microscale. It is believed that such macro- or microirregularities may improve the scale adherence by a mechanical keying effect.

Studies concerned with describing the influence of components other than oxygen in the gas and ash deposits on the adhesion of oxide barriers to alloys have not been performed. Experimental data show that chloride and sulfate deposits cause Al_2O_3 barriers to become less adherent. Procedures to combat such effects are not as yet available.

6 CONCLUDING REMARKS

The science of high-temperature corrosion reactions is not at a level where corrosion-resistant alloys can be designed without a substantial amount of experimental testing. The level of knowledge in this field is sufficient, however, to permit a coupling between the science and art. The techniques to be used in combination with empirical testing are summarized in Table II. It is first necessary to select a barrier to protect the alloy and then to attempt to have the barrier be developed on the alloy by modification of the alloy composition and alloy microstructure. To achieve this objective a substantial amount of empirical testing in the environment of interest will be required. Having successfully achieved the selective oxidation objective, the hot-corrosion resistance of the alloy can be determined and attempts made to improve it, if necessary. Finally, the corrosion resistance of the alloy can be increased to higher levels by striving to make the oxide barrier more adherent to the substrate.

Research is required to help in the design of corrosion-resistant alloys. There is a special need for studies concerned with examining the effects of other components in the environment in addition to oxygen on selective oxide-barrier formation and on barrier adhesion.

Table II. Techniques Useful for the Design of Corrosion-Resistant Alloys

Selective Oxidation	Hot-Corrosion Resistance	Scale Adhesion
Barrier Selection	Resistant to Acidic-Fluxing Propagation Mode	• Oxygen-active elements
Alloy Compositional Modifications	• Control of refractory metal content	• Oxide particles
• Secondary oxygen getter	• High chromium concentration	• Precious metals
• Oxygen-active elements		
• Refractory elements (double oxide formation)	Resistant to Basic Fluxing Sulfidation Propagation Mode	• Uniformity of composition and structure
	• High chromium concentration	
Alloy Microstructural Modifications	• Replace nickel with cobalt	
• Fine grains	• Close control of aluminum concentration	
• Oxide Dispersions		
• Homogeneity		

ACKNOWLEDGMENTS

The author wishes to acknowledge that much of the material presented in this paper was acquired and developed in cooperation with E. J. Felten, C. S. Giggins, and J. A. Goebel. The author desires to express thanks to C. E. Londin, V. Nevins, and A. R. Geary for their technical assistance.

REFERENCES

1. Salisbury, R. P., and Birks, N., *J. Iron and Steel Inst.*, **209**, 534 (1971).
2. Pettit, F. S., Goebel, J. A., and Goward, G. W., *Corrosion Sci.*, **9**, 903 (1969).
3. Bruckman, A., and Mrowec, S., *Werk. und Korr.*, **25**, 502 (1974).
4. Giggins, C. S., and Pettit, F. S., *Final Report on Oxide Scale Adherence Mechanisms Program*, for the Aerospace Research Laboratories, WPAFB, Ohio (Contract No. F33615-72-C-1702), October, 1975.
5. Caplan, D., and Sproule, G. I., *Oxid. of Metals*, **9**, 459 (1975).
6. Wagner, C., *J. Electrochem. Soc.*, **99**, 369 (1952).
7. Wagner, C., *J. Electrochem. Soc.*, **103**, 627 (1956).
8. Wagner, C., *J. Electrochem. Soc., ibid.*, p. 303.
9. Wagner, C., *Z. Elektrochem.*, **63**, 772 (1959).
10. Rapp, R. A., *Acta Met.*, **9**, 730 (1961).
11. Rapp, R. A., *Corrosion*, **21**, 382 (1965).
12. Kirkaldy, J. S., in *Oxidation of Metals and Alloys*, American Society for Metals (1971), p. 101.
13. Giggins, C. S., and Pettit, F. S., *J. Electrochem. Soc.*, **118**, 1782 (1971).
14. Wagner, C., *Corrosion Sci.*, **5**, 751 (1965).
15. Pickering, H. W., *J. Electrochem. Soc.*, **119**, 641 (1972).
16. Pettit, F. S., presented at Agard Conference Proceedings No. 120 on High Temperature Corrosion of Aerospace Alloys, Specialist Meeting held at the Technical University of Denmark, Lyngby, Denmark (1972).

17. Chatterji, D., Hampton, A. F., Graham, H. C., and Davis, H. H., in *Metal-Slag-Gas Reactions and Processes,* Z. A. Foroulis and W. W. Smeltzer (Eds.), Journal of the Electrochemical Society (1975), p. 369.
18. Giggins, C. S., and Pettit, F. S., *Met. Trans.,* **2**, 1071 (1971).
19. Stringer, J., and Wright, I. G., *Oxid. of Metals,* **5**, 59 (1972).
20. Graham, H. C., Hampton, A. F., and Davis, H. H., in *Metal-Slag-Gas Reactions and Processes,* Z. A. Foroulis and W. W. Smeltzer (Eds.), Journal of the Electrochemical Society (1975), p. 763.
21. Hancock, P., *Corrosion of Alloys at High Temperatures in Atmospheres Consisting of Fuel Combustion Products and Associated Impurities,* Her Majesty's Printing Office, London, England (1968).
22. Stringer, J., MCIC 72-08, Battelle, Columbus Laboratories, Columbus, Ohio.
23. Seybolt, A. U., *Trans. TMS-AIME,* **242**, 1955 (1968).
24. Bornstein, N. S., and DeCrescente, M. A., *Met. Trans.,* **2**, 2875 (1971).
25. Goebel, J. A., Pettit, F. S., and Goward, G. W., *Met. Trans.,* **4**, 261 (1973).
26. Goebel, J. A., and Pettit, F. S., *Final Report on Hot Corrosion of Cobalt-Base Alloys Program,* for the Aerospace Research Laboratories, WPAFB, Ohio (Contract No. F33615-72-C-1757), October, 1975.
27. Tien, J. K., and Pettit, F. S., *Met. Trans.,* **3**, 1578 (1972).
28. Wright, I. G., Wilcox, B. A., and Jaffee, R. I., *Final Report on Oxidation and Hot Corrosion of Ni-Cr and Co-Cr Base Alloys Containing Rare Oxide Dispersions,* Prepared under Contract N62269-74-C-0291, Naval Air Development Center by Battelle, Columbus Laboratories, Columbus, Ohio, May, 1975.

DISCUSSION on Paper by F. S. Pettit

SHEWMON: In coal-treatment plants, for example, coal gasification, metals will be exposed to "reducing" environments rich in CO and H_2, as well as to a significant sulfur potential. Sulfur, or H_2S, seems to be particularly aggressive to traditional stainless steels. What can be done to make alloys more resistant to sulfur-bearing gases in an atmosphere of low oxygen potential?

PETTIT: There are two approaches that can be used to influence the corrosion rates of alloys in sulfur-bearing gases of low oxygen potential. The first and most desirable approach is to attempt to develop continuous Cr_2O_3 or Al_2O_3 reaction-product barriers on the surfaces of the alloys. Even though the oxygen potential of the sulfur-bearing gases may be very low, this potential may still be large enough to permit the formation of such oxide barriers. In the event that the continuous oxide barriers are not stable on the surfaces of the alloys, then it is necessary to attempt to utilize sulfide barriers as a means of minimizing the corrosion rate. While sulfides are not as effective as oxides, some sulfides are better than others. For example, aluminum sulfide and chromium sulfide barriers appear to afford more protection than nickel, iron, or cobalt sulfide barriers.

STRINGER: "Reducing" is a relative term. Gasified coal has an oxygen activity of the order of 10^{-17} to 10^{-20} atm, which at the relevant temperatures, typically 500 to 1000 C, is sufficient to allow the growth of Cr_2O_3 and Al_2O_3. There is undoubtedly a severe problem, but it is associated with the simultaneous action of several oxidants, as Pettit indicated in the lecture. The atmosphere has a high carbon activity, and often a high sulfur activity—the presence of significant amounts of H_2S is a manifestation of the high sulfur activity—and the formation of carbides and sulfide may proceed prior to or simultaneously with the formation of oxide. This prevents (or may prevent) the establishment of a continuous protective oxide layer.

Fred, you mentioned scale adhesion, but don't you think that this is a particularly important topic with implications in many areas of technology—for example, exfoliation of scales from boiler tubes, breakaway corrosion of Zircaloy, oxidation of mild steel in CO_2 in the Magnox reactors, and so forth? Do you think that any of the approaches you mentioned could be used to help scale adhesion in these rather less exotic systems?

PETTIT: Scale adhesion plays a crucial role in determining the corrosion resistance of virtually all alloys. The studies concerned with the adhesion of oxide barriers, however, generally have been concerned with alloys for high temperatures (e.g., 900 C and above) and therefore the adhesion of Cr_2O_3 and α-Al_2O_3 barriers to alloys has received the most attention. The approaches used to improve the adhesion of Cr_2O_3 and Al_2O_3 on alloys would be expected to help scale adhesion in other systems providing the growth characteristics of these other scales were such that the conditions which developed at the scale-alloy interfaces were similar to those existing at Cr_2O_3-alloy and Al_2O_3-alloy interfaces.

HIRTH: Has there been effort with more complex alloys to promote continuous Al_2O_3 (rather than internal Al_2O_3) by time-temperature cycling; this should at least *incrementally* increase the oxygen pressure range for continuous films.

PETTIT: Time-temperature cycling can be used to promote the development of continuous Al_2O_3 scales on alloys. The technique of using a pretreatment to develop the desired protective barrier is not, however, of much practical value. In virtually all applications for high-temperature alloys, the oxide barriers are damaged during use. Therefore, it is necessary that high-temperature alloys possess the capability of developing the desired barriers under the conditions of use.

McCLINTOCK: Would you comment on allowable working levels of steady-state strain rate or cyclic-strain amplitude compared with typically allowable values for the base metal in a vacuum?

PETTIT: The working levels of steady-state strain rate or cyclic-strain amplitude allowable in Cr_2O_3 or Al_2O_3 oxide scales on alloys have not been measured and calculations based on values for the bulk oxides probably are not valid for the thin (i.e., ~1-4 μm) oxide scales. A comparison of such strain values with typically allowable values for the base metal in a vacuum may not be necessary, however, since cracking and spalling of these oxide barriers appears to result mainly from thermally induced stresses rather than from the external loads.

STRINGER: Hancock has shown, using an acoustic technique, that the criterion for cracking in an oxide layer associated with differential thermal expansion, is a strain criterion: once the strain in the oxide exceeds the fracture strain, cracking occurs. However, in many cases these cracks do not lead to catastrophic spalling of the scale—this appears to require propagation of cracks parallel to the metal/oxide interface, and we do not yet have a good understanding of this aspect of the scale failure mechanism.

TIEN: I wonder about the effectiveness of the very small (submicron) equiaxed particles as pegs in oxides to explain the observed adherence. Has any, however simple, stress analysis been done to evaluate whether these small particles can indeed "key" oxides?

PETTIT: To my knowledge, no stress analysis has been done to compare the effectiveness of macropegs, micropegs, or for that matter the lack of voids at the oxide-alloy interface as mechanisms to account for oxide-barrier adhesion. The basis for suggesting that these small particles may improve scale adhesion is as follows:

- Spalling of Al_2O_3 has been observed from a Ni-15Cr-6Al alloy even when no voids were observed at the Al_2O_3-alloy interface.

- Spalling of Al_2O_3 was not observed from this alloy when the small micropegs were present extending into the alloy even though no macropegs had been formed.

KEAR: Has anyone looked at the oxidation/hot-corrosion behavior of perfectly homogeneous alloys fabricated by rapid quenching from the melt? It seems to me that this new technique would be particularly useful for making iron, nickel or cobalt-base alloys that contain high concentrations of the oxygen-active elements (e.g., yttrium) that normally experience gross segregation during solidification.

PETTIT: The oxidation/hot corrosion properties of perfectly homogeneous alloys fabricated by rapid quenching from the melt have not been compared with those for equivalent alloys in other fabrication conditions. Alloys with high degrees of homogeneity frequently exhibit better resistance to the initiation of hot-corrosion attack than alloys with the same compositions but which ave compositional inhomogeneities. It has also been observed that the oxygen-active elements are most effective in improving the adhesion of oxide barriers when distributed as uniformly as possible throughout the alloy.

AGENDA DISCUSSION: STRUCTURE/ PROCESSING/PROPERTY PROOF— MECHANICAL PROPERTY LIMITED

Walter S. Owen

Massachusetts Institute of Technology
Cambridge, Massachusetts

Much of the discussion at this session was about the nature of the alloy design process. Most of the observations were broad generalizations of obscure origin and, consequently, it would be dangerous to attribute ideas to individuals. Only an account of some major impressions will be attempted here.

Most alloys are designed to meet the properties specified for a component of a structure or device. In this larger design system, metallic alloys compete with other materials. Until recent years, those components required to be strong and ductile were invariably made from alloys, but the advent of carbon- or boron-fiber composites introduced a competing class of materials. Other nonmetallic materials are now considered seriously, and the development of a ceramic gas turbine is within sight. One of the components which has been the subject of most intensive and sophisticated alloy design, the turbine blade, may be replaced by a nonmetallic material. Thus, even in the class of components which are "mechanical property limited", we should be concerned with the design of materials, not just alloys. Increasingly, components are designed from a composite of materials. It has been common practice for a long time in some designs to achieve bulk mechanical properties by the use of a suitably processed alloy and surface properties by a surface treatment or the application of a coating. Thus, even in this simple case, the designer is concerned with the design of a composite.

Within most large design systems, there are numerous possibilities for trade-offs between materials properties and other aspects of the total design such as cost, performance, reliability, and environmental properties. The term "cost" includes the cost of capital investment, energy, inspection, repair, maintenance, downtime, etc.—that is, the total life-cycle costs. A striking example of changes in design resulting from a reconsideration of the costs on a total-life basis is afforded by the analysis of the design of the domestic refrigerator described by Newton (George Newton, *Technology Review,* January 1976, pp. 56–63). There are many examples of radical changes in materials design for automobiles forced by new environmental or safety regulations. An aspect of materials design which is not yet of major importance, but which is likely to become so in the future, is materials substitution necessitated by changes in the relative availability, and thus the relative cost, of materials. Studies are already under way to find substitutes for chromium in steels, a change which will require careful study of a wide range of metallurgical technology and of the possibility of developing composites with equivalent mechanical and corrosion-resistant properties to, say, 18-8 stainless steel sheet. The competition between copper and aluminum alloys and composites for the transport of electric power is a much older alloy design problem, which is greatly influenced by the shifts in the relative energy and other costs of the two metals. Soon a microeconomic study of the feasibility of replacing aluminum, an energy-expensive metal, by magnesium alloys for some nonelectrical purposes will be undertaken. These are all examples of alloy design problems generated by an anticipated, or real, increase in the cost of a material. There is, of course, the possibility that a material may become much more abundant in the future. Manganese from deep-sea modules is such a case. Then there will be a need to use the cheap material in place of more expensive conventional materials. Although the substitution of manganese for part or all of the nickel in stainless or cryogenic steels is an old idea, the incentive to find technologically viable solutions to the problem will become much greater if there is a major change in the relative cost of nickel and manganese.

Much of the conference was concerned with possible ways of designing alloys with mechanical properties better than those of currently available alloys. This is much too narrow a view of the design function. Only in relatively new, advanced technologies such as the nuclear power industry is it usual to attempt to design an alloy from basic ideas. When this is done, it is usually a response to an unforeseen materials failure such as the void formation and swelling in canning materials. In longer established technologies such as the gas turbine or the subsonic airplane, improved mechanical properties are usually obtained by a process more accurately described as "alloy development"—that is, the alloy design evolves by successive modifications of the composition, thermomechanical treatment, and/or the fabrication process of an established alloy. Even in these circumstances, the improved alloy is adopted only if it can be produced and used cost effectively, either by reducing the manufacturing costs, by improving the service performance and/or the reliability, or by reducing energy consumption or adverse environmental impact. A good example of this kind of alloy development, which was described by a number of speakers at the conference, is the evolution of the superalloy turbine blade through coarse-grained cast, directionally solidified, and single-crystal processes. Often, the development of a structure requiring increases in the loading, or of a machine by increasing the power output, keeps pace with the evolution of alloys for key components. Designers attempting further developments of proven designs are understandably reluctant to introduce radical changes in the materials. When forced to do so, they insist on thorough testing of the new material, a long and expensive process. Several speakers said that it might take 10 years or more for a new alloy to be developed, tested, proved, and accepted. These speakers also emphasized the importance of specification and regulatory agencies in the design process. Since the primary concern of these agencies is safety, it is understandable that they should exert a conservative influence which ensures that, for structures such as airplanes and pressure vessels, materials are improved by a slow evolution with careful checks and testing at each stage in the development.

There is another kind of alloy design or development, exemplified by high-strength low-alloy steels for arctic pipelines, which was discussed briefly by Embury, but otherwise received little attention. That is the design of alloys for applications requiring large tonnages of material. The steels for gas pipelines are specified by the designers of the pipelines in terms of strength, a Charpy transition temperature, and weldability. In addition, the steel designer has to take into account formability and, perhaps, through-thickness strength. This set of properties can be achieved in many different ways. All steels designed for this application contain some microalloy (Cb, V, Ti, Mo) additions. The structure is developed by controlled rolling and cooling, and some form of sulfur reduction or sulfide particle shape control is used. Nevertheless, many different steels made by different processes have emerged and are being used. The steel developed or adopted by a steel producer is decided in large measure by the economics of capital investment in new equipment or the modification of existing equipment or processes. The role of the alloy designer in this situation is to develop alternative candidate alloys to be considered in the larger context of the total steelmaking, fabricating, and marketing processes.

There is a fourth reason why alloys may be designed: to exploit a newly found materials property. An example is the design of superconducting alloys which can be fabricated into composite wires or ribbons with sufficient strength to be used for the windings of large magnets. Another example is the design of shape-memory alloys with good memory retention, high strength, adequate ductility, elastic rigidity, and fatigue, erosion, and corrosion resistance for use as clamps, biomedical devices, damping alloys, or even as the thermal-mechanical converter in a heat engine. Here, unlike the cases considered earlier, the applications follow from the new set of properties developed by the new alloys. This is indeed an exciting, if rather uncommon, type of alloy design.

Whichever mode the alloy designer is operating in, the key to the design process is the structure-property relationships. First, the property must be identified and defined. It is not sufficient to define the property as "ductility", or even to specify how much ductility is needed. The anticipated mode of failure must be considered. A steel which in 3/4-inch plate fails with a ductile-tear fracture and a large shear lip might fail in 4-inch thick plate under plane strain with a brittle fracture. Further, when the ductility is terminated by plastic instability, creep rupture, or fatigue-crack propagation, the ductility property to be considered is different in each case. Next, the structure of the alloy which optimizes the property under consideration must be known or discovered by experiment. Finally, ways of producing the structure economically and reproducibly must be found. In practice, this sequence of steps is much more complicated than is suggested by these bare statements. All the structures are complex and seldom is there one structure which uniquely optimizes a particular property. To take another example from high-strength low-alloy steels, X65 steel (yield strength 65 ksi) can be achieved in steels with either a polygonal-ferrite (reduced-pearlite) or an acicular-ferrite (low-carbon bainite) structure by various combinations of strengthening mechanisms: fine ferrite grain size, substitutional solution, interstitial solution or dislocation substructure strengthening, strain aging, or dispersion hardening by carbonitrides. Further, almost always the structure has to be optimized with respect to a combination of properties. In addition to a yield strength of 65 ksi, pipeline steels must have a Charpy transition temperature of 0 F or less. If too large a contribution to the strength is expected from dispersion hardening, this ductility specification cannot be satisfied. Only by much reliance on ferrite grain refinement can the necessary ductility be achieved.

Alloy design was recognized to be a complex optimization process and, consequently, much of the attention of the conference was devoted to the question of whether our knowledge of the structure-property relationships involved is adequate for the needs of modern alloy design. As might have been expected, this question produced a mixed response. A small group of participants propounded the view that one can never know the answer to the question because improvements in alloy design are the result of the application of new understanding of the basic science of alloys and it is impossible to predict or plan for scientific discoveries. This point of view found little sympathy among the industrial metallurgists or even among some of the professors present. It was recognized by most people that alloy design is usually an evolutionary, not an inspirational, process. The question then becomes: Is our knowledge of the structure-property relationships sufficient to provide reliable guidelines? Stated in these terms, the question produced some fairly clear answers.

It was generally recognized that our knowledge of structural strengthening mechanisms is probably greater than we can use effectively at present. The practical yield stress is usually limited by considerations of hardenability or stress corrosion or other environmentally induced failure and not by structural limitations. Static ductility terminated by either cleavage or ductile fracture is, however, inadequately understood. The existing guidelines (the ductility increases with decreasing flow stress and increasing strain hardening) are too imprecise to be useful except in narrowly defined circumstances. McClintock and Hahn both presented analyses of the influences of structural variables on ductile and brittle fracture which offer hope that it will soon be possible to define these structure-property relationships with the same confidence as we can now define the strength parameters.

Ashby's steady-state and transient deformation maps for steels appear to provide reason to believe that guidelines for the design of steels to operate under conditions in which creep elongation is an important design consideration are close to being established. The situation with regard to stress-rupture failure is less satisfactory. Failure by fatigue-crack propagation is probably the least well understood of the common failure mechanisms. Only recently has any understanding of the influence of structural variables on this form of failure emerged. The realization that cyclic softening or hardening is a function of the efficiency of dislocation barriers during work hardening has been translated into practical terms by recent studies in which fatigue-crack propagation is compared in alloys in which precipitate particles can be sheared and in those in which they cannot. Further progress appears to await the development of working models of cyclic work hardening and softening—a subject on which much research effort has already been expended. Guidelines for alloy design often do not exist, or if they do they are of doubtful validity, in circumstances in which the anticipated failure mode is mixed, such as creep fatigue. When such failures are complicated by interactions with a hostile environment, such as in low-cycle, corrosion fatigue, alloys capable of operating only over a severely restricted range of stress, temperature, and strain rate can be designed.

Fuller discussion of chemical surface effects was scheduled for later sessions in the Colloquium.

If it is assumed that the optimum structure for an alloy designed for a particular application is known from laboratory studies, it is next necessary to decide whether the structure, or one close to it, can be obtained reproducibly in fabricated components. The primary purpose of the fabrication process is to shape the component as economically as possible. Of course, the fabrication process affects the structure; a cast alloy has a structure different from that of a forging of the same alloy. Thus, processing is a part of structure control and development. Further, the alloy may be modified in such a way as to improve its castability, forgeability, weldability, or deep-drawing properties, and this is also a consideration in alloy design.

The structure of an alloy in a finished product is a result of phase transformations and plastic deformations. Phase transformations occur during processing and during any additional thermal treatment, such as annealing or tempering, which may be applied. If these transformations are displacive, they may involve plastic flow which produces high densities of dislocations or twins in the crystal. In many cases, deformation also is applied at one or more stages of processing. Most phase transformations are changed profoundly by concomitant deformation. Thus, the development of structure during thermomechanical processing is complex, involving interacting changes, many of which when isolated are understood; but, as emphasized by Hornbogen in his discussion of recrystallization and grain growth in heterogeneous alloys, understanding of the development of structure by several simultaneous changes is primitive. Nevertheless, in recent years many advances have been made by designing the processing to act as thermomechanical treatments which develop a specific structure in the alloy at the same time as it is being shaped. Modern high-strength, low-alloy steels became possible only when hot-rolling schedules and steel compositions were designed to produce a fine grain size in the recrystallized austenite. Recently, some steels have been made by continuing the rolling to a temperature at which ferrite separates, with the result that the dislocation density in the ferrite is increased significantly, adding a useful increment to the strength in the rolling and cross-rolling directions but introducing some potential problem such as reduced through-thickness strength.

Few of the alloy structures with the most useful mechanical properties are equilibrium structures. In the last decade, many previously unattainable or unknown alloy structures, far removed from equilibrium, have been developed by radically new processing methods. Rapid quenching techniques for producing "metallic glasses" or extremely fine grained alloys, vapor deposition, sputtering, rheocasting and rheoforging, and compaction of ultrafine powders are examples of successful new processes which produce new structures. All produce very fine grained, nonequil'brium structures with unusual mechanical properties. They were discussed at length during the Colloquium. These new processes provide interesting challenges in alloy design. For the production of ribbon by rapid quenching, for example, if the solid alloy is to be amorphous, compositions must be selected which have a melting point at the bottom of a deep eutectic, and at the same time develop the excellent electrical and magnetic properties which, to date, are the basis of the most important applications of these alloys. To cite another example: one of the very attractive features of the ultrafine-powder technology which is developing so rapidly at present is that the ultrafine particle or dendritic, nonequilibrium structures which can be produced by these processes are often ductile, even in alloys that in the familiar cast, coarse-grained structure are unacceptably brittle owing to segregation of intergranular precipitates.

Much of the discussion focused on the adequacy of our understanding of phase transformations and other processes by which structure is developed and of the structure-property relationships of importance in real situations. The general conclusions reached by many, but certainly not all, of the participants can be summarized in a few simple statements. Thermally induced transformations involved in the older processing techniques, such as solidification and heat treatment, are sufficiently well understood for alloy design purposes, although there was much special pleading for further work on narrowly defined aspects of some transformations. The only major exception seems to be thermoelastic and mechanically reversible martensitic transformations of the kind which will be exploited in shape-memory alloys. However, there was general agreement that the complex changes occurring during thermomechanical processing, or during other new processing procedures such as splat cooling, are not at all well understood and require much attention in the future. The situation with regard to structure-property relationships was judged to be similar. For much-studied and well-defined properties, such as flow stress, the relationships are well established. As indicated earlier, the existing models relating the various forms of fracture failure to structure need refinement, particularly as they relate to stress corrosion and similar phenomena. However, it became clear during the course of the discussion that, just as understanding the development of structure by several interacting processes presented the greatest problems, understanding of the relationships between mixed modes of failure, such as creep-fatigue, and these structures requires the most attention.

AGENDA DISCUSSION: STRUCTURE/ PROCESSING/PROPERTY/PROOF— CHEMICAL PROPERTY LIMITED

*John Stringer**

*Electric Power Research Institute
Palo Alto, California*

*Ian G. Wright***

*Battelle-Columbus
Columbus, Ohio*

The Chairman opened the discussion by declaring that in his opinion, the alloy design process cannot be separated from the system design process. This becomes particularly obvious in the cases of high-temperature oxidation and corrosion and of stress corrosion cracking, since many of the problems encountered can be dealt with in a number of ways. As an instance, a high-temperature oxidation problem can be approached either by designing a better oxidation-resistant material or by removing the corrosive environment. In fact, the latter option is the one exercised whenever possi-

*Chairman.
**Secretary.

ble, so that cheap, abundant, and well-understood materials can be employed. In the face of such a problem, the first criterion is how to achieve the best conditions with the currently available materials; more exotic materials are in general considered only when the limits imposed by current materials render such a solution uneconomic.

Often there is a third option possible, that of using a protective coating, such as a paint coating on a metal subjected to low-temperature corrosion, or an aluminide diffusion coating on an alloy subjected to high-temperature oxidation. As a result, in this section we are very much concerned all the time with the concept of a material as part of an overall system. Exactly where one places the boundary of this system is, however, often difficult to specify.

After commenting on the usefulness of diagrams for conveying information, Stringer recalled with some awe the declared intention of Ashby to extend his deformation-mechanism maps to include the effect of environment. While such a convenient cataloging of environmental effects is eminently desirable, it is difficult to see how it could be achieved in such a concise form as that obtained for physical or mechanical properties. Data relating corrosion behavior to alloy and environmental variables are often quite complicated, and do not readily lend themselves to simple presentation. Corrosion rates are usually of less concern than are the occurrences of conditions leading to disastrous failures. For predicting corrosion behavior, in qualitative terms, our understanding in some areas of high-temperature oxidation and corrosion, and in some areas of low-temperature corrosion, is reasonably good, but in quantitative terms our understanding in most cases is poor. Therefore, from the point of view of being able to predict corrosion rates and events in a numerical sense, the subject of corrosion might be considered as being medieval.

If we were to proceed in an attempt to draw the corrosion-behavior equivalent of the Ashby-type diagram, we could certainly indicate the areas where various types of corrosion would be expected, but the declaration of regions where no corrosion would be expected is fraught with complications and dangers. At low temperatures, seemingly innocuous conditions may hold the danger of stress-corrosion cracking; at high temperatures, situations can exist where oxidation rates under laboratory conditions are acceptably low, but in the real system, the unavoidable presence of minor impurities can lead to a considerable enhancement in rate. The current problems with the British Magnox (CO_2-cooled) reactors provide a dramatic example of the latter case. In the design of systems to achieve corrosion resistance, therefore, due note must be taken that human factors are involved in the assemblage of structures, which are generally erected in areas where the components are subjected to some exposure to the ambient environment (i.e., weather) for unspecified periods before the final assembly.

In the design of materials for improved high-temperature corrosion resistance, there are two alternative criteria: either the whole alloy can have improved corrosion resistance, or only the region of the system in the vicinity of the surface need have improved corrosion properties. Where the whole alloy is required to exhibit the improved corrosion resistance, this requirement is usually added on after the usual strength criteria have been achieved, and the improved corrosion behavior has to be achieved with a minimal loss in the mechanical properties. Coatings, however, can be designed from the outset for corrosion resistance, and compatibility problems with the substrate can often be relegated to a later stage.

There are additional factors which should be borne in mind and which will serve to demonstrate the connection between corrosion considerations, and structural and strength aspects. The interaction of corrosion considerations with strength considerations is of particular importance. The simultaneous occurrence of deformation processes and corrosion may, for instance, have only a slight effect on the rates or behavior observed separately, or a strong synergistic effect may occur which leads to enhanced creep rates or corrosion effects.

One of the few studies made of the synergistic effect of high-temperature oxidation on creep was completed by McMahon and Sessions. This work was concerned with the effects of environment on the creep behavior of turbine blade alloys, in particular cast U700, and McMahon presented data showing the creep behavior in vacuum and air, indicated schematically in Fig. 1, and the resulting relationship of time to failure with the reciprocal of the grain size, as shown schematically in Fig. 2.

Interesting differences were also found in the behavior of directionally solidified U700 tested with the grain boundaries at different orientations to the tensile axis. In air, failure occurred by a mode analogous to stress corrosion, with cracks nucleating on the surface and propagating intergranularly. The surface regions were denuded of γ', and oxidation along the grain boundaries had also resulted in localized denudation of γ'. Oxygen had also apparently penetrated along the grain boundaries to oxidize preferentially

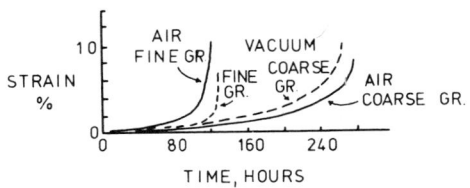

Figure 1. Creep behavior of cast U700 in vacuum and air.

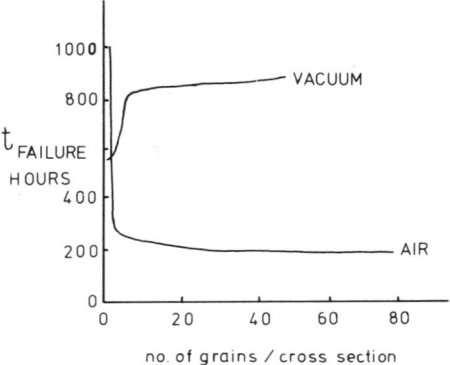

Figure 2. Relationship of time to failure with the reciprocal of the grain size.

chromium-rich phases at much greater depths than the crack lengths. The greatest oxygen penetration occurred when the grain boundaries were at 45 degrees to the stress axis. It was concluded that the enhanced oxidation was a result of increased oxygen diffusion in the moving grain boundaries (grain-boundary sliding). In vacuum, specimens failed through the classical nucleation and growth of creep cavities.

In the case of single crystals, during testing in air, surface oxidation leads to a softened, γ'-denuded zone at the alloy/oxide interface and also to the formation of surface cracks in areas of localized, more rapid, oxidation. It appears that these cracks do not propagate until they penetrate into the harder, nondenuded matrix beneath the γ'-free layer, and that this event depends on the relative rates of growth of the oxide cracks in the softened layer and the recession of the γ'-denuded layer. In vacuum, surface slip leads to the formation of surface offsets, which in turn lead fairly rapidly to shear crack formation, and the specimens fail by shearing off along the slip bands. In this case, then, the air environment has a beneficial effect.

Sims offered a further example of the properties of superalloys being changed by the simultaneous action of corrosion and stress in the "inert" environments now under consideration for gas-cooled reactors. In the HTGR concept, turbines will be operated in helium at 1000 psi and temperatures up to 1000 C, and component lives of 300,000 hours are required. Although chosen as a nonreactive environment, small amounts of oxygen and carbon are present in the helium which can react with the turbine alloys, but the oxygen level is too low to enable complete, protective oxide films to be produced. The result is internal oxidation and carburization of the reactive constituents of the alloys, to the extent that the stress-rupture capability of an alloy such as U700 is degraded by 15 to 25 percent at 900 C. Means of protecting present-day alloys, for example by preoxidizing, may have to be seriously considered until alloys with more inherent resistance are developed. In this latter category, molybdenum appears to possess thermochemical characteristics compatible with impure helium, and there is a report from Brown Boveri that molybdenum withstood stress-rupture tests in impure helium for 100,000 hours without attack.

Jaffee wondered if helium absorbed on the alloy surfaces might affect the adhesive forces between the metal atoms, but this was considered an unknown area. He also questioned whether there should be a maximum in the curve shown by McMahon of time to failure in vacuum (Fig. 2) where the number of grains becomes sufficient that the grain size effect can take over.

Frost commented that results of tests at ORNL in helium containing 1 ppm oxygen supported the points raised by Sims, but that the desirability of increasing the oxygen content to provide protective oxide films is countered by the increased reaction that would occur between the oxygen and the graphite in the core to give CO/CO_2. Sims replied that the trend in practical designs appeared to be toward a compromise atmosphere between the very pure helium required to ensure the life of the core and impurity levels, such that present-day materials could form some protection. He thought some effort should be made to evaluate the feasibility of controlled injection of water vapor to assist the maintenance of protective oxide films.

The Chairman pointed out that the oxygen activity, which determines whether or not an alloy can form an oxide film, depends not on the amounts of species present, but on the H_2O/H_2 or CO_2/CO ratios, so that simply stating the atmosphere impurity content in ppm can give a misleading impression of a lack of aggressiveness. It must be remembered that the lower the oxygen partial pressure, the lower is the mole fraction of chromium or aluminum required to form an external oxide layer, so that the low pressure of impurities cannot of itself account for the development of internal attack. The total amounts of oxidants present in the atmosphere may, however, determine whether complete coverage by protective films is possible in a reasonable time so that the possibility of kinetic limitations on oxidation behavior then arises. In the absence of complete coverage by a protective oxide, the oxidants can diffuse inwards and react internally with the more reactive minor constituents, as Sims had pointed out.

Pettit agreed that a simple reduction in the amount of oxygen present in the system, or in the oxygen partial pressure, would not cause a change from external to internal oxidation behavior. Rather, such a change would be expected to result from a mechanism such as the diffusion of carbon into the alloy to form carbides (in the absence of a complete protective oxide layer), followed by internal oxidation of the carbides.

Stevens wondered whether the oxygen partial pressure of the system ever was so low that alumina might be unstable, but several participants indicated that in real circumstances this was very unlikely.

Bement expressed concern that the transport of oxygen might be enhanced by potassium and sodium carried over in a coal gasification environment. A similar mechanism occurs in fast-breeder reactors where cesium is generated inside the fuel rods; the cesium transport along the grain boundaries enhances oxygen transport and enhances the internal oxidation and cracking. One also has to worry about the synergistic effects of fission-product transport in high-temperature gas-cooled reactors, and there are some instances where fission products also cause deleterious effects on these types of mechanisms.

Turning to more common practical atmospheres such as those in coal combustors and gasifiers, the Chairman described thermodynamic-phase-stability diagrams, also often called Pourbaix-Ellingham diagrams, which show regions of phase stability as a function of the activity of two oxidants. Consequently, provided the activities of the oxidants in the atmosphere are known, the phase in equilibrium can be specified, and in principle the sequence of phases in the corrosion system can be inferred. Using as an example the Fe-S-O stability diagram, he wondered whether it might be possible to use such reasoning in a practical design situation to define the operating conditions necessary to ensure that the materials of construction would be stable. Pettit thought this concept entirely feasible, but doubted that the conditions desirable for maximum alloy life would coincide with the most favorable operating conditions of the processes considered. Levy thought that such an approach would be inherently limited by operating transients, but it was agreed that in a well-regulated process, the conditions would be controlled and therefore known. Attempts have been made to apply the activity diagram approach in MHD reactors but, as Bement pointed out, the diagrams refer to equilibrium situations, and the activities in the non-equilibrium situations encountered in practice are largely unknown.

Tien pointed out that an area of importance in high-temperature corrosion which still is inadequately understood is scale adherence. He believed that scale spallation is a fracture problem, and he wondered whether anyone was looking at the state of stress in the growing scales. He doubted that the oxide particles produced by alloying additions commonly made to improve the resistance to scale spallation were large enough to act as pegs. Wilcox further questioned our quantitative understanding of stresses in oxide scales. Ansell remarked that most of the current work is concerned with the mechanical behavior of the oxides and not with the stress distributions set up by thermal cycling and loading. Studies are under way on the hot working of steels used for extrusion dies, where the stress concentrations are very high and the properties of the oxides and the variation in strength and structure of the oxide from the hot-worked steel through the surface becomes important. Results from this work have not yet been translated to the use of high-temperature alloys, but should be. The Chairman reported that some work is also being carried out at CERL to determine strains in the oxide arising from changes in temperature in terms of differential thermal expansion differences between the substrate and the oxide. Different alloy compositions have been used to produce changes in M_2O_3 and M_3O_4-type oxide compositions, and good correlation has been found between the macroscopic spalling tendencies of the oxide and the strains induced by the thermal-expansion differences. He also mentioned the work of Professor Hancock and his group who have demonstrated that oxides crack when the strain imposed by the difference in thermal expansion exceeds the (small) oxide fracture strain. However, in some cases, this cracking does not seem to result in a noticeable loss of protectiveness of the scale. Ansell thought, however, that the major concern was changes that could be effected in the plastic properties of the oxide itself, which could be brought about by alloying the oxide. In the ensuing discussion, it was agreed that only limited control of the compositions of thermally grown oxides is possible for thermodynamic reasons. There was some doubt expressed that any one property, such as the plasticity of the oxide, was responsible for scale adhesion.

Turning to stress-corrosion cracking, Bush presented some data to illustrate the interaction of the corrosion environment and stresses in the Peach Bottom BWR. A typical joint is one between a 28-inch pipe and a 4-inch pipe used in all BWR's. The weldolets or bypass lines are designed to 1 S_m (90 percent yield) at operating temperatures, whereas in practice stresses close to 6 S_m have been measured (linear extrapolation assuming the material is elastic). The yield stress of this material is approximately 16 ksi at 550 F. The residual stresses due to weld shrinkage on the inner surface in these systems are of the order of 40 ksi, and when the additional stresses arising from various sources are added, a total stress of 75 to 80 ksi is attained, which is quite substantial. The aggressive environment under operational conditions is 0.2 ppm oxygen. Upon shutdown, in full standby condition, the oxygen content rises to 8 to 9 ppm, which represents the saturation value. Chlorides are in the parts per billion range. Sensitization occurs essentially during the first three weld passes, in the sense that the system is nucleated, and each further pass increases the size of the precipitates, so that the conditions deteriorate with each repair weld. In the absence of stress, these precipitates (or misfits) are selectively removed, and in some locations a concentration cell can be set up at the bottom. Where these concentration cells are situated at or near grain boundaries, cracks can initiate and propagation has been observed at K_{ICSCC} of 9 ± 3 ksi $\sqrt{in.}$ Statistical evaluation of weld defects and failures suggests that after about 20,000 hours of operation the probability of failure in the 4-inch lines after a certain number of outages (>24 hours) approaches 1.

Frost commented that the failures described by Bush are in fact quite benign and relate to small leaks which in no way affect the safety of the reactors. McClintock expressed misgivings about the applicability of the codes set for nuclear materials. In reply to a question by Jaffee, Bush stated that present piping alloys can be used without experiencing failures when certain approved practices are followed in designing the system in which they are used.

A physical metallurgy view of stress-corrosion cracking was presented by Thomas. The relationship between microstructure and slip distribution in relation to the intergranular problem in precipitation-hardened alloys, e.g., Al-Zn-Mg type, and the transgranular problem in alloys such as stainless steels, was examined at Berkeley several years ago. He noted that Peter Swann at Imperial College is also very much involved with this approach and his *in situ* work in the high-voltage electron microscope should be very informative.

He suggested that one should pay attention to the mode of deformation, the intensity of slip, distribution of slip, and mechanical twinning (if any) rather than be concerned with so-called "planarity of slip". All the parameters presented by Slater which vary stress-corrosion sensitivity will also vary structure and (local) composition. Consequently, the slip pattern would be expected to change. If it is very heterogeneous and concentrated, for example, in PFZ regions adjacent to grain boundaries, then disruption of the protective film occurs. This effect would also occur in the matrix if cross slip is difficult (hence, one of the benefits of increasing nickel in austenitic alloys is to raise the stacking-fault energy). Not only may protective films be broken, thus allowing corrosion to occur on the fresh metal surface created at the slip step, but the sharp slip step produces a sharp notch so that the stress intensity rises in these regions. Cross slip near the surface would be effective in reducing both of these problems. Also, the deformation behavior (or fracture characteristics) of the films becomes important relative to the slip behavior of the underlying metal.

Thus, in summary, all the factors that localize slip into narrow intense regions would be expected to be detrimental in relation to stress corrosion. Stacking-fault energy, short-range order, solid-solution strengthening, etc., in the matrix as well as precipitate strength, size, and distribution in the matrix and the grain boundaries are among the important parameters in this regard. In steels, lower bainite and tempered dislocated martensite are similar except for the distribution of carbides. Thus, any difference in slip behavior and stress-corrosion behavior should be related to these differences.

Slater commented that Tennyson and Staehle were able to alter the nature of surface slip of nickel single crystals in an aqueous environment by changing the nature of the aqueous environment and the potential, so that the slip observed on filmed surfaces was different from that on clean surfaces when nickel was in the active region. Paxton commented that simple solutions to stress-corrosion problems are all too readily capable of circumvention during materials preparation and service, sometimes by human error. Thus, a second line of defense is advisable in critical applications. One such approach involves sacrificial coatings to provide cathodic protection. In some cases it might also be possible to design into materials, especially those of relatively high strength, the capacity to cause cracks to branch and thereby reduce K at the crack tips sufficiently to reduce velocity to a negligible level.

McMahon presented some data to compare inter- and transgranular cracking. In many cases it appears that intergranular is the less desirable since it can occur at low stress levels and propagate rapidly. Transgranular cracking alone might be tolerable. One hypothesis that can be tested is to suppose that intergranular cracking is an impurity effect. The paper by McMahon illustrated the dependence of threshold stress for hydrogen embrittlement of HY130 steel on the concentration of impurities at grain boundaries, and in fact in this case the effect of hydrogen and silicon was additive. Briant at the University of Pennsylvania found that the transition temperature increases as the grain boundary concentration of silicon increases, as shown in Fig. 3. In addition, crack growth curves obtained in hydrogen, shown schematically in Fig. 4, indicate that at a low transition temperature with a low concentration of silicon, a fairly high threshold and fairly low stage-two velocity are obtained, whereas for an embrittled alloy the threshold decreases and the velocity increases. What can be hoped for is that for perfectly clean grain boundaries, high thresholds and low velocities would result. The intention of this continuing work is to investigate the effects of grain boundary composition for different impurities on the stage-two velocity and on the threshold stress to establish the potencies of various impurities in HY130. For 4340 steel, the so-called one step temper embrittlement is represented by a minimum in the curve for room-temperature Char-

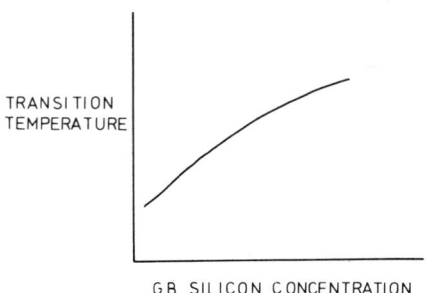

Figure 3. Increase in transition temperature with increase in grain-boundary concentration of silicon.

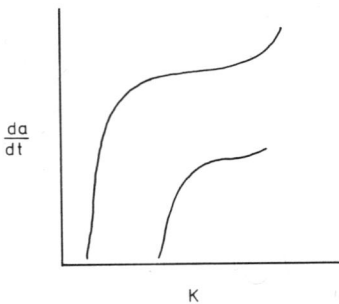

Figure 4. Crack growth curves obtained in hydrogen.

py energy versus temperature at 350 C, resulting from intergranular fracture. Auger spectra cannot determine the impurity levels, since silicon is
superimposed on iron and phosphorus is superimposed on molybdenum.
However, the embrittlement appears to be impurity related, because high-
purity 4340 does not exhibit the minimum at 350 C. Banerji at Pennsylvania
has found that this high-purity steel has a threshold much higher than those
of commercial-purity steels.

This one-step temper embrittlement is thought to be caused by
precipitation of carbides on grain boundaries during tempering, when insoluble elements such as phosphorus, sulfur, and silicon are rejected. This is
not an equilibrium segregation problem, and the alloy must be cleaned up to
eliminate it. Silicon can postpone this effect, and so has led to conflicting
views on the desirability of silicon additions. A similar behavior occurs in
AISI 304 during the sensitization treatment, where nickel is rejected from
the grain-boundary carbides. This leads to a large nickel gradient in the
vicinity of grain boundaries, and hence a chemical potential gradient for the
metalloids which causes them to diffuse towards the grain boundary. This
leads to an interesting point—would an ultrahigh-purity 304 alloy crack if
sensitized?

Levy remarked that in nuclear cladding applications, impurities (Cs,O)
are continuously generated in the grain boundaries, so that the purity of the
starting material would be less meaningful, but McMahon thought that
these were additional complications, the effect of which might be minimized if other impurities such as sulfur and phosphorus were removed. Bush
wondered why anyone would bother with ultrapure alloys when duplex
alloys containing more than 3 percent ferrite (which tends to mop up the impurities) have been used under the same conditions without any reputed
failures. Decker added that "clean" 304 produced by vacuum melting would
show improved performance but would still crack.

Zackay pointed out that Slater had stated in his paper that the susceptibility to stress corrosion in ultrahigh-strength steels tends to be ameliorated if the following factors are incorporated in the processing and structure of the steel: (a) the bainitic rather than the tempered martensitic structure is used; (b) that residual stresses are kept to a minimum; (c) that there be a maximum of crack branching; (d) that protective films be formed on the fresh metal surface created by the opening crack; and (e) that the carbon content be kept below a critical level. It is of interest that the new steels described in Zackay's paper incorporate many of these desirable features. By virtue of the heat treatment, isothermal holding above the M_s in the lower bainite region, the steels have low residual stresses, possess a microstructure largely of lower bainite and some tempered (low carbon) martensite, and films of retained austenite. Much of the carbon of the steel is in solid solution in this retained austenite. Further, McClintock has suggested that the high toughness of these steels may be due to the extensive branching of the propagating crack. Some preliminary tests have suggested that, as might be expected from the foregoing, these new steels have improved resistance to stress-corrosion cracking. Fracture toughness tests in liquid water showed that the new steels suffered no change in K_{Ic} (relative to ambient air atmospheres) while the same steels (without the extra silicon that characterizes the new ones and with a tempered martensitic microstructure, obtained by conventional heat treatment) exhibited a severe (factor of four) decrease in K_{Ic}. An extensive study of the apparent increase in resistance to stress-corrosion cracking of these new steels is currently under way.

Mayer described some results of Uhlig and others who found that steels with a very thin decarburized surface layer (3 to 4 mils) were virtually immune to stress-corrosion cracking.

Bush added that one microstructural effect that had not yet been mentioned was that the time to failure of sensitized 304 and 316 steels is often found to be ten times longer in the transverse direction than in the longitudinal direction. The grains are somewhat elongated, but some stress relief would have resulted from the sensitization treatment.

Owen commented that it has been said that the type of stress concentration at an oxide film or grain boundary found at the tip of a pileup in low-stacking-fault fcc metals is not found in bcc metals because dislocations cross slip out of any pileup. This is true. However, similar stress concentrations form at the tip of slip bands. These zones of plastic instability are clearly a function of macrostructure—distribution and coherency of carbides—and it is suspected that narrower and more stable slip bands form in bainites than in tempered martensite. This phenomenon should be studied much more carefully than it has been so far.

Hirth brought up the topic of compatibility effects. Elastic anisotropy under certain stress states would lead to stress singularity where a grain boundary hits a surface, which would have obvious mechanical effects, but could also have chemical effects such as affecting the entry and distribution of hydrogen. The same factors mentioned by Thomas would influence the plastic compatibility effects of grain boundaries, and the intensity of stresses in the region near, but not right at, the surface. These effects could also subtly interact with the effects mentioned by McMahon, in that gradients in solute concentrations would interact with dislocations and affect the degree to which pileups approach either the surface or a grain boundary. There is one other effect which is not a compatibility effect but which could influence the process; that is, it has been definitely established that the potential of the surface affects the plastic properties of insulators, so that there might be a subtle effect of electrical potential at the surface of a film influencing the plasticity of the film and thereby influencing the stress-corrosion process.

Hornbogen mentioned that the most effective way to get transcrystalline planar slip is the shear of very small particles, and in this case one can easily produce very concentrated slip in bcc structures, and the particles can be coherent or incoherent if they are small enough. Then tremendous concentrations of strain can occur in the slip bands in bcc lattices in spite of the high stacking-fault energy of the matrix. The second most important source of concentrated strain is grain boundaries, and in considering grain boundary embrittlement, it should not be forgotten that grain boundaries can be a source of preferred sliding if the grain boundary or its close environment are softer than the crystal. Then at a crack tip, especially in the overaged condition of precipitation-hardened alloys, one can get the grain boundary acting as a concentrated slip plane, which will lead to sliding and crack the protective layer, opening the layer at a plane where atoms can move in from the outside. Thus, one should not only consider an embrittled grain boundary, but also the grain boundary as a source of concentrated strain.

In response to the comments on stress-corrosion cracking, Slater emphasized that the desirability of designing materials which will promote crack branching is really only applicable to high-strength materials, since most cracks which occur in stress-corroded stainless steels are branched. It is fairly well known that very high purity stainless steels are relatively immune to stress-corrosion cracking, but it is also known that phosphorus plays a prominent role in the stress-corrosion cracking of austenitic steels in magnesium chloride, and that it causes transgranular stress-corrosion cracking. If a phosphorus-containing steel is sensitized and then stressed in magnesium chloride, transgranular, not intergranular, failure will occur. He also pointed out that alloys which are susceptible to stress-corrosion cracking are the high-strength steels and, in general, alloy-environment couples which show low corrosion rates (<5 to 10 mils per year). Good surface films are usually formed by these couples, which effectively preclude high corrosion rates, so that there is obviously some correlation between the filming tendency of the material and its susceptibility to stress-corrosion cracking in that environment.

In spite of the fact that such interesting topics as hydrogen effects and effects of trace impurities arising from fuels were not discussed, but in view of the advancing hour, the Chairman offered a few quotations from Lewis Carroll and brought the discussion to a close.

Part Five

CONCLUDING AGENDA DISCUSSION— CRITICAL ISSUES

CONCLUDING AGENDA DISCUSSION— CRITICAL ISSUES

J. P. Hirth*

The Ohio State University
Columbus, Ohio

A. H. Clauer**

Battelle-Columbus
Columbus, Ohio

The conference was wide ranging and hopefully covered most of the considerations that enter into alloy design. At times the very definition of alloy design was the topic of lengthy debate . . . does alloy design include only the process of developing new alloys for new or more severe applications, or does it also include careful modification of existing alloys to extend their service into applications hitherto closed or unknown to them?

---

*Chairman.
**Secretary.

Also, areas of needed work were suggested during the proceedings. The intent of the concluding discussion was to attempt to clarify these issues, to provide a summary of the conference, and to pinpoint areas requiring more attention in the future. Specifically, alloy design was discussed from four viewpoints: constraints imposed on the designers, properties needed by the designers, critical property limitations in the next decade, and possible advances in understanding fundamental properties.

1 CONSTRAINTS IMPOSED ON THE DESIGNERS

Hirth opened by stating that this area included "material design constraints other than the final properties to be achieved". He then presented a starting list of four design constraints: processing, codes, new processes, and strategic material alternatives. Hahn had presented one example of a processing constraint in his talk when he mentioned that increasing the fracture toughness in steels by lowering the carbon content resulted in the loss of the ability to draw tubing.

A more detailed example was presented by the chairman. This processing constraint is the roll-force temperature box for hot strip processing (shown in Fig. 1) which is representative of many specialty alloys that have desirable properties such as high creep or corrosion resistance, e.g., RA-333 nickel-base alloy*, suggested as a container for automobile emission control devices. This constraint is so severe that many steel companies are reluctant to fabricate these alloys, with the consequence that they are difficult to purchase. Figure 1 shows a roll-force temperature box limited at high forces by the possibility of roll breakage, at high temperatures by hot shortness, and at lower temperatures by the minimum force required to deform the material. The problem is that with the increased alloying elements (W, Mo, Co) introduced to enhance the creep and corrosion resistance, the necessary minimum forces increase (increasing alloy strength) and the hot-shortness-limiting temperature decreases (decreasing melting temperature). The result is a box of decreasing size (shown by the dashed arrows), which makes it more and more difficult to work within the restricted range of conditions required to successfully produce sound strip.

An awareness of these types of processing constraints by designers is important, particularly in the large-tonnage, low-cost materials as compared with the high-cost, low-tonnage specialty materials. The design of an alloy with better properties does not necessarily mean that it will be accepted in industry.

---

*This alloy has a nominal composition, in weight percent, of 45 Ni, 25 Cr, 18 Fe, 3 Mo, 3 W, 3 Co, 1.25 Si, 1.50 Mn, and 0.05 C.

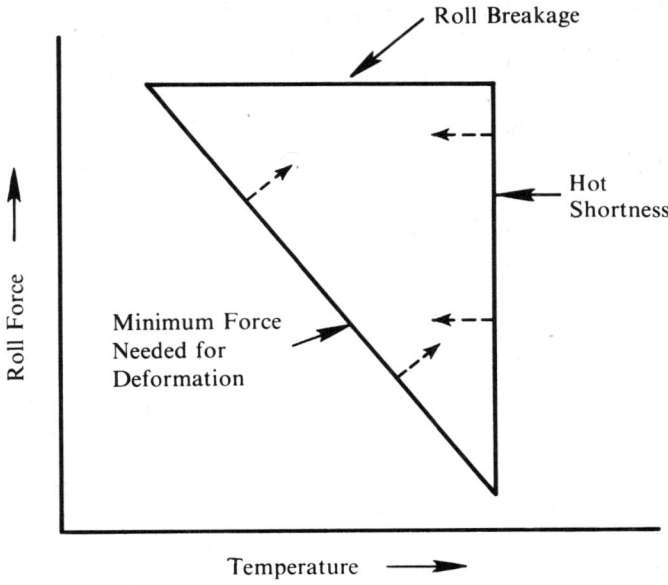

Fig. 1. Roll force-temperature box for hot processing of strip.

Earlier in the conference, both McClintock and Bush had expressed concern over the current trends in handling codes. Hirth commented on the length of time required to get code approval for a material and asked whether there are alternative ways to establish material reliability which would provide a time shortcut analogous to the use in nuclear technology of heavy-ion damage as an analogue of that produced by high neutron fluences. Bush then gave two examples. If one wanted to certify a new material for thick-wall vessels which would fit under the pressure-vessel code, either Section 8, 3, or perhaps 1, it would be particularly difficult to satisfy both the code and the nuclear aspects of the design. To place this in perspective, take as an example A543*, a material which has been considered and is actually covered in a code case. Although this material has been around for a long time it would still take 8 to 10 years to make it formally acceptable according to the code. Areas of investigation would include code variability, control of welding and the variability of the weldments, material variability, and radiation response. (Even though there are some data, the effect of residual elements and their control would need to be investigated, and possibly the fracture-mechanics curve would have to be duplicated.)

---

*Also known as HY 80.

The fracture toughness for this steel might easily be determined at −50 or −20 C, using small compact tension specimens, but the measurement of fracture properties at higher temperatures would probably require very large specimens to meet the ASTM criteria for valid fracture toughness measurements. Ten years ago the same problem was encountered in A533 and A508. It had been originally suggested that the best way to get these higher values was by linear extrapolation of the lower temperature results to higher temperatures. When this was done, the K_{Ic} at about 200 C was predicted to be only 15 to 20 ksi$\sqrt{\text{in}}$. Actually, the measured curve swept up when the nil-ductility transition temperature (NDTT) curve was exceeded, and a K_{Ic} value of 150 ksi $\sqrt{\text{in}}$. was obtained. Unfortunately, there is nothing in the theory that permits one to establish a priori what the plastic micromechanics of fracture properties are likely to be in 12-inch-thick sections. In the case of A543, the J-integral method could probably be used so that it would not now be necessary to go to 12-inch-thick specimens.

The second example is one that will have an impact on the LMFBR. This is Code Case 1539 on creep-fatigue interactions. This case went through eleven drafts as 1331 and has gone through three or four drafts as 1539. The latest draft is still not acceptable to the Nuclear Regulatory Commission. In the stainless steel vessel for the LMFBR, the creep-fatigue interaction will be an important consideration. How long it will take to get it in the code case, which is optional at this point since it is just a design parameter, and then get it into the code is difficult to say. It may still be at least 3 to 4 years away. Thus, it takes a long time to get the data and under the current mode of operation Bush does not see any shortcuts.

McClintock stated that as people learned more than was already included in the codes, they should include that information in their thinking and analysis in contrast to some situations he had observed. He was optimistic about the plastic micromechanics of fracture, pointing out that parameters predicted by such an approach were within a factor of three or so of the measured values.* Thus, although such fracture mechanics is still not precise enough for the code, it is good enough for a rough estimate of fracture behavior.

New processes have been touched on several times during the conference [amorphous metals, rapid powder quenching, and hot isostatic processing (HIP)]. Hirth asked how important these processes would become in the future and whether we were properly planning our strategy to effectively use these new processes.

Kear commented on new approaches in powder metallurgy. The first steps have already been made in powder processing to enable the

---

*See McClintock's paper in this volume.

microstructure to be controlled in a completely new way. Further advances are signalled by the development of amorphous metals. In all commercial powders, there is a wide distribution of powder sizes. Auxiliary cooling methods can be used to insure that all particle sizes have a small grain size so that the material will be superplastic in subsequent consolidation, and potentially superplastic in the as-consolidated condition. In a couple of years, this process should replace the current practice of rolling the powders and recrystallizing them before consolidation, thereby saving time, energy, and money. Another area of emphasis will be the controlled decomposition of amorphous or supersaturated solid solutions. One objective should be the production of wider material than the narrow ribbon now being produced. If, for example, sheet just 6 inches wide could be commercially produced, an "explosion" of metallurgical structural applications could ensue. To illustrate, if a superalloy could be taken directly from the melt into sheet form 12 inches wide, it would conceivably be superplastic if the cooling rates were sufficiently fast. Then complex structures such as honeycombs could be made from superalloys. He emphasized that no one yet knows how to make such wide sheet, but that it is a realistic objective which might be accomplished in 5 years. Similar things might be accomplished in tubes and more complex sections.

Hirth then mentioned microstructures which result from intermediate cooling rates, e.g., those included in the plots of Grant and others showing interdendritic arm spacing as a function of cooling rate. At the higher cooling rates, the interdendritic spacing decreases to tens of angstroms and in this condition the heat-treatment response could be very interesting, even in conventional materials. Kear agreed, saying that during rapid cooling, essentially a supersaturated solid solution could be obtained and the usual dendritic segregation could be prevented along with the phase separation that normally occurs there. In this instance, carbides and other constituents could be uniformly precipitated for the first time. Kear also mentioned in answer to an earlier question by McClintock that it should eventually be possible to fabricate a material containing a distribution of hard and soft particles by the appropriate blending together of selected powders.

Davis mentioned that the challenge in amorphous materials was in their production as the materials moved from the laboratory to the marketplace. Because of their unique combinations of properties—unusual corrosion resistance, strength, magnetic and electrical properties, and superconductivity—they could replace traditional materials in many areas. For example, precision resistors and thermometers could be designed.

Jaffee added that it would be interesting to speculate on what effect a revolutionary new process like electroslag melting would have on the design of steel for heavy sections. An immediate advantage is that the central zone-refining process permits larger forgings to be produced within the present mill capacity. But if the macrostructure of a large ingot can be modified to eliminate the detrimental impurities and inclusions which concentrate at the center, then certainly the same advantages would be conferred on conventionally melted ingots. This could permit us to go to compositions that cannot normally be made using conventional steelmaking processes.

On the topic of alternatives to strategic materials, Paxton presented comments from the viewpoint of the steel companies on materials now scarce or likely to be scarce within the next 15 years. The current "hot potato" is chromium. In a sense there is no replacement for chromium for oxidation and corrosion, although the carbide-strengthening contribution in chromium steels could probably be obtained in other ways. The only solutions today would appear to be to place more restrictions on nonessential uses of chromium, e.g., stainless steel sinks for households, and to make more use of cladding technology.

Manganese may also be in short supply in a few years. Currently it is obtained from Gabon, Brazil, India, South Africa, Australia, and Zaire, but no new ore bodies have been discovered. Ocean nodules show promise, but there are considerable technical and legal uncertainties. So far as aluminum is concerned, there is no shortage of bauxite, but the high-quality material is concentrated in a few countries, and these are trying to form an organization similar to the OPEC of the petroleum-producing countries. This has not happened yet, although bauxite prices have increased.

Tin supplies will be short, but other ways of providing surface protection for cans should be available by the time these supplies run out. In the short run, the production of tin-free steel by using chromium is a possibility. More likely, in the long run, organic thin-film coatings will be used to replace tin coatings.

There are still fairly rich nickel sulfide deposits, but the expected future sources are the oxide ores containing only 1-1/2 to 3 percent nickel. This low nickel content becomes significant in view of the increasing costs of energy required for separation. Also, much of this nickel laterite ore is located in developing countries, which makes supply uncertain.

Iron reserves in the U.S. and Canada are still large, but changing taxes may cause unpredictable shifts in the base in the next 10 to 15 years. Brazil has enormous reserves, but the economics of recovering the ore from the jungle make its development uncertain.

Two other strategic materials should be mentioned, for, although they are not metals, they are important in the recovery of metals. The first of these is fluorspar which is used to control the fluidity of slags and is important in the removal of sulfur and phosphorus from steel. It has been traditionally obtained from Mexico but the supply is getting short and the price is rising. While there are developments in South Africa, and some minor sources, its supply may become critical. A considerable amount of work is being done on the development of substitutes for fluorspar.

Another important mineral is bentonite. It is used as a binder in the manufacture of the pellets of taconite which are important for blast-furnace feed in high-capacity operations. The supply of bentonite is running low and it is lost during use. Since a large amount of this material is required to keep the pelletizing operations going*, substitutes are being sought.

New sources of minerals are appearing for the future. For example, recovery of nodules containing Mn, Ni, Cu, and Co from the ocean bottom is being actively pursued by many companies.

Sims suggested that since aluminum imparts good high-temperature oxidation resistance to metals as a result of Al_2O_3 scale formation and the ceramics SiC and Si_3N_4 show excellent oxidation resistance at high temperatures as a result of SiO_2 scale formation, perhaps it would be worthwhile to see whether aluminum and silicon (both elements are very common in the earth's crust) could be used to increase corrosion and oxidation resistance at intermediate temperatures. They could then be used to replace chromium.

Shewmon commented that as prices rise and supply goes down, metallurgists are going to be asked to design alloys to do the same job with less expenditure and fewer alloy additions. Hirth said that this area should be considered as a real challenge. In the future the compositional alternatives open to alloy designers may not be as broad as they are now. He went on to suggest that the Fe-Al-Si alloys might provide a useful substitute for stainless steels. A current drawback of these alloys is the ordering reaction and consequent embrittlement of alloys contianing more than about 3-1/4 percent silicon. However, powder-metallurgy processing could possibly get around this embrittlement problem and the substitution for stainless steels would accomplish savings in both strategic materials and cost. Jaffee suggested similar possibilities for magnesium alloys.

Stevens added that the consequences of a shortage of chromium and the development of suitable alternatives are being studied at the National Academy of Sciences under Earl Parker under the sponsorship of ERDA, NASA, GSA, and the Bureau of Mines.

2 PROPERTIES NEEDED BY THE DESIGNERS

Are the properties metallurgists are measuring and reporting the ones needed by designers? Hirth started a list of properties that designers had indicated were important to them. These were J_{Ic}, K_{Ic}, and K_{Iscc}; random

---

*In just one of U.S. Steel's plants, 18,000,000 tons of pellets a year are produced.

loading effects; environmental effects; constitutive equations; σ_{ij} and ϵ_{ij} (plastic); and elastic moduli. The elastic moduli were included in the list because some designers stated that such data were often not available for textured materials: Poisson's ratio in particular is difficult to find.

McClintock began the discussion by suggesting that J and K be changed to the plastic zone radius and crack-tip opening that would be derived from them. This conversion to length dimensions should make it easier to visualize what is going on. Also, the random loading effects and constitutive equations could together be called "history effects". An example of a simple case of a random loading effect is the Bauschinger effect from a simple load reversal. Another consideration is the homogeneity of slip or the mean free path of slip action which is a fundamental limitation on the constitutive equations; the local strain is not only a product of the local stress but also of the stress a short distance away. This size effect is important in such areas as hole growth on a fine scale and must be worked into the constitutive equations. Further, McClintock suggested that the mechanical researcher would benefit from more microstructural characterization, such as the typical sizes and spacings of the various microstructural constituents (to be compared with the previously mentioned fracture-mechanics dimensions), including inclusions as well as the hardening elements. More complete information in this area would permit more speculation on ideas which could ultimately shorten the time for code approval.

According to Begley, environmental effects are most difficult to describe because of the wide variety of possible service environments and their associated effects. Electrochemical techniques have provided some insight into material behavior. However, it is difficult to describe the environment in a piece of operating equipment, let alone to predict, a priori, what the response of the material will be in that environment. Hirth noted that ARPA is supporting a handbook program which will include what data are available on the effects of environments on engineering alloys. Begley commented on the caution required in using the handbook when it becomes available. For example, his group has found that a change in the caustic concentration is sufficient to completely change the fracture mode of steels from intergranular to transgranular. No reason could be found to predict when and in what form changes in behavior would occur. Thus, although handbook descriptions of behavior are useful, small changes in composition of the alloy or environment might produce significant changes in the material response.

Symenz indicated that aircraft design is in a comfortable state so far because it has developed good empirical relations over a 30-year period and changes have been made in small steps. Corrosion fatigue is usually not a problem. With the thin-gage material used, general corrosion failure would usually occur long before a corrosion fatigue failure.

It was noted by Jaffee that, in view of the cost of characterizing materials for Codes and the number and amounts of property data required by the aircraft industry, it will be necessary to be much more selective in the materials-development process. This does not mean that broad-ranging materials research should not be done, but it does mean that at each stage of material development for a given application, a decision should be made to eliminate those which probably will not measure up. Then the point will be reached as in steam turbines where only three steels are used and maximum testing and characterization can be performed on just these materials.

Campbell remarked that a real problem exists in characterizing microstructures. One has grain size and mean dislocation density, but these are clearly insufficient in view of the many different types of microstructures observed in alloys. It is difficult to see how one could develop constitutive equations except on a purely empirical basis, i.e., one could not go from a micrograph of the microstructure to a description of the macroscopic material behavior.

McClintock agreed wholeheartedly and went on to say that he thought plane-strain ductility tests and other tests that more closely approximate service conditions should be used more often. He mentioned the example of the double-grooved part-through cracks* which gave transition temperatures at least 30 C higher than that characteristic of the Charpy impact tests and emphasized that this was the sort of surprise that should be tested out in the laboratory before it is met in the field.

Levy commented that some of the listed quantities would be of value to the metallurgist measuring and describing the properties but would be of no use to the designer. Rather than concerning himself with particle spacings or microstructure he would want to see only the property curves. Hirth responded that ultimately the designer could be provided with a constitutive equation that contained these parameters, which he would then insert to calculate his property value. McClintock added that there now are also mechanicists who would want this information to interpret for the designer.

Thomas pointed out that, physically, microstructure can be defined to nearly the atomic scale with present-day techniques. The real difficulty now lies in characterizing the chemistry to a similar level. This is important in studying solute segregation, clustering, void formation, etc. It was pointed out that surface scientists had many tools that could be useful in this area but that they were not yet used to an appreciable extent by metallurgists.

Auger spectroscopy was brought up as a technique that addressed this problem. McMahon stated that this tool is only about 8 years old. The second generation equipment now has a resolution of about 5 microns, although techniques such as the use of field emission tips lower this to a

---

*Included in McClintock's paper in this volume.

few hundred angstroms. However, the sensitivity of present instruments is still insufficient for some important cases. In the 4340 steel which fractures intergranularly, the amounts of the elements segregated to the grain boundaries still cannot be detected. It is not clear whether ESCA will offer any improvements. It has better energy resolution, but may still have a high signal-to-noise ratio.

Thomas then mentioned a new scanning transmission electron microscope (STEM) which increases spatial resolution by decreasing the spot size to about 200 A. It amplifies the X-ray signal by differential analysis. This instrument, which is being considered for atomic resolution, may be useful to the metallurgist as a chemical analysis tool to supplement the Auger method.

Kear added that there is a new development with the field ion microscope, involving the use of a time-of-flight mass spectrometer to identify individual atom species desorbed and accelerated through an aperture. At Cambridge they have developed a gating system which enables them to obtain a concentration profile across a microstructural feature on the FIM image. This overcomes a limitation of the Auger method which requires that a microstructural feature must be on the fracture surface to be studied. Now, if one can get the feature of interest in the FIM image one can count the atoms and identify them as the interface is traversed. One problem which can possibly now be accomplished by this method is the determination of the concentration gradient across the γ-γ' interface in nickel-base alloys.

Frost commented that these basic research tools were not of any direct use to the designer, but were the province of the materials researcher. Ashby stated that although he agreed with Thomas that the techniques for measurement of the microstructure are available, they are not yet being applied quantitatively in the sense of determining the parameters that are wanted in design. Features such as grain size, particle size, and spacing are often measured in different ways by different investigators without mention of the method used. Also, often more than just one quantity should be measured, e.g., the distribution of dislocations in addition to the mean dislocation density, and the aspect ratio of nonequiaxed grains.

Bement called attention to the fact that the areas of yield surface measurements for anisotropic materials and the distortion of the subsequent flow surface by strain, particularly for multiaxial stress states, were in unsatisfactory states. Not only is it difficult to obtain constitutive equations for simple stress states, but there can be large strength differential effects, when comparing tensile and compressive behavior, which are somewhat unpredictable from current theories.

Suh pointed out that designers must have information on several additional properties. These properties include machinability and surface properties as they relate to bonding, friction, and wear. The problem of bonding and weldability is another aspect that should be considered in more detail by designers according to Bush. Weldability is often ignored or safety factors are used in design for it which have no basis in reality. Also, he had some reservations about the inspectability area. It is either ignored or is given consideration only during the construction and check-out phase. Because of gross environmental effects, inspection after the equipment goes into operation is often difficult and finding defects which may have developed is about like playing Russian roulette. This is an area that should be given more attention by designers.

Jaffee commented that reliability is an important property for an alloy. This property usually rests on two factors: the alloy is forgiving and has a large historical base. The use of alloys of high reliability is a characteristic of the power generation industry. Also, having these factors is the reason why an alloy which has mediocre properties, such as Ti-6Al-4V, is so widely used.

3 CRITICAL PROPERTY LIMITATIONS IN THE NEXT DECADE

What are the critical limitations of materials that metallurgists should be thinking of for important applications and what are possible solutions? Hirth opened with a starting list of limiting properties and applications. These included: the oxidation-strength limit of about 1100 C for high-temperature materials, the container for the fusion power system, strength-toughness, fatigue, corrosion, wear, and coatings. He then opened the discussion by commenting that one of the outputs of the conference should be a pinpointing of some questions which are regarded as crucial but have possible solutions within the next decade.

Decker indicated that he is concerned by the low national rate of effort being devoted to revolutionary alloy design, mentioned in an earlier Agenda Discussion by Owen, and to development of new alloy concepts. In the last 5 to 10 years, the area of process metallurgy has received a large share of attention. This has resulted in the upgrading of existing plant facilities and the construction of plants and equipment for new processes such as powder atomization and directional solidification. But these efforts are now becoming capital intensive and the rate of return will be decreasing in this area. Of course, new processing techniques do open up new opportunities for alloy design, as for glassy metals, but this is more often of an evolutionary nature in which the alloy compositions are modified to better adapt them to the process being used.

In contrast, the rate of effort now being spent on new alloy design appears to be a small fraction of what it was 10 years ago, perhaps at a 20-

year low. As a result we may find that within 5 to 10 years we will have a serious void in new alloy concepts. Therefore Decker would recommend that we increase the effort in this area. In response to a specific point on the above list, he felt there was a definite possibility that the strength-oxidation ceiling (1100 C) could be raised to higher temperatures.

Owen objected to the separation of alloy design from processing development. In many cases, processing is a part of alloy design; the alloy limits the available processing routes or the process is limited to certain types of alloys. Decker responded that this is a matter of emphasis and objective. The objective may not be to change properties, but rather to decrease costs or make a part in a different way. Owen countered that it would not be difficult to change the emphasis of much process design work today to transform it into alloy design. But Decker said that, often, different types of people would be involved. To change the research emphasis from process design to alloy design would require changing personnel.

Mayer pointed out that in the case of Metglas, revolutionary alloy design and processing were one and the same. This suggested that some new alloy concepts may require a more sophisticated processing approach than the more conventional processing routes used in the past.

Ansell, with respect to Decker's comments, stated that what is required to encourage new alloy development is an increase in the yield from processing or a decrease in the cost of production. For example, as the box in Fig. 1 grows smaller, as it often does with new, higher performance alloys, the product yield decreases and the costs of producing the alloy grow so large that they lower the cost/benefit ratio to the point where it is unattractive. Applications such as the nuclear industry could be exceptions. Here designers might be willing to pay the premium required when their applications are material-property limited for less expensive alloys. Developments such as the continuous casting of sheet bar would be required to overcome such yield problems. This would eliminate the capital and production costs required for breakdown of large ingots to make sheet products.

Sims added that there is a large effort to develop new ceramics for high-temperature applications. The effort has shown some promising results, and it may well be that significant new applications will emerge. Hirth mentioned the possible use of graphite in thermal power machines having reducing atmospheres. Frost pointed out that the understanding of the relations between microstructure and properties of ceramics is at a very early stage compared with that for metals. He suggested that if more attention were given to ceramics there could be a substantial increase in the materials applications above 1100 C.

Additional to the topic of fusion shells, Levy would include fission shells, i.e., fuel cladding. He feels that there are serious limitations in fuel cladding in the LMFBR. Bement made three points with regard to fusion shells and property instabilities in their design. Firstly, in the transition from fast breeder to fusion reactor, many more phenomena are involved. At present the national effort is broken up into groups dealing with each of these separately. But Bement feels that some major surprises may be in store when it is discovered that there are synergistic or combined effects between two or more of these phenomena. Some of these interactions or combined effects should be studied now. Secondly, the fusion reactors will be operated under cyclic conditions. Thus the time constants for the burn cycles should be compared with steady-state defect production and the downtimes should be compared with relaxation-recovery phenomena. Steady-state solutions to some of these problems are no longer adequate and the development of transient solutions is now required. Cyclic operation can have dramatic effects on void-size distributions and radiation-creep rates. Thirdly, during dynamic radiation creep, none of the conventional hardening laws apply. Some unique forms of instabilities are present. Under varying stress conditions, nearly the full initial transient response can be reinitiated rather than strain hardening being accumulated within the sample. More attention will have to be given to varying and multiaxial stress states in developing constitutive relations, and to the sources of instability and growth ratcheting which occur in these applications.

One of the concerns in the future will be recycling of materials, particularly the large tonnage materials. One of the problems here according to Bush will be the removal of certain trace elements that build up in the alloys during recycling, such as copper in steels. These elements in sufficient quantity can seriously degrade the alloy properties and result in structural failures if they are not carefully watched. Their removal might require plasma techniques or other costly processing.

Sherby suggested that a critical limitation to superplastic forming is that the highest limiting strain rate for superplasticity is so low, about 1% min^{-1}. If this ceiling could be raised to 100 or even 1000% min^{-1}, superplastic forming would become a much more attractive process.

4 POSSIBLE ADVANCES IN UNDERSTANDING FUNDAMENTAL PROPERTIES

What possible advances in fundamental understanding can be foreseen in the fields of metallurgy, chemistry, and physics which bear on alloy design? Here the starting list of items and examples included: theory of metals (electron theory), defects (dislocation core structures), solid solutions (ordering), transport (diffusion), phase transformations, constitutive relations (stress-strain from microstructures), elasticity-plasticity (nonlinear solutions), and corrosion and oxidation.

McMahon suggested adding to the list three items. First, he would add theory of cohesion of interfaces under theory of metals as a topic required to help explain observations such as hydrogen lowering the cohesion of interfaces within a material. Second, he cited the need for good thermodynamic data describing the stability of compounds such as those of the Groups IV through VI metalloids. Third, he felt that more understanding of the nature of surface films which form in the presence of the Groups IV through VI elements was required. This would perhaps help in understanding observations such as intergranular stress corrosion in temper-embrittled steel. One explanation is that this is caused by an inhibition of the passivating film caused by the presence of these elements. Hirth commented that decohesion is also important in developing a criterion for a particle decohering from the matrix in the plastic zone at the tip of a ductile crack. He also noted that there are almost no measurements of absolute surface energies at interfaces. Also connected with surface properties, there is a growing body of evidence that shows there are marked effects of surface electrostatic potentials on material behavior. Westwood has shown large effects, in machining and rock drilling, of small amounts of additives to surfaces which influence these potentials.

Also in connection with surfaces, Slater pointed out that, with the exception of copper and the noble metals, metals were protected by the formation of thin films, particularly in aqueous solutions. These films are still poorly understood. Such factors as the composition of the films, how the composition of the films is affected by the composition of the base material, and depletion of certain elements at the surface are very poorly understood. New techniques such as Auger spectroscopy and ion scattering spectrometry have contributed heavily in this area, but much is still unknown. For example, it is uncertain what molybdenum does in a 316 steel to make it more pitting resistant than a 304 steel. Until some of these effects are understood, new alloy design is going to be hindered. Before alloys can be designed for increased aqueous corrosion resistance without chromium, much more must be known about how other alloying elements affect filming behavior.

Haasen pointed out that one should differentiate between fundamental theory and modeling. Modeling can be very useful when the fundamental theory is difficult. For example, in alloy theory a fundamental description is difficult, but by the use of pairwise potentials, alloys can be described in a useful way with a reasonable number of parameters. The same is true of dislocation theory. Instead of waiting for fundamental dislocation theory to reach the point of developing constitutive equations to describe stress-strain curves, possibly three dislocation parameters could be used in a modeling experiment. These could be the mobile and immobile dislocation densities and the dislocation density tied up in arrays. One could then carry the model through and determine which parameters should be changed to more accurately describe alloy behavior. Full use has not been made of such possibilities and this is holding up the interaction between physical metallurgy and alloy design. There could be much more interaction in the fields of creep and fatigue.

McClintock would like more feedback from alloy designers as to what elastic-plastic or plastic solutions would be of most use to alloy developers. Several examples of needs brought up during the course of the conference were treatments of austenite films around martensite particles and of hard oxide films on a plastic substrate. But what are other situations like these where a mechanics solution would help to give insight? The people in applied mechanics often feel that their field is worked out, but looking at the materials area there is still a great deal that remains to be done. What is needed is identification of specific items for which solutions would be worth developing and testing out.

There is still no point defect theory which uses first principles to incorporate electron theory and size mismatch effects for calculating the binding energies between point defects and impurity atoms in the metal lattice. Bement remarked that this is very important in radiation-effects studies and must be important in other instances too. Until we know how to separate the contributions of these two effects to the binding energy, it is difficult to determine how to increase the binding energy or how to tailor it in specific instances.

Shewmon recalled that the plasticity of grain boundaries was brought up repeatedly in the conference. It is relied on heavily in processes using superplasticity, yet not enough is known about it.

Finally, Jaffee remarked that pair potentials had been mentioned for simulating bulk behavior and added that surface behavior can also be modeled by such potentials. This has been a very useful tool for successfully modeling such phenomena as the effect of hydrogen on the decohesion of iron.

Hirth concluded by stating that his view of the field of alloy design, which had not changed during the conference, is best described by Lewis Carroll's "Jabberwocky", which he then recited.

AUTHOR INDEX

663

SUBJECT INDEX